Ecological Methods

T.R.E. Southwood Kt, DSc, PhD, FRS
Lineacre Professor of Zoology, University of Oxford,
South Parks Road, Oxford

P.A. Henderson PhD
Pisces Conservation Limited, IRC House, The Square,
Pennington, Lymington, Hants

Third Edition

This book has an active website that provides additional
illustrations, details of equipment and computer programs
as well as references to work published since the manuscript was
completed.
The site is located at http://www.blackwell-science.com/
southwood

Blackwell
Science

© 2000 by T.R.E. Southwood and P.A. Henderson

BLACKWELL PUBLISHING
350 Main Street, Malden, MA 02148-5020, USA
9600 Garsington Road, Oxford OX4 2DQ, UK
550 Swanston Street, Carlton, Victoria 3053, Australia

First published 1966 by Methuen & Co Ltd
Second edition published 1978 by Chapman & Hall
Third edition published 2000 by Blackwell Science Ltd

7   2007

Library of Congress Cataloging-in-Publication Data

Southwood, Richard, Sir.
    Ecological methods / T.R.E. Southwood.   3rd ed. / revised by P.A. Henderson and T.R.E. Southwood.
        p. cm.
    Includes bibliographical references and index.
    ISBN 0-632-05477-8
    1. Ecology—Technique. 2. Animal populations. I. Henderson, P.A. II. Title.
QH541.28 S68   2000
577'.028—dc21   99-39859
                                                     CIP

ISBN-13: 978-0-632-05477-0

A catalogue record for this title is available from the British Library.

Set by Excel Typesetters Co., Hong Kong

For further information on
Blackwell Publishing, visit our website:
www.blackwellpublishing.com

Contents

Prefaces

Preface to the Third Edition

We have been encouraged to prepare this third edition by the continuing use of the earlier editions. In doing so, we have been struck by the contrast between the advances in some areas, especially data handling, and the enduring value of various other techniques. Ecology has continued its advance into the popular and political domain, though far from everything that has gained the 'eco-' prefix falls within our purview. The underlying paradigm of ecology has, however, shifted. In particular, the concept of the metapopulation is now recognized as central to the understanding of the distribution and abundance of animals and its exploration is aided by the accessibility of numerous data sets, often with large temporal or spatial scales. The availability of molecular techniques has encouraged the consideration of genetic and phylogenetic aspects and has permitted the growth of quantitative comparative analyses, described by P.H. Harvey and M.D. Pagel (1991) *The Comparative Method in Evolutionary Biology* (Oxford).

The extent to which we have felt that revision was necessary has varied greatly in different parts of the book. Where there has been little change in the method, we have retained the early references. We have done this on the bases that journal editors formerly permitted more detailed description of methods, that these papers will not be located by computer searches, and that these pioneers continue to deserve credit. Other portions have required considerable modification; we have deleted one chapter (12) and added a new one (15), as well as reorganizing the structure in some places. Although the primary focus remains on insects, which are in terms of species the dominant animals, we have taken the opportunity to explicitly expand the coverage to all major macroscopic groups.

The widespread availability of high-capacity personal computers (PCs), with software packages and access to the internet, has totally changed the speed and ease of handling (and sometimes accessing) data. We have therefore given references to some relevant software packages and websites, whilst eliminating many descriptions of time-consuming graphical methods. However, we believe that the advice given in the preface to the second edition is even more applicable today. The researcher who relies entirely on the output of a computer is in danger of drawing false conclusions and overlooking possible insights. It is essential to understand the features of the data (are there any outliers?), the assumptions of the methods, and the biological basis of the analysis and to acquire a feel for the capabilities and responses of the species under study.

The book has an active website that provides additional illustrations, details of equipment and computer programs and references to work published since the manuscript was completed. We aim to keep the revision up to date through this site; it would greatly assist us if ecologists could send us explanatory paragraphs and references to new work. These should be posted on http://www.blackwell-science.com/southwood. We would also be most grateful to have any errors drawn to our attention through the same route.

The interpretation of 'ecological methods' remains as described in the Preface to the First Edition, namely those methods peculiar to ecologists, either in their origin or in modification. Just as the measurement of physical factors, using the methods of the physical sciences, has always been outside the book's scope, so are the methods of molecular biology. These are described in a number of works, such as Snow A.A. & Parker P.G. (1998) Molecular markers for population ecology. *Ecology* **79**, 359–425.

We are most grateful to many ecologists who sent comments on the second edition. In particular, generous help has been given by Drs C. Henderson, D.J. Rogers, A.E. Magurran, W.D. Hamilton and G.R.W. Wint and Mr C. Hambler.

T.R.E. Southwood
P.A. Henderson
Oxford, October 1999

Preface to the Second Edition

In the twelve years since the First Edition was prepared there have been remarkable developments in ecology. The subject has changed its lay image, from a rather recondite branch of biology, to something that is widely considered 'good', but only vaguely understood. The public's focus on environmental problems and the insights into these that ecology can provide are a great challenge to ecologists to develop their subject: they need to be able to provide reliable quantitative inputs for the management of the biosphere. The enormous volume of work that it has been necessary to review for this edition is evidence of the extent to which ecologists are seeking to meet this challenge.

I believe that the theme of the first edition, the need for precise measurement and critical analysis, is equally valid today; although many recent studies show levels of sophistication that were beyond my wildest hopes when I embarked on the preparation of the first edition. In his review of the first edition Dr R.R. Sokal was kind enough to say it was an 'unusual book' for it covered both traps and mathematical formulae, topics that were usually of interest to different people. This, I am glad to say, is now no longer generally true. The computer and the electronic calculator have revolutionized the handling of ecological data, but neither can make a 'silk purse' of sound insight, out of a 'sow's ear' of unreliable raw data or confused analytical procedure. More than ever the ecologist needs to keep his biological assumptions in mind and remember the value of preliminary simple graphical analysis as a means of recognizing new patterns and gaining fresh insights.

It has been a gratifying, though exhausting, experience preparing the new edition! Progress has been so rapid in several areas that some chapters have been completely or largely rewritten (e.g. 11, 12, 13), whilst most have large new sections. It has been necessary to be highly selective in the additions to the bibliographies, even so there are nearly a thousand new entries and only a few older references could be deleted. I hope that, with the advent of *Ecological Abstracts*, the selective nature of the bibliographies will not handicap workers. As the mathematical side of ecology has grown, so have the problems of notation and it is now quite impossible (without extending far beyond the roman and greek alphabets!) to retain a unique notation throughout. Apart from widespread and generally accepted symbols, I have merely aimed to be consistent within a section.

I am most grateful to many ecologists who have helped me in this revision by sending me reprints of their papers or notes on difficulties and errors in the first edition. Detailed criticism, advice and help, including access to unpublished work has been generously given by M.H. Birley, P.F.L. Boreham, M.J.W. Cock, G.R. Conway, M.P. Hassell, R.M. May, A. Milne, S. McNeill, G. Murdie, G.A. Norton, S. Parry, P.M. Reader, G. Seber, N.E.A. Scopes, L.R. Taylor, R.A.J. Taylor, T.W. Tinsley, J.M. Webster, R.G. Wiegert and I.J. Wyatt. I am deeply indebted to them, and to others, especially Margaret Clements and my wife, who have assisted with patience and forbearance in the essential tasks associated with the preparation of the manuscript of the new edition.

T. R. E. Southwood
Imperial College
Silwood Park, Ascot.
October 1977

Preface to the First Edition

This volume aims to provide a handbook of ecological methods pertinent for the study of animals. Emphasis is placed on those most relevant to work on insects and other non-microscopic invertebrates of terrestrial and aquatic environments, but it is believed that the principles and general techniques will be found of value in studies on vertebrates and marine animals.

The term ecology is now widely used in the field of social, as well as biological, science; whilst the subject of ecology, covering as it does the relationship of the organism to its environment, has many facets. It is, in fact, true to say that the ecologist may have need of recourse to almost all of the methods of the biologist and many of those of the physical scientist: the measurement of the physical factors of the environment may be a particularly important part of an ecologist's work and he will refer to books such as R.B. Platt & J.E. Griffiths' (1964) *Environmental Measurement and Interpretation*.

There are, however, certain methods that are peculiar to the ecologist, those concerning the central themes of his subject, the measurement, description and analysis of both the population and the community. These are *ecological* methods (as opposed to 'methods for ecologists' which would need to cover everything from laboratory workshop practice to information theory); they are the topic of this book.

During the 10 years that I have been giving advanced and elementary courses on ecological methods at Imperial College, London, and at various Field Centres, the number and range of techniques available to the ecologist have increased enormously. It has been the comments of past students on the utility of these courses in helping to overcome the difficulties of coping with the scattered and growing literature that have encouraged me to attempt the present compilation. I am grateful to many former students for their criticisms and comments, as I am to the members of classes I was privileged to teach at the University of California, Berkeley, and at the Escuela Nacional de Agricultura, Mexico, whilst writing this book.

Although the general principles of most methods are of wide application, the study of a particular animal in a particular habitat may require certain special modifications. It is clearly impossible to cover all variants and therefore the reader is urged to consult the original papers that appear relevant to his problem. I am grateful to my publishers for agreeing to the publication of the extensive bibliographies, it is hoped that these will provide many leads on specific problems; they are, however, by no means exhaustive.

The present book is designed to be of use to those who teach the practical aspects of animal ecology in schools, training colleges and universities; insects, being numerically the dominant component of the macrofauna of terrestrial and many aquatic habitats, almost invariably come to the forefront of ecological field work. This volume is intended as an aid to all who need to measure and compare populations and communities of animals, not only for the research ecologist, but also for the conservationist and the economic entomologist. Population measurement is as necessary in the assessment of the effects of a pesticide and in the determination of the need for control measures, as it is in intensive ecological studies. It is frequently pointed out that ecological theories have outstripped facts about animal populations and I trust that it is not too presumptuous of me to hope that this collection of methods may encourage more precise studies and more critical analysis of the assembled data so that, in the words of O.W. Richards, we may have 'more light and less heat', in our discussions.

The topics have been arranged on a functional basis, that is, according to the type of information given by a particular method. As a result some techniques are discussed in several places, e.g. radiotracers will be found under marking methods for absolute population estimates (chapter 3), the measurement of predation and dispersal (chapter 9) and the construction of energy budgets (chapter 14). By its very nature ecology cannot be divided into rigid compartments, but frequent cross-references in the text, together with the detailed contents list and index, should enable the reader to find the information he needs. The sequence of chapters parallels, to a large extent, the succession of operations in a piece of intensive research.

It is a pleasure to express my great indebtedness to colleagues who have read and criticized various chapters in draft: Dr N.H. Anderson (ch. 6), Dr R.E. Blackith (ch. 2 & 13), Dr J.P. Dempster (ch. 1, 2 & 3), Mr G.R. Gradwell (ch. 10), Dr C.S. Holling (ch. 12), Mr S. Hubbell (ch. 14), Dr C.B. Huffaker (ch. 2), Dr G.M. Jolly, (section II of ch. 3), Dr C.T. Lewis, (section I of ch. 3), Dr R.F. Morris (ch. 10), Dr O.H. Paris (ch. 9 & 14), Mr L.R. Taylor (ch. 2 & 4), Professor G.C. Varley (ch. 10) and Dr N. Waloff (ch. 1, 2 & 10); frequently these colleagues have also made available unpublished material; they are of course in no way responsible for the views I have expressed or any errors. For access to 'in press' manuscripts, for unpublished data and for advice on specific points I am grateful to: Drs J.R. Anderson, R. Craig and D.J. Cross, Mr R.J. Dalleske, Drs W. Danthanarayana, H.V. Daly, E.A.G. Duffey, P.J.M. Greenslade, M.P. Hassell, P.H. Leslie, J. MacLeod, C.O. Mohr, W.W. Murdoch and F. Sonleitner, Mr W.O. Steel, Drs A.J. Thorsteinson, R.L. Usinger, H.F. van Emden, E.G. White, D.L. Wood and E.C. Young. Ecologists in all parts of the world have greatly helped by sending me reprints of their papers. I have been extremely fortunate too in the assistance I have received in translating; Mrs M. Van Emden has generously made extensive translations of works in German, and with other languages I have been helped by Dr F. Baranyovits (Hungarian), Dr T. Bilewicz-Pawinska (Polish), Mr Guro Kuno (Japanese), Mr P. Stys (Czechoslovakian) and Dr N. Waloff (Russian).

Much of the manuscript was prepared whilst I held a visiting professorship in the Department of Entomology and Parasitology of the University of California, Berkeley; I am indebted to the Chairman of that Department, Dr Ray F. Smith, for his interest and the many kindnesses and facilities extended to me and to the Head of my own Department, Professor O.W. Richards, F.R.S., for his support and advice. I wish to thank Mrs M.P. Candey and Mrs C.A. Lunn for assisting me greatly in the tedious tasks of preparing the bibliographies and checking the manuscript. My wife has encouraged me throughout and helped in many ways, including typing the manuscript.

T.R.E. Southwood
London, October 1965

1 Introduction to the Study of Animal Populations

Information about animal populations is sought for a variety of purposes; but the object of a study will largely determine the methods used and thus this must be clearly defined at the outset. Very broadly, studies may be divided into extensive and intensive (Morris, 1960). Extensive studies are carried out over larger areas than intensive studies. They are frequently used to provide information on distribution and abundance for conservation or management programmes (e.g. monitoring the status of mammal populations (Macdonald *et al.*, 1998)). Recent developments in remote sensing capability and geographical information system software have given great impetus to extensive studies in recent years (Chapter 15). For pests or parasites, they provide assessments of incidence or damage and may guide the application of control measures. Aphids, for example, are regularly monitored by the Rothamsted suction trap survey (Knight *et al.*, 1992). In extensive surveys, an area will often be sampled once or at the most a few times per study period. The timing of sampling in relation to the life cycle of the animal is obviously of critical importance and, for many species, only limited stages in the life cycle can be sampled. Extensive studies produce information about the spatial pattern of populations, and it is often possible to relate the level of the population to edaphic, oceanographic, or climatic factors.

Intensive studies involve the repeated observation of the population of an animal. Usually information is acquired on the sizes of the populations of successive developmental stages so that a life-table or budget may be constructed. Then, using this table, an attempt is made to determine the factors that influence population size and those that govern or regulate it (Varley & Gradwell, 1963; Sibley & Smith, 1998). It is important to consider at the start the type of analysis (see Chapter 11) that will be applied and so ensure that the necessary data are collected in the best manner. Intensive studies may have even more limited objectives, such as the determination of the level of parasitism, the amount of dispersal, or the overall rate of population change.

The census of populations and the stages at which mortality factors operate are necessary first stages in the estimation of the productivity (Chapter 14) of ecosystems. In survey and conservation work, the species make-up of the community and changes in its diversity associated with human activities are most frequently the features it is desired to measure. Special methods of analysis need to be used (Chapter 13), but difficulties usually arise because of the impossibility of recording the abundance of all the species living within a habitat with equal efficiency.

1.1 Population estimates

Population estimates can be classified into a number of different types; the most convenient classification is that adopted by Morris (1955), although he used the terms somewhat differently in a later paper (1960).

1.1.1 Absolute and related estimates

For large animals that are easily observed and have small, countable populations, such as rhinos, elephants, whales, and some birds, it may be possible to express the global or metapopulation size as a total number of individuals. However, for most animals, numbers will be expressed as a density per unit area or volume or density per unit of the habitat. Such estimates are given by distance sampling and related techniques (Chapter 9), marking and recapture (Chapter 3), sampling a known fraction of the habitat (Chapters 4–6), and removal sampling and random-walk techniques (Chapter 7).

1.1.1.1 Absolute population

This is defined as the number of animals per unit area. For planktonic animals, the number per unit volume can be more appropriate. It is almost impossible to construct a budget or to study mortality factors without the conversion of population estimates to absolute figures. This is because other measures of habitat are variable. For example, the amount of plant available to an insect is always changing; further, insects often move from the plant to the soil at different developmental stages. The importance of obtaining absolute estimates cannot be overemphasized.

1.1.1.2 Population intensity

This is the number of animals per unit of habitat, e.g. per leaf, per shoot, per plant, per host. Such a measure is often, from the nature of the sampling, the type first obtained (see also p. 148). When the level of the animal population is being related to habitat availability or plant or host damage, it is more meaningful than an estimate in absolute terms. It is also valuable in comparing the densities of natural enemies and their prey. However, the number of habitat units/area should be assessed, for differences in plant density can easily lead to the most intense population being the least dense in absolute terms (Pimentel, 1961). When dealing with different varieties of plants, differences in leaf area may account for apparently denser populations, in absolute terms, on certain varieties (Bradley, 1952). Thus the choice of the leaf or of the plant as the unit for expressing population intensity can affect the relative population estimate obtained (Broadbent, 1948; Fig. 1.1). Similarly, with litter fauna, owing to the effects of seasonal leaf fall, the intensity measure (in animals/weight of

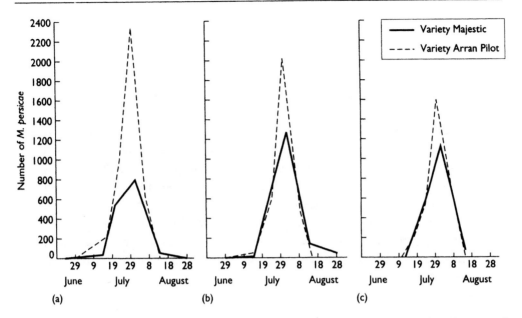

Fig. 1.1 The influence of habitat unit on relative population levels when these are measured in terms of population intensity: the populations of *Myzus persicae* on different varieties of potato. (a) Per 100 leaves; (b) per plant; (c) per plant, corrected for proportions of upper, middle, and lower leaves (after Broadbent, 1948).

litter) will give a different seasonal picture from an absolute estimate per square metre (Gabbutt, 1958). These examples also underline the importance of absolute estimates when interest lies primarily in the animal population.

1.1.1.3 Basic population
In some habitats, especially forests and orchards, it is often convenient to have an intermediate unit between that used for measuring intensity and absolute measures of ground area, e.g. 1 m² of branch surface (Morris, 1955) or branches of apple trees (Lord, 1968).

1.1.2 Relative estimates
These estimates, in which the number caught cannot be expressed as a density or intensity per area or habitat unit, only allow comparisons in space or time. They are especially useful in extensive work on species distributions, monitoring changes in species richness, environmental assessments, recording patterns of animal activity, and the investigation of the constitution of a polymorphic population. The methods employed are either the catch per unit effort type or various forms of trapping, in which the number of individuals caught depends on a number of factors besides population density (Chapter 7). There is no hard and fast line between relative and absolute methods, for absolute methods of sampling are seldom 100% efficient and relative methods can sometimes be corrected in various ways to give density estimates.

Relative methods are important in applied areas, such as fisheries or game management, where most of the information available may be derived from fishing or hunting returns. In fisheries research, catch per unit effort is often difficult to calculate from landing statistics because of changes in catch efficiency as fishing technology and the economy changes.

1.1.3 Population indices

The animals themselves are not counted, but their products (e.g. frass, webs, exuviae, tubes, nests) or effects (especially plant damage) are recorded.

Both population indices and relative estimates of population can sometimes be related to absolute population (if this is measured at the same time) by regression analysis. If such a study has been based on sufficient data, subsequent estimates from relative methods or indices can be converted to absolute terms, using various correction factors; such an approach is common in fisheries research (e.g. Beverton and Holt, 1957).

1.2 Errors and confidence

The statistical errors of various estimates can usually be calculated and are referred to as the fiducial limits (the estimate (x) being expressed as $x \pm y$, where y = fiducial limits). These are sometimes incorrectly referred to as 'confidence' limits, but the distinction between the two terms is in practice unimportant. The fiducial limits are calculated for a given probability level, normally 0.05, which means that there are only five chances in 100 that the range given by the fiducial limits does not include the true value (hence the expressions 5% probability level and 95% fiducial limits). If more samples are taken, the limits will be narrower, but the estimate may not move closer to the actual value. The biologist is often worried that some of the assumptions about sampling efficiency (e.g. does the 'knock-down' method really collect all the weevils on a tree?) may be incorrect. Intuitively, most believe, quite correctly, that this estimate should be compared with another method that has different assumptions. If the estimates are of the same order of magnitude, then the investigator can have much greater confidence that the result of his/her study is not misleading. It is therefore sound practice for the ecologist to contemporaneously estimate the population or other variables by more than one method.

When two estimates have been obtained from different sampling procedures, provided they are internally consistent (e.g. a t-test shows that the means are not significantly different), they may be combined to give a weighted mean, weighting each estimate inversely to its variance (Cochran, 1954). Under some circumstances, Bayes' theorem could be used (see p. 84) to give the combined estimate. Laughlin (1976) has suggested that the ecologist may be satisfied with a higher probability level (say 0.2) and thus narrower fiducial limits for estimates based on more than one method, because such estimates have a qualitative, biological assurance, in additional to that from the consistency of the data, that the true mean lies close to the estimate.

Most population studies are based on sampling and the values obtained are considered to have a generality that scales with the area from which the samples were drawn. All estimates have fiducial limits and the level of accuracy that should be aimed at is difficult to determine. Morris (1960) has aptly said that 'we are not likely to learn what precision is required by pessimistic contemplation of individual fiducial limits'. Excessive concerns about accuracy can always be used by those who favour the warmth of the hearth to being in the field. As the amount of time and labour that can be put into a problem is limited, it should always be borne in mind that the law of diminishing returns applies to the reduction of the statistical errors of sampling. In the long run, more knowledge of the ecology of the animal may be gained by studying other areas, by making other estimates, or by taking further samples than by straining for a very high level of accuracy in each operation. Against this must be set the fact that, when animals are being extracted from samples, the errors all lie below the true value, as animals will occasionally be missed. A number of very carefully conducted control samples may allow a correction factor to be applied, but the percentage of animals missed may vary with density; sometimes more are overlooked at the lowest densities (Morris, 1955).

An alternative to sampling is the continuous, or regularly repeated, study of a restricted cohort, e.g. the population of an aphid on a particular leaf or leaf-miners on a bough. Such studies have a very high level of accuracy, but they sacrifice generality. A combination of some cohort studies with larger-scale sampling often provides valuable insights.

References

Beverton, R.J.H. & Holt, S.J. (1957) *On the Dynamics of Exploited Fish Populations*. Ministry of Agriculture, Fisheries and Food, HMSO, London.

Bradley, R.H.E. (1952) Methods of recording aphid (Homoptera: Aphidae) populations on potatoes and the distribution of species on the plant. *Can. Entomol.* **84**, 93–102.

Broadbent, L. (1948) Methods of recording aphid populations for use in research on potato virus diseases. *Ann. Appl. Biol.* **35**, 551–66.

Cochran, W.G. (1954) The combination of estimates from different experiments. *Biometrics* **10**, 101–29.

Gabbutt, P.D. (1958). The seasonal abundance of some arthropods collected from oak leaf litter in S.E. Devon *Proc. X Int. Congr. Entomol.* **2**, 717.

Knight, J.D., Tatchell, G.M., Norton, G.A., &

Harrington, R. (1992) FLYPAST: an information management system for the Rothamsted Aphid Database to aid pest control research and advice. *Crop Prot.* **11**(5), 419–26.

Laughlin, R. (1976) Counting the flowers in the forest: combining two population estimates. *Aust. J. Ecol.* **1**, 97–101.

Lord, F.T. (1968) An appraisal of methods of sampling apple trees and results of some tests using a sampling unit common to insect predators and their prey. *Can. Entomol.* **100**, 23–33.

Macdonald, D.W., Mace, G., & Rushton, S. (1998) *Proposals for Future Monitoring of British Mammals*. Department of the Environment, London, 374 pp.

Morris, R.F. (1955) The development of sampling techniques for forest insect defoliators, with particular reference to the spruce budworm. *Can. J. Zool.* **33**, 225–94.

Morris, R.F. (1960) Sampling insect populations. *Ann. Rev. Entomol.* **5**, 243–64.

Pimentel, D. (1961) The influence of plant spatial patterns on insect populations. *Ann. Entomol. Soc. Am.* **54**, 61–9.

Sibley, R.M. & Smith, R.H. (1998) Identifying key factors using lambda contribution analysis. *J. Anim. Ecol.* **67**, 17–24.

Varley, G.C. & Gradwell, G.R. (1963) The interpretation of insect population changes. *Proc. Ceylon Assoc. Adv. Sci.* **18(D)**, 142–56.

2 The Sampling Programme and the Measurement and Description of Dispersion

2.1 Preliminary sampling

2.1.1 Planning and fieldwork

At the outset, the worker must be quite clear as to the problem he/she is proposing to investigate (Morris, 1960). As it is normally impossible to count and identify all the animals in a habitat, it is necessary to estimate the population by sampling. Naturally, these estimates should have the highest accuracy commensurate with the effort expended. This requires a plan that includes a sampling programme laying down the number of samples, their distribution, and their size. While the statistical principles for sampling are given in texts such as Hansen *et al.* (1953), Yates (1953), Stuart (1962), Cochran (1977), Elliott (1977), Green (1979), Seber (1982), Scheiner and Gurevitch (1993), and Underwood (1997), there is no universal sampling method. The importance of careful formulation of the hypothesis for test cannot be overstressed.

In community studies, preliminary work should consider species richness and potential problems with species identification. The appropriate degree of taxonomic discrimination must be decided, as it is important to maintain a consistent taxonomy. Sample sorting and species identification are often the most labour-intensive parts of a study and it may be useful to process a trial sample to assess the effort required. Saila *et al.* (1976) found that a single marine grab sample cost \$25 to collect and \$300 to sort. The planning of the timing of sampling requires knowledge of life cycles. Preliminary work will be necessary to gain some knowledge of the distribution of the animal(s). From these observations, an estimated cost per unit sample in terms of time, resources and money can be made.

The first decision concerns the scale of the universe to be sampled. Whether this is to be a single habitat (e.g. field, woodland, or pond) or representatives of the habitat type from a wide geographical area will depend on whether an intensive or an extensive study is planned. The correct definition of the target population or community is essential: if too small, it may not produce results representative of the structure as a whole; if too large, it will waste resources. A good example of this decision process for plankton is given in Omori and Ikeda (1984).

The second decision must be to define the accuracy or precision of the population estimates required. Accuracy measures how close an estimate is to the real value and precision measures reproducibility of the estimate. These decisions must be taken by considering both the objectives of the study and the variability of the system under study. For example, many species of insect pest exhibit a 10-fold or even 100-fold population change within a single season and recruitment in fish stocks can vary 10-fold between years (Cushing, 1992). For such species, an estimate of population density with a standard error of about 25% of the mean, enabling the detection of a doubling or halving of the population, is sufficiently accurate for damage control or stock assessment studies (Church & Strickland, 1954). For life-table studies, more especially on natural populations, a higher level of accuracy, frequently set at 10%, will be necessary. Conservationists often seek to build species inventories. Here completeness will replace accuracy as a measure of quality. While for large mammals or birds the aim might be to record all resident species, for high-diversity groups such as beetles the objective may be set at only 5–10%. Community studies often aim to generate summarizing statistics, such as measures of diversity or species richness (Chapter 13), which can be used to compare localities or changes through time. The accuracy and precision of these estimates must be carefully considered if changes are to be detected at the desired resolution.

In extensive work the amount of sampling in a particular locality will be limited and therefore a further decision concerning both the age-group to be studied and the timing of sampling will need to be made. For insects, the best stage for sampling may be that most closely correlated with the amount of damage (Burrage & Gyrisco, 1954), or, if the purpose of the survey is to assess the necessity for control, the timing should be such that it will give advanced information of an outbreak (Gonzalez, 1970). For aquatic organisms, timing is often determined by the reproductive cycle. For example, in temperate waters benthic surveys carried out in autumn will show a population dominated by recent recruits. The same survey carried out in the spring will show the resident community which has survived both competition for space and the rigours of the winter (Bamber & Spencer, 1984; Bamber, 1993). While the life-history stage to be sampled will depend on the objectives, the stage must not be one whose numbers change greatly with time and must be present for a period sufficient to allow the survey to be completed. Further, the easier the stage is to sample and count, the better. A problem with extensive surveys can be that different areas hold populations that differ in their position in the reproductive cycle. The reliability of samples in an extensive survey may be particularly sensitive to current weather conditions (Harris et al., 1972). The influence of latitudinal gradients on the timing of sampling can be important. In extensive surveys of marine plankton, the timing of blooms or larval production can vary by 1 month over 1° of latitude.

Although the preliminary sampling and the analysis of the assembled data will provide a measure of many of the variables, the actual decisions must still, in many cases, be a matter of judgement. Furthermore, as the density changes, so will many of the statistical parameters, and a method that is suitable at a higher density may be found inadequate if the population level drops. Shaw (1955) found that Thomas and

Jacob's (1949) recommendation for sampling potato aphids—one upper, middle, and lower leaf from each of 50 plants—was unsatisfactory in Scotland, in certain years, because of the lower densities.

Detailed and now classic accounts of the development of sampling programmes for various insects are given by Morris (1955, 1960), LeRoux and Reimer (1959), Harcourt (1961a, 1964), Lyons (1964), and Coulson et al. (1975, 1976), amongst others. Terrestrial sampling programmes for vertebrates, small mammals and birds are described by Caughley (1977), Smith et al. (1975), and Blondel and Frochet (1987), respectively. For aquatic systems, estuarine and coastal sampling programmes are described for many groups in Baker and Wolff (1987), sampling marine benthos by McIntyre et al. (1984) and zooplankton by Omori and Ikeda (1984).

The next step is to subdivide for sampling purposes. These subdivisions may be orientated with respect to an environmental gradient. If none is apparent, it is often convenient to divide the area into regions of regular shape, but this may be neither possible nor desirable. Ideally, the within-subdivision variability should be much less than the between-subdivision variability in habitat. The habitat must be considered from the biological angle and a decision made as to whether it might need further division. If it is woodland, for example, the various levels of the tree—upper, middle, and lower canopy and probably the tips and bases of the branches—would, on a priori grounds, be considered as potentially different divisions. The aspect of the tree might also be important. In herbage or grassland, if leaves or another small sampling unit is being taken, the upper and lower parts of the plants should be treated separately. Similarly, aquatic samples may be taken from different depths, from different positions with respect to flow, and from different plants or substrates.

It is often of value to take at least two different-sized sampling units (Waters & Hensen, 1959). In benthic surveys, a $0.1\,m^2$ grab sample suitable for sampling macroinvertebrates can itself be core-sampled for the meiofauna prior to processing (McIntyre & Warwick, 1984). In an insect study, one size of sample should be towards the smallest possible limit, e.g. a leaf blade or half a leaf. As a general principle, a higher level of reproducibility is obtained (for the same cost) by taking more smaller units than by taking fewer large ones.[1] Small sampling units may also enable precision to be increased by distinguishing between favourable and unfavourable microhabitats; Condrashoff (1964) found with a leaf miner that the upper and lower leaf surfaces should be considered as separate units. Two examples of any one size-sampling unit should be taken within each sampling plot or subsection. For example, for a field crop a preliminary plan could be as in Fig. 2.1, which gives a total of 160 samples.

In community studies, where it is essential to gain some idea of species diversity

---

[1] The only disadvantage of sampling by small units is the number of zeros that may result at low densities; this truncation may make analysis difficult and has led to the suggestion that larger-sized samples should be taken (Pradhan & Menon, 1945; Spiller, 1952). The decision must be related to the density of the animal, although moderate truncation can be overcome by suitable transformation; in other cases, it may be necessary to increase the size of the unit.

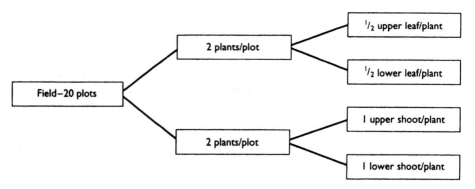

Fig. 2.1 Example sampling plan.

and the species–area relationship, a number of sampling methods may be needed. In benthic sampling, for example, a pilot study might use dredges to obtain a general idea of animal presence and distribution and a limited number of grab samples to estimate densities. During a pilot study, a record should be kept of the cost of each part of the sampling routine, normally expressed in man-hours.

2.1.2 Statistical aspects

Before the data gathered in the preliminary samples can be analysed, some aspects of statistics need to be considered.

2.1.2.1 The normal distribution and transformations

The most important and commonly used of the theoretical distributions is the normal or Gaussian distribution. This has a probability curve that is a symmetrical bell shape. The probability density of a normal variable $P(x)$ is:

$$P(x)\frac{1}{\sqrt{2\pi}\sigma}\exp\left[-\frac{1}{2}\left(\frac{x-\mu}{\sigma}\right)^2\right] \tag{2.1}$$

where μ is the mean and σ the standard deviation.

This distribution is symmetrical about the mean and the shape is determined by the standard deviation. It is important to ecologists because most of the common statistical tests assume the data to be normally distributed. While many distributions obtained from observation: e.g. the heights of humans—are approximately normal, the spatial distribution of individuals is seldom, if ever, normal. The distribution of a field population could approach normality only if the individuals were randomly scattered and the population was so dense or the size of sampling unit so large that considerable numbers were present in each sample. Therefore, in contrast to other distributions described below, the normal distribution is not of interest to ecologists as a means of describing dispersion. Its importance arises solely from the fact that, for most statistical methods to be applied to a set of data, the frequency distribution must be normal. The data should possess the associated properties that the variance is independent of the mean (or, more strictly, the variance is homogeneous, all errors coming from the same frequency curve) and its components are additive. Although analysis of variance is quite robust to deviations from normality—in some respects, more so than the χ^2 test (Reimer, 1959; Abrahamsen & Strand, 1970)—data whose frequency distribution is considerably skewed and with the variance closely related to the mean cannot be analysed without the risk of errors (Beall, 1942).

Ecological data are generally skewed. Such distributions are usually transformed by taking logarithms or square roots. For example, if the square-root transformation were applied to 9, 16, and 64, they would become 3, 4, and 8, and it will be observed that this tends to reduce the spread of the larger values. The interval between the second and third observations (16 and 64) is on the first scale nearly seven times that between the first and second observations; when transformed, the interval between the second and third observations is only four times that between the first and second. It is thus easy to visualize that a transformation of this type would tend to 'push' the long tail of a skew distribution in, so that the curve becomes more symmetrically bell-shaped. If all the observations (x) are positive and no a priori reason exists to choose between the many types of transformation possible, the Box–Cox empirical method can be used (Box & Cox, 1964). The general transformation function from x to x' is:

$$x' = \frac{(x^\gamma - 1)}{\gamma}, \quad \gamma \neq 0$$

$$x' = \ln x, \quad \gamma = 0 \tag{2.2}$$

The best transformation to normality is the value of exponent, γ, which maximizes the log-likelihood (L) equation:

$$L = \frac{v}{2}\ln s_{x'}^2 + (\gamma - 1)\frac{\gamma}{n}\sum \ln x \tag{2.3}$$

where $s_{x'}^2$ is the variance of the transformed distribution and v the degrees of freedom (number of observations − 1).

Examination of the general transformation function (2.2) will show that certain values of γ correspond to commonly used transformations; these are listed in Table 2.1. If the calculated value for γ falls close to that corresponding to a common transformation, then this transformation should be used. Thus, if $\gamma = 0.1$, a logarithmic transformation might be considered appropriate.

It must be stressed, however, that transformation does lead to difficulties, particularly in the consideration of the mean and other estimates (see below). It should not be undertaken routinely, but only when the conditions for statistical tests are grossly violated (LeRoux & Reimer, 1959; Finney, 1973).

The distribution of individuals in natural populations is such that the variance is

Table 2.1 Values of γ or $1 - 1/2b$ obtained by fitting either the Box–Cox transformation function or Taylor's power law and their corresponding transformations.

γ or $1 - \dfrac{1}{2b}$	Transformation
1	linear $y = ax + b$
0.5	square root $y = \sqrt{x}$
0	logarithmic
−0.5	reciprocal square root
−1	reciprocal $y = 1/x$

not independent of the mean. If the mean and variance of a series of samples are plotted, they tend to increase. This relationship has been shown by Taylor (1961, 1965) and Taylor *et al.* (1976, 1978) to fit a power law:

$$s^2 = a\bar{x}^b \tag{2.4}$$

where a and b are constants; a is largely a sampling factor, while b appears to be an index of aggregation characteristic of the species.

This relationship holds for a variety of distributions, from regular through random to highly aggregated. The same relationship between the mean $\bar{x}$ and the variance s^2 was discovered independently by Fracker and Brischle (1944) for quadrat counts of the currant, *Ribes*, and by Hayman and Lowe (1961) for counts of the cabbage aphid, *Brevicoryne brassicae*. The applicability of Taylor's power law is discussed by Rutledge and Swartz (1991). They held that it had no unique advantage over a quadratic of the form:

$$s^2 = a_1\bar{x} + a_2\bar{x}^2 \tag{2.5}$$

and noted that Taylor's power law was particularly poor in predicting the variance for means close to zero. If the goal is to minimize the squared errors in predicting the mean, then better methods are sometimes available. For example, if replicate counts are taken from a Poisson-distributed population, but sampling error allows the mean to vary from count to count, then it can be shown that the variance–mean relationship is:

$$\sigma^2 = \mu + c^2\mu^2 \tag{2.6}$$

where c is the coefficient of variation of the Poisson mean.

Taylor *et al.* (1978) had suggested a split domain model:

$$s^2 = a\mu^b, \quad a^{\frac{1}{1-b}} < \mu$$

and

$$s^2 = \mu, \quad \mu \le a^{\frac{1}{1-b}} \tag{2.7}$$

which would address the poor fit at low mean densities. However, the application of this model can be time-consuming and needs care (Perry & Woiwod, 1992). These authors compared a number of models, including the generalized form:

$$s^2 = \mu + e\mu^f, \quad e \geq 0 \tag{2.8}$$

They concluded that no approach was outstandingly superior, but noted that, with the development of generalized linear models, in which all additive effects and the relationship between mean and variance were specified separately, the need for variance-stabilizing transformations has been made unnecessary.

It seems that Taylor's power law, used with care, offers an approach to finding a general transformation which makes the variance independent of the mean. To do this, the series of means and variances necessary to calculate a and b may be obtained from several sets of samples from different areas, from sets of samples of different sizes, or by combining samples to form different-sized sampling units. The mean and the variance are calculated from the raw data by the usual methods. The constants a and b in equation (2.4) can then be found by non-linear regression, using computer statistics packages, such as Genstat or SAS. An approximate value of b can be found using linear regression to solve:

$$\log s^2 = \log a + b \log x \tag{2.9}$$

Having determined b each variable, x, is transformed to z using:

$$z = x^{1-\frac{1}{2b}} \tag{2.10}$$

An evaluation of this transformation and the determination of its parameters is given by Taylor et al. (1978).

As in the case of the Box–Cox method, an appropriate transformation for different values of $1 - 1/2b$ is given in Table 2.1. Healey and Taylor (1962) give tables for $1 - 1/2b = 0.2, 0.4, 0.6, 0.8$ and for the negative powers. Downing (1979), following an examination of benthic data, noted that b was frequently close to 1.5 and proposed a fourth-root transformation. This proposal was criticized by Taylor (1980).

These are precise transformations, but in practice, where sampling and other errors are fairly large, it will usually be found adequate to transform the data from a regular population by using squares, for a slightly contagious (aggregated) one by using square roots and for a distinctly contagious one by a logarithmic transformation.

In order to overcome difficulties with zero counts in log transformations, a constant (normally 1) is generally added to the original count (x); this is expressed as 'log(x + 1)'. Anscombe (1948) has suggested that a better transformation would be obtained by taking log($x + k/2$), where k is the dispersion parameter of the negative binomial (see below). As k is frequently in the region of 2, in many cases this refinement would have little effect. Andersen (1965) has shown that, if the mean and k are very small (less than 3 and approaching zero, respectively), then the variance will not be stabilized by $k^{1/2}$ or any of the common transformations. However, if independent samples are pooled or the size of the sampling unit increased, the data may be satisfactorily transformed.

It is customary to transform percentages to angles (arc sin), but Reimer (1959) has suggested this is only of value when the probability of finding the individual bearing the attribute (e.g. the leaf having a gall) is uniform within each area (for which a % infestation has been calculated) but varies considerably between the different classes (whose various % infestations one wishes to analyse). Even in such cases, it is necessary only when several of the % points lie outside the 20–80 range. When the various percentages are based on grossly unequal numbers of individuals, they will need to be weighted before the analysis of variance can be applied (Reimer, 1959).

The use of transformations can lead to problems and, as stressed above, transformation should not be routinely undertaken. If the fiducial limits are calculated from the transformed mean, this may be erroneous (Abrahamsen & Strand, 1970). The biological interpretation of estimates based on transformed data is often difficult and may lead to insurmountable problems in the construction of life budgets. There is indeed much to commend the use of the arithmetic mean (i.e. that based on the untransformed data) in population studies (van Emden et al., 1961; Lyons, 1964) and, if the distribution of the animal is random (see below), the fiducial limits are available in tables (Pearson & Hartley, 1958). If data have been transformed, the means of the untransformed and transformed values should be provided; back-transforms of, say, geometric means from logarithmic transformations are more difficult to interpret and contain biases unless the variances are small (Finney, 1973). Beauchamp and Olson (1973) suggest that the bias will be corrected by using the log-normal distribution for detransforming.

Checking the adequacy of the transformation. An adequate transformation should eliminate or considerably reduce two attributes of the data that are easily tested for. These are as follows.
1 Excess skewness of the frequency curve causing large deviations from normality, which can be detected either from a probit plot or by a statistical test (Snedecor & Cochran, 1989).
2 The dependence of the variance on the mean (instability or non-homogeneity of the variance), which may be shown graphically (Fig. 2.2).

Generally, as Hayman and Lowe (1961) have pointed out, 'as non-normality must be extreme to invalidate the analysis of variance it is better to concentrate on stabilising the variance of the samples'. A correct transformation for this property will also ensure the third property necessary for the analysis of variance: the additivity of the variance (Bliss & Owen, 1958). Indeed, all three properties are related and, for practical purposes, the distinction between transformation for normality and

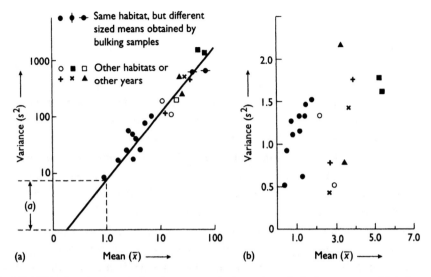

Fig. 2.2 (a) The plot of variance against the mean on a log/log scale to obtain the constant *a* of Taylor's power law, data from samples of olive scales, *Parlatoria oleae*, per twig. (b) The same data transformed to $x^{0.4}$, showing the relative independence of the variance from the mean.

that for stabilizing the variance need not be emphasized. The adequacy of a transformation in stabilizing the variance may be tested for graphically (Fig. 2.2b) or by calculating the correlation coefficient for the two terms (Harcourt, 1961b, 1965).

Alternatively, more precise methods may be used, as described by Forsythe and Gyrisco (1961), who used a non-parametric rank-correlation coefficient to test for non-normality. When normality has been established, the heterogeneity and non-additivity of the variance can be determined, using the tests of Hartley (1950) and Tukey (1949), respectively.

2.2 The sampling programme

2.2.1 The number of samples per habitat unit (e.g. plant, host, or puddle)

There are two aspects to consider: firstly, whether different regions of the unit need to be sampled separately and, secondly, the number of samples within each unit or subunit (if these are necessary) that should be taken for maximum efficiency. Although the habitat unit could, for example, be the fleece of a sheep, a bag of grain, a rock in a stream, or an area of seabed, for convenience, the word plant will, in general, be used in its place in the discussion below.

2.2.1.1 Subdivision of the habitat

If the distribution of the population throughout the habitat is biased towards certain subdivisions but the samples are taken randomly, what LeRoux and Reimer (1959) aptly term 'systematic errors' will arise. This can be overcome either by sampling so that the differential number of samples from each subdivision reproduces in the

samples the gradient in the habitat or by regarding each part separately and subsequently correcting. The question of the estimation of the area or volume of the plant
is discussed in Chapter 4. The amount of subdivision of the plant that various
workers have found necessary varies greatly. On apple, the eggs, larvae, and pupae of
the tortricid moth, *Archips argyrospilus*, were found for most of the year to be randomly distributed over the tree, so that only one level (the lower for ease) needed to
be sampled (Paradis & LeRoux, 1962). In contrast, on the same trees and in the same
years, the immature stages of two other moths showed marked differences between
levels at all seasons (LeRoux & Reimer, 1959). With the spruce budworm, *Choristoneura fumiferana*, Morris (1955) found that there were 'substantial and significant
differences from one crown level to another' and that there was a tendency for eggs
and larvae to be more abundant at the top levels, but there was no significant difference associated with different sides of the same tree. A similar variation with height
was found with the eggs of the larch sawfly, *Pristiphora erichsonii*, although here it was
concluded that, in view of the cost and mechanical difficulties of stratified sampling
at different heights, a reasonable index of the population would be obtained by sampling the mid-crown only (Ives, 1955). Studying all the organisms on aspen, *Populus
tremuloides*, Henson (1954) found it necessary to sample at three different levels of
the crown. Even with field crops, height often needs to be considered: Broadbent
(1948) recommended that potato aphids be estimated by picking three leaves—
lower, middle, and upper—from each plant and, when estimating the population of
the European corn-borer, *Ostrinia nubilalis*, Hudson and LeRoux (1961) showed
that the lower and upper halves of the maize stem needed to be considered separately, the former containing the majority of the larvae. The age of the plant may
also affect dispersion (Hirata, 1962; Peng & Brewer, 1994). In contrast, Teulon and
Cameron (1995) found few significant differences in the distribution of thrips eggs
and adults at different heights in sugar-maple trees.

Aspect is sometimes important; in Nova Scotia in the early part of the season, the
codling moth lays mostly on the south-east of the apple trees, but later this bias disappears (MacLellan, 1962). Aspect has also been found to influence the distribution
on citrus of the long-tailed mealy bug, *Pseudococcus adonidum* (Browning, 1959) and
the eggs of the oak leaf-roller moth, *Archips* (Ellenberger & Cameron, 1977) and of
three species of mite, each of which was most prevalent on a different side (Dean,
1959), but not that of the pine beetle, *Dendroctonus* (Dudley, 1971). Variations in the
spatial distribution of similar species in the same habitat, which complicates a sampling programme designed to record both, has also been recorded for two potato
aphids, by Helson (1958). Some insects are distributed without bias on either side of
·the mid-vein of leaves, so that they may be conveniently subsampled. Similarly,
Nelson *et al.* (1957) record that, when estimating populations of sheep keds, the
fleece of only one side need be sampled. Aspect can also be important in aquatic
systems. For example, in the northern hemisphere north-facing rocks in the intertidal zone offer the coolest, most shady habitat available, and thus tend to favour red
algae and its associated fauna (Baker & Crothers, 1987). Conversely, in Caithness, at

the northern limit of its range, the barnacle *Chthamalus* is more abundant on south-facing slopes. In aquatic habitats, orientation with respect to flow and the degree of exposure are important (Baker & Crothers, 1987). Intertidal communities were classified in terms of their exposure to wave action by Ballentine (1961). Plankton sampling may be undertaken at set depths. However, when a general measure of abundance throughout the water column is required a common practice is to collect what are termed double-oblique plankton-net hauls by lowering and raising the net while fishing (Omori & Ikeda, 1984). For example, Henderson *et al.* (1984) used samples collected in this way to estimate the growth and mortality of larval herring, *Clupea harengus*, which were known to be unevenly distributed with respect to depth (Henderson, 1987).

Occasionally, sampling may be restricted to one part of a plant upon which a large and constant proportion of the population of the animal lives. Wilson (1959) showed that, in Minnesota, 84% of the eggs of the spruce budworm, *C. fumiferana*, are laid on the tips of the branches and, if sampling is confined to these rather than entire branches, sampling time may be reduced by up to 40%.

A hierarchical design is one in which, for example, a number of plants are sampled from each of a number of plots from within a number of fields. If a certain number of samples are collected randomly within each strata this is often termed nested random sampling and is analysed by a nested analysis of variance. There may be two or three hierarchical levels, rarely more. Examples of such sampling programmes are in Bancroft and Brindley (1958), Harcourt (1961a), Buntin and Pedigo (1981), and Steffey *et al.* (1982).

2.2.1.2 The number of samples per subdivision

To determine the optimum number of samples per plant (or part of it) (n), the variance of within-plant samples (s_s^2) must be compared with the variance of the between-plant samples (s_p^2) and set against the cost of sampling within the same plant (c_s) or of moving to another plant and sampling within it (c_p):

$$n_s = \sqrt{\frac{s_s^2}{s_p^2} \times \frac{c_p}{c_s}} \qquad (2.11)$$

If the interplant variance, s_p^2, is the major source of variance and unless the cost of moving from plant to plant is very high, n will be of the order of 1 or less (which means 1 in practice). Interplant variance has been found to be much greater than within-plant variance in many insect populations, e.g. the spruce sawfly, *Diprion hercyniae* (Prebble, 1943), the lodgepole needle miner, *Recurvaria starki* (Stark, 1952), the cabbage aphid, *B. brassicae* (Church & Strickland, 1954), the spruce budworm, *C. fumiferana* (Morris, 1955), the diamondback moth, *Plutella maculipennis* (Harcourt, 1961a), the cabbage butterfly, *Pieris rapae* (Harcourt, 1962), the western pine beetle, *Dendroctonus brevicomis* (Dudley, 1971), the pine chermid, *Pineus pinifoliae* (Ford & Dimond, 1973), and the spider mite, *Panonychus ulmi* (Herbert & Butler, 1973). In most of these examples, the within-plant variance was small, so that only one sample was taken per plant or per stratum of that plant, although, of course, when this is done, within- and between-tree variances cannot be separated. However, with *Eoreuma loftini* on sugar cane, within-plant variance, from stalk to stalk, was found to be high, as was field-to-field variance, the lowest variance being found between samples from within a field (Meagher *et al.*, 1996). On apples, the within-tree variance of insect numbers may also be high, particularly at certain seasons, when as many as seven samples may be taken from a single tree (LeRoux & Reimer, 1959; LeRoux, 1961; Paradis & LeRoux, 1962).

Often, a considerable saving in cost without loss of accuracy in the estimation of the population, but with loss of information on the sampling error, may be obtained by taking randomly a number of subsamples which are bulked before sorting and counting. This is especially true where the extraction process is complex, as with soil samples; Jepson and Southwood (1958) bulked four random, 76.25 mm (3 inch) row samples of young oat plants and soil to make a single 305 mm (1 ft) row sample, which was then washed and the eggs of the frit fly, *Oscinella frit*, extracted. Such a process gave a mean as accurate as that obtained by washing all the 76.25 mm samples separately (at greater cost). Paradis and LeRoux (1962) sampled the eggs of a tortricid moth, *A. argyrospilus*, on apple by bulking 25 cluster samples.

2.2.2 The sampling unit, its selection, size, and shape

The criteria for the sample unit are broadly those of Morris (1955):

1 It must be such that all units of the universe have an equal chance of selection.

2 It must have stability (or, if not, its changes should be easily and continuously measured—as with the number of shoots in a cereal crop).

3 The proportion of the population using the sample unit as a habitat must remain constant.

4 The sampling unit must lend itself to conversion to unit areas.

5 The sampling unit must be easily delineated in the field.

6 The sampling unit should be of such a size as to provide a reasonable balance between the variance and the cost.

7 The sampling unit must not be too small in relation to the animal's size, as this would increase edge-effect errors.

8 The sampling unit for mobile animals should approximate to the average ambit of an individual. This 'condition' is particularly significant in studies on dispersion involving contiguous sampling units, Lloyd (1967) has suggested that a test of the appropriate size would be provided by several series of counts of animals in contiguous quadrats conforming to a Poisson series.

A sampling unit defined in relation to the animal's ambit or territory (e.g. gallery of a bark-beetle) will give different information from one defined in terms of the habitat (e.g. Cole, 1970). This re-emphasizes the need to be very clear as to objectives and hypothesis before commencing a sampling programme.

To compare various sampling units in respect of variance and cost, it is generally convenient to keep one or other constant. The same method of sampling must, of course, be used throughout. From preliminary sampling, the variances of each of the different units (s_u^2) can be calculated; these should then be computed to a common basis, which is often conveniently the size of the smallest unit. For example, if the smallest unit is 1 m of row, then the variance of a 2 m row unit will be divided by 2 and that of 4 m by 4. The costs will similarly be reduced to a common basis (C_u). Then the relative net cost for the same precision for each unit will be proportional to:

$$C_u s_u^2 \tag{2.12}$$

where C_u = cost per unit on a common basis and s_u^2 variance per unit on a common basis. Alternatively, the relative net precision of each will be proportional to the reciprocal of equation 2.12. The higher this reciprocal, the greater the precision for the same cost.

A full treatment of the methods of selecting the optimum-size sampling unit is given in Cochran (1977) and other textbooks, but as population density, and hence variance, is always fluctuating, too much stress should not be placed on a precise determination of the optimum size of the sampling unit. An example of the determination of optimal sample size is Zehnder's (1990) study of larval and adult Colorado potato beetle *Leptinotarsa decemlineata* (Say). He found that, while a five-stem sampling unit was most efficient, a three-stem unit would reduce costs with only a small reduction in sampling precision. Pieters (1977) compared the estimates of mean density and precision of arthropod samples collected from cotton using a D-Vac suction sample and found that mean estimates increased with reduced sample size while precision remained unchanged, leading to the conclusion that small sample sizes are better. Similarly, when sampling for benthic macrofauna, no advantage has been generally shown for $0.2\,m^2$ over $0.1\,m^2$ grab samplers (McIntyre *et al.*, 1984). With insects on plants, the nature of the plant usually restricts the possible sizes to, for example, half leaf, single leaf, or shoot (see p. 149). Because the sample size range initially chosen for examination is often arbitrary, it may not include the optimal sample size.

The shape of the sampling unit when this is of the quadrat type, rather than a biological unit, is theoretically of importance because of the bias introduced by edge effects. These are minimal with circles, maximal with squares and rectangles and intermediate with hexagons (Seber, 1982), because they are proportional to the ratio of sample unit boundary length to sampling unit area. If the total habitat is to be divided into numbered sampling units (for random number selection), then circular units are impractical, because of the gaps, and it is doubtful if the reduction of error from the use of hexagons normally justifies the difficulties of layout. Clearly, the larger the sampling unit, the proportionally smaller the boundary edge effect. The size of the organism will also influence this effect: the larger it is in relation to the sample size, the greater the chance of an individual lying across a boundary. This problem has been investigated for subcortical insects, where the damage to the edge individuals by the punch and the curved nature of the sampled substrate pose special problems (Safranyik & Graham, 1971). In general, edge effects can be minimized by a convention (e.g. of the animals crossing the boundaries, only those on the top and left-hand boundaries are counted).

2.2.3 The number of samples

Precision is measured by the scatter about the mean of the results obtained and is sometimes expressed as the coefficient of variation of the samples. This is different from the accuracy of a method, which measures how close the estimate is to the actual population. The total number of samples depends on the degree of precision required. This may be expressed either in terms of achieving a standard error of the mean within a predetermined magnitude or as a probability that the estimated mean is within a selected value of the mean (Karandinos, 1976; Ruesink, 1980). For many purposes, a standard error of 5% of the mean is satisfactory. Within a homogeneous habitat, the number of samples (n) required is given by:

$$n = \left(\frac{s}{E\bar{x}} \right)^2 \tag{2.13}$$

where s = standard deviation, $\bar{x}$ = mean, and E = the predetermined standard error as a decimal of the mean (i.e. for a standard error ±5%, then $E = 0.05$). This expression compares the standard deviation (s) of the observations with the standard error (E) acceptable for the contrasts we need to make; it will be noted from this equation, that, in any given situation, the value of the standard error will change with the square root of the number of samples: thus a large increase in n is necessary to bring about a small improvement in s.

Where sampling is necessary at two levels, e.g. a number of clusters per tree, the number of units (n) that need to be sampled at the higher level, e.g. trees (LeRoux & Reimer, 1959; Harcourt, 1961a), is given by:

$$n_t = \frac{\left(\dfrac{s_s^2}{n_s} \right) + s_p^2}{(\bar{x} + E)^2} \tag{2.14}$$

where n_t = the number of samples within the habitat unit (calculated as above), s_s^2 = variance within the habitat unit, s_p^2 = variance between the habitat unit (= interplant variance), $\bar{x}$ = mean per sample (calculated from the transformed data and given in this form), and E is as above.

If the dispersion of the population is well described by either the Poisson or negative binomial distribution, the desired number of samples is given, respectively, by:

$$N = \frac{1}{(E^2 \bar{x})}$$

and:

$$n = \frac{\frac{1}{\bar{x}} + \frac{1}{k}}{E^2} \tag{2.15}$$

where k is the dispersion parameter of the negative binomial (see p. 31).

The second approach uses confidence limits, so that the estimate is within a selected distance from the mean with a given probability. The general formula then becomes:

$$n = \left(\frac{ts}{D\bar{x}} \right)^2 \tag{2.16}$$

where t = 'Student's t' of standard statistical tables: its value depends on the number of samples and approximates to 2 for more than 10 samples at the 5% level — and D = the predetermined half-width of the confidence limits for the estimation of the mean expressed as a decimal (e.g. ±10% = 0.1). It will be seen that normally this gives a similar estimate to equation 2.13, provided the values of E and D are adjusted to accord with their meanings. The procedure is, perforce, somewhat approximate, depending on the preliminary estimates of the mean and standard deviation, and the inclusion of t does perhaps give it a slightly bogus air of precision! Additionally, there is normally a biological aspect to consider: as population characteristics change with time, so will the optimal number of samples (e.g. Bryant, 1976; Kapatos *et al.*, 1977).

If the distribution of the animal can be described by a Poisson or negative binomial distribution or if Taylor's power law applies, then s^2 can be substituted by other expressions (Buntin, 1994). For example, the formula when Taylor's power law applies is:

$$n = t^2 a \bar{x}^{b-2} D^{-2} \tag{2.17}$$

where parameters a and b are those in equation (2.4).

Equation 2.17 should be used with caution, because, as discussed in section 2.1.2, Taylor's power law may give a poor description of the variance–mean relationship at low densities; Riddle (1989) showed that equation (2.17) substantially underestimated the required sample size for low-density benthic animals. Shelton and Trumble (1991) concluded that equation 2.17 should only be used when $b < 2.0$. For convenience, Table 2.2 gives the number of samples for different levels of precision, mean density, and clumping, as calculated by Elliott and Drake (1981) using equation 2.17 with $a = 1$. Because the estimated number of samples required is the product of the tabulated value and the power law parameter, a, the value of which can itself change with sampling conditions and sample size (Taylor, 1984), the estimated sample number is sensitive to sample size, which must be chosen appropriately.

Another type of sampling programme concerns the measurement of the frequency of occurrence of a particular organism or event — for example, the frequency of occurrence of galls on a leaf or of a certain genotype in the population (Oakland, 1950; Cornfield, 1951; Henson, 1954; Cochran, 1977). Before an estimate can be

Table 2.2 Number (n) of sampling units per sample required for different levels of precision (95% confidence limits (CL) expressed as a percentage of the sample mean) for different values of the mean number caught per sample (x) and for different values of b (equation 2.4). The value of a in equation 2.17 is assumed to be 1. To find n, multiply the value of n found in the table by the actual value of a (after Elliott & Drake, 1981).

Precision D%	Sample mean (x)								Index of aggregation (b)
	0.5	1	5	10	20	50	100	1000	
10	800	400	80	40	22	10	6	1	1
20	200	100	22	12	7	3	1	1	1
40	50	26	7	4	1	1	1	1	1
60	24	13	3	1	1	1	1	1	1
80	14	8	1	1	1	1	1	1	1
100	10	6	1	1	1	1	1	1	1
10	566	400	178	126	90	56	40	15	1.5
20	140	100	45	32	24	16	12	5	1.5
40	35	26	13	10	8	6	5	1	1.5
60	17	13	7	5	4	2	1	1	1.5
80	11	8	5	3	2	1	1	1	1.5
100	8	6	2	2	1	1	1	1	1.5
10	400	400	400	400	400	400	400	400	2
20	100	100	100	100	100	100	100	100	2
40	26	26	26	26	26	26	26	26	2
60	13	13	13	13	13	13	13	13	2
80	8	8	8	8	8	8	8	8	2
100	6	6	6	6	6	6	6	6	2
10	283	400	894	1265	1789	2828	4000	12649	2.5
20	70	100	224	316	447	707	1000	3162	2.5
40	19	26	56	79	112	177	250	791	2.5
60	10	13	26	35	50	79	111	351	2.5
80	6	8	14	20	28	44	63	198	2.5
100	4	6	9	13	18	28	40	126	2.5

For definition of D, see equation 2.16.

made of the total number of samples required, an approximate value of the probability of occurrence must be obtained. For example, if it is found in a preliminary survey that 25% of the leaves of oak trees bear galls, the probability is 0.25. The number of samples (N) is given by:

$$N = \frac{t^2 p(1-p)}{D^2}$$
(2.18)

where p = the probability of occurrence (i.e. 0.25 in the above example) and D is as defined for equation (2.16).

If it is found that the leaves (or other units) are distributed differently in the different parts of the habitat, they should be sampled in proportion to the variances. For example, Henson (1954) found from an analysis of variance of the distribution of the leaf bunches of aspen that the level of the crown from which the leaves had been drawn caused a significant variation and, when this variance was portioned into levels, the values were: lower 112 993, middle 68 012, upper 39 436. Therefore leaf bunches were sampled in the ratio of 3 : 2 : 1 from these three levels of the crown. Note that, if the Taylor's power law constants a and b are substituted in equation (2.17), using the equalities $p = a$ and $1 - p = \bar{x}^{-b-2}$, we obtain equation (2.8).

Worked examples of the calculation of sample number for a variety of sampling programmes are given by Greenwood (1996).

2.2.4 The pattern of sampling

Again, it is important to consider the object of the programme carefully. If the aim is to obtain estimates of the mean density for use in, for example, life-tables, then it is desirable to minimize variance. But, if the dispersion (= distribution = pattern) of the animal is of prime interest, then there is no virtue in a small variance.

In order to obtain an unbiased estimate of the population, the sampling data should be collected at random, that is, so that every sampling unit in the universe has an equal chance of selection. In the simplest form—the unrestricted random sample—the samples are selected by the use of random numbers from the whole area (universe) being studied (random number tables are in many statistical works or may be generated using a computer or, failing these sources, the last two figures in the columns of numbers in most telephone books provide a substitute). The position of the sample site is selected on the basis of two random numbers giving the distances along two coordinates; the point of intersection is taken as the centre or a specified corner of the sample. If the size of the sample is large compared with the total area, then the area should be divided into plots, which will be numbered and selected using a single random number (e.g. Lloyd, 1967). Such a method eliminates any personal choice by the worker, whose bias in selecting sampling sites may lead to large errors (Handford, 1956).

However, a random-choice method is not an efficient way to minimize the variance, since the majority of the samples may turn out to come from one area of the field. The method of stratified random sampling is therefore to be preferred for most ecological work (Yates & Finney, 1942; Abrahamsen, 1969); here the area is divided up into a number of equal-sized subdivisions or strata and one sample is randomly selected from each stratum. Alternatively, if the strata are unequal in size, the number of units taken in each part is proportional to the size of the part; this is referred to as self-weighting (Wadley, 1952). Such an approach maximizes the accuracy of the estimate of the population, but an exact estimate of sampling error can only be obtained if additional samples are taken from one (or two) strata (Yates & Finney, 1942). The taking of one sample randomly and the other a fixed distance from it has been recommended by Hughes (1962) as a method of mapping aggregations. The fixed distance must be less than the diameter of the aggregations, which are assumed to be circular; the standard error cannot be calculated. However, the method has been found useful for soil and benthic faunas (Gardefors & Orrage, 1968).

When it is apparent that there are systematic variations in the density of the study animal across the study area, stratified sampling should be used. In general, the strata are chosen to minimize within-stratum variance. Individual strata need not form continuous patches within the study area. When the habitat is stratified, biological knowledge can often be used to eliminate strata in which few animals would be found. Such a restricted universe will give a greater level of precision for the calculation of a mean than an unrestricted and completely random sample with a wide variance. Prebble (1943) found that, with a pine sawfly, satisfactory estimates of the pupae were only obtained if sampling was limited to the areas around the bases of the trees. The variance of completely random sampling throughout the whole forest was too great, as many areas were included that were unsuitable pupation sites (see also Stark & Dahlsten, 1961).

The other approach is the systematic sample, taken at a fixed interval in space (or time). In general, such spatial data cannot be analysed statistically, but Milne (1959) has shown that, if the centric systematic area sample is analysed as if it were a random sample, the resulting statistics are 'at least as good, if not rather better' than those obtained from random sampling. The centric systematic sample is the one drawn from the exact centre of each area or stratum and its theoretical weakness is that it might coincide with some unsuspected systematic distribution pattern. As Milne points out, the biologist should, and probably would, always watch for any systematic pattern, either disclosing itself as the samples are recorded on the sampling plan or apparent from other knowledge. Such a sampling programme may be carried out more quickly than the random method and so has a distinct advantage from the aspect of cost (see also p. 19). Systematic sampling is often used in marine studies where the primary aim is to map distribution with respect to environmental gradients or suspected sources of pollution. In the absence of environmental gradients, marine benthic or plankton sampling is often undertaken at regularly distributed stations (McIntyre et al., 1984; Omori & Ikeda, 1984). This allows easier contour mapping of animal density, particularly if computers are used.

An example of an unbiased systematic method is given by Anscombe (1948). All the units (e.g. leaves) are counted systematically (e.g. from top to bottom and each stem in turn); then, every time a certain number (say, 50) is reached, that unit is sampled and the numbering is commenced again from 1; only one allocation of a random number is needed and that is the number (say, somewhere between 1 and 20) allotted to the first unit.

Biologists often use methods for random sampling that are less precise than the use of random numbers, such as throwing a stick or quadrant or the haphazard selection of sites. Such methods are not strictly random; their most serious objection is that they allow the intrusion of a personal bias and quite frequently marginal areas tend to be undersampled.

It may be worthwhile doing an extensive trial comparing a simple haphazard method with a fully randomized or systematic one, especially if the cost of the latter is high when compared with the former. Spiller (1952) found that scale insects on citrus leaves could be satisfactorily sampled by walking round the tree, clockwise and then anticlockwise, with the eyes shut and picking leaves haphazardly. For assessing the level of eggs of the red boll-worm, *Diparopsis castanea*, to determine the application of control measures, Tunstall and Matthews (1961) recommended two diagonal traverses across the field, counting the eggs at regular intervals.

Bias may intrude due to causes other than personal selection by the worker. Grains of wheat that contain the older larvae or pupae of the grain weevil, *Sitophilus granarius*, are lighter than uninfested grains. The most widespread method of sampling is to spread the grains over the bottom of a glass dish and then scoop up samples of a certain volume; as the lighter infected grains tend to be at the top, this method can easily overestimate the population of these stages (Howe, 1963). In contrast, for the earlier larval instars before they have appreciably altered the weight of the grain, such a simple method gives reliable results (Howe, 1963); it was undoubtedly this difference that led to Krause and Pedersen (1960) stressing the need for samples to contain a relatively high proportion of the same stage if good replication was to be obtained. The distribution of parasitized animals may be such that they are more vulnerable to capture. For example, roach infested with *Ligula intestinalis* swim close to the surface.

When choosing a sampling method, it is important to choose a method with a low and stable level of bias, for this cannot be reduced by increasing sample number (Dahlsten *et al.*, 1990).

2.2.5 The timing of sampling

In the absence of sampling constraints, the seasonal timing of sampling will be determined by the life cycle (e.g. Morris, 1955). Marine sampling is often constrained by climatic conditions, as fishing and bottom-sampling cannot be done in heavy seas. In extensive work, when only a single stage is being sampled, it is obviously most important that this operation should coincide with peak numbers (e.g. Edwards, 1962). This can sometimes be determined by phenological considerations (Unterstenhöfer, 1957), but the possibility of a control population in an outdoor cage (Harcourt, 1961a) to act as an indicator should be borne in mind. The faster the development rate, the more critical the timing. With intensive studies that are designed to provide a life-table, regular sampling will be needed throughout the season. It is not always realized that the time of day at which the samples are taken may also have a considerable effect. The diurnal rhythms of the insects may cause them to move from one part of the habitat to another, as Dempster (1957) found with the Moroccan locust, *Dociostaurus maroccanus* (Fig. 2.3).

Many grassland insects move up and down the vegetation, not only in response to weather changes, but also at certain times of the day or night (p. 268), and during the day quite a proportion of active insects may be airborne (see the observations of Southwood *et al.* (1961) on the numbers of adults of the frit fly, *O. frit*, on oats). There is a marked periodicity of host-seeking behaviour in many bloodsucking invertebrates (e.g. Camin *et al.*, 1971; Corbet & Smith, 1974). Diurnal changes in activity and distribution are common in aquatic organisms. Plankton may become

Fig. 2.3 The variation in the distribution of adults of the Moroccan locust *Dociostaurus maroccanus* at different times of the day: (a) early morning before 9.30 hours; (b) morning, after 9.30 hours and after noon. The histograms show the relative numbers on bare ground and areas with moderate and dense vegetation (after Dempster, 1957).

concentrated in surface waters at night and fish that remain hidden by day become active and vulnerable to trapping at night. Aquatic insects may emerge at a particular time of day. The ecologist may find that some of his/her sampling problems can be overcome, or at least additional information gained, if the work is done at night, dusk or dawn, rather than during conventional working hours.

2.3 Data processing

A sampling plan needs to consider the processes involved in data acquisition, organization, analysis, and presentation. Smooth and rapid progress along this chain is aided by the use of computers, but only if the data can be easily transferred between software. If different software products are used, then compatibility must be considered. The rapid improvement in the power of computers such as the IBM personal computer (PC) and the Apple Macintosh has allowed ecologists to undertake most of their computation on small, often portable, machines. These PCs can use powerful statistical and data manipulation packages, such as SAS, SPSS, Systat, or Genstat. For an ecologist, the statistics package of choice should handle non-linear models, time-series analysis, ordination techniques and a range of cluster analyses, as well as general statistical procedures.

Both software and hardware capability needs to be considered for each stage of the study. During data acquisition it is important to assess the data storage and processing requirements. Data have only been acquired when they have been processed into a usable form. Data processing and input rates are often different. An example of a common problem is from a study of salmon movement through a fish pass (P.A. Henderson, unpublished). Over a 10-week period, about 1680 h of video film was recorded of the water in the fish pass. It was originally planned to use an image-analysis program to identify and count the fish on the video. However, at the end of the fieldwork, it was found that the image-analysis software worked at less than one-tenth of real time, requiring >16 800 h of computer time to analyse the images. In practice, this meant that image analysis would only just be complete by the beginning of the next salmon migratory season. The only way to acquire the fish-count data within the required time was to employ three people for 6 weeks to view the video film.

Many software products are available for data organization. Those most frequently used by ecologists are termed spreadsheets. Examples of these programs are EXCEL and LOTUS 1-2-3. When choosing the product to use, consider carefully its capability to hold, manipulate, and convert the data. Ecological data sets of animal abundance can often be organized within a three-dimensional array of species (or other taxa) by station (locality) by time. This is the data structure spreadsheets are designed to handle. However, spreadsheets often have a limit on the number of rows or columns that can be used. Ecological data sets from high-diversity communities may easily exceed the upper limit of the system. If a spreadsheet has the capability to hold the data sets, then it is often the preferred software for data organization and elementary analysis, for spreadsheets allow rapid production of summary statistics, tables, and graphs. Specialist database applications, such as ACCESS, DB4, or PARADOX, can also be used to hold and organize data. These programs are better than spreadsheets if the worker needs to relate together different types of data and undertake operations such as sorting the data into different groups. They are usually inferior to spreadsheets in their ability to undertake calculations and plot graphs. Before committing to a particular database or spreadsheet application, the ability to move data into other programs must be considered. The three most common requirements are to export data to statistical, graphical, or word-processing packages. Manufacturers generally claim this is easier than it is. Large statistical software packages often offer their own spreadsheet capability. There is an advantage in using a group of programs all written for the same operating system, such as Windows 95 or 98. A common problem is backward compatibility. Most new software is designed to use data exported from existing programs. Often, however, data cannot be exported from recent to older programs.

2.4 Jackknife and bootstrap techniques

It is frequently difficult to estimate the accuracy of ecological indices because no equation exists for the variance of the statistic and the sampling distribution is unknown. Jackknife and bootstrap methods, which offer a non-parametric means of estimating parameters, are now widely used by ecologists to estimate the accuracy of population parameter estimates and indices for diversity or similarity. Both techniques are based on the repeated estimation of the parameters of interest on a sub-sample of the data, for which a computer is essential. Perhaps their most important application is for the estimation of error limits and confidence intervals for estimated parameters.

The jackknife method uses less computational effort than a bootstrap analysis and is thus often quicker and easier to undertake. However, it cannot be used to estimate confidence intervals. The general scheme is as follows.

1 Consider a situation where a parameter is to be estimated from n samples, e.g. a species diversity index from 10 kicknet samples from a stream.

2 Use all n samples to calculate the parameter of interest, E, the diversity index.

3 Now remove the samples one at a time and recalculate the parameter of interest, E_i.

4 Calculate the pseudovalues for the parameter of interest: $\phi_i = nE - (n-1)E_i$, where ϕ = pseudovalue for jackknife estimate i, n = the total number of replicates, E = the estimate for all n samples, E_i = the estimate with sample i removed.

5 Estimate the mean and standard error of the parameter of interest from the n pseudovalues. The jackknife estimate of bias is E minus the mean of the pseudovalues, and the estimate of the standard error of the sample is simply the standard error of the pseudovalues:

$$SE = \sqrt{\frac{\sum(\phi_i - \bar{\phi})^2}{n(n-1)}}$$

where $\bar{\phi}$ is the mean of the pseudovalues.

The bootstrap method differs from the jackknife in the means of sampling the original data set. The subsets of observations used to repeatedly calculate the parameter of interest are each selected at random, with replacement from the original data set of n observations and with each observation only included once. This procedure is repeated, often hundreds of times, so that many combinations of the original observations are generated. For example, if a sample comprised 10 species and 100 individuals, then each subset of data would be a randomly selected 100 individuals taken from a population that represented all the species according to their original frequency of observation. Each of these samples is then used to generate estimates of the parameter of interest (e.g. a diversity index), as for the jackknife method. The bootstrap estimate of the bias is simply found by subtracting the observed parameter value from the mean parameter obtained from the bootstrap replicates. The estimate of the standard error of the sample is simply the standard error of the replicate parameter estimates:

$$SE = \sqrt{\frac{\sum(\phi_i - \bar{\phi})^2}{b-1}}$$

where $\bar{\phi}$ is the mean of the replicates and b is the number of bootstrap samples taken. A number of methods can be used to estimate confidence intervals. The percentile bootstrap is the simplest and most commonly used. The 2.5 and 97.5 percentiles of the bootstrap distribution of the parameter are used as the 95% confidence intervals. For example, if 1000 bootstrap replicates have been taken, then these are arranged in order of magnitude and the 2.5 percentile calculated as the average of the 25th and 26th smallest values and the 97.5 percentile as the average of the 975th and 976th smallest values.

Bootstrap and jackknife estimates are used to estimate species richness (Chapter 13), population size (Chapter 3), and population growth rates (Chapter 12). Introductions to bootstrap and jackknife methods are given in Efron and Tibshirani (1986), Potvin and Roff (1993), and Scheiner and Gurevitch (1993).

2.5 Dispersion

The dispersion of a population — the description of the pattern of the distribution or disposition of the animals in space — is of considerable ecological significance. Not only does it affect the sampling programme and the method of analysis of the data, but it may be used to give a measure of population size (nearest-neighbour and related techniques) and, in its own right, is a description of the condition of the population. Changes in the dispersion pattern should be considered alongside changes in size when interpreting population dynamics. For example, if a mortality factor reduces the clumping of a sessile organism, this is an indication that it acts most severely on the highest densities, or, if the dispersion of a population becomes more regular, then intensification of competition should be suspected (Iwao, 1970a). An understanding of dispersion is vital in the analysis of predator–prey and host–parasite relationships (Crofton, 1971; Murdie & Hassell, 1973; Anderson, 1974; Hassell & May, 1974).

2.5.1 Mathematical distributions that serve as models

It is necessary to outline some of the mathematical models that have been proposed to describe the distribution of organisms in space; for a fuller treatment, reference should be made to early works, such as Anscombe (1950), Wadley (1950), Cassie (1962), and Katti (1996), to other papers cited in the text, to textbooks, e.g. Bliss and Calhoun (1954), and to Patil and Joshi's (1968) dictionary of distributions.

2.5.1.1 Binomial family

In the binomial distribution the variance is less than the mean, for the Poisson it is equal to the mean, and for the negative binomial it is greater than the mean. A full discussion of variance and its calculation will be found in most statistical textbooks (e.g. Bailey, 1959; Zar, 1984), but the estimated variance (S^2) of a distribution may be calculated using:

$$s^2 = \frac{\sum x^2 - \left[\dfrac{\left(\sum x\right)^2}{N}\right]}{N-1} \tag{2.19}$$

where Σ = the sum of, x = various values of the number of animal/sample, and N = number of samples.

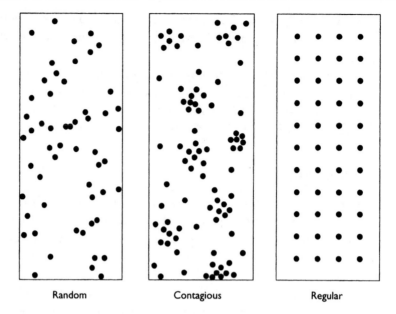

Random   Contagious   Regular

Fig. 2.4 Different types of spatial distribution.

The central place in this family is occupied by the Poisson series, which describes a random distribution (Fig. 2.4). It is important to realize that this does not mean an even or uniform distribution (Fig. 2.4), but that there is an equal probability of an organism occupying any point in space and that the presence of one individual does not influence the distribution of another. Because for the Poisson distribution the variance (S^2) is equal to the mean ($\bar{x}$), the probability of finding a certain number (x) of animals is described by one parameter, as follows:

$$p_x = e^{-\bar{x}} \frac{\bar{x}^x}{x!} \tag{2.20}$$

where e = base of natural (Napierian) logarithms. The goodness of fit of a set of data to the Poisson distribution may be tested by a χ^2 test on the observed and expected values.

Occasionally, it may be found that the variance is less than the mean; this implies a more regular (or uniform or even) distribution than that described by a Poisson series (Fig. 2.4).

Most commonly in ecological studies the variance will be found to be larger than the mean, that is, the distribution is contagious[2] (Fig. 2.4) — the population is

---

[2] The term 'contagious' is a mathematical one coined in connection with work on epidemiology and has certain implications that, to some extent, make its use in ecology inappropriate (Waters & Henson, 1959). An alternative is the term 'overdispersion', first introduced into ecology by Romell (1930), with its opposite — for more uniform spacing — 'underdispersion'; unfortunately, however, the use of these terms has been reversed by some ecologists. Therefore, the terms used here are contagious and regular, which are also those commonly used in plant ecology (Greig-Smith, 1978).

clumped or aggregated. Many contagious populations that have been studied can be adequately expressed by the negative binomial distribution. Examples of its use in insect and marine benthic populations are given in Bliss and Owen (1958), Lyons (1964) and Harcourt (1965). This distribution is described by two positive parameters, c and the exponent k, which is a measure of the amount of clumping and is often referred to as the dispersion parameter. The mean and variance of the negative binomial are given in terms of c and k by:

$$\mu = \frac{k}{c} \tag{2.21}$$

$$\sigma^2 = \frac{k}{c} + \frac{k}{c^2} \tag{2.22}$$

Defining for convenience $p = c/(c+1)$ and $q = 1 - p$, the probabilities of 0, 1, 2, 3, . . . events (animals per sample) are given by the terms of the expansion of:

$$p^k (1-q)^{-k} \tag{2.23}$$

for which the first few terms are

$$p^k \left\{ 1, kp, \frac{k(k+1)}{2!} q^2, k(k+1)\frac{(k+2)}{3!} q^3, \ldots \right\}. \tag{2.24}$$

Generally, for natural populations, values of k are in the region of 2; as $k \to \infty$, the distribution tends to the Poisson, whilst, as $k \to 0$, the distribution tends to the logarithmic series. The value of k is not a constant for a population, but often increases with the mean (Anscombe, 1949; Bliss & Owen, 1958; Waters & Hensen, 1959) (see p. 44). The truncated Poisson may be of value where the distribution is non-random, but the data are too limited to allow the fitting of the negative binomial (Finney & Varley, 1955).

Calculating k *of the negative binomial.* The value of k may be computed by several methods (Anscombe, 1949, 1950; Bliss & Fisher, 1953; Debauche, 1962; Legay, 1963). Three are presented here, two approximate and one more accurate. Another method is described by Katti (1966). This diversity of methods is because the maximum-likelihood estimation does not lead to a simple formula.
1 Having calculated the mean and the variance of the population using equation (2.19), k can be estimated using equations (2.21) and (2.22). The efficiency of this estimate for different values of k and x can be seen in Fig. 2.5. It is most useful for low-density populations.
2 k can be found iteratively by solving on a computer:

$$\log\left(\frac{N}{n_0}\right) = k\log\left(1+\frac{\bar{x}}{k}\right) \tag{2.25}$$

where N = total number of samples, n_0 = number of samples containing no animals, other parameters as above. A reasonable value of k to start with is 2. It will be seen from Fig. 2.5 that this method is efficient for most populations with very small

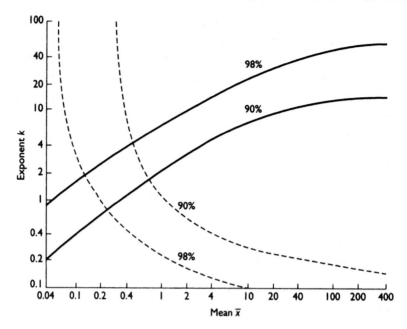

Fig. 2.5 Large sample efficiencies of the estimation of k of the negative binomial distribution by methods 1 (—) and 2 (- - -) (redrawn from Anscombe, 1950).

means, but for large ones only where there is extensive clumping. This can be expressed another way by saying that about one-third of the samples must be blank (empty) if the mean is below 10; with larger means, more blank samples must have been found for the method to be reasonably efficient.

3 The maximum-likelihood estimate of k is obtained by solving iteratively on a computer:

$$N \log_e\left(1 + \frac{\bar{x}}{k}\right) = \sum\left(\frac{Ax}{k+x}\right)$$

(2.26)

where $\log_e$ = natural logarithms, Ax = the sum of all frequencies of sampling units containing more than x individuals (e.g. $A_6 = \Sigma f_7 + f_8 + f_9$). Shenton and Wallington (1962) have shown that even this maximum-likelihood estimator of k will have a bias if the mean is small and k large.

Testing the fit of the negative binomial. When the value of k has been found, the agreement between the negative binomial series as a model and the actual distribution can be tested in three ways.

The expected frequencies of each value calculated using equation (2.24) may be compared by a χ^2 test with the actual values (Bliss & Fisher, 1953). An alternative expression for p_x using the gamma function $\Gamma(k)$ is:

$$p_x = \frac{\Gamma(k+x)}{x!\,\Gamma(k)}\left(\frac{\bar{x}}{\bar{x}+k}\right)^x\left(\frac{k}{k+\bar{x}}\right)^k$$

(2.27)

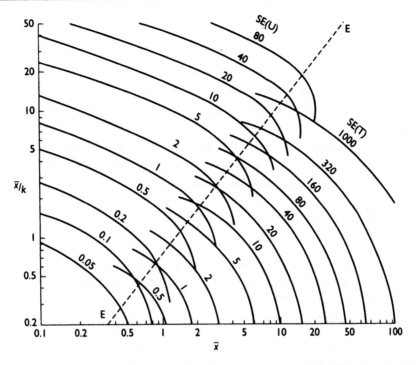

Fig. 2.6 Testing the fit of the negative binomial distribution — the standard errors of T and U for $N = 100$. For other values of N multiply the standard error by $10/\sqrt{N}$. The line E–E represents the contour along which the efficiencies of the methods are equal; the use of the second-moment test is preferable in the area to the left of the line and of the third-moment test in the area to the right (after Evans, 1953).

The χ^2 test has three fewer degrees of freedom than the number of comparisons that are made between expected and actual frequencies; those with small expectations are generally pooled, but Pahl (1969) considers that this is not necessary. Computer programs to estimate k by iteration and test the fit to a negative binomial model are widely available.

Such a comparison between actual and expected frequencies may be distorted by irregularities due to chance; there are two alternative tests based on the difference between the actual and expected moments (mean, variance, or skewness) compared with their standard errors (Anscombe, 1950; Bliss & Fisher, 1953). Which of these tests is most efficient will depend on the sizes of $\bar{x}$ and k. Evans (1953) has shown that with small means, unless k is large, the most efficient test is that based on the second moment (variance); for other values of $\bar{x}$ and k, the test based on the third moment (skewness) should be used. The actual dividing line between the two tests (E–E) is given in Fig. 2.6.

The test based on the second moment involves the calculation of the statistic U, which is the difference between the actual variance and the expected variance, given by:

$$U = s^2 - \left(\bar{x} + \frac{\bar{x}^2}{k} \right) \tag{2.28}$$

the value of k being derived by the second method above. If U is significantly less than its standard error ($\mathrm{SE}(U)$), most conveniently calculated from Fig. 2.6 (the exact formula is given by Evans (1953)), then the negative binomial may be taken as a satisfactory model.

The third-moment test involves the calculation of T, the difference between the actual third moment (skewness) of the data and its value predicted from the first two moments (mean and variance) of the same sample:

$$T = \left(\frac{\sum fx^3 - 3\bar{x} \sum fx^2 - 2\bar{x}^2 \sum fx}{N} \right) - s^2 \left(\frac{2s^2}{\bar{x}} - 1 \right) \tag{2.29}$$

This should be compared with its standard error (the square root of its large variance) and, if the negative binomial is a satisfactory model, T will be significantly smaller than the SE. The standard error $\mathrm{SE}(T)$ may be approximately read off from Fig. 2.6 or calculated using:

$$\mathrm{SE}(T) = \sqrt{ \frac{ 2\bar{x}(k+1)\dfrac{\bar{x}^2}{k^2}\left(1 + \dfrac{\bar{x}}{k}\right)^2 \left[2\left(3 + 5\dfrac{\bar{x}}{k}\right) + 3k\left(1 - \dfrac{\bar{x}}{k}\right) \right] }{N} } \tag{2.30}$$

where the symbols are as above and the value of k is found using equation (2.26). A large positive value of U or T means that the actual distribution is more skew than that described by the negative binomial, and a large negative value means that the actual distribution is less skew.

Calculating a common k. Samples may be taken from various fields or other units and each will have a separate k. The comparison of these and the calculation of a common k (if there is one) will be of value in transforming the data for the analysis of variance and for sequential sampling (see p. 53). The simplest method is the moment or regression method (Bliss, 1958; Bliss & Owen, 1958). Two statistics are calculated for each unit:

$$x' = \bar{x}^2 - \left(\frac{s^2}{N} \right) \tag{2.31}$$

$$y' = s^2 - \bar{x} \tag{2.32}$$

where $\bar{x}$ = the mean, s^2 = variance, and N = number of individual counts on which $\bar{x}$ is based. When y' is plotted against x' (Fig. 2.7) (including occasional negative or zero values of y'), the regression line of y' on x' passes through the origin and has the slope 1. An approximate estimate of the common k (k_c) is given by:

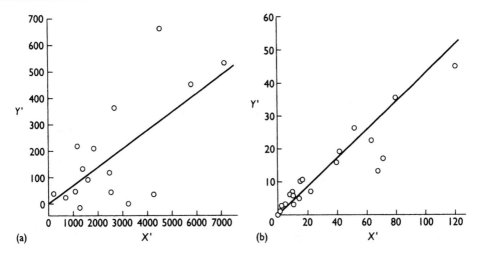

Fig. 2.7 Regression estimate of a common k for: (a) Colorado beetle, *Leptinotarsa*, in eight plots within each of 16 blocks; (b) wireworms (Coleoptera, Elateridae) in 175 sampling units in each of 24 irrigated fields (after Bliss & Owen, 1958).

$$\frac{1}{k_c} = \frac{\sum y'}{\sum x'} \tag{2.33}$$

It may be apparent from the plotting of y' on x' that a few points lie completely outside the main trend, and therefore, although their exclusion will mean that the resultant k is not common to the whole series of samples, it is doubtful if the k derived by including them would really be meaningful.

A further graphical test of the homogeneity of the samples is obtained by plotting $(1/k)$ (y'/x') against the mean, $\bar{x}$, for each sub-area or group of samples. If there is neither trend nor clustering (Fig. 2.8), we may regard the fitting of a common k as justified.

Fairly rough estimates of k_c are usually adequate, but in critical cases a weighted estimate of k_c should be obtained and whether or not this lies within the sampling error can be tested by the computation of χ^2. The method is given by Bliss (1958) and Bliss and Owen (1958). The latter authors describe another method of calculating a common k that is especially suitable for field experiments arranged in restricted designs (e.g. randomized blocks).

2.5.1.2 Logarithmic and other contagious models

A number of other mathematical models have been developed to describe various non-random distributions, and several of these may have more than one mode. Anscombe (1950) and Evans (1953) have reviewed these and they include the Thomas, Neyman's types A, B, and C, and the Polya–Aeppli distributions. The Thomas (1949) type is based on the assumption of randomly distributed colonies

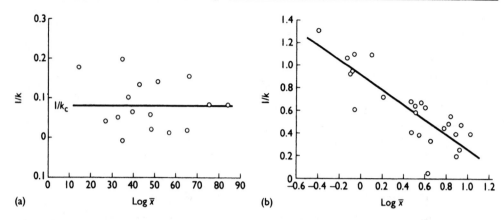

Fig. 2.8 The relation of $1/k$ to the mean for: (a) Colorada beetle; (b) wireworms, based on the same data as in Fig. 2.7 (after Bliss & Owen, 1958).

whose individual populations are values plus one, from a Poisson series. Neyman (1939) distributions are similar and are intended to describe conditions found soon after insect larvae hatched from egg batches; the modes are equally spaced. Skellam (1958) has shown that Neyman's type A is particularly applicable where the organisms occur in compact clusters, but that it can be used as an approximation in certain conditions when the clustering is less compact. The Polya–Aeppli distribution describes the situation when an initial wave of simultaneous invaders have settled and produced clusters of offspring; it may have one or two modes.

As mentioned above, the logarithmic model describes situations for which the negative binomial would give a very small value of k. The logarithmic series (Fisher *et al.*, 1943), which is derived from the negative binomial with k tending to zero and with the zero readings neglected, and the discrete and truncated (and censored) log-normal distributions (Preston, 1948; Grundy, 1952) have been found of most value in the description of the relationship between numbers of species and numbers of individuals and are discussed later (p. 471). Also they have been found to give a reasonable description of the distribution of the individuals of some insects (e.g. the citrus scale insect, *Aonidiella ornatum* (Spiller, 1952) or the eggs of the larch sawfly, *P. erichsonii* (Ives, 1955).

Working with plankton, with large mean values, Cassie (1962) suggested that the action of a series of environmental factors led to the population being distributed in a succession of Poisson series, the means of the series being themselves distributed according to the log-normal model. He called this the Poisson log normal; it differs from the negative binomial mainly in the left-hand flank, as ordinarily plotted, where it allows for fewer zero values.

2.5.1.3 Implications of the distribution models and of changes in the type of distribution

Neyman's distributions are based on precise models and, as Upholt and Craig (1940) found, if the biological assumptions underlying the distribution are not fulfilled, it will not adequately describe the dispersion of the population. It is perhaps for this reason that Neyman's distributions have been used relatively little by ecologists, most preferring the other distributions, which can be derived from a number of different hypotheses. The negative binomial can arise in at least five different ways (Anscombe, 1950; Waters & Hensen, 1959).

1 Inverse binomial sampling: if a proportion of individuals in a population possesses a certain character, the number of samples in excess of k that have to be taken to obtain k individuals with this character will have a negative binomial distribution with exponent k.

2 Heterogeneous Poisson sampling: if the mean of a Poisson distribution varies randomly between samples, under certain conditions a negative binomial results (Pielou, 1969). A biological example of this is the observation of Ito *et al.* (1962) that a series of counts of a gall-wasp on chestnut trees were distributed as a Poisson for each single tree, but when the counts from all trees were combined they were described by a negative binomial. The distribution of oribatid mites in the soil (Berthet & Gerard, 1965) and that of a tapeworm in fish (Anderson, 1974) also appear to arise according to this model.

3 Compounding of Poisson and logarithmic distributions: if a number of colonies are distributed as a Poisson but the number of individuals per colony follows a logarithmic distribution, the resulting distribution per unit area (i.e. independent of colonies) will be a negative binomial. Counts of bacteria (Quenouille, 1949) and the dispersion of eggs of the cabbage butterfly, *P. rapae* (Kobayashi, 1966) have been shown to satisfy this model.

4 Constant birth–death–immigration rates, the former two expressed per individual and the immigration rate per unit of time, will lead to a population whose size will form a negative binomial series.

5 True contagion: where the presence of one individual in a unit increases the chance that another will occur there also.

The logarithmic series (Fisher *et al.*, 1943) can also be derived from several modes of population growth (Kendall, 1948; Shinozaki & Urata, 1953).

It is clear, therefore, that, from mathematical considerations alone, it is unsound to attempt to analyse the details of the biological processes involved in generating a distribution from the mathematical model it can be shown to fit or, more often, approximately fit (Waters & Hensen, 1959). There is value in expressing the various possible mechanisms in biological terms. This has been done for parasite–host interactions by Crofton (1971) and Anderson (1974). These then provide alternative hypotheses, which may be tested by other observations.

The extent of clumping and the changes in it provide important evidence about the population. The uses and interpretation of the parameters will be discussed later (p. 42); here, changes in the actual type of distribution will be discussed. The main distinction lies between regular, random, and contagious distributions, with the respective implications that the animals compete (or at least tend to keep apart), have no effect on each other, or are aggregated or clumped. One must be careful to define the spatial scale over which the dispersion is described. The behavioural significance of the spacing pattern within the colony area (see below) is different from that between colonies.

The first difficulty is that the sampling method chosen by the experimenter may affect the apparent distribution (Waters & Hensen, 1959), as is shown by the example from the work of Ito *et al.* (1962) quoted above and that of Shibuya & Ouchi (1955), who found that the distribution of the gall-midge, *Asphondylia*, on soy bean was contagious if the plant was taken as the sampling unit, but random if the numbers per pod were considered. The contagion appeared to be due to there being more eggs on those plants with more pods. The effect of the size of the sampling unit on the apparent distribution has also been demonstrated for chafer-beetle larvae (Burrage & Gyrisco, 1954) and grasshopper egg pods (Putnam & Shklov, 1956). Careful testing of the size and pattern of sampling, as suggested above, will help to discover whether the contagion is an artefact or a reflection of the mixture of dispersion within 'colony areas' and the pattern of 'colony' distribution. Kennedy and Crawley (1967) aptly describe the dispersion of the sycamore aphid, *Drephanosiphon platanoides*, as 'spaced out gregarious'; they point out that, at close proximity—how close depending on the size of their 'reactive envelopes'—animals may repel each other (e.g. caddis larvae (Glass & Boubjerg, 1969)), but on a larger scale they will often show grouping or contagion. This may arise from patchiness of the habitat (Ferguson *et al.*, 1992), including differential predation (Waters & Hensen, 1959), or from the behaviour of the animals themselves, or a combination of both. The behaviour leading to aggregation, in the absence of special attractive areas in the habitat, may be of two types: inter-individual attraction or the laying of eggs (or young) in groups (Cole, 1946).

The dispersion of the initial insect invaders of a crop is often random; this randomness may be real or an artefact due to the low density relative to the sample size (see below). The distribution of aphids in a field during the initial phase of infestation is random, becoming contagious as each aphid reproduces (Sylvester & Cox, 1961; Shiyomi & Nakamura, 1964), although there may be differences between species (Kieckhefer, 1975; Badenhauser, 1994). The egg masses of many insects are randomly distributed (e.g. Chiang & Hodson, 1959) and the individual eggs and young larvae are clumped; however, the dispersion of the larvae may become random, or approach it, in later instars. Such changes in distribution with the age of the population have been observed in wireworms (Salt & Hollick, 1946), the rice stem borer, *Chilo simplex* (Kono, 1953), a chafer beetle, *Amphimallon majalis* (Burrage & Gyrisco, 1954), the cabbage butterfly, *P. rapae* (Ito *et al.*, 1960), the diamondback moth. *P. maculipennis* (Harcourt, 1961a), white grubs (*Phyllophaga*) (Guppy & Harcourt, 1970), leafhoppers (Nestel & Klein, 1995), and others. It is tempting to infer from such changes that either mortality or emigration or both are density-dependent; however, as Morisita (1962) has shown, this conclusion may not be justified with non-sedentary animals. Such changes could result from the alteration of the size of the area occupied by the colony relative to that of the sample or from the decrease in population density (see below). In contrast, the tendency to aggregate may be such that density-dependent mortality is masked, as in the pear lace-bug, *Stephanitis nashi*, where, even after the bugs have been artificially removed so as to produce a random distribution, it returns to non-randomness in a few days (Takeda & Hukusima, 1961), and, in the cabbage-root fly, *Erioschia brassicae*, k is similar for all immature stages (Mukerji & Harcourt, 1970).

Changes in the density of an animal often lead to changes, or at least apparent changes, in the distribution. When the population is very sparse, the chances of individuals occurring in any sampling unit are so low that their distribution is effectively random. The random distribution of low populations and the contagious distributions of higher ones have often been observed—for example, with wireworms (Finney, 1941), grasshopper egg pods (Davis & Wadley, 1949), a ladybird beetle, *Epilachna 28-maculata* (Iwao, 1956), the cabbage butterfly, *P. rapae* (Harcourt, 1961b), the pea aphid, *Acyrthosiphon pisum* (Forsythe & Gyrisco, 1963), and cotton insects (Kuehl & Fye, 1972). However, with the Nantucket pine-tip moth, *Rhyacionia frustrana*, as the population density increases still further, the distribution tends towards the Poisson again, i.e. the k value becomes high (Waters & Hensen, 1959). Populations of several other forest insects (Waters & Hensen, 1959) and of the periodical cicadas (Dybas & Davis, 1962) show a tendency to become more random at higher densities (p. 29). Indeed, one can envisage that, at even higher densities, the distribution would pass beyond the Poisson and become regular. Dispersions approaching a regular distribution have been but rarely observed; examples include the bivalve, *Tellina* (Holme, 1950), ants (Waloff & Blackith, 1962), aphids (Kennedy & Crawley, 1967), and a coral (Stimson, 1974).

In different habitats, the type of distribution may change, as Yoshihara (1953) found with populations of the winkle, *Tectarius granularis* (Mollusca), on rough and smooth rocks, although it is difficult to separate this habitat effect from associated changes in density.

Within the same habitat, different species will usually show different dispersion patterns, as has been found with leafhoppers (Homoptera: Auchenorrhyncha) on rice plants (Kuno, 1963). These differences can arise from several biological causes: one species may aggregate more than another in the same habitat because it disperses less, because it reproduces more, or because only certain parts of the habitat are suitable for it. However, Abrahamsen and Strand (1970) suggest that, for many soil organisms, there may be a relationship between k and x that can be expressed by the regression:

$$k = 0.27 + 0.334\sqrt{\bar{x}} \tag{2.34}$$

2.5.1.4 Comparison of aggregation indices

When a population is sampled, four basic bits of information are available.

1 The estimate (x) of the true mean, μ.

2 The estimate (s^2) of the true variance, σ^2.

3 The locality of the sample.

4 The size (unit).

The indices used for the description of animal populations are derived from various arrangements of this information. The simplest approach is the variance/mean ratio, $s^2/\bar{x}$, which, as described above, is unity for a Poisson (random) distribution. The index of clumping of David and Moore (1954) is simply:

$$I_{DM} = \frac{s^2}{\bar{x}} - 1 \qquad (2.35)$$

which gives a value of zero for a random (non-clumped) population.

A more general approach to the variance–mean relationship is given by Taylor's power law:

$$s^2 = a\bar{x}^b$$

This empirical relationship holds for many species (Taylor et al., 1978; Perry, 1981; Taylor, 1984) and has been found to be of considerable practical value in designing sampling programmes. Such a relationship can be shown to arise as a result of a variety of mechanisms, both stochastic demographic events and density-related behavioural processes (Anderson et al., 1982; Perry, 1988), and from its underlying mathematical basis (Tokeshi, 1995).

Pattern analysis in plant communities was developed by P. Greig-Smith (see Greig-Smith, 1978). The method is based on increasing quadrat size; where a peak arises in the plot of a measure of variance against quadrat size, this is considered as a 'clump area'. This method can only be applied to contiguous quadrats. Morisita (1954, 1959, 1962, 1964) developed an index, I_δ, that provided similar information but could be used on scattered quadrats. Zahl (1974) has suggested that Greig-Smith's approach can be made more robust by the use of the S-method of Scheffé.

With mobile animals, the degree of crowding experienced by an individual interested Lloyd (1967), who devised an index of mean crowding ($\dot{m}$):

$$\dot{m} = \mu + \left(\frac{\sigma^2}{\mu} - 1 \right) \qquad (2.36)$$

where μ and σ^2 are the true mean and variance, respectively. Thus $\dot{m}$ is the amount by which the variance mean ratio exceeds unity added to the mean. If the distribution conforms to a Poisson (random), then $\frac{\sigma^2}{\bar{x}} = 1$ and $\dot{m} = \mu$. With estimates based on samples, the index of mean crowding, $\dot{x}$, is given by:

$$\dot{x} \approx \bar{x} + \left(\frac{s^2}{\bar{x}} - 1 \right) \qquad (2.37)$$

For contagious dispersions described by the negative binomial (equations 2.25 and 2.26):

$$s^2 = \bar{x} + \frac{\bar{x}^2}{k} \qquad (2.38)$$

so equation (2.37) may be written:

$$\dot{x} = \bar{x} + \frac{\bar{x}}{k} \tag{2.39}$$

Rearranging equation (2.39):

$$\frac{\dot{x}}{\bar{x}} = 1 + \frac{1}{k} \tag{2.40}$$

one sees that the reciprocal of k is the proportion by which mean crowding exceeds mean density. Lloyd (1967) termed $\dfrac{\dot{x}}{\bar{x}}$ the 'patchiness'.

Clarification and unification of these various approaches were achieved by Iwao (1968, 1970a,b,c, 1972), who initially demonstrated that the relationship of mean crowding, $\dot{x}$, to mean density, $\bar{x}$, for a species could be expressed over a range of densities by a linear regression:

$$\dot{x} = \alpha + \beta\bar{x} \tag{2.41}$$

(Iwao used the expression $\dot{m}$–m method, but in reality one is normally working with estimates ($\dot{x}$ and $\bar{x}$), rather than the true mean and its derivatives. α and β are characteristic for the species and a particular habitat (Fig. 2.9) and their interpretation will be discussed later. Subsequently, he developed the ρ-index (Iwao, 1972):

$$\rho_1 = \frac{\dot{x}_i - \dot{x}_{i-1}}{\bar{x}_i - \bar{x}_{i-1}} \tag{2.42}$$

where $\dot{x}_i$ and $\bar{x}_i$ are the mean crowding index and mean density for the ith-sized quadrat sample. Thus, with a series of different-sized quadrats, values of ρ may be

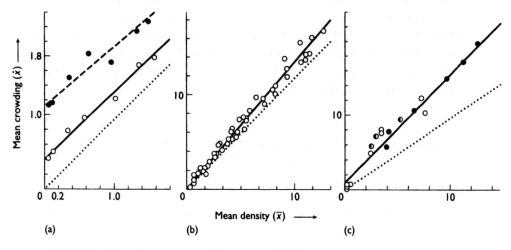

Fig. 2.9 Iwao's patchiness regression: (a) potato beetle, *Epilachna*, number of egg masses per plant (○) and of individual eggs (●); (b) spruce budworm, *Choristoneura*, numbers of larvae/twig; (c) pea aphid, *Macrosiphum*, total aphids (excluding first instar) per shoot (after Iwao, 1968).

plotted for the second and subsequent quadrat sizes (Fig. 2.10). Changes in the value of the ρ-index will indicate clump or territory size in a manner analogous, but theoretically preferable, to Greig-Smith's method.

$\dot{x}$ expresses 'crowding'. Lloyd (1967) derives it by estimating the mean number, less one, present in the area. Working with mites on leaves, Tanigoshi *et al.* (1975) restored the one to give an index they called 'mean colony size' ($\dot{C}$):

$$\dot{C} = \dot{x} + 1 \tag{2.43}$$

Lloyd (1967) termed this 'mean demand'—the mean number of individuals per quadrat per individual—and used it as an expression of trophic demand. The concept of mean colony size is useful, but is likely to be meaningful as formulated only when the sampling unit, as with spider mites on a leaf, contains an entire colony.

Thus, all the above indices are related to each other and to the simple concept of the comparison of the estimate of the variance (s^2) with that for the mean $\bar{x}$:

$$1 + \frac{s^2 - \bar{x}}{\bar{x}^2} = 1 + \frac{1}{k} = \frac{\dot{x}}{\bar{x}} = \frac{\dot{C} - 1}{\bar{x}} \approx I_\delta$$

Bringing in the third piece of information, quadrat size, the ρ-index is approximately the same as plotting the changes in I_δ with quadrat size; it is the proportion by which mean crowding changes against mean density with increased sample area. Essentially, in Greig-Smith's method, variance is plotted against sample area.

Perry and Hewitt (1991) criticized the above methods because there is no direct relationship between the mean and variance and the movements of the individual animals. Futher, they do not use the available spatial information. A methodology called spatial analysis by distance indices (SADIE) was developed (Perry & Hewitt, 1991; Aliston, 1994; Perry, 1994, 1995). The method involves the use of a computer simulation. The approach is to calculate the total distance by which the individuals need to be moved to make their distribution as crowded, regular, or random as possible. Perry (1994) gives examples where the method has been applied to a variety of animals and plants over a range of spatial scales; the software can be obtained at nominal cost (Perry, 1995).

2.5.2 Biological interpretation of dispersion parameters

2.5.2.1 Index of dispersion—the departure of the distribution from randomness
If the dispersion follows the Poisson distribution (p. 30), the mean will equal the variance and therefore departures of the coefficient of dispersion (or variation) $s^2/\bar{x}$ from unity will be a measure of the departure from the Poisson (from randomness). This is tested by calculating the index:

$$I_D = \frac{s^2(n-1)}{\bar{x}} \tag{2.44}$$

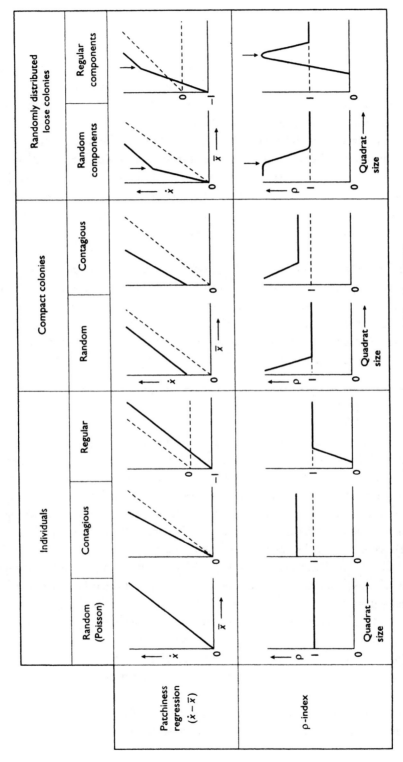

Fig. 2.10 The expected patchiness regressions and ρ-index plots for various patterns of dispersion (after Iwao, 1972).

where n = the number of samples and I_D is approximately distributed as χ^2 with $n-1$ degrees of freedom, so that, if the distribution is in fact Poisson, the value of I_D will not lie outside the limits (taken as 0.95 and 0.05) of χ^2 for $n-1$, as given in standard tables.[3] The index is therefore used as a test criterion for the null hypothesis that the pattern is random. The coefficient of variation or mean/variance ratio is a descriptive sample statistic that will approach zero for regularly distributed organisms, whilst a large value implies aggregation.

As Naylor (1959) has pointed out, the indices may be added if they are from the same-sized samples. Early examples of the use of this index are in Salt and Hollick's (1946) studies on wireworms, Naylor's (1959) on *Tribolium*, Overgaard Nielsen's (1963) on Culicoides and Milne's (1964) on the chafer beetle, *Phyllopertha*. In contagious populations, the value of this index is influenced by the density and size of the sampling unit (i.e. by the size of the mean).

2.5.2.2 'k' of the negative binomial — an index of aggregation in the population

If the negative binomial (p. 32) can be fitted to the data, the value of k gives a measure of dispersion; the smaller the value of k, the greater the extent of aggregation, whereas a large value (over about 8) indicates that the distribution is approaching a Poisson, i.e. is virtually random. This may be appreciated from the relationship of k to the coefficient of variation, cv:

$$cv = \frac{s}{\bar{x}} = \left[\frac{1}{k} + \frac{1}{\bar{x}}\right]^{\frac{1}{2}} \tag{2.45}$$

Clearly, the smaller the k, the larger the cv. The value of k is often influenced by the size of the sampling unit (Cole, 1946; Morris & Reeks, 1954; Waters & Hensen, 1959; Harcourt, 1961a; Ho, 1993) and therefore comparisons can be made only when using the same-sized unit. But, within this restriction, it does provide a useful measure of the degree of aggregation of the particular population, varying with the habitat and the developmental stage (Hairston, 1959; Waters & Henson, 1959; Harcourt, 1961a; Dybas & Davis, 1962). Examples of these variations are given in Table 2.3. As Waters and Henson (1959) have pointed out, the degree of aggregation of a population, which k expresses, could well affect the influence of predators and parasites.

The aggregation recognized by the negative binomial may be due either to active aggregation by the insects or to some heterogeneity of the environment at large (microclimate, soil, plant, natural enemies; p. 374). Dr R.E. Blackith has suggested that, if the mean size of a clump is calculated using Arbous and Kerrich's (1951) formula and this is found to be less than 2, then the 'aggregation' would seem to be due to some environmental effect and not to an active process. Sedentary animals, a large number of which are killed between settling and sampling, could be an excep-

---

[3] For values of n larger than those given in tables, χ^2 may be calculated (see p. 82).

Table 2.3 The variation k of the negative binomial with some factors, a, b, and c, with sampling unit, d, with developmental stage and change in density. a. *Eriophyes* leaf galls on *Populus tremuloides*. b. Nantucket pine-tip moth, *Rhyacionia frustrana* (from Waters & Henson, 1959). c and d. Cabbage white butterfly, *Pieris rapae* (from Harcourt, 1961b).

	Sampling unit	Mean number per unit	k
a	Single leaf	0.21 + 0.01	0.0611 + 0.0013
	1 leaf bunch	0.88 + 0.04	0.8830 + 0.0024
	2 leaf bunches	1.65 + 0.08	0.1150 + 0.0035
	5 leaf bunches	3.78 + 0.26	0.1740 + 0.0064
	Branch	14.23 + 1.68	0.2000 + 0.0119
b	Tip of shoot	0.49 + 0.03	0.449 + 0.021
	Branch whorl	2.36 + 0.36	0.214 + 0.035
	Tree	13.13 + 4.21	0.253 + 0.074

	Sampling unit	k for 1st-instar larvae	k for 4th-instar larvae
c	Quadrant of cabbage	1.38	1.96
	Half cabbage	2.28	4.24
	Whole cabbage	2.32	4.28

	Stage	Mean density	k
d	Egg	9.5	3.1
	1st instar	5.6	2.8
	2nd instar	4.4	2.8
	3rd instar	4.0	4.6
	4th instar	3.6	5.1
	5th instar	2.6	7.8
	Pupa	1.7	2.3

tion, in having a behavioural cause for an aggregation size of less than 2. Aggregations of two or more insects could be caused by either factor. The mean number of individuals in the aggregation is calculated by:

$$\lambda = \frac{\bar{x}}{2k} v \qquad (2.46)$$

where v is a function with a χ^2 distribution with $2k$ degrees of freedom and λ = the number of individuals in the aggregation for the probability level allocated to v. To find the mean size of the 'aggregate', the value at the 0.5 probability level is used. Usually, $2k$ degrees of freedom will be fractional, but an adequate χ^2 can be calculated by reference to graphs or by proportionality.

Let us take two examples from Table 2.3d, the eggs and pupae of *P. rapae*:

Eggs: $\lambda = \dfrac{9.5}{3.1 \times 2} \times 5.55 = 8.5$

Pupae: $\lambda = \dfrac{1.7}{2.3 \times 2} \times 3.54 = 1.3$

Fig. 2.11 The cause of contagion in the data—the plot of the mean aggregation (λ) of the two individuals for various values of the mean and k of the negative binomial. Populations whose value for the mean plotted against k lies below the line may be considered to exhibit contagion because of some environmental factor; in those populations whose value lies above the line, contagion could be due either to an active behavioural process or to the environment.

From this it can be concluded that the 'clumping' of the eggs could be due to a behavioural cause—the females tending to lay a number of eggs in close proximity or, because of environmental heterogeneity, only certain areas being suitable for oviposition. But 'clumping' of the pupae is due to environmental causes.

The values of the k and the mean for $\lambda = 2$ are plotted in Fig. 2.11, from which it can be ascertained for a particular population whose k and mean are known whether the mean 'aggregation' size is above or below 2. The negative binomial has been found useful in describing the pattern of predator attacks on patches of prey (Hassell, 1978). May (1978) has shown that the k of the attack pattern is the reciprocal of the coefficient of dispersion (or variation) (s/x) of the numbers of predators per patch. It has been argued that it is a useful model for the dynamics of animal aggregation and density dependence (Taylor *et al.*, 1979).

2.5.2.3 'b' of Taylor's power law—a coefficient of spatial variability for the species

The extent to which the parameter b of Taylor's power law (p. 12) is a constant for a species has been a matter of controversy. It may be concluded that it provides a description of the spatial structure of a species in a particular environment. There are many examples of it being relatively constant for a species (Taylor, 1961; Banerjee, 1976; Taylor *et al.*, 1978; Perry, 1988; Sawyer, 1989). However, as Riley *et al.* (1992) found when working on a weevil on peppers, whilst b did not vary significantly over time or with changes in population density, it was sensitive to the variety of the host plant and the type of sample. Bliss (1971) has pointed out that special techniques should be applied for the calculation of this regression, as the independent variate is not error-free. The same criticisms apply even more forcefully to Iwao's patchiness regression (see below), but in practice the more accurate approach is seldom required. As shown by Iwao (1970c), Taylor's power law will not fit certain theoretical distributions, although the biological frequency (and significance) of these is uncertain. It remains a simple and useful description of species distribution (Taylor, 1970; Usher, 1971; Egwuatu & Taylor, 1975; Kieckhefer, 1975; Croft et al., 1976; Ng et al., 1983; Pena & Baranowski, 1990; Steiner, 1990; Pringle & Giliomee, 1992; Peng & Brewer, 1994). High values of b show strong contagion. But, if the regression line crosses the Poisson line (slope $b = 1$), this shows a change to random distribution at lower densities. Hence, unless $a = 1$, the value of b in itself cannot be taken as a test of randomness, as shown by George (1974) for zooplankton. The constant a is largely a scaling factor related to sample size, but, if sample size is larger than the colony size and yet refers to a biologically meaningful entity (e.g. a leaf for spider mites), then a is also related to colony size. It would seem that the amount of scatter of the points around the line (see Fig. 2.2), which can be approximately estimated as the variance around the appropriate regression line (p. 12), is a measure of the effect of habitat variation on the extent of aggregation.

2.5.2.4 Lloyd's mean crowding and Iwao's patchiness regression indices for the population and species

The index of mean crowding $\dot{x}$ (Lloyd, 1967) is given approximately in equation (2.37). If the underlying distribution is known to fit the negative binomial, then equation (2.39) is applicable. This is an efficient estimate provided the number of samples (N) is large. When using equation (2.39), the most accurate method for estimating k (method 3, p. 321) should be used. A general formula is:

$$\dot{x} = \bar{x} + \left(\frac{s^2}{\bar{x}} - 1 \right)\left(1 + \frac{s^2}{N\bar{x}^2} \right)$$ (2.47)

the second term in parentheses correcting for sampling bias and k being calculated by moments (method 1, p. 31). The appropriate large-sample variance of the estimate is:

$$\mathrm{var}(\dot{x}) \approx \left[\frac{1}{N}\left(\frac{\dot{x}}{\bar{x}} \right)\left(\frac{s^2}{\bar{x}} \right)\left(\dot{x} + \frac{2s^2}{\bar{x}} \right) \right]$$ (2.48)

Lloyd (1967) discusses the problems of applying these to rare species and the statistical requirement—difficult for the biologist to meet—that one is a priori assured of encountering the species. The index $\dot{x}$ is the mean number of other individuals per sample unit per individual and is thus an expression of the intensity of interaction between individuals. In an extensive comparison of all the indices described hitherto applied to samples of Heliothis (corn-borer) eggs, Terry et al. (1989) found that $\dot{x}$ agreed least with the other indices, but it has been found useful with adult pear thrips, *Psylla pyricola*, which do not form colonies but have a contagious distribution (Burts & Brunner, 1981).

As will be seen from the first formula for the calculation of $\dot{x}$, when the population is randomly distributed the mean crowding index equals the mean density. Many examples have been given in the previous section of increased contagion with increased density. Iwao (1968) showed that $\dot{x}$ is related to the mean $(\overline{x})$ over a series of densities:

$$\dot{x} = \alpha + \beta \overline{x} \tag{2.49}$$

This is based on the changing relationship of Lloyd's index of patchiness, $\dfrac{\dot{x}}{\overline{x}}$, and may be termed Iwao's patchiness regression. The constant α indicates the tendency to crowding (+ve) or repulsion (−ve) and Iwao (1970b) termed it the 'index of basic contagion'. It is a property of the species and the mathematical basis suggests that it may warrant a biological interpretation (Tokeshi, 1995).

The coefficient β is related to the pattern in which the organism utilizes its habitat and Iwao named it the 'density contagiousness coefficient'. The theoretical effects of various underlying mathematical distributions on α and β and their application to many sets of field data are described by Iwao (1968, 1970a,b, 1972) and by Iwao and Kuno (1971) and they provide a measure of the numbers in the minimum clump in the sample unit, e.g. 38 for the individual eggs of *Epilachna* (Fig. 2.9) or 21.5 for *Macrosiphon* (excluding first instar) (Fig. 2.9). A negative value of α shows there is a tendency for the animals to repel each other, although, of course, there will not be actual points in this region (see Iwao & Kuno, 1971, Fig. 3). Coefficient β expresses the extent to which the colonies (as defined by β) are contagious at higher densities, e.g. in *Choristoneura* (Fig. 2.9b) or *Macrosiphum* (Fig. 2.9c). Although a single 'patchiness regression' can often be fitted for data from different areas, significantly different values of β could theoretically arise where the pattern of spatial heterogeneity differed between the two areas. Many comparative applications of Taylor's power law and Iwao's patchiness regression to insect sampling data have found the former to provide a more consistent fit (e.g. Pena & Duncan, 1992). Strong density-dependent mortality would influence the value of β. Density-independent mortality should alter the value of α but leave β unaffected (Iwao, 1970a,b).

2.5.2.5 Iwao's ρ index — a measure of colony area

The area occupied by a colony may be determined by the use of Iwao's (1972) index, given in equation (2.42). Clearly, this approach is only possible with animals living in habitats where sample size can be regularly increased. The index shows a sharp change in magnitude at the quadrat size corresponding to colony area (Fig. 2.10).

2.5.2.6 Breder's equations — a measure of the cohesion of aggregations

Working on fish schools, Breder (1954) found that the distance between individuals varied only slightly, and such uniform or regular distribution is indeed characteristic of the distribution of crowded individuals (p. 30). With such schools or social swarms, when the individuals are their normal distance (d) apart, the attraction (a) between individuals can be said to equal the repulsion (r), so that c, a measure of the cohesion in the equation $c = (a - r)/d^2$, is zero. Negative values of c will show 'over-crowding' and there will be a tendency for the individuals to move apart; positive values will show that normal aggregation has not been obtained — these will approach $+1$ asymptotically if the attractive force is taken as 1. If r_1 and r_2 are the repulsive actions of two animals on each other, then the equation becomes:

$$c = \frac{1 - r_1 r_2}{d^2}$$

(2.50)

or

$$r_1 r_2 = d_s^2$$

(2.51)

where d_s = the mean normal distance between individuals in the aggregation, because, by definition, $c = 0$ under these conditions.

The mean distance between individuals in a moving school may be measured from photographs and so the value for r found. It is then possible to plot c against d for a series of values (see Fig. 2.12). The angle at which the curves cross the 0 value is of interest: the steeper the line, the tighter the aggregation, angles of intersection of 40–50° being characteristic of fish that keep apart by a distance considerably less than their own length; smaller angles indicate looser aggregations, both spatially and, probably, Breder suggests, in their tendency to split up. The values of r given in Fig. 2.12 represents $\sqrt{r_1 r_2}$.

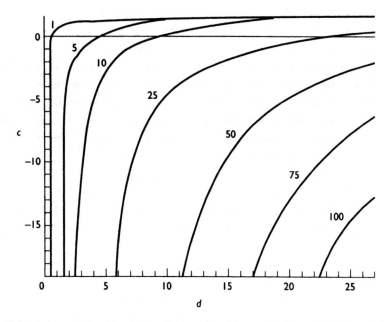

Fig. 2.12 Breder's curves for the cohesion of aggregations. The normal distance between individuals as a decimal fraction of the animal total length gives the value of d for $c = 0$, the appropriate curve for r ($= r_1 r_2$) may be read off (after Breder, 1954).

Some insects, such as locusts and army worms, form swarms, many of the characters of which would seem to parallel fish schools, and therefore the equations developed by Breder might well be used for comparisons with these insects. From Breder's equations, it would seem that the smaller the mean distance, in terms of their total lengths, between individuals, the more stable the aggregation. Extending this approach, various models have been developed for simulating the dynamics of fish schools and these may be used to analyse behaviour (Gotceitas & Colgan, 1991; Niwa, 1994).

2.5.2.7 Deevey's coefficient of crowding

This measure was used by Deevey (1947) to express the extent of crowding in barnacle populations, where the individuals actually impinge on one another as they grow, and is given by:

$$C_c = 2\pi r^2 N^2 \tag{2.52}$$

where r = the radius of the fully grown animal and N = the density of the animal per unit area. The units of r and N must of course be comparable (e.g. cm and cm^2) and then C_c will define the number of contacts per unit area. This coefficient could be used with any sedentary, approximately circular animal or with plant galls on leaves; when a realistic value can be given to r, for example the radius of a circular home range (p. 390), it could also be used with mobile forms.

2.5.3 Nearest-neighbour and related techniques: measures of departure from randomness of the distribution

There are basically two separate approaches: one may either select an individual at random and measure the distance between it and its nearest neighbour (true nearest-neighbour techniques) or one may select a point and measure the distance between this point and the nearest or nth-nearest individuals (sometimes referred to as closest-individual techniques). From each of these, either conclusions may be drawn about the departure of the distribution from random, if the population density is known from some other method, or alternatively, if one can assume the distribution to be random, its density can be estimated (see Chapter 9 for a discussion of the use of these methods for the estimation of density).

These techniques are most easily used with stationary, discrete, easily mapped organisms, e.g. trees, and have been used extensively by botanists (Greig-Smith, 1978). With animals, their mobility and the risk that one will fail to find the nearest neighbour limit their application (Turner, 1960). However, such methods have been applied to studies on grasshoppers (Blackith, 1958) and frogs (Turner, 1960) and, with more success, to studies of fairly conspicuous and relatively stationary animals (e.g. snails, Keuls *et al.*, 1963) or of well-marked colonies (e.g. ant mounds, Waloff & Blackith, 1962; Blackith *et al.*, 1963). Nearest-neighbour and related techniques could also be used with other relatively sedentary animals, such as barnacles, limpets, scale insects, tube-building animals, and gall-formers (especially on leaves).

The use of photography in mapping natural distributions of snails has been described by Heywood and Edwards (1961). This can be a valuable aid to the application of these techniques, not only to fairly sedentary organisms, but also, where the substratum is suitable, to far more mobile forms. The various dolichopodid and other flies on mud are an example of a community that might be studied in this way. There are, furthermore, a few other techniques, such as the 'squashing' or 'imprinting' method for estimating mite numbers on leaves (p. 155) or the deadheart method for estimating stem borers (p. 325), that do, or could easily, provide maps of the distribution of the individuals to which nearest-neighbour or closest-individual methods could be applied. Henson (1961) has used a combination of photography and nearest-neighbour methods to study the aggregations of a scolytid beetle in the laboratory (see also p. 151).

2.5.3.1 Nearest-neighbour methods

In this method, a point is selected at random, and then one searches around in tight concentric rings until an animal is found. The searching is then continued until its nearest neighbour is found and the distance between these two animals is measured. Strictly speaking, this is not a random method (Pielou, 1969); a truly random method would involve numbering all the animals in advance. The extent to which the method is robust in respect of this violation is not clear and thus, in general, the closest-individual method is to be preferred (Seber, 1982). There are two expressions; the simplest is due to Clark and Evans (1954):

$$D = \frac{1}{4\bar{r}^2} \tag{2.53}$$

where D = density per unit area and $\bar{x}$ = mean distance between nearest neighbours. As has been pointed out by Blackith (1958), this formula is based on the mean x squared $(\bar{x})^2$, whereas Craig's (1953) formula is based more correctly on the sum of x squared $(\Sigma x)^2$; the latter formula, however, is less easy to use and describes the density in a quarter segment.

If the density has been measured by some other method then the departure of the distribution from randomness will be given by the value of the numerator in the expressions (Clark & Evans, 1954; Waloff & Blackith, 1962):
uniform, hexagonal spacing competition:

$$\bar{x}^2 = \frac{1.154}{D} \tag{2.54}$$

random:

$$\bar{x}^2 = \frac{0.25}{D} \tag{2.55}$$

where $\bar{x}$ = mean distance between nearest neighbours and D = number of individuals per unit area. Smaller values of the numerator will indicate aggregation or clumping due to environmental or behavioural factors. These estimates should be based on large samples and, of course, the density and distance units of measurement must be the same.

The use of the second, third, . . . nth nearest neighbours, discussed by Morisita (1954) and Thompson (1956), enables not only more accurate density determinations to be made, but also the dispersion pattern to be detected over a larger area. However, as Waloff and Blackith (1962) pointed out, competition effects detectable between nearest neighbours may be masked at greater distances, because of heterogeneity of the habitat. Thompson (1956) shows that the statistic:

$$2\pi D \sum x_n^2 \tag{2.56}$$

where D = density and x = distance, is distributed as a χ^2 with $2N_n$ degrees of freedom (N = number of observations). This value may be calculated and compared with the expected value under the hypothesis of randomness (the χ^2 value for $2N$ is found from tables). A probability of χ^2 greater than 0.95 indicates significant aggregation, whilst a probability of less than 0.05 indicates significant regularity.

A simpler but—according to Thompson (1956)—less reliable comparison is given by the proportionality constants in the equation:

$$\bar{x}_n^2 = \frac{p}{D} \tag{2.57}$$

where p = the proportionality constant and the other symbols are as above. For a random distribution for $n = 1$ (i.e. the nearest neighbour), as indicated above, $p = 0.25$; for $n = 2, p = 0.87$; for $n = 3, p = 0.97$; for $n = 4, p = 1.05$.

2.5.3.2 Closest-individual or distance method

In this method, a point is selected at random and the distance between this and the nth-nearest individual measured. The main theoretical difficulty is the possibility of the random point being closer to the boundary than to an individual. Practically, it is more time-consuming than the nearest neighbour, because of the need to find the third-nearest or more distant individual. However, in spite of these problems, because of the difficulties of true random selection of an animal, referred to above (difficulties that apply less to botanical studies), this method is to be preferred.

The method has also been derived, apparently independently, by Keuls *et al.* (1963) and used for population estimations of the snail *Limnaea* (= *Galba*) *truncatula*. It depends on the population being randomly distributed, although, if the discontinuities are small, the errors that are made will tend to cancel each other out. The basic equation is:

$$D = \frac{r-1}{\pi} \cdot \frac{1}{a_n^2} \tag{2.58}$$

where r = the rank of the individual in distance from the randomly selected point, e.g. for the nearest neighbour $r = 1$, for the second nearest 2, a = the distance between the randomly selected point and the individual, and other symbols are as above. This basic equation is used in the ordered distance estimators in Chapter 9 to estimate density.

Various indices of non-randomness have been derived from equation 2.58 (for a fuller discussion, see Pielou, 1969; Seber, 1982). Two are as follows.

1 Pielou's index:

$$I_a = \bar{a}_i^2 D\pi \tag{2.59}$$

which requires a knowledge of the density per unit area (D), as well as the mean distance from a random point to the closest individual (a_i), $I_a = 1$ for random distributions, >1 for contagious, and <1 for regular. It has been used by Underwood (1976) with intertidal molluscs.

2 Eberhardt (1967) index:

$$I_E = \frac{\sum(\bar{a}_i^2)}{\left(\sum(\bar{a}_i)\right)^2} \tag{2.60}$$

This has the advantage that density need not be known, but the departure from randomness cannot easily be determined. $I_E = 1.27$ for a random distribution, being greater or less for contagious and regular, respectively.

2.6 Sequential sampling

In this type of sampling, the total number of samples taken is variable and depends on whether or not the results so far obtained give a definite answer to the question posed about the frequency of occurrence of an event (i.e. abundance of an insect). It is of particular value for the assessment of pest density in relation to control measures, when these are applied only if the pest density has reached a certain level.

2.6.1 Sampling numbers

Extensive preliminary work is necessary to establish the type of distribution (the relation of the mean and the variance) and the density levels that are permissible and those that are associated with extensive damage. From such data, it would be possible to plan to take a fixed number of samples that would enable one to be sure, in all instances, what the population level was. However, this would frequently lead to unnecessary sampling and a sequential plan usually allows one to stop sampling as soon as enough data have been gathered. For example, to take extremes, if 80 insects per shoot, or more, caused severe damage and no insects were found in the first five samples, common sense would tempt one not to continue for the additional, say 15, samples. Sequential sampling gives an exact measure (based on the known variance) so that with extremely high and extremely low populations very few samples need be taken and the expenditure of time and effort (cost) is minimal. As described later, this approach may also be used to obtain population estimates with a fixed level of precision (Kuno, 1963; Green, 1970).

As the distribution of most species can be fitted to the negative binomial, Iwao's patchiness regression, or Taylor's power law, these are the three types of distribution most commonly used to describe the population and formulae are given only for them; the principles are the same with other distributions. Although the negative binomial, being a probability model, is in principle to be preferred (Nyrop & Binns, 1991), variation in the value of k may indicate that the use of this distribution alone is inappropriate (Nyrop & Binns, 1991; Peng & Brewer, 1994; Nyrop et al., 1995). Many workers have found that Taylor's power law was the most robust (e.g. (Pena & Baranowski, 1990; Steiner, 1990; Pena & Duncan, 1992; Riley et al., 1992; Ho, 1993; Walker & Allsopp, 1993; Meagher et al., 1996)).

The method is described by Guenther (1997), Cho (1980), Nyrop and Binns (1991), and many textbooks. An early application was in Stark (1952) to normalized data for a needle miner (Lepidoptera). Whilst mostly applied to the monitoring of invertebrate, particularly insect, pests, it has also been applied to benthic community monitoring for environmental impact assessment (Resh & Price, 1984), the state of amphipod populations in acid-sensitive lakes (France, 1992), and stock assessment of fish (Weithman et al., 1979).

The first decision must be to fix the population level or levels related to the infestation classes. The most widely used plans have two levels: the higher one, where action is required, and the lower one, where it is definitely unnecessary. For example, we wish to distinguish between the hypothesis (H_1) that there are 200 or more egg masses per branch, sufficient to cause heavy damage, and the hypothesis (H_0) that there are 100 or fewer egg masses per branch, insufficient to cause damage. However, plans having only a single level have been developed (Nyrop et al., 1989; Legg et al., 1994), but this account will be based on the more traditional two-level plans.

The second decision concerns the level of probability of incorrect assessment one is prepared to tolerate; there are two types of error:

ψ = the probability of accepting H_1, when H_0 is the true situation

ω = the probability of accepting H_0 when H_1 is the true situation

Let us say that the same level is accepted for both, one false assessment in 20, i.e. a probability level of 0.05 for ψ and ω. Different levels may be necessary, because the costs of failing to apply control measures when an outbreak occurs may be much higher than those arising from the application of the treatment that is proved not to have been required.

For an animal whose distribution corresponds to the negative binomial, the values in Table 2.5 need to be calculated, the common k having been found already (p. 34) and the means being fixed as above (Oakland, 1950; Morris & Reeks, 1954).

The next aim is to plot the two lines (Fig. 2.13) that mark the 'acceptance' and 'rejection' areas (these terms have come from quality-control work, where sequential sampling was originally developed). The formulae for the lines are:

Table 2.4 Values needed for the assessment of which hypothesis to accept (see text for details).

	Infestation level	
	H_0	H_1
Mean = kp	kp_0 (e.g. 100)	kp_1 (e.g. 200)
$p = kp/k$	p_0	p_1
$q = 1 + p$	q_0	q_1
Variance = kpq	kp_0q_0	kp_1q_1

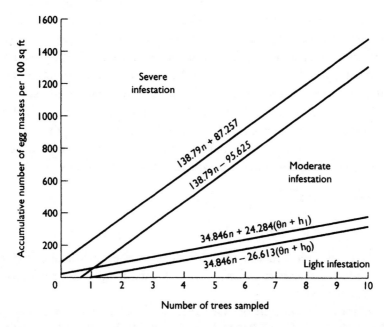

Fig. 2.13 A sequential sampling chart with two sets of acceptance and rejection lines (after Morris & Reeks, 1954).

$$d_0 = \theta n + b_0 \tag{2.61}$$

$$d_1 = \theta n + b_1 \tag{2.62}$$

where d = cumulative number of insects, n = number of samples taken, θ = the slope of the lines, given by:

$$\theta = k\frac{\log\dfrac{q_1}{q_0}}{\log\dfrac{p_1q_0}{p_0q_1}} \tag{2.63}$$

and b_0 and b_1 are the lower and upper intercepts, given by:

$$b_0 = \frac{\log\left[\dfrac{\omega}{(1-\varphi)}\right]}{\log\left(\dfrac{p_1 q_0}{p_0 q_1}\right)}$$ (2.64)

$$b_1 = \frac{\log\left[\dfrac{1-\omega}{(\varphi)}\right]}{\log\left(\dfrac{p_1 q_0}{p_0 q_1}\right)}$$ (2.65)

Once these calculations are made, a graph similar to that in Fig. 2.13 can be drawn up, and it will be clear that occasionally (especially when the true population lies between the population levels chosen, i.e. 100 and 200 in this example) the cumulative total continues to lie in the uncertain zone. In these circumstances, an arbitrary upper limit must be set to the number of samples taken. This limit (x), often known as the truncation mark, will be decided on considerations of accuracy and cost (Nyrop *et al.*, 1989). Normally, it will be laid down that, if after x samples the result still lies in the uncertain zone, treatment will be applied.

If Iwao's patchiness regression (p. 41) is used, the lines can be best calculated for successive values of n (number of samples taken) (Iwao, 1975):

$$d_1 = nx_c + t\sqrt{n\left[(\alpha+1)x_c + (\beta-1)x_c^2\right]}$$ (2.66)

$$d_0 = nx_c - t\sqrt{n\left[(\alpha-1)x_c + (\beta-1)x_c^2\right]}$$ (2.67)

where x_c = the critical density, α and β are the parameters from the patchiness regression, and t is taken from statistical tables for the acceptable error levels for φ and ω (for the d_1 equation).

Using the parameters of Iwao's regression and Taylor's power law, respectively, Kuno (1969) and Green (1970) devised methods that allowed population estimates to be obtained with a fixed level of precision. In Green's method, the cumulative number of animals to be sampled (T_n) is given by:

$$T_n^{b-2} = \frac{D_0^2}{\alpha} n^{b-1}$$ (2.68)

where α and b are the intercept and slope, respectively, in Taylor's power law (see p. 12), n is sample size, and D_0 is the chosen level of precision. This will be linear in its log form:

$$\log T_n = \frac{\log(D_0^2/\alpha)}{b-2} + \frac{b-1}{b-2}\log n$$ (2.69)

The decision lines can be generated by computer simulations and a number of algorithms have been described (Fowler & Lynch, 1987; Nyrop & Binns, 1991).

Table 2.5 A sequential sampling table, giving the uncertainty band for various numbers of samples (after Morris, 1954).

Sample tree	Moderate vs. severe infestation uncertainty band (cumulative total of insects)
1	42–225
2	183–366
3	323–506
4	460–643
5	599–782
6	738–921

Although Taylor's power law has been found to be a less biased predictor of variance than Iwao's regression, the parameters of these relationships are not as constant as theory suggests and vary from location to location, as well as in the same location with time. No method can be used to give plans that may be uncritically used at different times or in different places to give estimates with the required precision (Trumble *et al.*, 1989; Naranjo & Flint, 1995). Simulations may be undertaken and it is possible to determine operating characteristic curves, sample number curves, and the influence of the truncation mark for the different plans on the basis of information on the populations of the organism to be sampled. This will permit a choice of plan based on a consideration of the cost (number of samples) and benefit (accuracy). The actual fieldwork may be carried out based on the graph, a sequential table (see Table 2.5), or the data fed into a programmed computer.

General accounts of the utility of sequential sampling plans for insects have been given in reviews ranging over time from Ives (1954) to Pedigo and Buntin (1994). In addition to the papers already referred to, plans have been developed for around 200 species and/or situations; Fowler and Lynch (1987) and Pedigo and Buntin (1994) provide bibliographies.

2.6.2 Presence or absence sampling

It is sometimes very costly to estimate the actual numbers in a sample, but presence or absence can be easily assessed; this is true, for example, of aphids and spider mites, where the populations may be strongly clumped. The formulae used depend on assumptions concerning the distribution. This method provides an alternative approach for the development of sequential sampling plans as a basis for pest-control decisions.

If the dispersion can be described by the negative binomial distribution with a known k, the probability of a particular mean population per unit (N_y) being reached can be estimated (Gerrard & Chang, 1970; Wilson & Gerrard, 1971; Gerrard & Cook, 1972; Mailloux & Bostanian, 1989):

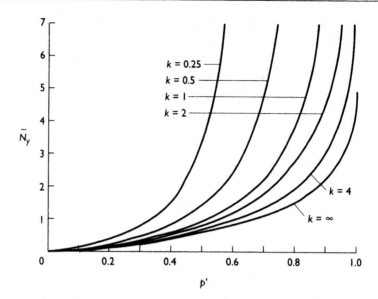

Fig. 2.14 Relationship between mean population density ($\bar{N}_y$) and the probability of cell occupancy (p') for selected values of k (after Gerrard & Cook, 1972).

$$\left(N_y\right) = k\left[\left(\frac{1}{1-p'}\right)^{\frac{1}{k}} - 1\right]$$
(2.70)

where p' = probability of the animal being present in the sampling unit determined on the basis of presence or absence sampling. The form of the relationship is shown in Fig. 2.14. It is reliable only if the critical density levels are related to values of p' less than about 0.8; above this, the uncertainty associated with the predictions is too great. Obviously, the sampling unit can be varied to achieve this relationship, but, for it to retain its value for rapid assessment, the new unit must be easily recognized. Furthermore, any major changes in the value of k with density would invalidate this approach.

Other approaches have been developed that do not depend on assuming a particular model for the dispersion. In a situation where the total number of potential sampling units (Φ) is meaningful and may be known, e.g. young trees in a plantation, the decision can be made (with a certain level of probability of it being untrue (α) that there is a zero or very low population (p_0) if a certain number of samples from which the organism is absent are drawn in succession. The number of such zero samples is given by (Kuno, 1991):

$$n_0 = \Phi\left(1 - \alpha^{1/\Phi p_0}\right)$$
(2.71)

However, for satisfactory levels of confidence the number of successive samples must be large in relation to Φ.

Of wider application is an approach that depends on a knowledge of the relationship between the number of samples where the animal is absent (or present in very low numbers) and the mean of the population/unit, the population intensity (N_y). In a general form, this is (Gerrard & Chang, 1970):

$$\ln\left(N_y\right) = a + b\ln(-\ln p_0)\qquad\qquad(2.72)$$

where a and b are constants for the animal in habitats of the type studied and p_0 is the probability of zero (or very low—if such a level has been chosen) numbers in a particular sampling unit. The values of a and b can be calculated from the regression of N_y on p_0 from preliminary field data. However, the estimate of the variance is less straightforward (Nachman, 1984; Kuno, 1986; Nyrop et al., 1989; Binns & Bostanian, 1990; Tonhasca et al., 1994) but is important for determining the relationship of the numbers of samples to be taken to the level of confidence in the decision. There is not only sampling error, but error in the estimation of a and b and 'error' which is a reflection of real biological variation in the dispersion, which would be reflected in changes in k of the negative binomial. Schaalje et al. (1991) provide a detailed review and recommendations based on simulations and field data.

2.6.3 Sampling a fauna

With faunal surveys, the problem of ensuring the detection of an adequate proportion of the species present must be considered. When sampling vegetation, Gleason (1922) pointed out that, as the area sampled increased, the numbers of hitherto unrecorded species added decreased, a concept referred to as the species–area curve, and one that has often been used and disputed in plant ecology (Goodall, 1952; Evans et al., 1955; Greig-Smith, 1978; Watt, 1964). The logarithmic series and log-normal distributions (p. 471) express the same basic assumptions another way.

One approach to this problem is, therefore, to take a number of preliminary samples and calculate the index of diversity based on the log series (p. 471). If one can make an estimate of the total number of individuals in the fauna, then the theoretical total number of species present can be arrived at from the index of diversity, and the total number of individuals that should be collected to find y of species present can be determined. A number of other estimates for the total number of species are presented in Chapter 13. The value of y will, of course, depend on the aim of the study. Unfortunately, as will be pointed out later (Chapter 13), this approach has a number of practical and theoretical limitations. The practical ones stem from the mosaic nature of most habitats (Hairston, 1964). Unless sampling is by a light trap or some other device that is independent of microhabitats, more new species are added by further samples than would be expected. The theoretical problems arise from the discovery that the log-normal distribution provides a better description of the phenomenon than the simpler logarithmic series (see p. 472 for a further discussion).

A number of non-parametric methods for estimating species richness have been developed. A particularly simple and effective estimator for the total number of species, S_T, in a locality is (Chao, 1984):

$$S_T = S_{obs} + \left(\frac{a^2}{2b} \right) \tag{2.73}$$

where S_{obs} is the observed number of species in a sample, a the number of species represented by only a single individual, and b the number of species represented by two individuals. Subsequently, Chao (1987) gave the variance of this estimate as:

$$\mathrm{var}(S_T) = b \left[\left(\frac{\frac{a}{b}}{4} \right)^4 + \left(\frac{a}{b} \right)^3 + \left(\frac{\frac{a}{b}}{2} \right)^2 \right] \tag{2.74}$$

If only presence–absence data are available, a similar approach can be taken (Chao, 1984):

$$S_T = S_{obs} + \left(\frac{L^2}{2M} \right) \tag{2.75}$$

where L is the number of species that occur in only one sample and M is the number of species that occur in only two samples. The variance is calculated using equation (2.74), with L and M substituting a and b, respectively. These estimates are probably best considered as the lower bound of total species number (see Colwell and Coddington, 1994), although P.A. Henderson and A.E. Magurran (pers. comm.) found this Chao estimator to be one of the most reliable available. In Chapter 13, other methods for estimating species number are presented; these require more data and thus are less useful during the planning stage.

Sometimes the mosaic nature of the habitat can be utilized to facilitate the planning of an optimal sampling programme. This is especially true where the effects of pollution are being studied in streams, but might be extended to other situations where a factor is acting simultaneously over a wide range of special habitats within a major habitat or community type—for example, the effect of aerial spraying on the fauna of a mixed woodland. If a fixed number of samples of the fauna are taken—say, from three different zones of a stream: riffles, pools, and marginal areas—it will be found that the first few samples will provide a more complete picture of the total observed fauna in one zone than in the other (Gaufin *et al.*, 1956). Because of the different diversities and microdistributions of the fauna in the different zones, and also possibly because of the differing efficiencies of the collecting methods under these conditions, the rates of accumulation of new species vary from zone to zone. A zone in which a high percentage of the total species taken in a large number of samples is recorded in the first few samples can have a smaller number of samples drawn from it than a zone in which the rate of accumulation is less rapid.

A method of calculating the average number of new species contributed by the Kth successive sample (when K can equal any number from 1 upwards) has been devised by Gaufin *et al.* (1956), who also provide a table of the coefficients necessary for up to a total of 10 samples. Their formula is:

$$S_K = \frac{1}{S} \sum_{i=1}^{n-k+1} a_i K S_i \qquad\qquad (2.76)$$

where S_K = the number of new species added by the Kth sample, n = the total number of samples taken in the preliminary survey, S = the total number of species taken in the n samples, S_i = the number of different species appearing in i out of n samples and a_i and K are coefficients by which successive values of S_i are multiplied in calculating the summation item. By summing a series of values of S_K (for $K = 1$ to a predetermined number K'), it is possible to compute the average number of species found in K' samples. A method of calculating the standard errors of such estimates has been developed by Harris (1957).

2.6.3.1 Biological and other qualitative aspects of sampling

Hitherto, stress has been laid on the statistical aspects of sampling; however, there are certain biological problems that are of cardinal importance. The ecologist should always remember that the computation of fiducial limits of estimates only tells him/her the consistency of the samples collected, and the statistical techniques cannot be blamed if these have been consistently excluding a major part of the population. This is indeed the most serious biological problem in all sampling work, namely to ensure that there is not a part of the population with a behavioural pattern or habitat preference such that it is never sampled (Macleod, 1958).

Less dangerous, because the phenomenon will be recorded, although it may well be misinterpreted, is the tendency for the behaviour of the animal to change and affect its sampling properties. The reactions of the fly *Meromyza* to weather conditions (Hughes, 1955) and of the larvae of different species of corixid water-bug to light (Teyrovsky, 1956) lead to errors when net sampling is carried out under different conditions (p. 247). Many, perhaps almost all, animals alter their behaviour with age, and this may lead to a confusion between a change in behaviour and death. For example, older females of the mirid grass bug, *Leptopterna dolobrata*, spend more time on the base of the grass shoots and less on the tops than do males. If a marking and recapture experiment is carried out and the bugs are collected by sweeping, the females will appear to be much shorter-lived than the males. However, this is because, as they age, the females enter the sampling zone less frequently; they are actually longer-lived than the males.

The variations in the distribution of an animal may well be linked with some character of the environment or its host plant. It is important to distinguish this variation from sampling variance; otherwise, the fiducial limits of the estimates may become so wide that no conclusions can be drawn (see Prebble, 1943). In other words, systematic errors will arise (p. 15). When population trend is eliminated, approximately 50% of the variance of the counts of populations of the sugar-cane froghopper, *Aeneolamia varia*, on sugar-cane stools could be accounted for in terms of the size of the stools (Fewkes, 1961). Hence, more 'accurate' estimates would be obtained by taking stool size into consideration and adjusting population estimates, or by allowing for it in the analysis. This is particularly important when comparisons of froghopper numbers are being made between insecticide-treated fields; the fields are of different ages and hence with stools of different sizes. Many other examples could be given of environmental factors influencing distribution; they are commonly microclimatic in nature, e.g. soil moisture affects the distribution of the cocoons of a sawfly (Ives, 1955). Even when the environment appears uniform, distribution may be systematically non-random; for example, the densest populations of the beetle *Tribolium* in flour are always adjacent to the walls of the container (Cox & Smith, 1957). Covariance analyses are often helpful in such situations.

References

Abrahamsen, G. (1969) Sampling design in studies of population densities in Enchytracidae (Oligochaeta). *Oikos* **20**, 54–66.

Abrahamsen, G. & Strand, L. (1970) Statistical analysis of population density data of soil animals, with particular reference to Enchytraeidae (Oligochaeta). *Oikos* **21**, 276–84.

Aliston, R. (1994) *Statistical analysis of animal populations*. PhD thesis, University of Kent, England.

Andersen, F.S. (1965) The negative binomial distribution and the sampling of insect populations. *Proc. XII Int. Congr. Entomol.* 395.

Anderson, R.M. (1974) Population dynamics of the cestode *Caryophyllaeus laticeps* (Pallas, 1781) in the bream (*Abramis brama* L.). *J. Anim. Ecol.* **43**, 305–21.

Anderson, R.M., Gordon, D.M., Crawley, M.J., & Hassell, M.P. (1982) Variability in abundance of animal and plant species. *Nature, Lond.* **296**, 245–8.

Anscombe, F.J. (1948) On estimating the population of aphids in a potato field. *Ann. Appl. Biol.* **35**, 567–71.

Anscombe, F.J. (1949) The statistical analysis of insect counts based on the negative binomial distribution. *Biometrics* **5**, 165–73.

Anscombe, F.J. (1950) Sampling theory of the negative binomial and logarithmic series distributions, *Biometrika* **37**, 358–82.

Arbous, A.G. & Kerrich, J.E. (1951) Accident statistics and the concept of accident-proneness. *Biometrics* **7**, 340–432.

Badenhauser, I. (1994) Spatial patterns of alate and apterous morphs of the *Brachycaudus helichrysi* (Homoptera: Aphididae) in sunflower fields. *Environ. Entomol.* **23**, 1381–90.

Bailey, N.T.J. (1959) *Statistical Methods in Biology.* English Universities Press, London.

Baker, J.M. & Crothers, J.H. (1987) Intertidal rock. In: Baker, J.M. & Wolff, W.J. (eds) *Biological Surveys of Estuaries and Coasts.* Cambridge University Press, Cambridge, New York, New Rochelle, 157–97.

Baker, J.M. & Wolff, W.J. (eds) (1987) *Biological Surveys of Estuaries and Coasts.* Estuarine and Brackish Water Sciences Association Handbook, Cambridge University Press, Cambridge.

Ballentine, W.J. (1961) A biologically-defined exposure scale for the comparative description of rocky shores. *Field St.* **1**, 1–19.

Bamber, R.N. (1993) Changes in the infauna of a sandy beach. *J. Exp. Mar. Biol. Ecol.* **172**, 93–107.

Bamber, R.N. & Spencer, J.F. (1984) The benthos of a coastal power station thermal discharge canal. *J. Mar. Biol. Assoc. UK* **64**, 603–23.

Bancroft, T.A. & Brindley, T.A. (1958) Methods for estimation of size of corn borer populations. *Proc. X Int. Congr. Entomol.* **2**, 1003–14.

Banerjee, B. (1976) Variance to mean ratio and the spatial distribution of animals. *Experientia* **32**, 993–4.

Beall, G. (1942) The transformation of data from entomological field experiments so that the analysis of variance becomes applicable. *Biometrika* **32**, 243–62.

Beauchamp, J.I. & Olson, I.S. (1973) Corrections for bias in regression estimates after logarithmic transformation. *Ecology* **54**, 1403–7.

Berthet, P. & Gerard, G. (1965) A statistical study of microdistribution of Oribatei (Acari). Part 1. The distribution pattern. *Oikos* **16**, 214–27.

Binns, M.R. & Bostanian, N.J. (1990) Robustness in empirically based binomial decision rules for integrated pest management. *J. Econ. Entomol.* **83**, 420–7.

Blackith, R.E. (1958) Nearest-neighbour distance measurements for the estimation of animal populations. *Ecology* **39**, 147–50.

Blackith, R.E., Siddorn, J.W., Waloff, N., & van Emden, H.F. (1963) Mound nests of the yellow ant, *Lasius flavus* L., on water-logged pasture in Devonshire. *Entomol. Monthly Mag.* **99**, 48–9.

Bliss, C.I. (1958) The analysis of insect counts as negative binomial distributions. *Proc. X Int. Congr. Entomol.* **2**, 1015–32.

Bliss, C.I. (1971) The aggregation of species within spatial units. In: Patil, G.P., Pielou, E.C., & Waters, W.E. (eds) *Statistical Ecology I (Spatial Patterns and Statistical Distributions).* Penn State University Press, Philadelphia, pp. 311–35.

Bliss, C.I. & Calhoun, D.W. (1954) *An Outline of Biometry.* Yale Co-operative Corporation, New Haven.

Bliss, C.I. & Fisher, R.A. (1953) Fitting the negative binomial distribution to biological data and note on the efficient fitting of the negative binomial. *Biometrics* **9**, 176–200.

Bliss, C.I. & Owen, A.R.G. (1958) Negative binomial distributions with a common *k*. *Biometrika* **45**, 37–58.

Blondel, J. & Frochet, B. (eds) (1987) *Bird Census and Atlas Studies Proceedings of the IXth International Conference on Bird Census and Atlas Work.* University of Dijon, Dijon.

Box, G.E.P. & Cox, D.R. (1964) An analysis of transformations. *J. Roy. Stat. Soc.* **26**, 211–52.

Breder, C.M. (1954) Equations descriptive of fish schools and other animal aggregations. *Ecology* **35**, 361–70.

Broadbent, L. (1948) Methods of recording aphid populations for use in research on potato virus diseases. *Ann. Appl. Biol.* **35**, 551–66.

Browning, T.O. (1959) The long-tailed mealybug, *Pseudococcus adonidum* L. in South Australia. *Aust. J. Agric. Res.* **10**, 322–37.

Bryant, D.G. (1976) Sampling populations of *Adelges piceae* (Homoptera: Phylloxeridae) on

balsam fir. *Abies balsamea. Can. Entomol.* **108**, 1113–24.

Buntin, G.D. (1994) Developing a primary sampling program. In: Pedigo, L.P. & Buntin, G.D. (eds) *Handbook of Sampling Methods for Arthropods in Agriculture.* CRC Press, Boca Raton, pp. 99–118.

Buntin, G.D. & Pedigo, L.P. (1981) Dispersion and sequential sampling of green cloverworm eggs in soybeans. *Environ. Entomol.* **10**, 980–5.

Burrage, R.H. & Gyrisco, G.G. (1954) Estimates of populations and sampling variance of European chafer larvae from samples taken during the first, second and third instar. *J. Econ. Entomol.* **47**, 811–17.

Burts, E.C. & Brunner, J.F. (1981) Dispersion statistics and sequential sampling plan for adult pear *Psylla. J. Econ Entomol.* **74**, 291–4.

Camin, J.H., George, J.F., & Nelson, V.E. (1971) An automatic tick collector for studying the rhythmicity of 'drop off' Ixodidae. *J. Med. Entomol* **8**, 394–8.

Cassie, R.M. (1962) Frequency distribution models in the ecology of plankton and other organisms. *J. Anim. Ecol.* **31**, 65–92.

Caughley, G. (1977) *Analysis of Vertebrate Populations.* John Wiley & Sons, New York.

Chao, A. (1984) Non-parametric estimation of the number of classes in a population. *Scand. J. Stat.* **11**, 265–70.

Chao, A. (1987) Estimating the population size for capture–recapture data with unequal matchability. *Biometrics* **43**, 783–91.

Chiang, H.C. & Hodson, A.C. (1959) Distribution of the first-generation egg masses of the European corn borer in corn fields. *J. Econ. Entomol.* **52**, 295–9.

Cho, C.-K. & Kuei, C. (1980) *An Introduction to Software Quality Control.* Wiley, New York.

Church, B.B. & Strickland, A.H. (1954) Sampling cabbage aphid populations on brussels sprouts. *Plant Pathol.* **3**, 76–80.

Clark, P.I. & Evans, F.C. (1954) Distance to nearest neighbor as a measure of spatial relationships in populations. *Ecology* **35**, 445–53.

Cochran, W.G. (1977) *Sampling Techniques*, 3rd edn. John Wiley & Sons, New York.

Cole, L.C. (1946) A theory for analysing contagiously distributed populations. *Ecology* **279**, 329–41.

Cole, W.E. (1970) The statistical and biological implications of sampling units for mountain pine beetle populations in lodgepole pine. *Res. Popul. Ecol.* **12**, 243–8.

Colwell, R.K. & Coddington, J.A. (1994) Estimating terrestrial biodiversity through extrapolation. *Phil. Trans. Roy. Soc. (Ser. B)* **345**, 101–18.

Condrashoff, S.F. (1964) Bionomics of the aspen leaf miner, *Phyllocnistis populiella* Cham. (Lepidoptera: Gracillariidae). *Can. Entomol.* **96**, 857–74.

Corbet, P.S. & Smith, S.M. (1974) Diel periodicities of landing of multiparous and parous *Aedes aegypti* (L.) at Dar es Salaam, Tanzania (Diptera, Culicidae). *Bull. Entomol Res.* **64**, 111–21.

Cornfield, J. (1951) The determination of sample size. *Am. J. Pub. Health* **41**, 654–61.

Coulson, R.N., Hain, F.P. Foltz, J.L., & Mayyasi, A.M. (1975) Techniques for sampling the dynamics of southern pine beetle populations. *Texas Agric. Exp. Sta. Misc. Pub.* **1185**, 3–18.

Coulson, R.N., Pulley, P.E., Foltz, J.L., & Martin, W.C. (1976) Procedural guide for quantitatively sampling within-tree populations of *Dendroctonus frontalis*. *Texas Agric. Exp. Sta. Misc. Pub.* **1267**, 3–26.

Cox, D.R. & Smith, W.L. (1957) On the distribution of *Tribolium confusum* in a container. *Biometrika* **44**, 328–35.

Craig, C.C. (1953) On a method of estimating biological populations in the field. *Biometrika* **40**, 216–18.

Croft, B.A., Welsh, S.M., & Dover, M.J. (1976) Dispersion statistics and sample size estimates for populations of the mite species *Panonychus ulmi* and *Amblyseius fallacis* on apple. *Environ. Entomol.* **5**, 227–34.

Crofton, H.D. (1971) A quantitative approach to parasitism. *Parasitology* **62**, 179–93.

Cushing, D.H. (1992) *Climate and Fisheries.* Academic Press, London.

Dahlsten, D.L., Rowney, D.L., Cooper, W.A., Tait, S.M., & Wenz, J.M. (1990) Long-term population studies of the douglas-fir moth in California. In: Watt, A.D., Leather, S.R., Hunter, M.D., & Kidd, N.A.C. (eds) *Population Dynamics of Forest Insects.* Intercept Press, Andover, England, pp. 45–58.

David, F.N. & Moore, G.P. (1954) Notes on contagious distributions in plant populations. *Ann. Bot. Lond. New Ser.* **18**, 47–53.

Davis, E.G. & Wadley, F. (1949) Grasshopper egg-pod distribution in the northern Great Plains and its relation to egg-survey methods. *USDA Circ.* **816**, 16 pp.

Dean, H.A. (1959) Quadrant distribution of mites on leaves of Texas grapefruit. *J. Econ. Entomol.* **52**, 725–7.

Debauche, H.R. (1962) The structural analysis of animal communities of the soil. In: Murphy, P.W. (ed.) *Progress in Soil Zoology.* Butterworths, London, pp. 10–25.

Deevey, E.S. (1947) Life tables for natural populations of animals. *Quart. Rev. Biol.* **22**, 283–314.

Dempster, J.P. (1957) The population dynamics of the Moroccan locust (*Dociostaurus maroccanus* Thunberg.) in Cyprus. *Anti-locust Bull.* **27**, 1–60.

Downing, J.A. (1979) Aggregation, transformation, and design of benthos sampling programs. *J. Fish, Res. Board Can.* **36**, 1454–63.

Dudley, C. (1971) A sampling design for the egg and first instar larval populations of the western pine beetle, *Dendroctonus brevicomis* (Coleoptera: Scolytidae). *Can. Entomol.* **103**, 1291–313.

Dybas, H.S. & Davis, D.D. (1962) A population census of seventeen-year periodical cicadas (Homoptera: Cicadidae: Magicicada). *Ecology* **43**, 432–44.

Eberhardt, L.L. (1967) Some developments in 'distance sampling'. *Biometrics* **23**, 207–16.

Edwards, R.L. (1962) The importance of timing in adult grasshopper surveys. *J. Econ. Entomol.* **55**, 263–4.

Efron, B. & Tibshirani, R. (1986) Bootstrap methods for standard errors, confidence intervals, and other measures of statistical accuracy. *Stat. Sci.* **1**, 54–77.

Egwuatu, R.I. & Taylor, T.A. (1975) Aspects of the spatial distribution of *Acanthomia tomentosicollis* Stal. (Heteroptera, Coreidae) in *Cajanus cajan* (pigeon pea). *J. Econ. Entomol.* **69**, 591–4.

Ellenberger, J.S. & Cameron, E.A. (1977) The spatial distribution of oak leaf roller egg masses on primary host trees. *Environ. Entomol.* **6**, 101–6.

Elliott, J.M. (1977) *Some Methods for the Statistical Analysis of Samples of Benthic Invertebrates.* Freshwater Biology Association, Ambleside, Cumbria.

Elliott, J.M. & Drake, C.M. (1981) A comparative study of seven grabs used for sampling macroinvertebrates in rivers. *Freshw. Biol.* **11**, 99–120.

van Emden, H.F., Jepson, W.F., & Southwood, T.R.E. (1961) The occurrence of a partial fourth generation of *Oscinella frit* L. (Diptera: Chioropidae) in southern England. *Entomol. Exp. Appl.* **4**, 220–5.

Evans, D.A. (1953) Experimental evidence

concerning contagious distributions in ecology. *Biometrika* **40**, 186–211.

Evans, F.C., Clark, P.I., & Brand, R.H. (1955) Estimation of the number of species present in a given area. *Ecology* **36**, 342–3.

Ferguson, C.S., Linit, M.J., & Krause, G. (1992) Dispersion and density of asiatic oak weevil (Coleoptera: Curculionidae) relative to oak density. *Environ. Entomol.* **21**, 247–52.

Fewkes, D.W. (1961) Stool size as a factor in the sampling of sugarcane froghopper nymph populations. *J. Econ. Entomol.* **54**, 771–2.

Finney, D.J. (1941) Wireworm populations and their effect on crops. *Ann. Appl. Biol.* **28**, 282–95.

Finney, D.J. (1973) Transformation of observations for statistical analysis. *Cotton Growers' Rev.* **50**, 1–14.

Finney, D.J. & Varley, G.C. (1955) An example of the truncated Poisson distribution. *Biometrics* **11**, 387–94.

Fisher, R.A., Corbet, A.S., & Williams, C.B. (1943) The relation between the number of species and the number of individuals in a random sample of an animal population. *J. Anim. Ecol.* **12**, 42–58.

Ford, R.P. & Dimond, J.B. (1973) Sampling populations of pine leaf chermid *Pirieus pinifoliae* (Homoptera: Chermidae) II. Adult Gailicolae on the secondary host. *Can. Entomol.* **105**, 1265–74.

Forsythe, H.Y. & Gyrisco, G.C. (1961) Determining the appropriate transformation of data from insect control experiments for use in the analysis of variance. *J. Econ. Entomol.* **54**, 859–61.

Forsythe, H.Y. & Gyrisco, G.G. (1963) The spatial pattern of the pea aphid in alfalfa fields. *J. Econ. Entomol.* **56**, 104–7.

Fowler, G.W. & Lynch, A.M. (1987) Sampling plans in insect pest management based on Wald's sequential probability ratio test. *Environ. Entomol.* **16**, 345–54.

Fracker, S.B. & Brischle, H.A. (1944) Measuring the local distribution of *Ribes*. *Ecology* **25**, 283–303.

France, R. (1992) Use of sequential sampling of amphipod abundance to classify the biotic integrity of acid-sensitive lakes. *Environ. Man.* **16**(2), 157–66.

Gardefors, D. & Orrage, L. (1968) Patchiness of some marine bottom animals: a methodological study. *Oikos* **19**, 311–21.

Gaufin, A.R., Harris, E.K., & Walter, H.J. (1956) A statistical evaluation of stream bottom sampling data obtained from three standard samples. *Ecology* **51**, 643–8.

George, D.C. (1974) Dispersion patterns in the zooplankton populations of a eutrophic reservoir. *J. Anim. Ecol.* **43**, 537–51.

Gerrard, D.I. & Chang, H.C. (1970) Density estimation of corn rootworm egg populations based upon frequency of occurrence. *Ecology* **51**, 237–45.

Gerrard, D.I. & Cook, R.D. (1972) Inverse binomial sampling as a basis for estimating negative binomial population densities. *Biometrics* **28**, 971–80.

Glass, L.W. & Boubjerg, R.V. (1969) Density and dispersion in laboratory populations of caddis fly larvae (Cheumatopsyche. Hydropsychidae). *Ecology* **50**, 1082–4.

Gleason, H.A. (1922) On the relation between species and area. *Ecology* **3**, 158–62.

Gonzalez, D. (1970) Sampling as a basis for pest management strategies. *Tall Timbers Conf. Ecol. Anim. Contr. Habitat Man.* **2**, 83–101.

Goodall, D.W. (1952) Quantitative aspects of plant distribution. *Biol. Rev.* **27**, 194–245.

Gotceitas, V. & Colgan, P. (1991) Assessment of patch profitability and ideal free distribution — the significance of sampling. *Behaviour* **119**, 65–76.

Green, R.H. (1970) On fixed precision level sequential sampling. *Res. Popul. Ecol.* **12**, 249–51.

Green, R.H. (1979) *Sampling Design and Statistical Methods for Environmental Biologists.* John Wiley & Sons, New York.

Greenwood, J.J.D. (1996) Basic techniques. In: Sutherland, W.J. (ed.) *Ecological Census Techniques.* Cambridge University Press, Cambridge, pp. 11–110.

Greig-Smith, P. (1978) *Quantitative Plant Ecology*, 3rd edn. Butterworths, London.

Grundy, P.M. (1952) The fitting of grouped truncated and grouped censored normal distributions. *Biometrika* **39**, 252–9.

Guenther, W.C. (1977) *Sampling Inspection in Statistical Quality Control.* Oxford University Press, New York.

Guppy, I.C. & Harcourt, D.G. (1970) Spatial pattern of the immature stages and teneral adults of *Phyllophaga* spp. (Coleoptera: Scarabaeidae) in a permanent meadow. *Can. Entomol.* **102**, 1354–9.

Hairston, N.G. (1959) Species abundance and community organisation. *Ecology* **40**, 404–16.

Hairston, N.G. (1964) Studies on the organization of animal communities. *J. Anim. Ecol.* **33**(suppl.), 227–39.

Handford, R.H. (1956) Grasshopper population sampling. *Proc. Entomol. Soc. BC* **52**, 3–7.

Hansen, M.H., Hurwitz, W.N. & Madow, W.G. (1953) *Sample Survey Methods and Theory*, Vol. 1. Wiley, New York.

Harcourt, D.G. (1961a) Design of a sampling plan for studies on the population dynamics of the diamondback moth, *Plutella maculipennis* (Curt.) (Lepidoptera: Plutellidae). *Can. Entomol.* **93**, 820–31.

Harcourt, D.G. (1961b) Spatial pattern of the imported cabbageworm, *Pieris rapae* (L.) (Lepidoptera: Pieridae), on cultivated Cruciferae. *Can. Entomol.* **94**, 945–52.

Harcourt, D.G. (1962) Design of a sampling plan for studies on the population dynamics of the imported cabbageworm, *Pieris rapae* (L.) (Lepidoptera: Pieridae). *Can Entomol.* **94**, 849–59.

Harcourt, D.G. (1964) Population dynamics of *Leptinotarsa decemlineata* (Say) in Eastern Ontario. II Population and mortality estimation during six age intervals. *Can. Entomol.* **96**, 1190–8.

Harcourt, D.G. (1965) Spatial pattern of the cabbage looper, *Trichoplusia ni*, on Crucifers. *Ann. Entomol. Soc. Am.* **58**, 89–94.

Harris, E.K. (1957) Further results in the statistical analysis of stream sampling. *Ecology* **38**, 463–8.

Harris, J.W.E., Collis, D.G., & Magar, K.M. (1972) Evaluation of tree-beating method for sampling defoliating forest insects. *Can. Entomol.* **104**, 723–9.

Hartley, H.O. (1950) The maximum F-ratio as a short-cut test for heterogeneity of variance. *Biometrika* **37**, 308–12.

Hassell, M.P. (1978) *The Dynamics of Arthropod Predator–Prey Systems*. Princeton University Press, Princeton, New Jersey.

Hassell, M.P. & May, R.M. (1974) Aggregation of predators and insect parasites and its effect on stability *J. Anim. Ecol.* **43**, 567–94.

Hayman, B.I. & Lowe, A.D. (1961) The transformation of counts of the cabbage aphid (*Brevicoryne brassicae* (L.). *NZ J. Sci.* **4**, 271–8.

Healey, M.J.R. & Taylor, L.R. (1962) Tables for power-law transformations. *Biometrika* **49**, 557–9.

Helson, G.A.H. (1958) Aphid populations: ecology and methods of sampling aphids *Myzus persicae* (Sulz.) and *Aulacorthum solani* (Kltb.). *NZ Entomol.* **2**, 20–3.

Henderson, P.A. (1987) The vertical and transverse distribution of larval herring in the River Blackwater estuary, Essex. *J. Fish Biol.* **31**, 281–90.

Henderson, P.A., Whitehouse, J.W. &

Cartwright, G. (1984) The growth and mortality of larval herring. *Clupea harengus* in the River Blackwater estuary, 1978–1980. *J. Fish Biol.* **24**, 613–22.

Henson, W.R. (1954) A sampling program for Poplar insects. *Can. J. Zool.* **32**, 421–33.

Henson, W.R. (1961) Laboratory studies on the adult behaviour of *Conopthorus coniperda* (Schwarz) (Coleoptera: Scolytidae). II. Thigmotropic aggregation. *Ann. Entomol. Soc. Am.* **54**, 810–19.

Herbert, H.J. & Butler, K.P. (1973) Sampling systems for European red mite, *Panonychus ulmi* (Acarina: Tetranychidae) eggs on apple in Nova Scotia. *Can. Entomol.* **105**, 1519–23.

Heywood, I. & Edwards, R.W. (1961) Some aspects of the ecology of *Potamopyrgus jenkinsi* Smith. *J. Anim. Ecol.* **31**, 239–50.

Hirata, S. (1962) Comparative studies on the population dynamics of important lepidopterous pests on cabbage. 2. On the habits of oviposition of *Pieris rapae crucivora*, *Plusia nigrisigna* and *Manestra* (*Barathra*) *brassicae* on cabbage plants. *Japan. J. Appl. Entomol. Zool.* **6**, 200–7.

Ho, C.-C. (1993) Dispersion statistics and sample size estimates for *Tetranychus kanzawai* (Acari: Tetranychidae) on mulberry. *Environ. Entomol.* **22**, 21–5.

Holme, N.A. (1950) Population dispersion in *Tellina tenuis* Da Costa. *J. Mar. Biol. Assoc. UK* **299**, 267–80.

Howe, R.W. (1963) The random sampling of cultures of grain weevils. *Bull. Entomol. Res.* **54**, 135–46.

Hudson, M. & LeRoux, E.J. (1961) Variation between samples of immature stages, and of mortalities from some factors, of the european corn borer, *Ostrinia nubilalis* (Hubner) (Lepidoptera: Pyralidae) on sweet corn in Quebec. *Can. Entomol.* **93**, 867–88.

Hughes, R.D. (1955) The influence of the prevailing weather on the numbers of *Meromyza variegata* Meigen (Diptera, Chloropidae) caught with a sweepnet. *J. Anim. Ecol.* **24**, 324–35.

Hughes, R.D. (1962) The study of aggregated populations. In: Murphy, P.W. (ed.) *Progress in Soil Zoology*. Butterworths, London, pp. 51–5.

Ito, Y., Gotoh, A., & Miyashita, K. (1960) On the spatial distribution of *Pieris rapae crucivora* population. *Japan. J. Appl. Entomol. Zool.* **4**, 141–5.

Ito, Y., Nakamura, M., Kondo, M., Miyashita, K., & Nakamura, K. (1962) Population dynamics of the chestnut gall-wasp, *Dryocosmus kuriphilus*

Yasumatsu (Hymenoptera: Cynipidae). 11. Distribution of individuals in bud of chestnut tree. *Res. Popul. Ecol.* **4**, 35–46.

Ives, W.G.H. (1954) Sequential sampling of insect populations. *For. Chron.* **30**, 287–91.

Ives, W.G.H. (1955) Effect of moisture on the selection of cocooning sites by larch sawfly, *Pristiphora erichsonii* (Hartwig). *Can. Entomol.* **87**, 301–11.

Iwao, S. (1956) The relation between the distribution pattern and the population density of the large twenty-eight-spotted lady beetle, *Epilachna 28-maculata* Motschulsky, in eggplant field: pattern of the spatial distribution of insect 6. *Japan. J. Ecol.* **5**, 130–5.

Iwao, S. (1968) A new regression method for analysing the aggregation pattern of animal populations. *Res. Popul. Ecol.* **10**, 1–20.

Iwao, S. (1970a) Analysis of contagiousness in the action of mortality factors on the western tent caterpillar population by using the m–m relationship. *Res. Popul. Ecol.* **12**, 100–10.

Iwao, S. (1970b) Analysis of spatial patterns in animal populations: progress of research in Japan. *Rev. Plant Prot. Res.* **3**, 41–54.

Iwao, S. (1970c) Problems of spatial distribution in animal population ecology. In: Patil, G.P. (ed.) *Random Counts in Biomedical and Social Sciences*. Penn State University Press, Philadelphia, pp. 117–49.

Iwao, S. (1972) Application of the *m–m method to the analysis of spatial patterns by changing the quadrat size. *Res. Popul. Ecol.* **14**, 97–128.

Iwao, S. (1975) A new method of sequential sampling to classify populations relative to a critical density. *Res. Popul. Ecol.* **16**, 281–8.

Iwao, S. & Kuno, E. (1971) An approach to the analysis of aggregation pattern in biological populations. In: Patil, G.P., Pielou, E.C., & Waters, W.E. (eds) *Statistical Ecology 1: Spatial Patterns and Statistical Distributions*. Penn State University Press, Philadelphia, 461–515.

Jepson, W.F. & Southwood, T.R.E. (1958) Population studies on *Oscinella frit* L. *Ann. Appl. Biol.* **46**, 465–74.

Kapatos, E., McFadden, M.W., & Pappas, S. (1977) Sampling techniques and preparation of partial life tables for the olive fly, *Dacus oleae* (Diptera: Trypetidae) in Corfu. *Ecol. Entomol.* **2**, 193–6.

Karandinos, M.G. (1976) Optimum sample size and comments on some published formulae. *Bull. Entomol.* **22**, 417–21.

Katti, S.K. (1966) Interrelations among generalized distributions and their components. *Biometrics* **22**, 44–52.

Kendall, D.G. (1948) On some modes of population growth leading to R.A. Fisher's logarithmic series distribution. *Biometrika* **35**, 6–15.

Kennedy, J.S. & Crawley, L. (1967) Spaced-out gregariousness in sycamore aphids *Drepanosiphum platanoides* (Schrank) (Hemiptera Callaphididae). *J. Anim. Ecol.* **369**, 147–70.

Keuls, M., Over, H.I., & De Wit, C.T. (1963). The distance method for estimating densities. *Stat. Neerland.* **17**, 71–91.

Kieckhefer, R.W. (1975) Field populations of cereal aphids in South Dakota spring grains. *J. Econ. Entomol.* **68**, 161–4.

Kobayashi, S. (1966) Process generating the distribution pattern of eggs of the common cabbage butterfly, *Pieris rapae crucivora*. *Res. Popul. Ecol.* **8**, 51–60.

Kono, T. (1953) Basic unit of population observed in the distribution of the rice-stem borer, *Chilo simplex*, in a paddy field. *Res. Popul. Ecol.* **2**, 95–105.

Krause, G.F. & Pedersen, J.R. (1960) Estimating immature populations of rice weevils in wheat by using subsamples. *J. Econ. Entomol.* **53**, 215–16.

Kuehl, R.O. & Fye, R.E. (1972) An analysis of the sampling distributions of cotton insects in Arizona. *J. Econ. Entomol.* **65**, 855–60.

Kuno, E. (1963) A comparative analysis on the distribution of nymphal populations of some leaf- and planthoppers on rice plant. *Res. Popul. Ecol.* **5**, 31–43.

Kuno, E. (1969) A new method of sequential sampling to obtain the population estimates with a fixed level of precision. *Res. Popul. Ecol.* **11**, 127–36.

Kuno, E. (1986) Evaluation of statistical precision and design of efficient sampling for the population estimation based on frequency of occurrence. *Res. Popul. Ecol.* **28**, 305–19.

Kuno, E. (1991) Verifying zero-infestation in pest control: a simple sequential test based on the succession of zero-samples. *Res. Popul. Ecol.* **33**, 29–32.

Legay, J.M. (1963) A propos de la répartition de la cécidomyie du Hêtre, *Mikiola fagi*. Un exemple de distribution binomiale négative. *Ann. Epiphytol. C* **14**, 49–56.

Legg, D.E., Nowierski, R.M., Feng, M.G. *et al.* (1994) Binomial sequential sampling plans and decision support algorithms for managing the Russian wheat aphid (Homoptera: Aphididae) in small grains. *J. Econ. Entomol.* **87**, 1513–33.

LeRoux, E.J. (1961) Variations between samples of fruit, and of fruit damage mainly from insect pests, on apple in Quebec. *Can. Entomol.* **93**, 680–94.

LeRoux, E.J. & Reimer, C. (1959) Variation between samples of immature stages and of mortalities from some factors of the eye-spotted bud moth, *Spilonota ocellana* (D & S) (Lepidoptera: Olethreutidae), and the pistol casebearer, *Coleophora serratella* (L.) (Lepidoptera: Coleophoridae), on apple in Quebec. *Can. Entomol.* **91**, 428–49.

Lloyd, M. (1967) Mean crowding. *J. Anim. Ecol.* **36**, 1–30.

Lyons, L.A. (1964) The spatial distribution of two pine sawflies and methods of sampling for the study of population dynamics. *Can. Entomol.* **96**, 1373–407.

McIntyre, A.D. & Warwick, R.M. (1984) Meiofauna techniques. In: Holme, N.A. & McIntyre, A.D. (eds) *Methods for the Study of Marine Benthos*. Blackwell Scientific Publications, Oxford, 217–44.

McIntyre, A.D., Elliott, J.M., & Ellis, D.V. (1984) Introduction: design of sampling programmes. In: Holme, N.A. & McIntyre, A.D. (eds) *Methods for the Study of Marine Benthos*. Blackwell Scientific Publications, Oxford.

MacLellan, C.R. (1962) Mortality of codling moth eggs and young larvae in an integrated control orchard. *Can. Entomol.* **94**, 655–66.

Macleod, J. (1958) The estimation of numbers of mobile insects from low incidence recapture data. *Trans. Roy. Entomol. Soc. Lond.* **110**, 363–92.

Mailloux, G. & Bostanian, N.J. (1989) Presence–absence sequential decision plans for management of *Lygus lineolaris* (Hemiptera: Miridae) on strawberry. *Environ. Entomol.* **18**, 827–34.

Manly, B.F.J. (1969) Some properties of a method of estimating the size of mobile animal populations. *Biometrika* **56**, 407–10.

May, R.M. (1978) Host–parasitoid systems in a patchy environment: a phenomenological study. *J. Anim. Ecol.* **47**, 833–43.

Meagher, R.L., Wilson, L.T., & Pfannenstiel, R.S. (1996) Sampling *Eoreuma loftini* (Lepidoptera: Pyralidae) on Texas sugarcane. *Environ. Entomol.* **25**, 7–16.

Milne, A. (1959) The centric systematic area-sample treated as a random sample. *Biometrics* **15**, 270–97.

Milne, A. (1964) Biology and ecology of the garden chafer, *Phyllopertha horticola* (L.), IX. Spatial distribution. *Bull. Entomol. Res.* **54**, 761–95.

Morisita, M. (1954) Estimation of population density by spacing method. *Mem. Fac. Sci. Kyushu Univ. E* **1**, 187–97.

Morisita, M. (1959) Measuring of the dispersion of individuals and analysis of the distributional patterns. *Mem. Fac. Sci. Kyushu Univ. E (Biol.)* **2**, 215–35.

Morisita, M. (1962) I_δ-index, a measure of dispersion of individuals. *Res. Popul. Ecol.* **49**, 1–7.

Morisita, M. (1964) Application of I_δ-index to sampling techniques. *Res. Popul. Ecol.* **69**, 43–53.

Morris, R.F. (1954) A sequential sampling technique for spruce budworm egg surveys. *Can. J. Zool.* **32**, 302–13.

Morris, R.F. (1955) The development of sampling techniques for forest insect defoliators, with particular reference to the spruce budworm. *Can. J. Zool.* **33**, 225–94.

Morris, R.F. (1960) Sampling insect populations. *Ann. Rev. Entomol.* **5**, 243–64.

Morris, R.F. & Reeks, W.A. (1954) A larval population technique for winter moth, *Operophtera brumata* (Linn.) (Lepidoptera: Geometridae). *Can. Entomol.* **86**, 433–8.

Mukerji, M.K. & Harcourt, D.G. (1970) Spatial pattern of the immature stages of *Hylemya brassicae* on cabbage. *Can. Entomol.* **102**, 1216–22.

Murdie, G. & Hassell, M.P. (1973) Food distribution, searching success and predator-prey models. In: Bartlett, M.S. & Hiorns, R.W. (eds) *The Mathematical Theory of the Dynamics of Biological Populations*. Academic Press, London, pp. 87–101.

Nachman, G. (1984) Estimates of mean population density and spatial distribution of *Tetranychus urticae* (Acarina: Tetranychidae) and *Phytoseiulus persimilis* (Acarina: Phytoseiidae) based upon the proportion of empty sampling units. *J. Appl. Ecol.* **21**, 903–13.

Naranjo, S.E. & Flint, H.M. (1995) Spatial distribution of adult *Bemisia tabaci* (Homoptera: Aleyrodidae) in cotton and development and validation of fixed-precision sampling plans for estimating population density. *Environ. Entomol.* **24**, 261–70.

Naylor, A.F. (1959) An experimental analysis of dispersal in the flour beetle, *Tribolium confusum*. *Ecology* **40**, 453–65.

Nelson, W.A., Slen, S.B., & Banky, E.C. (1957) Evaluation of methods of estimating populations of sheep ked, *Melopgagus ovinus* (L) (Diptera: Hippoboscidae), on mature ewes and young lambs. *Can. J. Anim. Sci.* **37**, 8–13.

Nestel, D. & Klein, M. (1995) Geostatistical analysis of leafhopper (Homoptera: Cicadellidae) colonisation and spread in deciduous orchards. *Environ. Entomol.* **24**, 1032–9.

Neyman, J. (1939) On a new class of 'contagious' distributions, applicable in entomology and bacteriology. *Ann. Math. Stat.* **10**, 35–57.

Ng, Y.-S., Trout, J.R., & Ahmad, S. (1983) Spatial distribution of the larval populations of the Japanese beetle (Coleoptera:Scarabaeidae) in turf grass. *J. Econ. Entomol.* **76**, 26–30.

Niwa, H.-S. (1994) Self-organizing model of fish schooling. *J. Theor. Biol.* **171**, 123–36.

Nyrop, J.P. & Binns, M. (1991) Quantitative methods for designing and analyzing sampling programs for use in pest management. In: Pimental, D. (ed.) *CRC Handbook of Pest Management*. CRC Press, Boca Raton, Florida, 67–132.

Nyrop, J.P., Agnello, A.M., Kovach, J., & Reissig, W.H. (1989) Binomial sequential classification sampling plans for European red mite (Acari: Tetranychidae) with special reference to performance criteria. *J. Econ. Entomol.* **82**, 482–90.

Nyrop, J.P., Villani, M.G., & Grant, J.A. (1995) Control decision rule for European chafer (Coleoptera: Scarabaeidae) larvae infesting turfgrass. *Environ. Entomol.* **24**, 521–8.

Oakland, G.B. (1950) An application of sequential analysis to whitefish sampling. *Biometrics* **6**, 59–67.

Omori, M. & Ikeda, T. (1984) *Methods in Marine Zooplankton Ecology*. John Wiley & Sons, New York.

Overgaard Nielsen, B. (1963) The biting midges of Lyngby Aamose (Culicoides: Ceratopogonidae). *Nat. Jutland.* **10**, 46 pp.

Pahl, P.J. (1969) On testing for goodness-of-fit of the negative binomial distribution when expectations are small. *Biometrics* **25**, 143–51.

Paradis, R.O. & LeRoux, E.J. (1962) A sampling technique for population and mortality factors of the fruit-tree leaf roller, *Archips argyrospilus* (WLK) (Lepidoptera: Tortricidae), on apple in Quebec. *Can. Entomol.* **94**, 561–73.

Patil, G.P. & Joshi, S.W. (1968) *A Dictionary and Bibliography of Discrete Distributions*. Oliver & Boyd, Edinburgh.

Pearson, E.S. & Hartley, H.O. (1958) *Biometrika Tables for Statisticians*. Cambridge University Press, Cambridge.

Pedigo, L.P. & Buntin, G.D. (eds) (1994) *Handbook of Sampling Methods for Arthropods in Agriculture*. CRC Press, Boca Raton, USA.

Pena, J.E. & Baranowski, R.M. (1990) Dispersion indices and sampling plans for the broad mite (Acari: Tarsonemidae) and the citrus rust mite (Acari: Eriophyidae) on limes. *Environ. Entomol.* **19**, 378–82.

Pena, J.E. & Duncan, R. (1992) Sampling methods for *Prodiplosis longifila* (Diptera: Cecidomyiidae) in limes. *Environ. Entomol.* **21**, 996–1001.

Peng, C. & Brewer, G.J. (1994) Spatial distribution of the red sunflower seed weevil (Coleoptera: Curculionidae) on sunflower. *Environ. Entomol.* **23**, 1101–5.

Perry, J.N. (1981) Taylor's power law for dependence of variance on mean in animal populations. *Appl. Stat.* **30**, 254–63.

Perry, J.N. (1988) Some models for spatial variability of animal species. *Oikos* **51**, 124–30.

Perry, J.N. (1994) Chaotic dynamics can generate Taylor's power law. *Proc. Roy. Soc. Lond. B* **257**, 221–6.

Perry, J.N. (1995) Spatial analysis by distance sampling. *J. Anim. Ecol.* **64**, 303–14.

Perry, J.N. & Hewitt, M. (1991) A new index of aggregation for animal counts. *Biometrics* **47**, 1505–18.

Perry, J.N. & Woiwod, I.P. (1992) Fitting Taylor's power law. *Oikos* **65**, 538–42.

Pielou, E.C. (1969) *An Introduction to Mathematical Ecology*. Wiley Interscience, New York, London.

Pieters, E.P. (1977) Comparison of sample-unit sizes for D-Vac sampling of cotton arthropods in Mississippi. *J. Econ. Entomol.* **71**, 107–8.

Potvin, C. & Roff, D.A. (1993) Distribution free and robust statistical methods: viable alternative to parametric statistics? *Ecology* **74**, 1617–28.

Pradhan, S. & Menon, R. (1945) Insect population studies. I. Distribution and sampling of spotted bollworm of cotton. *Proc. Nat. Inst. Sci. India* **6**(2), 61–73.

Prebble, M.L. (1943) Sampling methods in population studies of the European spruce sawfly, *Gilpinia hercyniae* (Hartig.), in eastern Canada. *Trans. Roy. Soc. Can. III, V* **37**, 93–126.

Preston, F.W. (1948) The commonness and rarity of species. *Ecology* **29**, 254–83.

Pringle, K.L. & Giliomee, J.H. (1992) Dispersion statistics and sample size estimates for monitoring mite populations in commercial apple orchards. *J. Appl. Ecol.* **29**, 143–9.

Putnam, L.G. & Shklov, N. (1956) Observations on the distribution of grasshopper egg-pods in Western Canadian stubble fields. *Can. Entomol.* **88**, 110–17.

Quenouille, M.H. (1949) A relation between the

logarithmic, Poisson, and negative binomial series. *Biometrics* **5**, 162–4.

Reimer, C. (1959) Statistical analysis of percentages based on unequal numbers, with examples from entomological research. *Can. Entomol.* **91**, 88–92.

Resh, V.R. & Price, D.G. (1984) Sequential sampling: a cost-effective approach for monitoring benthic invertebrates in environmental impact assessments. *Environ. Man.* **8**, 75–80.

Riddle, M.J. (1989) Precision of the mean and the design of benthos sampling. *Mar. Biol. (Berlin)* **103**(2), 225–30.

Riley, D.G., Schuster, D.J., & Barfield, C.S. (1992) Sampling and dispersion of pepper weevil (Coleoptera: Curculionidae) adults. *Environ. Entomol.* **21**, 1013–21.

Romell, L.G. (1930) Comments on Raunkiaer's and similar methods of vegetation analysis and the 'law of frequency'. *Ecology* **11**, 589–96.

Routledge, R.D. & Swartz, A. (1991) Taylor's Power Law re-examined. *Oikos* **60**, 107–12.

Ruesink, W.G. (1980) Introduction to sampling theory. In: Kogan, M. & Herzog, D.C. (eds) *Sampling Methods in Soybean Entomology*. Springer-Verlag, New York.

Safranyik, L. & Graham, K. (1971) Edge-effect bias in the sampling of sub-cortical insects. *Can. Entomol.* **103**, 240–55.

Saila, S.B., Pikanowski, R.A., & Vaughan, D.S. (1976) Optimum allocation strategies for sampling benthos in New York Bight. *Estuar. Coast. Mar. Sci.* **4**, 119–28.

Salt, C. & Hollick, F.S. (1946) Studies of wireworm populations. 11. Spatial distribution. *J. Exp. Biol.* **23**, 1–46.

Sawyer, A.J. (1989) Inconstancy of Taylor's *b*: simulated sampling with different quadrat sizes and spatial distributions. *Res. Popul. Ecol.* **31**, 11–24.

Schaalje, G.B., Butts, R.A., & Lysyk, T.J. (1991) Simulation studies of binomial sampling: a new variance estimator and density predictor, with special reference to the Russian wheat aphid (Homoptera: Aphididae). *J. Econ. Entomol.* **84**, 140–7.

Scheiner, S.M. & Gurevitch, J. (eds) (1993) *Design and Analysis of Ecological Experiments*. Chapman & Hall, New York.

Seber, G.A.F. (1982) *The Estimation of Animal Abundance and Related Parameters*. Griffin, London.

Shaw, M.W. (1955) Preliminary studies on potato aphids in north and north-east Scotland. *Ann. Appl. Biol.* **43**, 37–50.

Shelton, A.M. & Trumble, J.T. (1991) Monitoring insect populations. In: Pimentel, D. (ed.) *Handbook of Pest Management in Agriculture*, 2nd edn. CRC Press, Boca Raton, 45–55.

Shenton, L.R. & Wallington, P.A. (1962) The bias of moment estimators with an application to the negative binomial distribution. *Biometrika* **49**, 193–204.

Shibuya, M. & Ouchi, Y. (1955) Pattern of spatial distribution of soy bean pod gall midge in a soy bean field. *Oyo-Kontyu* **11**, 91–7.

Shinozaki, K. & Urata, N. (1953) Apparent abundance of different species and heterogeneity. *Res. Popul. Ecol.* **2**, 8–21.

Shiyomi, M. & Nakamura, K. (1964) Experimental studies on the distribution of the aphid counts. *Res. Popul. Ecol.* **6**, 79–87.

Skellam, J.G. (1958) On the derivation and applicability of Neyman's type A distribution. *Biometrika* **45**, 32–6.

Smith, M.H., Gardner, R.H., Gentry, J.B., Kaufman, D.W., & O'Farrell, M.H. (eds) (1975) *Density Estimations of Small Mammal Populations. Small Mammals, Their Productivity and Population Dynamics*. Cambridge University Press, Cambridge.

Snedecor, G.W. & Cochran, W.G. (1989) *Statistical Methods*. University of Iowa Press, Ames, Iowa.

Southwood, T.R.E., Jepson, W.F., & van Emden, H.F. (1961) Studies on the behaviour of *Oscinella frit* L. (Diptera) adults of the panicle generation. *Entomol. Exp. Appl.* **4**, 196–210.

Spiller, D. (1952) Truncated log-normal distribution of red scale (*Aonidella aurantiim* Mask.) on citrus leaves. *NZ J. Sci. Technol. (B)* **33**, 483–7.

Stark, R.W. (1952) Sequential sampling of lodgepole needle miner. *For. Chron.* **28**, 57–60.

Stark, R.W. & Dahlsten, D.L. (1961) Distribution of cocoons of a *Neodiprion* sawfly under open-grown conditions. *Can. Entomol.* **93**, 443–50.

Steffey, K.L., Tollefson, J.J., & Hinz, P.N. (1982) Sampling plan for population estimation of northern and western corn rootworm adults in Iowa cornfields. *Environ. Entomol.* **11**, 287–91.

Steiner, M. (1990) Determining population characteristics and sampling procedures for the western flower thrips (Thansanoptera: Thripidae) and the predatory mite *Amblyseius cucumeris* (Acari: Phytoseiidae) on greenhouse cucumber. *Environ. Entomol.* **19**, 1605–13.

Stimson, J. (1974) An analysis of the pattern of dispersion of the hermatypic coral, *Pocillopora meandrina* var *nobilis* Verrill. *Ecology* **55**, 445–9.

Stuart, A. (1962) *Basic Ideas of Scientific Sampling*. Griffin, London.

Sylvester, E. & Cox, E.L. (1961) Sequential plans for sampling aphids on sugar beets in Kern County, California. *J. Econ. Entomol.* **54**, 1080–5.

Takeda, S. & Hukusima, S. (1961) Spatial distribution of the pear lace bugs, *Stephanitis naski* Esaki et Takeya (Hemiptera: Tingitidae) in an apple tree and an attempt for estimating their populations. *Res. Bull. Fac. Agric. Gifu Univ.* **14**, 68–77.

Tanigoshi, L.K., Browne, R.W., & Hoyt, S.C. (1975) A study on the dispersion pattern and foliage injury by *Tetranychus medanieli* (Acarina: Tetranychidae) in simple apple ecosystems. *Can. Entomol.* **107**, 439–46.

Taylor, L.R. (1961) Aggregation, variance and the mean. *Nature* **189**, 732–5.

Taylor, L.R. (1965) A natural law for the spatial disposition of insects. *Proc. XII Int. Congr. Entomol.* 396–7.

Taylor, L.R. (1970) Aggregation and the transformation of counts of *Aphis fabae* Scop. on beans. *Ann. Appl. Biol.* **65**, 181–9.

Taylor, L.R. (1984) Assessing and interpreting the spatial distributions of insect populations. *Ann. Rev. Entomol.* **29**, 321–57.

Taylor, L.R., Kempton, R.A., & Woiwood, I.P. (1976) Diversity statistics and the log-series model. *J. Anim. Ecol.* **45**, 255–72.

Taylor, L.R., Wolwood, I.P., & Perry, J.N. (1978) The density-dependence of spatial behaviour and rarity of randomness. *J. Anim. Ecol.* **47**, 383–406.

Taylor, L.R., Woiwod, I.P., & Perry, J.N. (1979) The negative binomial as a dynamic ecological model for aggregation, and the density dependence of (*k*). *J. Anim. Ecol.* **48**, 289–304.

Taylor, W.D. (1980) Comment on aggregation, transformations, and the design of benthos sampling programs. *Can. J. Fish. Aquat. Sci.* **37**, 1328–9.

Terry, I., Bradley, J.R., & van Duyn, J.W. (1989) *Heliothis zea* (Lepidoptera: Noctuidae) eggs in soybeans: within field distribution and precision level sequential count plans. *Environ. Entomol.* **18**, 908–16.

Teulon, D.A.J. & Cameron, E.A. (1995) Within-tree distribution of pear thrips (Thysanoptera: Thrioidae) in sugar maple. *Environ. Entomol.* **24**, 233–8.

Teyrovsky, V. (1956) Fotopathie larey klestánek (Corixinae). *Acta Univ. Agric. Silv. Brunn.* **2**, 147–77.

Thomas, M. (1949) A generalization of Poisson's binomial limit for use in ecology. *Biometrika* **36**, 18–25.

Thomas, M. & Jacob, F.H. (1949) Ecology of potato aphids in north Wales. *Ann. Appl. Biol.* **30**, 97–101.

Thompson, H.R. (1956) Distribution of distance to *n*th neighbour in a population of randomly distributed individuals. *Ecology* **37**, 391–4.

Tokeshi, M. (1995) On the mathematical basis of the variance–mean power relationship. *Res. Popul. Ecol.* **37**, 43–8.

Tonhasca, A., Palumbo, J.C., & Byrne, D.N. (1994) Binomial sampling plans for estimating *Bemisia tabacci* populations in Cantaloupes. *Res. Popul. Ecol.* **36**, 181–6.

Trumble, J.T., Brewer, M.J., Shelton, A.M., & Nyrop, J.P. (1989) Transportability of fixed-precision sampling plans. *Res. Popul. Ecol.* **31**, 325–42.

Tukey, J.W. (1949) One degree of freedom for non-additivity. *Biometrics* **5**, 232–42.

Tunstall, J.P. & Matthews, G.A. (1961) Cotton insect control recommendations for 1961–2 in the Federation of Rhodesia and Nyasaland. *Rhodesia Agric. J.* **58**(5), 289–99.

Turner, F.B. (1960) Size and dispersion of a Louisiana population of the cricket frog, *Acris gryllus*. *Ecology* **41**, 258–68.

Underwood, A.J. (1976) Nearest neighbour analysis of spatial dispersion of intertidal Prosobrunel gastropods within two substrata. *Oecologia* **26**, 257–66.

Underwood, A.J. (1977) *Experiments in Ecology: Their Logical Design and Interpretation Using Analysis of Variance*. Cambridge University Press, Cambridge.

Unterstenhöfer, G. (1957) The basic principles of plant protection field tests. *Höfchen Briefe* **10**, 173–236.

Upholt, W.M. & Craig, R. (1940) A note on the frequency distribution of black-scale insects. *J. Econ. Entomol.* **33**, 113–14.

Usher, M.B. (1971) Properties of the aggregations of soil arthropods, particularly Mesostigmata (Acarina). *Oikos* **22**, 43–9.

Wadley, F.M. (1950) Notes on the distribution of insect and plant populations. *Ann. Entomol. Soc. Am.* **43**, 581–6.

Wadley, F.M. (1952) Elementary sampling principles in entomology. *USDA Pl. Quar. Bur. Entomol.* ET **302**, 17 pp.

Walker, P.W. & Allsopp, P.G. (1993) Sampling distributions and sequential sampling plans for *Eumargarodes laingi* and *Promargarodes* spp. (Hemiptera: Margarodidae) in Australian sugarcane. *Environ. Entomol.* **22**, 10–15.

Waloff, N. & Blackith, R.E. (1962) The growth and distribution of the mounds of *Lasius flavus* (Fabricius) (Hym.: Formicidae) in Silwood Park, Berkshire. *J. Anim. Ecol.* **31**, 421–37.

Waters, W.E. & Henson, W.R. (1959) Some sampling attributes of the negative binomial distribution with special reference to forest insects. *J. Econ. Entomol.* **52**, 1180–4.

Watt, A.S. (1964) The community and the individual. *J. Econ. Entomol.* **52**, 618–21.

Weithman, A.S., Reynolds, J.B., & Simpson, D.E. (1979) Assessment of structure of largemouth bass stocks by sequential sampling. *Proc. Ann. Conf. Southeastern Ass. Fish Wildlife Agencies* **33**, 415–24.

Wilson, L.F. (1959) Branch tip sampling for determining abundance of spruce budworm egg masses. *J. Econ. Entomol.* **52**, 618–21.

Wilson, L.F. & Gerrard, D.J. (1971) A new procedure for rapidly estimating european pine sawfiy (Hymenoptera: Diprionidae) population levels in young pine plantations. *Can. Entomol.* **103**, 1315–22.

Yates, F. (1953) *Sampling Methods for Censuses and Surveys.* Griffin, London.

Yates, F. & Finney, D.J. (1942) Statistical problems in field sampling for wireworms. *Ann. Appl. Biol.* **29**, 156–67.

Yoshihara, T. (1953) On the distribution of *Tectarius granularis. Res. Popul. Ecol.* **2**, 112–22.

Zahl, S. (1974) Application of the S-method to the analysis of spatial pattern. *Biometrics* **30**, 513–24.

Zar, J.H. (1984) *Biostatistical Analysis.* Prentice-Hall, Englewood Cliffs, New Jersey.

Zehnder, G.W. (1990) Evaluation of various potato plant sample units for cost-effective sampling on Colorado potato beetle (Coleoptera: Crysomelidae). *J. Econ. Entomol.* **83**, 428–33.

3 Absolute Population Estimates Using Capture–Recapture Experiments

When an absolute estimate of population size is required, these methods are the main alternative approach to those based on the count of animals within a fixed unit of the habitat. They offer the advantage that their accuracy does not depend on an assessment of the amount of habitat; their disadvantage is that their accuracy does depend on capturing a large proportion of the population. However, even an order-of-magnitude estimate may be satisfactory as a second, independent, check on population size; as stressed in Chapter 1, it is wise, whenever possible, to use more than one independent method. Seber (1982, 1986) reviews the statistical theory that underlies this approach. Originally developed to estimate population size, capture–recapture methods are now frequently used for the estimation of birth, death, and emigration rates. Lebreton *et al.* (1992) review methodologies for the estimation of survival rates. Populations can be classified as either open or closed. A closed population remains unchanged over the period of study, whereas an open population changes because of some combination of birth, death, and emigration. Reviews of methods applicable to closed and open populations are those of White *et al.* (1982) and of Pollock *et al.* (1990), respectively. Descriptions of the application of the methodology to particular animal groups are given by Nichols *et al.* (1981) for birds, by Burnham *et al.* (1987) for fish, and by Eberhardt *et al.* (1979) for marine mammals.

A mark–recapture method was first used for ecological study in 1896 by Petersen to estimate plaice, *Platichthys platessa*, populations; later, Lincoln (1930) independently developed the method to estimate waterfowl populations. The principle used to calculate population size in these early studies, in which the population was assumed closed, still applies. If a sample from a population is marked, returned to the original population, and then, after complete mixing, resampled, the number of marked individuals in the second sample will have the same ratio to the total numbers in the second sample as the total of marked individuals originally released have to the total population. A basic prerequisite for the use of these methods is a technique for marking the animals so that they can be released unharmed and unaffected into the wild and recognized again on recapture. Such techniques may also be used in other types of study, e.g. dispersal, longevity and growth, but for convenience all marking methodologies for animals will be discussed in the second half of this chapter.

3.1 Capture–recapture methods

Various assumptions underlie all methods of capture–recapture analysis. If the particular animal does not fulfil one or more conditions, it might be possible to allow for this to some extent, but a method of analysis should not be applied without ensuring, as far as is practicable, that its inherent assumptions are satisfied. Parsimony — the use of the smallest possible number of parameters to model a situation — is essential, so, though assumptions can be avoided by adding further parameters to a model, this inevitably leads to an increase in the variability of parameter estimates (Cormack, 1979; Begon, 1983). In the following account, models are presented in order of increasing complexity and decreasing number of underlying assumptions. The search for parsimony has led to the development of computer programs to identify and fit the most appropriate model from the wide range available (Crosbie & Manly, 1985; Lebreton & Clobert, 1986; Lebreton *et al.*, 1992). Table 3.1 lists the more readily available programs for analysis of mark–recapture experiments and tag–return data in general.

At the outset of study, neither the size of the population nor the correct assumptions to make about the capture probabilities are known. The ecologist therefore faces the task of both estimating the population parameters and identifying the most appropriate model for the purpose. Fieldwork should not be undertaken until the applicability of the method has been assessed. The effort that will be expended

Table 3.1 Computer software for the estimation of population size using mark–recapture data. All of the programs tabulated are for use on personal computers and most are available free of charge from the web from the addresses given.

Name	Description	Locality
CAPTURE	Calculates population size for closed population mark–recapture data	Patuxent software archive at http://www.mbr.nbs.gov/software.html
JOLLY	Computes population size, survival and capture probabilities for open population mark–recapture data.	Patuxent software archive at http://www.mbr.nbs.gov/software.html
JOLLYAGE	Computes population size, survival and capture probabilities for open population mark–recapture data with two age classes	Patuxent software archive at http://www.mbr.nbs.gov/software.html
MARK	Calculates population size and a range of other parameters from mark–recapture and other forms of re-encounter data.	http://www.cnr. colostate.edu/~gwhite/mark/mark.htm
POPAN	Computer package for managing and analysing mark–recapture data for open populations with multiple cohorts.	http://www.cs.umanitoba.ca/~popan/
SURGE	Widely used program for the analysis of mark–recapture data. It was not designed for use by those unfamiliar with the methods.	http://www.biol.sfu.ca/cmr/surge
POP/PRO	Population estimation using mark–recapture and removal sampling methods aimed mainly at fisheries workers.	http://comp.uark.edu/~tkwak/pp.html

on a capture–recapture study will be determined by resource availability and the percentage accuracy required (see Chapter 2). The latter is defined as:

$$A = \pm 100 \left(\frac{\text{Estimated} - \text{True}}{\text{True}} \right) \qquad (3.1)$$

Robson and Reiger (1964) suggest that A should range from ±50% for preliminary surveys, ±25% for management work, to ±10% for research work. Examination of published studies suggests that these ranges have rarely been achieved. Capture–recapture methods are labour-intensive and quite high proportions of the population must be marked if even 50% accuracy is to be gained. Having defined the required accuracy and made a guess at population size, it is helpful first to estimate the sampling effort that would be required for a simple Peterson–Lincoln index estimation (p. 83). All other methods will require greater effort. If the required effort is feasible, simulated data can then be explored to gain insight into the performance of the proposed models and their sensitivity to violations in the underlying assumptions.

3.1.1 Assumptions common to most methods
There are four.
1 The marked animals are not affected (either in behaviour or life expectancy) by being marked and the marks will not be lost.
2 The marked animals become completely mixed in the population.
3 The probability of capturing a marked animal is the same as that of capturing any member of the population; that is, the population is sampled randomly with respect to its mark status, age, and sex. Termed 'equal catchability', this assumption has two aspects: first, that all individuals of the different age-groups and of both sexes are sampled in the proportion in which they occur; secondly, that all the individuals are equally available for capture irrespective of their position in the habitat.
4 Sampling must be at discrete time intervals and the actual time involved in taking the samples must be small in relation to the total time.
The violation of these assumptions may invalidate the capture–recapture method, so care must be taken to test them, as will be discussed below.

3.1.1.1 Marking has no effect
Manly (1971a) has described a test to determine if the survival of individuals immediately after marking and release is different from that of animals that have been marked for some while (mortalities associated with marking are termed type I losses in fisheries science). It is thus a test of the effect of the marking process and not of the influence of bearing a mark. It depends on each animal bearing a unique mark and may be used in conjunction with the actual data derived from a series of, at least four, marking occasions. The difference (y) between the survival of animals newly exposed to marking and others is given by:

$$y = \ln\left(\frac{r_i'}{r_i} \times \frac{u_i}{R_i}\right) \tag{3.2}$$

where r_i' = the number of marked animals in the ith sample that are recaptured again later, r_i = the total marked animals in the ith sample, u_i = the total unmarked animals in the ith sample, and R_i = the total animals marked on the ith occasion and subsequently recaptured (see Table 3.6, p. 96). Now, if the process of marking has no effect on the animal, y will have an expected value of zero. The significance of any value of y can be determined by the statistic g, derived by dividing y by its variance:

$$g = \frac{y}{\left(\dfrac{1}{r_i'} + \dfrac{1}{R_i} - \dfrac{1}{r_i} - \dfrac{1}{u_i}\right)^{\frac{1}{2}}} \tag{3.3}$$

g will behave as a random normal variate with zero mean and variance. The one-tail test is normally justified because an increase in survival due to the marking process can be ruled out and thus, at the 5% level, $g < 1.64$ or, at the 10% level, $g < 1.29$ may be taken as showing that the marking process has no effect on survival. Robson (1969) has developed a similar test.

3.1.1.2 Marks will not be lost
Mark loss may be detected by double marking (Seber & Felton, 1981), where each marked individual carries two different marks. The total number of recaptures is given by:

$$r = r_a + r_b + r_{ab} \tag{3.4}$$

where r_a is the number of recaptures with only type a tag, r_b the number with only type b tag, and r_{ab} those with both a and b tags.

Seber (1982) shows that, if:

$$k = \frac{r_a r_b}{(r_a + r_{ab})(r_b + r_{ab})} \tag{3.5}$$

and

Table 3.2 Critical values for the X statistic.

Sampling periods, p	Observed number of negative differences							
	0	1	2	3	4	5	6	7
5	0.0035	0.0590	0.3056	0.6944	0.9410	0.9965	1.0000	1.0000
6	—	0.0012	0.0172	0.1052	0.3392	0.6608	0.8948	0.9828
7	—	—	0.0001	0.0020	0.0166	0.0627	0.2010	—
8	—	—	—	—	0.0001	0.0009	0.0323	0.1103

$$c = \frac{1}{1-k} \tag{3.6}$$

then the estimated number of recaptures in the absence of mark loss is given by:

$$\hat{r} = c(r_a + r_b + r_{ab}) \tag{3.7}$$

This estimate should be compared with r to determine the extent of the problem.

The probability of losing mark b or a is:

$$\frac{r_b}{r_b + r_{ab}} \tag{3.8}$$

and:

$$\frac{r_a}{r_a + r_{ab}} \tag{3.9}$$

respectively.

3.1.1.3 Equal catchability

The violation of this assumption, resulting in population over- or under-estimates, is common (Cormack, 1972; Bohlin & Sundstrom, 1977; Caughly, 1977). Eberhardt (1969) gives three general causes.

1 Differences in behaviour near traps.
2 Learning leading to trap addiction or shyness.
3 Unequal probability of capture because of trap positions.

Roff (1973) reviewed the statistical tests then available for the purpose of testing for equal catchability and, from simulations, concluded that under practical conditions they may be unreliable. An extreme example of such unreliability results if certain individuals cannot be caught by the chosen sampling method (Boonstra & Krebs, 1978). This warning is useful; the ecologist must use intuition and any biological indications for confirmation that the population is homogeneous (White, 1975). Reasons for and examples of unequal catchability are listed below, but, as shown by Krebs et al. (1994) in a study of feral house mice, there can be more than one reason for variable and low catchability and this can change over the course of a study.

1 Subcategories of the population are differently sampled because, by the nature of the habitat, only part of the population is available. This problem is particularly acute with subcortical, wood-boring, and subterranean insects. It is well exemplified by ants: the number and composition of the foragers of an ant colony are so variable that, unless one has a considerable knowledge of these aspects or can sample from within the nest (Stradling, 1970), the Petersen–Lincoln index and related methods cannot be used for population estimation (Ayre, 1962; Erickson, 1972). The marking and recapture of foraging ants may, however, be used to estimate the foraging area (Kloft et al., 1965) and, over a short period of time, the number of foragers (F) can be estimated using the following expression (Holt, 1955):

$$F = R_f \times (T_o + T_i) \tag{3.10}$$

where F = number of foragers; R_f = rate of flow of foragers per unit time, i.e. the average number of ants passing fixed points on all routes from a nest per minute; T_o = average time (in the same units as R_f) spent outside the nest (found by marking individuals on the outward journey and timing their return); and T_i = average time spent inside the nest (found by marking ants as they enter and timing their reappearance). In *Pogonomyrmex badius*, only about 10% of the total population forage above ground in any 2-week period (Golley & Gentry, 1964), whilst in the wood ant, *Formica*, a constant group of individuals forage, such that, if their number ($\hat{F}$) is estimated, colony size ($\hat{C}$) is given by (Kruk-de-Bruin *et al.*, 1977):

$$\log \hat{C} = \left(\log \hat{F} + 0.75\right)/1.01 \tag{3.11}$$

The estimation of the number of uncatchable individuals using double sampling is discussed by Aebischer (1986). The approach is to apply two different sampling methods, e.g. trapping and sighting. Some animals, which may be impossible to trap, may still be observed. For example, in a recapture experiment, ducks were caught on the nest and tagged; however, some ducks could not be caught because they ran off when the recorder approached. Using binoculars, the nesting ducks could be counted and the two types of information were later combined to obtain a population estimate.

2 Sub-categories of the population are differentially sampled because of differences due to sex and/or age or other causes. Lomnicki (1969), for example, found that land snails are very non-random in their movements, some individuals being more inclined to recapture than others. In many animals, especially when trapping techniques are used, differences will be linked to sex or age. Methods to adjust for changes linked to breeding behaviour are given in Nichols *et al.* (1994). It is indeed a good principle to enumerate and estimate males and females separately, at least in the first instance; many workers have found striking differences in their behaviour (e.g. Jackson, 1933a; Cragg & Hobart, 1955; MacLeod, 1958). Initially, different stages or age classes should also be considered separately and their survival rates determined using one of the formulae given below; if their survival rates differ significantly, one must continue to consider them separately. Indeed, every attempt should be made to ensure that the population estimated is as homogeneous as possible.

Provided there is no mortality, a χ^2 goodness-of-fit test may be applied to a table of the releases, recaptures, and non-recaptures of the different subclasses (White, 1975; Seber, 1982).

3 There is a periodicity in the availability of subcategories. This problem has been particularly encountered with mosquitoes, where an initial period of emigration (Sheppard *et al.*, 1969) and the gonotrophic and linked feeding cycles invalidate the equal-catchability assumption and may lead to fivefold overestimates (Conway *et al.*, 1974).

Other animals show periodicities in their behaviour and therefore, particularly when sampling is by a behavioural method (e.g. trapping) and the cycle is long compared with the sampling interval, workers should consider this problem. With the Fisher and Ford method (see p. 92), it may be allowed for by introducing additional terms into the model (Sheppard *et al.*, 1969; Conway *et al.*, 1974). This modification requires prior, independently gained, knowledge of the periodicities. The extent to which the incorporation of these additional terms improves the relationship of the expected and observed recaptures gives a measure of their value, but a precise statistical test is not available.

4 The processes of capturing and marking affect catchability. This can have two origins: the initial experience may alter catchability or it may be a cumulative effect. In the former case (more relevant to vertebrates), a regression method analogous to removal trapping (p. 255) may be used to estimate total population (Marten, 1970). The effect of repeated capture may be tested for by Leslie's test for random recaptures, demonstrated below (Orians & Leslie, 1958).

3.1.1.4 Statistical tests to detect variation in catchability
A number of tests are available.

Cormack's test. This method is appropriate if there are three sampling periods. Cormack (1966) gives two methods, the first for experiments where the capture technique is changed at each sampling occasion and the second where the same sampling methodology is used on all three occasions. Let:

n_1 = the number caught and marked in period 1
r_{10} = the number only caught in period 1
r_{12} = the number only caught in periods 1 and 2
r_{13} = the number only caught in periods 1 and 3
r_{123} = the number caught in periods 1, 2, and 3

For both methods 1 and 2, equal catchability is tested using the hypothesis that the coefficient of variation in catchability is zero.

For method 1, the test statistic is:

$$z = \frac{(r_{123}r_{10} - r_{12}r_{13})\sqrt{n_1}}{\sqrt{(r_{12} + r_{123})(r_{13} + r_{123})(n_1 - r_{12} - r_{123})(n_1 - r_{13} - r_{123})}} \qquad (3.12)$$

where z is the standard normal deviate. If z is negative, equal catchability can be assumed.

For method 2, first calculate estimates of the coefficient of variation (CV) and variance (s^2) of the catchability as follows:

$$\hat{CV} = \frac{(r_{123}r_{10}) - (r_{12}r_{13})}{(r_{12} + r_{123})(r_{13} + r_{123})} \qquad (3.13)$$

and

$$\hat{s}^2 = \frac{r_{123}^2 n_{12}}{(r_{12}+r_{123})^2 (r_{13}+r_{123})^2} \left[\frac{1}{r_{123}} - \frac{1}{n_1} - \frac{r_{12}+r_{13}}{(r_{12}+r_{123})(r_{13}+r_{123})} \right] \qquad (3.14)$$

Then use these values to derive the final estimates

$$CV = \frac{1-2\hat{CV}-\sqrt{1-8\hat{CV}}}{2(1+\hat{CV})} \qquad (3.15)$$

$$s^2 = \frac{(3-\sqrt{1-8\hat{CV}})^4 (\hat{s}^2)}{16(1-8\hat{CV})(1+\hat{CV})^4} \qquad (3.16)$$

The test statistic is

$$z = \frac{CV}{s^2} \qquad (3.17)$$

where z is the standard normal deviate. Unequal catchability is significant at the 5% level for $z > 1.64$. If the equations cannot be solved the test assumptions are violated, which indicates unequal catchability.

Zero-truncated Poisson test. This test is more powerful than Cormack's but requires more than three samples (Caughly, 1977). It is zero-truncated because there is no information on the animals which avoided capture. The expected frequency distribution, given equal catchability, is determined by the mean:

$$\bar{x} = \frac{\sum f_x x}{\sum f_x} \qquad (3.18)$$

where x is the number of times of capture and f_x the number of animals caught exactly x times.

Given p sampling periods, if $\bar{x}/s < 0.25$, then a Poisson distribution can be fitted to the data. When $\bar{x}/s > 0.25$, a binomial should be fitted (Seber, 1982). The mean for the Poisson distribution to be fitted has to be found by a numerical method. The observed and expected are compared using a χ^2 test.

Chapman's test. Chapman (1952) devised a non-parametric test for closed populations to test if tagging changed catchability. Let:

r_{ij} = the number of animals marked in period i and first recaptured in period j
n_i = the total number of animals caught in period i
p = the number of sampling periods
u_i = the number of unmarked animals caught in sample i

Construct an array where each row follows the fate of animals first marked at periods 1 to p − 1:

$$\begin{bmatrix} \dfrac{r_{12}}{u_1 n_2} & \dfrac{r_{13}}{u_1 n_3} & \dfrac{r_{14}}{u_1 n_4} & \cdots & \dfrac{r_{1p}}{u_1 n_p} \\[2ex] - & \dfrac{r_{23}}{u_2 n_3} & \dfrac{r_{24}}{u_2 n_4} & \cdots & \dfrac{r_{2p}}{u_2 n_p} \\[2ex] - & - & & \cdots & \vdots \\[2ex] - & - & - & & \dfrac{r_{p-1,p}}{u_{p-1} n_p} \end{bmatrix}$$

The test statistic, X, is defined using the sum of the differences between successive values in each row as follows: $X = D_1 + D_2 + D_3 + \ldots + D_{p-2}$ where D_i is the number of negative differences in the ith row. Table 3.2 (p. 76) gives the probabilities of obtaining X or fewer negative differences if catchability remains constant. Thus, for a study over 7 sampling periods, an X value of 3 or less would be expected with a probability of only 0.002, a result with would lead to the rejection of the null hypothesis of equal catchability.

Leslie's test. A comparison is made of the actual and expected variances of a series of recaptures of individuals known to be alive throughout the sampling period; individual marks must have been used. This test is best illustrated by a worked example taken from Leslie's appendix to Orians and Leslie (1958), based on the recaptures of shearwaters; with insects, of course, the recapture periods would be days or weeks rather than years.

Thirty-two individuals marked in 1946 were recovered for the last time in 1952; therefore they were available for recapture in the years 1947–51, inclusive, and two tables can be prepared (Tables 3.3 and 3.4).

The actual sum of squares:

$$\sum (x - \bar{x})^2 = \sum x^2 f(x) - \frac{\left[\sum x f(x)\right]^2}{N}$$
$$= 69 - 30.03 = 38.97$$

Table 3.3 Leslie's test: year by year analysis.

Year	No. of recaptures in each year (n_i)
1947	7
1948	7
1949	6
1950	4
1951	7
Total	31

Table 3.4 Leslie's test: all years analysis.

No. of recaptures for each individual x	Frequency of x, f(x)
0	15
1	7
2	7
3	2
4	1
5	0
	$N = \sum f(x) = 32$

The expected variance:

$$\sigma^2 = \frac{\sum n_i}{N} - \frac{\sum \left(n_i^2\right)}{N^2}$$

$$= \frac{31}{32} - \frac{199}{32^2} = 0.9688 - 0.1943 = 0.7745$$

Then:

$$X^2 = \frac{38.97}{0.7745} = 50.32$$

X^2 may be treated as equivalent to a χ^2 test and, for degrees of freedom (d.f.) $(N-1)$ of between 20 and 30, the probability of a value as great as or greater than this can be assessed from χ^2 tables; probabilities of less than 0.50 imply that capture is not random. With values over 30, use can be made of the fact that $\sqrt{2\chi^2} - \sqrt{2\text{d.f.} - 1}$ is approximately normally distributed about a mean of zero with unit standard deviation. In other words, one calculates the value for this expression and looks up its probability in the table of the normal deviate. In the present case:

$$\sqrt{2 \times 50.32} - \sqrt{(2 \times 31) - 1} = 10.03 - 7.81 = +2.22$$

The probability of a deviate as great as this is somewhere between 0.025 and 0.020 and therefore the shearwaters were not collected randomly. Leslie has suggested that the test should only be used when the number of individuals is 20 or more and the number of occasions on which recapture was possible is at least 3. This test will not distinguish whether the higher catchability of some individuals is due to catching effects or, as Lomnicki (1969) presumed with land snails, inherent individual differences. Carothers (1973) developed an extension of Leslie's test that is more appropriate when the number of occasions on which recapturing occurs is small.

3.1.2 Estimating closed populations

These methods assume that the population does not change over the period of study. It is therefore essential that this period is short compared with life expectancy. Further, individuals must not be able to enter or leave the study area. Methods in which the probability of capture is not assumed constant require a series of occasions (at least two) on which animals previously marked are recaptured, marked again, and released. We shall first describe the simple Petersen–Lincoln index which, as it uses only one period of recapture, assumes a constant probability of capture.

3.1.2.1 The Petersen–Lincoln estimator

Provided the four common assumptions (section 3.1.1) are satisfied, together with those of closure and constant probability of capture, it is legitimate to estimate the total population from the simple index used by Lincoln (1930):

$$\hat{N} = \frac{an}{r} \tag{3.19}$$

where $\hat{N}$ = the estimate of the number of individuals in the population, a = total number marked in the first sample, n = total number of individuals in the second sample, and r = total recaptures (individuals marked at time 1 and recaught at time 2).

When n is predetermined and approximately equal to a, the variance of this estimate is (Bailey, 1952):

$$\text{var}\,\hat{N} = \frac{a^2 n(n-r)}{r^3} \tag{3.20}$$

If the second sample consists of a series of subsamples and a large proportion of the population has been marked, then it is possible to utilize the recovery ratios ($= r/n$) in each to calculate the standard error of the estimated population (Welch, 1960):

$$\hat{N} = \frac{a}{R_T} \tag{3.21}$$

where R_T = the recovery ratio (r/n) based on the total animals in all of the samples. The variance is approximately:

$$\text{var}\left(\hat{N}\right) = \left(\frac{a}{R_T^2}\right)^2 \times R_T \frac{(1-R_T)}{y} \tag{3.22}$$

where y = total animals in subsamples. This approach is only valid if the marked individuals are distributed randomly in the subsamples, which may be tested by comparing the observed and theoretical variance.

Chapman (1951) shows that equation (3.19) is the best estimate of N as $N \to \infty$ but gives a large bias for small samples. When $a + n \geq N$, a better estimator is:

$$N^* = \frac{(a+1)(n+1)}{(r+1)} - 1 \tag{3.23}$$

Using a Poisson approximation to the hypergeometric distribution, Seber (1970) gives an approximate estimate of the variance:

$$\text{var}\left(N^*\right) = \frac{(a+1)(n_2+1)(a-r)(n_2-r)}{(r+1)^2(r+2)} \tag{3.24}$$

The above calculations are applicable to large samples where the value of r is fairly large; Bailey (1951, 1952) has suggested that, with small samples, a less biased estimate, based on a binomial approximation to the hypergeometric distribution, is:

$$\hat{N} = \frac{n(a+1)}{r+1} \tag{3.25}$$

and an approximate estimate of the variance is:

$$\text{var}(\hat{N}) = \frac{a^2(n+1)(n-r)}{(r+1)^2(r+2)} \tag{3.26}$$

Bonett (1988) compares the bias of the different estimates. The above methods are based on what is referred to as direct sampling, in which the size of n_2 is predetermined. If possible, inverse sampling, where the number of marked animals to be captured (r) is predetermined, has the advantage of giving unbiased estimates of population size ($\hat{N}$) and variance as follows (Bailey, 1952):

$$\text{var}(\hat{N}) = \frac{(a-r+1)(a+1)n(n-r)}{r^2(r+1)} \tag{3.27}$$

The sample size required to give a 95% probability of achieving a selected accuracy (A) (p. 75) can be obtained using computer programs. Examination of the results often shows that it is necessary to mark >50% of the population to achieve modest accuracy (Roff, 1973). It may well show that the proposed project is impossible.

Seber (1982) argues that, given two population estimates N_a and N_b with variances v_a and v_b, the null hypothesis H_0: $N_a = N_b$ can be tested using:

$$z = \frac{N_a - N_b}{\sqrt{(v_a^* + v_b^*)}} \tag{3.28}$$

which is approximately unit-normal when H_0 is true.

A potentially useful model for small r is a Bayesian model (Gaskell & George, 1972), which takes account of prior knowledge about the distribution of N. Gaskell and George's formula is essentially:

$$\hat{N} = \frac{an + 2N'}{r+2} \tag{3.29}$$

where N' is a prior estimate obtained, for example, from quadrat sampling or nearest-neighbour-type techniques. A Bayesian approach is also adopted by Zucchini and Channing (1986) and Nakai et al. (1995).

While simple, the Petersen–Lincoln index is still much used; some recent examples of applications are in Bozeman et al. (1985), Gaillard et al. (1986), Baker and Herman (1987), Gatz and Loar (1988), Menkens and Anderson (1988), Poljak et al. (1989), Law (1994), and Mazzotti (1995).

3.1.2.2 Other single-mark methods

A variant on the Petersen–Lincoln estimation is where a single marking period is followed by a sequence of samples, each comprising animals that are permanently removed from the population. Such a procedure is common in fisheries studies, where commercial landings yield the recaptures. Termed a mark–removal model, it has been studied in detail by Skalski and Robson (1982). Population size can be estimated using a least-squares method (Paloheimo, 1963). For fish populations, the Petersen–Lincoln method has also been combined with age–length data to estimate production (Newman & Martin, 1983).

Another variant, discussed by Minta and Mangel (1989), is a capture–resight experiment, where initially marked animals are noted but not captured during subsequent recording periods. The technique successfully estimated populations of badgers, bison, and crested porcupines.

3.1.2.3 Schnabel censuses

These are methods for closed populations using more than two periods of capture, s, (Schnabel, 1938; Darroch, 1958). The maximum-likelihood estimator of population size, N, is solved numerically using:

$$\left(1 - \frac{M}{N}\right) = \prod_{i=1}^{s}\left(1 - \frac{n_i}{N}\right) \tag{3.30}$$

where M is the total number of different animals captured and n_i is the number of animals captured in sample i.

When $s = 2$, equation (3.30) reduces to the simple Petersen–Lincoln index and, for $s = 3$, it is:

$$N^2(r_2 + r_3) - N(n_1 n_2 + n_1 n_3 + n_2 n_3) + n_1 n_2 n_3 = 0$$

where n_i and r_i are the total number of captures and recaptures, respectively, at time i. For $s > 3$, equation (3.30) needs to be solved iteratively.

A simpler approach is just an extension of the Petersen–Lincoln method to p sampling periods. First, estimate $\hat{N}$ for each sampling step i using:

$$\hat{N}_i = \frac{(M_i + 1)(n_i + 1)}{r_i + 1} - 1 \quad (i = 2, 3 \ldots p) \tag{3.31}$$

where $M_i = \Sigma n_i - r_i$. The variance v_i^* of this estimate is:

$$v_i^* = \frac{(M_i + 1)(n_i + 1)(M_i - r_i)(n_i - r_i)}{(r_i + 1)^2(r_i + 2)} \tag{3.32}$$

Then calculate the average population number:

$$\overline{N} = \frac{\Sigma \hat{N}_i}{(s - 1)} \tag{3.33}$$

If animals are sampled one at a time, we have a particular type of Schnabel census, which can be analysed by Craig's method (see below).

Given a series of samples of individuals that are uniquely tagged, some of the stringent assumptions underlying the Petersen–Lincoln index can be relaxed. A general approach to the selection and analysis of suitable models has been described by Otis *et al.* (1978) and White *et al.* (1982). They consider a family of models, which are systematically tested for applicability. Unlike the Petersen–Lincoln model, the assumption that the probability of capture (p) is constant can now be relaxed. The models they consider are as follows:

M_0—constant p;
M_t—time-varying p;
M_b—a behavioural response to capture changes p;
M_h—there is heterogeneity in p for unknown reasons;
Four mixed models: $M_{bh}, M_{th}, M_{tb}, M_{tbh}$.

By the use of the computer program CAPTURE (see p. 74), each of these models can be fitted to the data and the 'best' estimator identified (Pollock, 1974, 1975, 1981; White *et al.*, 1982). Log-linear models (Cormack, 1989), which are discussed further on p. 102, can also be applied to closed populations. Chao (1987) describes a point estimator and its associated confidence interval for the M_h model and subsequently (Chao, 1989) considered the application of the M_h and M_t models to sparse populations.

Examples of the application of the Schnabel census are in Poljak *et al.* (1989) and Bozeman *et al.* (1985).

3.1.2.4 Frequency-of-capture methods

The information gained from a series of samples may be considered another way. If the frequency of capture follows a particular distribution (f_1, animals captured once, f_2, animals captured twice, $f_3 \ldots f_x$), then the term f_0 represents those animals that have not been marked or captured at all and the sum of all the terms will represent the total population. Theoretically, problems do not arise from unequal catchability; a non-Poisson distribution may be fitted that describes the deviations from random in the probability of capture. Tanton (1965) fitted a zero-truncated negative binomial to the frequency of recapture of wood mice, *Apodemus*. Further details of such models, particularly useful in small-mammal work (e.g. Marten, 1970), are given by Seber (1982).

3.1.2.5 Craig's method

This method, devised for use with butterflies, assumes both that the population is closed and that each individual has a constant probability of capture. In practice, the butterflies, or other highly mobile but colonial animals, are collected randomly, marked, and immediately released, after recording the number of times, if any, the animal has previously been captured. The effective sample size is thus one. Craig

(1953) assumed that the frequency of recapture could be described by two mathematical models: the truncated Poisson (p. 31) and the Stevens distribution function. On the basis of these two models, six different methods can be used to estimate the size of the population, three based on moments and three on maximum likelihood; Craig found, however, that they gave similar results. The first[1] of the two simplest, based on the Poisson, is:

$$\hat{N} = \frac{\sum x f_x^2}{\left(\sum x^2 f_x - \sum x f_x\right)} \tag{3.34}$$

where $\hat{N}$ = the estimate of population, x = the number of times an individual had been marked, f_x = the frequency with which individuals marked x times had been caught, and thus $\Sigma x f_x$ = the total number of different times animals were captured (namely, 1 × the number caught once + 2 × the number caught twice + 3 × the number caught thrice, etc.). This method is simple, allowing a direct solution, but is subject to greater sampling error than equation (3.35). It is useful for obtaining a trial value for use in solving equation (3.35) (and in the other methods given by Craig).

The second is:

$$\log \hat{N} - \log\left(\hat{N} - \sum f_x\right) = \frac{\sum x f_x}{N} \tag{3.35}$$

where the symbols are as above, $\Sigma x f$ being the total different individual animals caught (namely, the number caught once + the number caught twice + the number caught thrice, etc.). Craig (1953) gives formulae for the variances of these estimates:

For equation 3.34:

$$\mathrm{var}\left(\hat{N}\right) = \frac{2N}{\left(\sum x f_x\right)^2} \tag{3.36}$$

For equation 3.35:

$$\mathrm{var}\left(\hat{N}\right) = \frac{N}{\left(e^{\sum x f_x} - 1 - \sum x f_x\right)} \tag{3.37}$$

It is clear that this method is based on different assumptions from the Lincoln index. It demands that the animals be very mobile so that their chances of recapture are effectively random immediately after release and yet they must not leave the habitat. It has been used with butterflies and might be applied to other large, conspicuous, flying or very mobile animals under certain circumstances. Phillips and

---

[1] The actual numbering of these methods is the reverse of that given by Craig, but it is most logical here to give the simplest method first.

Campbell (1970) showed that the Schnabel census method is more robust when portions of the population are inaccessible.

3.1.2.6 Change-in-ratio methods (Kelker's selective removal)

This method uses the natural marks of a population, normally the difference between the sexes, but theoretically any other recognizable distinction could be used (e.g. the different morphs of a polymorphic species). The proportion of the different forms or components is determined, a known number of one form is removed ('selective removal'), and the new ratio is found; from the change in the ratio, the total population can be estimated:

$$\hat{N} = \frac{K_\alpha}{\left[D_{\alpha 1} - \dfrac{\left(D_{\beta 1} D_{\alpha 2}\right)}{D_{\beta 2}} \right]} \qquad (3.38)$$

where α and β are the components (e.g. sexes) in a population, α is the component part of which is removed during the interval between times 1 and 2, and β is the component (or components) no part of which is removed; then K_α = number of component α that are killed; D = the proportion of the population represented by a component at time 1 or 2; thus $D_{\alpha 1}$ = the proportion of α in the population as a decimal before any were removed; $D_{\beta 1}$ = the proportion of the population as a decimal represented by β after K individuals of α were removed.

Such an approach is clearly most applicable with game, where individuals of a certain sex or age class are culled (e.g. Kelker, 1940; Hanson, 1963); it is sometimes referred to as the dichotomy method (Chapman, 1954).

The mathematical background has been developed by Chapman (1955), who compared the results from this method with those given by the Petersen–Lincoln-type capture–recapture analysis; he showed that a capture–recapture estimation procedure will yield more information for equal effort. A further objection to the use of this method with insect populations is that the selective removal and the resulting atypical unbalance could seriously prejudice any further study of the population; in populations of game animals, such removal might be one of the 'normal' mortality factors.

This method does not depend on the assumption that initial capture does not alter the probability of subsequent recapture, and this is an advantage over those methods that utilize artificial marks and might indicate conditions for their use, perhaps only as a test of estimates derived by Lincoln-type methods. Chapman (1955) gives a technique for combining the two approaches. Seber (1982) provides a comprehensive review of this approach.

3.1.3 Estimations for open populations

Open populations change under some combination of birth, death, immigration, and emigration during the course of the capture–recapture study. Of necessity, such populations must be subject to at least three periods of sampling.

3.1.3.1 Review of method development

If the animals are marked on a series of two or more occasions, then an allowance may be made for the loss of marked individuals between the time of the initial release and the time when the population is estimated. These methods have been reviewed by Seber (1982, 1986). Their history and development are given below. A simple, non-algebraic development of the main formulations is provided by Cormack (1973), who shows how the estimates are based on the largest 'known groups'.

Jackson (1933a, 1937, 1953), Dowdeswell *et al.* (1940), and Fisher and Ford (1947) were the first to devise methods for estimating the population using the data from a series of marking and recapture occasions. Besides assumptions 1–4 listed above (p. 75), they also assumed a constant probability of capture and survival rate; even though this was known to be an approximation, it was considered necessary for an algebraic solution.

Jackson developed two methods for his work on the tsetse fly, where he calculated a theoretical recapture immediately after release by either the 'positive method', where the loss ratio was calculated for that day over all release groups, or the 'negative method', where the loss ratio was calculated from the subsequent daily percentage losses for the two release groups. It is possible that the negative method might still be found useful in a situation such as that provided by the use of radioactive isotopes, where marking is carried out on a limited number of occasions followed by a long series of recaptures.

Fisher's method is often referred to as the 'trellis method', as the data are initially set out in a trellis diagram; details of the way it works are given by Dowdeswell (1959). The survival rate has to be determined by trial and error. With the use of computers, this comparatively robust method has enjoyed a revival. A useful discussion and summary of these early methods are given by MacLeod (1958). Although these methods are of historical interest and have been widely used, more modern methods are to be recommended.

Bailey (1951, 1952), who introduced maximum-likelihood techniques into capture–recapture analysis and so was able to calculate the variances of his estimates, made an important advance. The equations in Bailey's triple-catch method are simple and may be solved directly to provide estimates of various population parameters.

Another significant step was made at about the same time by Leslie and Chitty (1951) and Leslie (1952), who compared three different methods of classifying the animals according to their marks. In Leslie's method A, mentioned above, the animals were classified according to the occasion on which they were marked, and thus an animal bearing several marks would give several entries in the recapture table. Jackson, Fisher, and Bailey all also used this method of classification in drawing up their recapture tables, but Leslie and Chitty (1951) showed that it leads to loss of information. Leslie's method B, in which the animals were classified according to the date on which they were last marked (ignoring all earlier marks), was completely efficient under the assumptions he made. Leslie (1952) gives a full account of his method and a worked example, but his formulae require solution by iterative methods when more than three sampling occasions are considered. Jolly (1963) simplified the calculations by providing formulae for explicit solutions (i.e. direct, not iterative). This method is particularly applicable if a fairly large number of individuals have been recaptured several times (multiple recaptures) in a long series of samples, a situation that often arises with work on mammals.

Working on blowflies, where the incidence of recapture was very low, MacLeod (1958) found that none of the then known methods were usable, and he derived two formulae (based on the Lincoln index) which gave estimates, albeit rather approximate, of the total population. For both methods, the mortality rate had to be ascertained independently—in his case, in the laboratory—and for one method the percentage of flies not immigrating over a given time was measured by a separate experiment.

All the above methods are based on deterministic models that assume that the survival rate over an interval is an exact value, whereas it would be more correct to state that in nature an animal has a probability of surviving over the interval. This probability is well expressed by a stochastic model, but initially it was thought that the computations arising from a stochastic model would be too complex. Darroch (1958, 1959) showed that, under certain conditions, a fully stochastic model, giving explicit solutions for the estimation of population parameters, was possible, and Jolly (1965) and Seber (1965) have, independently, extended this method to cover situations in which there is both loss (death and emigration) and dilution (births and immigration). Their methods give similar solutions, except that Jolly's makes allowance for the common experience of animals dying after capture.

The Jolly–Seber method efficiently groups the data and is fully stochastic; however, its reliability strictly depends on the assumption that every marked animal has the same probability of survival. This requires that the probability of any animal surviving through any period is not affected by its age at the start of the period. Manly and Parr (1968) pointed out that, with short-lived adult insects (e.g. many Lepidoptera) studied over their emergence period, the animals marked on the ith occasion are likely, at that time, to be younger than those marked on the first occasion. If all deaths were independent of age, this in itself would not invalidate the Jolly–Seber method, but this cannot be assumed. They therefore devised a method free of this assumption, based on intensity of sampling, but depending on a relatively high frequency of multiple recapture. The term 'multiple recapture' has been used extensively and is appropriate with mark-grouping methods (e.g. Bailey's), where the number of marks an animal bears (or the number of occasions on which it has been captured) affects its classification. However, it will be noted that in the date-grouping methods no significance is attached to any mark other than the last and hence, as Jolly (1965) points out, the term multiple recapture should not be used in connection with these methods.

The Jolly–Seber method was developed further by Buckland (1980) for the analysis of tags returned from dead animals. He also modified the method to situations where cohorts differ in catchability and provided methods for estimating the survival curve (Buckland, 1982; Buckland *et al.*, 1983). Trap shyness was discussed in Buckland and Hereward (1982). Generalizations to cope with survival and capture probabilities, which vary with capture history, were developed by Robson (1969) and Pollock (1975), and age-dependent capture and survival were considered by Pollock (1981). Pollock and Mann (1983) considered this problem in fisheries research. Pollock *et al.* (1990) reviews the Jolly–Seber model and give details of computer software. A contemporaneous development has been the application of log-linear models by Cormack (1989), which may be solved using the computer package GLIM.

3.1.3.2 Choice of method

Comparative studies on actual or simulated data have been made by several workers, including Parr (1965), Sheppard *et al.* (1969), Manly (1970), Bishop and Sheppard (1973), Roff (1973), and Schwarz *et al.* (1993). Two approaches to method choice can be taken.

The systematic approach is to consider the 'goodness of fit' of each member of a family of models which represent all the different assumptions possible (like the approach of Otis *et al.* (1978) and White *et al.* (1982) for closed populations). In practice, this is not always possible because of problems of fitting some models and measuring the goodness of fit. The log-linear formulation (Cormack, 1989) is useful for this approach, as all the standard models can be represented using a consistent formulation. Each candidate model may then be solved using GLIM, which yields a deviance that measures lack of fit. The model that gives the best combination of fit and parsimony can then be selected. Table 3.1 lists computer programs that aid model selection.

The second approach is to identify the assumptions that can be made prior to the experiment, select a parsimonious model, and then design the experiment to collect the data required to meet the desired accuracy. This approach will require preliminary experimentation or prior knowledge. It is prudent to apply more than one method and compare the results. Robust methods which use a mix of models should also be considered when undertaking longer-term studies (see p. 545). The following general notes on the merits of the various methods may be of use.

The Fisher and Ford (1947) trellis method is robust for small samples and, provided the survival rate remains relatively constant, gives population estimates not dissimilar to those of the Jolly–Seber method. It has the advantage that periodicity of availability for sampling, known from biological knowledge, may be incorporated in the trellis model (see Conway *et al.*, 1974).

Bailey's triple catch and the Jolly–Seber method both provide estimates of the variance; however, these have been shown to be related to the population estimate and are particularly unreliable for small samples. The Jolly–Seber method population estimate is usually reliable when 9% or more of the population is sampled and the survival rate is not less than 0.5 (Bishop & Sheppard, 1973). It is less sensitive to age-dependent variations in the mortality rate than Fisher and Ford's method, but is not of course independent of them. Except when there is only a very limited number of sampling occasions, the Jolly–Seber method is superior to Bailey's.

The Jolly–Seber method may seriously overestimate the survival rate (Bishop & Sheppard, 1973): it remains, however, probably the most useful and gives good estimates if a high proportion of the population can be sampled. Manly and Parr's (1968) method is not affected by age-dependent mortality but requires the sampling of a relatively high proportion of the population (>25% for populations under 250, >10% for larger populations (Manly, 1970). When these conditions can be achieved, it should be used if mortality is thought to be related to age, particularly if it is high early in the lifespan studied.

The above methods utilize batch marks; when it is desired to model individual behaviour, then individual marks need to be used and a log-linear formulation may be used (Cormack, 1981, 1989).

3.1.3.3 The Fisher–Ford method

The data from a series of recaptures are set out in a 'trellis diagram' — a constant survival rate that best fits the data is found by a trial-and-error process, and one essentially makes a series of Lincoln index estimates, working backwards in time from day to day from when the releases were originally made. Then the population estimate is:

$$\hat{N}_t = \frac{n_t a_i \Phi_{i-t}}{r_{ti}} \qquad (3.39)$$

where n_t = total sample at time t, a_i = total marked animals released at time i, Φ_{i-t} = the survival rate over the period $i - t$ and r_{ti} = the recaptures at time t of animals marked at time i. Details of the method are given by Fisher and Ford (1947) and in an expository form by Dowdeswell (1959) and Parr (1965).

As already mentioned, one of the strengths of this robust model is that additional variables, such as the periodicity of availability, may be incorporated in the model and the computer used to find the best fit to the various parameters (Sheppard *et al.*, 1969; Conway *et al.*, 1974). A model for the mosquito, *Aedes aegypti*, assuming a 4-day feeding (and hence availability) cycle, with some secondary feeding on the second and third days, is shown in Table 3.5. The total recaptures for any day will have components bearing the marks of different days. For example, r_4 may contain recaptures from the first, second, and third marking occasions, and the values of these will, if the model is appropriate, fit the expressions $p_4 a_1 \Phi^3 \beta_3$, $p_4 a_2 \Phi^2 \beta_2$, and $p_4 a_3 \Phi \beta_1$, respectively. The sampling intensity (catch rate), p, will vary from occasion

Table 3.5 Population model for mark and recapture with constant survival (ϕ), but cycling proportionate availability (β) and varying sampling intensity (catch rate) (p): the modification of the Fisher & Ford model by Conway et al. (1974).

1 $n_1 a_1$						
2 $n_2 a_2$	$p_2 a_1 \phi \, \beta_1$					
3 $n_3 a_3$	$p_3 a_1 \phi^2 \beta_2$	$p_3 a_2 \phi \, \beta_1$				
4 $n_4 a_4$	$p_4 a_1 \phi^3 \beta_3$	$p_4 a_2 \phi^2 \beta_2$	$p_4 a_3 \phi \, \beta_1$			
5 $n_5 a_5$	$p_5 a_1 \phi^4 \beta_0$	$p_5 a_2 \phi^3 \beta_3$	$p_5 a_3 \phi^2 \beta_2$	$p_5 a_4 \phi \, \beta_1$		
6 $n_6 a_6$	$p_6 a_1 \phi^5 \beta_1$	$p_6 a_2 \phi^4 \beta_0$	$p_6 a_3 \phi^3 \beta_3$	$p_6 a_4 \phi^2 \beta_2$	$p_6 a_5 \phi \, \beta_1$	
7 $n_7 a_7$	$p_7 a_1 \phi^6 \beta_2$	$p_7 a_2 \phi^5 \beta_1$	$p_7 a_3 \phi^4 \beta_0$	$p_7 a_4 \phi^3 \beta_3$	$p_7 a_5 \phi^2 \beta_2$	$p_7 a_6 \phi \, \beta_1$
8 $n_8 a_8$	$p_8 a_1 \phi^7 \beta_3$	$p_8 a_2 \phi^6 \beta_2$	$p_8 a_3 \phi^5 \beta_1$	$p_8 a_4 \phi^4 \beta_0$	$p_8 a_5 \phi^3 \beta_3$	$p_8 a_6 \phi^2 \beta_2$ $p_8 a_7 \phi \, \beta_1$

to occasion; here it is the value for the fourth occasion. The marked animals (from a_i) available for recapture are modified by the survival rate, Φ, and availability proportion, β. As it is a 4-day cycle, β itself will have four values and each represents the different proportions available (in this case feeding) on each day of the cycle. The value of β_0 is 1 (all survivors are available on the fourth day); the other β values, given by Conway et al. (1974), are somewhat complex algebraically because in their situation the periodicity of availability was complicated by some mosquitoes feeding on the second and third days of the cycle, as well as on the first.

However, if we simplify and generalize the model and assume that any cohort of the animals in question is only available for sampling every fourth day, $\beta_0 = 1$ and β_1, β_2, and $\beta_3 = 0$, and certain terms will then disappear from the trellis (Table 3.5). The survival rate and β values for the trellis may be determined by the normal least-squares method and then the sampling intensities for each catch date determined.

In the simplified general case, with synchronized availability, the total population for the ith occasion is then estimated by:

$$\hat{N} = \frac{n_i(1 - \Phi^x)}{p_i \Phi^y (1 - \Phi)} \tag{3.40}$$

where x = the period of the cycle of availability (e.g. 4 days) and y = the period between birth and becoming available for capture (e.g. for mosquitoes 1 day). If y is zero, then the denominator becomes $p_i (1 - \Phi)$. Full details for the special case of *A. aegypti* are given by Conway et al. (1974); but the availability of a number of insects (including the rhinoceros beetle) to trapping methods shows a marked periodicity related to age, and the method is therefore of wider application.

Recent applications of the Fisher–Ford method are in Rosewell and Shorrocks (1987) and Wilkin and Scofield (1991).

3.1.3.4 Bailey's triple-catch method

As indicated above, this method is based on a deterministic model of survival and uses an inefficient method of grouping the data, although the latter is not a serious fault if very few animals are recaptured more than once (Sonleitner & Bateman, 1963). The formulae for the calculation of the various parameters are relatively simple; Bailey (1951) gives the mathematical background and Bailey (1952) and MacLeod (1958) describe it. It was found particularly appropriate by Coulson (1962) in a study on craneflies, *Tipula*, with very short life expectancies, but a series of triple-catch estimates may be used to estimate a population over a longer period (e.g. Iwao *et al.*, 1963). It seems that the results are most reliable when large numbers are marked and recaptured (Parr, 1965). It has been used with a variety of insects and other invertebrates (e.g. snails: Woolhouse & Chandiwana, 1990; and lobster: Evans & Lockwood, 1994).

For the purposes of illustration, the sampling is said to take place on days 1, 2, and 3. In practice, the intervals between these sampling occasions may be of any length, provided they are long enough to allow for ample mixing of the marked individuals with the remainder of the population and not so long that a large proportion of the marked individuals have died. With large samples, the population on the second day is estimated:

$$\hat{N}_2 = \frac{a_2 n_2 r_{31}}{r_{21} r_{32}} \tag{3.41}$$

where a_2 = the number of newly marked animals released on the second day, n_2 = the total number of animals captured on the second day, and r = recaptures, with the first subscript representing the day of capture and the second the day of marking; thus r_{21} = the number of animals captured on the second day that had been marked on the first, r_{31} = the number of animals captured on the third day that had been marked on the first. It is clear that we are really concerned with the number of marks and hence the same animal could contribute to r_{31} and r_{32}. The logic of the above formula may be seen as follows:

$$\hat{N}_2 = \frac{\hat{a}_1 n_2}{r_{21}} \tag{3.42}$$

which is the simple Lincoln index with $\hat{a}_1$ = the estimate of the number of individuals marked on day 1 that are available for recapture on day 2. Now, if the death-rate is constant:

$$\frac{\hat{a}_1}{a_2} = \frac{r_{31}}{r_{32}}$$

Substituting for $\hat{a}_1$ in the above equation, we have the estimate of population size given in equation (3.41). The large-sample variance of the estimate is:

$$\text{var } \hat{N}_2 = \hat{N}_2^2 \left(\frac{1}{r_{21}} + \frac{1}{r_{32}} + \frac{1}{r_{31}} - \frac{1}{n_2} \right) \tag{3.43}$$

Where the numbers recaptured are fairly small, there is some advantage in using 'Bailey's correction factor', i.e. the addition of 1 so that:

$$\hat{N}_2 = \frac{a_2(n_2+1)r_{31}}{r_{21}r_{32}} \tag{3.44}$$

with approximate variance:

$$\text{var } \hat{N}_2 = \hat{N}_2^2 - \frac{a_2^2(n_2+1)(n_2+2)(r_{31}-1)r_{31}}{(r_{21}+1)(r_{21}+2)(r_{32}+1)(r_{32}+2)} \tag{3.45}$$

The loss rate, which is compounded of the numbers actually dying and the numbers emigrating, is given by:

$$\gamma = \ln\left(\frac{a_2 r_{31}}{a_1 r_{32}}\right)^{\frac{1}{t_1}} \tag{3.46}$$

where t_1 = the time interval between the first and second sampling occasions. The dilution rate, which is the result of births and immigration, is given by:

$$\beta = \ln\left(\frac{r_2 n_3}{n_2 r_{31}}\right)^{\frac{1}{t_2}} \tag{3.47}$$

where t_2 = the time interval between the second and third sampling occasions. Both these are measures of a rate per unit of time (the unit being the units of t). Bailey (1952) gives formulae for their variance.

3.1.3.5 Jolly–Seber stochastic method

The advantages of the stochastic model on which this method is based have already been discussed and its general utility has been indicated (p. 92). The basic equation in Jolly's method is:

$$\hat{N}_i = \frac{\hat{M}_i n_i}{r_i} \tag{3.48}$$

where $\hat{N}_i$ = the estimate of population on day i, $\hat{M}_i$ = the estimate of the total number of marked animals in the population on day i (i.e. the counterpart of 'a' in the simple Lincoln index), r_i = the total number of marked animals recaptured on day i, and n_i = the total number captured on day i.

   The procedure may be demonstrated by the following example, taken from Jolly (1965). (The notation is slightly modified to conform to the rest of this section.) It would be normal practice to undertake these calculations on a computer.
1 The field data are tabulated as in Table 3.6 according to the date of initial capture (or mark) and the date on which the animal was last captured. The columns are then

Table 3.6 The tabulation of recapture data according to the data on which the animal was last caught for analysis by Jolly's method (after Jolly, 1965).

Day of capture i	Total captured n_i	Total released a_i	Day when last captured (j)												
1	54	54	*1*												
2	146	143	10	*2*											
3	169	164	3	34	*3*										
4	209	202	5	18	33	*4*									
5	220	214	2	8	13	30	*5*								
6	209	207	2	4	8	20	43	*6*							
7	250	243	1	6	5	10	34	56	*7*						
8	176	175	0	4	0	3	14	19	46	*8*					
9	172	169	0	2	4	2	11	12	28	51	*9*				
10	127	126	0	0	1	2	3	5	17	22	34	*10*			
11	123	120	1	2	3	1	0	4	8	12	16	30	*11*		
12	120	120	0	1	3	1	1	2	7	4	11	16	26	*12*	
13	142		0	1	0	2	3	3	2	10	9	12	18	35	*13*
$R_i =$				80	70	71	109	101	108	99	70	58	44	35	

Table 3.7 Calculated table of the total number of marked animals recaptured on a given day bearing marks of the previous day or earlier (after Jolly, 1965).

Day i	Day i – 1											
	1											
2	**10**	*2*										
3	3	**37**	*3*									
4	5	23	**56**	*4*								
5	2	10	23	**53**	*5*							
6	2	6	14	34	**77**	*6*						
7	1	7	12	22	56	**112**	*7*					
8	0	4	4	7	21	**40**	**86**	*8*				
9	0	2	6	8	19	31	**59**	**110**	*9*			
10	0	0	1	3	6	11	28	**50**	**84**	*10*		
11	1	3	6	7	7	11	19	31	**47**	**77**	*11*	
12	0	1	4	5	6	8	15	19	30	**46**	**72**	*12*
13	0	1	1	3	6	9	11	21	30	42	**60**	**95**	*13*
$Z(i-1)+1 = 14$	57	71	89	121	110	132	121	107	88	60	
	Z_2	Z_3	Z_4	Z_5	Z_6	Z_7	Z_8	Z_9	Z_{10}	Z_{11}	Z_{12}	

summed to give the total number of (animals released on the ith occasion = s_i of Jolly), subsequently recaptured (R_i), e.g. for day 7, $R = 108$.

2 Another table is drawn up (Table 3.7) giving the total number of animals recaptured on day i bearing marks of day j or earlier (Jolly's a_{ij}); this is done by adding each row in Table 3.6 from left to right and entering the accumulated totals. The number marked before time i which are not caught in the ith sample, but are caught subsequently (Z_i), is found by adding all but the top entry (printed in bold) in each column. Thus Z_7 is given by the figures enclosed in Table 3.6. The figures above the

line, i.e. the top entry in each column, represent the number of recaptures (r_i) for the day on its right, e.g. $r_7 = 112$.

3 Then the estimate of the total number of marked animals at risk in the population on the sampling day may be made:

$$\hat{M}_i = \frac{a_i Z_i}{R_i} + r_i \qquad (3.49)$$

Thus:

$$\hat{M}_7 = \frac{243 \times 110}{108} + 112 = 359.50$$

Similarly for other $\hat{M}_i$, the results being entered in Table 3.8, in which other population parameters are entered as calculated.

4 The proportion of marked animals in the population at the moment of capture on day i is found and entered in the final table:

$$\alpha_i = \frac{r_i}{n_i}$$

Thus:

$$\alpha_7 = \frac{112}{250} = 0.448$$

5 The total population is then estimated for each day (Table 3.9) (equations 3.48 and 3.49):

$$\hat{N} = \frac{\hat{M}_i}{\alpha_i}$$

6 The probability that an animal alive at the moment of release of the ith sample will survive till the time of capture of the $(i + 1)$th sample is found:

$$\hat{\Phi}_i = \frac{M_{i+1}}{\hat{M}_i - r_i - a_i}$$

Survival rates estimates slightly over 1 may arise from sampling effects, but 'rates' greatly above this indicate a major error! Frequently it will be found that the marks of one occasion have been lost or were not recognized. This survival rate may be converted to a loss rate (the effect of death and emigration):

$$\hat{\gamma}_i = 1 - \hat{\Phi}_i$$

7 The number of new animals joining the population in the interval between the ith and $(i + 1)$th samples and alive at time $i + 1$ is given by:

$$\hat{B}_i = \hat{N}_{i+1} - \hat{\Phi}_i \left(\hat{N}_i - n_i + a_i \right)$$

Table 3.8 The final table for a Jolly type mark and recapture analysis (after Jolly, 1965).

Day	Proportion of recaptures	No. marked animals at risk	Total population	Survival rate	No. of new animals	Standard errors			Standard errors due to errors in the estimation of parameter itself	
i	$\hat{\alpha}_i$	$\hat{M}_i$	$\hat{N}_i$	$\hat{\phi}_i$	$\hat{B}_i$	$\sqrt{\{v(\hat{N}_i)\}}$	$\sqrt{\{v(\hat{\phi}_i)\}}$	$\sqrt{\{v(\hat{B}_i)\}}$	$\sqrt{\{v(\hat{N}_i/N)\}}$	$\sqrt{\left\{v(\hat{\phi}_i) - \dfrac{\hat{\phi}_i^2(1-\hat{\phi}_i)}{\hat{M}_{i+1}}\right\}}$
1	—	0	—	0.649	—	—	0.114	—	—	0.093
2	0.0685	35.02	511.2	1.015	263.2	151.2	0.110	179.2	150.8	0.110
3	0.2189	170.54	779.1	0.867	291.8	129.3	0.107	137.7	128.9	0.105
4	0.2679	258.00	963.0	0.564	406.4	140.9	0.064	120.2	140.3	0.059
5	0.2409	227.73	945.3	0.836	96.9	125.5	0.075	111.4	124.3	0.073
6	0.3684	324.99	882.2	0.790	107.0	96.1	0.070	74.8	94.4	0.068
7	0.4480	359.50	802.5	0.651	135.7	74.8	0.056	55.6	72.4	0.052
8	0.4886	319.33	653.6	0.985	13.8	61.7	0.093	52.5	58.9	0.093
9	0.6395	402.13	628.8	0.686	49.0	61.9	0.080	34.2	59.1	0.077
10	0.6614	316.45	478.5	0.884	84.1	51.8	0.120	40.2	48.9	0.118
11	0.6260	317.00	506.4	0.771	74.5	65.8	0.128	41.1	63.7	0.126
12	0.6000	277.71	462.8	—	—	70.2	—	—	68.4	—
13	0.6690	—	—	—	—	—	—	—	—	—

This may be converted to the dilution rate β:

$$\frac{1}{\beta} = 1 - \frac{\hat{B}_i}{\hat{N}_{i+1}}$$

8 If desired, the standard errors (the square roots of the variances) are obtained from:

$$\text{var}(\hat{N}_1) = \hat{N}_i(\hat{N}_i - n_i)\left[\frac{\hat{M}_i - r_i + a_i}{\hat{M}_i}\left(\frac{1}{R_i} - \frac{1}{a_i}\right) + \frac{1 - \alpha_i}{r_i}\right] + \hat{N}_i - \sum_{j=0}^{i=1}\frac{\hat{N}_i^2(j)}{\hat{B}_j} \quad (3.50)$$

The special problems involved in the computation of the summation term are discussed by Jolly (1965).

$$\text{var}(\hat{\Phi}_i) = \Phi_i^2\left[\frac{(\hat{M}_{i+1} - r_{i+1})(\hat{M}_{i+1} - r_{i+1} + a_{i+1})}{\hat{M}_{i+1}^2}\left(\frac{1}{R_{i+1}} - \frac{1}{a_{i+1}}\right)\right.$$
$$\left. + \frac{\hat{M}_i - r_i}{\hat{M}_i - r_i + a_i}\left(\frac{1}{R_i} - \frac{1}{a_i}\right) + \frac{1 - \Phi_i}{\hat{M}_{i+1}}\right] \quad (3.51)$$

$$\text{var}(\hat{B}_i) = \frac{\hat{B}_i^2(\hat{M}_{i+1} - r_{i+1})(\hat{M}_{i+1} - r_{i+1} + a_{i+1})}{\hat{M}_{i+1}^2}\left(\frac{1}{R_{i+1}} - \frac{1}{a_{i+1}}\right)$$
$$+ \frac{\hat{M}_i - r_i}{\hat{M}_i - r_i + a_i}\left[\frac{\Phi_i a_i(1 - \alpha_i)}{\alpha_i}\right]^2\left(\frac{1}{R_i} - \frac{1}{a_i}\right)$$
$$+ \frac{[(\hat{N}_i - n_i)(\hat{N}_{i+1} - \hat{B}_i)](1 - \alpha_i)(1 - \Phi_i)}{\hat{M}_i - r_i + a_i}$$
$$+ \hat{N}_{i+1}(\hat{N}_{i+1} - n_{i+1})\frac{1 - \alpha_{i+1}}{r_{i+1}} + \Phi_i^2\hat{N}_i(\hat{N}_i - n_i)\frac{1 - \alpha_i}{r_i} \quad (3.52)$$

Jolly (1965) also gives terms for the covariance.

9 Jolly shows that the variances of the population and survival rate estimates given above contain an error component, due to the real variation in population numbers, apart from errors of estimation. If only errors of estimation are required, the formulae are:

$$\text{var}\left(\frac{\hat{N}_i}{N_i}\right) = \hat{N}_i(\hat{N}_i - n_i)\left\{\frac{\hat{M}_i - r_i + a_i}{\hat{M}_i}\left(\frac{1}{R_i} - \frac{1}{a_i}\right) + \frac{1 - \alpha_i}{r_i}\right\} \quad (3.53)$$

$$\text{var}\left(\frac{\hat{\Phi}_i}{\Phi_i}\right) = \text{var}(\hat{\Phi}_i) - \frac{\hat{\Phi}_i^2(1 - \Phi_i)}{\hat{M}_{i+1}} \quad (3.54)$$

The errors that arise in this method may be thought of as having two causes: (i) 'small-sample biases' inherent in the method; and (ii) biases due to the violation of the assumptions. The former will be minimal when the number of recaptures is large. Therefore, Jolly suggests that, when the cost of marking is high, it may be an advantage if the sampling for recaptures is separated from the sampling and marking of further animals to provide the next value of a_i. Sampling for recaptures could be carried out extensively by less experienced staff than are needed to mark. Manly (1971b) and Roff (1973) have shown that these errors often express themselves as underestimates of the variance, because of the correlation between the estimates and their variances. Some improvement is obtained if N and Φ are transformed to logarithms before making the calculations (Manly, 1971b).

Buckland (1980) discusses a solution to the frequently encountered problem of the Jolly–Seber model, which is that survival and birth rate estimates are found to be impossible (e.g. a negative birth rate). Jolly (1982) presents modifications which, by assuming constant survival and/or capture rates, give a more parsimonious model, leading to more efficient estimators. Brownie et al. (1986) presented further modifications, and also indicate how to carry out goodness-of-fit tests that utilize individual capture-history information.

Recent examples of the use of the Jolly–Seber method, often with comparisons with other methods, are: with insects (Gouteux & Buckland, 1984; Komazaki & Sakagami, 1989; Nelemans et al., 1989); other invertebrates (Nakai et al., 1995; Wilkin & Scofield, 1991); fish (Schwarz et al., 1993; Law, 1994); amphibians (Freeland, 1986); reptiles (Le Gall et al., 1986; Woodward et al., 1987); birds (Loery & Nichols, 1985; Loery et al., 1987; Spendelow & Nichols, 1989; Boano & Cucco, 1991); diving mammals (Baker and Herman, 1987); and terrestrial mammals (Baker and Herman, 1987; Paradis et al., 1993).

3.1.3.6 Robust design

While the Jolly–Seber method has been the traditional approach to the estimation of demographic parameters in long-term capture–recapture studies of wildlife and fish species, it can give biased estimates of population size and recruitment. Pollock (1982) proposed a sampling scheme in which a series of closely spaced samples were separated by longer intervals, such as a year. For this 'robust design', Pollock suggested a flexible *ad hoc* approach that combines the Jolly–Seber estimators with closed-population estimators, to reduce bias caused by unequal catchability and to provide estimates for parameters that are unidentifiable by the Jolly–Seber method alone. Kendall et al. (1995) describe the analysis of data obtained using the robust design, which has been recommended for small-mammal studies by Nichols et al. (1984) and Nichols and Pollock (1990).

3.1.3.7 Manly and Parr's method

The assumption that survival probability is equal is often quite unjustified in entomological studies (Iwao et al., 1966). To overcome this Manly and Parr (1968) devised a method based on the intensity of sampling, each animal being considered to have the same chance of capture on the ith occasion, so that:

$$\hat{N}_i = \frac{n_i}{p_i}$$

where $p_i =$ the sampling intensity. In practice:

$$\hat{p}_i = \frac{r_i}{\hat{M}_i}$$

using the nomenclature of the Jolly–Seber method. Thus:

$$\hat{N}_i = \frac{an_i}{r_i}$$

the basic Lincoln–Petersen index. The animals must be individually marked or have date-specific marks. Manly and Parr prepared an individual animal table, allocating the symbol x for the first or last occurrence of an individual mark, y for the intermediate occasion when an individual mark was captured, and z for occasions when it was there, but not recaptured, i.e. the blanks left between the two x values after the y values have been inserted. A greatly curtailed example is given in Table 3.9. The sampling intensity will clearly be the ratio of those animals captured known to be present before and after, to the total number available known to be present before and after. For any day, these are given by Σy_i and $(\Sigma y_i + \Sigma z_i)$, respectively. Then:

$$\hat{p}_i = \frac{\sum y_i}{\left(\sum y_i + \sum z_i\right)}$$

so:

$$\hat{N}_i = \frac{n_i}{p_i} = \frac{n_i\left(\sum y_i + \sum z_i\right)}{\sum y_i} \tag{3.55}$$

and survival may be estimated as:

$$\Phi_{i-(i+1)} = \frac{r_{i,i+1}}{n_i b_{i+1}} \tag{3.56}$$

where $r_{i,i+1} =$ the animals caught in both ith and $(i + 1)$th samples. The births — number of new animals entering the population — are estimated by:

$$\hat{B}_{i-(i+1)} = \hat{N}_{i+1} - \hat{\Phi}_i \hat{N}_i$$

Manly (1969) gives the variance of the population estimate as:

$$\text{var}\left(\hat{N}_i\right) = \frac{\hat{N}_i\left(\hat{N}_i - n_i\right)\left[\hat{N}_i - \left(\sum y_i + \sum z_i\right)\right]}{n_i\left(\sum y_i + \sum z_i\right)} \tag{3.57}$$

As mentioned, this method is not robust unless a fairly large sample is taken (Manly, 1970). Seber (1982) shows how the necessary data may be tabulated in a form similar to that of the Jolly–Seber method; this would undoubtedly be more convenient with large sets of data, but Manly and Parr's original arrangement is simpler for exposition.

Table 3.9 Manly and Parr's method of recording data for their method of estimation (obviously one would normally mark far more than three animals on each occasion).

	Sampling days					
	1	2	3	4	5	6
1	x	y	z	x	—	—
2	x	z	y	z	x	—
3	x	—	—	—	—	—
4		x	z	x	—	—
5		x	y	z	y	x
6		x	z	y	x	—
7			x	y	z	x
8			x	z	x	—
9			x	y	z	x
$\Sigma y_i =$		1	2	3	1	
$\Sigma z_i =$		1	3	3	2	

3.1.3.8 Cormack's log-linear method

When each individual is uniquely tagged the log-linear formulation (Cormack, 1981, 1989) can describe many of the models proposed for the estimation of population parameters by capture–recapture. All the animals in the population are considered to be multinomially distributed between 2^s capture histories, where s is the number of samples. The expected number of individuals showing each capture history is given by the product of the probability of the event(s) and the size of the population. By taking logs, these equations are made linear and can then be solved using the readily available GLIM computer package. Examples of its use are given in Cormack (1989) and Baker (1990).

3.2 Methods of marking animals

Marking techniques are often classified as either group (batch) or individual (tagging) techniques. While methods such as fin clipping, branding, or staining are often used in a manner that precludes individual identification, this need not always be so. For example, a range of coloured dyes may be applied to different areas of the body. The number of unique marks available (C) is given by:

$$C = (x+1)^n - 1 \tag{3.58}$$

where x is the number of colours or types of mark and n the number of sites to hold the mark or tag. Batch methods enable a large number of animals to be marked in the same way and are perfectly adequate for most capture–recapture population estimations and dispersal studies. Almost all methods are capable of one or two variants, so that two or three groups may be marked differently.

  If each individual can be separately marked, additional information can be obtained on longevity and dispersal and, if they can be aged and sexed initially, survival can be related to these and other characters. Birth- and death-rates may be more easily calculated and the excessive handling of animals recaptured more than once, which arises from remarking, may be avoided (Dobson & Morris, 1961). Individual marks may also, for example, allow the assessment of the rate of oviposition in the field by weighing individual females (Richards & Waloff, 1954), and they provide a method for assessing the randomness of recapture. The extent to which the higher cost, in terms of effort, of marking animals individually is justified will depend on the percentage of the marked individuals that are recovered: high recovery rates justify elaborate marking programmes; with low recovery rates individual marking is seldom justified.

Below, we review the principal techniques available. Marking methods have also been reviewed in Stonehouse (1977), and a number of guides applicable to particular animal groups, e.g. birds (Calvo & Furness, 1992) and fish (Laird & Stott, 1978), have been published.

The chosen method must seek to meet the following list of requirements.

1 Marking should not affect growth, longevity, or behaviour of the animals. An attempt should always be made to confirm that this is true in the particular case under investigation. For example, although the pigments used in most markers may be non-toxic, the solvents are often toxic. This may be checked in the laboratory or field cage by keeping samples of living marked and unmarked individuals and comparing longevity (e.g. Stobow *et al.*, 1992; Kideys & Nash, 1993; Morgan & Walsh, 1993; Montgomery *et al.*, 1995) or in the field by comparing the longevity of individuals bearing differing numbers of marks (e.g. Richards & Waloff, 1954; Dobson & Morris, 1961). Newly emerged insects may be more sensitive to the toxic substances used in markers than older insects (Jackson, 1948; Dobson *et al.*, 1958), and the attachment of labels to their wings, which has no effect on old insects, may cause distortion due to interference with the blood circulation (Waloff, 1963). Marking can also cause disease, often in unexpected ways. For example, ear tags on the white-footed mouse (*Peromyscus leucopus*) were found by Ostfeld *et al.* (1993) to result in increased tick infestations, probably because they interfered with grooming. External tagging and branding can cause wounds, leading to infections. Calvo and Furness (1992) in a review of the effects of marking on birds points out that very few studies tested for harmful effects or data biases caused by marking, but, where assessments have been made, it is clear that all methods of marking can have adverse effects.

2 It should not affect vulnerability to predation. An animal's natural camouflage may be lost, making it more liable or, as Hartley (1954) observed with marked snails, less liable to predation. This effect is difficult to assess; it can to some extent be avoided by marking in inconspicuous places or by the use of fluorescent powders (Pal, 1947), dyes in powder form (Quarterman *et al.*, 1954), phenolphthalein solution (Peffly & Labrecque, 1956), tetracycline (Wastle *et al.*, 1994), or internal tags, such as coded wire tags or radioactive tracers, whose presence is only detectable by the use of a special technique after recapture. The effect on predation of a conspicuous, but convenient, marking method could be measured by marking further

individuals with one of these invisible methods and comparing longevity. The effect of a conspicuous mark can be checked through choice experiments in the laboratory: Buckner (1968) confirmed that small mammals are not influenced if their prey is stained with vital dyes. Special care should be taken to avoid atypical predation when animals are released (see p. 106).

3 The mark must be durable. Some paints, particularly cellulose lacquers, may flake off; adhesives can lose strength (e.g. Treble *et al.*, 1993); student's oil-paints and powdered dyes wash off; some fluorescent powders may lose this property on exposure to sunlight (Polivka, 1949) or be abraded during collection (Dow, 1971); tags may be shed or get ejected (Hampton & Kirkwood, 1990; Wisniewolski & Nabialek, 1993); and a radioactive isotope could decay and/or be excreted by the animal. Animals will lose marks on their cuticle when they moult. Laboratory tests of durability are not always reliable; Blinn (1963) found that cellulose lacquers would remain on the shells of land snails for 2 years in the laboratory, but they only lasted about 1 year in the field, and the rate at which a radioactive isotope is lost may depend upon the animal's diet (p. 367) and other factors (p. 516).

4 The mark must be easily detected. However, a conspicuous mark on animals sampled by a method that relies on the sight of the collector may result in the over-reporting of marked individuals (Edwards, 1958).

5 The objectives must justify the cost. The amount of effort that can be put into a marking programme and the choice of marking method need to be related to the percentage of recoveries that can be expected. A high cost per individual marked will be justified where the recovery rate is high or the animals have a special significance.

6 Generally, marks must be easily applied in the field and without the need for anaesthetic. Ethical aspects of the distress and harm caused to the animals need to be considered (Putman, 1995).

Consideration may also have to be given to the effect of preservatives on the mark and its toxicity, appearance, and safety if the marked animals can be seen, caught, or eaten by the public.

After selecting an appropriate marking technique, it is important to consider the handling and release of the animals. During a marking campaign animals must be captured, marked, held for recovery, and released. At each stage in this chain, undue stress must be minimized and this requires planning. When badly undertaken, it results in the death or injury of many animals, a result which is probably commoner than many biologists would hope.

3.2.1 Handling techniques

Handling can be stressful, even fatal, and should be carefully considered. Most experimental designs assume that the marked animals behave normally, but, if handling is poor, this will not be so. For example, to radiotag freshwater dolphins, a standard method is to trap the animals in nets and then harpoon them, an unpleasant procedure which may change their behaviour, at least when in the presence of humans. Netting can be highly stressful to fish and it may only be possible to capture fish at set localities or times in the tidal cycle. This has resulted in the capture of salmon in freshwater pools followed by their release in estuarine water; the behavioural consequences of such an abrupt change cannot be known. Capture and marking can also affect the behaviour of invertebrates. Singer and Wedlake (1981), for example, demonstrated that the behaviour of the papilionid butterfly, *Graphium sarpedon* L. changed following capture and release. The trapping techniques that may be used are considered in Chapters 4, 5, and 6. There are methods described below where the animal is marked without capture; clearly, these should avoid these problems.

If the animals are to be marked by spraying or dusting, this can be done whilst they are still active, either in the field or in small cages. But, for marking with paints, especially precise spotting, and for most methods of labelling and mutilation, it is necessary for the animal to be still. Insects, crustaceans, molluscs, birds, and small mammals can often be held in the hand or between the fingers; where this is impossible the animal must be held by some other method, anaesthetized, or chilled. Fish can be placed between two boards fixed to give a V-shaped cross-section. Invertebrates may be held still under a net or with a hair. One end of the hair is fixed and the other has a piece of Plasticine on it; the animal is placed underneath and the hair may be tightened under a stereoscopic microscope by pressure on the Plasticine (Banks, 1955). More robust insects may be held immobile on top of a cork by a piece of Terylene net, which is kept taut by a ring (of about 5 cm diameter, but of course depending on the size of the animal). This ring is pushed down into the surrounding Plasticine (Fig. 3.1b) and the animal can be marked through the holes of the net (Murdoch, 1963). Conway *et al.* (1974) held mosquitoes for spot marking between diaphragms of nylon mesh and stockinet. These devices may be placed under a microscope.

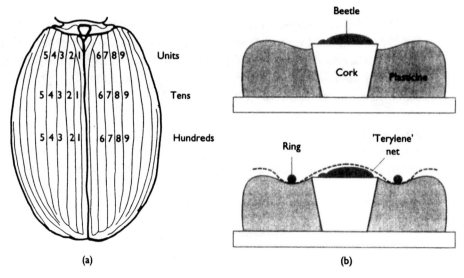

Fig. 3.1 (a) A 'mutilation system' for marking carabid beetles individually by scraping certain positions on the elytra, applied to *Agonum fuliginosum*. (b) A device for holding a hard-bodied animal whilst it is marked (modified from Murdoch, 1963).

Insects can also be held by suction (Hewlett, 1954). Muir (1958) found that a mirid bug was conveniently held by the weak suction of a pipette, formed from the ground-down point of a coarse hypodermic needle, connected to a water pump. The picking up of the bugs by this weak suction was facilitated by coating the inside of the glass dish in which they were held by an unsintered dispersion of Fluon GPI (polytetrafluoroethylene, PTFE), which presented an almost frictionless surface (Muir, 1958; Radinovsky & Krantz, 1962). Butterflies may be held in a clamp (Horn, 1976).

If none of these methods can be used or if the insects are so active that they cannot even be handled and counted, they may have to be immobilized. Chilling at a temperature between 1 and 5°C is probably the best method and may be done in a lagged tank surrounded by an ice–water mixture (MacLeod & Donnelly, 1957). Alternatively, an anaesthetic may be employed: carbon dioxide is often used and is easily produced from 'dry ice' (Caldwell, 1956) or bought in cartridges designed to carbonate drinks (Dobson, 1993; Killick-Kendrick, 1993). Caldwell (1956) gives a table of recovery times of houseflies from carbon dioxide anaesthesia at 21°C (Table 3.10).

Working on the earwig, *Forficula*, Lamb and Wellington (1974) found that, although activity may be quickly resumed after carbon dioxide anaesthesia of less than 1 min, normal behaviour could be affected for many hours; they recommend a 24 h recovery period. Dalmat (1950) found that female blackflies (Simulidae) laid more eggs than normal after being subjected to carbon dioxide, but Edgar (1971) found this did not affect the development of the spider *Lycosa lugubris*. Other anaesthetics are ether, chloroform, nitrogen, and nitrous oxide. Honey-bees are prematurely aged by these (Ribbands, 1950; Simpson, 1954). For insects, use of anaesthetics should, wherever possible, be avoided in ecological and behavioural studies.

The use of anaesthetic while marking fish is advised by Laird and Stott (1978). MS 222 and benzocaine are the two most commonly used. Wedermeyer (1970) suggested that benzocaine was better than MS 222 at reducing the stress experienced by fish. Drug immobilization with dart guns is often used for the capture of large mammals and arboreal animals (Eltringham, 1977).

3.2.2 Release

The release of the animals after marking is an operation that is too often casually undertaken. Some methods allow animals to be marked without capture, while

Table 3.10 Recovery times (min) of houseflies from carbon dioxide anaesthesia (from Caldwell, 1956).

Exposure time	Recovery time
Up to 5	1–2
5–30	3–5
30–60	5–10

others allow them to be marked in the field and immediately released. However, often the animal is incarcerated, handled, and much disturbed before being released. It should not be surprising if they show a high level of activity immediately after release; indeed, Greenslade (1964) recorded with individually marked ground beetles that there was far greater movement on the day after release than subsequently. Two approaches can be used to minimize this effect.

If the animal has a marked periodicity of movement (i.e. is strictly diurnal or nocturnal), then it should be released during its inactive period; e.g. radioactively tagged individuals of the frit fly (*Oscinella frit*) were released at dusk (its period of activity is from dawn to late afternoon) (Southwood, 1965). Animals that are active at most times of the day may be restrained from flying immediately after release by covering them with small cages (Evans & Gyrisco, 1960). The release sites should be chosen carefully. It is especially important to avoid the release of small flying insects in the middle of the day, when their escape flights may carry them beyond the shelter of the habitat into winds or thermals that can transport them for miles. Of course, only apparently healthy, unharmed individuals should be released.

The release points should be scattered throughout the habitat, as it is essential that the marked animals mix freely with the remainder of the population; e.g. Muir (1958) returned the arboreal mirids he had marked to all parts of the tree. Very sedentary animals may, indeed, invalidate the use of the capture–recapture method if they do not move sufficiently to remix after marking, as Edwards (1961) found, surprisingly, with a population of the grasshopper, *Melanoplus*. The extent of the remixing may be checked, to some degree, by a comparison of the ratio of marked to unmarked individuals in samples from various parts of the habitat; the significance of the difference may be tested by a χ^2 (Iwao et al., 1963).

The release of disorientated or unadapted animals can also result in high predation losses. Frequently, marked crustaceans and fish have adapted their coloration to the conditions within the holding tanks so that they have no natural camouflage when released. A general feature of predatory fish is that they recognize disorientation and quickly attack. Thus, a common feature of fish-tagging studies is a high mortality rate shortly after release. Fish-tagging in the Amazon has proved difficult because of piranha predation. Fish should be released as near as possible to their place of capture and shoaling species should be released in a group.

3.2.3 Surface marks using paints and solutions of dyes

The hard cuticle or shell of arthropods and molluscs allows the use, in terrestrial environments, of a wide range of paints and surface treatments. In aquatic or moist habitats the range is much more limited.

Artist's oil-paint is an extensively used marking material; it can, of course, be obtained in a variety of colours and has been used successfully for marking moths (Collins & Potts, 1932), tsetse flies (Jackson, 1933b), bedbugs (Mellanby, 1939), locusts and grasshoppers (Richards & Waloff, 1954), mirids (Muir, 1958), flies (Cragg & Hobart, 1955; Dobson et al., 1958), beetles (Mitchell, 1963), and others.

However, Davey (1956) found them toxic to certain locusts, although this may have been the effect of the dilutant and they are, of course, slow-drying. Artist's poster paints have been used to mark mosquitoes (Gillies, 1961; Slooff & Herath, 1980) as have Humbrol enamel paints (Trpis & Hausermann, 1986). More recently, various paint markers have been used (e.g. Sugimoto *et al.*, 1994).

Nitrocellulose lacquers or paints (e.g. model-aircraft dope) and alkyl vinyl resin paints are quick-drying and have been used by a number of workers—on snails, where they were applied to a small area on the underside of the shell, from which the periostracum had been scraped (Sheppard, 1951), and on grasshoppers (Richards & Waloff, 1954), where they were found to be less satisfactory than artist's oil-paints. Other arthropods to which they have been applied include ants (Holt, 1955), lace-bugs (Southwood & Scudder, 1956), dragonflies (Corbet, 1952), mites (Hunter, 1960), tipulid flies (Freeman, 1964), various beetles (Greenslade, 1964; Malter, 1996; Ohguski & Sawada, 1998), and mosquitoes (Sheppard *et al.*, 1969). Fluorescent lacquer enamels or fluorescent pigments with gum arabic glue plus a trace of detergent have been used to mark tsetse flies (Jewell, 1956, 1958; McDonald, 1960), chafer beetles (Evans & Gyrisco, 1960), and lepidopterous cater-pillars (Wood, 1963). Animals marked in these ways may be spotted after dark in the field at distances of up to 8–10 m by the use of a beam of ultraviolet (UV) light pro-duced by a battery-powered lamp (McDonald, 1960). Fluorescent dye solutions, in alcohol or acetone, principally rhodamine B, have been used to mark mosquitoes (Chang, 1946), *Drosophila* (Wave *et al.*, 1963), and the tick *Argus* (Medley & Ahrens, 1968). Felt-tipped pens provide a convenient way for marking some large insects (Iwao *et al.*, 1966), and Porter and Jorgensen (1980) marked ants with coloured fluo-rescent ink extracted from 'magic marker' pens. Wada *et al.* (1975) used an aqueous fluorescent solution to stain mosquitoes, which comprised 1.0% yellow 8G, 1.0% Kaycal BZ, 0.1% rhodamine 6G, and 0.5% crystal violet. Reflecting paints may also be used to mark animals for detection at night and have the advantage that they can be seen with a small hand torch for up to 10 m. Nail varnish has been found effective for gastropods (Fenwick & Amin, 1983; Woolhouse, 1988; Gosselin, 1993).

White aluminium paint adheres well to scraped areas of the elytra of carabid beetles (Murdoch, 1963). Solutions of various stains in alcohol, such as eosin, orange G, and Congo red, have been used to mark adult Lepidoptera (Nielsen, 1961); petroleum-based inks (e.g. Easterbrook Flowmaster) have been found particularly useful, staining the integument below the scales (Wolf & Stimmann, 1972). Working on houseflies, Peffly and Labrecque (1956) used a 6% solution of phe-nolphthalein in acetone; the marked flies were identified on recapture by placing them in 1% sodium hydroxide solution, whereupon they became purple. Fales *et al.* (1964) used waterproof inks to mark face flies.

An interesting recent idea is marking the mirid, *Lygus hesperus*, adults externally by submersion in rabbit immunoglobulin G (IgG) solution, which will presumably be undetectable to other animals while allowing detection using an immulogical test even from samples obtained from within the guts of predators (Hagler *et al.*, 1992).

The distribution of some insects has been studied by feeding the adults dyes, such as rhodamine B, which give the eggs a distinctive colour (e.g. Narayanaswamy et al., 1994).

Henderson and Holmes (1987) marked the carapace of the common shrimp, Crangon crangon, using cyanoacrylate instant glue mixed with a coloured power or dye. This mixture hardened and bonded to the surface even underwater. The mark was found to last at least 1 month or until the shrimp moulted.

Commercial hair dye and lighteners can be used to stain feathers (Ellis & Ellis, 1975). Bird feathers can be marked by dipping in dyes. However, application of paint or dyes to the skin, fur, or feathers of a vertebrate rarely gives a reliable, lasting mark. Marking methods for birds are discussed by Calvo and Furness (1992). The feathers of incubating herring gulls (Larus argentatus) can be marked by applying a dye (rhodamine B, malachite green or picric acid) to their eggs (Belant & Seamans, 1993). The authors discuss the merits of oil-based silica gel, petroleum jelly, and vegetable shortening as a carrier for the dyes and acetic acid, isopropyl alcohol, and propylene glycol as fixatives. Lizards (Hain, 1965), nile crocodiles, Crocodylus niloticus (Pooley, 1962), and snakes (Parker, 1976) have been given temporary marks using enamel paints, lacquers, and other quick-drying materials.

3.2.3.1 Application methods

When the paints or solutions are in their most concentrated form, they are most conveniently applied to insects by the use of an entomological pin, a sharpened matchstick, a single bristle, or even a fine dry grass stem (Jackson, 1933b; Corbet, 1952; Muir, 1958; Hunter, 1960). With quick-drying cellulose lacquer, it may be necessary to dilute slightly with acetone or another solvent; if this is not done, a fine skin may form over the droplet on the pin and it will not adhere firmly to the animal. With artist's oil-paints, dyes in solution, or diluted lacquers, a camel-hair brush may be used (Wood, 1963), although generally this method has no advantage over the use of a pin and frequently leads to the application of too large a mark. If the mark covers any of the sense organs or joints, the specimen will have to be discarded. Slooff and Herath (1980) applied poster paints to mosquitoes with a tiny loop of 0.0024-gauge wire. The loop was dipped into the paint and then into 96% ethanol, which made the paint form a small blob prior to application on the scutum of the mosquito. Freeman (1964) found that a fine syringe was suitable for applying cellulose lacquers to tipulid flies.

The paints or solutions may be further diluted with acetone, dilute alcohol, or other solvents and sprayed on. This may be done with a hand-atomizer (e.g. a nasal spray) while the insects are contained in a small wire cage (Leeuwen, 1940; Evans & Gyrisco, 1960). Mortality during marking by this method can be reduced if, immediately after spraying, the insects are quickly dried in the draught from an electric fan (Leeuwen, 1940). Large numbers of moths may be rapidly marked, and this technique has been extended to field marking of locusts (Davey, 1956) and butterflies (Nielsen, 1961). With the use of a spray gun, individuals can be marked at a distance of 5 m or more. Davey showed that, for the same cost (labour and time), nearly 10 times as many locusts could be marked by this method than by that involving the capture and handling of each individual.

If small labels can be attached to the insects' wings (see below), there is no problem in marking each individual, and Nielsen (1961) records that in the laboratory he marked butterflies individually on the wing with a rubber stamp. With most small invertebrates, individual marks have to be obtained by a combination of spots in various positions, the numerical range of the coding often being increased by the use of various colours. It is generally wise to follow the policy of Michener *et al.* (1955) and ensure that all individuals bear the same number of marks, so that, if one is lost, number 121 doesn't become, say, 21, but is immediately recognized as an 'unreadable' mark. It is also desirable to use the minimum number of colours in marking any individual insect, as the change in colours increases the handling time. Obviously a change in colour between, say, the ranges 1–99 and 100–199 is less of a practical problem than the use of two colours in a single mark. The actual pattern will depend on the size and shape of the insect, the number of colours available, and the number of individual marks required. Patterns for bodies and wings have been devised for dragonflies (Borror, 1934; Parr *et al.*, 1968), bedbugs (Mellanby, 1939), tsetse flies (Jackson, 1953), bees (Frisch, 1950), grasshoppers (Richards & Waloff, 1954; White, 1970; Nakamura *et al.*, 1971), snails (Blinn, 1963), crane-flies (Freeman, 1964), mosquitoes (Sheppard *et al.*, 1969; Conway *et al.*, 1974), and Lepidoptera (Ehrlich & Davidson, 1960; Brussard, 1971; Dempster, 1971). The system of Richards and Waloff (1954) is logical and versatile, enabling up to 999 individuals to be marked with continuous numbering (Fig. 3.2a). On the right of the thorax is a spot that represents the unit, on the left one that represents the tens, and the head spot represents the hundreds; 1–5 and 10–50 are white spots, 6–0 and 60–00 are red; ten different colours are used for the head mark, including the zero class (i.e. for numbers 1 to 99). With this system the addition of a further marking site would allow another 10 000 individuals to be numbered. With smaller insects, the number of spots can be reduced by increasing the number of colours used, as Richards and Waloff did with the hundred mark; however, the practical problem of switching quickly between more than three colours is serious. The rapid reading of the mark is another advantage, and Richards and Waloff's code is particularly clear in this respect, because the same colours and corresponding positions represent the same figures in units and tens.

Richards and Waloff's (1954) code may be considered a decimal code, based on tens. Ehrlich and Davidson (1960) developed a 1–2–4–7 marking system, modified by Brussard (1971) to mark up to 1000 individuals with one colour (Fig. 3.4b). A binomial system for marking mosquitoes was developed by Sheppard *et al.* (1969); up to 255 individuals can be marked with a single colour (Fig. 3.4c) but, as with Brussard's method, the number of marks is variable and the quick and accurate reading of the mark a matter of experience.

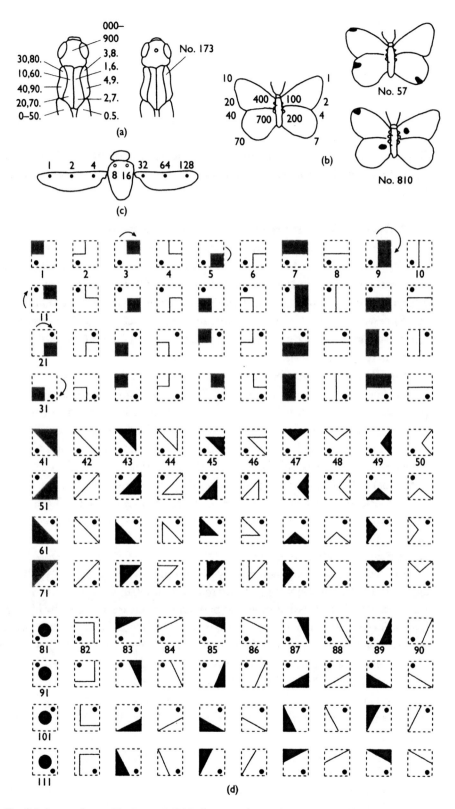

Fig. 3.2 Systems for marking insects individually using colour and position codes. (a) Richards and Waloff's (1954) decimal system. (b) Brussard's (1971) modification of Ehrlich and Davidson's (1960) 1–2–4–7 system. (c) Sheppard *et al.*'s (1969) binomial system. (d) White's (1970) 'shape code' (see text for further explainations).

A more elaborate code of shapes has been devised by White (1970) (Fig. 3.2d). The system is self-checking (if marks are lost), a single colour may be used up to 160, but, as White points out, rapid reading becomes more difficult beyond 120.

If the body and/or wings are unsuitable for spotting, the legs may be marked, but the removal of such marks by the animals during cleaning movements is a real risk. Kuenzler (1958) was able to number lycosid spiders by marking their legs with white enamel paint and Corbet (1956) hippoboscid flies by marking both the body and the legs.

3.2.4 Dyes and fluorescent substances in powder form

Hairy insects or vertebrates may be marked by dusting them with various dyes in powder form; this is most easily done by applying the dusts from a powder dispenser (insufflator) or by producing a dust storm in a cage with a jet of air (e.g. produced by a bicycle pump). Frankie (1973) describes a suitable apparatus and Williams *et al.* (1979) describes a unit for marking large numbers of stable flies, *Stomoxys calcitrons*. Only a very small quantity of powder is necessary. Such methods have been applied mostly to terrestrial insects.

Non-fluorescent dyes that have been found useful are the rotor and waxoline group. The marked insects are recognized by laying them on a piece of white filter-paper and dropping acetone on to them, when a coloured spot or ring forms beneath those that have been marked. As the testing involves the killing of insects, this method is not suitable for recapture work, and laboratory tests showed that, with blowflies, the mark may only last for 1 week and seldom for more than 2 (MacLeod & Donnelly, 1957). However, dyes of two different colours may be applied to the same insect and distinguished in the spotting. This method has been used for calypterate flies by Quarterman *et al.* (1954) and MacLeod and Donnelly (1957). It is possible that, under certain circumstances, it could be used for marking large aggregated populations in the field.

Fluorescent substances, detected by placing the animals under a UV lamp, have also been used extensively for marking (e.g. Foott, 1976; Wheye & Ehrlich, 1985; Aars *et al.*, 1995). Day-Glo fluorescent pigments, which are activated by blue visible light and UV, are used to mark insects. Service (1995) found that 'series A' pigments gave good adhesion. Examples of their use are by Sheppard *et al.* (1973), with taban-ids, and Isaacs and Byrne (1998), with whiteflies; the latter achieved large-scale group marking (for dispersal studies) by spraying the dust over a whole field with a tractor-mounted boom. Other fluorescent powders designed to mark insects are available from USR Optrix, Hackettstown, New Jersey, USA. Multiple marking is possible, although combinations sometimes produce distinct fluorescences (Moth & Barker, 1975), and, because they may be detected without killing the insect, they are suitable for use in capture–recapture population estimation (Crumpacker, 1974). Although with the stable fly, *Stomoxys*, field cage experiments suggested that they could adversely affect longevity (Labrecque *et al.*, 1975), this was not detected in trials with *Drosophila* (Moth & Barker, 1975). For some powders, better adhesion can be obtained by mixing one part of the dye with six parts of gum arabic, adding water until a paste is formed, and then drying the paste and pulverizing it in a mortar. This powder is applied to the insects, which are then placed in high humidity; the gum arabic particles absorb sufficient moisture to make them adhere to the insect. This method has been used for marking mosquitoes by Reeves *et al.* (1948) and Sinsko and Craig (1979). The movements of foraging bees have been studied by marking them with a fluorescent powder as they leave the hive; this is conveniently done by forcing them to walk between two strips of velveteen liberally dusted with the marker (Smith & Townsend, 1951); tabanids caught in a canopy trap (p. 290) have been self-marked in a similar way (Sheppard *et al.*, 1973). Bees have also been marked when visiting flowers by dusting these with a mixture of the fluorescent powder and a carrier, such as talc or lycopodium dust. The bees leave a trail of powder, which can be detected after dark with a UV lamp, on the other flowers they have visited (Smith, 1958; Johansson, 1959). Fluorescein and rhodamine B have been found useful in this work; all the bees leaving the hive are marked and the marks last for weeks (Smith & Townsend, 1951). The possibility that unmarked individuals may bear a few particles that will fluoresce under UV light should be remembered when using these markers. Unmarked mosquitoes caught in the wild have been found with fluorescent blue, purple, green, white, yellow, and orange spots (Reeves *et al.*, 1948); therefore the use of rhodamine B, which fluoresces red, was recommended. Fluorescent-dyed melamine copolymer resins were found effective for marking adult lesser grain borers, *Rhyzopertha dominica* (Dowdy & McGaughey, 1992). Fluorescent labels can be used to detect night-foraging animals such as bats (Griffin, 1970). Nocturnal observation techniques in general are reviewed by Hill and Clayton (1985).

An ingenious self-marking method for newly emerged calypterate flies has been devised by Norris (1957). The principle is that the soil or medium which contains the fly puparia is covered with a mixture of about 40 parts to one of fine sand and fluorescent powder or the puparia are coated with dye; as the flies emerge a small quantity of the dust adheres to the ptilinum and, when, after emergence, the ptilinum is retracted, the dust becomes lodged in the ptilinal suture. Sometimes, in an examination of the faces of such marked flies in UV light, this suture will be seen to shine vividly, but a more reliable technique is to crush the whole head on a filter-paper, at the same time adding a small amount of the appropriate solvent; the mark may be seen on the paper, using UV if necessary (Steiner, 1965). Under laboratory conditions, Cook and Hain (1992) showed that the same principle could be applied to the bark beetles *Dendroctonus frontalis* and *Ips grandicollis*, which were self-marked with fluorescent powders upon emergence from treated logs. Niebylski and Meek (1989) describe a self-marking device for mosquitoes emerging from ditches. The apparatus holds folds of cheesecloth impregnated with fluorescent powder which attaches to emerging adults as they push past. A trap for marking newly emerged blackflies, *Simulium venustum* and *Stegopterna mutata*, with fluorescent dust is described by Dosdall *et al.* (1992).

3.2.5 Marking formed by ingestion or absorption of dyes

Arthropods lose external marks when they moult. Therefore, the marking of the tissues of invertebrates with a vital dye incorporated in the food is a valuable tool; its significance is enhanced if it can be detected without killing the animal and if the mark is retained from larval to adult life (see below). Many workers have screened a wide range of dyes, but few have been found satisfactory: the majority are either rapidly excreted or prove toxic (Reeves *et al.*, 1948; Zacharuk, 1963; Daum *et al.*, 1969; Hendricks, 1971). Calco oil red (N-1 700, American Cyanamid Co.), if fed to immature stages, has been found to mark adults and sometimes the resultant eggs (but not the larvae) of several cotton insects (*Heliothis, Platyedra, Anthonomus*) (Daum *et al.*, 1969; Hendricks & Graham, 1970; Graham & Mangum, 1971). The dye is fed in a natural oil (e.g. cottonseed) with the larval diet at a concentration of about 0.01% dye/unit diet; it is detected in the adult by crushing the abdomen. Similar results have been obtained with oil-soluble Deep Black BB and Blue II (BASF Corp., USA) (Hendricks, 1971).

Sawfly larvae have been marked by feeding them, in the last instar, on foliage treated with solutions of rhodamine B (series 4; 3.7 g/L) and Nile blue sulphate (series 5; 0.4 g/L) (Heron, 1968). The cocoons, adults, and eggs were all marked and the dyes were visible externally, rhodamine B fluorescing bright yellow-orange under UV. Mosquitoes (Reeves *et al.*, 1948), *Drosophila* (Wave *et al.*, 1963), and the eyes of a small fish (O'Grady & Hoy, 1972) have also been marked with rhodamine B. The eggs of the uzi fly, *Exorista bombycis*, were stained pink when the adults were fed with 0.15% rhodamine B, with no affect on adult longevity (Narayanaswamy *et al.*, 1994). Houseflies have been marked with thiorescein (Shura-Bura & Grageau, 1956) and fluorescein (Zaidenov, 1960); but in some cases only a small proportion of the insects could be induced to feed on the solution. South (1965) marked slugs by feeding them on agar jelly containing 0.2% neutral red; the digestive gland became deeply stained and the colour was easily visible through the foot. Bloodsucking Diptera have been marked by allowing them to feed on a cow which had had 200 ml of an aqueous solution containing 4 g of trypan blue, intravenously administered over 20 min. The dyestuff can be detected by a paper-chromatographic technique, in which the gut contents of the fly is mixed with 0.1 N sodium hydroxide solution and applied to a narrow strip of Whatman no. 1 chromatographic filter-paper, which is developed in 0.1 N sodium hydroxide solution; the trypan blue remains at the origin, while other marks due to the gut contents move away (Knight & Southon, 1963). Haematophagous animals may also be marked with specific agglutinins (Cunningham *et al.*, 1963). These dyes or stains may often be detected in the faeces of predators that have fed on marked animals (Hawkes, 1972).

Much less work has been undertaken on aquatic organisms. However, I. Walker (pers. comm.) found that copepod guts could be stained so as to be visible while alive by feeding them cooked beetroot. Young fish have been marked by adding vital stains to their food (Loab, 1966) or to the food of the mother (Bagenal, 1967).

Various compounds can permanently stain bone, teeth, spines, scales, shell, or otoliths as they are formed, creating a mark which can be used to measure the subsequent growth rate of the structure. These techniques are termed date banding or time stamps. Because of their properties of localization in hard tissues, such as bone, low toxicity, and fluorescence, tetracycline antibiotics have been used with success to mark many different animal groups. A bibliography of the marking of fish with tetracycline and its effects has been produced by Wastle *et al.* (1994). Some recent examples of oxytetracycline application by injection are: Pacific sailfish, black marlin (Speare, 1992), larval southern brook lamprey, *Icthyomyzon gagei* (Medland & Beamish, 1991), and subadult red drums, *Sciaenops ocellatus* (Bumgaurdner, 1991). Examples of batch immersion marking for larval or young fish are with: whitefish, *Coregonus lavaretus* (Beltran *et al.*, 1995), shad, *Alosa sapidissima* (Hendricks *et al.*, 1991), grayling, *Thymallus thymallus* (Nagiec *et al.*, 1995), striped bass, *Morone saxatilis* (Secor *et al.*, 1991), and walleye, *Stizostedion vitreum* (Brooks *et al.*, 1994). Survival of walleye (*S. vitreum*) fry following immersion in tetracycline has been shown to be high (Peterson & Carline, 1996). Acute toxicities of oxytetracycline hydrochloride and calcein to juvenile striped bass, *M. saxatilis*, following immersion marking are reported by Bumguardner and Kin (1996). Successful marking of otilith and bone within the egg is reported for Arctic char, *Salvelinus alpinus* (Rojas Beltran *et al.*, 1995), and coregonids (Ruhle & Winecki Kuhn, 1992).

Oxytetracycline can also be administered in the food, e.g. for juvenile cod, *Gadus morhua* (Pedersen & Carlsen, 1991; Nordeide *et al.*, 1992), and, for red fox, *Vulpes vulpes*, striped skunks, *Mephitis mephitis*, and raccoon, *Procyon lotor*, it has been assessed as a method of marking individuals that have taken bait containing the rabies vaccine (Nunan *et al.*, 1994). Tetracycline administered orally as a single 300 mg dose marked white-tailed deer for at least 150 days (Van Brackle *et al.*, 1994), but they found that tetracycline treatment of bait apparently reduced its palatability.

Tetracycline staining can also be applied to the skeletal structures of lower animals, such as the molluscs (Ekaratne & Crisp, 1982). Examples are: abalone, *Haliotis rubra* (Day *et al.*, 1995), whelk, *Buccinum undatum* (Kideys & Nash, 1993), bryozonans (Barnes, 1995), and the sponge, *Clathrina cerebrum* (Bavestrello *et al.*, 1993). Abelone, *H. rubra*, has also been marked by injection, which Pirker and Schiel (1993) found caused more stress to the animal than immersion.

Other florescent compounds can also be used. The otoliths of cod, *G. morhua*, have been marked by immersing the eggs, larvae, and juveniles in solutions of alizarin complexone (AC) (Blom *et al.*, 1994). Strontium, which is chemically similar to calcium and can substitute for this element in bony tissues of fishes, has been advocated as an inexpensive and efficient means of chemically marking juvenile salmonid scales (Snyder *et al.*, 1992).

3.2.6 Marking by injection, Panjet, or tattooing

For vertebrates, such as fish, dyes can be applied by Panjet inoculator, injection, or tattooing machine. An individual marking scheme, using a Panjet and alcian blue applied to large numbers of brown trout, is described by Bridcut (1993). Care must be taken to ensure that the force applied to the skin is not sufficient to cause internal injuries. Ducklings have been dyed by injecting the egg prior to hatching (Evans, 1951).

Tattooing has been frequently used. For small mammals, individual marking for life can be achieved by tattooing the ear, using tattooing tongs (Klimisch, 1986). Honma *et al.* (1986) considered ear tattooing superior to clipping, tagging, and tattooing for marking small mammals. Joly and Miaud (1989) noted that, although tattooing had been little used for urodeles, it offered advantages over other methods. Tattoos are not always permanent; on sturgeon, *Acipenser transmontanus*, they did not last 1 year (Rien *et al.*, 1994).

It has been found possible to mark crayfish by injecting a small amount of 'Bates numbering-machine ink' into the venter of the abdomen; the black, blue, and red inks were found to be non-poisonous (Slack, 1955; Black, 1963); similarly Indian ink may be used in fish (O'Grady & Hoy, 1972). This method might be applied to large arthropods that have an area of almost transparent cuticle. Latex injections have been used for crayfish (Slack, 1955) and fish (Davis, 1955). Coloured monofilament tags have been injected between the integument and the muscle on the ventral surface of prawns (Teboul, 1993). When applied to fish, such marks are often lost by ejection through the skin (Arnold, 1966).

Coded wire tags injected into crustaceans and fish are commonly used (e.g. Morrison, 1990; Fitz & Wiegert, 1991; Nass & Bocking, 1992; Bergstedt *et al.*, 1993; Ingram, 1993; Frith & Nelson, 1994). These may be detected magnetically. An alternative is visible implant tags, which are injected just under the skin, where they can be seen and read. However, these are susceptible to shedding (e.g. Niva, 1995), although coded wire tags injected into trout so as to be visible were found to have excellent retentions of 96% in postocular tissue and 99% in adipose fins after 238 days (Oven & Blankenship, 1993).

More recently, passive integrated transponder (PIT) tags only a few millimetres long have been developed for individually tagging aquatic animals. They are now extensively used on fish, and other applications include flatback turtles, *Natator depressus* (Parmenter, 1993), and red king crab, *Paralithodes camtschatica* (Donaldson *et al.*, 1992). The presence of these tags is detected electronically. The small size of PIT and coded wire tags suggests they have potential for studies on the larger terrestrial invertebrates.

3.2.7 External tags
Bands and rings are used extensively in work on birds, reptiles, and mammals. Ear tags, originally developed for domestic animals, are used on large mammals (e.g. seals: Prestrud, 1992; Testa & Rothery, 1992). A great variety of external tags are commercially available for fish (see Laird & Stott, 1978) and these can also be used for other animals groups (e.g. rattlesnakes: Smith, 1994). Rice and Taylor (1993) describe a waist band for anurans. A permanent method for tagging growing macro-

crustaceans is described by Schmalbach *et al.* (1994); these animals are often problematic because they moult even as adults. A nylon-filament tag attached to the prawn *Macrobrachium rosenbergii* did not interfere with moulting or growth over a 17-month period. Molluscs can also be tagged, e.g. abalone (*H. rubra*) were individually marked, using a nylon rivet to attach a numbered disc to the trema (Prince, 1991).

The small size of most insects often precludes these convenient methods. Butterflies and locusts have, however, been marked by attaching small labels with a word or a code written in waterproof black ink to part of their wings; in the Lepidoptera, the area should first be denuded of scales. Earlier workers used paper or cellophane stuck on with an adhesive, such as Durofix, which was later replaced by Sellotape. Freilich (1989) found that plastic tags attached with cyanoacrylate instant glue gave reliable results for benthic macroinvertebrates, and stoneflies retained tags for 6 months in a circular laboratory stream tank and up to 81 days under field conditions in a turbulent stream. Klock *et al.* (1953) devised a machine that glued lengths of coloured thread to anaesthetized flies. Punched ferrous labels have been used on bees; they may be magnetically removed (Gary, 1971).

Small (1.0–1.5 mm diameter), lightweight, plastic tags, produced using photographic film with a minimum of equipment, provide a means for individually marking insects without reliance on complex colour-coding or other schemes. Rubink (1988) describes their use in mark–recapture studies of native screw-worm, *Cochliomyia hominivorax*, in Central America. Sullivan *et al.* (1990) attached numbered, coloured discs, measuring 2.5 mm in diameter, to the shells of the planorbid snail *Helisoma duryi* with a waterproof epoxy, which gave 100% tag retention for over 5 months without significant mortality.

A recent development has been miniature radar tags for flying insects (Riley, 1996; Roland *et al.*, 1996). These tags, which reflect harmonic radar, were originally fitted to walking beetles (Mascanzoni & Wallin, 1986). The tags used by Roland (1996) weighed 0.4 mg and comprised a Schottky low-barrier diode mounted in the centre of an 8 cm dipole of very fine alminium wire. Mounted with rubber cement to the thorax or abdomen with the dipole trailing behind the insect, they were successfully fitted to the Apollo butterfly, *Parnassius sminthus*, the common alpine butterfly, *Erebia epipsodea*, the forest tent moth, *Malacosoma disstria*, the tachinid fly *Patelloa pachypyga* and the sarcophagid fly *Arachnidomyia aldrichi*. The transmitter–receiver used to track the insects can be carried in a backpack.

3.2.8 Branding

Hot branding is commonly used for large vertebrates. Hot-wire branding can also be used for fish but is rarely the method of choice. Freeze-branding has been applied to a variety of animals, including fish (Peterson & Key, 1992) and diving mammals — but not always with success — and small mammals. In mammals, it produces a white mark by killing the pigment-producing cells. Peterson and Key (1992) compared survivorship, tag retention, and growth of juvenile walleyes (*S. vitreum*) hand-tagged with binary-coded wire microtags (BCWMTs) or cold-branded with liquid nitrogen. They concluded that BCWMTs were superior to cold-branding. There are reports of tumours linked to freeze-branding in cattle (Yeruham *et al.*, 1993).

3.2.9 Mutilation

This method is also more widely used with vertebrates, especially fish, amphibians, and reptiles (Laird & Stott, 1978), than with the smaller insects, for a mark, in order to be easily visible, may be proportionally so large as to affect the insect's behaviour. Lepidoptera have been marked by clipping their wings (Querci, 1936) and beetles by damaging their elytra in various ways—incising the edges (Grüm, 1959), punching or burning small holes (Skuhravy, 1957; Schjøtz-Christensen, 1961), or scraping away the surface of the elytra between certain striae (Murdoch, 1963). Individual marks can be given to beetles by cutting small triangular and square notches in particular positions on the elytra using a pair of scissors (Goldwasser *et al.*, 1993). Crabs have been marked by cutting some of the teeth on the carapace (Edwards, 1958) and orthopteroids by notching the pronotum and amputating tegmina (Gangwere *et al.*, 1964). Fin-clipping is commonly applied to fish (Laird & Stott, 1978) but can be problematic, particularly in tropical waters, where regeneration is fast, natural fin damage common, and infection rife. Rien *et al.* (1994), working with the white sturgeon (*A. transmontanus*), compared a variety of tag and mutilation marks and concluded that lateral scute removal made a mark that lasted for at least 2 years.

An interesting variant on these procedures was the use of skin autografts by Rafinski (1977) to mark amphibians. Grafts were successful in more than 95% of cases and individuals could still be recognized 3 years later. For snakes, Spellerberg and Prestt (1977) considered carefully executed scale clipping to be the most humane and reliable permanent marking method, while Hutton and Woolhouse (1989) marked crocodiles by coded toe clipping or tail-scute notching.

3.2.10 Natural marks, parasites, and genes

Higher animals especially may be individually recognized by their unique markings or scars, e.g. characteristic fluke markings used to identify sperm whales (Dufault & Whitehead, 1995). Fish populations have been distinguished by the incidence of parasitic infections (Margolis, 1963; Templeman & Fleming, 1963; Bamber & Henderson, 1985) or by meristic characteristics, such as gill-raker form or number, otolith shape, or scale circuli.

Dispersal may be studied by the use of mutant genes (Peer, 1957; Levin, 1961; Hausermann *et al.*, 1971) or, when various genotypes are clearly distinct, by their different proportions in adjacent colonies (e.g. Sheppard, 1951; Richards & Waloff, 1954; Goodhart, 1962). However, it should be remembered that selection may operate differentially in the different colonies—perhaps on young stages before the genotype becomes identifiable. Different sexes and age classes also provide naturally marked groups.

3.2.11 Rare elements

Many elements, when exposed to a source of neutrons, become radioactive and emit a characteristic spectrum of gamma rays, a process termed neutron activation. Various workers have marked animals by incorporating rare elements in them and subsequently recognizing the mark by neutron activation and gamma spectroscopy. The great advantage of the method is that, theoretically at least, the mark may be retained from larval to adult life and self-marking is possible. For example, if a small quantity of a rare element is mixed into a mosquito's breeding pool, any adults emerging should be detectable at any time in their life and the contributions of different breeding sites to a population could be determined. The disadvantages are that the equipment is extremely expensive (but it is often available in nuclear physics or engineering centres); the procedure for gamma spectroscopy for rare elements is fatal for the animals (so it cannot be used in recapture studies); and there are often difficulties, undoubtedly of a basic physiological nature, in getting an adequate quantity of the rare element absorbed into the body and retained in the tissues (rather than just held in the gut).

Dysprosium, first used by Riebartsch (1963), has been found to be a lifelong marker for some Lepidoptera (Jahn *et al.*, 1966) and *Drosophila* (Richardson *et al.*, 1969). Rubidium has marked cabbage looper, *Trichoplusia ni*, larva (Berry *et al.*, 1972), bollworm, *Helicoverpa zea* (Stadelbacher, 1991), tobacco budworm, *Heliothis virescens* (Stadelbacher, 1991), gypsy moth, *Lymantria dispar* (Fleischer *et al.*, 1990), and the dipterans *Culicoides variipennis* (Holbrook *et al.*, 1991) and *A. aegypti* (Reiter *et al.*, 1995). Other elements used are europium for Lepidoptera (Jahn *et al.*, 1966; Ito, 1970), manganese for the fruit fly, *Ceratitis* (Monro, 1968), and lutetium, samarium, and europium for the larvae of clams, barnacles, and polychaetes (Levin *et al.*, 1993).

Caesium, disprosium, rubidium, and strontium, applied as chlorides to the foliage of host plants, have been found to mark phytophagous insects, including those feeding on flower-heads (e.g. Fleischer *et al.*, 1990; Stadelbacher, 1991; Dempster *et al.*, 1995). Berry *et al.* (1972) found a straight-line relationship between the concentrations of rubidium in foliar spray and in the male moths developing from the larvae that had been reared on the sprayed plants, but with the pea aphid, *Acyrthosiphon pisum*, the biological half-life was only about 1 day and detectable quantities only remained for 4 days after the aphid had left the plant (Frazer & Raworth, 1974). Ito (1970) showed that incorporation of europium (44 or 80 p.p.m.) into the diets of the larvae of the moths *Hyphantria* and *Spodoptera* did not affect growth and survival. The tick *Dermacentor variabilis* was marked by injecting the host mice with rubidium chloride (RbCl) (Burg, 1994).

Alternatively, the rare elements may be applied topically, essentially being used as labels. Cerium (50 μg $CeCl_3$/insect in an alcoholic solution), which is absorbed into the cuticle of insects, has been used by Rahalkar *et al.* (1971) to mark the elytra of weevils (*Rhynchophorus*), and gold was used similarly by Bate *et al.* (1974) to mark the elm bark beetle. Emerging calypterate diptera may self-mark by contamination of the ptilinal suture. Various dyes are often used for this purpose (p. 112); rare elements may be even more effective. Haisch *et al.* (1975) marked emerging *Rhagoletis* with dysprosium and samarium, mixed with fine sand and silica gel (1 : 1 by weight) at concentrations of 0.1% by weight. Provided the pupae were 3 cm deep, the error in self-marking was about 2%.

The trace elements within an animal can vary with the habitat, potentially allowing the origin of individuals to be identified by their 'chemoprint'. Identification of breeding site or place of origin has been applied to fish, birds, and insects. Dempster *et al.* (1986) applied this approach to the identification of brimstone butterfly, *Gonepteryx rhamni*, populations, but found that differences related to the place of origin faded with age.

3.2.12 Radioactive isotopes

Radioactive labelling, which was most popular soon after radioisotopes became readily available, is now less used to mark animals for the study of population size or dispersal. It has been largely replaced by safer techniques, such as rare-element tracers and fluorescent powders. In many countries, there are health and safety restrictions on the use of radioisotopes in the field and these must be checked before use. Useful bibliographies of the early literature on the entomological use of radiotracers are given in Anon. (1963) and Jenkins (1963); methodological details for the use of radioisotopes are given in Knoche (1991).

The radioactive isotopes of elements are unstable and they disintegrate, emitting radiations and forming other, usually non-radioactive, isotopes; the rate of disintegration is characteristic for each isotope and is described in terms of its half-life. Of the three types of radiation, entitled alpha, beta, and gamma, the isotopes used in ecological work produce only the two last-named. Beta particles have less power of penetration than gamma rays; however, the actual energies (expressed as million electron volts (MeV)) of the radiations differ for the different isotopes. Radioactivity is measured in units of becquerels (1 Bq = one disintegration/s) and the actual mass of the chemical element having a certain level of radioactivity will vary depending on the half-life of the isotope. The specific activity of any material or solution gives the relationship of the amount of radioisotope to the total element content and is expressed in terms such as Bq/g. Further basic information on isotopes may be found in Knoche (1991). Table 3.11 gives the characters of the principal isotopes used in ecological work. The use of isotopes in ecology is by no means confined to marking; they may also be used in studies on predation and energy flow (see Chapters 10 and 14). Great care should be exercised in all work with radioactive isotopes and precautions taken not to contaminate the environment (and the operators!). For this reason, long half-life isotopes should normally be avoided in fieldwork.

There are basically two methods of marking animals with isotopes, although the dividing line is not sharp. The isotopes may be used as a label outside the animal

Table 3.11 Some characters of the principal isotopes used in ecological research.

Isotope	Symbol	Half-life (approximate)	Radiations and energies (MeV) Beta (maximum)	Gamma
^{14}Carbon	C	5760 years	0.16	
^{137}Caesium	Cs	30 years	0.31	0.66
^{60}Cobalt*	Co	5.27 years	0.31	1.2
^{65}Zinc*	Zn	245 days	0.33	1.1
^{195}Gold	Au	185 days		0.1
^{45}Calcium	Ca	165 days	0.25	
^{182}Tantalum	Ta	115 days	0.51	Range, max. 1.2
^{35}Sulphur	S	87 days	0.17	
^{46}Scandium	Sc	84 days	0.36	0.89
^{192}Iridium	Ir	74 days	0.67	0.6
^{89}Strontium	Sr	55 days	1.50	
^{59}Iron	Fe	45 days	0.46	1.2
^{32}Phosphorus	P	14.2 days	1.71	
^{131}Iodine	I	8 days	0.61	0.36

*Used as labels.

or alternatively they may be fed to the animal and incorporated in its tissues. Which method and which isotope will depend on the precise nature of the work. If it is desired to trace the animal's movements, in the soil or in wood, or to locate it from a distance, then it is the penetrating gamma radiations that will have to be detected.

If the object is to mark part of the population in a way that can be recognized after the animal has been recaptured, then a wider range of isotopes is available, including those of carbon, calcium, phosphorus, and sulphur, which, if fed to the animal, are readily incorporated into its tissues. This marking method is preferable to the use of radioactive labels, in that animals can be tagged with the minimum amount of manipulation and the mark is invisible to predators. It is also almost unique (see above), as the mark will not be shed when the insect moults; this enables the method to be used for the estimation of population size and mortality in immature stages (Cook & Kettlewell, 1960) and the isotope may be passed on to the next generation (Beard, 1965). It is necessary, however, to choose an isotope (e.g. sulphur) with a reasonably long half-life (unless the insect's life cycle is very short) and to carry out laboratory tests on the extent to which the radioactive isotope atoms are excreted by the animal and replaced by normal ones (see below and p. 517).

If there is a risk of the mark being transferred naturally to other individuals (secondary tagging), then radioactive isotopes cannot be used in a mark–recapture protocol. Secondary tagging occurs with ants, as they pass food to other members of their colonies by trophallaxis, whilst some gregarious insects (e.g. *Triatoma*) feed intra-integumentally on the gut contents of others. Whether or not secondary tagging occurs, isotopes may be used to determine the foraging area of an ant colony (Erickson, 1972). Stradling (1970) has shown that, if the ants are starved for 4 days after marking, before return to the nest, phosphorus-32 (^{32}P) becomes incorporated into the tissues and secondary tagging does not occur.

3.2.12.1 Labels

Besides acting as a suitable gamma source, radioactive labels need to be non-toxic and durable. They may be attached externally, when corrosion or loss are major problems, or inserted into the body, when toxicity or disturbed behaviour present difficulties. Among the elements used are cobalt (^{60}Co), tantalum (^{182}Ta), iron (^{59}Fe), and iridium (^{192}Ir). For protection, iron may be used in stainless-steel wire and iridium in a platinum–iridium wire (Baldwin & Cowper, 1969), and cobalt can be gold-plated (Fredericksen & Lilly, 1955). A benzene-soluble DeKhontinsky-type cement has been found to be a good glue for attaching labels (Sanders & Baldwin, 1969). Alternatively, cobalt, in the form of the nitrate, can be made up with a water-resistant resin glue (e.g. Bond Fast) at a strength of about 6×10^7 Bq/ml and applied to the surface of the insect (Green et al., 1957); likewise, ^{192}Ir, as the chloride, mixed with some cellulose paint and rubber solution (van der Meijden, 1973), or zinc-65 (^{65}Zn), as the oxide, mixed with cellulose paint and applied with a special applicator (Mitchell et al., 1973), can be used. Minute labels (about 0.05 mm × 0.16 × 0.23–0.46 mm) of metallic tantalum have been used to mark coccinellid larvae and pentatomoid bugs, being attached by Seccotine glue or cellulose paint (Banks et al., 1961), and to mark earthworms, being introduced into the coelomic cavity (Gerard, 1963). If the initial specific activity of the labels is too high, there may be marked adverse effects; a specific activity of about 4.8×10^4 Bq was found appropriate for fieldwork. Workers using cobalt have applied labels with much higher specific activities, and Sullivan (1961) has shown that, if this is greater than 1.8×10^6 Bq, the survival of the weevil Pissodes strobi is affected; this does not, however, become apparent for several months. But we are doubtful of the validity of the implied claim of some workers that, if the dosage is just small enough for the mortality to be negligible, then the animals' behaviour is normal. It is a good principle to use the minimum radiation dose consistent with detection, as was done by van der Meijden (1973), who showed that the pupation success of moth larvae was not affected by labels containing about 1×10^5 Bq of ^{192}Ir; these labels could be detected from about 60 cm.

Submerging the animals in a solution containing the radioactive isotope is another method that may be classified as labelling, although it is possible that a small amount of the isotope could become incorporated in the tissues. This method was first developed by Roth and Hoffman (1952), who marked houseflies, wasps, leafhoppers, grasshoppers, and Coleoptera by dipping them for 1 min into a solution containing ^{32}P at the concentration of 1.8×10^5 Bq/ml, together with a wetting agent. The marker did not penetrate the body of the animal, and yet it could be removed only by prolonged washing. Large numbers of insects may be marked

Table 3.12 Submersion of insects in radioactive isotope solutions.

Animal	Isotope	Strength of solution (Bq/ml)	Authors
Boll weevil, *Anthonomus*	^{60}Co	1.8×10^4	Babers *et al.* (1954)
Bug, *Eurygaster*	^{60}Co	4.5×10^8	Rakitin (1963)
Pine weevil, *Pissodes*	^{46}Sc	c. 2.8×10^8	Godwin *et al.* (1957)
Ant, *Lasius*	^{32}P	c. 3.7×10^4	Odum & Pontin (1961)
Tick, *Amblyomma*	^{32}P	3.7×10^5	Knapp *et al.* (1956)
Tick, *Amblyomma*	^{59}Fe	3.7×10^5	Smittle *et al.* (1967)

quickly by this technique and it has frequently been used; some examples are given in Table 3.12.

The strengths of the solution applied varied greatly; it is possible that in some instances activity levels well above the minimum necessary were used. Roth and Hoffman (1952) found that the isotopes on their insects could be detected for about 18 days and Godwin *et al.* (1957) could recognize their marked weevils for 5 months, the difference largely stemming from the half-lives of the isotopes. The marking can conveniently be done in a covered funnel, with a constricted neck or straining plate to prevent the insects being washed out (Knapp *et al.*, 1956). More delicate animals cannot be marked by this method and, because of the violence involved, it is doubtful if it should be used even with robust species where an alternative technique is available.

A few instances are recorded of marking insects by spraying them with radioactive isotopes; this, however, is extremely hazardous for the operator and cannot be recommended.

Radioactive techniques for small-mammal marking are reviewed by Linn (1977). The general approach is similar to that adopted for insects.

3.2.12.2 Incorporation in tissues

Some of the advantages of this method have been alluded to above. The natural food of either the larva or the adult may be made radioactive or, more conveniently, the animal is offered a sugar solution containing the isotope—a large number of species of animal will take up fluid from such solutions. The actual distribution of the isotope in the body will depend on the distribution of the appropriate element and not on the method of feeding (see Hoffman *et al.*, 1951).

Plants may be made radioactive in a variety of ways (Sudia & Linck, 1963; Wiegert & Lindeborg, 1964; Krall & Simmons, 1977); two of the simplest are to grow them in a culture solution containing an isotope or to soak the seeds for a day in isotope-containing solution (Monk, 1967). All the parts of a large coniferous tree, including the pollen, may be tagged by introducing ^{32}P into the branches and trunk (Graham, 1957). Examples of phytophagous insects that have been marked through their host plant are turnip fly with ^{32}P (Oughton, 1951), plum weevil with ^{32}P and strontium-89 (^{89}Sr) (Rings & Layne, 1953), mealy bugs with ^{32}P (Cornwell, 1956), lepidopterous larvae with sulphur-35 (^{35}S) (Cook & Kettlewell, 1960), and a mirid, *Orthotylus*, with ^{32}P (Lewis & Waloff, 1964). The specific activity of the solution depends on the method of detection it is planned to use (see below), on the isotope, on the animal, and on the plant. For example, with ^{32}P, Cornwell (1956) used 1×10^6 Bq/ml and detected marked mealy bugs using a Geiger counter, but Lewis and Waloff (1964), with an activity level of 1.8×10^4 Bq/ml, used autoradiography to identify marked mirids. Hubbell *et al.* (1965) found that the quantity of radioisotope assimilated from a food source would vary under different conditions (see p. 517). With the gall-midge, *Dasyneura brassicae*, marked by treating its galls with ^{32}P, concentrations of over 2×10^5 Bq/ml were necessary to provide a clear autoradiographic image; concentrations greater than 9×10^6 Bq/ml (6×10^4 Bq/g host plant) reduced midge size and fecundity, but emergence was not affected by higher levels of activity (Sylven & Lonsjo, 1970). Zinc-65 is effective label for fungal host plants (Coleman, 1968).

Bloodsucking animals have been marked through their hosts (e.g. mosquitoes with ^{32}P (Yates *et al.*, 1951), blackflies with ^{32}P (Bennett, 1963), and female ticks and their eggs with carbon-14 (^{14}C), ^{144}Ce, and other isotopes (Sonenshine, 1970).

Aquatic larvae and the resulting adults become marked if reared in solutions containing isotopes (e.g. Garby *et al.*, 1966).

The sensitivity of the insect to radiation clearly needs to be taken into account in deciding the dose rate. The problem has been thoroughly studied by Robertson (1973) in a scale insect, and she has related dose to body size. The effect of the level of beta radiation from an isotope in the tissues is cumulative, and corresponds to the effects of a short-term exposure to beta radiation of about one-third the dose. The reproductive female is most sensitive (median lethal dose (LD$_{50}$) of 42 Gy); high sensitivity is also shown in the period just prior to moulting. Robertson recommends a level of 7×10^5 Bq/g (of insect body) for insects to survive and reproduce, and 5.5×10^6 Bq/g of insect body may be permissible for short-term work; these results are probably a general guide for small insects.

The incorporation of an isotope in an artificial food, bait, or drinking-water is a technique that has been widely used (e.g. Fay *et al.*, 1963). A convenient method for the mass marking of large numbers of small, delicate insects has been developed by Lewis and Waloff (1964) for the mirid *Orthotylus*, and has been used extensively with small Diptera, *Oscinella*, and *Drosophila*. The basic principle is that the insects drink an isotope-containing sugar solution, which is presented to them on a piece of saturated lens paper in a polystyrene Petri dish (Fig. 3.3). This has three metal studs (rivets) stuck to the base for legs, and holes are made in the upper and lower portions with a hot metal tube. A plastic stopper fits into the top hole and the lower one is covered with gauze (stuck on with acetone or a similar solvent). Most small insects soon die in the laboratory from desiccation and, as marking takes about 24 h, the dishes are stacked under bell-jars whose atmosphere is kept humid by a tube of an appropriate glycerine solution with a filter-paper wick (Johnson, 1940); 10% solution was used in this work, giving 98% relative humidity (r.h.).

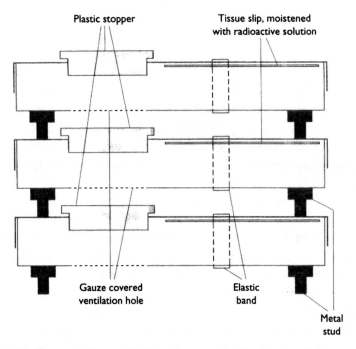

Fig. 3.3 Stacked polystyrene dishes modified for the presentation of radioactive solutions to small insects (after Lewis & Waloff, 1964).

The procedure is as follows. The lens tissue-paper slip is placed on the lid and moistened with the radioactive solution (about 0.2 ml), which should also contain the appropriate quantity (i.e. 10% in the work mentioned above) of glycerine; the paper will then adhere to the lid by surface tension. The two halves of the dish should then be put together and held by an elastic band. A known number of insects can then be tipped straight from a specimen tube in which they have been collected into the dish through the hole in the lid and the plastic stopper is inserted. The dishes are then stacked under a bell-jar; after about a day, they may be taken into the field at an appropriate time (see p. 106) and released by removing the lids.

When working with very low concentrations of an isotope, a significant proportion of the ions will become adsorbed on to the fibres of the lens tissue-paper and become unavailable to the insects (Lewis & Waloff, 1964). This can be overcome by raising the concentration of non-radioactive ions, i.e. of the 'carrier', normally an orthophosphate or sulphate, so that the final concentration of this salt in the solutions applied to the dishes is not less than 0.1%. This is conveniently done by diluting the stock isotope solution with an omnibus 5% sugar, 0.1% carrier, and 10% glycerine solution. The specific activity of the ^{32}P solution used by Lewis and Waloff (1964) was 1.8×10^4 Bq/ml. Because of the low energy of ^{35}S radiations, higher concentrations of this isotope must be used; Lewis and Waloff found a specific activity of 7.4×10^5 Bq/ml adequate.

3.2.12.3 Detection

There are two simple methods of detection adequate for mark–recapture studies: by a Geiger–Müller tube and by autoradiography. The former is indicated if the animal is to be traced in the field (e.g. van der Meijden, 1973). The latter demands that the animals be recaptured and killed, but has the advantage that very low levels of radioactivity can be used for marking, so one can discount the possibility that, although the animal may appear normal, its behaviour is affected by the radiations. A method for continuously locating small mammals in a limited arena is described by Graham and Ambrose (1967); it could also be used with large carabid beetles or similar insects.

The most convenient autoradiographic method for small insects is to stick them between two strips of Sellotape (Gillies, 1958). They may, if necessary, be examined under the microscope and retained for permanent reference. The strips are applied to the surface of an X-ray film; the exposure time will vary with the specific activity of the insects and the speed of the film — it is usually between 2 and 10 days. Autoradiographic discrimination between ^{32}P- and ^{35}S-marked insects is possible, owing to the different energies of these two isotopes. The images they produce on X-ray film are distinct: those due to ^{35}S give a clear silhouette of the marked insect; those due to ^{32}P are stronger but the shape of the insect is indistinct. However, it would be unwise to rely on the different form of the image. The short half-life of ^{32}P can be used to identify it; if the test strip containing the insects is exposed to film twice, at about a 14-day interval, the ^{32}P images could be much weaker on the second than on the first film — those from ^{35}S would alter little. However, as shown in Fig. 3.4, it is possible to distinguish the two isotopes simultaneously (Lewis & Waloff, 1964).

In studies on energy or nutrient flow using isotopes (see Chapter 14), it is often necessary to use another method: a scintillation counter. The quantification of the level of radioactivity provided by the scintillation counter may be useful; Smittle *et al.* (1973) showed that successive egg rafts of mosquitoes tagged as larvae had significantly different levels of counts.

Fig. 3.4 Double-exposure autoradiographic methods of discriminating between animals marked with ^{32}P and ^{35}S — diagrammatic sections of arrangement of test strip and films in cassette. (a) Method of Gillies (1958). (b) Method of Lewis and Waloff (1964).

The length of time a radioactively marked animal remains detectable will depend on the initial dose and on the effective half-life; the latter is the resultant of the half-life of the isotope and the rate of replacement of the isotope atoms in the tissues by inactive ones, i.e. on the biological half-life. Jensen and Fay (1951) found with various calypterate flies marked with ^{32}P that the rate of replacement was much faster when they were fed on a diet rich in phosphorus (milk) than when on one poor in this element (honey solution). If maintained on sugar after feeding on solutions containing 9×10^5–15×10^6 Bq/ml ^{32}P, eye gnats, *Hippelates*, remained detectable by a Geiger counter for just over 15 days (Dow, 1959); Jensen and Fay's (1951) results were similar. Autoradiography, of course, will detect lower levels, and Gillies (1961) records that *Anopheles* marked with ^{32}P as larvae were readily detectable 8 weeks after emerging as adults. Although each species under study should be tested, present evidence would suggest that ^{32}P will remain detectable for at least 2–3 weeks (Bennett, 1963) and ^{35}S considerably longer (Cook & Kettlewell, 1960), the half-life of the isotope being the most important factor.

3.2.13 Radio and sonic tags

Radio tags transmit at radio frequencies and thus can be used in both terrestrial and aquatic environments. Sonic tags emit ultrasound and are only useful in water, where the signal is detected with a hydrophone. These tags are used to study movement, the animal's immediate environment, and its physiological activity. In addition to transmitting a signal, which can be used to track the individual, they can also transmit information on, for example, temperature, salinity, heart rate, and depth of dive (pressure) (Amlaner & Macdonald, 1980). The key problem with these tags is their size and weight, which create difficulties of attachment and interference with the animal's movement. The batteries, which must be included within the tag, limit miniaturization. Alternative technologies for simply tracking animals are the much smaller PIT (p. 116) or radar tags (p. 117), which do not need an *in situ* power supply.

References

Aars, J., Andreassen, H.P., & Ims, R.A. (1995) Root voles: litter sex ratio variation in fragmented habitat. *J. Anim. Ecol.* **64**(4), 459–72.

Aebischer, N.J. (1986) Estimating the proportion of uncatchable animals in a population by double-sampling. *Biometrics* **42**(4), 973–80.

Amlaner, C.J. & Macdonald, D.W. (eds) (1980) *A Handbook on Biotelemetry and Radio Tracking.* Pergamon Press, Oxford.

Anon. (1963) Radioisotopes and ionizing radiations in entomology. *Bibl. Ser. Int. Atomic Energy Ag.* **9**, 414 pp.

Arnold, D.E. (1966) Marking fish with dyes and other chemicals. *U.S. Fish. Wildl. Serv. Tech. Pap.* **10**, 44 pp.

Ayre, G.L. (1962) Problems in using the Lincoln index for estimating the size of ant colonies (Hymenoptera: Formicidae). *J. NY Entomol. Soc.* **70**, 159–66.

Babers, F.H., Roan, C.C., & Walker, R.L. (1954) Tagging boll weevils with radioactive cobalt. *J. Econ. Entomol.* **47**, 928–9.

Bagenal, T.B. (1967) A method of marking fish eggs and larvae. *Nature* **214**, 113.

Bailey, N.T.J. (1951) On estimating the size of mobile populations from recapture data. *Biometrika* **38**, 293–306.

Bailey, N.T.J. (1952) Improvements in the interpretation of recapture data. *J. Anim. Ecol.* **21**, 120–7.

Baker, C.S. & Herman, L.M. (1987) Alternative population estimates of humpback whales (*Megaptera novaeangliae*) in Hawaiian waters (USA). *Can. J. Zool.* **65**(11), 2818–21.

Baker, S.G. (1990) A simple EM algorithm for capture–recapture data with categorical covariates. *Biometrics* **46**(4), 1193–200.

Baldwin, W.F. & Cowper, G. (1969) The use of radioactive platinum–iridium wire (Ir 92) as an internal tag for tracing insects. *Can. Entomol.* **101**, 151–2.

Bamber, R.N. & Henderson, P.A. (1985) Diplostomiasis in the sand smelt from the Fleet, Dorset and its use as a population indicator. *J. Fish Biol.* **26**, 223–9.

Banks, C.J. (1955) The use of radioactive tantalum in studies of the behaviour of small crawling insects on plants. *Br. J. Anim. Behav.* **3**, 158–9.

Banks, C.J., Brown, E.S., & Dezfulian, A. (1961) Field studies of the daily activity and feeding behaviour of Sunn Pest, *Eurygaster integriceps* Put. (Hemiptera: Scutelleridae) on wheat in North Iran. *Entomol. Exp. Appl.* **4**, 289–300.

Barnes, D.K.A. (1995) Seasonal and annual growth in erect species of Antarctic bryozoans. *J. Exp. Mar. Biol. Ecol.* **188**(2), 181–98.

Bate, L.C., Lyon, W.S., & Wollerman, E.H. (1974) Gold tagging of elm bark beetles and identification by neutron activation analysis. *Radiochem. Radioanal. Lett.* **17**, 77–85.

Bavestrello, G., Cattaneo Vietti, R., Cerrano, C., & Sara, M. (1993) Rate of spiculogenesis in *Clathrina cerebrum* (Porifera: Calcispongiae) using tetracycline marking. *J. Mar. Biol. Assoc. UK* **73**(2), 457–60.

Beard, R.L. (1965) Competition between DDT-resistant and susceptible house flies. *J. Econ. Entomol.* **58**, 584.

Begon, M. (1983) Abuses of mathematical techniques in ecology: applications of Jolly's capture–recapture method. *Oikos* **40**, 155–8.

Belant, J.L. & Seamans, T.W. (1993) Evaluation of dyes and techniques to color-mark incubating herring gulls. *J. Field Ornithol.* **64**(4), 440–51.

Beltran, R.R., Champigneulle, A., & Vincent, G. (1995) Mass-marking of bone tissue of *Coregonus lavaretus* L. and its potential application to monitoring the spatio-temporal distribution of larvae, fry and juveniles of lacustrine fishes. *Hydrobiologia* **301**, 399–407.

Bennett, G.F. (1963) Use of P32 in the study of a population of *Simulium rugglesi* (Diptera: Simuliidae) in Algonquin Park, Ontario. *Can. J. Zool.* **411**, 831–40.

Bergstedt, R.A., Swink, W.D., & Seelye, J.G. (1993) Evaluation of two locations for coded wire tags in larval and small parasitic phase sea lampreys. *N. Am. J. Fish. Manag.* **13**(3), 609–12.

Berry, W.L., Stimmann, M.W., & Wolf, W.W. (1972) Marking of native phytophagous insects with rubidium: a proposed technique. *Ann. Entomol. Soc. Am.* **65**, 236–8.

Bishop, J.A. & Sheppard, P.M. (1973) An evaluation of two capture–recapture models using the technique of computer simulation. In: Bartlett, M.S. & Hiorns, R.W. (eds) *The Mathematical Theory of the Dynamics of Biological Populations*. Academic Press, London, pp. 235–52.

Black, J.B. (1963) Observations on the home range of stream-dwelling crawfishes. *Ecology* **44**, 592–5.

Blinn, W.C. (1963) Ecology of the land snails *Mesodon thyroidus* and *Allogona profunda*. *Ecology* **44**, 498–505.

Blom, G., Nordeide, J.T., Svasand, T., & Borge, A. (1994) Application of two fluorescent chemicals, alizarin complexone and alizarin red S, to mark otoliths of Atlantic cod, *Gadus morhua* L. *Aquacult. Fish. Manag.* **25** (suppl. 1), 229–43.

Boano, G. & Cucco, M. (1991) Annual survival rates of a marsh warbler, *Acrocephalus palustris*, breeding population in northern Italy. *Riv. Ital. Ornitol.* **61**(1–2), 10–18.

Bohlin, T. & Sundstrom, B. (1977) Influence of unequal catchability on population estimates using the Lincoln index and the removal method applied to electrofishing. *Oikos* **28**, 123–9.

Bonett, D.G. (1988) Small sample bias and variance of Lincoln–Petersen and Bailey estimators. *Biometr. J.* **30**(6), 723–7.

Boonstra, R. & Krebs, C.J. (1978) Pitfall trapping of *Microtus townsendii*. *Ecology* **60**, 567–73.

Borror, D.J. (1934) Ecological studies of *Argia moesta* Hagen (Odonata: Coenag-Rionidae) by means of marking. *Ohio J. Sci.* **34**, 97–108.

Bozeman, E.L., Helfman, G.S., & Richardson, T. (1985) Population size and home range of American eels (*Anguilla rostrata*) in a Georgia (USA) tidal creek. *Trans. Am. Fish. Soc.* **114**(6), 821–5.

Bridcut, E.E. (1993) A coded alcian blue marking technique for the identification of individual brown trout, *Salmo trutta* L.: an evaluation of its use in fish biology. *Biol. Environ.* **93B**(2), 107–10.

Brooks, R.C., Heidinger, R.C., & Kohler, C.C. (1994) Mass-marking otoliths of larval and juvenile walleyes by immersion in oxytetracycline, calcein, or calcein blue. *N. Am. J. Fish. Manag.* **14**(1), 143–50.

Brownie, C., Hines, J.E., & Nichols, J.D. (1986) Constant-parameter capture–recapture models. *Biometrics* **42**(3), 561–74.

Brussard, P.F. (1971) Field techniques for investigations of population structure in a ubiquitous butterfly. *J. Lepidop. Soc.* **25**, 22–9.

Buckland, S.T. (1980) A modified analysis of the Jolly–Seber capture–recapture model. *Biometrics* **36**, 419–35.

Buckland, S.T. (1982) A mark–recapture survival analysis. *J. Anim. Ecol.* **51**, 833–47.

Buckland, S.T. & Hereward, A.C. (1982) Trap-shyness of yellow wagtails (*Motacilla flava flavissima*) at a premigratory roost. *Ringing and Migration* **4**, 15–23.

Buckland, S.T., Rowley, I., & Williams, D.A. (1983) Estimation of survival from repeated sightings of tagged galahs. *J. Anim. Ecol.* **52**, 563–73.

Buckner, C.H. (1968) Reactions of small mammals to vital dyes. *Can. Entomol.* **100**, 476–7.

Bumguardner, B.W. (1991) Marking subadult red drums with oxytetracycline. *Trans. Am. Fish. Soc.* **120**(4), 537–40.

Bumguardner, B.W. & Kin, T.L. (1996) Toxicity of oxytetracycline and calcein to juvenile striped bass. *Trans. Am. Fish. Soc.* **125**(1), 143–5.

Burg, J.G. (1994) Marking *Dermacentor variabilis* (Acari: Ixodidae) with rubidium. *J. Med. Entomol.* **31**(5), 658–62.

Burnham, K.P., Anderson, D.R., White, G.C., Brownie, C., & Pollock, K.H. (1987) *Design and Analysis Methods for Fish Survival Experiments Based on Release–Recapture*. Am. Fish. Soc. Monograph, Bethesda, Maryland.

Caldwell, A.H. (1956) Dry ice as an insect anaesthetic. *J. Econ. Entomol.* **49**, 264–5.

Calvo, B. & Furness, R.W. (1992) A review of the use and the effects of marks and devices on birds. *Ringing and Migration* **13**, 129–36.

Carothers, A.D. (1973) The effects of unequal catchability on Jolly–Seber estimates. *Biometrics* **29**, 79–100.

Caughley, G. (1977) *Analysis of Vertebrate Populations*. John Wiley & Sons, New York.

Chang, H.T. (1946) Studies on the use of fluorescent dyes for marking *Anopheles quadrimaculatus* Say. *Mosquito News* **6**, 122–5.

Chao, A. (1987) Estimating the population size for capture–recapture data with unequal catchability. *Biometrics* **43**(4), 783–92.

Chao, A. (1989) Estimating population size for sparse data in capture–recapture experiments. *Biometrics* **45**(2), 427–38.

Chapman, D.G. (1951) Some properties of the hypergeometric distribution with applications to zoological sample censuses. *Univ. Calif. Publ. Stat.* **1**(7), 131–60.

Chapman, D.G. (1954) The estimation of biological populations. *Ann. Math. Stat.* **25**, 1–15.

Chapman, D.G. (1995) Population estimation based on change of composition caused by a selective removal. *Biometrika* **42**, 279–90.

Coleman, D.C. (1968) A method for intensity labelling of fungi for ecological studies. *Mycologia* **60**, 960–1.

Collins, C.W. & Potts, S.F. (1932) Attractants for the flying gypsy moths as an aid to locating new infestations. *USDA Tech. Bull.* **336**, 43 pp.

Conway, G.R., Tripis, M., & McClelland, G.A.H. (1974) Population parameters of the mosquito *Aedes aegypti* (L.) estimated by mark–release–recapture in a suburban habitat in Tanzania. *J. Anim. Ecol.* **43**, 289–304.

Cook, L.M. & Kettlewell, H.B.D. (1960) Radioactive labelling of lepidopterous larvae: a method of estimating larval and pupal mortality in the wild. *Nature* **187**, 301–2.

Cook, S.P. & Hain, F.P. (1992) The influence of self-marking with fluorescent powders on adult bark beetles (Coleoptera: Scolytidae). *J. Entomol. Sci.* **27**(3), 269–79.

Corbet, G.B. (1956) The life-history and host relations of a hippoboscid fly *Ornithomyia fringillina* Curtis. *J. Anim. Ecol.* **25**, 403–20.

Corbet, P.S. (1952) An adult population study of *Pyrrhosoma nymphula* (Sulzer) (Odonata: Coenagrionidae). *J. Anim. Ecol.* **21**, 206–222.

Cormack, R.M. (1966) A test for equal catchability. *Biometrika* **22**, 330–42.

Cormack, R.M. (1972) The logic of capture–recapture estimates. *Biometrics* **28**, 337–43.

Cormack, R.M. (1973) Commonsense estimates from capture–recapture studies. In: Bartlett, M.S. & Hiorns, R.W. (eds) *The Mathematical Theory of the Dynamics of Biological Populations*. Academic Press, London, pp. 225–34.

Cormack, R.M. (1979) Models for capture–recapture. In: Cormack, R.M. & Patil, G.P. (eds) *Sampling Biological Populations*. International Cooperative Publishing House, Fairfield, Maryland, pp. 217–55.

Cormack, R.M. (1981) Log-linear models for capture–recapture experiments on open populations. In: Hiorns, R.W. & Cooke, D. (eds) *The Mathematical Theory of the Dynamics of Biological Populations*. Academic Press, London, pp. 197–216.

Cormack, R.M. (1989) Log-linear models for capture–recapture. *Biometrics* **45**(2), 395–414.

Cornwell, P.B. (1956) Some aspects of mealybug behaviour in relation to the efficiency of measures for the control of virus diseases of cacao in the Gold Coast. *Bull. Entomol. Res.* **47**, 137–66.

Coulson, J.C. (1962) The biology of *Tipula subnodicornis*, Zetterstedt, with comparative observations on *Tipula paludosa* Meigen. *J. Anim. Ecol.* **31**, 1–21.

Cragg, J.B. & Hobart, J. (1955) A study of a field population of the blowflies *Lucilia caesar* (L.) and *L. serricata* (MG.). *Ann. Appl. Biol.* **43**, 645–63.

Craig, C.C. (1953) On the utilisation of marked specimens in estimating populations of flying insects. *Biometrika* **40**, 170–6.

Crosbie, S.F. & Manly, B.F.J. (1985) A new approach to parsimonius modelling of capture–mark–recapture experiments. *Biometrics* **41**, 385–98.

Crumpacker, D.W. (1974) The use of micronized fluorescent dusts to mark adult *Drosophila pseudoobscura. Am. Midl. Nat.* **91**, 118–29.

Cunningham, M.P., Harley, J.M., & Grainge, E.B. (1963) The labelling of animals with specific agglutinins and the detection of these agglutinins in the blood meals of *Glossina. Rep. E. Afr. Tryp. Res. Org.* **1961**, 23–4.

Dalmat, H.T. (1950) Studies on the flight range of certain Simuliidae, with the use of aniline dye marker. *Ann. Entomol. Soc. Am.* **43**, 537–45.

Darroch, J.N. (1958) The multiple recapture census: I Estimation of a closed population. *Biometrika* **45**, 343–51.

Darroch, J.N. (1959) The multiple-capture census II. Estimation when there is immigration or death. *Biometrika* **46**, 336–51.

Daum, R.J., Gast, R.T., & Davich, T.B. (1969) Marking adult boll weevils with dyes fed in a cottonseed oil bait. *J. Econ. Entomol.* **62**, 943–4.

Davey, J.T. (1956) A method of marking isolated adult locusts in large numbers as an aid to the study of their seasonal migrations. *Bull. Entomol. Res.* **46**, 797–802.

Davis, C.S. (1955) The injection of latex solution as a fish marking technique. *Invest. Indiana Lakes Streams* **4**, 111–16.

Day, R.W., Williams, M.C., & Hawkes, G.P. (1995) A comparison of fluorochromes for marking abalone shells. *Mar. Freshw. Res.* **46**(3), 599–606.

Dempster, J.P. (1971) The population ecology of the cinnabar moth, *Tyria jacobaeae* L. (Lepidoptera, Arctiidae). *Oecologia* **7**, 26–67.

Dempster, J.P., Atkinson, D.A., & Cheesman, O.D. (1995) The spatial population dynamics of insects exploiting a patchy food resource I. Population extinctions and regulation. *Oecologia* **104**, 340–53.

Dempster, J.P., Lakhani, K.H., & Coward, P.A. (1986) The use of chemical composition as a population marker in insects: a study of the brimstone butterfly. *Ecol. Entomol.* **11**, 51–65.

Dobson, J.R. (1993) A portable carbon dioxide dispenser and its uses in field entomology. *Antenna* **17**, 10–13.

Dobson, R.M. & Morris, M.G. (1961) Observations on emergence and life-span of wheat bulb fly, *Leptohylemyia coarctata* (Fall.) under field-cage conditions. *Bull. Entomol. Res.* **51**, 803–21.

Dobson, R.M., Stephenson, J.W., & Lofty, J.R. (1958) A quantitative study of a population of wheat bulb fly, *Leptohylemyia coarctata* (Fall.) in the field. *Bull. Entomol. Res.* **49**, 95–111.

Donaldson, W.E., Schmidt, D., Watson, L., & Pengill, Y.D. (1992) Development of a technique to tag adult red king crab, *Paralithodes camtschatica* (Tilesius, 1815), with passive integrated transponder tags. *J. Shellfish Res.* **11**(1), 91–4.

Dosdall, L.M., Galloway, M.M., & Gadawski, R.M. (1992) New self-marking device for dispersal studies of black flies (Diptera: Simuliidae). *J. Am. Mosquito Contr. Assoc.* **8**(2), 187–90.

Dow, R.P. (1959) Dispersal of adult *Hippelates pusio*, the eye gnat. *Ann. Entomol. Soc. Am.* **52**, 372–81.

Dow, R.P. (1971) The dispersal of *Culex nigripalpus* marked with high concentrations of radiophosphorus. *J. Med. Entomol.* **8**, 353–63.

Dowdeswell, W.H. (1959) *Practical Animal Ecology*. Methuen, London.

Dowdeswell, W.H., Fisher, R.A., & Ford, E.B. (1940) The quantitative study of populations in the lepidoptera. 1. *Polyommatus icarus* Rott. *Ann. Eugen.* **10**, 123–36.

Dowdy, A.K. & McGaughey, W.H. (1992) Fluorescent pigments for marking lesser grain borers (Coleoptera: Bostrichidae). *J. Econ. Entomol.* **85**(2), 567–9.

Dufault, S. & Whitehead, H. (1995) An assessment of changes with time in the marking patterns used for photoidentification of individual sperm whales, *Physeter macrocephalus. Mar. Mammal Sci.* **11**(3), 335–43.

Eberhardt, L.L. (1969) Population estimates from recapture frequencies. *J. Wildlife Management* **33**, 28–39.

Eberhardt, L.L., Chapman, D.G., & Gilbert, J.R. (1979) A review of marine mammal census methods. *Wildlife Monographs* **63**, 1–46.

Edgar, W.D. (1971) The life-cycle, abundance and seasonal movement of the wolf spider, *Lycosa (Pardosa) lugubris*, in central Scotland. *J. Anim. Ecol.* **40**, 303–22.

Edwards, R.L. (1958) Movements of individual members in a population of the shore crab, *Carcinus maenas* L., in the littoral zone. *J. Anim. Ecol.* **27**, 37–45.

Edwards, R.L. (1961) Limited movement of individuals in a population of the migratory grasshopper, *Melanoplus bilituratus* (Walker) (Acrididae) at Kamloops, British Columbia. *Can Entomol.* **93**, 628–31.

Ehrlich, P.R. & Davidson, S.E. (1960) Techniques for capture–recapture studies of Lepidoptera populations. *J. Lepidop. Soc.* **14**, 227–9.

Ekaratne, S.U.K. & Crisp, D.J. (1982) Tidal micro-growth bands in intertidal gastropod shells, with an evaluation of band-dating techniques. *Proc. Roy. Soc. B* **214**, 305–23.

Ellis, D.H. & Ellis, C.H. (1975) Color marking Golden eagles with human hair dyes. *J. Wildlife Management* **39**, 446–7.

Eltringham, S.K. (1977) Methods of capturing wild animals for marking purposes. In: Stonehouse, B. (ed.) *Animal Marking*. Macmillan Press, London, pp. 13–23.

Erickson, J.M. (1972) Mark–recapture techniques for population estimates of *Pogonomyrmex* and colonies: an evaluation of the ^{32}P technique. *Ann. Entomol. Soc. Am.* **65**, 57–61.

Evans, C.D. (1951) A method of color-marking young waterfowl. *J. Wildlife Management* **15**, 101–3.

Evans, C.R. & Lockwood, A.P.M. (1994) Population field studies of the Guinea chick lobster (*Panulirus guttatus* Latreille) at Bermuda: abundance, catchability, and behavior. *J. Shellfish Res.* **13**(2), 393–415.

Evans, W.G. & Gyrisco, G.G. (1960) The flight range of the European chafer. *J. Econ. Entomol.* **53**, 222–4.

Fales, J.H., Bodenstein, P.F., Mills, G.D., & Wessel, L.H. (1964) Preliminary studies on face fly dispersion. *Ann. Entomol. Soc. Am.* **57**, 135–7.

Fay, R.W., Kilpatrick, J.W., & Baker, I.T. (1963) Rearing and isotopic labelling of *Fannia canicularis*. *J. Econ. Entomol.* **56**, 69–71.

Fenwick, A. & Amin, M.A. (1983) Marking snails with nail varnish as a field experimental technique. *Ann. Trop. Med. Parasitol.* **77**, 387–90.

Fisher, R.A. & Ford, E.B. (1947) The spread of a gene in natural conditions in a colony of the moth *Panaxia dominula* L. *Heredity* **1**, 143–74.

Fitz, H.C.W. & Wiegert, R.G. (1991) Tagging juvenile blue crabs, *Callinectes sapidus*, with microwire tags: retention, survival, and growth through multiple molts. *J. Crustac. Biol.* **11**(2), 229–35.

Fleischer, S.J., Ravlin, F.W., Delorme, D., Stipes, R.J., & McManus, M.L. (1990) Marking gypsy moth (Lepidoptera: Lymantriidae) life stages and products with low doses of rubidium injected or implanted into pin oak. *J. Econ. Entomol.* **83**(6), 2343–8.

Foott, W.H. (1976) Use of fluorescent powders to monitor flight activities of adult *Glischrochilus quadrisignatus* [Coleoptera: Nitidulidae]. *Can. Entomol.* **108**, 1041–4.

Frankie, G.W. (1973) A simple field technique for marking bees with fluorescent powders. *Ann. Entomol. Soc. Am.* **66**, 690–1.

Frazer, B.D. & Raworth, D.A. (1974) Marking aphids with rubidium. *Can. J. Zool.* **529**, 1135–6.

Fredericksen, C.F. & Lilly, J.H. (1955) Measuring wireworm reactions to soil insecticides by tagging with radioactive cobalt. *J. Econ. Entomol.* **48**, 438–42.

Freeland, W.J. (1986) Populations of cane toad, *Bufo marinus*, in relation to time since colonization. *Aust. Wildl. Res.* **13**(2), 321–30.

Freeman, B.E. (1964) A population study of *Tipula* species (Diptera, Tipulidae). *J. Anim. Ecol.* **33**, 129–40.

Freilich, J.E. (1989) A method for tagging individual benthic macroinvertebrates. *J. N. Am. Benthol. Soc.* **8**(4), 351–4.

Frisch, K.V. (1950) *Bees, Their Vision, Chemical Senses, and Language*. Comstock, Ithaca, New York.

Frith, H.R. & Nelson, T.C. (1994) Abundance, age, size, sex and coded wire tag recoveries for chinook salmon escapements of Campbell and Quinsam Rivers, 1993. *Can. Manuscr. Rep. Fish. Aquat. Sci.* **2251**, 1–59.

Gaillard, J.M., Boisaubert, B., Boutin, J.M., & Clobert, J. (1986) Estimation of population numbers from capture–mark–recapture data: application to roe deer. *Gibier Faune Sauvage* **3**, 143–58.

Gangwere, S.K., Chavin, W., & Evans, F.C. (1964) Methods of marking insects, with especial reference to Orthoptera (Sens. Lat.). *Ann. Entomol. Soc. Am.* **57**, 662–9.

Garby, L., Yasuno, M., & Phuriuethaya, V. (1966) Labelling mosquitoes with radio-active iodine 1. *Trans. Roy. Soc. Trop. Med. Hyg.* **60**, 136.

Gary, N.E. (1971) Magnetic retrieval of ferrous labels in a capture–recapture system for honey bees and other insects. *J. Econ. Entomol.* **64**, 961–5.

Gaskell, T.L. & George, B.J. (1972) A Bayesian modification of the Lincoln index. *J. Appl. Ecol.* **9**, 377–84.

Gatz, A.J.J.R. & Loar, J.M. (1988) Petersen and removal population size estimates: combining methods to adjust and interpret results when assumptions are violated. *Environ. Biol. Fish.* **21**(4), 293–308.

Gerard, B.M. (1963) An earthworm labelled with radioactive tantalum. *Nature* **200**, 486–7.

Gillies, M.T. (1958) A simple autoradiographic method for distinguishing insects labelled with phosphorus-32 and sulphur-35. *Nature* **182**, 1683–4.

Gillies, M.T. (1961) Studies on the dispersion and survival of *Anopheles gambiae* Giles in East Africa, by means of marking and release experiments. *Bull. Entomol. Res.* **52**, 99–127.

Godwin, P.A., Jaynes, H.A., & Davis, J.M. (1957) The dispersion of radio-actively tagged white pine weevils in small plantations. *J. Econ. Entomol.* **50**, 264.

Goldwasser, L., Schatz, G.E., & Young, H.J. (1993) A new method for marking Scarabaeidae and other Coleoptera. *Coleop. Bull.* **47**, 21–6.

Golley, F.B. & Gentry, J.B. (1964) Bioenergetics of the southern harvester ant, *Pogonomyrmex badius*. *Ecology* **45**, 217–25.

Goodhart, C.B. (1962) Thrush predation on the snail *Cepaea hortensis*. *J. Anim. Ecol.* **279**, 47–57.

Gosselin, L.A. (1993) A method for marking small juvenile gastropods. *J. Mar. Biol. Assoc. UK* **73**(4), 963–6.

Gouteux, J.P. & Buckland, S.T. (1984) Tsetse fly ecology in the preforested area of Ivory Coast: 8. Population dynamics. *Cahiers ORSTOM Ser. Entomol. Med. Parasitol.* **22**, 19–34.

Graham, B.F. (1957) Labelling pollen of woody plants with radioactive isotopes. *Ecology* **38**, 156–8.

Graham, H.M. & Mangum, C.L. (1971) Larval diets containing dyes for tagging pink bollworm moths internally. *J. Econ. Entomol.* **64**, 376–9.

Graham, W.I. & Ambrose, H.W. (1967) A technique for continuously locating small mammals in field enclosures. *J. Mammal.* **48**, 639–42.

Green, G.W., Baldwin, W.F., & Sullivan, C.R. (1957) The use of radioactive cobalt in studies of the dispersal of adult females of the European pine shoot moth *Rhyacionia buoliana* (Schifr.). *Can. Entomol.* **89**, 379–83.

Greenslade, P.J.M. (1964) The distribution, dispersal and size of a population of *Nebria brevicollis* (F.), with comparative studies on three other Carabidae. *J. Anim. Ecol.* **33**, 311–33.

Griffin, D.R. (1970) Migrations and homing of bats. In: Wimsatt, W.A. (ed.) *Biology of Bats*. Academic Press, London.

Grüm, L. (1959) Seasonal changes of activity of the Carabidae. *Ekol. Polska A* **7**, 255–68.

Hagler, J.R., Cohen, A.C., Bradley Dunlop, D., & Enriquez, F.J. (1992) New approach to mark insects for feeding and dispersal studies. *Environ. Entomol.* **21**(1), 20–5.

Hain, M.L. (1965) Ecology of the lizard *Uta mearnsi* in a desert canyon. *Copeia* **1965**, 78–81.

Haisch, A., Stark, H., & Forester, S. (1975) Markierung von Fruchtfliegen und ihre Erkennung durch Indikatoraktivierung. *Entomol. Exp. Appl.* **18**, 31–43.

Hampton, J.K. & Kirkwood, G.P. (1990) Tag shedding by southern bluefin tuna *Thunnus maccoyii*. *US Nat. Mar. Fish. Serv. Fish. Bull.* **88**(2), 313–22.

Hanson, W.R. (1963) Calculation of productivity, survival and abundance of selected vertebrates from sex and age ratios. *Wildl. Monogr.* **9**, 1–60.

Hartley, P.H.T. (1954) Back garden ornithology. *Bird Study* **1**, 18–27.

Hausermann, W., Fay, R.W., & Hacker, C.S. (1971) Dispersal of genetically marked female *Aedes aegypti* in Mississippi. *Mosquito News* **31**, 37–51.

Hawkes, R.B. (1972) A fluorescent dye technique for marking insect eggs in predation studies. *J. Econ. Entomol.* **65**, 1477–8.

Henderson, P.A. & Holmes, R.H.A. (1987) On the population biology of the common shrimp *Crangon crangon* (L.) (Crustacea: Caridea) in the Severn Estuary and Bristol Channel. *J. Mar. Biol. Ass. UK* **67**, 825–47.

Hendricks, D.E. (1971) Oil-soluble blue dye in larval diet marks adults, eggs, and first stage F, larvae of the pink bollworm. *J. Econ. Entomol.* **64**, 1404–6.

Hendricks, D.E. & Graham, H.M. (1970) Oil-soluble dye in larval diet for tagging moths, eggs, and spermatophores of tobacco budworms. *J. Econ. Entomol.* **63**, 1019–20.

Hendricks, M.L., Bender, T.R.J.R., & Mudrak, V.A. (1991) Multiple marking of American shad otoliths with tetracycline antibiotics. *N. Am. J. Fish. Manag.* **11**(2), 212–19.

Heron, R.J. (1968) Vital dyes as markers for behavioural and population studies of the larch sawfly, *Pristiphora erichsonii* (Hymenoptera: Tenthredinidae). *Can. Entomol.* **1009**, 470–5.

Hewlett, P.S. (1954) A micro-drop applicator and its use for the treatment of certain small insects with liquid insecticide. *Ann. Appl. Biol.* **41**, 45–64.

Hill, S.B. & Clayton, D.H. (1985) *Wildlife after Dark: a Review of Nocturnal Observation Techniques*. Occasional Paper 17, James Ford Bell Museum of Natural History, University of Minnesota.

Hoffman, R.A., Lindquist, A.W., & Butts, J.S. (1951) Studies on treatment of flies with radioactive phosphorus. *J. Econ. Entomol.* **44**, 471–3.

Holbrook, F.R., Belden, R.P., & Bobian, R.J. (1991) Rubidium for marking adults of *Culicoides variipennis* (Diptera: Ceratopogonidae). *J. Med. Entomol.* **28**(2), 246–9.

Holt, S.J. (1955) On the foraging activity of the wood ant. *J. Anim. Ecol.* **24**, 1–34.

Honma, M., Iwaki, S., Kast, A., & Kreuzer, H. (1986) Experiences with the identification of small rodents. *Exp. Anim.* **35**(3), 347–52.

Horn, H.S. (1976). A clamp for marking butterflies in capture–recapture studies. *J. Lepidop. Soc.* **30**(2), 145–6.

Hubbell, S.P., Sikora, A., & Paris, H. (1965) Radiotracer, gravimetric and calorimetric studies of ingestion and assimilation rates of an isopod. *Health Phys.* **11**, 1485–501.

Hunter, P.E. (1960) Plastic paint as a marker for mites. *Ann. Entomol. Soc. Am.* **53**, 698.

Hutton, J.M. & Woolhouse, M.E.J. (1989) Mark–recapture to assess factors affecting the proportion of a Nile crocodile population seen during spotlight counts to monitor crocodile abundance. *J. Appl. Ecol.* **26**, 381–95.

Ingram, B.A. (1993) Evaluation of coded wire tags for marking fingerling golden perch, *Macquaria ambigua* (Percichthyidae), and silver perch, *Bidyanus bidyanus* (Teraponidae). *Aust. J. Mar. Freshw. Res.* **44**(6), 817–24.

Isaacs, R. & Byrne, D.N. (1998) Aerial distribution, flight behaviour and eggload: their inter-relationship during dispersal by the sweetpotato whitefly. *J. Anim. Ecol.* **67**, 741–50.

Ito, Y. (1970). A stable isotope, europium-151 as a tracer for field studies of insects. *Appl. Entomol. Zool.* **5**, 175–81.

Iwao, S., Mizuta, K., Nakamura, H., Oda, T., & Sato, Y. (1963) Studies on a natural population of the large 28-spotted lady beetle, *Epilachna vigintioctomaculata* Motschulsky. 1. Preliminary analysis of the overwintered adult population by means of the marking and recapture method. *Japan. J. Ecol.* **13**, 109–17.

Iwao, S., Kiritani, K., & Hokyo, N. (1966) Application of a marking and recapture method for the analysis of larval–adult populations of an insect, *Nezara viridula* (Hemiptera: Pentatomidae). *Res. Popul. Ecol.* **8**, 147–60.

Jackson, C.H.N. (1933a) On the true density of tsetse flies. *J. Anim. Ecol.* **2**, 204–9.

Jackson, C.H.N. (1933b) On a method of marking tsetse flies. *J. Anim. Ecol.* **2**, 289–90.

Jackson, C.H.N. (1937) Some new methods in the study of *Glossina morsitans*. *Proc. Zool. Soc. Lond.* **1936**, 811–96.

Jackson, C.H.N. (1948) The analysis of a tsetse-fly population. III. *Ann. Eugen.* **14**, 91–108.

Jackson, C.H.N. (1953) A mixed population of *Glossina morsitans* and *G. swynnertoni*. *J. Anim. Ecol.* **22**, 78–86.

Jahn, E., Lippay, H., Weidinger, N., & Schwach, G. (1966) Untersuchungen uber die Usbreitung von Nonnenfaltem durch Markierung mit Seltenen erden. *Anz. Schadlingskunst* **39**, 17–22.

Jenkins, D.W. (1963) Use of radionuclides in ecological studies of insects. In: Schultz, V. & Klement, A.W. (eds) *Radioecology*. Rheinhold, New York, pp. 431–43.

Jensen, J.A. & Fay, R.W. (1951) Tagging of adult horse flies and flesh flies with radioactive phosphorus. *Am. J. Trop. Med.* **31**, 523–30.

Jewell, G.R. (1956) Marking of tsetse flies for their detection at night. *Nature* **178**, 750.

Jewell, G.R. (1958) Detection of tsetse fly at night. *Nature* **181**, 1354.

Johansson, T.S.K. (1959) Tracking honey bees in cotton fields with fluorescent pigments. *J. Econ. Entomol.* **52**, 572–7.

Johnson, C.G. (1940) The maintenance of high atmosphere humidifiers for entomological work with glycerol–water mixtures. *Ann. Appl. Biol.* **27**, 295–9.

Jolly, G.M. (1963) Estimates of population parameters from multiple recapture data with both death and dilution—a deterministic model. *Biometrika* **50**, 113–28.

Jolly, G.M. (1965) Explicit estimates from capture–recapture data with both death and immigration—stochastic model. *Biometrika* **52**, 225–47.

Jolly, G.M. (1982) Mark–recapture models with parameters constant in time. *Biometrics* **38**, 301–21.

Joly, P. & Miaud, C. (1989) Tattooing as an individual marking technique in urodeles. *Alytes* **8**(1), 11–16.

Kelker, G.H. (1940) Estimating deer populations by a differential hunting loss in the sexes. *Proc. Utah Acad. Sci. Arts Lett.* **17**, 65–9.

Kendall, W.L., Pollock, K.H., & Brownie, C. (1995) A likelihood-based approach to capture–recapture estimation of demographic parameters under the robust design. *Biometrics* **51**(1), 293–308.

Kideys, A.E. & Nash, R.D.M. (1993) A note on the mortality associated with fluorescent marking in a gastropod. *Turk. J. Zool.* **17**(2), 175–8.

Killick-Kendrick, R. (1993) 'Sodastream' CO_2 dispenser for killing insects. *Antenna* **17**, 115–16.

Klimisch, J.J. (1986) Ear tattooing technique for individual identification of white small rodents. *Zeitschr. Versuchstierkunde* **28**(6), 277–81.

Klock, J.W., Pimentel, D., & Stenburg, R.L.

(1953) A mechanical fly-tagging device. *Science* **118**, 48–9.

Kloft, W., Holldobler, B., & Haishch, A. (1965) Traceruntersuchungen zur Abgrenzurg von Nestarealen Holzzerstorender Rossameisen (*Camponolus herculeanus* L. und *C. ligniperda* Satr.). *Entomol. Exp. Appl.* **8**, 20–6.

Knapp, S.E., Farinacci, C.J., Herbert, C.M., & Saenger, E.I.. (1956) A method for labelling the lone star tick with a radioactive indicator (P32). *J. Econ. Entomol.* **49**, 393–5.

Knight, R.H. & Southon, H.A.W. (1963) A simple method for marking haematophagous insects during the act of feeding. *Bull. Entomol. Res.* **54**, 379–82.

Knoche, H.W. (1991) *Radioisotopic Methods for Biological and Medical Research*. Oxford University Press, Oxford.

Komazaki, S. & Sakagami, Y. (1989) Capture-recapture study on the adult population of the white spotted longicorn beetle, *Anoplophora malasiaca* (Thomson) (Coleoptera: Cerambycidae), in a citrus orchard. *Appl. Entomol. Zool.* **24**(1), 78–84.

Krall, J.H. & Simmons, G.A. (1977) Tree root injection of phosphorus-32 for labelling defoliating insects. *Environ. Entomol.* **6**, 159–60.

Krebs, C.J., Singleton, G.R., & Kenney, A.J. (1994) Six reasons why feral house mouse populations might have low recapture rates. *Wildl. Res.* **21**(5), 559–67.

Kruk-de-Bruin, M., Rost, L.C.M., & Draisama, F.G.A.M. (1977) Estimates of the number of foraging ants with the Lincoln-index method in relation to the colony size of *Formica polyctena*. *J. Anim. Ecol.* **46**, 457–70.

Kuenzler, E.J. (1958) Niche relations of three species of lycosid spiders. *Ecology* **39**, 494–500.

Labrecque, G.C., Bailey, D.L., Meifert, D.W., & Weidhaas, D.E. (1975) Density estimates and daily mortality rate evaluations of stable fly *Stomoxys calcitrans* (Diptera: Muscidae) populations in field cages. *Can. Entomol.* **107**, 597–600.

Laird, M.L. & Stott, B. (1978) Marking and tagging. In: Bagenal, T. (ed.) *Methods for Assessment of Fish Production in Fresh Waters*. Blackwell Scientific Publications, Oxford, pp. 84–100.

Lamb, R.I. & Wellington, W.G. (1974) Techniques for studying the behavior and ecology of the European earwig, *Forficula auricularia* (Dermaptera: Forficulidae). *Can. Entomol.* **106**, 881–8.

Law, P.M.W. (1994) Simulation study of salmon carcass survey capture–recapture methods. *Calif. Fish Game* **80**, 14–28.

Lebreton, J.D. & Clobert, J. (1986) *User's Manual*

for the Program Surge, Version 2.0. CEPE/CNRS, Montpellier.

Lebreton, J.D., Burnham, K.P., Clobert, J., & Anderson, D.R. (1992) Modeling survival and testing biological hypotheses using marked animals: a unified approach with case studies. *Ecol. Monogr.* **62**(1), 67–118.

Leeuwen, E.R.V. (1940) The activity of adult codling moths as indicated by captures of marked moths. *J. Econ. Entomol.* **33**, 162–6.

Le Gall, J.Y., Bosc, P., Chateau, D., & Taquet, M. (1986) An estimation of the number of adult female green turtles *Chelonia mydas* per nesting season at Tromelin and Europa (Indian Ocean) (1973–1985). *Oceanogr. Trop.* **21**(1), 3–22.

Leslie, P.H. (1952) The estimation of population parameters from data obtained by means of the capture–recapture method II. The estimation of total numbers. *Biometrika* **39**, 363–88.

Leslie, P.H. & Chitty, D. (1951) The estimation of population parameters from data obtained by means of the capture–recapture method. 1. The maximum likelihood equations for estimating the death-rate. *Biometrika* **38**, 269–92.

Levin, L.A., Huggett, D., Myers, P., Bridges, T., & Weaver, J. (1993) Rare earth tagging methods for the study of larval dispersal by marine invertebrates. *Limnol. Oceanogr.* **38**(2), 346–60.

Levin, M. (1961) Distribution of foragers from honey bee colonies placed in the middle of a large field of alfalfa. *J. Econ. Entomol.* **54**, 431–4.

Lewis, C.T. & Waloff, N. (1964) The use of radioactive tracers in the study of dispersion of *Orthotylus virescens* (Douglas & Scott) (Miridae, Hetcroptera). *Entomol. Exp. Appl.* **7**, 15–24.

Lincoln, F.C. (1930) Calculating waterfowl abundance on the basis of banding returns. *USDA Circ.* **118**, 4 pp.

Linn, I.J. (1977) Radioactive techniques for small mammal marking. In: Stonehouse, B. (ed.) *Animal Marking*. Macmillan Press, London.

Loab, H.A. (1966) Marking brown trout fry with the dye Sudan black. *NY Fish Game J.* **13**, 109–18.

Loery, G. & Nichols, J.D. (1985) Dynamics of a black-capped chickadee (*Parus atricapillus*) population, 1958–1983. *Ecology* **66**(4), 1195–203.

Loery, G., Pollock, K.H., Nichols, J.D., & Hines, J.E. (1987) Age-specificity of black-capped chickadee survival rates: analysis of capture–recapture data. *Ecology* **68**(4), 1038–44.

Lomnicki, A. (1969) Individual differences among adult members of a snail population. *Nature* **223**, 1073–4.

McDonald, W.A. (1960) Nocturnal detection of

tsetse flies in Nigeria with ultra-violet light. *Nature* **185**, 867–8.

MacLeod, J. (1958) The estimation of numbers of mobile insects from low-incidence recapture data. *Trans. Roy. Entomol. Soc. Lond.* **110**, 363–92.

MacLeod, I. & Donnelly, J. (1957) Individual and group marking methods for fly population studies. *Bull. Entomol. Res.* **48**, 585–92.

Malter, S.F. (1996) Interpatch movement of the red milkweed beetle, *Tetraopes tetraphthalmus*: individual responses to patch size and isolation. *Oecologica* **105**, 447–53.

Manly, B.F.J. (1970) A simulation study of animal population estimation using the capture–recapture method. *J. Appl. Ecol.* **7**, 13–39.

Manly, B.F.J. (1971a) Estimates of a marking effect with capture–recapture sampling. *J. Appl. Ecol.* **8**, 181–9.

Manly, B.F.J. (1971b) A simulation study of Jolly's method for analysing capture–recapture data. *Biometrics* **27**(2), 415–24.

Manly, B.F.J. & Parr, M.J. (1968) A new method of estimating population size survivorship, and birth rate from capture–recapture data. *Trans. Soc. Br. Entomol.* **189**, 81–9.

Margolis, L. (1963) Parasites as indicators of the geographical origin of sockeye salmon *Oncorhynchus nerka* (Walbaum) occuring in the North Pacific Ocean and adjacent seas. *Bull. Int. N. Pacif. Fish. Commn.* **11**, 101–56.

Marten, G.G. (1970) A regression method for mark–recapture estimates with unequal catchability. *Ecology* **51**, 291–5.

Mascanzoni, D. & Wallin, H. (1986) The harmonic radar: a new method of tracing insects in the field. *Ecol. Entomol.* **11**, 387–90.

Mazzotti, S. (1995) Population structure of *Emys orbicularis* in the Bardello (Po Delta, Northern Italy). *Amphib. Reptil.* **16**(1), 77–85.

Medland, T.E. & Beamish, F.W.H. (1991) Lamprey statolith banding patterns in response to temperature, photoperiod, and ontogeny. *Trans. Am. Fish. Soc.* **120**(2), 255–60.

Medley, J.G. & Ahrens, E.H. (1968) Fluorescent dyes for marking and recovering fowl ticks in poultry houses treated with insecticides. *J. Econ. Entomol.* **61**, 81–4.

van der Meijden, E. (1973) Experiments on dispersal, late-larval predation, and pupation in the cinnabar moth (*Tyria jacobaeae* L.) with a radioactive label (^{111}Ir). *Netherl. J. Zool.* **23**, 43–5.

Mellanby, K. (1939) The physiology and activity of the bed-bug (*Cimex lectularius* L.) in a natural infestation. *Parasitology* **31**, 200–11.

Menkens, G.E.J.R. & Anderson, S.H. (1988)

Estimation of small-mammal population size. *Ecology* **69**(6), 1952–9.

Michener, C.D., Cross, E.A., Daly, H.V., Rettenmeyer, C.W., & Wille, A. (1955) Additional techniques for studying the behaviour of wild bees. *Insectes Soc.* **2**, 237–246.

Minta, S. & Mangel, M. (1989) A simple population estimate based on simulation for capture–recapture and capture–resight data. *Ecology* **70**(6), 1738–51.

Mitchell, B. (1963) Ecology of two carabid beetles, *Bembidion lampros* (Herbst.) and *Trechus quadristriatus* (Schrank.). *J. Anim. Ecol.* **32**, 377–92.

Mitchell, H.C., McGovern, W.L., Cross, W.H., & Mitlin, N. (1973) Boll weevils: tagging for hibernation and field studies. *J. Econ. Entomol.* **66**, 563–4.

Monk, C.D. (1967) Radioisotope tagging through seed soaking. *Bull. Georgia Acad. Sci.* **25**, 13–17.

Monro, J. (1968) Marking insects with dietary manganese for detection by neutron activation. *Ecology* **49**, 774–6.

Montgomery, S.S., Brett, P.A., Blount, C., *et al.* (1995) Loss of tags, double tagging and release methods for eastern king prawns, *Penaeus plebejus* (Hess) laboratory and field experiments. *J. Exp. Mar. Biol. Ecol.* **188**, 115–31.

Morgan, M.J. & Walsh, S.J. (1993) Evaluation of the retention of external tags by juvenile American plaice (*Hippoglossoides platessoides*) using an aquarium experiment. *Fish. Res.* **16**(1), 1–7.

Morrison, J.K., Coyle, C.L., & Bertoni, S.E. (1990) Histological effect of tagging chum and coho salmon fry with coded wire tags. *Progr. Fish Cult.* **52**(2), 117–19.

Moth, I.I. & Barker, J.S.F. (1975) Micronized fluorescent dusts for marking drosophila adults. *J. Nat. Hist.* **91**, 393–6.

Muir, R.C. (1958) On the application of the capture–recapture method to an orchard population of *Blepharidopterus angulatus* (Fall.) (Hemiptera–Heteroptera, Miridae), *Rep. E. Malling Res. Sta.* **1957**, 140–7.

Murdoch, W.W. (1963) A method for marking Carabidae (Col.). *Entomol. Monthly Mag.* **99**, 22–4.

Nagiec, M., Czerkies, P., Goryczko, K., Witkowski, A., & Murawska, E. (1995) Mass-marking of grayling, *Thymallus thymallus* (L.), larvae by fluorochrome tagging of otoliths. *Fish. Manag. Ecol.* **2**(3), 185–95.

Nakai, K., Musashi, T., Inoguchi, N., Saido, T., Kishida, T., & Matsuda, H. (1995) Estimation of survival rate in the early stage of released artificial young abalone by the capture–recapture data. *Nippon Suisan Gakkaishi* **59**(11), 1845–50.

Nakamura, K., Ito, Y., Nakamura, M., Matsumoto, T., & Hayakawa, K. (1971) Estimation of population productivity of *Parapleurus alliaceus* Germar (Orthoptera: Acridiidae) on a *Miscanthus sinensis* Anders. grassland, I Estimation of population parameters. *Oecologia (Berlin)* **7**, 1–15.

Narayanaswamy, K.C., Kumar, P., Manjunath, D., & Datta, R.K. (1994) Determination of flight range of the uzi fly, *Exorista bombycis* (Louis) (Diptera: Tachinidae) through marking technique by adding dye to the adult diet. *Ind. J. Sericult.* **33**(1), 40–3.

Nass, B.L. & Bocking, R.C. (1992) Abundance, age, size, sex and coded wire tag recoveries for chinook salmon escapements of Kitsumkalum River, 1989–1990. *Can. Manuscr. Rep. Fish. Aquat. Sci.* **2147**, 1–61.

Nelemans, M.N.E., Den Boer, P.J., & Spee, A. (1989) Recruitment and summer diapause in the dynamics of a population of *Nebria brevicollis* (Coleoptera: Carabidae). *Oikos* **56**(2), 157–69.

Newman, R.M. & Martin, F.B. (1983) Estimation of fish production rates and associated variances. *Can. J. Fish. Aqu. Sci.* **40**, 1729–36.

Nichols, J.D. & Pollock, K.H. (1990) Estimation of recruitment from immigration versus *in situ* production using Pollock's robust design. *Ecology* **71**(1), 21–6.

Nichols, J.D., Hines, J.E., Pollock, K.H., Hinz, R.L., & Link, W.A. (1994) Estimating breeding proportions and testing hypotheses about costs of reproduction with capture–recapture data. *Ecology* **75**(7), 2052–65.

Nichols, J.D., Noon, B.R., Stokes, S.L., & Hines, J.E. (1981) Remarks on the use of mark–recapture methodology in estimating avian population size. *Studies in Avian Biology* **6**, 121–36.

Nichols, J.D., Pollock, K.H., & Hines, J.E. (1984) The use of a robust capture–recapture design in small mammal population studies: a field example with *Microtus pennsylvanicus*. *Acta Theriol.* **29**(26–36), 357–65.

Niebylski, M.L. & Meek, C.L. (1989) A sef-marker device for emergent adult mosquitos. *J. Am. Mosquito Contr. Assoc.* **5**, 86–90.

Nielsen, E.T. (1961) On the habits of the migratory butterfly, *Ascia monuste* L. *Biol. Meddr. Dan. Vid. Selsk.* **23**, 1–81.

Niva, T. (1995) Retention of visible implant tags by juvenile brown trout. *J. Fish Biol.* **46**(6), 997–1002.

Nordeide, J.T., Holm, J.C., Ottera, H., Blom, G., & Borge, A. (1992) The use of oxytetracycline as a marker for juvenile cod (*Gadus morhua* L.). *J. Fish Biol.* **41**(1), 21–30.

Norris, K.R. (1957) A method of marking Calliphoridae (Diptera) during emergence from the puparium. *Nature* **180**, 1002.

Nunan, C.P., Macinnes, C.D., Bachmann, P., Johnston, D.H., & Watt, I.D. (1994). Background prevalence of tetracycline-like fluorescence in teeth of free ranging red foxes (*Vulpes vulpes*), striped skunks (*Mephitis mephitis*) and raccoons (*Procyon lotor*) in Ontario, Canada. *J. Wildl. Dis.* **30**(1), 112–14.

Odum, E.P. & Pontin, A.J. (1961) Population density of the underground ant, *Lasius flavus*, as determined by tagging with P32. *Ecology* **42**, 186–8.

O'Grady, I.I. & Hoy, I.B. (1972) Rhodamine-B and other stains as markers for the mosquito fish, *Gambusia affinis*. *J. Med. Entomol.* **9**, 571–4.

Ohgushi, T. & Sawada, H. (1998) What changed the demography of an introduced population of an herbivorous lady beetle? *J. Anim. Ecol.* **67**, 679–88.

Orians, G.H. & Leslie, P.H. (1958) A capture–recapture analysis of a shearwater population. *J. Anim. Ecol.* **27**, 71–86.

Ostfeld, R.S., Miller, M.C., & Schnurr, J. (1993) Ear tagging increases tick (*Ixodes dammini*) infestation rates of white footed mice (*Peromyscus leucopus*). *J. Mammal.* **74**(3), 651–5.

Otis, D.K., Burnham, K.P., White, G.C., & Anderson, D.R. (1978) Statistical inference from capture data on closed animal populations. *Wildl. Monogr.* **62**, 1–135.

Oughton, J. (1951) Tagging root maggot flies by means of radioactive phosphorus. *Ann. Rep. Entomol. Soc. Ont.* **81**, 91–2.

Oven, J.H. & Blankenship, H.L. (1993) Benign recovery of coded wire tags from rainbow trout. *N. Am. J. Fish. Manag.* **13**(4), 852–5.

Pal, R. (1947) Marking mosquitoes with fluorescent compounds and watching them by ultra-violet light. *Nature* **160**, 298–9.

Paloheimo, J.E. (1963) Estimation of catchabilities and population sizes of lobsters. *J. Fish Res. Board Can.* **20**, 59–88.

Paradis, E., Guedon, G., & Pradel, R. (1993) Estimation of sex- and age-related survival rates in a microtine population. *J. Wildl. Manag.* **57**(1), 158–63.

Parker, W.S. (1976) Population estimates, age structure, and denning habits of whipsnakes, *Masticophis taeniatus taeniatus*. *Copeia* **1972**, 892–5.

Parmenter, C.J. (1993) A preliminary evaluation of the performance of passive integrated transponders and metal tags in a population study of the flatback sea turtle (*Natator depressus*). *Wildl. Res.* **20**(3), 375–81.

Parr, M.J. (1965) A population study of a colony of

imaginal *Ischnura elegans* (van der Linden) (Odonata: Coenagriidae) at Dale, Pembrokeshire. *Field St.* **2**(2), 237–82.

Parr, M.J., Gaskell, T.I., & George, B.J. (1968) Capture–recapture methods of estimating animal numbers. *J. Biol. Educ.* **2**, 95–117.

Pedersen, T. & Carlsen, B. (1991) Marking cod (*Gadus morhua* L.) juveniles with oxytetracycline incorporated into the feed. *Fish. Res.* **12**(1), 57–64.

Peer, D.F. (1957) Further studies on the mating range of the honey bee, *Apis mellifera* L. *Can. Entomol.* **89**, 108–10.

Peffly, R.L. & Labrecque, G.C. (1956) Marking and trapping studies on dispersal and abundance of Egyptian house flies. *J. Econ. Entomol.* **49**, 214–17.

Peterson, D.L. & Carline, R.F. (1996) Effects of tetracycline marking, transport density, and transport time on short-term survival of walleye fry. *Progr. Fish Cult.* **58**(1), 29–31.

Peterson, M.S. & Key, J.P. (1992) Evaluation of hand tagging juvenile walleyes with binary coded wire microtags. *N. Am. J. Fish. Manag.* **12**(4), 814–18.

Phillips, B.F. & Campbell, N.A. (1970) Comparison of methods of estimating population size using data on the whelk *Dicathais aegrota* (Reeve). *J. Anim. Ecol.* **399**, 753–9.

Pirker, J.G. & Schiel, D.R. (1993) Tetracycline as a fluorescent shell-marker in the abalone *Haliotis iris. Mar. Biol.* **116**(1), 81–6.

Polivka, J. (1949) The use of fluorescent pigments in a study of the flight of the Japanese bettle. *J. Econ. Entomol.* **42**, 818–21.

Poljak, A., Huber, D., & Greguric, J. (1989) Estimation of street pigeon (*Columba livia*) population size in Zagreb (Yugoslavia). *Larus* **42**, 141–50.

Pollock, K.H. (1974) *The Assumption of Equal Catchability of Animals in Tag–Recapture Experiments*. Cornell University, Ithaca, New York.

Pollock, K.H. (1975) A *k*-sample tag–recapture model allowing for unequal survival and catchability. *Biometrika* **62**, 577–83.

Pollock, K.H. (1981) Capture–recapture models allowing for age-dependent survival and capture rates. *Biometrics* **37**, 521–9.

Pollock, K.H. (1982) A capture–recapture design robust to unequal probability of capture. *J. Wildlife Management* **46**, 752–7.

Pollock, K.H. & Mann, R.H.K. (1983) Use of an age-dependent mark–recapture model in fisheries research. *Can. J. Fish. Aqu. Sci.* **40**, 1449–55.

Pollock, K.H., Nichols, J.D., Brownie, C., & Hines, J.E. (1990) Statistical inference for capture–recapture experiments. *Wildl. Monogr.* **107**, 1–98.

Pooley, A.C. (1962) The nile crocodile (*Crocodylus niloticus*). *Lammergeyer* **2**, 1–55.

Porter, S.D. & Jorgensen, C.D. (1980) Recapture studies of the harvester ant *Pogonomyrmex owyheei* Cole, using a fluorescent marking technique. *Ecol. Entomol.* **5**, 263–9.

Prestrud, P. (1992) Denning and home range characteristics of breeding arctic foxes in Svalbard. *Can. J. Zool.* **70**(7), 1276–83.

Prince, J.D. (1991) A new technique for tagging abalone. *Austr. J. Mar. Freshw. Res.* **42**(1), 101–6.

Putman, R.J. (1995) Ethical considerations and animal welfare in ecological field studies. *Biodivers. Conserv.* **4**(8), 903–15.

Quarterman, K.D., Mathis, W., & Kilpatrick, J.W.X. (1954) Urban fly dispersal in the area of Savannah, Georgia. *J. Econ. Entomol.* **47**, 405–12.

Querci, O. (1936) Aestivation of Lepidoptera. *Entomol. Rec.* **48**, 122.

Radinovsky, S. & Krantz, G.W. (1962) The use of Fluon to prevent the escape of stored-product insects from glass containers. *J. Econ. Entomol.* **55**, 815–16.

Rafinski, J. (1977) Autotransplantation as a method for permanent marking of urodele amphibians. *J. Herpetol.* **11**, 241–7.

Rahalkar, G.W., Mistry, K.B., Harnalkar, M.R., Bharathan, K.G., & Gopalayengar, A.R. (1971) Labelling adults of red palm weevil (*Rhynchophorus ferrugineus*) with cerium for detection by neutron activation. *Ecology* **52**, 187–8.

Rakitin, A.A. (1963) The use of radioisotopes in the marking of *Eurygaster integriceps* Put. [In Russian.] *Entomol. Obozr.* **42**, 39–48. (Trans. *Entomol. Rev.* **42**, 20–38.)

Reeves, W.C., Brookman, B., & Hammon, W.M. (1948) Studies on the flight range of certain *Culex* mosquitoes, using a fluorescent-dye marker, with notes on *Culiseta* and *Anopheles*. *Mosquito News* **8**, 61–9.

Reiter, P., Amador, M.A., Anderson, R.A., & Clark, G.G. (1995) Short report: dispersal of *Aedes aegypti* in an urban area after blood feeding as demonstrated by rubidium-marked eggs. *Am. J. Trop. Med. Hyg.* **52**(2), 177–9.

Ribbands, C.R. (1950) Changes in behaviour of honeybees, following their recovery from anaesthesia. *J. Exp. Biol.* **27**, 302–10.

Rice, T.M. & Taylor, D.H. (1993) A new method for making waistbands to mark anurans. *Herpetol. Rev.* **24**(4), 141–2.

Richards, O.W. & Waloff, N. (1954) Studies on the biology and population dynamics of British grasshoppers. *Anti-locust Bull.* **17**, 1–182.

Richardson, R.H., Wallace, R.J., Gage, S.J., Bouchey, G.D., & Dennell, M. (1969) XII. Neutron activation techniques for labelling *Drosophila* in natural populations. In: Wheeler, M.R. (ed.) *Studies in Genetics*, vol. V, pp. 171–86. Publication 6918, University of Texas, Houston.

Riebartsch, K. (1963) Inaktive Markierung von Insekten mit Dysprosium. *Nachrbl. Deutsch. Pflschut.-dienst. Stuttg.* **15**, 154–7.

Rien, T.A., Beamesderter, R.C.P., & Foster, C.A. (1994) Retention, recognition, and effects on survival of several tags and marks for white sturgeon. *Calif. Fish Game* **80**(4), 161–70.

Riley, J.R. (1996) Tracking bees with harmonic radar. *Nature* **379**, 29–30.

Rings, R.W. & Layne, G.W. (1953) Radioisotopes as tracers in plum curculio behaviour studies. *J. Econ. Entomol.* **46**, 473–7.

Robertson, G. (1973) The sensitivity of scale insects to α-particles and to γ-rays. *Int. J. Radiat. Biol.* **23**, 313–23.

Robson, D.S. (1969) Mark–recapture methods of population estimation. In: Johnson, N.L. & Smith, H. (eds) *New Developments in Survey Sampling*. Wiley, New York.

Robson, D.S. & Regier, H.A. (1964) Sample size in Petersen mark–recapture experiments. *Trans. Am. Fish. Soc.* **93**, 215–26.

Roff, D.A. (1973) On the accuracy of some mark–recapture estimators. *Oecologia* **12**, 15–34.

Rojas Beltran, R., Gillet, C., & Champigneulle, A. (1995) Immersion mass-marking of otoliths and bone tissue of embryos, yolk-sac fry and fingerlings of Arctic charr *Salvelinus alpinus* (L.). *Nord. J. Freshw. Res.* **71**, 411–8.

Roland, J., McKinnon, G., Backhouse, C., & Taylor, P.D. (1996) Even smaller radar tags on insects. *Nature* **381**, 120.

Rosewell, J. & Shorrocks, B. (1987) The implications of survival rates in natural populations of *Drosophila*: capture–recapture experiments on domestic species. *Biol. J. Linn. Soc.* **32**(4), 373–84.

Roth, A.R. & Hoffman, R.A. (1952) A new method of tagging insects with P32. *J. Econ. Entomol.* **45**, 1091.

Rubink, W.L. (1988) A photographic technique for producing high-quality insect tags for mark/release/recapture studies. *Entomol. News* **99**(3), 167–71.

Ruhle, C. & Winecki Kuhn, C. (1992) Tetracycline marking of coregonids at the time of egg fertilization. *Aquat. Sci.* **54**(2), 165–75.

Sanders, C.I. & Baldwin, W.F. (1969) Iridium-192 as a tag for carpenter ants of the genus *Camponotus* (Hymenoptera: Formicidae). *Can. Entomol.* **101**, 416–18.

Schjøtz-Christensen, B. (1961) For plantnga biologien hos *Amara infirma* Dft. og *Harpalus neglectus* Serv. *Flora Fauna* **67**, 8–12.

Schmalbach, A.E., Quackenbush, L.S., & Melinek, R. (1994) A method for tagging the Malaysian prawn *Macrobrachium rosenbergii*. *Aquaculture* **122**(23), 147–59.

Schnabel, Z.E. (1938) The estimation of the total fish population of a lake. *Am. Math. Monogr.* **45**, 348–52.

Schwarz, C.J., Bailey, R.E., Irvine, J.R., & Dalziel, F.C. (1993) Estimating salmon spawning escapement using capture–recapture methods. *Can. J. Fish. Aquat. Sci.* **50**(6), 1181–97.

Seber, G.A.F. (1965) A note on the multiple-recapture census. *Biometrika* **52**, 249.

Seber, G.A.F. (1970) Estimating time-specific survival and reporting rates for adult birds from band returns. *Biometrika* **57**, 313–18.

Seber, G.A.F. (1982) *The Estimation of Animal Abundance and Related Parameters*. Griffin, London.

Seber, G.A.F. (1986) A review of estimating animal abundance. *Biometrics* **42**, 267–92.

Seber, G.A.F. & Felton, R. (1981) Tag loss and the Petersen mark–recapture experiment. *Biometrika* **68**, 211–19.

Secor, D.H., White, M.G., & Dean, J.M. (1991) Immersion marking of larval and juvenile hatchery-produced striped bass with oxytetracycline. *Trans. Am. Fish. Soc.* **120**(2), 261–6.

Service, M.W. (1995) *Mosquito Ecology: Field Sampling Methods*, 2nd edn. Chapman & Hall, London, 988 pp.

Sheppard, D.C., Wilson, G.H., & Hawkins, J.A. (1973) A device for self-marking of Tabanidae. *J. Environ. Entomol.* **2**, 960–1.

Sheppard, P.M. (1951) Fluctuations in the selective value of certain phenotypes in the polymorphic land snail *Cepaea nemoralis* (L.). *Heredity* **5**, 125–54.

Sheppard, P.M., MacDonald, W.W., Tonn, R.I., & Grab, B. (1969) The dynamics of an adult population of *Aedes aegypti* in relation to dengue haemorrhagic fever in Bankok. *J. Anim. Ecol.* **38**, 661–702.

Shura-Bura, B.L. & Grageau, U.L. (1956) Fluorescent analysis studies of insect migration. [In Russian.] *Entomol. Obozr.* **35**, 760–3.

Simpson, J. (1954) Effects of some anaesthetics on honeybees: nitrous oxide, carbon dioxide, ammonium nitrate, smoker fumes. *Bee World* **35**, 149–55.

Singer, M.C. & Wedlake, P. (1981) Capture does affect probability of recapture in a butterfly species. *Ecol. Ent.* **6**, 215–16.

Sinsko, M.J. & Craig, G.B. (1979) Dynamics of an isolated population of *Aedes triseriatus* (Diptera: Culicidae) I population size. *J. Med. Entomol.* 15, 89–98.

Skalski, J.R. & Robson, D.S. (1982) A mark and removal field procedure for estimating population abundance. *J. Wildlife Management* 46, 752–60.

Skuhravy, V. (1957) Studium pohybu některych Střevlikovitych značkováním jedincu. *Acta Soc. Entomol. Bohem.* 53, 171–9.

Slack, K.V. (1955) An injection method for marking crayfish. *Prog. Fish Cult.* 17, 36–8.

Slooff, A. & Herath, P.R.J. (1980) Ovarian development and biting frequency in *Anopheles calicifacies* Giles in Sri Lanka. *Trop. Geog. Med.* 32, 306–11.

Smith, C. (1994) Fish tags for observing free ranging rattlesnakes. *Herpetol. Rev.* 25(2), 58.

Smith, M.V. (1958) The use of fluorescent markers as an aid in studying the forage behaviour of honeybees. *Proc. X Int. Congr. Entomol.* 4, 1063.

Smith, M.V. & Townsend, G.F. (1951) A technique for mass-marking honeybees. *Can. Entomol.* 83, 346–8.

Smittle, B.J., Hill, S.O., & Philips, F.M. (1967) Migration and dispersal patterns of ^{59}Fe-labelled lone star ticks. *J. Econ. Entomol.* 60, 1029–31.

Smittle, B.J., Lowe, R.E., Ford, H.R., & Weidhaas, D.E. (1973) Techniques for ^{32}P labelling and assay of egg rafts from field-collecting *Culex pipiens quingquefoaseiatus* Say. *Mosquito News* 33, 215–20.

Snyder, R.J., McKeown, B.A., Colbow, K., & Brown, R. (1992) Use of dissolved strontium in scale marking of juvenile salmonids: effects of concentration and exposure time. *Can. J. Fish. Aquat. Sci.* 49(4), 780–2.

Sonenshine, D.E. (1970) Fertility and fecundity of *Dermacentor variabilis* reared from radioisotope-tagged larvae. *J. Econ. Entomol.* 63, 1675–6.

Sonleitner, F.I. & Bateman, M.A. (1963) Mark–recapture analysis of a population of queensland fruit-fly, *Dacus tryoni* (Frogg.) in an orchard. *J. Anim. Ecol.* 32, 259–69.

South, A. (1965) Biology and ecology of *Agriolimax reticulatus* (Müll.) and other slugs: spatial distribution. *J. Anim. Ecol.* 34, 403–17.

Southwood, T.R.E. (1965) Migration and population change in *Oscinella frit* L. (Diptera) on the oatcrop. *XII Int. Congr. Entomol.* 4, 420–1.

Southwood, T.R.E. & Scudder, G.G.E. (1956) The bionomics and immature stages of the thistle lace bugs (*Tings ampliata* H.-S. and *T. cardui* L.; Hem., Tingidae). *Trans. Soc. Br. Entomol.* 12, 93–112.

Speare, P. (1992) A technique for tetracycline injecting and tagging billfish. *Bull. Mar. Sci.* 51(2), 197–203.

Spellerberg, I.F. & Prestt, I. (1977) Marking snakes. In: Stonehouse, B. (ed.) *Animal Marking*. Macmillan Press, London, 133–41.

Spendelow, J.A. & Nichols, J.D. (1989) Annual survival rates of breeding adult roseate terns. *Auk* 106(3), 367–74.

Stadelbacher, E.A. (1991) Bollworm and tobacco budworm (Lepidoptera: Noctuidae): labeling of adults produced in wild geranium, *Geranium dissectum*, treated with rubidium or strontium chloride. *J. Econ. Entomol.* 84(2), 496–501.

Steiner, L.F. (1965) A rapid method for identifying dye-marked fruit flies. *J. Econ. Entomol.* 589, 374–5.

Stobow, T., Fowler, G.M., & Sinclaire, A.F. (1992) Short term tagging mortality of laboratory held juvenile Atlantic herring (*Clupea harengus harengus*). *J. NW Atlantic Fish. Sci.* 12, 27–33.

Stonehouse, B. (ed.) (1977) *Animal Marking: Recognition Marking of Animals in Research*. Macmillan Press, London.

Stradling, D.J. (1970) The estimation of worker ant populations by the mark–release–recapture method: an improved marking technique. *J. Anim. Ecol.* 39, 575–91.

Sudia, T.W. & Linck, A.J. (1963) Methods for introducing radionuclides into plants. In: Schultz, V. & Klement, A.W.E. (eds) *Radioecology*. New York, Rheinhold, pp. 417–23.

Sugimoto, T., Sakuratani, Y., Setokuchi, O. *et al.* (1994) Using the mark-and-release method in the estimation of adult population of sweet potato weevil, *Cylas formicarius* (Fabricius) in a sweet potato field. *Appl. Entomol. Zool.* 29(1), 11–19.

Sullivan, C.R. (1961) The survival of adults of the white pine weevil, *Pissodes strobi* (Peck), labelled with radioactive cobalt. *Can. Entomol.* 93, 78–9.

Sullivan, J.J., Bishop, H.S., Schneider, K.R., & Rodrick, G.E. (1990) Marking snails with numbered, colored discs: a technique for identifying individual specimens. *Trop. Med. Parasitol.* 41(3), 289–90.

Sylven, E. & Lonsjo, H. (1970) Studies on a method of radioactive tagging of a gall midge, *Dasyneura brassicae* Winn. (Diptera, Cicedomyiidae). *Entomol. Scand.* 1, 291–6.

Tanton, M.T. (1965) Problems of live-trapping and population estimation for the wood mouse. *J. Anim. Ecol.* 38, 511–29.

Teboul, D. (1993) Internal monofilament tags used to identify juvenile *Penaeus monodon* tag retention and effects on growth and survival. *Aquaculture* 113, 167–70.

Templeman, W. & Fleming, A.M. (1963) Distribution of *Lernaeocera branchialis* L. on the cod as an indicator of cod movements in the Newfoundland area. *Spec. Publs. Int. Commn. NW. Atlant. Fish.* **4**, 381–22.

Testa, J.W. & Rotherly, P. (1992) Effectiveness of various cattle ear tags as maker for Weddell seals. *Mar. Mammal Sci.* **8**(4), 344–53.

Treble, R.J., Day, R.W., & Quinn, T.J. (1993) Detection and effects on mortality estimates of changes in tag loss. *Can. J. Fish. Aquat. Sci.* **50**(7), 1435–41.

Trpis, M. & Hausermann, W. (1986) Dispersal and other population parameters of *Aedes aegypti* in an African village and their possible significance in epidemiology of vector-borne diseases. *Am. J. Trop. Med. Hyg.* **35**, 1263–79.

Van Brackle, M.D., Linhart, S.B., Creekmore, T.E., Nettles, V.F., & Marchinton, R.L. (1994) Oral biomarking of white-tailed deer with tetracycline. *Wildl. Soc. Bull.* **22**(3), 483–8.

Wada, Y., Suenaga, O., & Miyagi, I. (1975) Dispersal experiment of *Aedes togoi*. *Trop. Med.* **16**, 137–46.

Waloff, Z. (1963) Field studies on solitary and transiens desert locusts in the red sea area. *Anti-locust Bull.* **40**, 1–93.

Wastle, R.J., Babaluk, J.A., & Decterow, G.M. (1994) A bibliography of marking fishes with tetracyclines including references to effects on fishes. *Can. Tech. Rep. Fish. Aquat. Sci.* **1951**, 1–26.

Wave, H.E., Henneberry, T.J., & Mason, H.C. (1963) Fluorescent biological stains as markers for *Drosophila*. *J. Econ. Entomol.* **56**, 890–1.

Wedermeyer, G. (1970) Stress of anaesthesia with MS-222 and benzocaine in rainbow trout (*Salmo gairdneri*). *J. Fish. Res. Bd. Can.* **27**, 909–14.

Welch, H.E. (1960) Two applications of a method of determining the error of population estimates of mosquito larvae by the mark and recapture technique. *Ecology* **41**, 228–9.

Wheye, D. & Ehrlich, P.R. (1985) The use of fluorescent pigments to study insect behaviour: investigating mating patterns in a butterfly population. *Ecol. Entomol.* **10**, 231–4.

White, E.G. (1970) A self-checking coding technique for mark–recapture studies. *Bull. Entomol. Res.* **60**, 303–7.

White, E.G. (1975) Identifying population units that comply with capture–recapture assumptions in an open community of alpine grasshoppers. *Res. Popul. Ecol.* **16**, 153–87.

White, G.C., Anderson, D.R., Burnham, K.P., & Otis, D.L. (1982) *Capture–Recapture and Removal Methods for Sampling Closed Populations*. Los Alamos National Laboratory, Los Alamos, New Mexico.

Wiegert, R.G. & Lindeborg, R.G. (1964) A 'stem well' method of introducing radioisotopes into plants to study food chains. *Ecology* **45**, 406–10.

Wilkin, P.J. & Scofield, A.M. (1991) The structure of a natural population of the medicinal leech, *Hirudo medicinalis*, at Dungeness, Kent (England, UK). *Freshw. Biol.* **25**(3), 539–46.

Williams, D.F., Patterson, R.S., & LaBrecque, G.C. (1979) Marking large numbers of stable flies *Stomoxys calcitrans* (L.) for a sterile release program. *Mosquito News* **39**, 146–8.

Wisniewolski, W. & Nabialek, J. (1993) Tag retention and survival of fish tagged in controlled pond experiments. *Aquat. Sci.* **55**(2), 143–52.

Wolf, W.W. & Stimmann, M.W. (1972) An automatic method of marking cabbage looper moths for release–recovery identification. *J. Econ. Entomol.* **65**, 719–22.

Wood, G.W. (1963) The capture–recapture technique as a means of estimating populations of climbing cutworms. *Can. J. Zool.* **41**, 47–50.

Woodward, A.R., Hines, T.C., Abercrombie, C.L., & Nichols, J.D. (1987) Survival of young American alligators on a Florida lake (USA). *J. Wildl. Manag.* **51**(4), 931–7.

Woolhouse, M.E.J. (1988) A mark–recapture method for ecological studies of schistosomiasis vector snail populations. *Ann. Trop. Med. Parasitol.* **82**, 485–97.

Woolhouse, M.E.J. & Chandiwana, S.K. (1990) Population biology of the freshwater snail *Bulinus globosus* in the Zimbabwe highveld. *J. Appl. Ecol.* **27**, 41–59.

Yates, W.W., Gjullin, C.M., & Lindquist, A.W. (1951) Treatment of mosquito larvae and adults with radioactive phosphorus. *J. Econ. Entomol.* **44**, 34–7.

Yeruham, I., Perl, S., Yakobson, B., Ben Yosef, H., & Lampert, M. (1993) Skin tumors in cattle following tattooing by liquid nitrogen. *Israel J. Vet. Med.* **48**(1), 38–40.

Zacharuk, R.Y. (1963) Vital dyes for marking living elaterid larvae. *Can. J. Zool.* **41**, 991–6.

Zaidenov, A.M. (1960) Study of housefly (Diptera, Muscidae) migrations in Chita by means of luminescent tagging. [In Russian.] *Entomol. Obozr.* **39**, 574–84. (Trans. *Entomol. Rev.* **39**, 406–14.)

Zucchini, W. & Channing, A. (1986) Bayesian estimation of animal abundance in small populations using capture–recapture information. *S. Afr. J. Sci.* **82**(3), 137–40.

4 Absolute Population Estimates by Sampling a Unit of Habitat: Air, Plants, Plant Products, and Vertebrate Hosts

Sampling a unit of habitat is one of four approaches to the absolute population estimate, the other three methods being: (i) distance or nearest neighbour (p. 344); (ii) mark–recapture (p. 73); and (iii) removal trapping (p. 255). In this approach, the habitat is sampled, together with the animals it contains. Hence, two separate measurements have to be made: the total number of animals in the unit of the habitat sampled and the total number of these units in the whole habitat of the population being studied. The second may involve the use of the techniques of the botanist, forester, surveyor, or hydrologist and cannot be considered in detail here. The first concerns the extraction of animals from the samples and sometimes the taking of samples; this and the next two chapters will be concerned mainly with these problems in five habitats, two biotic (plants and vertebrate hosts) and three physical (air, soil, and water).

It is important to remember that, if sampling is unbiased, the errors in the population estimates obtained by this method will normally lie below the true value. This is because extraction efficiency can never be more than 100% — one cannot normally find more animals than are present! As the basis of the approach is to multiply the mean count per sampling unit by the typically large number of such units in the habitat, the underestimation of total number may, in terms of individuals, be large.

4.1 Sampling from the air

In some ways, air is the simplest of the five habitats. It is homogeneous in all environments and permits a universal solution, which stems from the work of C.G. Johnson and L.R. Taylor. They developed suction traps, standardized them, and measured their efficiency so precisely that aerial populations can be assessed with a greater level of accuracy than those in most other habitats. The only environmental variable for which allowance needs to be made is wind speed. There is no evidence that insects are in any way attracted or repelled by the traps (Taylor, 1962a), although precautions must be taken to avoid the development of a charge of static electricity on the traps (Maw, 1964). Only a very small proportion of the insects collected in suction traps are damaged by them.

The basic features of the suction trap are an electric fan that pulls or drives air through a fine gauze cone; this filters out the insects, which are collected in a jar or cylinder. The trap may be fitted with a segregating device, which separates the catch according to a predetermined time interval, and this provides information, not only about the numbers flying, but also about the periodicity of flight. Few insects are active equally throughout 24 h (Lewis & Taylor, 1965) and therefore the daily catch divided by the daily intake of air of the fan does not present as true a picture of aerial density as when each hour is considered separately.

Other traps for flying insects — sticky and water traps and suspended nets — are greatly influenced by wind speed, their efficiencies being low at low wind speeds. Although Taylor (1962b) has shown how it is possible to convert sticky-trap and suspended-net catches into absolute densities if the wind speeds are known, these methods are used in general for the relative estimation of populations and are discussed in Chapter 7.

4.1.1 Sampling apparatus

4.1.1.1 Exposed-cone (Johnson–Taylor) suction trap

Originally developed in the 1950s (Johnson, 1950; Taylor, 1951; Johnson & Taylor, 1955a), the air (with the insects) is drawn through the fan first and subsequently via a cone mesh (c. 10.25/cm), where the insects are impinged and then drawn down to a collecting tube at the apex of the cone (Fig. 4.1). The trap is often fitted with a disc-dropping mechanism to segregate the catch temporally. Goodenough et al. (1983) describe a method for the segregation of the catch when it is preserved in liquid. Both 9-in (23 cm) and 12-in (46 cm) Johnson–Taylor insect suction traps are commercially produced by the Burkard Manufacturing Co., Rickmansworth, England (web address: http://www.burkard.co.uk). In these models, the sample is collected in a plastic bottle with side overflow ports and in standard models the catch can be segregated at pre-set time intervals of 5 min–1 h. Exposed-cone samplers using Vent-Axia axial fans (Vent-Axia Ltd., Fleming Way, Crawley, West Sussex, RH10 2NN, England) can be easily built.

The efficiency of Johnson–Taylor suction traps changes with wind speed and they are most suitable for sheltered positions between plants. For example, using a 9-in trap, crosswinds of more than 22.4 km/h gave significant reductions in the air intake of the mouth of the fan (Johnson, 1950; Taylor, 1955). If the trap is standing amongst a crop, so that the cone is sheltered from the effects of the crosswinds, which only impinge on the mouth of the fan, the delivery loss is only 0.5% per km/h crosswind (Taylor, 1955).

Johnson–Taylor traps have been used extensively for the study of aphids and other small, weakly flying insects. Examples of their application to other groups are: aerial spiders (Dean & Sterling, 1985, 1990); pollen beetles in rape fields (Sedivy, 1993); Psocoptera (Locatelli & Limonta, 1993, 1994); lesser grain borers, *Rhyzopertha dominica* (Leos Martinez et al., 1986).

Numerous small suction traps have been designed. An inexpensive and easily constructed suction trap suitable for monitoring the flights of aphids is described by

Fig. 4.1 Cross-section of the exposed-cone or Johnson–Taylor suction trap.

Allison and Pike (1988). For sites lacking mains electricity, Belding *et al.* (1991) describe a small suction trap powered by a deep cycle battery, recharged by a photovoltaic panel. This trap, suitable for sampling weakly flying insects, such as aphids, has four 12 V, brushless, direct current (DC) fans, which are turned on and off by a timer under photoelectric control. Wainhouse (1980) describes a portable, battery-driven suction trap for sampling small insects.

Because suction traps can catch large numbers of insects, Hobbs and Hodges (1993) describe an optical method for the automatic recording and classification of the insect catches. This is based on the principle that, when the sample is illuminated against a dark background, the amount of scattered light can be used to measure insect presence and size.

4.1.1.2 Enclosed-cone types of suction trap, including the Rothamsted 12 m trap

In these traps, the fan unit is at the bottom of a metal cylinder and the metal-gauze collecting cone opens into the mouth of the trap (Fig. 4.2), so that the insects are filtered out before the air passes through the fan. Because the cone and mouth of the fan are protected from crosswinds, they have a better performance than exposed-cone designs at higher wind speeds. Johnson and Taylor (1955a,b) originally

Fig. 4.2 Cross-section of the 18-in propeller enclosed-cone type of suction trap.

developed various traps in this class, including a large but lightweight model. The 18-in propeller trap has proved a versatile and robust model, which is relatively easily constructed from widely available components. It is powered by a longitudinally mounted, 400 W electric motor, which drives an 18-in (46 cm) diameter propeller fan mounted in a galvanized-iron cylindrical duct. The air-filtering cone is of 32-mesh (=12 mesh/cm) copper gauze and at its base the glass collecting jar screws into a socket; the jar contains 70% alcohol. To empty the trap, the cone is removed and the jar unscrewed (Fig. 4.2).

The Rothamsted 12 m trap is a centrifugal-flow trap, which has proved valuable for continuous monitoring of insect abundance (Taylor & Palmer, 1972; Taylor, 1974) because they operate uniformly over a wide range of conditions. In the original design, the cylinder enclosing the cone could be extended vertically to sample at heights of up to 12.2 m, with a narrow, high-velocity inlet and an expansion chamber to reduce air speed before filtering. The collecting cone is at the base of the trap, so that it may be emptied at ground level. The traps were powered by a 1.5 brake horse-power (bhp), 1440 rev/min motor, with direct drive to a 30.5 cm diameter horizon-tal-discharge, forward-bladed centrifugal fan; the air-filtering cone was 15.74 mesh/cm gauze and the diameter of the cylinder 25.4 cm. A network of 12.2 m suction traps has been operated by the Rothamsted insect survey since 1964 to monitor the aerial movement of aphids and other insects. Many variants on this design have evolved in other countries. A standard design for a 12 m suction trap is described by Macaulay *et al.* (1988). In the USA, networks of Allison–Pike traps (which only sample 20% of the volume of the Rothamsted trap) have been established for the same purpose (Allison & Pike, 1988).

4.1.2 Rotary and other traps

The rotary or whirligig trap consists of a gauze net that is rotated at a speed of about 16 km/h on the end of an arm (Fig. 4.3); this type of trap has been developed a number of times, sometimes with a single net and sometimes with a net at each end of the arm (Williams & Milne, 1935; Chamberlin, 1940; Stage *et al.*, 1952; Nicholls, 1960; Vité & Gara, 1961; Gara & Vité, 1962; Prescott & Newton, 1963). The great-est efficiency will be obtained if the net is built so as to sample isokinetically—that is, the air-flow lines in it are straight; Taylor (1962a) has described such a net (Fig. 4.3). Rotary traps sample more or less independently of wind speed, until the wind speed exceeds that of the trap (Taylor, 1962a). Experiments indicate that they sample 85% of the flying population, but Juillet (1963) has suggested that some insects may be able to crawl out and escape; this can be prevented by adequate baffles inside the net. It is also probable that strong flyers with good vision may avoid the area of the moving trap. It is not possible to time-segregate the catches of rotary traps. A large rotary trap designed to measure aerial spider density is described by Topping *et al.* (1992). A comparison of the catch of spiders and other invertebrate groups between this trap's catch and that of a suction trap showed that the rotary trap was operating at a greater efficiency for most groups. However, the two traps caught equal numbers of spiders and non-staphylinid beetles.

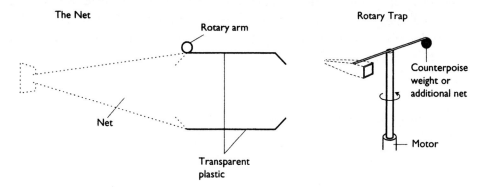

Fig. 4.3 A rotary trap, with isokinetic net.

4.1.3 Comparison and efficiencies of the different types of suction traps

Three factors influence the choice of the type of suction trap: the density of the insects being studied, the wind speeds in the situation where sampling is proposed and the necessity or otherwise for information on periodicity. If information is required on periodicity, then a trap with a catch-segregating mechanism must be used, and this rules out the 18-in propeller trap. The other two factors generally give the same indications, for, the sparser the insects, the larger the desired air intake (so as to sample an adequate number) and, the stronger the winds, the stronger (larger) the trap should be. Now, at ground level and up to 1–1.5 m amongst vegetation, insect populations are usually dense and wind speeds seldom exceed 10 km/h); under these conditions, the 9-in exposed-cone type is generally used. In more exposed situations (e.g. above short turf) or on towers or cables several metres above the ground, the enclosed-cone traps should be used; the greater the height, the more powerful the winds and the sparser the insects and, therefore, the stronger the fan required. Where information on periodicity can be foregone, the 18-in propeller trap provides a robust, simple, and relatively inexpensive piece of equipment.

The average air deliveries of the traps given by Taylor (1955; L.R. Taylor, pers. comm.) are: 9-in Vent-Axia, 7.84 m^3/min; 18-in propeller trap, 77.37 m^3/min; centrifugal-flow traps, 97.6 m^3/min. There is, however, a certain amount of individual variation, which over a long period could lead to significant differences in catches. Taylor (1962a) has suggested that, unless the actual deliveries of the traps have been checked in the field, differences in catch should exceed 6% for 9-in traps or 10% for 18-in traps before they are attributed to population differences.

The effects of wind speed on the absolute efficiencies of the traps vary according to the size of the insects, as well as the performance of the fan. Taylor (1962a) has shown that the efficiency of a trap for a particular insect is given by the formula:

$$E = (W + 3)(0.0082\, CE - 0.123) + (0.104 - 0.159 \log i)$$

where E = the log efficiency of the trap, W = wind speed in miles/h, CE = coefficient of efficiency of the trap, and i = insect size in 1/3 mm^2. The coefficient of efficiency of a trap = (35.3 × volume sampled in m^3/h)$^{1/3}$/(0.39 inlet diameter of the fan in cm)$^{1/2}$; those for the above traps are: 9-in, 8.7; 12-in, 12.1; 18-in, 12.1; and 30-in, 13.4. Sizes (i.e. length × wing-span) of various insects in mm^2 given by Taylor (1962a) are given in Table 4.1.

Table 4.1 Sizes (length × wing-span) of some insects (in mm^2) (from Taylor, 1962a).

Minute Phoridae and Sphaeroceridae	1–3
Oscinella, Psychodidae and small Aphididae	3–10
Drosophila, Chlorops, large Aphididae	10–30
Fannia, Musca	30–100
Calliphora, Lucilia	100–300
Sarcophaga, Tabanus	300–1000

Estimates of aphid density determined by radar have been compared with simultaneous catches from both a Rothamsted 12.2 m and an aerofoil trap (Schaefer *et al.*, 1985). The 12.2 m trap effectiveness was found to decrease by a factor of two for each 2.4 m/s (about 5 m.p.h.) increase in average wind speed. The two trap sensitivities did not differ significantly in their catches, but the results are significantly different ($P < 0.001$) from those calculated using the above formula, which is based upon a comparison of catches from suction traps with rotary (whirligig) trap and tow-net catches. Absolute calibration of the aerofoil trap was achieved using a remote-sensing IRADIT infrared system to measure the aerial density of aphid-sized insects near to the trap inlet in very light winds, when the effectiveness was about unity, and the 12.2 m trap was predicted to perform similarly.

4.1.4 Conversion of catch to aerial density

The number of insects caught is divided by the volume of air sampled; the actual crude delivery figures, given above, may be used for suction traps close to the ground if the wind speed seldom exceeds 5 m.p.h. (2.4 m/s); in other situations, appropriate corrections should be made (Taylor, 1955). This figure is then corrected for the efficiency of extraction, of the particular insect, by use of the formula for efficiency given immediately above, though this may need modification in the light of the findings of Schaffer *et al.* (1985). This formula gives a negative value in logs and this is the proportion by which the actual catch is less than the real density.

Where catches have been segregated at hourly intervals, the tables of conversion factors may be used (Table 4.2); Taylor (1962a) also gives conversion factors for 12-in and 30-in traps. The conversion factors are given in logs so that:

Log catch/h + conversion factor = log density/10^6 ft^3 of air

To convert to log density per m^3, divide by 28 317.

4.1.5 Conversion of density to total aerial population

Johnson (1957) has shown that the density of insects (f) at particular height (z) is given by:

$$f_z = C(z + z_e)^{-\lambda}$$

where f_z = density at height z, C = a scale factor dependent on the general size of the population in the air, λ = an index of the profile and of the aerial diffusion process, and z_e = a constant added to the actual height and possibly related to the height of the boundary layer — the height at which the insects' flight speed is exceeded by the wind speed (Taylor, 1958). This may be solved by iteration using a computer or by the graphical method described by Johnson (1957).

The total number of insects between heights z_1 and z_2 (n) may then be integrated as follows:

$$n = \frac{C}{1 - \lambda}\left[(z_2 + z_e)^{1-\lambda} - (z_1 + z_e)^{1-\lambda}\right]$$

Table 4.2 Conversion factors, log F, for suction trap catches (after Taylor, 1962).

9-IN VENT-AXIA TRAP

Insect size (mm^2)	Wind speed (miles/h)				
	0–2	2–4	4–6	6–8	8–10
1–3	1.89	2.00	2.12	2.23	2.34
3–10	1.97	2.08	2.20	2.31	2.42
10–30	2.05	2.16	2.28	2.39	2.50
30–100	2.13	2.24	2.36	2.47	2.58
100–300	2.21	2.32	2.44	2.55	2.66
300–1000	2.29	2.40	2.52	2.63	2.74

18-IN PROPELLER TRAP

Insect size (mm^2)	Wind speed (miles/h)						
	0–2	2–4	4–6	6–8	8–10	10–14	14–20
1–3	0.85	0.89	0.95	1.01	1.07	1.15	1.29
3–10	0.93	0.97	1.03	1.09	1.15	1.23	1.37
10–30	1.01	1.05	1.11	1.17	1.23	1.31	1.45
30–100	1.09	1.13	1.19	1.25	1.31	1.39	1.53
199–300	1.17	1.21	1.27	1.33	1.39	1.47	1.61
300–1000	1.25	1.29	1.35	1.41	1.47	1.55	1.69

The symbols are as above.

With individual species of insect, the profiles may be found to be rather irregular, and approximate integrations may then be made graphically (Johnson *et al.*, 1962).

4.2 Sampling from plants

In many ways, this is the most difficult habitat from which to sample invertebrates. It differs from the soil and the air both in being much more heterogeneous and in continually changing. It is frequently convenient to take a part of the plant as the sampling unit; the resulting population estimate is not an absolute one but a measure of population intensity. The distinction between these terms has already been discussed (p. 2), but the point is so vital that some further elaboration here is justified. If the amount of damage to the plant is the primary concern, it may seem reasonable to make all estimates in terms of numbers of, say, mites per leaf. This is the intensity of mites that the tree has to withstand and, if mite populations are being related to some index of the health of the tree, an estimate of 'population intensity' is probably the most relevant. But, if the study is concerned with the changes in the numbers of mites in a season, a series of estimates of population intensity could easily be misleading. If the number of mites per leaf fell throughout the summer, this could be due to an actual reduction in the mite population or to an increase in the numbers of leaves. In order to determine which of these explanations is correct, the number of leaves per branch would have to be counted on each sampling date and the mite population could then be expressed in terms of numbers per branch. This is sometimes called a 'basic population estimate', but it differs from the measure of population intensity only in degree and not fundamentally, for from year to year the number of branches or indeed the number of trees in the forest or orchard will alter. It is only when these units are also counted and the whole converted into numbers per unit surface area of soil, commonly per square metre or hectare, that it is possible to make a valid comparison between populations differing in time (and space), which is the essence of many population studies.

The labour of obtaining the additional data on the density of the plant unit is often comparatively slight, and it cannot be stressed too strongly that, whenever possible, this should always be done, as it makes the resulting data more meaningful and useful for testing a wider range of hypotheses.

4.2.1 Assessing the plant

The simplest condition is found in annual crops, where the regular spacing and uniform age of the plants make the estimation of the number of shoots or plants per length of row a relatively easy matter; the variance of even a few samples is often small and the determination of the number of row length units per unit of area (e.g. per m^2) is also straightforward. With natural herbaceous vegetation, there will usually be far more variability, both in the number of individual host plants per unit area and in their age and form (see also p. 18). Although the number of plants per unit area is sometimes an adequate measure, the variation in size and age often means that the weight of the plant or the number of some part of the plant frequented by the insect (e.g. the flower-head) gives a more adequate measure. Occasionally, botanical measures, such as frequency, as measured by a point quadrat (Greig-Smith, 1964; Mueller-Dombois & Ellenburg, 1974), are useful.

The most complex situation is found when one attempts to assess the habit of arboreal insects. The leaf is often, but not always, the sampling unit (e.g. Richards & Waloff, 1961); its age, aspect, and height above the ground need to be considered, as well as the age of the tree. Frequently, a stratified random-sampling plan minimizes the variance of the population estimate. For example, to sample the citrus mealy bug, Browning (1959) found the variance was minimal if each aspect of the tree was sampled in turn and individual leaves picked haphazardly rather than sampling the mealy bugs on all the leaves on a single branch.

The precise estimation of the number of leaves per branch is often a minor piece of research on its own. Actual numbers may be counted or dry weight used as the measure. In some trees, the actual weight of the foliage may continue to increase after the leaves have ceased elongating (Ives, 1959). The diameter of the trunk may be taken as a cruder measure of foliage weight and of shoot number in young growing trees, especially conifers (Harris, 1960).

4.2.2 Determining the numbers of invertebrates

Irrespective of habitat, it is difficult to devise an 'all-species method', as the more mobile animals have a greater tendency to escape, small species are more easily missed, and large or firmly attached species are more difficult to detach or remove.

4.2.2.1 Direct counting

With large, conspicuous animals, including insects, it is sometimes possible to count all the individuals; for example, Moore (1964) was able to make a census of the male dragonflies over a pond using binoculars. Smaller insects on distant foliage, e.g. the upper parts of trees, may be counted using a powerful tripod-mounted telescope, capable of short-range work and with a wide aperture (e.g. Questar); a transect may be made through the canopy of a tree by systematically turning the elevation adjustment of the telescope (L.E. Gilbert, pers. comm.). This should prove a very useful method for the survey of forest insects. Direct counts may also be used to follow a particular cohort (see p. 5), from which we can obtain true estimates of population; the difficulties of determining the true mean and other parameters, which may be severe in highly aggregated populations, are avoided.

More often, a part of the habitat is delimited and the animals within the sample are counted. This is easily accomplished for relatively immobile ground-living animals, such as the snail *Theba pisana* (Baker & Vogelzang, 1988; Baker & Hawke, 1990), where quadrats may be placed upon the ground and the animals immediately counted. With care, quadrats can also be used to estimate fairly mobile, but relatively large, insects, such as grasshoppers (Richards & Waloff, 1954) or froghoppers (cercopids). These quadrats are made of wire and it is often useful to attach white marker ribbons. They are placed in the field and are left undisturbed for some while before a count is made. Each quadrat is approached carefully and the numbers of insects within it counted as they fly or hop out. Alternatively, a high-sided, box quadrat, possibly driven into the soil, can be used to retain less mobile insects while collecting or counting is in progress (Balogh & Loksa, 1956). In such cases and with other relatively immobile large insects, e.g. the wart-biter cricket, *Decticus verrucivorus* (Cherrill & Brown, 1990), the sunn pest, *Eurygaster integriceps* (Banks & Brown, 1962), or lepidopterous larvae and pupae (Arthur, 1962), a count is made as soon as the quadrat is in position.

Where the insects are restricted to particular plants, then these or particular parts of them, such as fruits or flowers, may be collected or examined and the numbers on each sample counted. Care must be exercised, as many insects can be dislodged as the samples are taken (Satchell & Mountford, 1962) or small animals can be overlooked (Condrashoff, 1967). Forest and orchard pests are often assessed as numbers per shoot and crop insects as numbers per plant or per leaf. Aphids on crops such as wheat may be sampled *in situ* by inspecting a set number of tillers (e.g. Cannon, 1986). It will commonly be found that the position of the shoot or leaf—whether it is in the upper, middle, or lower part of the tree or plant—influences the numbers of insects upon it, and due allowance should be made for this stratification by a subdivision of the habitat (p. 7). When small abundant insects are actually being counted, it is desirable that the sampling unit should be as small as possible; Shands *et al.* (1954) showed that various aphids on potatoes could be adequately estimated by counting, not every aphid on a selected leaf, but only those on the terminal and two basal leaflets; even a count based on a half of each of those leaflets could be satisfactory.

Photographs (colour transparencies) have been found to be a practical and rapid method of recording aphids on rose shoots; they are taken from four aspects at a distance of about 40 cm and those aphids seen against the shoot (not silhouettes) are counted by viewing under a dissecting microscope (Maelzer, 1976). Noldus *et al.* (1985) estimated the density of the scale insect *Trialeurodes vaporariorum* on beans using enlarged photographs of the leaves under a grid of 5 × 5 mm squares and then counting the individuals in each quadrat. Video cameras can also be used to record insects. The potential of photography and video for direct sampling (and recording) of insect populations has probably not been fully developed.

Insects in stored products, especially cacao, are often estimated by 'snaking'. The bag is opened and the contents tipped on to a floor in a long wavy line; the insects may be counted directly and a large number will often be found in the last piece tipped out, known as the 'tail'. This method, although technically providing an absolute type of sample, is really more properly regarded as a relative method and is often necessary to establish the basis of a presence-or-absence sampling programme (p. 57) for extensive work.

4.2.2.2 The separation of exposed small animals from the foliage on which they are living

Knock-down — by chemicals, jarring, and heat. Knock-down by chemicals and jarring are discussed more fully below, as they are more generally used when the whole twig or tree is the sampling unit. Here we are concerned with their use with animals on herbaceous plants and with animals on trees when the leaf is the sampling unit. Some insects can be made to drop off foliage by exposing them to the vapours of certain chemicals. With the wide range of organic substances available, this approach could undoubtedly be extended; some recognized repellents (Jacobson, 1966) might be screened. Aphids can be made to withdraw their stylets and will mostly drop off the plant if exposed to the vapour of methyl isobutyl ketone (Gray & Schuh, 1941; Helson, 1958) or γ-BHC (Way & Heathcote, 1966). Aphids that do not cover the host plant with a thick deposit of honeydew and are fairly active may be separated from the foliage by exposing them to this vapour and then shaking them in the type of sampling can described by Gray and Schuh (1941) (Fig. 4.4a). With the pea aphid, *Acyrthosiphon pisum*, these authors shook the can 50 times, following a 5 min interval, and found that after one such treatment an average of 89.3% of the aphids were extracted, after 10 min (and 100 shakes) 97% and after 15 min 99.1%. In the six samples tested, the lowest extraction rate was 98.5% after 15 min; the numbers extracted from each sample in the can were between 3000 and 7000. Walker *et al.* (1972) found with *Schizaphis* that, because of entanglement with

(a)

(b)

Fig. 4.4 (a) Aphid sampling can (modified from Gray & Schuh, 1941). (b) Sampling tray for flea-beetle, weevils, and others on small crop plants (after La Croix, 1961).

honeydew, extractions were in the region of 85% after 15 min exposure; shaking more than 50 times did not increase the recovery efficiency (and the effect from 25 to 50 was only marginal). Laster and Furr (1962) found that the fauna of sorghum heads could be killed by insecticide and collected in a paper cone strapped round the stem and stabilized with a little sand in the base.

Thrips may be extracted from flowers and grass heads, as well as foliage, by exposing them to turpentine vapour (Lewis, 1960). A convenient apparatus has been described by Lewis, who found the method was over 80% efficient for adults but confirmed Taylor and Smiths (1955) conclusion that it was unsatisfactory for larvae. Lewis and Navas (1962) obtained thrips and some other insects that were over-wintering in bark crevices by breaking the bark sample into small pieces and sieving; this dislodges the majority, but, to obtain 97–100% extraction efficiencies, the fragments were exposed to turpentine vapour.

Heat was used by Hughes (1963) to stimulate *Brevicoryne brassicae* to leave cabbage leaves, and Hoerner's (1947) onion thrips extractor worked on the same principle; these techniques can be considered modifications of the Berlese funnel (p. 229). An ultrasonic vibrator (about 20 kHz, 100 W output) was found to dislodge eriophyid mites from grass (Gibson, 1975). The foliage was placed in 50 or 75% alcohol (ethanol) and exposed for between 45 and 60 s (maximum). (See also p. 165.)

Insects that fall from their host plant when disturbed, e.g. some beetles, can be sampled from young field crops by enclosing the base of the plant in a hinged metal tray (Fig. 4.4b). The plant is then tapped (jarred) and beetles from it may be counted as they fall on to the tray. This method was developed by La Croix (1961) for flea beetles, *Podagrica*, on cotton seedlings; they must be counted quickly before they hop off the tray. This is facilitated by painting the tray white.

Brushing. Mites, and apparently other small animals apart from insects, are removed from leaves if these are passed between two spiral brushes revolving at high speed in opposite directions (Henderson & McBurnie, 1943; Chant, 1962). The animals are collected in a receptacle in which an adhesive may be used to secure the specimens. Clear varnish was the original adhesive, but various surface-active agents have been found more satisfactory if the mites are to be examined at once: e.g. Emcol 5100, an alkanolamine condensate (Morgan *et al.*, 1955), Tween 20 (polyoxyethylene sorbitan monolaurate), and Span (sorbitan monolaurate) (Cleveland, 1962). For permanent storage, R.C. Muir (cited in Chant, 1962) recommends replacing the glass plate with a cardboard disc repeatedly coated with varnish. Adult mites are usually removed slightly more efficiently than the eggs; in general, the efficiency for fruit-tree mites on apple leaves has been found to be between 95 and 100% (Morgan *et al.*, 1955), but it is much less on peach leaves because of their complex curvature (Putman, 1966).

Washing. Small animals, principally mites, aphids, and thrips, can be washed off herbaceous plants or single leaves with various solutions. This approach can be more efficient, in terms of both accuracy and cost, than direct counting. With aphids, the extraction will be facilitated if they are first exposed to the vapour of methyl isobutyl ketone (= 4-methyl 2-pentanone) (Pielou, 1961). Dilute soap, detergent, or alcohol solutions are often adequate for washing (e.g. Taylor & Smith, 1955), but with eggs a solvent must be used that will dissolve the cement. Benzene heated almost to boiling-point removes many of the eggs of mites on fruit-tree twigs (Morgan et al., 1955), as does hot sodium hydroxide solution (Kobayashi & Murai, 1965). Benzene is not only highly inflammable, but also carcinogenic; such procedures should be undertaken in a fume cupboard with appropriate saftey precautions. If conifer foliage with moth eggs is soaked for 30 h in a 1.5% solution of sodium hydroxide, a 96% recovery rate may be achieved (Condrashoff, 1967), and petrol has been found satisfactory for washing eggs of the corn ear worm, *Heliothis zea*, from the silks of maize (Connell, 1959). Thrips may be washed off foliage with ethanol (Le Pelley, 1942), the solution filtered, and the thrips separated from the other material by shaking in benzene and water; the thrips pass into the benzene, where they may be counted (see also p. 222) (Bullock, 1963). Moth eggs may be freed from bark and moss by soaking for 45 min in 2% bleach solution (sodium hypochlorite solution (5.25% free chlorine) diluted with 98 parts water); the mixture is then agitated and passed through a sieve, which separates the larger debris (Otvos & Bryant, 1972). Wardlow and Jackson (1984) extracted apple-rust mite, *Aculus schlechtendali*, from buds using methanol.

The animals removed by washing are often so numerous that they require automatic counting methods. An alternative approach is to count a sample; this may be done by means of a counting grid or disc (Fig. 4.5) (Strickland, 1954; Morgan et al., 1955). The animals are mixed with 70% alcohol or some other solution, agitated in a Petri dish, and allowed to settle. The dish is stood above a photographically produced counting disc (on celluloid) and illuminated from below. Strickland (1954) found that the percentage error of the count for aphids using the disc in Fig. 4.5 ranged from 5 (with about 3000 aphids/dish) to 15 (with fewer than 500 aphids/dish).

If the number of animals is too great for the counting grid to be used directly, the 'suspension' of the animals in a solution may be divided into aliquots (Newell, 1947). The diluted samples may then be estimated with a counting disc or volumetrically or the two methods combined. In the volumetric method, the animals are allowed to settle or are filtered out of the solution; they are then estimated as, for example, so many small specimen tubefuls and the numbers in a few of the tubes are counted partially by the use of a counting disc or *in toto* (Banks, 1954; Walker et al., 1972). If the population consists of very different-sized individuals, the volumetric method may lead to inaccuracies unless special care is taken. This variation can, however, be used to aid estimation; Pielou (1961) showed that, if apple aphids, *Aphis pomi*, were shaken with alcohol and the solution placed in Imhoff sedimentation cones (wide-mouthed and graduated), the sediment stratifies into layers: first the larger living aphids, then

Fig. 4.5 A counting grid (after Strickland, 1954). All animals that are wholly within the clear areas or overlap the left-hand or outer margins should be counted and these will be one-sixth of the total.

the smaller living aphids, and finally dead aphids and cast skins; the volumes of these can easily be read off and calibrated for the various categories. A method of separating the instars based on the same principle has been developed by Kershaw (1964). A routine for subsampling when collections vary enormously in size is described by Corbet (1966).

Imprinting. This technique has been used only with mites and their eggs. The infested leaves are placed between sheets of glossy absorbent paper and passed between a pair of rubber rollers (e.g. a household wringer); where each mite or its egg has been squashed, a stain is left (Venables & Dennys, 1941; Austin & Massee, 1947). Summers and Baker (1952) used the same method to record the mites beaten from almond twigs, and the advantage of this method is its speed and the provision of a permanent record; furthermore, with certain precautions, nearest-neighbour techniques (p. 344) might be applied to the resulting marks. However, Chant and Muir (1955) found with the fruit-tree red spider mite, *Panonychus ulmi*, that the number recorded by the imprinting method was significantly less than the number recorded by the brushing method (see above), which is therefore to be preferred in general. Trumble (1994) describes the imprinting procedure for the two-spotted spider mite, *Tetranychus urticae*, on strawberry. The leaves are crushed on filter discs impregnated with bromophenol blue, which turns green in the presence of protein from the mites.

4.2.2.3 The expulsion of animals from tall vegetation

Jarring or beating. This is a collector's method and originally the tree was hit sharply with a stick and the insects collected in an umbrella held upside down under the tree! The umbrella is now replaced by a beating tray, which is basically a cloth-covered frame, flat or slightly sloping towards the centre, which is large enough to collect all the insects that drop off the tree. The colour of the cloth should produce the maximum contrast with that of the insect being studied, and the insects are rapidly collected from the tray by an aspirator or pooter (see Fig. 7.6). In general, this is only a relative method, but with some insects, such as leaf beetles (chrysomelids), many weevils, and lepidopterous larvae, which fall, rather than fly, from the host plant when disturbed, a sufficiently high proportion may be collected for it to be regarded as an absolute method (e.g. Richards & Waloff, 1961; White, 1975). Weather, strength of the blow, and time of day can all change efficiency (Harris *et al.*, 1972). As very small animals, like mites, may be overlooked on a beating tray and active ones may escape, more accurate counts can sometimes be obtained by fastening a screen over a large funnel. The twigs are tapped on the screen and mites and insects funnelled into a container (Steiner, 1962). The beating tray and funnel methods may be combined. Wilson (1962) attached a removal jar below a hole in the centre of the tray and Coineau (1962) placed a grid above a funnel in the centre of a net tray; this grid helped to separate twigs and other debris from the insects. Arthropods on row crops (e.g. soy beans) may be collected by quietly unrolling a length of oilcloth groundsheet (or polythene) between two adjacent rows and briskly shaking the plants over it; the method is time-consuming but seems to provide absolute estimates of large, easily dislodged, species (e.g. pentatomids) (Rudd & Jensen, 1977).

Although the tree is generally hit sharply with a stick, Legner and Oatman (1962) suggest the use of a rubber mallet, and Richards and Waloff (1961) found that some host plants (e.g. *Sarothamnus*) could be damaged by such violence (so leading to a change in the habitat); they recommended vigorous shaking by hand.

Chemical knock-down. The most widespread approach to the sampling of arboreal insects is now based on this use of insecticides (Stork *et al.*, 1997). The pyrethrin group is particularly useful, as it affects most arthropods and initially causes enhanced activity, which can lead to the insect falling from the tree. The pioneering work was undertaken in plantations and orchards (Collyer, 1951; Gibbs *et al.*, 1968), when the tree was shrouded and the total fauna collected from large sheets spread on the ground below. These early studies were usually concerned with one or two pest species, which could be rapidly picked out from the mass of material.

Around 1980, the method was used to obtain information on the total fauna of trees, but, if all the specimens are saved, sampling a whole tree produces a very large number of individuals (and a substantial task) but gives no measure of intra-tree variance (see p. 15). Sampling efficiency was therefore increased by taking a limited sample from under any individual tree or any small area of forest; a small number (normally five to 10) of sampling sheets or trays were either suspended or placed on the ground under each tree (Moran & Southwood, 1982; Southwood *et al.*, 1982; Erwin, 1983; Stork, 1987; Chey *et al.*, 1998). Where the ground cover is low and sparse or when it can be removed, the required samples may be collected on cloth sheets, either spread directly on the ground or stretched on a metal frame with short legs. This design in similar to the 'frass collector' (see Fig. 8.1); it keeps the cloth above the vegetation and provides a precise area. A square metre is a convenient size and insects may be removed (with a pooter) virtually as soon as they have fallen. With tall trees, particularly in rain forest, it is necessary to suspend the collecting devices below the canopy (Erwin, 1983; Stork & Hammond, 1997); these should retain the specimens that fall, as they cannot be continuously tended, and funnel-trap designs (see Fig. 10.4b) have been found effective. Washing the sides down with ethanol from a wash-bottle may clear them. It is always important to seek to avoid disturbing the ground-cover fauna before sampling; otherwise its members may fly into the trees and provide an atypical element in the subsequent sample.

If the trees to be sampled are less than about 15 m in height, the spray may be applied from a standard backpack sprayer, using an extension lance if necessary. In rain forests, it may be necessary to use an even finer spray from a 'mist-blower' or 'fogger' (e.g. Swingfog SN50) and allow this to drift and penetrate the canopy; this process may be repeated at short intervals.

Paarmann and Kerck (1987) found that, if natural pyrethrum (without a synergist, 1% diluted in diesel oil) was used as the knock-down agent, many insects would recover and often survive to develop and breed normally. This enables biological investigations to be made on species that are not otherwise easily collected. It also has the advantage that sampling will have less effect on the population because the considerable part of the population that falls outside the area of the sampling devices may recover. Recovery rates differ between groups, so there is a risk of another type of bias if further studies are to be made in the same locality. Most synthetic pyrethrins are more toxic than the natural compound and few insects recover, but knock-down may be more complete, especially if, as with neo-pybuthrin, this is quick.

The proportion of the fauna sampled is influenced by many factors: the method, the type of insect (the highly mobile and the highly cryptic, e.g. borers, are under-represented), the type of tree, and the weather conditions. However, most studies have found that around 75% of the total population fell within the first hour; this figure may be increased if the tree can be repeatedly shaken or jarred towards the end of the period or shortly afterwards (Southwood *et al.*, 1982; Adis *et al.*, 1997).

In Britain, where the fauna is relatively well known, the entomofauna of a tree species as revealed by knock-down sampling was compared by Southwood *et al.* (1982) with that obtained by collation from faunal lists; they concluded that the two approaches gave comparable data. Majer and Recher (1988) compared this method with branch clipping (a direct count) and found that it sampled a more diverse fauna, although it was less efficient for sessile insects.

Collection of naturally descending animals. Many arboreal larvae pupate in the soil and these may be collected as they descend to provide a population estimate of the total numbers at that particular developmental stage (see p. 350). Cone (1963) found that the vine weevil (*Otiorrhynchus sulcatus*) dropped off its host plant at daybreak and the population could therefore be determined by placing funnel-type traps under the grapevines.

4.2.2.4 The extraction of animals from herbage and debris

Here one is concerned with animals living in the herbage layer above the soil and the problem is not, as in the above section, to make them fall down, but to get them to come up, either artificially or naturally.

Suction apparatus. Based on the principles of domestic vacuum cleaners, a number of types of suction sampler for animals living on or amongst the herbage down to the soil surface have been developed. Today the three most widely used purpose-built commercial units are the D-Vac, Univac, and Vectis. Cheap machines can be made by converting those designed for other material, such as garden litter (Wright & Stuart, 1992), and designs have been published which offer advantages of cost, weight, and sampling efficiency over the established commercial models (e.g. Stewart & Wright, 1995). Suction sampling is only really effective in dry, upright vegetation less than about 15 cm high.

The D-Vac (Riverside, California, USA) was developed by Dietrick *et al.* (1959) and uses a 20 cm hose (Fig. 4.6). The efficiency of this sampler is greatly affected by the mode of operation; the nozzle must be moved vertically into the foliage and down to the ground surface, although even when this is done, if the ground is subsequently searched carefully, a number of beetles, isopods, and other 'heavy' arthropods will usually be found to have been missed. If the nozzle is swept horizontally across the foliage, very small catches are obtained (Richmond & Graham, 1969). Several comparisons of the D-Vac with other methods have shown efficiencies of around 50–70%. Some of these studies have been concerned with coccinellid beetles and other predators on cotton (Turnbull & Nicholls, 1966; Leigh *et al.*, 1970; Shepard *et al.*, 1972; Smith *et al.*, 1976). Dewar and Dean (1982) compared D-Vac, direct counting, and washing sampling methods for sampling aphids on wheat and concluded that the efficiency of the D-Vac declined with aphid density and varied with species and life-history stage. The D-Vac cannot be recommended for grassland Coleoptera sampling, for which reported sampling efficiencies range between only 7–27% (Duffey, 1980) and 1–17% (Morris & Rispin, 1987) of that obtained using heat extraction. However, Törmälä (1982) found suction sampling to be the most effective method for grassland Auchenorhyncha (Hemiptera), but efficiency can vary seasonally (Duffy, 1980) because larvae are not caught as efficiently as adults (Andrezejewska, 1965). Perfect *et al.* (1989) discuss the precision of population estimates obtained for planthoppers and leafhoppers on rice and Hand (1986) determined the capture efficiency for aphids on grasses and cereals. While effective, the

Back-pack mounted
3-horsepower engine

Flexible air hose

Collection cone. Screen sieve bag
retains arthropods and small debris.

Fig. 4.6 The D-Vac suction sampler.

D-Vac is heavy and difficult to handle. In order to increase efficiency, Turnbull and Nicholls (1966) reduced nozzle and hose diameters, getting a nozzle speed approaching 90 miles/h. The D-Vac sampler overcomes the problem of adhesion to the hose by having the collecting bag incorporated in the widened nozzle (see Fig. 4.6), and a squirrel-cage type of fan produces a strong air current, so long as it is not impeded.

The recently introduced Vortis sampler (Arnold, 1994), manufactered by the Burkard Manufacturing Co., Rickmansworth, England, unlike the D-Vac sampler does not have an animal-collecting bag through which the main airstream must pass (Fig. 4.7). As the collected material does not impede air flow, the Vectis has the advantage that the flow rate will not decline as material is sucked up. The air is sucked in above ground level and is turned through 180° as it lifts loose particles up the 10 cm diameter main tube (collecting area 0.2 m²). A series of vanes create a vortex, so that, when the insects reach the expansion chamber, they fall freely into an externally attached collecting pot or bag. Comparative studies of the efficiency of the Vortis have yet to be published. The Vortis has the advantage that air flow is constant and cannot be impeded, so that a less powerful, lighter motor can be used.

Owing to the formation of a water film, in which animals may become trapped, inside the suction hose, it is difficult to sample damp herbage with the narrow

Fig. 4.7 The Vortis suction sampler (after Arnold, 1994).

suction-hose types of apparatus, but use of a clear hose enables a check to be kept on this potential source of error. Arnold *et al.* (1973) used Vacuflex hose, which, although of clear plastic, is reinforced with a spiral wire. When sampling arable habitats with the Univac, it may be an advantage to use two bags, one of nylon net, which retains the insects, and outside this a larger cloth bag to prevent the soil passing through the fan. The narrow suction-hose type of sampler, with its high nozzle wind speed and a fan whose efficiency does not fall off when impeded, is particularly suited for work with the fauna of the lower parts of vegetation and the soil surface.

In conclusion, the efficiency of suction samplers is affected by several factors, but of particular importance are nozzle wind speed (for heavy and/or tenacious species) and speed of enclosure (for active flying species). The Univac and Vortis are excellent on the latter count, but the nozzle wind speed of the D-Vac is, with several types of collecting head, on the borderline of efficient extraction. Leigh *et al.* (1970) consider that the D-Vac needs to be standardized for the particular conditions before it may be considered as an extraction method for absolute estimates. Low efficiencies for the D-Vac occur if the animals are very active (escaping as the sampler approaches) or if they can cling close to the soil surface, when the nozzle speeds of the normal D-Vac (*c.* 60 km/h) are inadequate to dislodge them. Other samplers seem more efficient under a wider range of conditions (Johnson *et al.*, 1957; Heikinheimo & Raatikainen, 1962). The method is more satisfactory than sweeping (p. 267), which can give biased results. Larger models seem efficient; Kirk and Bottrell (1969) devised a tractor-mounted model and the McCoy Insect Collector, which incorporated blowing (a principle first used by Santa (1961)) as well as sucking, in a 'lorry-sized' motor vehicle. Its efficiency is high, and McCoy and Lloyd (1975) show that it can detect two boll weevils in 10 acres (4 ha) of cotton. These vehicular machines, which sample from such a large proportion of the habitat, overcome one of the main difficulties with smaller samplers, i.e. the high variance of the samples due to the patchy nature of many, especially natural, habitats. Theoretically, the answer lies in taking more samples, but the worker easily becomes overburdened with material to sort. For samples from large vehicular machines, mechanized and computerized methods of sorting have been developed.

Small suction samplers have been used to obtain invertebrates from rodents' nests (Levi *et al.*, 1959) and flowers (Kennard & Spencer, 1955).

Cylinder or covering method. The basis of this method is the enclosure of an area of herbage within a covered cylinder; the animals are collected by some method, often after being knocked down by an insecticide. Kretzschmar (1948) used a modification of the Romney (1945) technique for sampling soy-bean insects; he placed plastic transparent base plates beneath the plants the day before sampling. These greatly facilitate the subsequent collection of the insects, but tests should be made to ensure that their presence does not influence the density of insects on the plants above. In order to prevent the escape of active insects, Balogh and Loksa (1956) attached the sampling cylinder to the end of a long pole and rapidly brought this down on to the crop. These workers killed the insects with powerful fumigants, such as carbon disulphide and hydrogen cyanide, and then removed them by hand. As it is difficult to reach and search the base of a tall cylinder, Skuhravý *et al.* (1959) and Skuhravý and Novák (1961) used two cylinders and removed the tall outer one after the insects had been knocked down. Cherry *et al.* (1977) combined this method with the 'tent' technique (see below). Smaller plants have been sampled by enclosing them in an impermeable plastic tent, which is then flooded with carbon dioxide.

The combination of the 'covering method' with a suction apparatus for collection probably provides the most efficient method for sampling the fauna of herbage. The most appropriate cylinder or tent (see below) and suction apparatus will depend on the animal, the vegetation, and the availability of services (electricity, easy carriage of heavy equipment). Turnbull and Nicholls (1966) devised a quick trap (Fig. 4.8), which they used in conjunction with a modified D-Vac sampler. The quick trap is put in position several hours prior to sampling, with the net portion folded (the dotted outline in Fig. 4.8). When the sample is to be taken, the suspension cord, which can (and should) extend some distance from the trap, is pulled, which jerks it out of the slot at the top of tripod leg A. The tension springs expand the trap into position, but the operator should immediately check that the base ring is in close

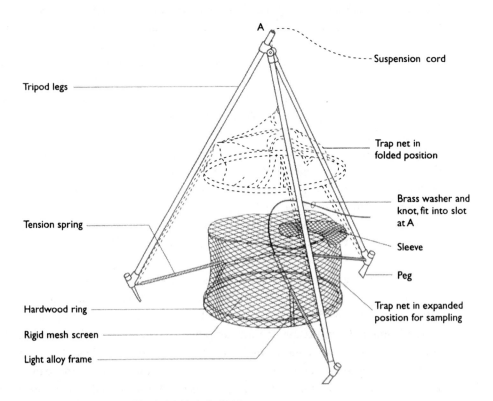

Fig. 4.8 'Quick trap' (after Turnbull & Nicholls, 1966).

contact with the ground. The suction-sampler hose is inserted through the top of the trap and the area systematically worked over. The mixture of small arthropods and debris can be sorted by hand or by a combination of funnel extractors and flotation (Dondale *et al.*, 1971; see also Chapter 6, pp. 229 and 218). Turnbull and Nicholls (1966) made most useful comparisons of this method with several others (Fig. 4.9). The extent of the advantage compared with the sweep net or normal D-Vac sampler varied between taxa.

Tents for sampling strongly phototactic animals. A large muslin-covered cage or 'tent' (Fig. 4.10) is quickly put into position when the animals are least active, usually at dusk. When they are active again, the following morning, the animals (if large) may be directly collected from the sides of the tent; alternatively, it may be covered with a black shroud, leaving only the muslin-topped, celluloid collecting cylinder exposed at the apex. After some time—about 15 min in bright sunshine—the majority of the animals will be in the cylinder, which can be quickly removed and is conveniently carried back to the laboratory by pushing its base into appropriately sized rings of Plasticine on a metal tray. If possible, someone should then get inside the 'tent' and remove the remaining animals from the inside of the walls; on a dull day, quite a large proportion of the total population may fail to enter the collecting cylinder. This method has been used successfully with blowflies (*Lucilia*) (MacLeod & Donnelly, 1957), mosquitoes (Chinaev, 1959), and frit flies, *Oscinella frit* (Southwood *et al.*, 1961). A miniature tent for use with mites is described by Jones (1950), and McGovern *et al.* (1973) have devised a model that they used to separate phototactic (and resilient) animals (e.g. boll weevils) from the debris in suction-apparatus samples.

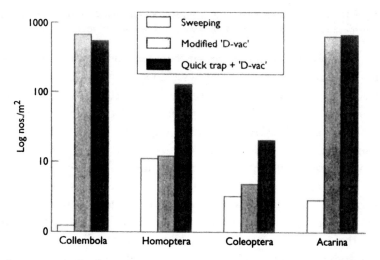

Fig. 4.9 Comparison of population estimates for different taxa using three methods of sampling: sweep net, modified D-Vac sampler, and quick-trap covering method plus D-Vac (after Turnbull & Nicholls, 1966).

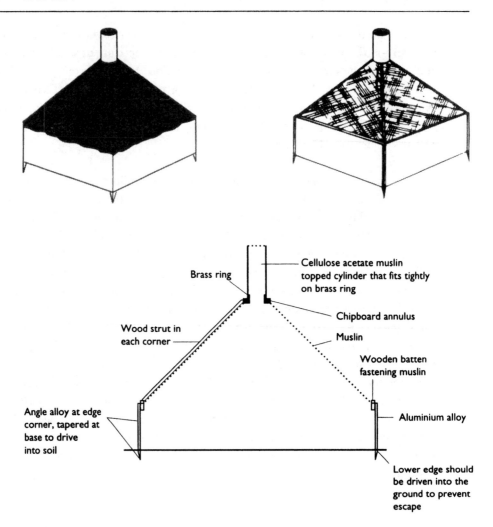

Fig. 4.10 A 'tent' for sampling strongly phototactic animals from herbage. A sagittal section through one corner and a side, with sketches of its appearance shrouded and unshrouded.

An apparatus that is intermediate between a cylinder and a tent has been devised by Wiegert (1961) for sampling grasshoppers, cercopids, and other insects that readily leave the vegetation if disturbed. It consists of a truncated nylon-covered strap-iron cone, topped with a metal funnel and a collecting jar. As with the tent or cylinder, this is placed rapidly over the vegetation, which is then agitated. The sampling cone is then gradually moved on to a flat sheet, of wood or other substance, which is placed adjacent to it. Eventually, most of the vegetation can be slid out of the cone, whilst the active insects remain; it is then inverted and the insects are jarred down into the collecting container. Wiegert considered the efficiency for the froghopper *Philaenus* to be greater than 90%. The leafhopper *Empoasca* was efficiently collected by Cherry *et al.* (1977) in a rather similar device; a black plastic dust bin, with a collecting cylinder (as used in tents (Fig. 4.10)) in its base, was quickly lowered upside down, suspended from the end of a long pole.

Extraction by heat drying and/or flotation. A further extension of the use of the animal's reactions to certain physical conditions leads logically to the removal of samples of herbage or debris to the laboratory, where they are subjected to drying by heat. This is the basis of the Berlese/Tullgren funnel. This is principally a method for soil and litter and hence its many modifications are discussed in Chapter 6. It is frequently utilized to extract animals from herbage samples, especially the mixtures that result from suction sampling (see above) (e.g. Dietrick *et al.*, 1959; Dondale *et al.*, 1971). Also, adult beetles and late-instar larvae from grain samples may be extracted using this approach (Smith, 1977). It is probable that it would be appropriate for many foliage insects that are moderately mobile; special attention would need to be given to the design of the funnel if they are strongly phototactic. Harris (1971) found the method fully efficient, compared with hand-sorting, for the larvae of the pear psyllid, *Psylla pyricola*, provided the heat source was not too powerful. Flotation methods (Chapter 6) may also be useful for the material from a suction sampler (Dondale *et al.*, 1971). With animals that rest in litter, such as weevils and leaf beetles, it may be possible (given a plentiful supply of water) to float the animals off the sample in the field. Wood and Small (1970) did this by enclosing the area in a solid metal frame and floating off the beetles and litter, through an overflow, on to a fine sieve.

4.2.2.5 Methods for animals in plant tissues

Although leaf-miners may be directly counted (p. 325), it is seldom possible to estimate so simply the number of eggs embedded in plant tissue or larval or adult insects boring into the stems of herbaceous plants or trees. The main impetus for the development of techniques for such animals has been provided by work on insect pests of stored grain and timber. Doubtless, considerable advances will be made when the approaches developed in this field are applied and extended to insects in the tissues of herbaceous plants.

Dissection. This method is widely applicable, if tedious. Sometimes, it is possible to limit the dissection of stems, fruits, or other parts to those that are obviously damaged (see Chapter 8).

Bleaching and/or selective staining. The plant tissue may be rendered transparent by treatment in eau-de-Javelle, lactophenol (Koura, 1958; Carlson & Hibbs, 1962), or 10% sodium hydroxide solution (Apt, 1950), so that the insects become visible; this method has been used with grains and lentils. Plant-parasitic nematodes are assessed by simultaneously bleaching the plant tissues and staining the worms in a mixture of lactophenol and cotton blue (Goodey, 1957). The eggs of various Hemiptera in potato leaves can be detected by bleaching the leaves (boiled in water until limp and then in 95% alcohol over a water-bath) and then staining in saturated solution of methyl red and differentiating in a slightly alkaline solution (containing sodium hydroxide) of the same dye; the leaf tissues become orange or yellow, but the eggs remain bright red and may be counted by transmitted light (Curtis, 1942). However, Carlson and Hibbs (1962) found that the same eggs could be counted after merely boiling for 1 min in lactophenol, when the leaf tissues become bleached and the egg proteins coagulated. Insect-egg plugs in grains have been found to be stained selectively by gentian violet (when moist) (Goosens, 1949) or by various alkaloids, derberine sulphate, chelidonium, and primuline, which fluoresce yellow, orange, and light blue under ultraviolet light (Milner *et al.*, 1950). If the grain is crushed, the insect protein may be recognized by the ninhydrin test (Dennis & Decker, 1962).

X-rays. In 1924, Yaghi suggested that X-rays could be used in the detection of wood-boring insects, but it is only since the 1950s that the principle has been extensively applied, more particularly to insects in homogeneously textured materials: in grain (Milner *et al.*, 1950, 1953; Pedersen & Brown, 1960) and other seeds (Simak, 1955). It is also of wide applicability to the assessment of insects in moist plant tissues: for wood borers (Berryman & Stark, 1962; Bletchly & Baldwin, 1962; Amman & Rasmussen, 1969), stem borers (Goodhue & van Emden, 1960), and boll-worms (Graham *et al.*, 1964). The cavity made by the larva or adult is much more easily detected in the radiographs than the difference between animal and plant tissues. Where one has to rely on the latter distinction, as with the blackcurrant gall mite, *Cecidophyopsis ribis*, then the value of the method may be limited, for long exposures to maximize the distinction will lead to blurred images (Smith, 1960). It has generally been found that 'soft' X-rays (5–35 kV) are best for the work; the thicker the plant material, the greater the voltage required for the same exposure time or the greater the exposure time, which needs to be increased as the water content of the plant tissue rises (Amman & Rasmussen, 1969). Small portable X-ray units (e.g. Picker) with a thin (*c.* 2 mm) beryllium window are particularly suitable for ecological work. Further technical details are given in the papers cited above. It has been shown by De Mars (1963) that X-ray detection of a bark beetle, *Dendroctonus*, is virtually as efficient as dissection, eight times faster, and less than a quarter the cost, even when allowance is made for materials. Wickman (1964) reached similar conclusions for its use with Siricidae. The interpretation of X-rays of bark insects is discussed by Berryman (1964). Appropriate safety procedures should be followed when using X-rays. No doubt, medical scanners, ultrasound and nuclear magnetic resonance (NMR) could be used in certain circumstances, but their costs generally rule them out.

Methods based on the different mass of the whole and the infested material. Where the plant material is of fairly uniform weight, e.g. grain, advantage may be taken of the fact that material in which insects have, or are, feeding will be lighter. The lighter grain can be made to float (e.g. in 2% ferric nitrate solution (Apt, 1952)) or to fall more rapidly when projected (Katz *et al.*, 1954). Such an approach naturally leads on to flotation techniques, which are discussed in Chapter 6.

Acoustic detection. Insects living in plant material often make sounds, normally from their feeding. The aural detection of this sound, particularly in grain-infesting species, was first proposed by Adams *et al.* (1953) and Street (1971) used a modified gramophone cartridge transducer. Recent advances allowing the construction of electronic acoustic detection devices capable of identifying the acoustic signature of feeding or some other activity are now under development. Such detectors may greatly facilitate non-destructive sampling in the future.

4.2.3 Special sampling problems with animals in plant material

4.2.3.1 The marking of turf samples

Van Emden (1963) has described a method whereby small areas of turf may be marked and subsequently found and recognized without impairing mowing or grazing or attracting the attention of vandals. Wire rings, marked with a colour code, were sunk into the soil around the samples and here they could remain for weeks or even years. Their approximate positions were recorded by reference to a grid and a diagonal tape, and their condition or other relevant ecological information noted. After the desired period, the rings may be rediscovered and the ecological changes in the precise area observed, by plotting the approximate position with reference to the grid and finding the actual ring with a mine-detector.

4.2.3.2 The sampling of bulk grain

When studying insect populations in stored grain, it is often desired to draw samples from known depths, ensuring that they are not contaminated by insects or grain from the outer layers. Burges (1960) has developed a method for doing this: a hollow spear with a lateral aperture is forced into the required position and then the desired sample sucked out through its shaft by a domestic vacuum cleaner. Other types of aspirator for sampling grain have been described by Chao and Peterson (1952) and Ristich and Lockard (1953).

4.2.3.3 The sampling of bark

A circular punch that cuts $0.1\,\text{ft}^2$ $(92.9\,\text{cm}^2)$ samples from bark was devised by Furniss (1962). The punch is made from a segment of a 4-in (10 cm) diameter steel pipe fitted with a central handle so that it can be rapidly and symmetrically hammered into the bark. Very thick bark may need to have the outer layer shaved off first if Scolytidae are being sampled and the phloem layer is required, but not, of course, if the crevice fauna is being studied (Lewis & Navas, 1962). There is also a problem of estimating the actual area sampled, which will vary with the curvature of the trunk or bough (Dudley, 1971).

4.3 Sampling from vertebrate hosts

There are two important variables to be considered when attempting to obtain complete samples of vertebrate ectoparasites from their hosts: first, the readiness or otherwise with which they will leave their hosts and, secondly, whether the host can be killed or not. Furthermore, if the host is killed, can the skin be destroyed whilst obtaining the parasites or must it be retained in good condition? Sampling from living animals will generally be more likely to provide absolute population counts of those parasites that are readily dislodged from the host (e.g. fleas, hippoboscids) than for those that are more firmly attached (e.g. lice, mites); to estimate these, the host will often need to be killed and even the skin destroyed. After collection, each host should be kept isolated by being placed into individually sealed bags or containers, in order to avoid cross-infection, which can rapidly occur. Clayton and Walter (1997) provide a detailed review of the application of methods to birds.

4.3.1 Sampling from living hosts
Care must be taken in handling the host, for Stark and Kinney (1962) found that many fleas would rapidly leave a struggling or agitated host.

4.3.1.1 Searching
This method is usually satisfactory only with relatively large parasites (e.g. ticks), active ones (e.g. hippoboscids (Ash, 1952)), or those that sit on exposed and hairless parts of the body. In the rare cases where the host is almost completely hairless, minute parasites can be counted by this method, e.g. on humans (Johnson & Mellanby, 1942). Nelson *et al.* (1957) compared various relative sampling techniques for the sheep ked, *Melophagus ovinus*, with the 'total live counts' (MacLeod, 1948) and with the 'picked-off' count; in the latter method, every ked that can be found on the sheep is picked off. They showed that the 'total live count', in which the living keds are counted as the fleece is parted, underestimated the total population, the actual mean estimates for different breeds of sheep ranging from 61 to 81% of the 'picked-off' count. This experiment emphasizes the importance of confirming that apparently absolute methods are really giving estimates of the total population. Ash (1960) sampled lice from living birds by removing a proportion of the feathers from the area frequented; the lice on these were determined and the total feathers in the area counted. Clayton and Walther (1997) consider visual examination of birds to be suitable for some ticks and feather mites, but fumigation (p. 169) yields a higher proportion of the parasite load (Walther & Clayton, 1997). Buxton (1947) determined the head-louse populations of men by shaving the scalp and dissolving the hair.

4.3.1.2 Combing
A fine-toothed comb may be used to remove the ectoparasites on living mammals, but, with lice, only a small part of the population is removed (see below). It is often an advantage if the host is anaesthetized with ether (Mosolov, 1959), but this also affects the parasites and is strictly a combination of combing and fumigation (see below).

4.3.1.3 Fumigation

This technique is particularly successful for Hippoboscidae and fleas (Siphonaptera) and is used for obtaining these parasites from birds at ringing stations and elsewhere during ecological studies on vertebrates. The hosts are collected from a trap and immediately placed in a white cloth (linen) bag for transportation to the laboratory. Here they may be exposed to an anaesthetic, such as ether (Mosolov, 1959; Janion, 1960), chloroform (Williamson, 1954), ethyl acetate, or carbon dioxide, which will dislodge the parasites. Janion merely sprinkled the bags containing the hosts (mice in his work) with ether and, after a few seconds of gentle squeezing, the mouse was released, the fleas remaining in the bag. Mosolov removed the rodents from the bag and shook and combed them with forceps and obtained a wide range of parasites. He points out that, as larval ectoparasites are often pale and the adults dark, they are most readily seen if they are shaken on to coloured paper. The 'Fair Isle apparatus' described by Williamson (1954) was designed to limit exposure of the host to the anaesthetic (Fig. 4.11). Basically, the bird is placed in a jar from which its head protrudes and into which the anaesthetic can be introduced. Fowler and Cohen (1983) describe a modification in which a rubber diaphram placed in the lid of the chamber has a slit cut into it through which the head of the bird can be pushed. The body of the bird within the chamber is then exposed to the aneasthetic. Bear (1995) describes further improvements to the apparatus, and Walther and Clayton (1997) note that a higher recovery rate is acheived by ruffling the feathers when the bird is removed. This apparatus removed up to 80% of ectoparasites (Fowler & Cohen, 1983; Poiani, 1992), but it does not sample the head!

A simple and effective method for birds is to dust with insecticidal power and then ruffle the feathers over a coloured surface (Clayton & Walther, 1997). Malcomson (1960) dusted the plumage with pyrethrum powder and allowed the bird to flutter for about 5 min under an inverted paper cone. Pyrethrum produces a rapid knock-down and has no side-effects on birds or mammals (Casida, 1973; Jackson, 1985). However, it is not 100% effective and superior results may be obtained using commercial formulations sold for treating pets, which combine pyrethrin with a pyrethrin derivative and piperonyl butoxide. These are sold as dust and in aerosol form. Clayton and Walther (1997) found the latter worked better on birds. The silicon aerogel Dri-Die 67, which kills arthropods by desiccation, was considered by Clayton and Walther (1997) to be 100% effective within 3 h, but it is long-lasting and can remove oil from feathers or fur, resulting in the host suffering from exposure during wet weather. Chlordimeform causes ticks to detach from their hosts (Gladney et al., 1974). Kalamarz (1963) found that lice were efficiently removed from the host after fumigation if it was vacuum-cleaned with a special brush nozzle.

4.3.1.4 Trapping

Ectoparasites that leave the host after feeding may be trapped. Bird ticks have been

Fig. 4.11 The Fowler–Cohen design for a chamber to collect ectoparasites from birds.

sampled by placing the host in a cage over a pan of water, into which the parasites fall after feeding (Krantz, 1978; Sonenshine, 1993). Castro (1973) used a funnel trap to collect mites emerging from the feathers of house sparrows.

4.3.2 Sampling from dead hosts

4.3.2.1 Searching and combing

Searching and combing, combined with brushing in all directions with a stiff brush, will provide many specimens from mammal pelts (Spencer, 1956), but often this is really only a relative method. With birds, however, Ash (1960) considered that a feather-by-feather search was the only efficient method for lice. Janzen (1963) used a novel principle, which may have a wider application, to obtain almost absolute samples of the beaver beetle (*Platypsyllus castoris*). The host's pelts were first frozen and then brought into a warm environment; the beetles would move away from the chilled pelts and the majority were found by searching, while most of the remainder were extracted by combing.

4.3.2.2 Fumigation

If the mammals are trapped live, but can be sacrificed, fumigation with hydrogen cyanide (produced from calcium cyanide) for about 15 min, followed by combing, will yield most parasites (Ellis, 1955; Murray, 1957). Bird ecto parasites have been sampled by fumigation with methyl bromide (Harshbarger & Raffensperger, 1959) and ethyl actetate (Clayton *et al.*, 1992), followed by ruffling the feathers.

4.3.2.3 Dissolving

The pelt of the host is dissolved, most conveniently by incubating with the proteolytic enzyme trypsin for 2 days, followed by boiling in 10% caustic potash for some minutes (Cook, 1954; Clayton & Walther, 1997). After such treatment, virtually only the ectoparasites remain and even minute listrophorid mites can be recovered. This method is generally associated with Hopkins (1949), who records the following comparison with searching and beating: the latter method had yielded 31 lice (larvae and adults) from three pelts; when these were dissolved, a further 1208 specimens were recovered. Although this method is very useful for mammal ectoparasites, differing results have been reported for bird ectoparasites, Ash (1960) found it unsatisfactory, but Clayton *et al.* (1992) obtained recovery rates for adult lice of between 91 and 100%.

4.3.2.4 Cleaning

The pelt is shaved and the hair placed in lysol or some other cleaning medium, when the lice and their eggs may be counted under a microscope (Murray, 1957). If the exact distribution of the parasites is to be determined, Murray recommends stunning the animals, soaking their coats with ether or chloroform to kill all lice *in situ*, and then placing the animal in a closed jar until dead. The hair is then shaved from each area (these may be delimited by a grid) and mounted in Berlese's mounting medium.

4.3.2.5 Washing

Large numbers of ectoparasites, especially mites, can be removed by washing the pelt or the animal in a solution (<5%) of detergent (Lipovsky, 1951). Optimal results are obtained by placing the host body in a container with 1–2% detergent solution and using a mechanical shaker, such as a paint mixer. Henry and McKeever (1971) found this method, when applied to rats, to be >90% efficient for mites, fleas, and lice but only 66% efficient for ticks, while McGroarty and Dobson (1974) found efficiencies of >95% for lice and >85% for feather mites from the bodies of house sparrows.

4.3.3 Sampling from vertebrate 'homes'

Birds' and rodents' nests, bat roosts, and other vertebrate 'homes' may usually be treated as litter and their fauna extracted by modified Berlese funnels (Sealander

& Hoffman, 1956) and the other methods described in Chapter 6; however, Woodroffe (1953) concluded that there was no absolute method—he found a combination of warming and sieving most efficient. Drummond (1957) and Wasylik (1963) collected continuous samples of the mites in mammal and avian nests by placing funnels or gauze-covered tubes below the nests; such traps basically resemble pitfall traps (p. 276) and provide relative estimates.

References

Adams, R.E., Wolfe, J.E., Milner, M., & Shellenberger, J.A. (1953) Aural detection of grain infested internally with insects. *Science* **118**, 163–4.

Adis, J., Paarmann, W., De Fonseca, C.R.V., & Rafael, J.A. (1997) Knockdown efficiency of natural pyrethrum and survival rate of living arthropods obtained by canopy fogging in central Amazonia. In: Stork, N.E., Adis, J., & Didham, R.K. (eds) *Canopy Arthropods*. Chapman & Hall, London, pp. 67–84.

Allison, D. & Pike, K.S. (1988) An inexpensive suction trap and its use in an aphid monitoring network. *J. Agric. Entomol.* **5**(2), 103–8.

Amman, G.D. & Rasmussen, L.A. (1969) Techniques for radiographing and the accuracy of the X-ray method for identifying and estimating numbers of the mountain pine beetle. *J. Econ. Entomol.* **62**, 631–4.

Andrzejewska, L. (1965) Stratification and its dynamics in meadow communities of Auchenorhyncha (Homoptera). *Ekologia Polska, Seria A* **13**, 685–715.

Apt, A.C. (1950) A method for detecting hidden infestation in wheat. *Milling Prod.* **15**(5), 1.

Apt, A.C. (1952) A rapid method of examining wheat samples for infestation. *Milling Prod.* **17**(5), 4.

Arnold, A.J. (1994) Insect suction sampling without nets, bags or filters. *Crop Protection* **13**, 73–6.

Arnold, A.J., Needham, P.H., & Stevenson, J.H. (1973) A self-powered portable insect suction sampler and its use to assess the effects of azinphos methyl and endosulfan on blossom beetle populations on oil seed rape. *Ann. Appl. Biol.* **75**, 229–33.

Arthur, A.P. (1962) A skipper, *Thymelicus lineola* (Ochs.) (Lepidoptera: Hesperiidae) and its parasites in Ontario. *Can. Entomol.* **94**, 1082–9.

Ash, J.S. (1952) Records of Hippoboscidae (Dipt.) from Berkshire and Co. Durham in 1950, with notes on their bionomics. *Entomol. Monthly Mag.* **88**, 25–30.

Ash, J.S. (1960) A study of the Mallophaga of birds with particular reference to their ecology. *Ibis* **102**, 93–110.

Austin, M.D. & Massee, A.M. (1947) Investigations on the control of the fruit tree red spider mite (*Metatetranychus ulmi* Koch) during the dormant season. *J. Hort. Sci.* **239**, 227–53.

Baker, G.H. & Vogelzang, B.K. (1988) Life-history, population-dynamics and polymorphism of *Theba pisana* (Mollusca, Helicidae) in Australia. *J. Appl. Ecol.* **25**, 867–87.

Baker, G.H. & Hawke, B.G. (1990) Life-history and population-dynamics of *Theba pisana* (Mollusca, Helicidae) in a cereal pasture rotation. *J. Appl. Ecol.* **27**, 16–29.

Balogh, I. & Loksa, I. (1956) Untersuchungen über die Zoozönose des Luzernenfeldes. Strukturzönologische Abhandlung. *Acta Zool. Hung.* **2**, 17–114.

Banks, C.J. (1954) A method for estimating populations and counting large numbers of *Aphis fabae* Scop. *Bull. Entomol. Res.* **45**, 751–6.

Banks, C.J. & Brown, E.S. (1962) A comparison of methods of estimating population density of adult sunn pest, *Eurygaster integriceps* Put. (Hemiptera, Scutelleridae) in wheat fields. *Entomol. Exp. Appl.* **5**, 255–60.

Bear, A. (1995) An improved method for collecting bird ectoparasites. *J. Field Ornithology* **66**, 212–14.

Belding, M.J., Isard, S.A., Hewings, A.D., & Irwin, M.E. (1991) Photovoltaic-powered suction trap for weakly flying insects. *J. Econ. Entomol.* **84**(1), 306–10.

Berryman, A.A. (1964) Identification of insect inclusions in X-rays of *Ponderosa* pine bark infested by western pine beetle, *Dentroctonus brevicomis* Le Conte. *Can. Entomol.* **96**, 883–8.

Berryman, A.A. & Stark, R.W. (1962) Radiography in forest entomology. *Ann Entomol. Soc. Am.* **55**, 456–66.

Bletchley, J.D. & Baldwin, W.J. (1962) Use of X-rays in studies of wood boring insects. *Wood* 27, 485–8.

Browning, T.O. (1959) The long-tailed mealybug *Pseudococcus adonidium* L. in South Australia. *Aust. J. Agric. Res.* 10, 322–39.

Bullock, J.A. (1963) Extraction of Thysanoptera from samples of foliage. *J. Econ. Entomol.* 569, 612–14.

Burges, H.D. (1960) A spear for sampling bulk grain by suction. *Bull. Entomol. Res.* 51, 1–5.

Buxton, P.A. (1947) *The Louse*, 2nd edn. Arnold, London, 164 pp.

Cannon, R.J.C. (1986) Summer populations of the cereal aphid *Metopolophium dirhodum* (Walker) on winter-wheat. 3. Contrasting years. *J. Appl. Ecol.* 23, 101–14.

Carlson, O.V. & Hibbs, E.T. (1962) Direct counts of potato leafhopper, *Empoasca fabae*, eggs in *Solanum* leaves. *Ann. Entomol. Soc. Am.* 55, 512–15.

Casida, J.E. (1973) *Pyrethrum, the Natural Insecticide*. Academic Press, New York.

Castro, S.D. (1973) A method for collection of the quill mites, *Syringophiloides minor* (Berlese) (Prostigmata: Syringophilidae), from living birds. *J. Medical Entomology* 10, 524.

Chamberlin, I.C. (1940) A mechanical trap for the sampling of aerial insect populations. *USDA Bur. Entomol. Plant Quar. ET* 163, 12.

Chant, D.A. (1962) A brushing method for collecting mites and small insects from leaves. *Progr. Soil Zool.* 1, 222–5.

Chant, D.A. & Muir, R.C. (1955) A comparison of the imprint and brushing machine methods for estimating the numbers of the fruit tree red spider mite, *Metaletranychus ulmi* (Koch), on apple leaves. *Rep. E. Malling Res. Sta. (A)* 1954, 141–5.

Chao, Y. & Peterson, A. (1952) A new type of aspirator. *J. Econ. Entomol.* 45, 751.

Cherrill, A.J. & Brown, V.K. (1990) The habitat requirements of adults of the wart-biter *Decticus verrucivorus* (L) (Orthoptera, Tettigoniidae) in southern England. *Biological Conservation* 53, 145–57.

Cherry, R.H., Wood, K.A., & Ruesink, W.G. (1977) Emergence trap and sweep net sampling for adults of the potato leafhopper from alfalfa. *J. Econ. Entomol.* 70, 279.

Chey, V.K., Holloway, J.D., Hambler, C., & Speight, M.R. (1998) Canopy knockdown of arthropods in exotic plantations and natural forest in Sabah, north-east Borneo, using insecticidal mist-blowing. *Bull. Entomol. Res.* 88, 15–24.

Chinaev, P.P. (1959) Methods in quantitative sampling of bloodsucking mosquitoes (Diptera,

Culicidae). [In Russian.] *Entomol. Obozr.* 38, 757–65. (Transl. *Entomol. Rev.* 38, 679–86.)

Clayton, D.H., Gregory, R.D., & Price, R.D. (1992) Comparative ecology of neotropical bird lice. *J. Anim. Ecol.* 61, 781–95.

Clayton, D.H. & Walther, B.A. (1997) Collection of Arthropod parasites of birds. In: Clayton, D.H. & Moore, J. (eds) *Host Parasite Evolution*. Oxford University Press, Oxford, pp. 419–40.

Cleveland, M.L. (1962) Adhesives for holding mites to glass plates. *J. Econ. Entomol.* 55, 570–1.

Coineau, Y. (1962) Nouvelles méthodes de prospection de la faune entomologique des plantes herbacées et ligneuses. *Bull. Soc. Entomol. Fr.* 67, 115–19.

Collyer, E. (1951) A method for the estimation of insect populations of fruit trees. *Rep. E. Malling Res. Sta.* 1949–50, 148–51.

Condrashoff, S.F. (1967) An extraction method for rapid counts of insect eggs and small organisms. *Can. Entomol.* 99, 300–3.

Cone, W.W. (1963) The black vine weevil, *Brachyrhinus sulcatus*, as a pest of grapes in south central Washington. *J. Econ. Entomol.* 56, 677–80.

Connell, W.A. (1959) Estimating the abundance of corn ear worm eggs. *J. Econ. Entomol.* 52, 747–9.

Cook, E.F. (1954) A modification of Hopkins' technique for collecting ectoparasites from mammal skins. *Entomol. News* 15, 35–7.

Corbet, P.S. (1966) A method for sub-sampling insect collections that vary widely in size. *Mosquito News* 26, 420–4.

Curtis, W.E. (1942) A method of locating insect eggs in plant tissues. *J. Econ. Entomol.* 35, 286.

Dean, D.A. & Sterling, W.L. (1985) Size and phenology of ballooning spiders at 2 locations in eastern Texas (USA). *J. Arachnol.* 13(1), 111–20.

Dean, D.A. & Sterling, W.L. (1990) Seasonal patterns of spiders captured in suction traps in eastern Texas (USA). *Southw. Entomol.* 15(4), 399–412.

De Mars, C.J. (1963) A comparison of radiograph analysis and bark dissection in estimating numbers of western pine beetle. *Can. Entomol.* 95, 1112–16.

Dennis, N.M. & Decker, R.W. (1962) A method and machine for detecting living internal insect infestation in wheat. *J. Econ. Entomol.* 55, 199–203.

Dewar, A.M. & Dean, G.J. (1982) Assessment of methods for estimating the numbers of aphids (Hemiptera: Aphididae) in cerials. *Bull. Entomol. Res.* 72, 672.

Dietrick, E.J., Schlinger, E.I., & Bosch, R.V.D. (1959) A new method for sampling arthropods using a suction collecting machine and modified Berlese funnel separator. *J. Econ. Entomol.* **52**, 1085–91.

Dondale, C.D., Nicholls, C.F., Redner, J.H., Semple, R.B., & Turnbull, A.L. (1971) An improved Berlese–Tüllgren funnel and a flotation separator for extracting grassland arthropods. *Can. Entomol.* **103**, 1549–52.

Drummond, R.O. (1957) Observations on fluctuations of acarine populations from nests of *Peromyscus leucopus*. *Ecol. Monogr.* **27**, 137–52.

Dudley, C.O. (1971) A sampling design for the egg and first instar larval populations of the western pine beetle, *Dendroctonus brevicomis* (Coleoptera: Scolytidae). *Can. Entomol.* **103**, 1291–313.

Duffey, E. (1980) The efficiency of the Dietrick vacuum sampler (D-vac) for invertebrate population studies in different types of grassland. *Bull. Ecol.* **11**, 421–31.

Ellis, L.L. (1955) A survey of the ectoparasites of certain mammals in Oklahoma. *Ecology* **36**, 12–18.

van Emden, H.F. (1963) A technique for the marking and recovery of turf samples in stem borer investigations. *Entomol. Exp. Appl.* **6**, 194–8.

Erwin, T.L. (1983) Beetles and other insects of tropical rainforest canopies at manaus, Brazil, sampled by insecticidal fogging. In: Sutton, S.L., Whitmore, T.C., & Chadwick, A.C. (eds) *Tropical Rain Forest Ecology and Management*. Blackwell Scientific Publications, Oxford, pp. 59–75.

Fowler, J.A. & Cohen, S. (1983) A method for the quantitative collection of ectoparasites from birds. *Ringing and Migration* **4**, 185–9.

Furniss, M.M. (1962) A circular punch for cutting samples of bark infested with beetles. *Can. Entomol.* **94**, 959–63.

Gara, R.I. & Vité, J.P. (1962) Studies on the flight patterns of bark beetles (Coleoptera: Scolytidae) in second growth *Ponderosa* pine forests. *Contrib. Boyce Thompson Inst.* **21**, 275–90.

Gibbs, D.G., Pickett, A.D., & Leston, D. (1968) Seasonal population changes in cocoa capsids (Hemiptera, Miridae) in Ghana. *Bull. Entomol. Res.* **58**, 279–93.

Gibson, R.W. (1975) Measurement of eriophyid mite populations on ryegrass using ultrasonic radiation. *Trans. Roy. Entomol. Soc. Lond.* **127**, 31–2.

Gladney, W.J., Ernst, S.E., & Drummond, R.O. (1974) Chlordimeform: a detachment-stimulating chemical for three-host ticks. *J.*

Med. Entomol. **11**, 569–72.

Goodenough, J.L., Jank, P.C., Carroll, L.E. *et al.* (1983) Collecting and preserving airborne arthropods in liquid at timed intervals with a Johnson–Taylor-type suction trap. *J. Econ. Entomol.* **76**(4), 960–3.

Goodey, J.B. (1957) *Laboratory Methods for Work with Plant and Soil Nematodes*, 3rd edn. Technical Bulletin 2, Ministry of Agriculture, HMSO, London.

Goodhue, R.D. & van Emden, H.F. (1960) Detection of stem borers in Gramineae by X-rays. *Plant Pathol.* **9**, 194.

Goosens, H.J. (1949) A method for staining insect egg plugs in wheat. *Cereal Chem.* **26**(5), 419–20.

Graham, H.M., Robertson, O.T., & Martin, D.F. (1964) Radiographic detection of pink bollworm larvae in cottonseed. *J. Econ. Entomol.* **57**, 419–20.

Gray, K.W. & Schuh, J. (1941) A method and contrivance for sampling pea aphid populations. *J. Econ. Entomol.* **34**, 411–15.

Greig-Smith, P. (1964) *Quantitative Plant Ecology*. Butterworths, London.

Hand, S.C. (1986) The capture efficiency of the Dietrick vacuum insect net for aphids on grasses and cereals. *Ann. Appl. Biol.* **108**, 233.

Harris, J.W.E., Collis, D.G., & Magar, K.M. (1972) Evaluation of the tree-beating method for sampling defoliating forest insects. *Can. Entomol.* **104**, 723–9.

Harris, M. (1971) Sampling pear foliage for nymphs of the pear *Psylla*, using the Berlese–Tullgren funnel. *J. Econ. Entomol.* **64**, 1317–18.

Harris, P. (1960) Number of *Rhyacionia buoliana* per pine shoot as a population index, with a rapid determination method of this index at low population levels. *Can. J. Zool.* **38**, 475–8.

Harshbarger, J.C. & Raffensperger, E.M. (1959) A method for collecting and counting populations of the shaft louse. *J. Econ. Entomol.* **52**, 1215–16.

Heikinheimo, O. & Raatikainen, M. (1962) Comparison of suction and netting methods in population investigations concerning the fauna of grass leys and cereal fields, particularly in those concerning the leafhopper, *Calligypona pellucida* (F.). *Valt. Maatalousk. Julk. Helsingfors* **191**, 31 pp.

Helson, G.A.H. (1958) Aphid populations: ecology and methods of sampling aphids *Myzus persicae* (Sulz.) and *Aulacorthum solani* (Kltb). *NZ Entomol.* **2**, 20–3.

Henderson, C.F. & McBurnie, H.V. (1943) Sampling technique for determining populations of the citrus red mite and its predators. *USDA Circular* **671**, 11 pp.

Henderson, I.F. & Whitaker, T.M. (1976) The efficiency of an insect sampler in grassland. *Ecol. Entomol.* **2**, 57–60.

Henry, L.G. & McKeever, S. (1971) A modification of the washing technique for quantitative evaluation of the ectoparasite load of small mammals. *J. Med. Ent.* **8**, 504–5.

Hobbs, S.E. & Hodges, G. (1993) An optical method for automatic classification and recording of a suction trap catch. *Bull. Entomol. Res.* **83**(1), 47–51.

Hoerner, J.L. (1947) A separator for onion thrips. *J. Econ. Entomol.* **40**, 755.

Hopkins, G. (1949) The host associations of the lice of mammals. *Proc. Zool. Soc. Lond.* **119**, 387–604.

Hower, A.A. & Ferguson, W. (1972) A square-foot device for use in vacuum sampling alfalfa insects. *J. Econ. Entomol.* **65**, 1742–3.

Hughes, R.D. (1963) Population dynamics of the cabbage aphid *Brevicoryne brassicae* (L.). *J. Anim. Ecol.* **32**, 393–424.

Ives, W.G.H. (1959) A technique for estimating tamarack foliage production, a basis for detailed population studies of the larch sawfly. *Can. Entomol.* **91**, 513–19.

Jackson, J.A. (1985) On the control of parasites in nest boxes and the use of pesticides near birds. *Sialia* **7**, 17–25.

Jacobson, M. (1966) Chemical insect attractants and repellants. *Ann. Rev. Entomol.* **11**, 403–22.

Janion, S.M. (1960) Quantitative dynamics in fleas (Aphaniptera) infesting mice of Puszcza Kampinoska Forest. *Bull. Acad. Pol. Sci. II* **8**(5), 213–18.

Janzen, D.H. (1963) Observations on populations of adult beaver-beetles, *Platypsyllus castoris* (Platypsyilidae: Coieoptera). *Pan. Pacif. Entomol.* **32**, 215–28.

Johnson, C.G. (1950) A suction trap for small airborne insects which automatically segregates the catch into successive hourly samples. *Ann. Appl. Biol.* **37**, 80–91.

Johnson, C.G. (1957) The distribution of insects in the air and the empirical relation of density to height. *J. Anim. Ecol.* **26**, 479–94.

Johnson, C.G. & Mellanby, K. (1942) The parasitology of human scabies. *Parasitology* **34**, 285–90.

Johnson, C.G. & Taylor, L.R. (1955a) The development of large suction traps for airborne insects. *Ann. Appl. Biol.* **43**, 51–61.

Johnson, C.G. & Taylor, L.R. (1955b) The measurement of insect density in the air. *Lab. Pract.* **4**, 187–92, 235–9.

Johnson, C.G., Southwood, T.R.E., & Entwistle, H.M. (1957) A new method of extracting arthropods and molluscs from grassland and herbage with a suction apparatus. *Bull. Entomol. Res.* **48**, 211–18.

Johnson, C.G., Taylor, L.R., & Southwood, T.R.E. (1962) High altitude migration of *Oscinella frit* L. (Diptera: Chioropidae). *J. Anim. Ecol.* **31**, 373–83.

Jones, B.M. (1950) A new method for studying the distribution and bionomics of trombiculid mites (Acarina: Trombidiidae). *Parasitology* **40**, 1–13.

Juillet, J.A. (1963) A comparison of four types of traps used for capturing flying insects. *Can. J. Zool.* **41**, 219–23.

Kalamarz, E. (1963) Badania nad biologia Mallophaga IV. Nowe methody zbierania ektopasozytów. *Ekol. Polska B* **9**, 321–5.

Katz, R, Farrell, E.P., & Milner, M. (1954) The separation of grain by projection I. *Cereal Chem.* **31**, 316–25.

Kennard, W.C. & Spencer, J.L. (1955) A mechanical insect collector with high manoeuvrability. *J. Econ. Entomol.* **48**, 478–9.

Kershaw, W.J.S. (1964) Aphid sampling in sugar beet. *Plant. Pathol.* **13**, 101–6.

Kirk, I.W. & Bottrell, D.G. (1969) A mechanical sampler for estimating boll weevil populations. *J. Econ. Entomol.* **62**, 1250–1.

Kobayashi, F. & Murai, M. (1965) Methods for estimating the number of the Cryptomeria red mite, especially with the removal by solutions. *Res. Popul. Ecol.* **7**, 35–42.

Koura, A. (1958) A new Transparency method for detecting internal infestation in grains. *Agric. Res. Rev.* **36**, 110–13.

Krantz, G.W. (1978) *A Manual of Acarology*. Oregon State University Book Stores, Corallis, USA.

Kretzschmar, G.P. (1948) Soy bean insects in Minnesota with special reference to sampling techniques. *J. Econ. Entomol.* **41**, 586–91.

La Croix, E.A.S. (1961) Observations on the ecology of the cotton-flea-beetles in the Sudan Gezira and the effect of sowing date on the level of population in cotton. *Bull. Entomol. Res.* **52**, 773–83.

Laster, M.L. & Furr, R.E. (1962) A simple technique for recovering insects from sorghum heads in insecticide tests. *J. Econ. Entomol.* **55**, 798.

Legner, E.F. & Oatman, E.R. (1962) Foliage-feeding Lepidoptera on young non-bearing apple trees in Wisconsin. *J. Econ. Entomol.* **55**, 552–4.

Leigh, T.F., Gonzalez, D., & Van Den Bosch, R. (1970) A sampling device for estimating absolute insect populations on cotton. *J. Econ. Entomol.* **63**, 1704–6.

Leos Martinez, J., Granovsky, T.A., Williams,

H.J., Vinson, S.B., & Burkholder, W.E. (1986) Estimation of aerial density of the lesser grain borer (*Rhyzopertha dominica*) (Coleoptera: Bostrichidae) in a warehouse using dominicalure traps. *J. Econ. Entomol.* **79**(4), 1134–8.

Le Pelley, R.H. (1942) A new method of sampling thrips populations. *Bull. Entomol. Res.* **33**, 147–8.

Levi, M.I., Chernov, S.G., Labunets, N.F., & Kosminskii, R.B. (1959) Aspiration method for the collection of fleas from rodents' nests. [In Russian.]. *Med. Parazitol.* **28**, 64–9.

Lewis, T. (1960) A method for collecting Thysanoptera from Gramineae. *Entomologist* **939**, 27–8.

Lewis, T. & Navas, D.E. (1962) Thysanopteran populations overwintering in hedge bottoms, grass litter and bark. *Ann. Appl. Biol.* **50**, 299–311.

Lewis, T. & Taylor, L.R. (1965) Diurnal periodicity of flight by insects. *Trans. Roy. Entomol. Soc. Lond.* **116**, 393–469.

Lipovsky, L.J. (1951) A washing method of ectoparasite recovery with particular reference to chiggers. *J. Kansas Entomol. Soc.* **24**, 151–6.

Locatelli, D.P. & Limonta, L. (1993) Catches of Psocoptera by a suction trap and notes for a list of Italian species. *Boll. Zool Agrar. Bachicolt.* **25**(2), 131–41.

Locatelli, D.P. & Limonta, L. (1994) Psocoptera captured by using a suction trap in Valtellina during 1992–1993. *Boll. Zool Agrar. Bachicolt.* **26**(2), 279–82.

Macaulay, E.D.M., Tatchell, G.M., & Taylor, L.R. (1988) The Rothamsted insect survey '12-metre' suction trap. *Bull. Entomol. Res.* **78**(1), 121–30.

McCoy, J.R. & Lloyd, E.P. (1975) Evaluation of airflow systems for the collection of boll weevils from cotton. *Econ. Entomol.* **68**, 49–52.

McGovern, W.L., Leggett, J.E., Johnson, W.C., & Cross, W.H. (1973) Techniques for separating boll weevils and other small insects from samples taken with insect collecting machines. *J. Econ. Entomol.* **66**, 1332.

McGroarty, D.L. & Dobson, R.C. (1974) Ectoparasite populations on house sparrows in northwestern Indiana. *American Midland Naturalist* **91**, 479–86.

MacLeod, J. (1948) The distribution and dynamics of ked populations, *Melophagus ovinus* Linn. *Parasitology* **39**, 61–8.

MacLeod, J. & Donnelly, J. (1957) Some ecological relationships of natural populations of calliphorine blowflies. *J. Anim. Ecol.* **26**, 135–70.

Maelzer, D.A. (1976) A photographic method and

a ranking procedure for estimating numbers of the rose aphid, *Macrosiphum rosae* (L.) on rose buds. *Aust. J. Ecol.* **1**, 89–96.

Majer, J.D. & Recher, H.F. (1988) Invertebrate communities on Western Australian eucalyptus: a comparison of branch clipping and chemical knockdown procedures. *Austr. J. Ecol.* **13**, 269–78.

Malcomson, R.O. (1960) Mallophaga from birds of North America. *Wilson Bull.* **72**(2), 182–97.

Maw, M.G. (1964) An effect of static electricity on captures in insect traps. *Can. Entomol.* **96**, 1482.

Milner, M., Barney, D.L., & Shellenberger, J.A. (1950) Use of selective fluorescent stains to detect egg plugs on grain kernels. *Science* **112**, 791–2.

Milner, M., Katz, R., Lee, M.R., & Pile, W.B. (1953) Application of the Polaroid–Land process to radiographic inspection of wheat. *Cereal Chem.* **30**, 169–70.

Moore, N.W. (1964) Intra- and interspecific competition among dragonflies (Odonata). *J. Anim. Ecol.* **33**, 49–71.

Moran, V.C. & Southwood, T.R.E. (1982) The guild composition of arthrpod communities in trees. *J. Anim. Ecol.* **51**, 289–306.

Morgan, C.V.G., Chant, D.A., Anderson, N.H., & Ayre, G.L. (1955) Methods for estimating orchard mite populations, especially with the mite brushing machine. *Can. Entomol.* **87**, 189–200.

Morris, M.G. & Rispin, W.E. (1987) Abundance and diversity of the coleopterous fauna of a calcareous grassland under different cutting regimes. *J. Appl. Ecol.* **24**, 451–65.

Mosolov, L.P. (1959) A method of collecting the ectoparasites of rodents, without destroying the host population. [In Russian, English summary.] *Med. Parazitol.* **28**, 189–92.

Mueller-Dombois, D. & Ellenberg, H. (1974) *Aims and Methods of Vegetation Ecology.* J. Wiley & Sons, New York, 547 pp.

Murray, Y.F. (1957) An ecological appraisal of host–ectoparasite relationships in a zone of epizootic plague in central California. *Am. J. Trop. Med. Hyg.* **6**, 1068–86.

Nelson, W.A., Slen, S.B., & Banky, E.C. (1957) Evaluation of methods of estimating populations of the sheep ked, *Melophagus ovinus* (L.) (Diptera: Hippoboscidae), on mature ewes and young lambs. *Can. J. Anim. Sci.* **37**, 8–13.

Newell, I.M. (1947) Quantitative methods in biological and control studies of orchard mites. *J. Econ. Entomol.* **40**, 683–9.

Nicholls, C.F. (1960) A portable mechanical insect trap. *Can. Entomol.* **92**, 48–51.

Noldus, L.P.J.J., Rumei, X., & van Lenteren, J.C.

(1985) The parasite–host relationship between *Encarsia formosa* Gahan (Hymenoptera, Aphelinidae) and *Trialeurodes vaporariorum* (Westwood) (Homoptera, Aleyrodidae). 17. Within-plant movement of adult greenhouse whiteflies. *J. Appl. Ent* **100**, 494–503.

Otvos, I.S. & Bryant, D.G. (1972) An extraction method for rapid sampling of eastern hemlock looper eggs, *Lambdina fiscellaria* (Lepidoptera: Geometridae). *Can. Entomol.* **104**, 1511–14.

Paarmann, W. & Kerck, K. (1997) Advances in using canopy fogging technique to collect living arthropods from tree-crowns. In: Stork, N.E., Adis, J., & Didham, R.K. (eds) *Canopy Arthropods*. Chapman & Hall, London, pp. 53–66.

Pedersen, J.R. & Brown, R.A. (1960) X-ray microscope to study behaviour of internal-infesting grain insects. *J. Econ. Entomol.* **53**, 678–9.

Perfect, T.J., Cook, A.G., & Ferrer, E.R. (1989) Population sampling for planthoppers, leafhopers (Hemiptera: Delphacidae & Cicadellidae) and their predators in flooded rice. *Bull. Entomol. Res.* **73**, 345–55.

Pielou, D.P. (1961) Note on a volumetric method for the determination of numbers of apple aphid, *Aphis pomi* Deg., on samples of apple foliage. *Can. J. Plant Sci.* **41**, 442–3.

Poiani, A. (1992) Ectoparasitism as a possible cost of social life: a comparitive analysis using Australian passerines (Passeriformes). *Oecologia* **92**, 429–41.

Prescott, H.W. & Newton, R.C. (1963) Flight study of the clover root *Curculio*. *J. Econ. Entomol.* **56**, 368–70.

Putman, W.L. (1966) Sampling mites on peach leaves with the Henderson–McBurnie machine. *J. Econ. Entomol.* **59**, 224–5.

Richards, O.W. & Waloff, N. (1954) Studies on the biology and population dynamics of British grasshoppers. *Anti-locust Bull.* **17**, 1–184.

Richards, O.W. & Waloff, N. (1961) A study of a natural population of *Phytodecta olivacea* (Forster) (Colcoptera, Chrysomeloidea). *Phil. Trans. Roy. Soc.* **244**, 204–57.

Richmond, C.A. & Graham, H.M. (1969) Two methods of operating a vacuum sampler to sample populations of the cotton fleahopper on wild hosts. *J. Econ. Entomol.* **62**, 525–6.

Ristich, S. & Lockard, D. (1953) An aspirator modified for sampling large populations. *J. Econ. Entomol.* **46**, 711–12.

Romney, V.E. (1945) The effect of physical factors upon catch of the beet leafhopper (*Eutettix tenellus* (Bak.) by a cylinder and two sweep methods. *Ecology* **26**, 135–47.

Rudd, W.G. & Jensen, R.L. (1977) Sweep net and

ground cloth sampling for insects in soybeans. *J. Econ. Entomol.* **70**, 301–4.

Santa, H. (1961) A method for sampling plant and leaf-hopper density in winter and early spring. [In Japanese.] *Plant Prot. Tokyo* **8**, 353–5.

Satchell, J.E. & Mountford, M.D. (1962) A method of assessing caterpillar populations on large forest trees, using a systemic insecticide. *Ann. Appl. Biol.* **50**, 443–50.

Schaefer, G.W., Bent, G.A., & Allsopp, K. (1985) Radar and opto-electronic measurements of the effectiveness of Rothamsted insect survey suction traps. *Bull. Entomol. Res.* **75**(4), 701–16.

Sealander, J.A. & Hoffman, C.E. (1956) A modified Berlese funnel for collecting mammalian and avian ectoparasites. *Southw. Nat. Dallas* **1**, 134–6.

Sedivy, J. (1993) Variation in the population density of pollen beetles (*Meligethes aeneus* F.) in winter rape. *Ochrana Rostlin* **29**(1), 9–15.

Shands, W.A., Simpson, G.W., & Reed, L.B. (1954) Subunits of sample for estimating aphid abundance on potatoes. *J. Econ. Entomol.* **47**, 1024–7.

Shepard, M., Sterling, W., & Walker, J.K. (1972) Abundance of beneficial arthropods on cotton genotypes. *Environ. Entomol.* **1**, 117–21.

Simak, M. (1955) Insect damage on seeds of Norway spruce determined by X-ray photography. *Medd. Stat. Skogsforsknings-Inst. Uppsala* **41**, 299–310.

Skuhravý, V. & Novák, V. (1961) The study of field crop entomocenoses. [In Russian]. *Entomol. Obozr.* **41**, 807–14. (Transl. *Entomol. Rev.* **41**, 454–8.)

Skuhravý, V., Novák, K., & Starý, P. (1959) Entomofauna jetele (*Trifolium pratense* L.) a jeji vývoj. *Rozpr. Čsl. Akad. Věd.* **69**(7), 3–82.

Smith, B.D. (1960) Population studies of the black current gall mite (*Phytoptus ribis* Nal.). *Rep. Agric. Hort. Res. Sta. Bristol* **1960**, 120–4.

Smith, J.W., Stadelbacher, E.A., & Gantt, C.W. (1976) A comparison of techniques for sampling beneficial arthropod populations associated with cotton. *Environ. Entomol.* **5**, 435–44.

Smith, L.B. (1977) Efficiency of Berlese–Tüllgren funnels for removal of the rusty grain beetle, *Cryptolestes ferrugineus*, from wheat samples. *Can. Entomol.* **109**, 503–9.

Sonenshine, D.E. (1993) *Biology of Ticks*. Oxford University Press, Oxford, England.

Southwood, T.R.E., Jepson, W.F., & van Emden, H.F. (1961) Studies on the behaviour of *Oscinella frit* L. (Diptera) adults of the panicle generation. *Entomol. Exp. Appl.* **4**, 196–210.

Southwood, T.R.E., Moran, V.C., & Kennedy, C.E.J. (1982) The assessment of arboreal insect fauna: comparisons of knockdown sampling and faunal lists. *Ecol. Entomol.* **7**, 331–40.

Spencer, G.J. (1956) Some records of ectoparasites from flying squirrels. *Proc. Entomol. Soc. BC* **52**, 32–4.

Stage, H.H., Gjullin, C.M., & Yates, W.W. (1952) Mosquitoes of the north-western states. *USDA Agric. Handb.* **46**, 95.

Stark, H.E. & Kinney, A.R. (1962) Abandonment of disturbed hosts by their fleas. *Pan. Pacif. Entomol.* **38**, 249–51.

Steiner, H. (1962) Methoden zur Untersuchung der Populationsdynamik in Obstanlagen (inc. Musternahme und Sammeln). *Entomophaga* **7**, 207–14.

Stewart, A.J.A. & Wright, A.F. (1995) A new inexpensive suction apparatus for sampling arthropods in grassland. *Ecol. Entomol.* **20**(1), 98–102.

Stork, N.E. (1987) Guild structure of Bornean rainforest trees. *Ecol. Entomol.* **12**, 69–80.

Stork, N.E., Adis, J., & Didham, R.K. (eds) (1997) *Canopy Arthropods.* Chapman & Hall, London, 567 pp.

Stork, N.E. & Hammond, P.M. (1997) Sampling arthropods from tree-crowns by fogging with knockdown insecticides: lessons from studies of oak tree beetle assemblages in Richmond Park (UK). In: Stock, N.E., Adis, J., & Didham, R.K. (eds) *Canopy Arthropods.* Chapman & Hall, London, 567 pp.

Street, M.W. (1971) A method for aural monitoring of in-kernel insect activity. *J. Georgia Entomol. Soc.* **6**, 72–5.

Strickland, A.H. (1954) An aphid counting grid. *Plant Pathol.* **3**, 73–5.

Summers, F.M. & Baker, G.A. (1952) A procedure for determining relative densities of brown almond mite populations on almond trees. *Hilgardia* **21**, 369–82.

Taylor, E.A. & Smith, F.F. (1955) Three methods for extracting thrips and other insects from rose flowers. *J. Econ. Entomol.* **48**, 767–8.

Taylor, L.R. (1951) An improved suction trap for insects. *Ann. Appl. Biol.* **38**, 582–91.

Taylor, L.R. (1955) The standardization of air flow in insect suction traps. *Ann. Appl. Biol.* **43**, 390–408.

Taylor, L.R. (1958) Aphid dispersal and diurnal periodicity. *Proc. Linn. Soc. Lond.* **1699**, 67–73.

Taylor, L.R. (1962a) The absolute efficiency of insect suction traps. *Ann. Appl. Biol.* **50**, 405–21.

Taylor, L.R. (1962b) The efficiency of cylindrical sticky insect traps and suspended nets. *Ann. Appl. Biol.* **50**, 681–5.

Taylor, L.R. (1974) Monitoring change in the distribution and abundance of insects. *Rep. Rothamsted Exp. Sta.* **1973**, 202–39.

Taylor, L.R. & Palmer, J.M.P. (1972) Aerial sampling. In: van Emden, H. (ed.) *Aphid Technology.* Academic Press, London, New York, pp. 189–234.

Topping, C.J., Sunderland, K.D., & Bewsey, J. (1992) A large improved rotary trap for sampling aerial invertebrates. *Ann. Appl. Biol.* **121**(3), 707–14.

Törmälä, T. (1982) Evaluation of five methods of sampling field layer arthropods, particularly the leafhopper community in grassland. *Annales Entomologici Fennici* **48**, 1–16.

Trumble, J.T. (1994) Sampling arthropod pests in vegetables. In Pedigo, L.P. & Buntin, G.D. (eds) *Handbook of Sampling Methods for Arthropods in Agriculture.* CRC Press, Boca Raton, pp. 603–26.

Turnbull, A.L. & Nicholls, C.F. (1966) A 'quick trap' for area sampling of arthropods in grassland communities. *J. Econ. Entomol.* **59**, 1100–4.

Ulrich, G.W. (1986) A new larval fish trap: the quatrefoil trap. *Prog. Fish. Cult.* **46**, 216–19.

Venables, E.P. & Dennys, A.A. (1941) A new method of counting orchard mites. *J. Econ. Entomol.* **34**, 324.

Vité, J.P. & Gara, R.I. (1961) A field method for observation on olfactory responses of bark beetles (Scolytidae) to volatile materials. *Contrib. Boyce Thompson Inst.* **219**, 175–82.

Wainhouse, D. (1980) A portable suction trap for sampling small insects. *Bull. Entomol. Res.* **70**(3), 491–4.

Walker, A.L., Cate, J.R., Pair, S.D., & Bottrell, D.G. (1972) A volumetric method for estimating populations of the greenbug on grain sorghum. *J. Econ. Entomol.* **65**, 422–3.

Walther, B.A. & Clayton, D.H. (1997) Dust-ruffling: a simple method for quantifying the ectoparasite loads of live birds. *J. Field Ornithology* **68**, 509–18.

Wardlow, L.R. & Jackson, A.W. (1984) Comparison of laboratory methods for assessing numbers of apple rust mite (*Aculus schlechtendali*) overwintering on apple. *Plant Pathol.* **33**, 57.

Wasylik, A. (1963) Metoda analizy ciaglej roztoczy gniazd ptasich. *Ekol. Polska B* **9**, 219–24.

Way, M.T. & Heathcote, G.D. (1966) Interactions of crop density of field beans, abundance of *Aphis fabae* Scop., virus incidence and aphid control by chemicals. *Ann. Appl. Biol.* **57**, 409–23.

White, T.C.R. (1975) A quantitative method of beating for sampling larvae of *Selidosema suavis* (Lepidoptera: Geometridae) in plantations in New Zealand. *Can. Entomol.* **107**, 403–12.

Wickman, B.E. (1964) A comparison of

radiographic and dissection methods for measuring siricid populations in wood. *Can. Entomol.* **96**, 508–10.

Wiegert, R.G. (1961) A simple apparatus for measuring density of insect population. *Ann. Entomol. Soc. Am.* **54**, 926–7.

Williams, C.B. & Milne, P.S. (1935) A mechanical insect trap. *Bull. Entomol. Res.* **26**, 543–51.

Williamson, K. (1954) The Fair Isle apparatus for collecting bird ectoparasites. *Br. Birds* **47**, 234–5.

Wilson, L.F. (1962) A portable device for mass-collecting or sampling foliage inhabiting arthropods. *J. Econ. Entomol.* **55**, 807–8.

Wood, G.W. & Small, D.N. (1970) A method of sampling for adults of *Chlamisus cribripennis. J. Econ. Entomol.* **63**, 1361–2.

Woodroffe, G.E. (1953) An ecological study of the insects and mites in the nests of certain birds in Britain. *Bull. Entomol. Res.* **44**, 739–72.

Wright, A.F. & Stuart, A.J.A. (1992) A study of the efficiency of a new inexpensive type of suction apparatus in quantitative sampling of grassland invertebrate populations. *Bull. Br. Ecol. Soc.* **23**, 116–20.

Yaghi, N. (1924) Application of the Roentgen ray tube to detection of boring insects. *J. Econ. Entomol.* **17**, 662–3.

5 Absolute Population Estimates by Sampling a Unit of Aquatic Habitat

Many of the methods described in this chapter have comparatively low efficiencies, so that the emphasis must be on the use of a unit of habitat, rather than on the absoluteness of the estimate. Sampling costs and difficulties frequently dictate the use of a low-efficiency method, which must be calibrated by a subsidiary study to obtain an efficiency conversion factor, by which the results can be adjusted to give an absolute estimate. The choice of method is determined by the nature of the habitat, the size and mobility of the animals, and the scale of the study. Marine studies frequently require larger, heavier, more complex samplers than those needed for fresh waters, but similar methodological approaches are applicable. As the primary determinant of the method is the type of habitat, methods will be classified under the following headings: (i) open water: inhabited by surface dwellers, swimming macrofauna, and planktonic animals; (ii) vegetation: animals living on or around submerged or floating plants; and (iii) bottom or benthic fauna: animals living on or in the substrate.

Methods for the study of the microfauna of inland or marine waters, which are not covered here, are introduced in works such as Edmondson and Winberg (1971), Baker and Wolff (1987), Ford (1993), and Pepper *et al.* (1995). In contrast to terrestrial habitats, major difficulties in making absolute estimates of the fauna lie in actually taking a sample of known size. As the problems of extraction, where this further operation is needed, are similar to those encountered with terrestrial samples, these are discussed together in Chapter 6.

5.1 Open water

5.1.1 Nets

5.1.1.1 Plankton

Nets allow quantitative sampling for average density; other methods (p. 183) must be used to study small-scale variation in plankton density. Typical plankton nets comprise a nylon-gauze cone attached to a metal frame (Fig. 5.1). The mouth may be fitted with a collar or mouth-reducing cone and the cod-end should have a sampling bucket. A flowmeter is attached in front of the mouth to measure the volume of water sampled. The main causes of sampling error are: (i) mesh penetration of

Fig. 5.1 Typical plankton nets. (a) A Birge cone tow-net suitable for fresh waters with macrophyte beds. (b) A modified working party 2 net used for vertical tows from depths of less than 200 m to the surface. (c) A Gulf III net used in marine environments for horizontal or oblique tows at speeds up to 3 m/s.

animals; (ii) net avoidance; and (iii) changes in filtration efficiency. These errors are minimized by the correct choice of net type, mesh size, and sampling speed. Care must be taken to ensure that the net is not moved too fast so that it 'pushes aside' some water and the animals in it; this is particularly important with fine-mesh nets (Ricker, 1938; Fujita, 1956); however, if the net is moved too slowly, the more agile animals with good vision may be able to avoid it. Filtration efficiency is defined as the volume swept out by the net, divided by the volume of water filtered. It is advisable to use a net with a mesh 75% of the minimum dimension of the animal to be sampled (Vannucci, 1968). Fine nets are easily clogged, and Omori and Ikeda (1984) advise the use of mesh sizes greater than $100\,\mu m$ and $200\,\mu m$ in ocean and coastal waters, respectively. For meso- and macroplankton, a $330\,\mu m$ mesh is commonly used. The open area ratio of a net is given by:

$$R = \frac{aP}{A}$$

where a is the surface area of the net, A the mouth area, and P the mesh porosity. Tranter and Smith (1968) found a $330\,\mu m$, 3.2 open area ratio net towed at 0.7–1.0 m/s had a filter efficiency >0.85 until it had filtered $49\,m^3$ of coastal water with abundant plankton. The volume of water sampled at >0.85 efficiency increased to $300\,m^3$ when the open area ratio was increased to 6.4. Omori and Ikeda (1984) state that a net should have an open area ratio >3.5 and preferably >6.0 if it is to be used with high plankton densities.

A plankton net may be towed vertically, horizontally, or obliquely (Fig. 5.2). If a sample is required from a set depth, the net must be fitted with a closing or opening–closing mechanism. An example is the Clarke–Bumpus sampler (Clarke & Bumpus, 1940; Tonolli, 1971), which is a normal cone-shaped plankton net; however, in its mouth, there is a metal cylinder with a propeller blade (which records the throughput of water), two stabilizing vanes, and a shutter mechanism, which is controlled by messenger which is released to slide down the tow line. Bongo (McGowan & Brown, 1966), MTD (Motoda, 1971), and ORI nets have also been fitted with opening and closing mechanisms.

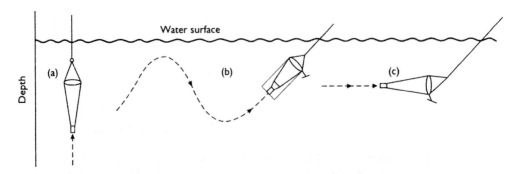

Fig. 5.2 Sampling profiles for plankton nets. (a) Vertical. (b) Vertical–oblique. (c) Horizontal.

Net avoidance is a serious problem when sampling larger plankton, which swim well, and has been much discussed with respect to larval fish, which can dart clear of the mouth of a net. The only solution is to use a large-mouthed net that can be trawled at high speed. The Isaacs–Kidd midwater trawl (Isaacs & Kidd, 1953), which can be towed at 2.5 m/s, has been widely used. High-speed samplers, such as the Gulf III (Gehringer, 1952), MTD underway plankton sampler V (Motoda, 1959), Miller sampler (Miller, 1961), and jet net (Clarke, 1964), have been developed for deploying from ships moving at up to 7 m/s. While, for some plankton, net avoidance is decreased using high-speed nets, the small-mouth diameters of these nets can increase avoidance in swift-swimming groups.

Specialist nets have been developed to sample close to the water surface (e.g. David, 1965; Matsuo *et al.*, 1976; Ellertson, 1977) and the sea- or lake-bed (Wickstead, 1953; Omori, 1969; Rothlisberg & Pearcy, 1977). In fresh waters, where nets can become quickly clogged, the Birge cone net, as modified by Wolcott (1901) (see Fig. 5.1), is a convenient tow-net; the anterior wire-mesh cone ensures that water-weed and other large debris do not enter the net, which is easily emptied by the removal of the bottom cap, conveniently made from a metal screw cap.

5.1.1.2 Particular methods for fish

The near-shore open beach is one of the few habitats where netting can quantitatively sample open-water fish. A beach seine, comprising a curtain of net stretching from the surface to the bed, is used to encircle a body of water and trap its fish, which are landed by pulling the net on to a beach. To be successful, water depth must be less than two-thirds of the depth of the net; otherwise, fish will escape under the lead line. The method is only applicable on beaches that shelve gently, lack underwater obstructions, such as fallen trees, and have little plant growth. The area swept by a seine can be increased by allowing it to drift with the flow parallel to the bank; however, it is unlikely that the sample will reflect the total area swept. Some fish, such as mullet, escape by jumping the float line; this loss can be avoided by simultaneously fishing two seines, one inside the other and kept sufficiently close together for the fish that jump the first net to have insufficient space to accelerate to jump the second. Beach seines range in size from those used to sample young fish, which may be only 2 m long, to large commercial nets 200 m long. On gently sloping beaches, drop or lift nets have been used to estimate fish densities. While these methods are some of the few that can give absolute estimates of species, such as young flatfish, their main use has been to calibrate more easily used techniques, such as beam trawling.

Electric fishing, which is normally undertaken within a limited stretch of stream defined by stop nets and uses a removal trapping approach, is described in Chapter 7 (p. 264). The use of beam and other trawls is discussed on p. 197.

5.1.2 Pumps

Pumps are often used when the objective is to sample plankton from a point locality or defined depth in shallow water. They are also advantageous in turbulent water, where the flowmeter in a plankton net will no longer work. The disadvantages of this technique are that the precise depth from which the water is drawn is unknown, some animals may react to the current and so avoid capture, and delicate animals may suffer mechanical damage and become lost. Miller and Judkins (1981) discuss pumping-system design, which can be classified into systems that use either submersible or deck pumps.

5.1.3 Water-sampling bottles

A known volume of surface water may be collected with a bucket. For subsurface water, one of the most frequently used samplers is the Van Dorn bottle (Van Dorn, 1957), which will collect 5–20 L samples from a selected depth (Fig. 5.3). The device consists of a plastic cylinder, the ends of which can be sealed by stoppers. The sampler is lowered on a cable with the stoppers open so that water can flow through the cylinder. When correctly positioned, sending a messenger down the cable closes the stoppers and the sample is hauled to the surface. It is normal to deploy a number of bottles on a single cable. The low volume of the samples limits its usefulness for population estimation.

Fig. 5.3 A Van Dorn water-sampling bottle.

5.1.4 The Patalas–Schindler volume sampler

This sampler, which is particularly useful for small-scale limnological studies (e.g. Hardy, 1992), is a hybrid between a water sampler and a plankton net. The main body is a perspex box with a capacity of 12 L, with a large flap valve on the upper and lower surfaces and a net attached to a side-wall (Fig. 5.4). The sampler is lowered to the desired depth and then raised swiftly so that the valves close, retaining the water sample. At the surface, the box is turned on its side and its water drained via the side net. This procedure can be repeated a number of times to filter a larger volume before the sample is collected from the sample pot.

5.1.5 Particular methods for insects

No absolute quantitative method has been devised for estimating surface-dwelling insects. The larger forms may be counted directly *in situ*; their numbers and dispersion can also be studied by photography and possibly nearest-neighbour techniques used in their estimation (p. 345). Pond dipping with a long-handled ladle is usually regarded as a relative method (p. 247), but Croset *et al.* (1976) found that, provided

Fig. 5.4 A cross-section of the Patalas–Schindler sampler. The sampler comprises a clear Perspex box, with valves on the upper and lower surfaces and a plankton net attached to the side. The sampler is lowered into the water, causing the valves to open. Once at the desired depth, the sampler is pulled swiftly to the surface, causing the valves to close. Once at the surface, the water in the sampler is filtered by running it through the attached plankton net and the sample removed.

the ladle was large enough (so that no more than 5% of samples were empty), this method could be regarded as sampling from a unit of the habitat (the volume of the ladle). Such samples provided population estimates of culicid larvae that were comparable to those from mark–recapture or removal trapping.

Because mosquito larvae dive to the bottom of a pond when disturbed and then gradually make their way up to the surface again, sampling by dipping with a net or strainer will give variable results, depending on the skill of the collector. Welch and James (1960) have, however, used this habit in the Belleville mosquito sampler (Fig. 5.5). The sampler consists of a cylinder, a cone, a concentrator, and a bucket. The cylinder is placed in the pond, its base firmly pushed on to the substrate, the cone slipped inside, and the apparatus left for 20 min. The concentrator and bucket are then fixed to the top and the whole is rapidly reversed. Most of the larvae are collected in the bucket as the water drains out through it and the concentrator; a few may remain stuck on the sides, so the apparatus should be rinsed.

The angle of the cone may have to be adjusted for different species; for *Aedes*, Welch and James found the minimum was 33°. The greater the angle of the cone, the greater its height; as the water must always be deeper than the height of the cone, this value will limit the depth of water in which the sampler can be used. Since the samplers need to be left for some time, a practical set of equipment for fieldwork would consist of 10 cylinders and cones with a single concentrator and bucket.

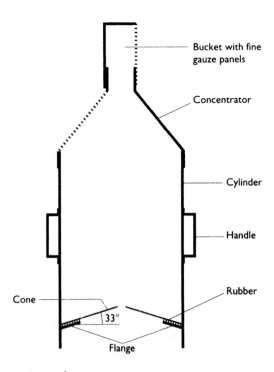

Fig. 5.5 Belleville mosquito sampler.

Laboratory tests gave disappointingly low efficiencies—in the region of 30–40%—and increasing the 'rising time' from 5 to 20 min did not markedly increase the number of larvae extracted. Welch and James suggest that the sampler be standardized by comparing the catch for a 24 h period with that for a shorter time. The series of catches may then be corrected to give an absolute estimate. A number of samples from a restricted habitat could be treated as a removal-trapping experiment to obtain an absolute population estimate (p. 255). With small bodies of water, e.g. mosquito-breeding pools, it may be possible to empty the complete pond, carefully separating the animals by a series of sieves, the contents of each sieve being sorted in a large pan. As developed by Christie (1954), the method allows the mosquito larvae to be returned to the pond after enumeration.

5.2 Vegetation

Besides the more general situation presented by rooted, but submerged, vegetation, particular solutions to the sampling problem have been developed for use in emergent rooted vegetation (e.g. reed beds) and with floating vegetation. Floating meadows, composed of grasses, are particularly extensive in tropical South America. The methods described below are loosley grouped according to habitat, but may be applicable in other situations, depending on the nature of the organisms being sampled.

Animals among floating vegetation are frequently sampled by dipping with a pond net or a strainer, but such methods usually give only relative estimates of population density. For absolute estimates, it is necessary to enclose a unit volume of the vegetation and associated water. When this has been strained, the organisms are usually separated by hand-sorting; however, some of the methods described in Chapter 4 might be found useful (e.g. clearing and staining for eggs in water plants). Extraction by spraying the material with a fine mist, as used in plant nematode work, might also be found applicable (p. 238).

5.2.1 Emergent vegetation

Perhaps the simplest practicable sampling device for zooplankton in vegetated littoral areas is the littoral sampling tube of Pennak (1962). This consists of a length of 64 mm diameter light-weight tubing (Pennak used a tube made of rubber-impregnated cloth with spiral reinforcing to maintain a constant diameter). To one end is attached some plankton net mesh, which can be removed, and to the open end a recovery line (Fig. 5.6). The sample is taken by lowering the open end of the tube vertically to the chosen depth, ensuring that the mesh end remains in the air. Holding the mesh end steady, the recovery line is used to raise the open end to the surface, when the mesh end is released. The tube is then raised and the plankton retained on the mesh washed into a sample holder. Tubes up to 9 m long can be constructed and the method is applicable to habitats other than amongst aquatic vegetation, e.g. from under ice.

Fig. 5.6 (a) Taking samples with a littoral sampling tube; (b) schematic view of the multiple sampling device: 1. Handle; 2. Clear sampling tube; 3. Tube spacer; 4. Tube clamp; 5. Clamp strap; 6. Spring.

A number of simple devices have been used to sample amongst flooded rushes and grasses. Swanson (1978) used a column sampler constructed from a 135 cm long, 6.5 cm diameter acrylic tube, with a disc cemented at one end with a 1.5 cm hole for a rubber bung (Fig. 5.6). The open end was sharpened to give a cutting edge. Handles were attached near the centre of the tube. Working in water <1 m deep, samples were collected by first pushing the tube into the substrate. The tube was then closed with a bung and the water column raised, a hand placed over the open end as soon as possible, and the tube up-ended. The multiple-tube sampler (Fig. 5.6), described by Euliss *et al.* (1992), for collecting both water-column and benthic invertebrates in shallow water, works on a similar principal. This simple multiple corer was used because it could sample the benthic and water-column fauna simultaneously. This allowed superior estimates of invertebrate density to those obtained when using separate devices for the benthos and water column, which both tended to catch animals such as chironomids and oligochaetes, which live at the sediment–water interface and thus, when added together, gave a biased estimate.

The Wisconsin trap and Gerking sampler (see below) may also be used with emergent vegetation.

5.2.2 Submerged vegetation

Sampling cages may be used to enclose a column of water of a known volume (James & Nicholls, 1961). It is a screen cage that is pushed down into the mud; the enclosed water is hand-sieved, and then the substrate is dredged up and sorted on sorting trays attached to the sides of the cage. There is always the danger with this method that some animals will be missed in the hand-sieving. Bates (1941) and Goodwin and Eyles (1942) describe earlier versions of this sampling device.

5.2.2.1 Wisconsin trap

As described by Welch (1948), this is simply a canvas and gauze net with a closable mouth (Fig. 5.7). The trap is lowered over the vegetation; the jaws of the trap are closed just above the substrate, the plants being uprooted or cut off. The whole is then raised from the water and drained.

5.2.2.2 The Gerking sampler

Designed by Gerking (1957) for the sampling of littoral macrofauna, this equipment would seem to be satisfactory, apart from the labour involved, for obtaining absolute

Fig. 5.7 (a) Gerking sampler: sketch of the outer shell and cross-section. (b) Hess sampling cylinder for floating vegetation. (c) Wisconsin trap: sketch of jaws and cross-section.

samples of all but the most active animals. It consists of two galvanized-iron shells, each with a square cross-section; they may be nearly 1 m high if necessary (Fig. 5.7). The two shells of the sampler are placed in position, the lower edges of the outer shell being forced into the mud. One side of the base of the outer shell is open (see sketch, Fig. 5.7), and through this opening the stems of the water plants are quickly cut with a pair of grass shears. The metal and screen sliding door is then inserted and this effectively closes the inner shell, which may be slowly raised. The outer cylinder is left in position and Gerking used an Ekman dredge to remove the substrate from within the same area; alternatively, it could be scooped out. If it is not desired to sample the substrate simultaneously, then only a single shell need be used; it should correspond in shape to the outer shell, but have a number of gauze panels.

5.2.2.3 McCauley's samplers

McCauley (1975) described two samplers for the macroinvertebrates on aquatic vegetation. The general model (Fig. 5.8) for submerged and floating vegetation can

Fig. 5.8 (a) Hess vegetation sampling cylinder. (b) McCauley's sampler for submerged and floating vegetation (after McCauley, 1975).

be conveniently operated from the back of a boat. The lower plate of the sampler has a cutting edge, it is moved into position as close to the bottom as possible. When this has been done, the sampling cylinder is released (triggered) and the pair of powerful springs force it down on to the lower plate. The plants are removed and the contents pumped out, using a hand-operated plastic bilge pump, and filtered through a fine-mesh sieve. The sampler may then be removed and the inside washed down to remove any remaining animals.

McCauley's sampler for rushes consists of a length of acrylic tubing, which encloses the sample, and a sharpened brass plate, which is released by a trigger and cuts the bases of the rushes, rather like the sliding plate of the Gerking sampler.

5.2.3 Sampling floating vegetation

One of the simplest devices is the sampling cylinder, described by Hess (1941), which consists of a stout galvanized cylinder with copper-mesh screening (Fig. 5.8). This is lowered under the vegetation, moved into position, raised, and any plant stems crossing its edge cut by striking the edges of the cylinder with a wooden paddle. If necessary, this could be aided by sharpening the upper edges of the cylin-der. The cylinder is then raised from the water and the plants and animals are retained within it.

The McCauley sampler (see above) may be used with floating vegetation, but such methods are only practical when the layer of vegetation is thin and easily cut.

Floating vegetation can form thick mats in some habitats, such as tropical lakes. The fish and macrocrustaceans living in floating-grass, *Paspalum repens*, meadows of Amazonian flood-plains can be quantitatively sampled using a seine net to encircle an area of 2–12 m^2 of the vegetation and then passing the lead line under the root mat to close the bag prior to lifting the entire mass into a boat for sorting (Henderson & Hamilton, 1995; Henderson & Crampton, 1997). This method works because, at high water, the roots are detached from the substrate and the associated fauna is loath to flee into open water when disturbed. For insects and other invertebrates in the same habitat, Junk (1973) used a device, attached to the front of a boat, which could be used to take 'bites' from the outer edge of a floating meadow. A hand-net of 0.1 m^2 mouth area with a powerful frame, so that it could be thrust under the roots of floating plants and levered swiftly to the surface, was used for insects, crustaceans, and small fish by Henderson and Crampton (1997).

5.3 Bottom fauna

A large number of methods have been developed for sampling the bottom-fauna aquatic habitats; one early reviewer has commented that the number of samplers is nearly proportional to the number of investigators (Cummins, 1962)! This fact arises because five variables affect the choice of sampler: (i) the animal; (ii) the nature of the bottom substrate—soft or hard; (iii) the current; (iv) the depth of the water; (v) the object of the study, e.g. a survey for pollution, an assessment of the food potentially available for fish, or an intensive ecological investigation for the development of a life-table for a single species.

A few generalizations are possible. Fast-flowing water has the advantage that the current may be used to carry animals, disturbed from the substrate, into a sampler (as with the Surber sampler, p. 194); however, it limits the use of devices that enclose a unit area, for, as these are lowered, the increased current immediately beneath them may scour the organisms from the very area that is to be sampled.

After the sample has been taken, the problem remains of separating the animals from the substrate. If the sample is too large or contains too many organisms, it may be divided and subsampled (see also p. 154). The material needs to be thoroughly agitated for satisfactory subsampling. Huckley (1975) describes a sample-splitter that uses compressed air for mixing. Hand-sorting is the most widely used method (Frost, 1971); its accuracy may be checked by staining any remaining animals with rhodamine B and viewing under ultraviolet (UV) light (Eckblad, 1973). Methods for sorting animals from sediment are discussed in Chapter 6.

Benthic sampling is often undertaken in open water with no absolute visual points of reference, so it is important at the outset to consider how the position at which the samples are taken is to be determined. If repeat sampling is planned, then the accuracy and repeatability of the method to return to the same location must also considered. Global positioning systems (GPS) have made the determination of absolute position to ±100 m generally available. Older methods for position fixing and also those for depth determination are reviewed by Holme and Willerton (1984). When sampling in the littoral zone, it is particularly important to find the level, as this determines the time the site will be exposed. Benthic communities are partially determined by sediment type, which is often collected as part of a benthic sample; methods for sediment analysis are reviewed by Buchanan (1984).

Eleftheriou and Holme (1984) and Baker and Wolff (1987) review methods for marine benthic sampling, and Elliott and Tullet (1978) present a bibliography of benthic sampling methods for fresh waters. Sampling efficiencies for grabs and air-lifts are assessed by Elliott and Drake (1981) and Drake and Elliott (1982), respectively, while sampling in stony streams has been reviewed by Macan (1958). Baker *et al.* (1987) discuss the type of sampling programme and pattern. Because benthic animals are usually particular in their substrate requirements, it is important to characterize the substrate at the sampling locality. Kajak (1963) discusses the problems of sample-unit size and the number of samples (see Chapter 2). While some marine habitats comprising a single sediment type can show little spatial variability in their fauna, other habitats, such as streams, have tremendous variability of microhabitats and the variance of a series of samples is often extremely large.

5.3.1 Hand-net sampling of forest litter
In shallow waters, a hand net may be the most practical way to lift a portion of a sub-

strate composed of plants or forest litter. Henderson and Walker (1986) used shallow hand-nets of $0.1\,m^2$ area to quantitatively sample the insect and fish infaunal community of Amazonian stream litter banks. The net was rapidly thrust into the side of a submerged litter bank and rapidly raised vertically to the surface.

5.3.2 Lifting stones

The majority of the animals in fast-flowing streams will be underneath the stones on the bottom. Scott and Rushforth (1959) have investigated from the mathematical angle the influence of stone size and spacing on the area of the stream bed covered by stones. They propose the symbol C, for this parameter and discuss its value. An estimate of the absolute population of animals may be obtained by picking up the stones from an area of the bottom; a net should be held on the downstream side to catch those animals washed off (Macan, 1958; Ulfstrand, 1968). If an estimate of population intensity (p. 2) is to be made, it will be necessary to make an estimate of the effective surface area of the stones. Calow (1972) showed that this could be calculated if the longest perimeter of the stone (x) was found; then:

surface area $= 2.22 + 0.26x$

Some animals may be especially difficult to remove from the stones and Britt (1955) found that these would become active if placed in a very weak acid–alcohol solution (2–5% alcohol, 0.03–0.06% hydrochloric acid). A flotation solution (p. 221) might serve the same purpose and separate the animals as well.

Alternatively, a net may be fixed in position, the stones and remainder of the substrate are disturbed from a unit area upstream, and the animals caught in the net. The simplest approach is the 'kicking technique', when the stones immediately upstream are 'kicked'. Frost et al. (1971) found that the first kick disturbed about 60% of the fauna yielded by 10 kicks; this is essentially a catch per unit effort (p. 262). The same principle, but with a more precise delimitation of the area, is used in the Surber sampler (Surber, 1936) (Fig. 5.9). The variability of this technique, with operator and stream conditions, and its efficiency have been tested by Needham and Usinger (1956). As some stones will lie partly inside and partly outside the frame, the selection and rejection of these is a matter of personal judgement, and Usinger and Needham (1954) found that one operator continually sampled half as much area again as the other four operators, but the latter were consistent. Comparison of the results from the Surber sampler with absolute counts of animals from buried trays (see above) showed that it caught only about a quarter of the population. The exact proportion will vary from site to site, but it is obvious that the Surber sampler cannot be used for absolute population estimates from a shallow stream without a careful test of its efficiency to determine what correction factor should be applied (Kroger, 1972, 1974).

5.3.3 The planting of removable portions of the substrate

One of the most accurate methods of sampling the bottom fauna is to place a bag,

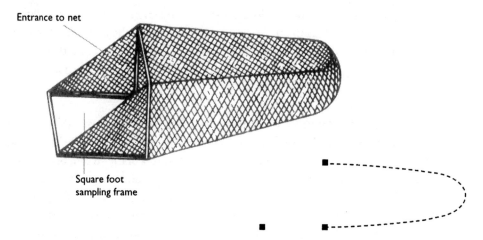

Fig. 5.9 Surber sampler: sketch and cross-section.

tray, or box in the stream or pond bed and either replace the substrate or allow the sediment to accumulate naturally (Usinger & Needham, 1956; Ford, 1962). An elaborate apparatus of this type is the box of Ford. This has two fixed wooden sides and a bottom and is placed in a hole in the bed of the stream, the two sides being parallel to the course of the water. After a suitable time (Ford left it for 6 weeks), the other two sides, which are made of Perspex, are slid into position; the box is then made watertight and may be lifted from the stream bed with the sample undisturbed. In Ford's model, it was possible to divide the sample horizontally so that each stratum could be separately analysed. One of the wooden sides had a series of slots in it; the egress of mud and water through these was prevented by rubber flaps on the inside and the other sides were grooved at the corresponding level; Perspex sheets could then be pushed through these slots, dividing the sample. Simpler 'basket samplers' have been widely used and are generally found to give satisfactory assessments of the macroinvertebrate fauna (Mason *et al.*, 1973; Crossman & Cairns, 1974). A larger framed net, which may be rapidly pulled to the surface, has been designed for use in the shallow waters of the Florida Everglades (Higer & Kolipinski, 1967).

The fauna of rocks or concrete substrates may be studied by placing easily removable blocks or plates of similar composition. Britt (1955) placed blocks of concrete, heavily scored on the undersurface, on a rubble and gravel bottom in deep water. The blocks can be easily located and raised with the help of a buoy and cord, but some animals may be lost as the block is raised. This loss is overcome in Mundie's (1956) method, devised for the study of Chironomidae on the sloping slides of artificial reservoirs. The artificial substrate is a plate of asbestos-cement composition (Fig. 5.10); attached to the centre of the plate is a guide wire, which must be attached to a line above the water surface. The plates are left for the period of time necessary for them to be indistinguishable from the surrounding substrate (1–3 months) and then retrieved. Mundie's ingenious method of retrieval was as

Fig. 5.10 Artificial substrate and sampling apparatus (after Mundie, 1956). (a) Artificial substrate plate with central pipe, to which wire is attached. (b) Retrieving cone. (c) Gauze-bottomed landing bucket.

follows: a retrieving cone (Fig. 5.10) is slid down the guideline; this cone has flap-valves on the upper surface and these open as it descends, ensuring that there is no surge of water to disturb the sample as the cone settles; rubber flanges around these valves and the base of the cone protect the sample from disturbance as the plate is gradually hauled up by the guide wire. Before the sample breaks the surface, a gauze-bottomed bucket (Fig. 5.10) is placed below it; this retains any organisms washed out as the water in the sampler drains through the flap valves of the cone.

Completely artificial substrates may also be used, but these give only relative estimates, as settlement on them will be different from that on the natural substrate. These standardized artificial substrate units (SASUs) have been found particularly useful for estimating relative changes in the numbers of blackflies (Simuliidae) (Wolfe & Peterson, 1958; Williams & Obeng, 1962) and Disney (1972) points out that the numbers obtained will depend on: (i) the absolute population of the black-flies; (ii) the relative area of other suitable substrata; (iii) the length of time the artificial substrate units are exposed; (iv) the intensity and nature of the factors that cause the larval blackflies to move from the substrate where they are already established; and (v) the nature of the substrate and its acceptability. The latter effect will vary

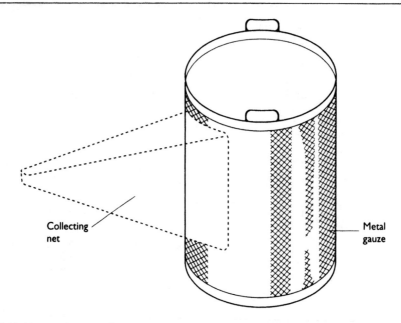

Collecting
net

Metal
gauze

Fig.5.11 Hess circular square-foot sampler.

with the animal and the substrate (Glime & Clemons, 1972; Mason *et al.*, 1973; Benfield *et al.*, 1974).

5.3.4 Cylinders and boxes for delimiting an area

As pointed out above, when a box or cylinder is lowered into flowing water, as it approaches the bottom it is likely to cause this to be scoured. This problem is least serious with the Hess circular square-foot sampler (Hess, 1941) (Fig. 5.11), which is made of a fairly coarse mesh and may be rapidly turned so as to sink a little way into the bottom. The smaller organisms may be lost through the coarse grid. If a finer grid is used, scouring becomes progressively more serious; on the other hand, drifting animals may pass through a coarse net and add cumulatively to the sample. Therefore Waters and Knapp (1961) designed a circular sampler with fine-mesh screening; the emptying of the collecting bag was facilitated by attaching it to the sleeve with a zip fastener. The use of these samplers is restricted to streams, as a current is necessary to carry the organisms into the collecting bag. The Wilding sampler (Wilding, 1940; see also Welch, 1948) may be used in still or moving water. It consists of two cylinders; the outer cylinder has finely perforated sides and a band of saw-teeth along the lower edge to aid penetration into the bed of the stream or lake (these are not essential). This cylinder delimits the area, and the larger organisms, rocks, and other debris are removed; then the whole is stirred and the inner cylinder (of slightly smaller diameter) lowered inside. The inner cylinder has a rotary valve at the bottom through which virtually all the water in the outer cylinder enters it (Fig. 5.12a,b). When it is resting on the bottom, the rotary valve is closed and the inner cylinder, containing all the water and smaller

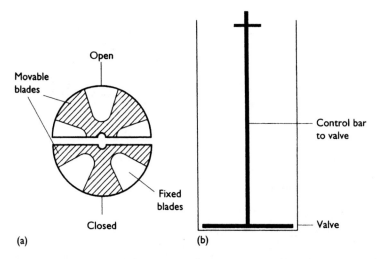

Fig. 5.12 Simplified diagram of the Wilding sampler. (a) Diagram of rotary valve, showing open (upper half) and closed (lower half) positions. (b) Section of cylinder.

animals, removed. Its contents can be passed through a nest of sieves by opening the valve slowly.

The Gerking sampler and cages described above (p. 189) may also be used for delimiting an area of bed for sampling, and simple metal frames and cylinders were used for this purpose by Scott (1958) and Dunn (1961).

5.3.5 Trawls, bottom sledges, and dredges

Marine epifauna may be quantitatively sampled by beam, Agissiz and otter trawls (Fig. 5.13); however, their efficiency is generally low and they are species-selective (Mason *et al.*, 1979), so they are probably best considered relative methods. Numerous adaptations have been made to increase efficiency, including the use of additions, such as tickler chains, which scare animals burrowing into the substrate up into the mouth of the net. For mobile animals, efficiency can change greatly between day and night. An additional problem is the measurement of the area swept, for which the commonest solution is to fit a measuring wheel to the trawl or dredge frame (see Eleftheriou & Holme, 1984).

Dredges are heavy devices designed to scrape organisms off hard surfaces while not digging deeply into the substrate (Fig. 5.13). Examples include the naturalist's or rectangular dredge, rock dredge, and sand dredge. Like trawls, dredges have low sampling efficiency. Dickie (1955) found a specialized scallop dredge to range in efficiency from 5 to 12%. For epifauna in general, Richards and Riley (1967) consider efficiency to be about 10%. Given their low efficiencies, trawls and dredges are at best semiquantitative. They are, however, extremely useful for preliminary surveys to determine substrate and community composition.

Bottom sledges have been designed to glide over the substrate while sampling the epibenthic animals and proximobenthic plankton. They often include opening

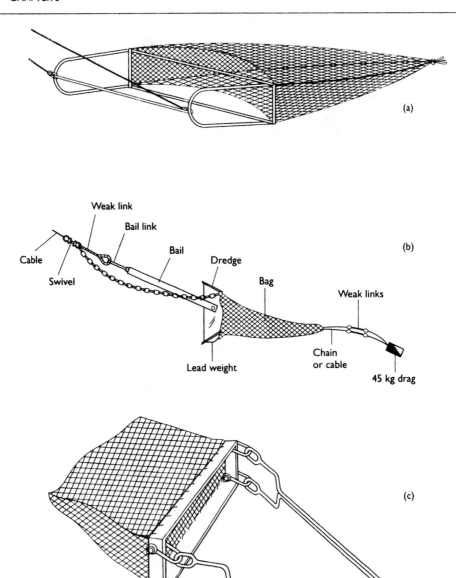

Fig. 5.13 Trawls and dredges for epifauna. (a) Agassiz trawl. (b) Rock dredge (from Nalwalk *et al.*, 1962). (c) Naturalist's or rectangular dredge. Note weak link of twine joining one arm to the ring.

and closing mechanisms, to control when the sample is taken, and water flowmeters. A typical example is the 'suprabenthic' sampler (Brunel *et al.*, 1978), which is a modification of the sledge design used by Macer (1967). Colman and Segrove (1955) describe a hand-towed sledge net for shallow-water sampling.

In fresh waters, the use of the Surber sampler and the various sampling cylinders is limited to shallow waters (the depth of an arm's length or less); in deeper waters, either a moving net (drag, scoop, shovel) or a metal sampling box (dredge, grab, etc. — see below) must be used. Several nets that may be pushed or pulled have been described (see Welch, 1948; Macan, 1958; Albrecht, 1959). One of the most robust models is that devised by Usinger and Needham (1954, 1956) (Fig. 5.14) for sampling stone or gravel bottoms (none of these movable nets are really suitable for soft mud). The tines on the mouth disturb the substrate, but prevent large stones and debris from entering the bag, and the weights and the heavy steel frame ensure that the tines really scour the bottom. The bag is attached by means of a brass zip-fastener and protected from tearing by a canvas sleeve. Usinger and Needham (1954) found that, under the conditions they were working in, this drag caught about one-quarter of the animals from the area it traversed. Therefore, unless standardized, this is only a relative method. A modification of this drag, particularly suitable for sampling unionid mussels, has the tines removed, an adjustable blade attached to the lower edge, and a vertical handle (in addition to the tow-rope) fixed to the top leading edge (Negus, 1966).

5.3.6 Grabs

Samplers with closing jaws are called grabs. They are often the method of choice for quantitative infaunal sampling. The basic principle is to lower the grab to the substrate surface, where the grab is activated so that the jaws cut into and enclose a sample of the substrate, which can then be raised to the surface (Fig. 5.15). The fundamental problems in grab design are to ensure, first, that it rests correctly on the substrate, so that when the jaws are closed the designed volume of substrate is taken, and, secondly, that the jaws fully close to retain the sample during recovery. Wigley (1967) used motion pictures to determine the effectiveness of closing of various designs of Petersen's grab.

More than 60 types of grab have been described (Elliott & Tullett, 1978). These tend to be variants of five basic types: Peterson, van Veen, Birge–Ekman, Smith–McIntryre, and orange-peel (Fig. 5.16). The choice of grab will depend on depth, flow, substrate, weather (sea conditions), and the type of craft or platform from which it will be operated. Eleftheriou and Holme (1984) and Elliott and Tullett (1978) give extensive bibliographies of comparative sampling efficiency. Efficiency of capture of animals is only one criterion that needs to be considered. Others include substrate penetration characteristics, sample size, reliability of operation, and ease of handling.

In marine habitats with large boats and mechanical winches, heavy grabs such as Petersen (Petersen & Boysen Jensen, 1911), Okean (Lisitsyn & Udintsev, 1955), van Veen (van Veen, 1933), Ponar (Powers & Robertson, 1967), Hunter (Hunter & Simpson, 1976), Smith–McIntyre (Smith & McIntyre, 1954), Day, orange-peel (Brida & Reys, 1966), Baird (Baird, 1958), Hamon, and Holme (Holme, 1949), can be used. Grab sampling at sea requires experience, and specialist advice should be sought; a general introduction to the subject is given in Eleftheriou and Holme

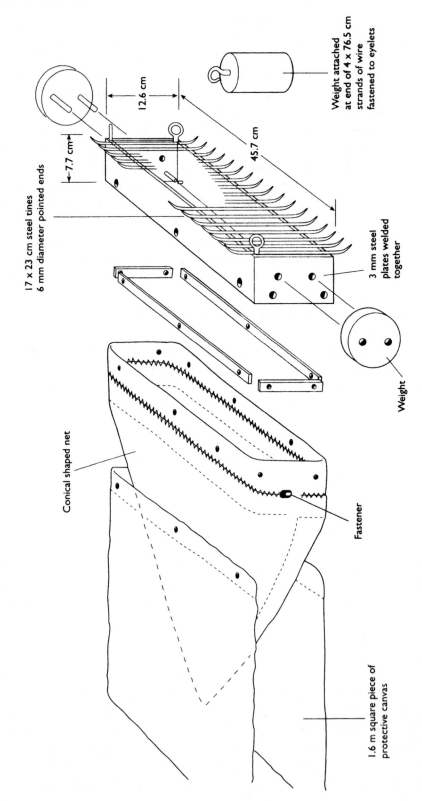

Fig. 5.14 Usinger and Needham's drag for freshwater sampling (after Usinger & Needham, 1956).

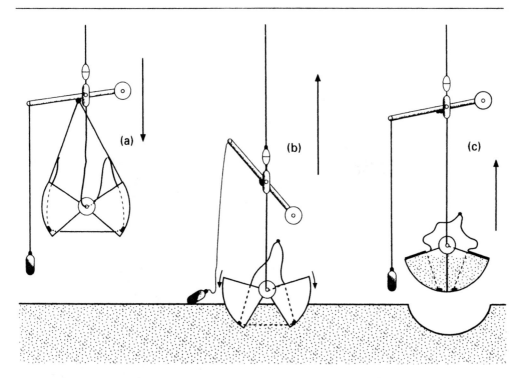

Fig. 5.15 The operation of a Petersen grab, showing a method of closure that uses a counterpoised weight (after Lisitsyn & Udintsev, 1955).

(1984). Marine grabs can rarely penetrate deeper than 10–15 cm and thus will not sample deep-burrowing animals, which can be sampled with a Foster anchor dredge, box-corer, or suction sampler. Further, they will rarely capture faster-moving epibenthic animals, such as crabs, which should be quantified using photographic, video, or trawl methods.

Lighter grabs for use from small craft in ponds, rivers, and lakes include van Veen (21 kg), Ponar (weighted 23 kg, unweighted 14 kg), Birge–Ekman (7 kg), Friedinger (8 kg), Dietz–La Fond (21 kg), and Allan (8 kg). In water up to 3 m deep, the Birge–Ekmann and Allen grabs can be pole-operated. Elliott and Drake (1981) compared the performance of all these grabs for freshwater work and concluded that the Ponar grabs performed best with mud or fine gravel bottoms and were the only design to sample at >3 cm depth when small stones (8–16 mm largest dimension) were present. When stones >16 mm (largest dimension) were present, only the weighted Ponar sampled to a depth of 2 cm. No grab performed well when the substrate was predominantly stones. Ponar, Birge–Ekman, and Allen grabs performed well in muddy substrates (particle size 0.004–0.06 mm). Freshwater-grab capture efficiency for non-mobile objects in a fine gravel substrate is tabulated in Table 5.1.

The Ekman grab has been compared with various corers for sampling chironomid larvae in mud; Milbrink and Wiederholm (1973) found little difference, but

Fig. 5.16 The basic types of grab design. (a) Peterson. (b) van Veen. (c) Birge–Ekman. (d) Smith–McIntyre.

Karlsson *et al.* (1976) found that coring followed by separation by flotation (p. 221) recovered more larvae.

5.3.7 Dendy (1944) inverting sampler

Freshwater substrates comprising mud with fine gravel or small sticks may be

Table 5.1 Efficiency of small grabs suitable for freshwater work measured as the percentage of pellets collected from a fine gravel substrate (from Elliott & Drake, 1981).

Grab	Efficiency at the surface	Efficiency at c. 3 cm	Area sampled per sample (m^2)
Ponar	100	70	0.056
Van Veen	87	56	0.1
Birge–Ekman	73	37	0.023
Allan	51	36	0.035
Friedinger	59	7	0.03
Dietz–Le Frond	22	26	0.016

sampled by the Dendy (1944) inverting sampler (Fig. 5.17). Sticks and twigs are a particular problem for grabs, as they stop the jaws closing. This apparatus consists of a brass cylinder of about 8 cm diameter, its top covered with a piece of brass gauze. The cylinder is on the end of a long handle (this limits the depth of operation), which is used to drive it into the substrate; the pull cord is then tugged and the cylinder inverted and lifted to the surface. Only material that can pass out through the gauze is lost in the ascent (apart from a small amount from the very bottom of the sample). The small size of this sampler relative to the Ekman grab may indicate its use in preference to the latter from cost and sampling-pattern considerations (see p. 20). But it does have the disadvantage of disturbing the light layer on the surface of the bottom and, for this reason, Kajak (1971) recommends against its use.

5.3.8 Box samplers and corers

Samples from sand or mud beaches are obtained by driving a high-sided square or round cross-section frame into the substrate. Typical areas for square-sided frames are 0.1–0.25 m² and the sample is taken by digging out the substrate to the desired depth. Round cross-section samplers, which can simply be a length of pipe with the rim sharpened at one end, usually take a smaller sample, which can be lifted undisturbed, although Grange and Anderson (1976) used a corer of 0.1 m² cross-section to a depth of 0.1 m on some beaches. To aid the retention of the sample as the corer is withdrawn from the substrate, the tube may be fitted with a bung. Divers can also use corers (Gamble, 1984).

In large-scale marine studies, box corers (Reineck, 1963; Bouma & Marshall, 1964) are designed to push a corer into the substrate, and the corer is then closed prior to retrieval. For macrofaunal studies, corers must be large, heavy devices (400–750 kg), which can only be operated from large vessels. When available and practicable, they are considered amongst the most reliable and efficient samplers of soft sediments. Large-diameter geological corers may be suitable for biological sampling but have been little used (Eleftheriou & Holme, 1984).

Piston-type samplers have been developed for small-scale, generally freshwater, use. These enable a vertical core to be taken from the bottom and are ideal for the

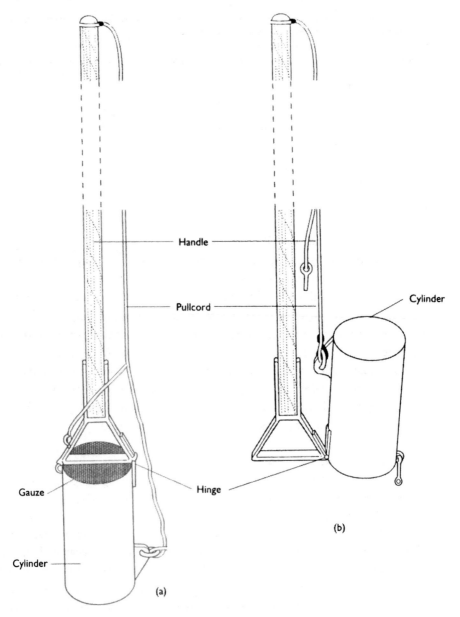

Fig. 5.17 Dendy inverting sampler (after Dendy, 1944). (a) Position when entering mud. (b) Inverted, ready to raise to the surface.

study of stratification of micro-organisms and for the assessment of the physical and chemical properties of the various layers. One of the earliest was the Jenkin surface-mud sampler (Jenkin & Mortimer, 1938; Mortimer, 1942; Goulder, 1971; Milbrink, 1971; Fig. 5.18).

Elgmork (1962) modified the Freidinger water sampler to take samples of soft mud; after sampling, a piston was inserted and the sample slowly pushed out.

Fig. 5.18 Modified Jenkin surface-mud sampler (after Goulder, 1971), with horizontal steel plates in position to isolate slices of sediment.

The stratification of *Chaoborus punctipennis* was studied with this apparatus. The Kajak–Brinkhurst sampler (Kajak, 1971) is very suitable for soft sediments; the lid may be closed by a messenger after the sample is taken. It has been found to be nearly as efficient as using a diver (Holopainen & Sarvala, 1975). A multiple corer has been developed by Hamilton *et al.* (1970).

Livingston (1955) described a piston sampler, particularly well adapted for obtaining the lower layers of sediment from a muddy bottom; Vallentyne (1955) and Rowley and Dahl (1956) introduced modifications that made the construction simpler and less costly, and Brown (1956) has modified it to sample the mud–water interface as well and, by the use of a clear plastic tube, to allow the inspection of the intact core.

A different approach is that of Shapiro (1958), who surrounded the sampling tube (corer) with a tapered jacket. Just before lowering, the jacket is filled with crushed solid carbon dioxide and *n*-butyl alcohol. It is lowered quickly through the water, allowed to settle, left for about 5 min, and retrieved. The freezing mixture is replaced with water; the frozen core may soon be slid out and divided and handled in the solid state. As pointed out by Kajak (1971), soil corers, especially O'Connor's (p. 214), are very useful for hard bottoms in shallow water.

5.3.9 Airlift and suction devices

Airlifts, which are used in marine archaeology, mining etc., use the principle that, when compressed air is discharged under water, an air–water mixture is formed which rises to the surface. A typical application is the diver-operated airlift sampler described by Barnett and Hardy (1967). This works by sucking material up a long vertical tube into a collecting sieve by the injection into the tube of compressed air (Fig. 5.19). This airlift can be used to remove the sediment from within a core manually pushed into the substrate. Mackey (1972) describes a simple apparatus, essentially for taking core samples. Compressed air is fed to the centre of the base of a length of plastic drainpipe, the top of the pipe being angled and opening into a collecting net or sieve. A more elaborate device, in which the sampling area is delimited by a rectangular frame and the compressed air is liberated through a series of openings (Fig. 5.19), was developed by Pearson *et al.* (1973). They compared its efficiency with the Suber sampler and the Allen grab; overall, there were not great differences, but particular animals were sampled in greater numbers by one or other method. The airlift sampled significantly larger numbers of *Polycelis*, *Ephemerella*, *Sphaerium*, and *Caenis*. Drake and Elliott (1982) review three airlift samplers for use in rivers.

A number of samplers employ suction either to force a coring tube into the substrate or to draw the sediment into a collector. The Knudsen sampler (Knudsen, 1927; Barnett, 1969), which operates by sucking a coring tube into the substrate, using a cable-operated pump, was the first sampler of this type designed. For shallow waters, Kaplan *et al.* (1974) and Thayer *et al.* (1975) describe samplers that work on the same principle. More recently, a number of samplers use pumped water. The

Fig. 5.19 Air-lift suction sampler (after Barnett & Hardy, 1967). (a) Diagram of the apparatus. (b) Method of use, with a diver extracting the contents of a sampling cylinder pushed into the substrate.

benthic suction sampler of True *et al.* (1968) sucks a $0.1\,\mathrm{m}^2$ core into the substrate, while the substrate is drawn up into a collecting basket. Many other suction samplers have been developed (e.g. Emig & Lienhart, 1971; Mulder & Arkel, 1980; Grussendorf, 1981) (Fig. 5.20). Suction samplers have also been designed to be slowly towed along the bottom. Allen and Hudson (1970) used a vacuum sledge to quantify buried pink shrimp.

5.4 Poisons and anaesthetics used for sampling fish in rock pools and small ponds

Chemofishing has long been practised by humans and has been effectively used by ecologists seeking quantitative estimates of fish number in small, isolated bodies of

Fig. 5.20 Suction sampler of Emig (1977). (a) In profile; (b) from below. Air introduced through the small tube produces suction in the central tube, through which sediment fauna is drawn up. The five compensating tubes (e) and the gap between cylinder and cone (g) are provided not only to enable the tube to dig into the substatum but also to help bring the sediment into suspension so that it is more easily collected. Diameters of various tubes (in cm) are shown.

water, such as rock pools and small ponds. In the past, rotenone, which almost inevitably killed the fish, was widely used; more recently, fish anaesthetics, such as MS 222, have been the method of choice. While, in theory, these compounds should allow the fish to recover when placed in clean water, this is far harder to achieve in practice. For example, it can be difficult to capture fish after exposure, as they may not float to the surface, resulting in them receiving a lethal exposure. Further, recovered fish returned to the wild are often highly vulnerable to predation from birds and other fish.

References

Albrecht, M.-L. (1959) Die quantitative Untersuchung der Bodenfauna fliessender Gewässer Untersuchungsmethoden und Arbeitsergebnisse). *Z. Fisch. (NF)* **8**, 481–550.

Allen, D.M. & Hudson, J.H. (1970) *A Sled-mounted Suction Sampler for Benthic Organisms.* Special Report—Fisheries, United States Fish and Wildlife Service.

Baird, R.H. (1958) *A Preliminary Report on a New Half Square Metre Bottom Sampler.* CM 1958/70, Shellfish Committee, International Council for the Exploration of the Sea.

Baker, J.M. & Wolff, W.J. (eds) (1987) *Biological*

Surveys of Estuaries and Coasts. Cambridge University Press, Cambridge.

Baker, J.M., Hartley, J.P., & Dicks, B. (1987) Planning biological surveys. In: Baker, J.M. & Wolff, W.J. (eds) *Biological Surveys of Estuaries and Coasts.* Cambridge University Press, Cambridge, pp. 1–26.

Barnett, P.R.O. (1969) A stabilizing framework for the Knudsen bottom sampler. *Limnol. Oceanogr.* **14**, 648–9.

Barnett, P.R.O. & Hardy, B.L.S. (1967) A diver-operated quantitative bottom sampler for sand macrofaunas. *Helgoland. wissensch. Meeresuntersuch.* **15**, 390–8.

Fig. 5.20 Suction sampler of Emig (1977). (a) In profile; (b) from below. Air introduced through the small tube produces suction in the central tube, through which sediment fauna is drawn up. The five compensating tubes (e) and the gap between cylinder and cone (g) are provided not only to enable the tube to dig into the substatum but also to help bring the sediment into suspension so that it is more easily collected. Diameters of various tubes (in cm) are shown.

water, such as rock pools and small ponds. In the past, rotenone, which almost inevitably killed the fish, was widely used; more recently, fish anaesthetics, such as MS 222, have been the method of choice. While, in theory, these compounds should allow the fish to recover when placed in clean water, this is far harder to achieve in practice. For example, it can be difficult to capture fish after exposure, as they may not float to the surface, resulting in them receiving a lethal exposure. Further, recovered fish returned to the wild are often highly vulnerable to predation from birds and other fish.

References

Albrecht, M.-L. (1959) Die quantitative Untersuchung der Bodenfauna fliessender Gewässer Untersuchungsmethoden und Arbeitsergebnisse. *Z. Fisch. (NF)* **8**, 481–550.

Allen, D.M. & Hudson, J.H. (1970) *A Sled-mounted Suction Sampler for Benthic Organisms*. Special Report—Fisheries, United States Fish and Wildlife Service.

Baird, R.H. (1958) *A Preliminary Report on a New Half Square Metre Bottom Sampler*. CM 1958/70, Shellfish Committee, International Council for the Exploration of the Sea.

Baker, J.M. & Wolff, W.J. (eds) (1987) *Biological*

Surveys of Estuaries and Coasts. Cambridge University Press, Cambridge.

Baker, J.M., Hartley, J.P., & Dicks, B. (1987) Planning biological surveys. In: Baker, J.M. & Wolff, W.J. (eds) *Biological Surveys of Estuaries and Coasts*. Cambridge University Press, Cambridge, pp. 1–26.

Barnett, P.R.O. (1969) A stabilizing framework for the Knudsen bottom sampler. *Limnol. Oceanogr.* **14**, 648–9.

Barnett, P.R.O. & Hardy, B.L.S. (1967) A diver-operated quantitative bottom sampler for sand macrofaunas. *Helgoland. wissensch. Meeresuntersuch.* **15**, 390–8.

Fig. 5.18 Modified Jenkin surface-mud sampler (after Goulder, 1971), with horizontal steel plates in position to isolate slices of sediment.

The stratification of *Chaoborus punctipennis* was studied with this apparatus. The Kajak–Brinkhurst sampler (Kajak, 1971) is very suitable for soft sediments; the lid may be closed by a messenger after the sample is taken. It has been found to be nearly as efficient as using a diver (Holopainen & Sarvala, 1975). A multiple corer has been developed by Hamilton *et al.* (1970).

Livingston (1955) described a piston sampler, particularly well adapted for obtaining the lower layers of sediment from a muddy bottom; Vallentyne (1955) and Rowley and Dahl (1956) introduced modifications that made the construction simpler and less costly, and Brown (1956) has modified it to sample the mud–water interface as well and, by the use of a clear plastic tube, to allow the inspection of the intact core.

A different approach is that of Shapiro (1958), who surrounded the sampling tube (corer) with a tapered jacket. Just before lowering, the jacket is filled with crushed solid carbon dioxide and *n*-butyl alcohol. It is lowered quickly through the water, allowed to settle, left for about 5 min, and retrieved. The freezing mixture is replaced with water; the frozen core may soon be slid out and divided and handled in the solid state. As pointed out by Kajak (1971), soil corers, especially O'Connor's (p. 214), are very useful for hard bottoms in shallow water.

5.3.9 Airlift and suction devices

Airlifts, which are used in marine archaeology, mining etc., use the principle that, when compressed air is discharged under water, an air–water mixture is formed which rises to the surface. A typical application is the diver-operated airlift sampler described by Barnett and Hardy (1967). This works by sucking material up a long vertical tube into a collecting sieve by the injection into the tube of compressed air (Fig. 5.19). This airlift can be used to remove the sediment from within a core manually pushed into the substrate. Mackey (1972) describes a simple apparatus, essentially for taking core samples. Compressed air is fed to the centre of the base of a length of plastic drainpipe, the top of the pipe being angled and opening into a collecting net or sieve. A more elaborate device, in which the sampling area is delimited by a rectangular frame and the compressed air is liberated through a series of openings (Fig. 5.19), was developed by Pearson *et al.* (1973). They compared its efficiency with the Suber sampler and the Allen grab; overall, there were not great differences, but particular animals were sampled in greater numbers by one or other method. The airlift sampled significantly larger numbers of *Polycelis*, *Ephemerella*, *Sphaerium*, and *Caenis*. Drake and Elliott (1982) review three airlift samplers for use in rivers.

A number of samplers employ suction either to force a coring tube into the substrate or to draw the sediment into a collector. The Knudsen sampler (Knudsen, 1927; Barnett, 1969), which operates by sucking a coring tube into the substrate, using a cable-operated pump, was the first sampler of this type designed. For shallow waters, Kaplan *et al.* (1974) and Thayer *et al.* (1975) describe samplers that work on the same principle. More recently, a number of samplers use pumped water. The

Fig. 5.19 Air-lift suction sampler (after Barnett & Hardy, 1967). (a) Diagram of the apparatus. (b) Method of use, with a diver extracting the contents of a sampling cylinder pushed into the substrate.

benthic suction sampler of True *et al.* (1968) sucks a 0.1 m² core into the substrate, while the substrate is drawn up into a collecting basket. Many other suction samplers have been developed (e.g. Emig & Lienhart, 1971; Mulder & Arkel, 1980; Grussendorf, 1981) (Fig. 5.20). Suction samplers have also been designed to be slowly towed along the bottom. Allen and Hudson (1970) used a vacuum sledge to quantify buried pink shrimp.

5.4 Poisons and anaesthetics used for sampling fish in rock pools and small ponds

Chemofishing has long been practised by humans and has been effectively used by ecologists seeking quantitative estimates of fish number in small, isolated bodies of

Bates, M. (1941) Field studies of the anopheline mosquitoes of Albania. *Proc. Entomol. Soc. Wash.* **43**, 37–58.

Benfield, E.F., Hendricks, A.C., & Cairns, J. (1974) Proficiencies of two artificial substrates in collecting stream macroinvertebrates. *Hydrobiologia* **45**, 431–40.

Bouma, A.H. & Marshall, N.F. (1964) A method for obtaining and analysing undisturbed oceanic bed samples. *Mar. Geol.* **2**, 81–99.

Brida, C. & Reys, J.P. (1966) Modifications d'une benne 'orange peel' pour les prélèvements quantitatifs du benthos de substrats meubles. *Recueil Trav. Sta. Mar. Endoume* **41**, 117–21.

Britt, N.W. (1955) New methods of collecting bottom fauna from shoals or rubble bottoms of lakes and streams. *Ecology* **36**, 524–5.

Brown, S.R. (1956) A piston sampler for surface sediments of lake deposits. *Ecology* **37**, 611–13.

Brunel, P., Besner, M., Messier, D., *et al.* (1978) Le traineau suprabenthique Macer–GIROQ: appareil amélioré pour l'échantillonage quantitatif étage de la petite faune nageuse au voisinage du fond. *Int. Rev. Gasamten Hydrobiol.* **63**, 815–29.

Buchanan, J.B. (1984) Sediment analysis. In: Holme, N.A. & McIntyre, A.D. (eds) *Methods for the Study of Marine Benthos*. Blackwell Scientific Publications, Oxford, pp. 41–65.

Calow, P. (1972) A method for determining the surface area of stones to enable quantitative density estimates of littoral stone dwelling organisms to be made. *Hydrobiologia* **40**, 37–50.

Christie, M. (1954) A method for the numerical study of larval populations of *Anopheles gambiae* and other pool-breeding mosquitoes. *Ann. Trop. Med. Parasitol. Liverpool* **48**, 271–6.

Clarke, G.L. & Bumpus, D.F. (1940) The plankton sampler—an instrument for quantitative plankton investigations. *Spec. Publ. Am. Soc. Limnol. Oceanogr.* **5**, 1–8.

Clarke, W.D. (1964) The jet net, a new high speed plankton sampler. *J. Mar. Res.* **22**, 284–7.

Colman, J.S. & Segrove, F. (1955) The tidal plankton over Stoupe Beck Sands, Robin Hood's Bay (Yorkshire, North Riding). *J. Anim. Ecol.* **24**, 445–62.

Croset, H., Papierok, B., Rioux, J.A. *et al.* (1976) Absolute estimates of larval populations of culicid mosquitoes: comparison of capture–recapture removal and dipping methods. *Ecol. Entomol.* **1**, 251–6.

Crossman, I.S. & Cairns, J. (1974) A comparative study between two different artificial substrate samplers and regular sampling techniques. *Hydrobiologia* **44**, 517–22.

Cummins, K.W. (1962) An evaluation of some techniques for the collection and analysis of benthic samples with special emphasis on lotic waters. *Am. Midl. Nat.* **67**, 477–503.

David, P.M. (1965) A neuston net: a device for sampling the surface fauna of the oceans. *J. Mar. Biol. Assoc. UK* **45**, 313–20.

Dendy, J.S. (1944) The fate of animals in stream drift when carried into lakes. *Ecol. Monogr.* **14**, 333–57.

Dickie, L.M. (1955) Fluctuations in abundance of the giant scallop *Placopecten magellanicus* (Gmelin) in the Digby area of the Bay of Fundy. *J. Fish. Res. Board Can.* **12**, 797–857.

Disney, R.H.L. (1972) Observations on sampling pre-imaginal populations of blackflies (Dipt., Simuliidae) in West Cameroon. *Bull. Entomol. Res.* **61**, 485–503.

Drake, C.M. & Elliott, J.M. (1982) A comparative study of three air-lift samplers used for sampling benthic macro-invertebrates in rivers. *Freshw. Biol.* **12**, 511–33.

Dunn, D.R. (1961) The bottom fauna of Llyn Tegid (Lake Bala), Merionethshire. *J. Anim. Ecol.* **31**, 267–81.

Eckblad, J.W. (1973) Population studies of three aquatic gastropods in an intermittent backwater. *Hydrobiologia* **41**, 199–219.

Edmondson, W.T. & Winberg, G.G. (eds) (1971) *A Manual on Methods for the Assessment of Secondary Productivity in Freshwaters*. IBP Handbook 17, Blackwell, Oxford, London.

Eleftheriou, A. & Holme, N.A. (1984) Macrofauna techniques. In: Holme, N.A. & McIntyre, A.D. (eds) *Methods for the Study of Marine Benthos*. Blackwell Scientific Publications, Oxford, pp. 140–216.

Elgmork, K. (1962) A bottom sampler for soft mud. *Hydrobiologia* **20**, 167–72.

Ellertson, B.A. (1977) A new apparatus for sampling surface fauna. *Sarsia* **63**, 113–14.

Elliott, J.M. & Drake, C.M. (1981) A comparative study of seven grabs used for sampling benthic macroinvertebrates in rivers. *Freshw. Biol.* **11**, 99–120.

Elliott, J.M. & Tullett, P.A. (1978) *A Bibliography of Samplers for Benthic Invertebrates*. Freshwater Biological Association, Occasional Publication **4**, 61 pp.

Emig, C.C. (1977) Un nouvel aspirateur sous-marins, à air comprimé. *Marine Biology* **43**, 379–80.

Emig, C.C. & Lienhart, R. (1971) Principe de l'aspirateur sous-marin automatique pour sédiments meubles. *Vie Milieu* **22**, 573–8.

Euliss, N.H., Swanson, G.A., & Mackay, J. (1992) Multiple tube sampler for benthic and pelagic invertebrates in shallow wetlands. *J. Wildl. Manag.* **56**, 186–91.

Ford, I.B. (1962) The vertical distribution of larvae Chironomidae (Dipt.) in the mud of a stream. *Hydrobiologia* **19**, 262–72.

Ford, T.E. (1993) *Aquatic Microbiology: an Ecological Approach.* Blackwell Science, Oxford.

Frost, S. (1971) Evaluation of a technique for sorting and counting stream invertebrates. *Can. J. Zool.* **49**, 878–83.

Frost, S., Huni, A., & Kershaw, W.E. (1971) Evaluation of a kicking technique for sampling stream bottom fauna. *Can. J. Zool.* **49**, 167–73.

Fujita, H. (1956) The collection efficiency of a plankton net. *Res. Popul. Ecol.* **3**, 8–15.

Gamble, J.C. (1984) Diving. In: Holme, N.A. & McIntyre, A.D. (eds) *Methods for the Study of Marine Benthos.* Blackwell Scientific Publications, Oxford.

Gehringer, J.W. (1952) An all metal plankton sampler (model Gulf III). *Spec. Scient. Rep. US Fish. Wildl. Serv. Fish.* **88**, 7–12.

Gerking, S.D. (1957) A method of sampling the littoral macrofauna and its application. *Ecology* **38**, 219–26.

Glime, J.M. & Clemons, R.M. (1972) Species diversity of stream insects on *Fontinalis* spp. compared to diversity on artificial substrates. *Ecology* **53**, 458–64.

Goodwin, M.H. & Eyles, D.E. (1942) Measurements of larval populations of *Anopheles quadrimaculatus*, Say. *Ecology* **23**, 376.

Goulder, R. (1971) Vertical distribution of some ciliated protozoa in two freshwater sediments. *Oikos* **22**, 199–203.

Grange, K.R. & Anderson, P.W. (1976) A soft-sediment sampler for the collection of biological specimens. *Rec. NZ Oceanogr. Inst.* **3**, 9–13.

Grussendorf, M.J. (1981) A flushing–coring device for collecting deep-burrowing infaunal bivalves in intertidal sand. *Fish. Bull. Nat. Fish. Serv. NOAA Seatle* **79**, 383–5.

Hamilton, A.L., Burton, W., & Flannagan, J.F. (1970) A multiple corer for sampling profundal benthos. *J. Fish. Res. Board Can.* **27**, 1867–9.

Hardy, E.R. (1992) Changes in species composition of cladocera and food availability in a floodplain lake. Lago Jacaretinga, central Amazon. *Amazoniana* **12**, 155–68.

Henderson, P.A. & Crampton, W.G.R. (1997) A comparison of fish diversity and abundance between nutrient rich and poor lakes in the Upper Amazon. *J. Trop. Ecol.* **13**, 175–98.

Henderson, P.A. & Hamilton, H.R. (1995) Standing crop and distribution of fish in drifting and attached floating meadow within an Upper Amazonian varzea lake. *J. Fish. Biol.* **47**, 266–76.

Henderson, P.A. & Walker, I. (1986) On the leaf-litter community of the Amazonian blackwater stream Tarumazinho. *J. Trop. Ecol.* **2**, 1–17.

Hess, A.D. (1941) New limnological sampling equipment. *Limnol. Soc. Am. Spec. Publ.* **6**, 1–15.

Higer, A.L. & Kolipinski, M.C. (1967) Pull-up trap: a quantitative device for sampling shallow-water animals. *Ecology* **48**, 1008–9.

Holme, N.A. (1949) A new bottom-sampler. *J. Mar. Biol. Assoc. UK* **28**, 323–32.

Holme, N.A. & Willerton, P.F. (1984) Position fixing of ship and gear. In: Holme, N.A. & McIntyre, A.D. (eds) *Methods for the Study of Marine Benthos.* Blackwell Scientific Publications, Oxford, pp. 27–40.

Holopainen, I.J. & Sarvala, J. (1975) Efficiencies of two corers in sampling soft-bottom invertebrates. *Ann. Zool. Fenn.* **12**, 280–4.

Huckley, P. (1975) An apparatus for subdividing benthos samples. *Oikos* **26**, 92–6.

Hunter, W. & Simpson, A.E. (1976) A benthic grab designed for easy operation and durability. *J. Mar. Biol. Assoc. UK* **56**, 951–7.

Isaacs, J.D. & Kidd, L.W. (1953) *Isaacs–Kidd Midwater Trawl: Final Report.* Ref. 53–3, Scripps Institute of Oceanography, University of California.

James, H.G. & Nicholls, C.F. (1961) A sampling cage for aquatic insects. *Can. Entomol.* **939**, 1053–5.

Jenkin, B.M. & Mortimer, C.H. (1938) Sampling lake deposits. *Nature* **142**, 834.

Junk, W.J. (1973) Investigations on the ecology and production-biology of the floating meadows (*Paspalum echinochloa*) on the Middle Amazon. *Amazoniana* **4**, 9–102.

Kajak, Z. (1963) Analysis of quantitative benthic methods. *Ekol. Polska A* **11**, 1–56.

Kajak, Z. (1971) Benthos of standing water. In: Edmondson, W.T. & Winberg, G.G. (eds) *IBP Handbook 17.* Blackwell, Oxford, London, pp. 25–65.

Kaplan, E.H., Welker, J.R., & Krause, M.G. (1974) A shallow-water system for sampling macrobenthic infauna. *Limnol. Oceanogr.* **19**, 346–50.

Karlsson, M., Bohlin, T., & Stenson, J. (1976) Core sampling and flotation: two methods to reduce costs of a chironomid population study. *Oikos* **27**, 336–8.

Knudsen, M. (1927) A bottom sampler for hard bottom. *Meddel. Komission. Havundersogelser Fisk* **8**, 1–4.

Kroger, R.L. (1972) Underestimation of standing crop by the Surber sampler. *Limnol. Oceanogr.* **17**, 475–8.

Kroger, R.L. (1974) Invertebrate drift in the Snake River, Wyoming. *Hydrobiologia* **44**, 369–80.

Lisitsyn, A.P. & Udintsev, G.B. (1955) New model dredges. [In Russian.] *Trudy Vses. Gidrobiol. Obsch.* **6**, 217–22.

Livingston, D.A. (1955) A lightweight piston sampler for lake deposits. *Ecology* **36**, 137–9.

Macan, T.T. (1958) Methods of sampling the bottom fauna in stony streams. *Mitt. int. Verein. theor. angew. Limnol.* **8**, 1–21.

McCauley, V.J.E. (1975) Two new quantitative samplers for aquatic phytomacrofauna. *Hydrobiologia* **47**, 81–9.

Macer, T.C. (1967) A new bottom-plankton sampler. *J. Cons. Perm. Int. Explor. Mer* **31**, 158–63.

McGowan, J.A. & Brown, D.M. (1966) *A New Opening–Closing Paired Zooplankton Net*. Ref. 66–23, Scripps Institute of Oceanography, University of California.

Mackey, A.P. (1972) An air-lift for sampling freshwater benthos. *Oikos* **23**, 413–15.

Mason, J., Chapman, C.J., & Kinnear, J.A.M. (1979) Population abundance and dredge efficiency studies on the scallop, *Pecten maximus* (L.). *Rapp. Proc. Verb. Reunions Cons. Expl. Mer* **175**, 91–6.

Mason, W.T., Weber, C.I., Lewis, P.A., & Julian, E.C. (1973) Factors affecting the performance of basket and multiplate macro-invertebrate samplers. *Freshw. Biol.* **3**, 409–36.

Matsuo, Y., Nemoto, T., & Marumo, R.A. (1976) A convertible neuston net for zooplankton. *Bull. Plankton Soc. Japan* **23**, 26–30.

Milbrink, G. (1971) A simplified tube bottom sampler. *Oikos* **22**, 260–3.

Milbrink, G. & Wiederholm, T. (1973) Sampling efficiency of four types of mud bottom sampler. *Oikos* **24**, 479–82.

Miller, D.A. (1961) A modification of the small Hardy plankton sampler for simultaneous high-speed plankton hauls. *Bull. Mar. Ecol.* **5**, 165–72.

Miller, D.A. & Judkins, D.C. (1981) Design of pumping systems for sampling zooplankton, with descriptions of two high-capacity samplers for coastal studies. *Biol. Oceanogr.* **1**, 29–56.

Mortimer, C.H. (1942) The exchange of dissolved substances between mud and water in lakes. III and IV. *J. Ecol.* **30**, 147–201.

Motoda, S. (1959) Devices of simple plankton sampling. *Mem. Fac. Fish. Hokkaido Univ.* **7**, 73–94.

Motoda, S. (1971) Devices of simple plankton apparatus, V. *Bull. Fac. Fish. Hokkaido Univ.* **22**, 101–6.

Mulder, M. & Arkel, M.A.V. (1980) An improved system for quantitative sampling of benthos in shallow water using the flushing technique. *Netherl. J. Sea Res.* **14**, 119–22.

Mundie, J.H. (1956) A bottom sampler for inclined rock surfaces in lakes. *J. Anim. Ecol.* **259**, 429–32.

Nalwalk, A.J., Hersey, J.B., Reitzel, J.S., & Edgerton, H.E. (1962) Improved techniques of deep-sea dredging. *Deep-Sea Research* **8**, 301–2.

Needham, P.R. & Usinger, R.L. (1956) Variability in macrofauna of a single riffle in Prosser creek, California, as indicated by the Surber sampler. *Hilgardia* **24**, 383–409.

Negus, C.L. (1966) A quantitative study of growth and production of unionid mussels in the River Thames at Reading. *J. Anim. Ecol.* **35**, 513–32.

Omori, M. (1969) A bottom-net to collect zooplankton living close to the sea-floor. *J. Oceanogr. Soc. Japan* **25**, 291–4.

Omori, M. & Ikeda, T. (1984) *Methods in Marine Zooplankton Ecology*. John Wiley & Sons, New York.

Pearson, R.G., Litterick, M.R., & Jones, N.V. (1973) An air-lift for quantitative sampling of the benthos. *Freshw. Biol.* **3**, 309–15.

Pennak, R.W. (1962) Quantitative zooplankton sampling in littoral vegetation areas. *Limnol. Oceanogr.* **7**, 487–9.

Pepper, I.L., Gerba, C.P., & Brendecke, J.W. (eds) (1995) *Environmental Microbiology: a Laboratory Manual*. Academic Press, London.

Petersen, C.G.J. & Boysen Jensen, P. (1911) Valuation of the sea. I. Animal life of the sea bottom, its food and quantity. *Report from the Danish Biological Station* **20**, 81 pp.

Powers, C.F. & Robertson, A. (1967) Design and evaluation of an all-purpose benthos sampler. *Great Lakes Res. Div. Spec. Rep.* **30**, 126–31.

Reineck, H.E. (1963) Der Kastengreifer. *Natur Museum* **93**, 102–8.

Richards, S.W. & Riley, G.A. (1967) The benthic epifauna of Long Island Sound. *Bull. Bingham Oceanogr. Collec.* **19**, 89–135.

Ricker, W.E. (1938) On adequate quantitative sampling of the pelagic net plankton of a lake. *J. Fish. Res. Board Can.* **4**, 19–32.

Rothlisberg, P.C. & Pearcy, W.G. (1977) An epibenthic sampler used to study the ontogeny of vertical migration of *Pandalus jordani* (Decapoda, Caridea). *Fish. Bull. US* **74**, 994–7.

Rowley, J.R. & Dahl, A. (1956) Modifications in design and use of the Livingstone piston sampler. *Ecology* **37**, 849–51.

Scott, D. (1958) Ecological studies on the Trichoptera of the River Dean, Cheshire. *Arch. Hydrobiol.* **54**, 340–92.

Scott, D. & Rushforth, J.M. (1959) Cover on river bottoms. *Nature* **183**, 836–7.

Shapiro, J. (1958) The core-freezer—a new sampler for lake sediments. *Ecology* **39**, 758.

Smith, W. & McIntyre, A.D. (1954) A spring-loaded bottom sampler. *J. Mar. Biol. Assoc. UK* **33**, 257–64.

Surber, E.W. (1936) Rainbow trout and bottom fauna production in one mile of stream. *Trans. Am. Fish. Soc.* **66**, 193–202.

Swanson, G.A. (1978) A water column sampler for invertebrates in shallow wetlands. *J. Wildl. Manag.* **42**, 670–1.

Thayer, G.W., Williams, R.B., Price, T.J., & Colby, D.R. (1975) A large corer for quantitatively sampling benthos in shallow water. *Limnol. Oceanogr.* **20**, 474–80.

Tonolli, V. (1971) Methods of collection: zooplankton. In: Edmondson, W.T. & Winberg, G.G. (eds) *IBP Handbook 17*. Blackwells, Oxford, London, pp. 1–20.

Tranter, D.J. & Smith, P.E. (1968) Loss of organisms through meshes. In: Tranter, D.J. & Fraser, J.H. (eds) *Zooplankton Sampling: Monographs on Oceanographic Methodology*. UNESCO, Paris.

True, M.A., Reys, J.-P., & Delauze, H. (1968) Progress in sampling the benthos: the benthic suction sampler. *Deep Sea Res.* **15**, 239–42.

Ulfstrand, S. (1968) Benthic animal communities in Lapland streams. *Oikos* **105** (suppl.), 1–120.

Usinger, R.L. & Needham, P.R. (1954) A plan for the biological phases of the periodic stream sampling program. Mimeographed, Final Report to Calif. St. Wat. Pollution Cont. Bd.

Usinger, R.L. & Needham, P.R. (1956) A drag-type riffle-bottom sampler. *Progr. Fish Cult.* **18**, 42–4.

Vallentyne, J.R. (1955) A modification of the Livingstone piston sampler for lake deposits. *Ecology* **36**, 139–41.

Van Dorn, W.G. (1957) Large volume water sampler. *Trans. Am. Geophys. Union* **37**, 682–4.

Vannucci, M. (1968) Loss of organisms through meshes. In: Tranter, D.J. & Fraser, J.H. (eds) *Zooplankton Sampling: Monographs on Oceanographic Methodology*. UNESCO, Paris.

van Veen, J. (1933) Onderzoek naar het zandtransport von rivieren. *De Ingenieur* **48**, 151–9.

Waters, T.F. & Knapp, R.J. (1961) An improved stream bottom fauna sampler. *Trans. Am. Fish. Soc.* **90**, 225–6.

Welch, H.E. & James, H.G. (1960) The Belleville trap for quantitative samples of mosquito larvae. *Mosquito News* **20**, 23–6.

Welch, P.S. (1948) *Limnological Methods*. McGraw-Hill, New York.

Wickstead, J. (1953) A new apparatus for the collection of bottom plankton. *J. Mar. Biol. Assoc. UK* **32**, 347–55.

Wigley, R.L. (1967) Comparative efficiencies of van Veen and Smith–McIntyre grab samplers as revealed by motion pictures. *Ecology* **48**, 168–9.

Wilding, J.L. (1940) A new square-foot aquatic sampler. *Limnol. Soc. Am. Spec. Publ.* **4**, 1–4.

Williams, T.R. & Obeng, L. (1962) A comparison of two methods of estimating changes in *Simulium* larval populations, with a description of a new method. *Ann. Trop. Med. Parasitol. Liverpool* **56**, 359–61.

Wolcott, R.H. (1901) A modification of the Birge collecting net. *Joppl. Microsc. Lab. Meth.* **4**, 1407–9.

Wolfe, L.S. & Peterson, D.G. (1958) A new method to estimate levels of infestations of black-fly larvae (Diptera: Simulidae). *Can. J. Zool.* **36**, 863–7.

6 Absolute Population Estimates by Sampling a Unit of Soil or Litter Habitat: Extraction Techniques

The extraction methods described here may be used not only with soils and aquatic sediments but also with plant and animal debris, litter and dung, plant material collected by suction apparatus or other means, and the nests of vertebrates. The actual methods for obtaining the samples from these other habitats are discussed in Chapters 4 and 5. In general, the matrix holding the animals will be termed soil if of terrestrial origin and sediment if of aquatic or marine-sample origin.

Much of the work in soil zoology was originally aimed at the extraction of a large segment of the fauna by a single method. However, most workers have now concluded that a method that will give an almost absolute estimate of one species or group will give, at most, a rather poor relative estimate for another. The efficiency of extraction varies, not only with the animal, but also with the soil (see Table 6.1), the physical properties and size distribution of its particles, its water content, and the amount of vegetable matter in it. Therefore, although with certain animals under certain conditions each of these methods will give absolute population estimates, none of them will provide such data under all conditions. Further information on ecological methods in soil zoology is given in Phillipson (1970), Dindal (1990), and Gorny and Grum (1993). Techniques for sorting animals from benthic samples are reviewed by Eleftheriou and Holme (1984), McIntyre and Warwick (1984), and Hartley et al. (1987).

Bees, cicindelid and *Bledius* larvae, certain crabs, and other comparatively large animals that make holes or casts in bare ground or the seabed may be counted directly *in situ*. In most studies, however, it is necessary both to take a sample and to extract the animals.

6.1 Sampling

The number and size of the samples and the sampling pattern in relation to statistical considerations have been discussed in Chapter 2. The soil samples are usually taken with a corer; golf-hole borers or metal tubing sharpened at one end make simple corers. It has been suggested that some animals may be killed by compression when the core is forced from such 'instruments', and furthermore it is highly desirable to keep the core undisturbed (especially for extraction by behavioural methods), so more elaborate corers have been developed. In general, the larger the animal and the sparser its population, the bigger the sample (e.g. Frick (1962) took 8-in (20 cm) diameter cores for *Nomia* bees). The depth to which it is necessary to sample varies with the animal and the condition of the soil; it will be particularly deep in areas with marked dry seasons (Price, 1975); many soil animals have seasonal and diel vertical migrations (Erman, 1973). Termites and ants present special problems, due to their patchy distribution and the difficulty of taking adequately deep samples before the disturbance has caused them to escape (Brian, 1970; Sands, 1972).

With the O'Connor (1957) split corer (Fig. 6.1), the risk of compressing the sample by forcing it out of the corer is avoided. After the core has been taken, the clamping band can be loosened, the two aluminium halves of the cover separated,

Fig. 6.1 Soil corers. (a) O'Connor split corer showing compartments (after O'Connor, 1957). (b) O'Connor split corer assembled. (c) Soil corer with sample tubes (after Dhillon & Gibson, 1962). (d) Soil corer for canister extractor (after MacFadyen, 1961).

and the sample exposed. Furthermore, the sample can then be easily divided into the different soil layers: litter, humus, upper 2 cm, etc. Plastic or metal rings may be inserted inside the metal sheath of the corer, just behind the cutting ring (MacFadyen, 1961; Dhillon & Gibson, 1962; Vannier & Vidal, 1965; Fig. 6.1); these will enable the core to be extracted with its natural structure intact (see p. 229).

In order to penetrate hard tropical or frozen tundra soils, it may be necessary to have equipment of the types described by Belfield (1956) and Potzger (1955). In contrast, in soft humus-rich situations, such as manure heaps, it is difficult to take an undisturbed sample; von Törne (1962b) has devised a sampler for these habitats, consisting of two concentric tubes, each with cutting teeth. The inner tube is continuously pushed down firmly and held still, thereby protecting the sample, while the outer one is rotated and cuts through the compost. Tanton (1969) described the construction of a corer of this type, with a series of internal plastic rings that can separate the sample into subsamples. Vannier and Alpern (1968) described another corer that allows subsamples to be obtained from different depths.

Semifluid substances, such as the contents of water-filled tree holes, may be sampled with a device that includes an auger (worm bit) inside a normal corer; the turning of the auger takes the sample into the corer and retains it (Kitching, 1971).

Many pests of field crops lay their eggs on the bases of the plants and on the soil; the number of eggs usually falls off very rapidly with distance away from the plant (e.g. Lincoln & Palm, 1941; Abu Yaman, 1960). The scissor type of sampler described by Webley (1957) can take suitable samples of young plants and the soil around them.

Fallen leaves and other debris are usually sampled with quadrats, such as a metal box with top and bottom missing and the lower edge sharpened (Gabbutt, 1959).

6.2 Bulk staining

The staining of samples to facilitate sorting is occasionally used, particularly with marine and freshwater benthic samples. Suitable stains include rose bengal, rhodamine B, eosin, or Lugol's iodine. It is common practice to preserve benthic samples using formalin with 4 g/L rose bengal, so that the organisms become stained during storage. For some animals, rose bengal may obscure diagnostic features. After sieving, the animals appear bright red in the sorting tray and are easily picked from retained detritus. Rhodamine-B-stained organisms fluoresce under ultraviolet (UV) light (Hamilton, 1969). For samples with a high detritus content, Williams and Williams (1974) advocate a counterstaining method, where a high contrast is achieved by using rose bengal or Lugol's iodine and chlorazol black E.

6.3 Mechanical methods of extraction

Mechanical methods have the advantages that theoretically they extract all stages, mobile and sedentary, and are not dependent on the behaviour of the animal or the condition of the substrate; samples for mechanical extraction may be stored frozen for long periods before use. Their disadvantages are that, compared with behavioural methods, the operator must expend a great deal of time and energy on each sample, that sometimes they damage the animals, and that, as mobile and immobile animals are extracted, it may be difficult to distinguish animals that were dead at the time of sampling from those that were alive.

There are a number of distinct mechanical processes that can be used to separate animals from the soil and vegetable material: sieving, flotation, sedimentation, elutriation, and differential wetting. The following account, however, is intended to be functional, rather than classificatory or historical. Only the main types of processes will be outlined and it must be stressed that, although there are already many different combinations and variants, others will need to be developed for particular animals and substrates.

6.3.1 Dry sieving

This method may be of use for the separation of fairly large (and occasionally small) animals from friable soil or fallen leaves. Its disadvantages are that small specimens are often lost and a considerable amount of time needs to be spent in hand-sorting the sieved material. The Reitter sifter (see Kevan, 1962) is a simple device, but this is really more a collector's tool than a means of estimating populations. M. Lane and E. Shirck made the first mechanical sieve with a to-and-fro motion, although it was worked by hand. Motorized sieve shakers are commercially available.

Molluscs have been extracted by dry-sieving methods (Jacot, 1935), but these have been shown to be inaccurate, some snails remaining in the leaves (Williamson, 1959). Dry sieving has also been used to separate mosquito eggs from mud (Stage *et al.*, 1952). A modified grain drier is the basis of this method; the samples are dried until almost dusty and passed through a mesh sieve and then into the top of four shaker sieves of the grain drier. Coarser particles are removed by the first three sieves, but the eggs and similar-sized soil particles are held on the last one (80-mesh/in; 31/cm); they then fall through an opening on to a 60-mesh/in (24/cm) roll screen, a carefully adjusted air current blowing away the lighter particles during the fall. Because of their spindle shape, the eggs, with only a very small quantity of soil, eventually pass 'end-on' through the roll screen into the 'catch pan'. Such a degree of mechanization is possible because only a single organism with a constant and regular shape and a uniform weight was required.

The active red-legged earth mite, *Halotydeus destructor*, lives amongst the grass and plant debris in Australian pastures; Wallace (1956) has shown that it can be very accurately sampled by taking shallow cores, which are then inverted and rotated, still within the corer, over a sieve in a funnel. The sides of the metal corer are tapped a number of times and the mites fall through the sieve and may be collected in tubes below the funnel.

6.3.2 Wet sieving

This technique is most often used in conjunction with other methods (see below); on its own, it is of particular value where the organisms are much smaller than the parti-cles of the substrate (e.g. small snails amongst freshly fallen leaves) or much larger (e.g. many mud-dwellers), and hence separation by size alone is sufficient. The sieves may be of three basic designs: flat, revolving, or 'three-dimensional'. The method is particularly important for marine benthic samples.

Flat sieves may be built into a tower (see Fig. 6.3a) and a series of sieves was the basis of early uses of this method by N.A. Cobb and H.M. Morris. Sieving is often used in studies on the bottom animals of aquatic habitats for the separation of the mud from the organisms and stones that are retained by the sieve (Eleftheriou & Holme, 1984). The mesh size to be used is determined by the size of the organism being sampled. In marine studies, macrofauna is usually defined as animals retained by a 0.5–1.0 mm mesh. The exact mesh size to be used will be dictated by the particle size distribution, as the aim is to pass the majority of the sediment.

Gastropod molluscs and possibly other small animals may be efficiently sepa-rated from fresh leaf litter by wet sieving, but here it is the organisms that pass through the sieve and the unwanted material that is retained (Williamson, 1959). The leaves are placed on a coarse sieve (mesh size 1 cm), above a fine cloth-covered grid, the whole is immersed in a vessel of water (Fig. 6.2a), and the leaves are stirred from time to time; after about 15 min, the vessel can be drained from below and the molluscs will be found on the cloth sheet, which is then inspected under the microscope. Clearly this method is not satisfactory for molluscs amongst soil or humus, as much of these materials will pass through the upper sieve; here a combi-nation of wet sieving and flotation may be used (see p. 218).

Simple revolving sieves were used by McLeod (1961) for the extraction of sawfly cocoons. These consisted of wooden boxes, hinged in the middle and with screen tops and bottoms (Fig. 6.2b); they were revolved once or twice while a jet of water from a hose was played on to them. More elaborate revolving models were designed by Horsfall (1956) and Read (1958); these are discussed below.

The so-called 'three-dimensional' or rod sieve was designed by Stewart (1974) to facilitate the separation, undamaged, of tipulid larvae or other large animals from vegetable material in grassy soil samples, prior to flotation. Stewart's apparatus

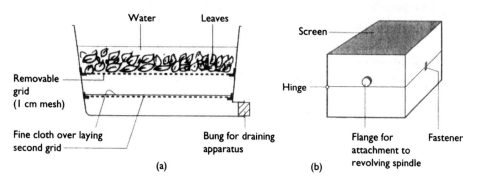

Fig. 6.2 (a) Tank for separating molluscs from leaf litter by wet sieving, based on Williamson's design. (b) simple sieve box for wet-sieving cocoons, based on McLeod's design.

consists of 13 tiers of parallel galvanized-steel rods set in an open-ended box. The rods are progressively closer in each tier (2.5 cm closing to 1.3 cm apart); the side of the box may be removed (for cleaning, a process assisted by a sliding panel). The samples is placed on the top row of rods and the box moved up and down in a tank of water; the inside of the tank is lined with a normal screen sieve. The vegetation will pass down between the rods, freeing insects and mineral material, which remain in the tank. The three-dimensional sieve is removed (and if necessary cleaned) and the sieve inside the tank removed and the normal flotation procedure followed (see below).

6.3.3 Soil washing and flotation

When the organisms to be extracted and the rest of the sample are of different particle size, sieving alone is sufficient for extraction; however, in most situations, after sieving, the animals required remain mixed with a mass of similar-sized mineral and vegetable matter. As the specific gravity of mineral and biotic materials is frequently different, the extraction may be taken a stage further by flotation of the animals; unfortunately, plant material usually floats equally well. The combination of flotation, devised by A. Berlese many years before, with soil washing is usually associated with Ladell (1936). This approach, much modified and improved by Salt and Hollick (1944) and Raw (1962), is the basis of one of the most widely used, versatile, and efficient techniques; it consists of four stages, three of which will be discussed here.

6.3.3.1 Pretreatment

In order to disperse the soil particles—particularly important if there is a high clay content—the cores may be soaked in water and deep-frozen. Chemical dispersion may also be used, the core being soaked in solutions of sodium citrate (d'Aguilar *et al.*, 1957) or sodium oxalate (Seinhorst, 1962). For the heavy clay soils, it may be necessary to combine chemical and physical methods; the core is gently crumbled into a plastic container and covered with a solution of sodium hexametaphosphate (50 g) and sodium carbonate (20 g) in 1 L of water; the whole is then placed under reduced pressure in a vacuum desiccator for a time. After the restoration of atmospheric pressure, the samples is frozen for at least 48 h and may be stored in this condition (Raw, 1955).

6.3.3.2 Soil washing

The sample is placed in the upper (and coarsest) sieve of the washing apparatus (Fig. 6.3a); it may be washed through by slow jets of water or single sieves 'dunked' in the settling can. The material retained in the sieves must be carefully teased apart and thoroughly washed; a few large animals may be removed at this stage. The mesh of the lowest sieve is such as to allow the animals required to pass through into the settling can, which is pivoted and then tipped into the 'Ladell can'. A 'three-dimensional' sieve may be necessary for the above process if there is a lot of grass in

Fig. 6.3 Soil-washing apparatus (modified from Salt & Hollick, 1944). (b) Ladell can and associated equipment during the agitation phase of flotation (diagrammatic).

the sample and large animals are being separated (see above). The Ladell, which resembles an inverted, bottomless, paraffin can, has a fine phosphor-bronze sieve and its lower opening immersed in the drainage tank; this tank should be arranged so that the level of water maintained in it, and hence in the Ladell, is slightly above the sieve of the latter—this minimizes blockage of the sieve. The Ladell is allowed to drain. This, often tedious, process may be aided by tapping the side of the can with the hand from time to time. The standard Ladell has a fine sieve (mesh size about 0.2 mm); for many animals this may be much finer than is necessary and, to overcome this, Stephenson (1962) has designed a Ladell with a set of interchangeable sieve plates of various mesh sizes.

In conclusion it must be warned that soil washing is an invariably wet operation and is best carried out in a room with a concrete floor, the operator wearing rubber boots and suitable apparel.

6.3.3.3 Flotation

The Ladell is removed from the tank to a stand (Fig. 6.3) and allowed to drain completely, and then the process of flotation is begun. The lower opening of the Ladell is closed with a bung and the flotation liquid introduced until the Ladell is about two-thirds full (Fig. 6.3b). Concentrated magnesium sulphate solution (specific gravity *c.* 1.2) is the most usual flotation liquid, but solutions of sodium chloride (Golightly, 1952), potassium bromide (d'Aguilar *et al.*, 1957), or zinc chloride (Sellmer, 1956) may be used. Air is then bubbled up from the bottom of the Ladell; this agitation is continued for 2–3 min and serves to free any animal matter that may have been trapped on the sieve. More of the flotation solution is introduced from below until the liquid and 'float' passes over the lip, and the latter is retained on the collecting tube, which usually consists of a glass tube with a piece of bolting silk held in place with a rubber band. A wash-bottle may be used to help direct the float out of the Ladell. When the surface and sides of the Ladell are completely clean, the animal and plant material will all be in the collecting tube and the process is complete. The flotation liquid is then drained from the Ladell, for further use, by appropriate manipulations of the reservoir and pinch-clips (see Fig. 6.3).

The animals may now be separated from the plant material by direct examination under the microscope or it may be necessary to carry out a further separation based on differential wetting and/or centrifuging (see below). However, before these are discussed other washing and flotation techniques must be reviewed.

The apparatus of d'Aguilar *et al.* (1957) first removes the smallest silt particles by agitating the sample with a water current in a cylinder with a fine gauze side; subsequently the material is passed through a series of sieves, followed by flotation. Edwards *et al.* (1970) have developed a fully mechanized soil-washing process, which, in general, is as efficient as the Salt–Hollick process described above.

The eggs and puparia of the cabbage root fly (*Erioischia brassicae*) may be separated from the soil by sieving and flotation in water (Abu Yaman, 1960; Hughes, 1960). With such insects, therefore, there is no need for the Ladell—the final sieve should be fine enough to retain them; this is immersed in water and the insects float to the surface. A certain amount of foam often develops in soil washing and hinders examination of the float; it may be dispersed with a small quantity of caprylic acid (Abu Yaman, 1960). To separate relatively large insects (root maggots), Read (1958) used a cylindrical, aluminium, screen sieve, which was sprayed with water, rotated, and half-submerged in a tank; all the fine soil passed out of the sieve, which could then be opened and the floating insects removed. Gerard (1967) found a combination of wet sieving and simple magnesium-sulphate flotation efficient for earthworm sampling, although Nordstrom and Rundgren's (1972) studies suggest that the additional labour involved, compared with chemical extraction, is not always justified (see p. 238).

A revolving sieve was also used by Horsfall (1956) for mosquito eggs. Here the function of the drum, which consisted of three concentric sieves, was to disintegrate the sample and retain the larger materials, the eggs passing out and being collected in the finest of a series of sieves.

6.3.4 Flotation

If the organisms differ in specific weight from the material in which they are dispersed, then separation by flotation may be possible. The normal approach is to mix the sample into a liquid of specific gravity sufficient to allow the animals to float and the mineral particles to sink. Because organic debris also floats, such methods are inappropriate for samples of sediment rich in organic detritus. A large variety of liquids have been used, including magnesium sulphate (Ladell, 1936), sodium chloride (salt) (Lyman, 1943), carbon tetrachloride (Birkett, 1957; Whitehouse & Lewis, 1966), heptane (Walter et al., 1987), sugar solutions (Lackey & May, 1971; Higgins, 1977), and zinc chloride (Sellmer, 1956). With organic solvents, such as carbon tetrachloride, care must be taken over ventilation. At least for meiofaunal samples, these solutions have been largely replaced by Ludox, a colloidal silica polymer. De Jonge and Bouwman (1977) extracted nematodes by mixing sediment samples with a 25% v/v solution of Ludox–TM[1] (specific gravity 1.39 g). The surface of the mixture was then covered with water to prevent gel formation and left for 16h, after which the surface layer was sieved and the organisms collected. Ludox can also be used with centrifugation (see p. 224). Eleftheriou and Holme (1984) state that this method has also been used successfully for marine macrofauna. Sellmer (1956) used a 75% solution of zinc chloride (specific gravity 2.1) to float *Gemma*, a small bivalve mollusc, from marine sediment. Laurence (1954) separated insect larvae from dung by merely stirring the samples in 25% solution of magnesium sulphate, and Iversen (1971) obtained 75–92% of mosquito eggs in samples from temporary woodland pools by this method, though Service (1968) added centrifugal flotation (see p. 224). The efficiencies of this method, for the extraction of all stages of the midge *Leptoconops* from sand, were determined by Davies and Linley (1966) as: eggs 18%; first-instar larvae 29%; second-instar larvae 72%; third- and fourth-instar larvae 95%; and pupae 85%. Sodium chloride solution or brine has been widely used for arthropods (e.g. Dondale et al., 1971) and calcium chloride for larval and pupal midges (Sugimoto, 1967); sucrose solution (specific gravity 1.12) has been found effective for stream animals (Pask & Costa, 1971).

For semi-arid soils, where behavioural extractors can be inefficient, Walter et al.

---

[1] Obtainable from Du Pont de Nemours & Co., Wilmington, Delaware, USA.

(1987) recommend heptane flotation. The basic protocol is as follows (Walter *et al.*, 1987; Kethley, 1991).

1 Pour the sample into a stoppered measuring cylinder and add 1 L of 50% ethyl alcohol and about 10 ml of heptane.

2 Replace stopper; without shaking, invert cylinder and allow the heptane to rise. Repeat this step twice.

3 Allow the cylinder to stand until the fine sediment has settled—a minimum of 4 h.

4 With minimal disturbance of sediment, decant the heptane layer into a sieve.

5 Rinse the sieve with 95% ethyl alcohol to remove the heptane and wash the sample into a sorting dish.

Slugs can be separated by wet sieving and flotation (Pinder, 1974). If the sample is first dried, flotation alone may be used for snails (Vágvölgyi, 1953) and nematode cysts (Goodey, 1957). Vágvölgyi's method has been modified by Mason (1970), who found it 84% efficient. The sample is put into hot water and stirred. Dead shells and litter float and are removed; badly broken shells, soil, and the originally live snails, now killed by heat, sink. The sediment is then dried at 120°C for about 16 h, when the contents of the originally live snails contract; thus, when the sample is carefully crumbled into a weak solution of detergent, these now float and may be removed by careful searching. Badly broken shells, soil, and other debris again sink. With nematode cysts, drying, dry sieving, and flotation in water is the usual sequence; various details and modifications are given by Goodey (1957) and in papers in Murphy (1962a).

A flotation technique for separating fine plant material from insect eggs is described by King (1975). After rotary sieving, the mixture of eggs, fine plant debris, and sand was added to water containing one part per hundred of 6% hydrogen peroxide. Oxygen, produced from the hydrogen peroxide when it comes into contact with the plant debris, causes this to float to the surface. After removal of the plant debris, the eggs may be floated off in magnesium sulphate and glycerine solution (specific gravity 1.2).

6.3.5 The separation of plants and insects by differential wetting

A mixture of animal and plant material results from soil washing and flotation and also from various methods of sampling animals on vegetation (Chapter 4).

There are three possible approaches to their separation, one depending on the flotation effect of oxygen derived from decomposing hydrogen peroxide (see above) and two dependent on the lipoid and waterproof cuticle of arthropods. Either the arthropod cuticle can be wetted by a hydrocarbon 'oil' or the plant material can be waterlogged so that it will sink in aqueous solutions.

6.3.5.1 Wetting the arthropod cuticle

If arthropods and plant material are shaken up in a mixture of petrol or other hydrocarbon and water and then allowed to settle, the arthropods, whose cuticles are wetted by the petrol, will lie in the petrol layer above the water and the plant material in the water. As Murphy (1962b) has pointed out, this separation will be imperfect if the plant material contains much air or if the specific gravities of the two phases are either too dissimilar or too close. The former may be overcome by boiling the suspension in water first (Cockbill *et al.*, 1945), the second by the choice of appropriate solutions. Ethyl alcohol (60%) may be used for the aqueous phase with a light oil. This concept was originally introduced into soil faunal studies by Salt and Hollick (1944). The float from the collection tube (see Fig. 6.3) is placed in a wide-necked vessel with a little water, benzene is added, and the whole is shaken vigorously and allowed to settle. The arthropods will be in the benzene phase, which may be washed over into an outer vessel by adding further water below the surface from a pipette or wash-bottle. Raw (1955) introduced the idea of freezing the benzene (m.p. 50°C); the plug plus animals may then be removed and the benzene evaporated in a sintered-glass crucible. (The fumes should be carried away by an exhaust fan, as they are a cumulative poison.) A combination of decahydronophalene (Dekalin) + carbon tetrachloride (specific gravity 1.2) and zinc sulphate solution (specific gravity 1.3) has been found particularly effective, the arthropods floating off in the organic solvent mixture and the organic debris in the zinc sulphate solution, run off in a separating funnel (Heath, 1965). Xylene and paraffin have also been used for the oil phase and gelatin, which solidifies on cooling, for the aqueous phase; in the latter case, it is the plant material that is removed in a plug. If the cuticle is not easily wetted by the benzene, extraction is improved if a solution with a specific gravity of about 1.3 is used instead of water and the initial mixture exposed to a negative pressure (Sugimoto, 1967).

6.3.5.2 Waterlogging the plant material

This process, by boiling under reduced pressure (Hale, 1964) or vacuum extraction, may be used as part of the above technique. Repeated freezing and thawing also serve to impregnate the vegetable matter with water.

Danthanarayan (1966) has devised a method for the separation of the eggs of the weevil *Sitona* from the float by waterlogging the plant material by repeated freezing, filtering, and transferring the whole float to a tube containing saturated sodium chloride solution; this is then centrifuged for 5 min (1500 r.p.m.) and the eggs and other animal matter come to the surface, whilst the plant material sinks. The eggs of some insects (e.g. the wireworm, *Ctenicera destructor*) sink, rather than float, in sodium chloride (salt) solution, because they are covered with small soil particles, which have adhered to the coating from the female's glands. Doane (1969) found that this could be overcome by centrifuging in 20% alcohol solution and cleaning in 5% bleach (sodium hypochlorite) solution, followed by flotation and centrifuging in salt solution.

6.3.5.3 Grease-film separation

Using the same principle as above—namely, the resistance of arthropod cuticle to wetting by water and its lipophilic nature—Aucamp (1967) devised a 'grease-film machine', in which greased glass slides are revolved in an aqueous suspension of the litter or soil sample; the arthropods adhere to the grease. However, small stones remove the grease, thus impairing the efficiency of the greased surface and providing an alternative surface. Shaw (1970) devised an apparatus to overcome this problem, in which the soil sample is enclosed in a cylinder of coarse mesh in the centre of a plastic box; the lid of the box is greased with anhydrous lanolin and the box filled with water containing a little sodium hexametaphosphate. Several such boxes may be revolved on a wheel, after which the grease and arthropods can be washed from the lid into a dish with hot soapy water or other solvent. Fragments and many whole arthropods may be recovered on a large scale from litter by a Speight (1973) 'greased-belt machine' (Fig. 6.4). The litter is poured in aqueous suspension on to a moving (9–12 m/min), fine nylon, bolting-cloth belt ($350 \, \text{cm} \times 15 \, \text{cm}$; 1.6 mm mesh), coated with petroleum jelly slightly thinned with liquid paraffin. At its lowest point, the belt passes through a water trough, when mineral and plant debris falls off. Subsequently, the belt, with the arthropod material now on its undersurface passes over a special rebated roller, under a powerful jet of water which washes the arthropods off into a tray. Speight found that extraction efficiency for whole arthropods or fragments was 70–90%, except for whole organisms with diameters of less than 0.75 mm or greater than 10.0 mm, for which it is unsatisfactory. About 1 L of sample material can be extracted in 1 h; the belt can run for 10 h before 'regreasing' is necessary. The apparatus is particularly appropriate for dead organisms and fragments. To remove fine organic matter, Speight treated all samples with formalin, before wet sieving (1 mm mesh) and adding to the feeder tank.

6.3.6 Centrifugation

This technique is little used for microfaunal studies, because only small samples can be treated at any time. Müller (1962) found that Acarina and Collembola could be more efficiently extracted from soil by centrifugal flotation in saturated salt (sodium chloride) solution than by certain funnels and, after initially separating mosquito

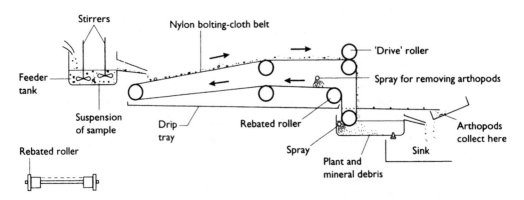

Fig. 6.4 Speight's greased-belt machine for the separation of arthropods from litter (after Speight, 1973).

eggs from soil by washing and flotation in magnesium sulphate (see above), Service (1968) completed the extraction by this method. A similar technique has been used to separate nematode cysts from plant debris. Murphy (1962b) gives a useful summary, in tabular form, of early work using centrifugal flotation to separate animals from foodstuffs, soil, and excreta.

For the extraction of marine meiofauna, McIntyre and Warwick (1984) describe a technique which comprises the following steps.

1 Remove large particles and fine silt by decanting (see below) and sieve separation (p. 216).

2 Wash the contents of the sieve into a centrifuge tube.

3 Add about a heaped teaspoon of kaolin to every 250 ml of sample, shake vigorously, and spin for 7 min at 6000 r.p.m.

4 Pour off supernatant, add Ludox–TM (specific gravity 1.15) and resuspend sediment.

5 Spin again at 6000 r.p.m. for 7 min.

6 Pour supernatant into a sieve and wash with tap water before placing in a sorting dish.

7 Repeat steps 4–6 twice more.

6.3.7 Sedimentation

Sedimentation is of little value for terrestrial and marine macrofaunal studies; however, live marine meiofauna, comprising groups such as nematodes, copepods, ostracods, and gastrotrichs, which will pass through a 1 mm sieve, can be separated by decantation (McIntyre & Warwick, 1984). The basic procedure has the following steps.

1 Wash the sample into a stoppered measuring cylinder and make up to 800 ml. The size of the sample should give a sedimentation depth of about 30 cm.

2 Invert the cylinder several times then leave to settle for 60 s.

3 Decant the supernatant through a sieve of pore size $\leq 63 \mu$m.

4 Repeat steps 1–3 three times.

5 Wash material from the sieve for sorting and counting.

To stop the animals attaching to sediment particles and sinking, magnesium chloride or alcohol can be added to the sample. This technique, while ideally applied to live samples, can also be used for dead material. For groups such as nematodes, elutriation (see below) may be more efficient.

Seinhorst (1962) describes a sedimentation method ('two Erlenmeyer method') for soil nematodes, in which the flask containing the suspension was moved across a series of collecting vessels, the first of which was also inverted and the sediment in it further precipitated. The efficiency of the method, however, was only 60–75%. Another sedimentation method for nematodes is described by Whitehead and Hemming (1965), but their 'tray method' (p. 235) was generally more efficient.

6.3.8 Elutriation

Elutriation is the separation of organisms by washing a sample in a constant stream of water adjusted so that the animals are carried away and the heavier sediment remains behind. The technique is particularly appropriate for marine benthic samples from sand habitats. In some methods, both water and air jets are used (Lauff *et al.*, 1961; Worswick & Barbour, 1974). For macrofauna, an apparatus which is easily constructed is Barnett's fluidized sand-bath (Fig. 6.5), for which efficiencies of 98–100% have been claimed. Water is forced vertically via a scintered sheet into a chamber holding sand, which becomes fluidized. The sample is tipped into this chamber and the animals rise to the water surface, where they are retained by a sieve on the overflow. This apparatus can be used for samples as large as 20 L and separation can be completed in 10 min. For meiofauna, an elutriation system can be easily built in the laboratory from standard glassware. The method was first used for meiofauna by Boisseau (1957); the general design of the apparatus is shown in Fig. 6.5. Hockin (1981) gives a simple design using Quickfit glassware. Small terrestrial arthropods, such as Pauropoda and some Collembola, which float in water, may be extracted by von Törne's (1962a) elutriator and sieving process.

Fig. 6.5 Barnett's fluidized sand-bath. Water enters at the bottom, passes upwards, and overflows at the top. Large animals can be removed from the surface of the sand and small animals filtered from the outflowing water (after Eleftheriou & Holme, 1984).

Elutriation is incorporated in two widely used methods for the extraction of soil nematodes: the Oostenbrink and Seinhorst elutriators. In the Oostenbrink model (Fig. 6.6a), the nematodes are washed out of the sample and carried in suspension by the opposing water currents out of the overflow and through a series of sieves (Oostenbrink, 1954; Goffart, 1959; Murphy, 1962b).

The Seinhorst elutriator is also part of a combined elutriation and sieving process. The soil is first passed through a coarse sieve and the suspension retained in a flask. This is inverted on top of the elutriator and the cork removed *in situ* (Fig. 6.6b); the upward flow of water (45 ml/min), coupled with the narrow sections A_2 and B_2, ensures that the nematodes are retained in sections A and B. After about half an hour, the flask and the sections A and B may be drained through stopcocks A_3 and B_3. This material and that collected from the overflow is then wet-sieved. A few large worms may pass into the soil-collecting container and this should be re-elutriated, but only for a brief period so that they will still be retained in vessel A

Fig. 6.6 Elutriators. (a) Oostenbrink's model. (b) Seinhorst's model (after Murphy, 1962a).

(Goodey, 1957; Seinhorst, 1962). This method might well be adapted for single-species studies on other organisms; those of a comparatively uniform size and mass, e.g. eggs, would be easily separated at a single level, whereas healthy, parasitized, and dead individuals having different masses would separate at different levels.

In a comparative study of methods for the extraction of nematodes from leaf litter, Schouten and Kmarp (1991) found the blender–elutriation–cotton-wool method to be the most efficient. The sample of litter is first homogenized in a blender. The nematodes are then extracted, using a funnel elutriator (as described above), the trapped suspension is sieved through a 40 μm sieve, and finally the sieve washings are filtered through a double cotton-wool filter.

6.3.9 Sectioning

Small animals may be examined and counted in soil sections. Haarløv and Weis-Fogh (1953, 1955) devised a method of fixing and freezing a soil core and then impregnating it with agar and sectioning. Gelatin is probably a more suitable substance for impregnation (Minderman, 1957; Anderson & Healey, 1970) and an automatic method for this is described by Vannier and Vidal (1964). Although the method is valuable for giving qualitative information on feeding sites and micro-distribution, it is less suitable than other methods for population studies (Pande & Berthet, 1973).

6.4 Behavioural or dynamic methods

In these methods, the animals are made to leave the substrate under some stimuli, e.g. heat, moisture (lack or excess), or a chemical. Their great advantage is that, unlike the mechanical methods, once the extraction has been set up, it may usually be left, virtually unattended, and thus large quantities of material may be extracted simultaneously in batteries of extractors. Another important advantage is the ability to extract animals from substrates containing a large amount of vegetable material. The disadvantage is that, being based on the animal's behaviour, the extraction efficiency will vary with the condition of the animals and be influenced by changes in climate, water content, etc., experienced before and after sampling, as well as by variations in these conditions in the apparatus itself. If samples have to be retained for several days, polythene bags may be suitable containers (Rapoport & Oros, 1969), leaf litter may be held in cloth bags (e.g. cotton pillowcases) for up to 4–5 days, as long as they are turned frequently and not exposed to excess moisture or extreme temperatures. Obviously, eggs and other immobile stages cannot be extracted by this method.

The exact behavioural mechanism of extraction is not fully understood. Temperature gradients are established from the start, but in some types of funnel the humidity gradient is not clearly marked in the lower part of the sample (Brady, 1969). The pattern of egress of the mites is irregular and there is often a marked 'flush' towards the end of the extraction (Brady, 1969), this has been associated with the moisture content reaching about 20%, which triggers positive geotaxis (Nef, 1970, 1971). It seems that oribatid mites are positively geotactic in dry conditions and negatively geotactic when the soil is moist; the adaptive significance of such responses is obvious and they are probably an important part of the mechanism of the vertical movements of soil fauna (p. 214). Whether all extractors act in this way, by the animal responding to a definite stimulus, rather than moving along a gradient, is uncertain. Many workers have sought to maximize the gradients (e.g. MacFadyen, 1961; Kempson *et al.*, 1963) and these extractors have been found efficient (e.g. Edwards & Fletcher, 1970). There can be no doubt that, for certain groups, it is essential that the stimuli (usually heating and drying) are not applied too quickly (Lasebikan, 1971), but, with the diversity of soil organisms, it is not surprising that it is impossible to construct an extractor that is equally efficient for all groups. Therefore, although dry funnels, in particular, have been used extensively for community studies, their efficiency varies with soil type and animal group (MacFadyen, 1961; Satchell & Nelson, 1962; Block, 1966; Edwards & Fletcher, 1970). Edwards (1991) has reviewed most modifications of behavioural extractors and provides details of preferred methods for different soil types (see p. 240).

6.4.1 Dry extractors

The basic apparatus is the Berlese–Tullgren funnel, the heated-copper funnel designed at the turn of the century by the Italian entomologist, A. Berlese, and subsequently modified by the Swede, A. Tullgren, who used a light-bulb as a heat source. The funnel has been considerably modified and improved by many workers, as described in the reviews of MacFadyen (1962), Murphy (1962a), and Edwards (1991). Some of the most important innovations have been the demonstration by Hammer (1944) that, when soil is being extracted, the core should be retained intact and inverted, thereby enabling the animals to leave the sample by the natural passageways, and the discovery by Haarløv (1947) that serious losses could result from the animals becoming trapped in condensation from the core on the sides of the funnel. Haarløv (1947) recommended that the core should never touch the sides of the funnel and subsequent workers have sometimes referred to the space between the core and the sides of the funnel as the 'Haarløv passage'. When litter is being extracted and the 'passage' is difficult to maintain, it may be helpful to increase air circulation with wide plastic piping (Paris & Pitelka, 1962). Large numbers of small funnels were first grouped together by Ford (1937), and this approach has been much developed and improved by MacFadyen (1953, 1955, 1961, 1962), who also introduced the concepts of steepening the heat gradient and arranging a humidity gradient. Murphy (1955), Newell (1955), Dietrick *et al.* (1959), and Kempson *et al.* (1963) have all introduced devices to reduce the fall of soil into the sample, thereby ensuring a cleaner extraction. A folding Berlese funnel for use on expeditions has been developed by Saunders (1959) and a simpler model, in which the funnels are constructed of oilcloth and the animals 'forced down' by the vapour from naphthalene held in cheesecloth bags above the sample, has been devised by Brown (1973).

For large volumes of litter, Norton and Kethley (1988) describe a lightweight, collapsible, and easily transportable Berlese–Tullgren funnel that is simple to construct.

There are many variants of the dry funnel in use, but most of them approximate to one of the following types.

6.4.1.1 Large Berlese funnel

This is used for extracting large arthropods, e.g. Isopoda, Coleoptera, from bulky soil or litter samples (MacFadyen, 1961) and also for the extraction of insects from suction apparatus and other samples containing much vegetation (Clark *et al.*, 1959; Dietrick *et al.*, 1959; Dondale *et al.*, 1971). It can also be used to extract microarthropods from materials such as moss (Smrz, 1992). Desirable features are an air circulation system, introduced by MacFadyen, which may be opened to ensure a rapid drying for the extraction of desiccation-resistant animals, such as beetles and ants, or partly closed for a slower 'wet regime' (but avoid condensation) for beetle larvae, *Campodea*, and other animals that are susceptible to desiccation. MacFadyen recommends testing the humidity below the sample by cobalt thiocyanate papers; for the resistant animals, humidities down to 70% are tolerable, but for the others it should not fall below 90%. As phototactic animals, e.g. halticine beetles, are often attracted upwards towards the light-bulb of the funnel and so fail to be extracted, this is replaced as a heat source by a Nichrome or similar wire-grid heating element—this also heats the sample more uniformly (Clark *et al.*, 1959). The extraction of positively phototactic animals has been further improved by Dietrick *et al.* (1959), who placed a 75 W spotlight below the collecting jar; this was switched on intermittently. An apparatus incorporating these and other features is illustrated in Fig. 6.7. As MacFadyen and others have frequently stressed, for most animals the heat should be applied slowly and therefore it may be useful to be able to control the voltage of the heating element via a rheostat (Dietrick *et al.*, 1959). Optimal temperature and humidity levels over the course of an extraction can be achieved using a computerized control system. Extraction time will vary from a few days to over a month (e.g. Park & Auerbach, 1954), depending on the substrate and the animal.

Most recent enhancements to the Berlese funnels have been designed to produce steep temperature and humidity gradients. Crossley and Blair (1991) describe a low-cost design that can be used in a cool room, with Christmas-tree lights used to produce a 20°C gradient.

6.4.1.2 Horizontal extractor

Designed by Duffey (1962) for extracting spiders from grass samples, this extractor (Fig. 6.8) would probably serve equally well for any rapidly moving litter animal. It has the advantages that steep gradients are built up and that debris cannot fall into the collecting trough; it is also relatively compact and built in paired units. Heat is provided by a 500 W domestic heating element. Air passes in through the ventilation holes just above the aqueous solution of the collecting gutters; this helps to prevent

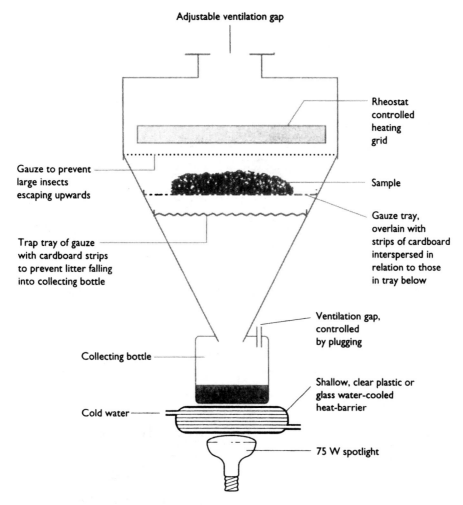

Fig. 6.7 Large Berlese funnel with modifications.

too rapid desiccation. Duffey recommended a 0.001% solution of phenylmercuric acetate (a fungicide and bactericide), with a few drops of a surface-active agent (detergent). The samples are held in wire trays 6 in (10 cm) wide. The extraction takes only a day or two.

6.4.1.3 High-gradient (multiple-canister) extractor

Collembola, mites, and other small animals may be obtained from small, compact, soil samples in this apparatus, designed by MacFadyen (1961). The samples are taken with the special corer (see Figs 6.1d and 6.9a) and removed, surrounded by the plastic ring; a sieve plate is fitted to each ring. The collecting canisters are cooled in a water-bath and 95% humidity is maintained below the samples during extraction (Fig. 6.9b). Heat is provided by two 160 W elements, giving 10 W per sample, and the temperature of the upper surface of the sample reaches 35°C; the temperature of

Fig. 6.8 A pair of horizontal extractors, based on Duffey's (1962) design. (a) Sketch. (b) Sectional view.

Fig. 6.9 High-gradient extractor (after MacFadyen, 1961). (a) Canister, core, and sieve plate. (b) Whole apparatus.

the lower surface is determined by the water-bath (e.g. 15°C), so that a very steep gradient is set up over the 3 cm deep sample. Extraction may be completed in 5 days, the maximum heat being applied after the first 24 h. MacFadyen recommended the use of a 0.3 solution of the fungicide Nipagin (which contains streptomycin and ter-ramycin) in the collecting canisters, but the tests of Kempson *et al.* (1963) (see below) indicate that a saturated aqueous solution of picric acid, with a little photo-graphic wetting agent, might be preferable. Constructional details are given by MacFadyen (1961), who has also (MacFadyen, 1962) described variants suitable for use under expedition conditions, the energy source then being kerosene or bottled gas, and the effect of various modifications on the physical conditions in the extractor (MacFadyen, 1968). Block (1966) found that this apparatus extracted 76% of the Acarina from mineral soils.

6.4.1.4 The Kempson bowl extractor (Fig. 6.10)

Developed by Kempson et al. (1963), this method is suitable for the extraction of mites, Collembola, Isopoda, and many other arthropods in woodland litter. Extraction rates of 90–100% are recorded for groups other than the larvae of holometabolous insects, for which it is not efficient. The apparatus consists of a box or shrouded chamber, containing an ordinary light-bulb and an infrared lamp (250 W), which is switched on in pulses, at first only for a few seconds at a time and gradually increased until it is on for about a third of the time. The pulsing of the lamp is controlled by a time switch. The extraction bowls are sunk into the floor of the chamber. Each bowl contains a preservative fluid and is immersed in a cold water-bath. A saturated aqueous solution of picric acid, with some wetting agent, was used by Kempson et al. (1963), but Sunderland et al. (1976) have found an 80 g/L solution of trisodium orthophosphate ($Na_3PO_4 \cdot 12H_2O$) to be a cheap, safe, and suitable alternative. The litter is placed in a plastic tray above the bowl and is supported by two pieces of cotton fillet net laid on a coarse plastic grid. The top of the tray is covered by a fine, black, nylon screen. Kempson et al. (1963) give full instructions for the construction of this apparatus. Its great advantages, besides its

Fig. 6.10 Kempson bowl extractor (after Kempson et al., 1963).

comparatively simple construction, are the high humidity maintained on the lower surface of the sample, the lack of debris among the extracted material (but the fillet net does not appear to present any barrier to the animals), the prevention by the black screen of animals escaping upwards, and the simple mechanism of gradually increasing the amount of heat to which the sample is subjected. Extractions with this apparatus generally take about a week.

6.4.2 Wet extractors

The principle of this method is similar to that of the dry extractors, the animals being driven out of their natural substrate under the influence of a stimulus, possibly heat or, as the observations of Williams (1960a) would suggest for some cases, reduced oxygen tension. As the substrate is flooded with water in all variants of the method, there is no risk of desiccation and the method is particularly successful with those groups that are but poorly extracted by a Berlese–Tullgren funnel—namely, nematode and enchytraeid worms, insect larvae, and various aquatic groups (Williams, 1960a). The method, which is faster than dry extraction, seems to have been discovered independently at least three times.

6.4.2.1 Baermann funnel

As originally designed, this consisted of a glass funnel with a piece of metal gauze or screen resting in the funnel and a piece of rubber tubing with a pinch-cock on the stem (Fig. 6.11a); the sample, which must not be very deep, is contained in a piece of muslin and partly flooded with warm water. Nematode worms leave the sample and fall to the bottom of the funnel, where they collect in the stem. They may be drawn off, in a little water, by opening the pinch-cock (Peters, 1955). A number of refinements of this method have been introduced.

Nielsen (Overgaard, 1947–48) ran a battery of Baermann funnels within a box

Fig. 6.11 (a) Simple Baermann funnel. (b) Sand extractor.

with a lamp above; this heated the surface of the water in the funnels, which was initially cold. Nematodes and rotifers could be extracted from soil and moss by this method, but it was not satisfactory for tardigrades. In order to extract Enchytraeidae, O'Connor (1957, 1962) used larger polythene funnels, spread the sample directly on to the gauze, without muslin, and submerged it completely; a powerful shrouded lamp heated the surface of the water to 45°C in 3 h. Dry samples must be moistened before being placed in the funnel and, if the soil temperature is low, samples must be gradually warmed to room temperature for a day. Extraction usually takes about 3 h. In a series of tests, O'Connor (1962) and Peachey (1962) have shown this method to be as efficient as the Nielsen inverted extractor (see below) for grassland soils and more efficient for peat and woodland soils. It is also very efficient for second-instar and later tipulid larvae in moorland soils (Hadley, 1971), but Kiritani and Matuzaki (1969) found that only 10% of the larvae of the cotton root-knot eelworm, *Meloidogyne incognita*, were extracted after a week. Didden *et al.* (1995) compared the efficiency of extraction of Enchytraeidae by wet-funnel techniques and concluded that all of the methods tried gave good estimates. However, the omission of heating and a long extraction time gave the highest rates of extraction. These results suggest that, given sufficient time, heat should not be used to extract enchytraeids.

Whitehead and Hemming (1965) introduced a modification that has been termed the 'Whitehead tray'. The soil sample is placed in a small (23 cm × 33 cm), shallow tray of 8 mesh/cm phosphor-bronze wire cloth, the inside of the tray being lined with paper tissue and the outside supported by a plastic-coated wire basket. This stands in a plastic photographic tray or similar vessel and is flooded with water until the surface reaches the underside of the soil (i.e. it is not inundated as much as in the Baermann funnel (Fig. 6.11a)). After 24 h at room temperature, the wire basket is carefully lifted out and the water in the tray agitated and poured into vessels resembling separating funnels. After about 4 h, the nematodes collect at the tapered part of the funnel, which may be separated off by a rubber bung or diaphragm on a handle; the stopcock is opened and the nematodes are drawn off. The process may be repeated to concentrate them further. In comparative tests with clay, loam, and sand and different nematodes, Whitehead and Hemming (1965) showed that this method usually extracted more efficiently than Seinhorst's 'two Erlenmeyer method', elutriation, or a sedimentation method they developed.

6.4.2.2 Hot-water extractors

These have been developed by Nielsen (1952–53) for Enchytraeidae and by Milne *et al.* (1958) for tipulid larvae. Nielsen's apparatus involved the heating of the lower surface of the sample, contained in an earthenware vessel, by hot water; the animals moved up and entered a layer of sand, kept cool by a cold-water coil, which was placed above the sample. The only advantage of this method over O'Connor's funnel (see above) is that it seems to extract more of the young worms (Peachey, 1962); under conditions with a large amount of humus, it is less efficient (see above).

The larvae of many insects, notably of some Coleoptera and Diptera, earthworms, molluscs, and most motile animals seem to be efficiently and rapidly extracted from soil samples if these are heated from below. Schjøtz-Christensen (1957) recorded 85% extraction of Elateridae using Nielsen's apparatus, modified by heating more strongly and omitting the cooling coil. Such an apparatus is not very different from that of Milne et al. (1958) (Fig. 6.12), which was found to be virtually 100% efficient for active larvae and pupae. Essentially, their device consists of two galvanized boxes, one within the other; the space between is a water-bath heated by a thermostatically controlled immersion heater. The water temperature should not rise above 90°C. The inner box has a wire-gauze base and holds the sample. The initial mode of operation is to fill the water-bath until the level is just above the base of the sample, the heater is switched on until the temperature registered by the thermometer in the centre of the turf is 40°C, and the water level is then raised to the level where it would just, eventually, flood the surface of the sample.

Heating and observation are continued, the insects being picked up as they leave the turf (but not before or they will retreat). The light above the extraction apparatus should not be too bright; Milne and co-workers recommended as a maximum, a 4 ft (122 cm), 40 W, natural fluorescent tube 3 ft (91.5 cm) above the sample; however, other species of insects might be more or less sensitive. Extraction times varied from 1 h in a light soil to nearly 3 h in heavy peat-covered clay. The chief disadvantage of this method as at present designed is that the apparatus must be watched continuously and the animals removed as they appear; however, equivalent mechanical methods usually take longer.

6.4.2.3 Sand extractors
Extremely simple in design, consisting of a metal can, a piece of wire gauze, and some sand (Fig. 6.11b), these devices seem to be remarkably efficient for extracting

Fig. 6.12 Hot-water extractor (after Milne et al., 1958). (a) Sketch with segment 'removed'. (b) Sectional diagram.

aquatic and semi-aquatic insect larvae and other animals from mud and debris. The method was described by Bidlingmayer (1957) for the extraction of ceratopogonid midge larvae from salt marshes and has been improved and extended by Williams (1960a,b), who has found that the following freshwater invertebrates could be extracted from their soil substrate by this technique: Coelenterata, Turbellaria, Nematoda, Oligochaeta, Ostracoda, Acarina, dipterous larvae, Gastropoda, and Plecopoda. The sample, about 5 cm thick, is cut so as to fill the container completely; it is then covered with 5 cm of dry, clean sand and the whole flooded with water. After 24 h, the sand may be scooped off and will contain large numbers of animals, but for 100% extraction of *Culicoides* larvae it should be left for 40 h (Williams, 1960b) and for other groups tests would need to be made. The animals may be separated from the sand, as Bidlingmayer and Williams recommend, by stirring in a black photographic tray, when the pale moving objects easily show up against the dark background. If the promise of this method is justified, it would clearly repay further development. The final separation might be facilitated by the adoption of a sedimentation, elutriation, or flotation method, aided perhaps by the replacement of the sand by fine, washed gravel, as used by Nielsen (1952–53) in his extractor (see above), which really combines this method with the hot-water process).

The chief disadvantages of the method in its present form is that it is difficult to make a clean separation of the sand from the soil after extraction. The placing of a piece of wide-mesh wire gauze between the two substrates might aid this operation, or, if the extraction is done in a square vessel, a plate could be pushed across between the layers at the end of extraction and the top layer poured off.

6.4.2.4 Cold-water extractor

South (1964) found that slugs could be efficiently extracted from 1 ft (30.5 cm) deep turfs by slowly immersing these in water. Initially, they are stood in 1 in (2.5 cm) of water; after about 17 h, the water level is raised to half the depth of the turf (i.e. 6 in (15.2 cm)); after 2 days, the water level is again raised, this time in several stages, until, at the end of another 24 h, it is within $\frac{1}{2}$ in (1.2 cm) of the top of the turf. This approach has also been used to extract the collembolan *Hyopogastrura* (van der Kraan, 1973) and overwintering rice water weevils, *Lissorhoptrus* (Gifford & Trahan, 1969). The latter authors used a technique basically similar to the 'van Emden biscuit tin' (Fig. 10.4, p. 356), but constructed from plastic bleach bottles, in which the sample is flooded. Such a technique is, of course, only suitable for robust insects.

6.4.2.5 Low-salinity extractor

Uhlig (1966, 1968) developed a method for marine meiofauna using melting seawater ice. The sample is placed in a tube with a gauze base. A layer of cotton wool, followed by crushed sea ice, is placed above it and the tube suspended in a dish of sea water (Fig. 6.13). As the ice melts, the flow of cold, low-salinity water forces the fauna into the dish. Uhlig considered that it was the salinity rather than the

Fig. 6.13 Cold-water extractor for marine interstitial fauna (after Uhlig, 1966, 1968).

temperature gradient that forced the animals to move. This method can only be applied to sandy substrates.

6.4.2.6 Mistifier
This is essentially a method of extracting active nematodes from plant tissue. The plant material is spread on a tray, broken up, and continually sprayed with very fine jets of water ('mists') (Webster, 1962). Combined with a cotton-wool filter to collect the nematodes, this method can be much more efficient than other techniques for certain eelworms (de Guiran, 1966; Oostenbrink, 1970).

6.4.2.7 Chemical extraction
Chemical fumes may be used to drive animals from vegetation (p. 151); the extension of such methods to soil and litter animals has been successful only with aphids and thrips (MacFadyen, 1953) (MacFadyen used dimethyl phthalate and 2-cyclohexyl-4,6-dinitrophenol). Chemicals in a liquid form, notably potassium permanganate solution, orthodichlorobenzene, and formalin, often with a 'wetter', have been use for the extraction of insect larvae and earthworms (Svendsen, 1995; Milne *et al.*, 1958; Raw, 1959). Formalin should be applied four times, with 15 min intervals, at the rate of 75 ml of 4% formaldehyde per 3.5 L water per square metre of ground

(Raw, 1959; Nordstrom & Rundgren, 1972). Many workers have found these methods to be inefficient for earthworms (Satchell, 1955; Svendsen, 1955); for tipulids in the grasslands, studied by Milne *et al.* (1958), the efficiency was 85%. However, the method can be efficient, depending on the type of soil, the species of animal, and the season (which will influence the activity level of the animal). The more porous the soil (Boyd, 1958) and the more active the earthworm, the more satisfactory the method. Among earthworms, it is as efficient as hand-sorting for *Lumbricus terrestris* during active periods; on sandy soils, it seems suitable for *Lirabellus* and *Dendrobaena octaedra*, but is seldom reliable for *Allobophora* spp. Where it is efficient, it is more reliable for small worms than hand-sorting (Nordstrom & Rundgren, 1972). Daniel *et al.* (1992) found little difference in the efficiency of extraction of juvenile *L. terrestris* using formaldehyde and chloroacetophenone (CN).

6.4.2.8 Electrical extraction

By discharging a current from a water-cooled electrode driven into the soil, Satchell (1955) was able to expel large numbers of earthworms; but, as he has pointed out, this method suffers from the disadvantage that the exact limits of the volume of soil treated are unknown (Nelson & Satchell, 1962) and the efficiency of extraction seems directly related to the soil pH (Edwards & Lofty, 1975). This method has been used to collect earthworms for pollution studies (Pizl & Josens, 1995).

6.5 Summary of the applicability of the methods

It has already been pointed out that the efficiency of the different methods varies greatly with the animal group and the substrate, as is shown, for example, by the studies of Raw (1959, 1960), Nef (1960), MacFadyen (1961), O'Connor (1962), and Satchell and Nelson (1962). The separation methods recommended by Edwards (1991) for different soil arthropod groups collected from woodland, pasture, and arable land are given in Table 6.1. The comments made here should be taken as only general indications; more specific information is given under each method.

6.5.1 Substrate type

Because of the difficulties of separating plant and animal material, the mechanical methods have generally been considered unsuitable for litter and for soils, pond mud, and other substrates containing a large amount of organic matter. For these media, behavioural methods are likely to be the most satisfactory, and the hot-water and sand extraction techniques may well repay further development. There are, however, some exceptions; flotation is suitable for dung-dwelling insects (Laurence, 1954), Williamson's wet sieving (which may contain a behavioural element) works well for snails in fresh litter, and Satchell and Nelson (1962) found that flotation was more efficient than dry funnels for the extraction of scutacarid mites from a moder soil.

Table 6.1 Recommended methods for extraction of soil arthropods (after Edwards, 1991).

| | Soil type/system | | | | | | | | |
| | Peat | | | Clay/Loam | | | Sand | | |
Group	W	P	A	W	P	A	W	P	A
Isopoda	AC	ABC	ABC	AC	ABC	ABC	All	All	All
Pauropoda	ABC	ABC	ABC	ABC	ABC	All	All	All	All
Symphyla	ABC	ABC	ABC	All	ABC	All	All	All	All
Diplopoda	ABC	ABC	ABC	ABC	ABC	All	All	All	All
Chilopoda	ABC	ABC	ABC	All	ABC	All	All	All	All
Acarina	ABC	ABC	ABC	ABCF	ABCDE	ACD	ABCF	ABC	ABC
Pseudoscorpionida	ABC	ABC	ABC	ABC	ABC	ABC	All	All	All
Araneae	ABC	ABC	ABC	ABC	ABC	ABC	ABC	ABC	All
Collembola	ABC	ABC	ABC	All	ABCD	ABCD	ABCD	ABCD	ABCD
Protura	ABC	ABC	ABC	ABC	All	All	All	All	All
Psocoptera	ABC	ABC	ABC	All	All	All	All	All	All
Thysanoptera	ABC	ABC	ABC	DE	DE	All	All	All	All
Hemiptera	ABC	ABC	ABC	All	All	All	All	All	All
Hymenoptera	ABC	ABC	ABC	All	All	All	All	All	All
Coleoptera	ABC	ABC	ABC	All	All	CDEF	CDEF	All	All
Diptera	ABC	ABC	ABC	All	All	All	All	CDEF	CDEF
Eggs/pupae	DEF	DEF	DEF	DEF	DEF	DEF	DEF	DEF	DEF

All, all methods suitable; A, Tullgren funnel; B, Kempson extractor; C, air-conditioned funnel; D, Salt and Hollick flotation; E, other flotation; F, grease-film extractor; W, woodland; P, pasture; A, arable.

With friable and sandy soils or aquatic sediments, mechanical methods are generally to be preferred, and techniques as simple as wet sieving or flotation, perhaps using a heavy organic solvent, may be satisfactory for the study of a single species. Marine benthic macrofauna are usually preserved in the field, where a stain such as rose bengal can be added; later they can be sorted by wet sieving followed by manual picking of individuals from the debris. Occasionally, flotation or elutriation is used, possibly after initial sieving.

6.5.2 Animal type

Obviously, immobile stages can only be extracted by mechanical methods. Comparatively large and robust animals can often be estimated simply by sieving. When the specific gravity of the animal is very different from that of the substrate particles, possible techniques include elutriation and centrifugal separation, as well as the more widely used flotation. The two first-named methods might also be useful for the separation of parasitized and unhealthy individuals from healthy ones and for the extraction of the eggs of a single species. Broadly speaking, Acarina and Collembola seem to be more efficiently extracted by an appropriate funnel technique and insects, except on peat soils, by a flotation method (Edwards & Fletcher, 1970). Marine benthic meiofauna in coarse sediments can often be separated by simple decantation or flotation and from muddy substrates by sieving and centrifugation.

Behavioural methods must naturally be adapted to fit the behaviour of the animal. Fast-moving animals that are comparatively resistant to desiccation need to be extracted by the horizontal extractor or the large Berlese funnel; the last-named, when modified as described here, is also the only method really suitable for positively phototactic species, such as lygaeid bugs and phytophagous beetles. Groups such as Collembola and most mites need a slower extraction in the Kempson or multiple-canister extractors. Those animals that are very sensitive to desiccation (e.g. nematode and enchytraeld worms, dipterous and probably many coleopterous larvae) should be separated in wet extractors, which is also, of course, the behavioural method appropriate for animals in aquatic sediments. The degree of complexity for a single group is well illustrated by the earthworms, for, while *L. terrestris*, but not other species, may be efficiently extracted from orchard soils by a surface application of formalin and several species from shallow soils by wet sieving and flotation, in other habitats even hand-sorting fails (Raw, 1959, 1960; Nelson & Satchell, 1962).

6.5.3 Cost

The time required to undertake an extraction is the most important factor determining the total cost of the procedure. Methods differ greatly in labour and time required. When using behavioural methods, there is often a direct relationship between cost (time) and extraction efficiency. It has generally been found that behavioural methods work more quickly when the samples are exposed to higher gradients of environmental conditions, such as temperature, salinity, or humidity. However, using weaker gradients and taking longer over the extraction often, but by no means always, produce higher rates of extraction.

References

Abu Yaman, I.K. (1960) Natural control in cabbage root fly populations and influence of chemicals. *Meded. Landb. Hoogesch. Wageningen* **60**, 1–57.

Anderson, J.M. & Healey, I.N. (1970) Improvements in the gelatine-embedding technique for woodlands soil and litter samples. *Pedobiologia* **8**, 108–20.

Aucamp, J.L. (1967) Efficiency of the grease film extraction technique in soil micro-arthropod survey. In: Graff, O. & Satchell, J.E. (eds) *Progress in Soil Biology*. North-Holland, Amsterdam, pp. 515–24.

Belfield, W. (1956) The arthropoda of the soil in a west African pasture. *J. Anim. Ecol.* **259**, 275–87.

Bidlingmayer, W.L. (1957) Studies on *Culicoides furens* (Poey) at Vero Beach. *Mosqito News* **17**, 292–4.

Birkett, L. (1957) Flotation technique for sorting grab samples. *J. Comm. Perm. Int. Explor. Mer.* **22**, 289–92.

Block, W. (1966) Some characteristics of the MacFadyen high gradient extraction for soil micro-arthropods. *Oikos* **17**, 1–9.

Boisseau, J-P. (1957) Technique pour l'étude quantitative de la faune interstitielle des sables. *Comptes Rendus du Congrés des Societes Savantes de Paris et des Départements* **1957**, 117–19.

Boyd, J.M. (1958) Ecology of earthworms in cattle-grazed machair in Tiree, Argyll. *J. Anim. Ecol.* **27**, 147–57.

Brady, J. (1969) Some physical gradients set up in Tullgren funnels during the extraction of mites from poultry litter. *J. Appl. Ecol.* **6**, 391–402.

Brian, M.V. (1970) Measuring population and energy flow in ants and termites. In: Phillipson, J. (ed.) *Methods of Study in Soil Ecology*. UNESCO, Paris, pp. 231–4.

Brown, R.D. (1973) Funnel for extraction of leaf litter organisms. *Ann. Entomol. Soc. Am.* **66**, 485–6.

Clark, E.W., Williamson, A.L., & Richmond, C.A. (1959) A collecting technique for pink bollworms and other insects using a Berlese funnel with an improved heater. *J. Econ. Entomol.* **52**, 1010–12.

Cockbill, G.F., Henderson, V.E., Ross, D.M., & Stapley, J.H. (1945) Wire-worm populations in relation to crop production. I. A large-scale flotation method for extracting wireworms from soil samples and results from a survey of 600 fields. *Ann. Appl. Biol.* **32**, 148–63.

Crossley, D.A. & Blair, J.M. (1991) A high efficiency, 'low technology' Tullgren-type extractor for soil microarthropods. *Agric. Ecosys. Environ.* **34**, 187–92.

d'Aguilar, J., Benard, R., & Bessard, A. (1957) Une méthode de lavage pour l'extraction des arthropodes terricoles. *Ann. Epiphytol. C* **8**, 91–9.

Daniel, O., Jager, P., Cuendet, G., & Bieri, M. (1992) Sampling of *Lumbricus terrestris* (Oligocheata, Lumbricidae). *Pedobiologia* **36**, 213–20.

Danthanarayan, A.W. (1966) Extraction of arthropod eggs from soil. *Entomol. Exp. Appl.* **9**, 124–5.

Davies, J.B. & Linley, J.B. (1966) A standardised flotation method for separating *Leptoconops* (Diptera: Ceratopogonidae) and other larvae from sand samples. *Mosquito News* **26**, 440.

de Guiran, G. (1966) Infestation actuelle et infestation potentielle du sol par nématodes phytoparasites du genre *Meloidogyne*. *Comptes Rendus Hebd. Séances Acad. Sci. (Paris)* **262**, 1754–6.

De Jonge, V.N. & Bouwman, L.A. (1977) A simple density separation technique for quantitative isolation of meiobenthos using colloidal silica Lodox-TM. *Marine Biology* **82**, 379–84.

Dhillon, B.S. & Gibson, N.H.E. (1962) A study of the Acarina and Collembola of agricultural soils. 1. Numbers and distribution in undisturbed grassland. *Pedobiologia* **1**, 189–209.

Didden, W., Born, H., Domm, H., *et al.* (1995) The relative efficiency of wet funnel techniques for the extraction of Enchytraeidae. *Pedobiologia* **39**, 52–7.

Dietrick, E.J., Schlinger, E.I., & Van Den Bosch, R. (1959) A new method for sampling arthropods using a suction collecting machine and modified Berlese funnel separator. *J. Econ. Entomol.* **52**, 1085–91.

Dindal, D.L. (1990) *Soil Biology Guide*. Wiley Interscience, New York.

Doane, J.F. (1969) A method for separating the eggs of the prairie grain wireworm, *Ctenicera destructor*, from soil. *Can. Entomol.* **101**, 1002–4.

Dondale, C.D., Nicholls, C.F., Redner, J.H., Semple, R.B., & Turnbull, A.L. (1971) An

improved Berlese–Tullgren funnel and a flotation separator for extracting grassland arthropods. *Can. Entomol.* **103**, 1549–52.

Duffey, E. (1962) A population study of spiders in limestone grassland: description of study area, sampling methods and population characteristics. *J. Anim. Ecol.* **31**, 571–99.

Edwards, C.A. (1991) The assessment of populations of soil-inhabiting invertebrates. *Agric. Ecosys. Environ.* **34**, 145–76.

Edwards, C.A. & Fletcher, K.E. (1970) Assessment of terrestrial invertebrate populations. In: Phillipson, J. (ed.) *Methods of Study in Soil Ecology*. UNESCO, Paris, pp. 57–66.

Edwards, C.A. & Lofty, J.R. (1975) The invertebrate fauna of the Park Grass Plots. 1. Soil fauna. *Rep. Rothamsted Exp. Sta.* **1974**, *Part* 2, 133–54.

Edwards, C.A., Whiting, A.E., & Heath, G.W. (1970) A mechanized washing method for separation of invertebrates from soil. *Pedobiologia* **10**, 141–8.

Eleftheriou, A. & Holme, N.A. (1984) Macrofauna techniques. In: Holme, N.A. & McIntyre, A.D. (eds) *Methods for the Study of Marine Benthos*, 2nd edn. Blackwell Scientific Publications, Oxford, pp. 140–216.

Erman, D.C. (1973) Invertebrate movements and some diel and seasonal changes in a Sierra Nevada peatland. *Oikos* **24**, 85–93.

Ford, J. (1937) Fluctuations in natural populations of Collembola and Acarina. *J. Anim. Ecol.* **6**, 98–111.

Frick, K.E. (1962) Ecological studies on the alkali bee, *Nomia melanderi*, and its bombyliid parasite, *Heterostylum robustum* in Washington. *Ann. Entomol. Soc. Am.* **559**, 5–15.

Gabbutt, P.D. (1959) The bionomics of the wood cricket, *Nemobius sylvestris* (Orthoptera: Gryllidae). *J. Anim. Ecol.* **28**, 15–42.

Gerard, B.M. (1967) Factors affecting earthworms in pastures. *J. Anim. Ecol.* **36**, 235–52.

Gifford, J.R. & Trahan, G.B. (1969) Apparatus for removing overwintering adult rice water weevils from bunch grass. *J. Econ. Entomol.* **62**, 752–4.

Goffart, H. (1959) Methoden zur Bodenuntersuchung auf nichtzystenbildende Nematoden. *Nachr. Bl. deutsch. Pfl. Sch. Dienst. Stuttgart* **11**, 49–54.

Golightly, W.H. (1952) Soil sampling for wheat-blossom midges. *Ann. Appl. Biol.* **39**, 379–84.

Goodey, J.B. (1957) *Laboratory Methods for Work with Plant and Soil Nematodes*. HMSO, London.

Gorny, M. & Grum, L. (eds) (1993) *Methods in Soil Zoology*. Elsevier, Amsterdam.

Haarløv, N. (1947) A new modification of the Tullgren apparatus. *J. Anim. Ecol.* **16**, 115–21.

Haarløv, N. & Weis-Fogh, T. (1953) A microscopical technique for studying the undisturbed texture of soils. *Oikos* **4**, 44–7.

Haarløv, N. & Weis-Fogh, T. (1955) A microscopical technique for studying the undisturbed texture of soils. In: Kevan, D.K.M. (ed.) *Soil Zoology*. School of Agriculture, University of Nottingham, Nottingham, pp. 429–32.

Hadley, M. (1971) Aspects of the larval ecology and population dynamics of *Niolophilus ater* Meigen (Diptera: Tipulidae) on Pennine moorland. *J. Anim. Ecol.* **40**, 445–66.

Hale, W.G. (1964) A flotation method for extracting Collembola from organic soils. *J. Anim. Ecol.* **33**, 363–9.

Hamilton, A.C. (1969) A method of separating invertebrates from sediments using longwave ultraviolet light and florescent dyes. *J. Fish. Res. Bd. Can.* **26**, 1667–72.

Hammer, M. (1944) Studies on the oribatids and collemboles of Greenland. *Medd. Grønland* **141**, 1–210.

Hartley, J.P., Dicks, B., & Wolff, W.J. (1987) Processing sediment macrofauna samples. In: Baker, J.M. & Wolff, W.J. (eds) *Biological Surveys of Estuaries and Coasts*. Cambridge University Press, Cambridge, pp. 131–9.

Heath, G.W. (1965) An improved method for separating arthropods from soil samples. *Lab. Pract.*, April 1965.

Higgins, R.P. (1977) Two new species of Echinodered (Kinorhyncha) from South Carolina. *Trans. Am. Microscopical Soc.* **96**, 340–54.

Hockin, D.C. (1981) A simple elutriator for extracting meiofauna from sediment matrices. *Marine Ecology Progress Series* **4**, 241–2.

Horsfall, W.R. (1956) A method for making a survey of floodwater mosquitoes. *Mosquito News* **16**, 66–71.

Hughes, R.D. (1960) A method of estimating the numbers of cabbage root fly pupae in the soil. *Plant Pathol.* **9**, 15–17.

Iversen, T.M. (1971) The ecology of the mosquito population (*Aedes communis*) in a temporary pool in a Danish beech wood. *Arch. Hydrobiol.* **69**, 309–32.

Jacot, A.P. (1935) Molluscan populations of old growth forests and rewooded fields in the Asheville basin of N. Carolina. *Ecology* **16**, 603–5.

Kempson, D., Lloyd, M., & Ghelardi, R. (1963) A new extractor for woodland litter. *Pedobiologia* **3**, 1–21.

Kethley, J. (1991) A procedure for extraction of microarthropods from bulk soil samples with emphasis on inactive stages. *Agric. Ecosys. Environ.* **34**, 193–200.

Kevan, D.K.McE. (1962) *Soil Animals*. Witherby, London.

King, A.B.S. (1975) The extraction, distribution and sampling of the eggs of the sugarcane froghopper, *Aeneolamia varia saccharina* (Dist.) (Homoptera, Cercopidae). *Bull. Entomol. Res.* **65**, 157–64.

Kiritani, K. & Matuzaki, T. (1969) On the extraction rate of nematodes by the Baermann funnel. *Bull. Kochi Inst. Agric. For. Sci.* **2**, 25–30.

Kitching, R.L. (1971) A core sampler for semi-fluid substances. *Hydrobiologia* **37**, 205–9.

van der Kraan, C. (1973) Populations okologische untersuchungen an *Hypogastrura vintica* Tullb. 1872 (Collembola) anf Schiermonnikoog. *Faun-okol. Mitt.* **4**, 197–206.

Lackey, R.T. & May, B.E. (1971) Use of sugar flotation and dye to sort benthic samples. *Trans. Am. Fish. Soc.* **100**, 794–7.

Ladell, W.R.S. (1936) A new apparatus for separating insects and other arthropods from the soil. *Ann. Appl. Biol.* **23**, 862–79.

Lasebikan, B.A. (1971) The relationship between temperature and humidity and the efficient extraction of Collembola by a dynamic-type method. *Rev. Ecol. Biol. Sol* **8**, 287–93.

Lauf, G.M., Cummins, K.W., Enksen, C.H., & Parker, M. (1961) A method for sorting bottom fauna samples by elutriation. *Limnol. Oceanogr.* **6**, 462–6.

Laurence, B.R. (1954) The larval inhabitants of cow pats. *J. Anim. Ecol.* **23**, 234–60.

Lincoln, C. & Palm, C.E. (1941) Biology and ecology of the alfalfa snout beetle. *Mem. Cornell Univ. Agric. Exp. Sta.* **236**, 3–45.

Lyman, F.E. (1943) A pre-impoundment bottom fauna study of Watts Bar Reservoir area (Tennessee). *Trans. Am. Fish. Soc.* **72**, 52–62.

MacFadyen, A. (1953) Notes on methods for the extraction of small soil arthropods. *J. Anim. Ecol.* **22**, 65–78.

MacFadyen, A. (1955) A comparison of methods for extracting soil arthropods. In: Kevan, D.K.McE. (ed.) *Soil Zoology*. School of Agriculture, University of Nottingham, Nottingham, pp. 315–32.

MacFadyen, A. (1961) Improved funnel-type extractors for soil arthropods. *J. Anim. Ecol.* **30**, 171–84.

MacFadyen, A. (1962) Soil arthropod sampling. *Adv. Ecol. Res.* **1**, 1–34.

MacFadyen, A. (1968) Notes on methods for the extraction of small soil arthropods by the high gradient apparatus. *Pedobiologia* **8**, 401–6.

McIntyre, A.D. & Warwick, R.M. (1984) Meiofauna techniques. In: Holme, N.A. & McIntyre, A.D. (eds) *Methods for the Study of Marine Benthos*, 2nd edn. Blackwell Scientific Publications, Oxford, pp. 217–44.

McLeod, J.M. (1961) A technique for the extraction of cocoons from soil samples during population studies of the Swaine sawfly, *Neodiprion swainei* Midd. (Hymenoptera: Diprionidae). *Can. Entomol.* **91**, 888–90.

Mason, C.F. (1970) Snail populations, beech litter production and the role of snails in litter decomposition. *Oecologia (Berlin)* **5**, 215–39.

Milne, A., Coggins, R.E., & Laughlin, R. (1958) The determination of the number of leatherjackets in sample turves. *J. Anim. Ecol.* **27**, 125–45.

Minderman, G. (1957) The preparation of microtome sections of unaltered soil for the study of soil organisms *in situ. Plant Soil* **8**, 42–8.

Müller, G. (1962) A centrifugal–flotation extraction technique and its comparison with two funnel extractors. In: Murphy, P.W. (ed.) *Progress in Soil Zoology*. Butterworths, London, pp. 207–11.

Murphy, P.W. (1955) Notes on processes used in sampling, extraction and assessment of the meiofauna of heathland. In: Kevan, D.K.McE. (ed.) *Soil Zoology*. School of Agriculture, University of Nottingham, Nottingham, pp. 338–40.

Murphy, P.W. (ed.) (1962a) *Progress in Soil Zoology. Papers from a Colloquium on Research Methods Organised by the Soil Zoology committee of the International Society of Soil Science held at Rothamsted Experimental Station Hertfordshire, 4 July 1958*. Butterworths, London.

Murphy, P.W. (1962b) Extraction methods for soil animals. II. Mechanical methods. In: Murphy, P.W. (ed.) *Progress in Soil Zoology*. Butterworths, London, pp. 115–55.

Nef, L. (1960) Comparaison de l'efficacité de différentes variantes de l'appareil de Berlese–Tullgren. *Z. angew. Entomol.* **46**, 178–99.

Nef, L. (1970) Réactions des acariens à une dessiccation lente de la litière. *Rev. Ecol. Biol. Sol* **7**, 381–92.

Nef, L. (1971) Influence de l'humidité sur le géotactisme des Oribates (Acarina) dans l'entracteur de Berlese–Tullgren. *Pedobiologia* **11**, 433–45.

Nelson, J.M. & Satchell, J.E. (1962) The extraction of Lumbricidae from soil with special reference to the hand-sorting method. In: Murphy, P.W. (ed.) *Progress in Soil Zoology*. Butterworths, London, pp. 294–9.

Newell, I. (1955) An autosegregator for use in collecting soil-inhabiting arthropods. *Trans. Am. Microsc. Soc.* **74**, 389–92.

Nielsen, C.O. (1952–53) Studies on Enchytraeidae. 1. A technique for extracting Enchytraeidae from soil samples. *Oikos* **4**, 187–96.

Nordstrom, S. & Rundgren, S. (1972) Methods of sampling lumbricids. *Oikos* **239**, 344–52.

Norton, R.A. & Kethley, J.B. (1988) A collapsible, full-sized Berlese-funnel system. *Entomological News* **99**, 41–7.

O'Connor, F.B. (1957) An ecological study of the Enchytraeid worm population of a coniferous forest soil. *Oikos* **8**, 162–99.

O'Connor, F.B. (1962) The extraction of Enchytraeidae from soil. In: Murphy. P.W. (ed.) *Progress in Soil Zoology*. Butterworths, London, pp. 279–85.

Oostenbrink, M. (1954) Een doelrnatige methode voor het toetsen van aaltjesbestrijdingsmiddelen in grond met *Hoplolaimus uniformis* als proefdier. *Meded. Landb. Hoogesch. Gent* **19**, 377–408.

Oostenbrink, M. (1970) Comparison of techniques for population estimation of soil and plant nematodes. In: Phillipson, J. (ed.) *Methods of Study in Soil Ecology*. UNESCO, Paris, pp. 249–55.

Overgaard, C. & Nielsen, C.O. (1947–48) An apparatus for quantitative extraction of nematodes and rotifers from soil and moss. *Nat. Jutl.* **1**, 271–8.

Pande, Y.D. & Berthet, P. (1973) Comparison of the Tullgren funnel and soil section methods for surveying oribatid populations. *Oikos* **24**, 273–7.

Paris, O.H. & Pitelka, F.A. (1962) Population characteristics of the terrestrial Isopod *Armadillidium vulgare* in California grassland. *Ecology* **43**, 229–48.

Park, O. & Auerbach, S. (1954) Further study of the tree-hole complex with emphasis on quantitative aspects of the fauna. *Ecology* **35**, 208–22.

Pask, W.M. & Costa, R.M. (1971) Efficiency of sucrose flotation in recovering insect larvae from benthic steam samples. *Can. Entomol.* **103**, 1649–52.

Peachey, J.E. (1962) A comparison of two techniques for extracting Enchytraeidae from moorland soils. In: Murphy, P.W. (ed.) *Progress in Soil Zoology*. Butterworths, London, pp. 286–93.

Peters, B.G. (1955) A note on simple methods of recovering nematodes from soil. In: Kevan. D.K.McE. (ed.) *Soil Zoology*. School of

Agriculture, University of Nottingham, Nottingham, pp. 373–4.

Phillipson, J. (ed.) (1970) *Methods of Study in Soil Ecology*. UNESCO, Paris.

Pinder, L.C.V. (1974) The ecology of slugs in potato crops, with special reference to the differential susceptibility of potato cultivars to slug damage. *J. Appl. Ecol.* **11**, 439–51.

Pizl, V. & Josens, G. (1995) The influence of traffic pollution on earthworms and their heavy metal contents in an urban ecosystem. *Pediobiologia* **39**, 442–53.

Potzger, J.E. (1955) A borer for sampling in permafrost. *Ecology*, **36**, 161.

Price, D.W. (1975) Vertical distribution of small arthropods in a Californian pine forest soil. *Ann. Entomol. Soc. Am.* **68**, 174–80.

Rapoport, E.H. & Oros, E. (1969) Transporte y manipuleo de las muestras de suclo y su efecto sobre la micro y mesofauna. *Rev. Ecol. Biol. Soc.* **6**, 31–9.

Raw, F. (1955) A flotation extraction process for soil micro-arthropods. In: Kevan, D.K.McE. (ed.) *Soil Zoology*. School of Agriculture, University of Nottingham, Nottingham, pp. 341–6.

Raw, F. (1959) Estimating earthworm populations by using formalin. *Nature* **184**, 1661–2.

Raw, F. (1960) Earthworm population studies: a comparison of sampling methods. *Nature* **187**, 257.

Raw, F. (1962) Flotation methods for extracting soil arthropods. In: Murphy, P.W. (ed.) *Progress in Soil Zoology*. Butterworths, London, pp. 199–201.

Read, D.C. (1958) Note on a flotation apparatus for removing insects from soil. *Can. J. Plant Sci.* **38**, 511–14.

Salt, G. & Hollick, F.S.J. (1944) Studies of wireworm populations. 1. A census of wireworms in pasture. *Ann. Appl. Biol.* **31**, 53–64.

Sands, W.A. (1972) Problems in attempting to sample tropical subterranean termite populations. *Ekol. Polska* **20**, 23–31.

Satchell, J.E. (1955) An electrical method of sampling earthworm populations. In: Kevan, D.K.McE. (ed.) *Soil Zoology*. School of Agriculture, University of Nottingham, Nottingham, pp. 356–64.

Satchell, J.E. & Nelson, J.M. (1962) A comparison of the Tullgren-funnel and flotation methods of extracting Acarina from woodland soil. In: Murphy, P.W. (ed.) *Progress in Soil Zoology*. Butterworths, London, pp. 212–16.

Saunders, L.G. (1959) Methods for studying *Forcipomyia* midges, with special reference to

cacao-pollinating species (Diptera, Ceratopogonidae). *Can. J. Zool.* **27**, 33–51.

Schjøtz-Christensen, B. (1957) The beetle fauna of the Corynephoretum in the ground of the Mols Laboratory, with special reference to *Cardiophorus asellus* Er. (Elateridae). *Nat. Jutl.* **6–7**, 1–120.

Schouten, A.J. & Kmarp, K. (1991) A comparative study in the efficiency of extraction methods for nematodes from different forest litters. *Pedobiologia* **35**, 393–400.

Seinhorst, J.W. (1962) Extraction methods for nematodes inhabiting soil. In: Murphy, P.W. (ed.) *Progress in Soil Zoology*. Butterworths, London, pp. 243–56.

Sellmer, G.P. (1956) A method for the separation of small bivalve molluscs from sediments. *Ecology* **37**, 206.

Service, M.W. (1968) A method for extracting mosquito eggs from soil samples taken from oviposition sites. *Ann. Trop. Med. Parasitol.* **62**, 478–80.

Shaw, G.G. (1970) Grease-film extraction of an arthropod: a modification for organic soils. *J. Econ. Entomol.* **63**, 1323–4.

Smrz, J. (1992) The ecology of the microarthropod community inhabiting the moss cover of roofs. *Pedobiologia* **36**, 331–40.

South, A. (1964) Estimation of slug populations. *Ann. Appl. Biol.* **53**, 251–8.

Speight, M.C.D. (1973) A greased-belt technique for the extraction of arthropods from organic debris. *Pedobiologia* **13**, 99–106.

Stage, H.H., Gjullin, C.M., & Yates, W.W. (1952) *Mosquitoes of the Northwestern States*. Agricultural Handbook 46, USDA, 95 pp.

Stephenson, J.W. (1962) An improved final sieve for use with the Salt and Hollick soil washing apparatus. In Murphy, P.W. (ed.) *Progress in Soil Zoology*. Butterworths, London, pp. 202–3.

Stewart, K.M. (1974) A three-dimensional net sieve for extracting Tipulidae (Diptera) larvae from pasture soil. *J. Appl. Ecol.* **11**, 427–30.

Sugimoto, T. (1967) Application of oil–water flotation method to the extraction of larvae and pupae of aquatic midges from a soil sample. *Japan. J. Ecol.* **17**, 179–182.

Sunderland, K.D., Hassell, M., & Sutton, S.L. (1976) The population dynamics of *Philoscia muscorum* (Crustacea, Oniscoidea) in a dune grassland ecosystem. *J. Anim. Ecol.* **45**, 487–506.

Svendsen, J.A. (1955) Earthworm population studies: a comparison of sampling methods. *Nature* **175**, 864.

Tanton, M.T. (1969) A corer for sampling soil and litter arthropods. *Ecology* **5**, 134–5.

von Törne, E. (1962a) An elutriation and sieving apparatus for extracting microarthropods from soil. In Murphy, P.W. (ed.), *Progress in Soil Zoology*. Butterworths, London, pp. 204–6.

von Törne, E. (1962b) A cylindrical tool for sampling manure and compost. In: Murphy, P. W. (ed.) *Progress in Soil Zoology*. Butterworths, London, pp. 240–2.

Uhlig, G. (1966) Untersuchungen zur Extraktion der vagilen Mikrofauna aus marinen Sedimenten. *Zoologischer Anzeiger* 29, 151–7.

Uhlig, G. (1968) Quantitative methods in the study of interstitial fauna. *Trans. Am. Microscopical Soc.* 87, 226–32.

Vágvölgyi, J. (1953) A new sorting method for snails, applicable also for quantitative researches. *Ann. Hist. Nat. Mus. Nat. Hung.* 44(N.S. 3), 101–4.

Vannier, G. & Alpern, I. (1968) Techniques de prélèvements pour l'étude des distributions horizontales et verticales des microarthropodes du sol. *Rev. Ecol. Biol. Sol* 5, 225–35.

Vannier, G. & Vidal, P. (1964) Construction d'unappareil automatique pour couper les inclusions de sol dans la gelatine. *Rev. Ecol. Biol. Sol* 1, 575–86.

Vannier, G. & Vidal, P. (1965) Sonde pédologique pour l'échantillonnage des microarthropodes. *Rev. Ecol. Biol. Sol* 2, 333–7.

Wallace, M.M.H. (1956) A rapid method of sampling small free-living pasture insects and mites. *J. Aust. Inst. Agric. Sci.* 22, 283–4.

Walter, D.E., Kethley, J., & Moore, J.C. (1987) A heptane flotation method for recovering microarthropods from semiarid soils, with comparison to the Merchant–Crossley high-gradient extraction method and estimates of microarthropod biomass. *Pediobiologia* 30, 221–32.

Webley, D. (1957) A method of estimating the density of frit fly eggs in the field. *Plant Pathol.* 6, 49–51.

Webster, J.M. (1962) The quantitative extraction of *Ditylenchus dipsaci* (Kuhn) from plant tissues by a modified Seinhorst mistifier. *Nematologica* 8, 245–51.

Whitehead, A.G. & Hemming, J.R. (1965) A comparison of some quantitative methods of extracting small vermiform nematodes from soil. *Ann. Appl. Biol.* 55, 25–38.

Whitehouse, J.W. & Lewis, B.G. (1966) The separation of benthos from stream samples by flotation with carbon tetrachloride. *Limnol. Oceanogr.* 11, 124–6.

Williams, D.D. & Williams, N.E. (1974) A counterstaining technique for use in sorting benthic samples. *Limnol. Oceanogr.* 19, 152–4.

Williams, R.W. (1960a) A new and simple method for the isolation of fresh water invertebrates from soil samples. *Ecology* 41, 573–4.

Williams, R.W. (1960b) Quantitative studies on populations of biting midge larvae in saturated soil from two types of Michigan bogs (Diptera: Ceratopogonidae). *J. Parasitol.* 46, 565–6.

Williamson, M.H. (1959) The separation of molluscs from woodland leaf-litter. *J. Anim. Ecol.* 28, 153–5.

Worswick, J.M. & Barbour, M.T. (1974) An elutriation apparatus for macroinvertebrates. *Limnol. Oceanogr.* 19, 538–40.

7 Relative Methods of Population Measurement and the Derivation of Absolute Estimates

Most relative methods require comparatively simple equipment and, as they often serve to concentrate the animals, they provide impressive collections of data from situations where few animals will be found by absolute methods. This makes them particularly appropriate for initial faunal surveys and, from a statistical viewpoint, plentiful data will appear to be preferable to the hard-won, often scanty, information obtained from unit-area sampling. Most traps will collect specimens continuously, providing a relatively large return for the amount of time spent working with them; the cost (see p. 18) of the data is low. With all these apparent advantages, it is hardly surprising that these methods have been extensively used and developed. There are probably more accounts of their design and use in the literature than references to all the other topics in this book (therefore the list at the end of this chapter is highly selective).

7.1 Factors affecting the size of relative estimates

The biological interpretation of relative population estimates (p. 3) is extremely difficult. Their size is influenced by the majority or all of the following factors.

1 Changes in actual numbers: population changes.
2 Changes in the numbers of animals in a particular 'phase'.
3 Changes in activity following some change in the environment.
4 Sexual or species-specific trap response.
5 Changes in the efficiency of the traps or the searching method.

It is clear, therefore, that the estimation of absolute population by relative methods is difficult; what one is really estimating is the proportion of those members of the population that are vulnerable to the trap under the prevailing climatic conditions and the current level of trapping efficiency. The influence of factors 2–5 on these relative methods must be considered further.

Theoretically, a trap may catch a sufficiently large proportion of the population to appreciably reduce the population being assessed. In practice, with insects this is rarely the case. The failure of early attempts to control insect populations by light traps (e.g. Stahl, 1954) confirm the conclusions of Williams *et al.* (1955) that, in general, the previous night's trapping has little effect. However, in isolated communities, e.g. small oceanic islands, ponds, and streams, or when the trap contains a powerful attractant (Petruska, 1968), or with many vertebrates, a high proportion of the population may be captured. In these circumstances the removal trapping method (p. 255) may be used to estimate population.

7.1.1 The 'phase' of the animal

The susceptibility of an animal to capture or observation will alter with age if behavioural attributes or responses are age-dependent. Many relative methods rely, to some extent, on the animal's movements. Southwood (1962) divided insect movements into migratory or trivial, a classification that can also be applied to other animals. For many terrestrial species, migratory movements occur mainly early in adult life or between reproductive periods (Johnson, 1960, 1963; Kennedy, 1961). The situation is often different for aquatic organisms, such as fish, where dispersal and migration often occur early in life, e.g. the migration of European eel, *Anguilla anguilla*, leptocephalus larvae from the Sargasso Sea to European waters.

For insects, trivial movements, during which they will be especially responsive to bait, occur mostly in later life. The effects of these phenomena on insect-trap catches had in fact been recorded before they themselves were recognized; Geier (1960) showed with the codling moth that the majority of the females taken in light traps were in the pre-reproductive phase, whilst bait-trap catches were predominantly mature (egg-laying) or post-reproductive females. Similarly, Spradbery and Vogt (1993) showed age-dependent trapping bias in the Old World screw-worm fly, *Chrysomya bezziana*, to vary with the stage of ovarian development. Another example is the fall-off in numbers of *Culicoides* midges caught on sticky traps which occurs before the actual population starts to decline (Nielsen, 1963). The reaction of animals to stimuli, important in many trapping techniques, also varies greatly with phase: this is particularly true of bloodsucking insects (Gillies, 1974).

Small mammals often vary in their vulnerability to trapping because of changes in their exploratory behaviour and response to novel objects, which are in turn influenced by factors such as sex, age, and social status (e.g. Carley & Knowlton, 1968; Myllymaki *et al.*, 1971; Adler & Lambert, 1997). Noting that feral house-mouse populations have low recapture rates (0–20%) in live-trapping studies, Krebs *et al.* (1994) studied the movement of radio-collared mice. While low recapture rates during the breeding season were due to low trappability, during the non-breeding period this was caused by nomadic movements away from the study area.

Fish often vary in their vulnerability to light traps with age (often inferred from size distributions). Following light-trap sampling for coral-reef fish, Choat *et al.* (1993) caught almost no larvae <5 mm long, and Thorrold (1992) reported that the pomacentrids, lethrinids, clupeids, mullids, and scombrids were exclusively late-stage larvae and pelagic juveniles. In a study of North American freshwater fish, Gregory and Powles (1985) found that common carp and bluntnose minnow entered traps almost exclusively as yolk-sac larvae, Iowa darter and pumpkinseed only as yolk-sac and postlarvae, and yellow perch when from 5 to 33 mm total length. There are no published studies that have examined individual differences in the vulnerability of adult fish to traps, though Bagenal (1972) suggested that the reproductive status of perch, *Perca fluviatilis*, might be influencing catch rate. Fish traps, such as bag-nets and salmon butts, are usually positioned along the banks of channels, where they intercept fish undertaking seasonal, tidal, or migratory movements and give catches biased with respect to age, sex, and reproductive condition.

7.1.2 The activity of the animal

The level of activity of an animal follows a diurnal cycle, e.g. some insects fly by day and others at night, some fish and mammals may forage at dawn and dusk and others by night or day. The expression of this activity will be conditioned by the prevailing climatic conditions. The separation of changes in trap catch due to climate from those reflecting population change has long exercised entomologists. Williams (1940) approached the problem by taking running means of the catches; the variation in these running means (i.e. the long-term variation) reflected population changes, and the departures of the actual catch from them reflected the influence of climate. Working with groups of species, such as the larger Lepidoptera, Williams was able to demonstrate both the long-term effects of climate through population change and the short-term effects on activity. When studying the populations of airborne aphids, Johnson (1952) found that the running means reflected current climatic conditions and the deviations population trends; this was due to the relatively short period of time any given aphid spends in the air. Thus the running-mean technique is unsafe on biological grounds, being influenced by the relative frequency of population and climatic change; furthermore, its use places severe restrictions on the number of degrees of freedom available for the calculation of significance levels. Even if no attempt is made to separate population trend from activity, when the regression of the actual catch of a group of animals on temperature is calculated, a highly significant result is frequently obtained, thereby emphasizing the role of temperature in the determination of catch size (e.g. Williams, 1940; Southwood, 1960).

The activity of terrestrial vertebrates is greatly influenced by weather. Small mammals show changes of activity with respect to variables such as temperature, cloud cover, and rain, which change the number trapped (Mystowska & Sidorowicz, 1961; Lockart & Owings, 1974; Vickery & Bider, 1978; Plesner Jensen & Honess, 1995). Reptile activity is greatly influenced by temperature. Changes in physical factors, such as water temperature, flow, light, tide, or oxygen concentration, greatly alter the activity of fish, although little quantification has been undertaken of their influence on the efficiency of capture. However, Bagenal (1972) found that captures of perch, *P. fluviatilis*, declined when water temperature rose from 10 to 20°C.

Occasionally, it is possible to obtain a significant regression of catch size on temperature for a single species, particularly for groups such as fish, which increase their swimming activity with temperature. However, activity may be determined by thresholds, e.g. insects may have upper and lower temperature thresholds for flight (Taylor, 1963). If activity is known to occur between thresholds, then, once these have been determined, fluctuations in numbers between the lower and upper thresholds may be considered as due to other causes. The thresholds may be determined by classifying each trapping period as either 1, when one or more insects of the particular species were caught, or 0, when none were caught. Then the trapping periods are grouped according to the prevailing temperature and the percentage of occasions, with flight plotted against temperature (Fig. 7.1). For example, if there were 25 trapping periods when the temperature was 16°C and one or more specimens were collected on 10 occasions, a point would be entered at the 40% flight occurrence (Fig. 7.1). Thresholds for light and other physical conditions may be determined in the same way. The transition from 0% flight occurrence to 100% is not sharp, presumably because of the variation in the individual animals and the microclimate of the sites from which they have flown. Thresholds for flight may appear to vary with the trapping method; Esbjerg (1987) found that turnip moth, *Agrotis segetum*, would fly to sex traps in wind speeds where none would be caught in a light trap.

Taylor (1963) elegantly demonstrates that, when mixed populations of several species are considered, the series of thresholds will lead to an apparent regression of activity on temperature of the type demonstrated by Williams and others; he concludes, therefore, that regression analysis as a means of interpreting the effect of temperature on insect flight should be limited to multispecific problems. Flight thresholds for temperature and wind velocity have been demonstrated in water beetles, *Helophorus* (Landin, 1968; Landin & Stark, 1973), and for temperature and light in aphids (Halgren & Taylor, 1968; Dry & Taylor, 1970).

7.1.3 Differences in the response between species, sexes, and individuals
Our ability to observe or catch different species varies greatly and this must be taken into account when choosing methods for faunal surveys. For example, small mammal species differ in their response to novel objects and those with a higher

Fig. 7.1 The graphical determination of the flight threshold of a species (after Taylor, 1963).

aversion are caught less frequently (Myllymaki *et al.*, 1971). For example, Feldhamer and Maycroft (1992) found that the mean number of captures per individual for golden mice, *Ochrotomys nuttalli*, was significantly less than that for white-footed mice, *Peromyscus leucopus*. Fish species vary greatly in their trappability. Pelagic species may avoid structures and thus never enter traps, while some species that live within floating meadows or weed beds may be so adept at navigating maze-like environments that they can escape. Species also differ greatly in their ability to detect and avoid gill-nets, which are particularly ineffective against electric fish and other elongate, non-spiny forms which do not use sight for navigation. Individual fish differ in their willingness to explore new objects and thus become trapped. There are reports that the presence of a fish alarm substance, which is released from injured skin, will deter fish, such as fat-head minnows, *Pimephales promelas*, from entering traps (Mathis & Smith, 1992). While in this case the existence of pheromones has been questioned (Magurran *et al.*, 1996; Henderson *et al.*, 1997), the general principle that traps should be kept clean and give no signal of their prior use is sound practice. Eastop (1955) and Heathcote (1957a) have compared the ratios of the numbers of different species of aphids caught in yellow sticky traps to the numbers caught in suction traps, presumably the true population. Eastop, for example, found ratios varying from 31 to 0.5 and, even within a genus (*Macrosiphum*), they ranged from 14.7 to 0.8. Analogous observations on the relative numbers of different species of mosquitoes caught in light traps compared with the results from a rotary trap have been made by Love and Smith (1957), and they found ratios (which they called the 'index of attraction') from over 7 down to 0.24.

In many groups, significantly more of one sex or the other are caught in traps (for light traps, see Williams, 1939; Masaki, 1959). For example, mirid bug males make up the majority of light-trap catches and there is some evidence that male mirids do engage in significantly more trivial movement, 'flits', than females (Southwood, 1960; Waloff & Bakker, 1963), but the excess of male Miridae is greater in ultraviolet than in tungsten-filament light traps (Southwood, 1960), so that at least a proportion of this predominance must be due to a selective effect of the traps between the sexes, rather than a real difference in flight activity. The positioning of traps can influence the sex-ratio bias; pheromone traps for the oblique-banded leaf-roller, *Choristoneura rosaceana*, placed in the tops of trees always caught higher numbers of male moths than those in lower- or middle-canopy positions (Agnello *et al.*, 1996). An interesting case of a difference between sexes is the large number of male relative to female tsetse flies usually taken on 'fly rounds' (Glasgow & Duffy, 1961). The biological interpretation of this seems to be that newly emerged female flies usually feed on moving prey and the early pairing, desirable in this species, is achieved by numbers of males following moving bait (Bursell, 1961). For mammals, sex differences are also common; mature males are often caught more easily than females (e.g. Andreassen & Bondrup Nielsen, 1991; Drickamer *et al.*, 1995), but the reverse can be the case. For example, Adler and Lambert (1997) found that adult females of the Central American spiny rat, *Proechimys semispinosus*, were more trappable than males.

Individual differences in capture rates have been related to genetics. Gerard *et al.* (1994) found that heterozygous house mice, *Mus domesticus*, formed by the crossing of two chromosomal strains, had a lower trappability than the homozygous wild types. Drickamer *et al.* (1995) also reported lower captures of heterozygous females. Diseases and parasites can also affect the trap response and should always be considered as a potential source of bias in a survey of infection rates, as it is frequently easier to trap sick animals. Webster *et al.* (1994) found that *Toxoplasma gondii* changed avoidance behaviour in wild brown rats, *Rattus norvegicus*, resulting in the more frequent capture of infected individuals.

7.1.4 The efficiency of the trap or searching method

The efficiency of a method of population estimation is the percentage of the animals actually present that are recorded. The efficiency of a searching method depends both on the skill of the observer and on the habitat, e.g. any observer is likely to see tiger beetles (Cicindellidae) far more easily on lacustrine mud-flats than on grass-covered downs.

Several types of traps—sticky, water, and flight—catch insects that are carried into or on to them by the wind, and Johnson (1950) and Taylor (1962) have shown that the efficiency of such traps varies with wind speed, for which the results can be corrected for the size of the insect (Table 7.1, see also p. 146). Other variables noted by Martin *et al.* (1994) as influencing sweep-net and sticky-trap sampling efficiency for three species of blackflies, *Simulium truncatum* complex, *Prosimulium mixtum*, and *Stegopterna mutata*, were wind speed, light, temperature, saturation deficit, and time of day. A similar list would be found for many insects.

Light traps are also affected by wind (e.g. Gregg *et al.*, 1994) and generally catch fewer specimens of most insects on nights when the moon is full (e.g. Taylor, 1986). This is considered to be due to a fall in efficiency of the traps (Williams *et al.*, 1956; Miller *et al.*, 1970; Bowden & Church, 1973). However, from time series obtained from tower-mounted light traps in northern New South Wales, Gregg *et al.* (1994) found no correlation with the phase of the moon. The efficiency of baits varies from many causes, e.g. the ageing and fermentation of artificial or non-living baits (Kawai & Suenaga, 1960) affecting their 'attractiveness', whilst living baits may differ in

Table 7.1 The percentage efficiency of a cylindrical sticky trap for different insects at different wind speeds (after Taylor, 1962).

	Wind speed (miles/h)				
	1	2	6	10	20
Small aphid (*Jacksonia*)	20	45	64	66	68
Small fly (*Drosophila*)	35	58	72	73	74
Housefly (*Musca*)	45	64	76	76	76
Bumble-bee (*Bombus*)	52	68	78	78	78

unexplained ways—Saunders (1964) found that two apparently identical black zebu oxen trapped different numbers of tsetse flies. Field biologists often find that colleagues differ greatly in their attractiveness to biting flies and some can be employed as a useful decoy. The effectiveness of a given bait may vary from habitat to habitat (Starr & Shaw, 1944).

Fish traps and gill-nets vary in efficiency with changes in water turbidity, turbulence, light, flow, temperature, and depth. Bottom-trawl efficiency changes with the nature of the substrate, and open-water fish and crustaceans may be able to avoid open-water trawls and nets when visibility is good. Efficiency of almost all towed nets varies with speed and this should be standardized. The avoidance response of temperate fish tends to be faster at higher temperatures, so that nets and suction samplers are often more effective in winter than in summer.

The efficiency of the various methods will be discussed further below; the examples already indicate that variation in efficiency is a limitation on the value of relative estimates even for comparative purposes.

7.2 The uses of relative methods

It is apparent from the above section that the actual data from relative methods should be used and interpreted with far more caution than has often been shown. Comparisons of different species and different habitats are particularly fraught with danger.

7.2.1 Measures of the availability

This is the most direct approach; the availability of the population of an animal is the result of the response to the stimuli, the activity, and the abundance, i.e. the product of factors 1–3 and 5 in the list above (p. 247); it may be defined as the ratio of total catch to total effort. Thus, assuming the efficiency of the trap or search does not change, the raw data of catch per unit time or effort will provide a measure of availability, or what Heydemann (1961) terms 'Aktivitatsdicht'.

True availability is meaningful in many contexts: it is most easily interpreted with natural 'baits', e.g. the availability of bloodsucking insects to their normal prey (in a bait trap) is a measure of their 'biting level', and the availability of flying insects to colonize or oviposit on a trap host plant is a measure of these parameters. Extrapolations from more artificial situations can only be made in the light of additional biological knowledge. The availability of codling moth females to a light trap gives a convenient indication of the magnitude and phenology of emergence and oviposition, and this may be used to time the application of control measures. But the peak of availability of the females to the bait traps indicates that the main wave of oviposition has passed (Geier, 1960), and therefore in warm climates, where the eggs will hatch quickly, this information may be too late for effective control measures.

In general, measures of availability may be used for the immediate assessment of the 'attacking' or colonization potential of a population and its phenology. Over a long period of time, the changes in the species composition of the catches of the same trap in the same position may be used to indicate changes in the diversity of the fauna (see p. 462). It is only under exceptional circumstances that the trapping seems to affect population size, i.e. when the traps are powerful and the population restricted.

7.2.2 Indices of absolute population

When the efficiency of the trap and the responsiveness of the animal to it can be regarded as constant and if the effects of activity can be corrected for, then the resulting value is an index of the size of the population in that particular phase. The effects of activity due to temperature or other physical factors on the catch size of sticky, flight, pitfall, and similar traps may be eliminated by the determination of the thresholds, using Taylor's (1963) method (see above). With net catches, the regression technique of Hughes (1955) serves the same purpose (p. 267). The animals' diel periodicity cycle may need to be known before these corrections can be made and the index derived. Such an index may be used in the place of actual absolute population estimates in damage assessment and in studies on the efficiency of pest-control measures. The value of independent estimates of population size has been stressed (p. 4) and a series of such indices may be compared with actual population estimates; if the ratio of one to the other is more or less constant, the reliability of the estimates has been confirmed. Indeed, these indices may be used in place of absolute population estimates for any comparative purpose, but are of course of no value in life-table construction.

7.2.3 Estimates of absolute population

It is possible to derive estimates of absolute population from relative estimates by three approaches: (i) by 'calibration' with absolute estimates; (ii) by estimating efficiency; and (iii) from the rate by which trapping reduces the sizes of successive samples (removal trapping). Distance-sampling methods also provide absolute estimates; this is achieved by fitting a function that describes the decreasing efficiency of sighting an animal or its sign (e.g. nest) with distance from the observer. Such methods, described in Chapter 9, are only applicable to large or easily spotted animals.

7.2.3.1 'Calibration' by comparison with absolute estimates

When a series of indices of absolute population has been obtained simultaneously with estimates of absolute population by another method, the regression of the index on absolute population may be calculated (e.g. Zoebisch *et al.* (1993) used a linear regression to relate sticky-trap catches of the fly *Liriomyza trifolii* to population density). This can then be used to give estimates of population directly from the indices, but such 'corrections' should only be made under the same conditions that held during the initial series of comparisons.

7.2.3.2 Correcting the catch to allow for variations in trap efficiency

This approach is really a refinement of that above; trap catches and absolute estimates are made simultaneously under a variety of conditions that are known to affect trap efficiency. A table of the correction terms for each condition can then be drawn up and will give much greater precision than a regression coefficient for all conditions. A valuable approach along these lines has been made by Taylor (1962), who used the data from Johnson's (1950) experiment to determine the efficiency for sticky traps and tow-nets, using a calibrated suction trap (see p. 142) and these gave a direct measure of the absolute density of the aerial population. Taylor was therefore able to construct a table of correction terms (Table 7.1, p. 252); they are of course only applicable to the particular traps Johnson used and it is important to note that the sticky trap was white, not yellow. Taylor also computed the expected fall in efficiency of the traps, allowing for the different volumes of air passing at different winds and the effect of this on the impaction of the insects on the sticky trap, assuming that they behave as inert particles; he found that a large part of the fall in efficiency at low wind speeds could be accounted for by these factors.

7.2.3.3 Removal trapping or collecting

The principle of removal trapping or collecting is that the number of animals in a closed, finite population will decline, as will the catch per unit effort, if captured animals are removed each time the population is sampled. Although Le Pelley (1935) first demonstrated that this fall of catch is geometric in observations on the hand-picking of the coffee bug, *Antestiopsis*, the method has been developed by mammalogists, starting in 1914 (Seber, 1982). Examples of recent studies on mammals are Rickart *et al.* (1991), Woodman *et al.* (1995), and Christensen (1996). These methods are particularly appropriate to small, isolated populations, for which the probability of capturing each individual on each trapping occasion can be made >0.2. Removal methods are also applicable to the estimation of fish populations in streams and ponds which are sampled by electric fishing (see p. 264).

If this approach is to be used in life-table studies, it is necessary to use a method of catching that does not kill the animals; they will have to be kept captive and then released at the end of the estimation.

The basic procedure is to undertake s distinct periods of trapping and to record the number caught for the first time on each occasion, u_s. The various approaches for calculating N, the total population size, are based on either maximum likelihood or regression methods. Given constant sampling effort (e.g. the same number of traps on each occasion), maximum-likelihood methods are superior (White *et al.*, 1982; Seber, 1986) and will be described in detail below. These estimates of N for a constant probability of capture model were first published by Moran (1951) and developed by Zippin (1956); a more general model, allowing variable probabilities of capture, was presented by Otis *et al.* (1978). The suggestion (Seber, 1986) that the jackknife estimator for the M_{bb} model presented by Pollock and Otto (1983) should be tried has received little attention, probably because most ecologists find the maximum-likelihood methods satisfactory. Regression methods are still used in studies where the sampling effort is variable, not being under the control of the researcher. For example, in fisheries research, the number of fish landed varies with the activity of the fishing fleet. Methods for fisheries research are given in Ricker (1975) and Seber (1982) and will not be considered further here. When designing a removal-trapping experiment, the aim should be to maintain constant sampling effort.

Maximum-likelihood method with constant probability of capture. For the application of the method the following conditions must be satisfied (Moran, 1951).
1 The catching or trapping procedure must not lower (or increase) the probability of an animal being caught. For example, the method will not be applicable if the insects are being caught by the sweep-net and, after the first collection, the insects drop from the tops of the vegetation and remain around the bases of the plants, or if the animals are being searched for and, as is likely, the most conspicuous ones are removed first (Kono, 1953).
2 The population must remain stable during the trapping or catching period; there must not be any significant natality, mortality (other than by the trapping), or migration. The experimental procedure must not disturb the animals so that they flee from the area. As Glasgow (1953) has shown, if the trapping is extended over a period of time, immigration is likely to become progressively more significant as the population falls.
3 The population must not be so large that the catching of one member interferes with the catching of another. This is seldom likely to be a problem with insects, where each trap can take many individuals, but may be significant in vertebrate populations, where one trap can only hold one animal.
4 The chance of being caught must be equal for all animals. This is the most serious limitation in practice. Some individuals of an insect population, perhaps those of a certain age, may never visit the tops of the vegetation and so will not be exposed to collection by a sweep-net. In vertebrates, 'trap-shyness' may be exhibited by part of the population. In electric fishing, where the method is used extensively, smaller individuals are more difficult to stun and individuals occupying territories under banks or other obstructions may be particularly difficult to catch.

Zippin (1956, 1958) has considered some of the specific effects of failures in the above assumptions. If the probability of capture falls off with time, the population will be underestimated and if the animals become progressively more susceptible to capture, the population will be overestimated. Changes in susceptibility to capture will arise not only from the effect of the experiment on the animal, but also from changes in behaviour associated with weather conditions or a diel periodicity cycle.

The expected number captured on each sampling occasion $E(u_s)$ is:

$$E(u_s) = N(1-p)^{s-1} p \qquad (7.1)$$

where p is the probability of capture on each sampling occasion. Thus, for the first sampling occasion, the expected number caught is Np and for the second $N(1 - p)p$ and so on.

Numerical iterative techniques are required to find the maximum-likelihood value for N and are best undertaken on a computer, using programs such as CAPTURE (see Table 3.1, p. 74). The maximum-likelihood estimates for N and p are given by solving the equations:

$$\hat{N} = \frac{T}{\left(1 - \hat{q}^s\right)}. \tag{7.2}$$

and:

$$\frac{\hat{q}}{\hat{p}} - \frac{k\hat{q}^s}{\left(1 - \hat{q}^s\right)} = \frac{\sum\limits_{i=1}^{s}(i-1)u_i}{T} = R \tag{7.3}$$

where T is the total caught over all k samples and $q = 1 - p$. First R is calculated and then equation (7.3) is solved numerically for q. This value is then used in equation (7.2) to estimate N.

Following Zippin (1956, 1958), we show below how the calculations can be undertaken for $k = 3, 4, 5, 6,$ or 7 without a computer. Zippin's procedure is best illustrated using as a worked example the results presented in Table 7.2.

The total catch, T, is calculated:

$$T = 65 + 43 + 34 + 18 + 12 = 172$$

Then:

$$\sum_{i=1}^{k}(i-1)u_i = (1-1)65 + (2-1)43 + (3-1)34 + (4-1)18 + (5-1)12 = 213$$

is found.

Next the ratio R is determined: $R = 213/172 = 1.238$ and the estimate of the total population is solved using equations (7.2) and (7.3). However, for $s = 3, 4, 5, 6,$ or 7 these last steps may be circumvented by the using of Zippin's charts (Figs 7.2 and 7.3). Thus, in the present example, for $s = 5$ and $R = 1.24$ the value of $(1 - \hat{q}^s)$ is read off Fig. 7.2 as 0.85, so that:

Table 7.2 Example data from a removal-trapping experiment.

	Sampling occasion					
	1st (u_1)	2nd (u_2)	3rd (u_3)	4th (u_4)	5th (u_5)	Total
Number caught	65	43	34	18	12	172
Expected number for $p = 0.33$	66.66	44.66	29.92	20.04	13.43	174.71
Chi-squared						1.06

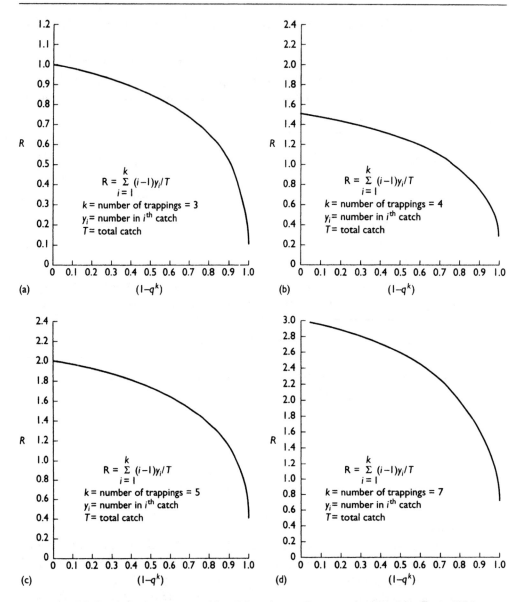

Fig. 7.2 Graphs for the estimation of $(1 - q^s)$ from the ratio R in removal trapping (after Zippin, 1956).

$$\hat{N} = \frac{172}{0.85} = 202$$

The standard error (SE) is given by:

$$SE(\hat{N}) = \sqrt{\frac{\hat{N}(\hat{N} - T)T}{T^2 - \hat{N}(\hat{N} - T)\left[(sp)^2 / (1 - p)\right]}}$$

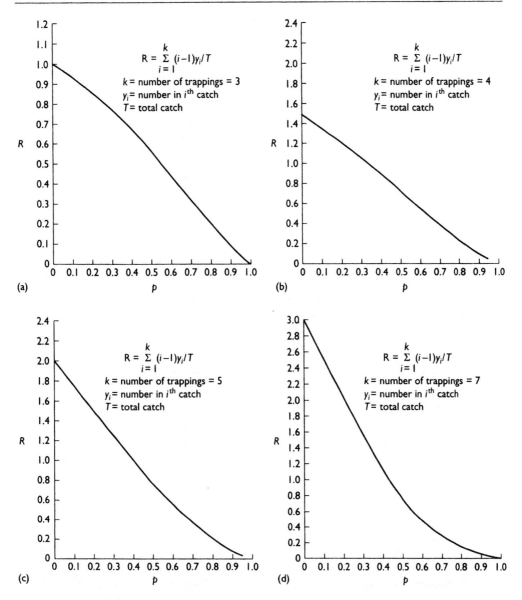

Fig. 7.3 Graphs for the estimation of p from the ratio R in removal trapping (after Zippin, 1956).

where the notation is as above and p is read from Fig. 7.3. In our example SE(N) = 14.46. Therefore the 95% confidence limits of the estimate are approximately:

$$202 \pm 2 \times 14.46 = 202 \pm 28.9$$

If the lower confidence interval is less than the total number of captures, T, then T should be taken as the lower confidence interval.

It has been shown by Zippin (1956, 1958) that a comparatively large proportion of the population must be caught to obtain reasonably precise estimates. His conclusions are presented in Table 7.3, from which it may be seen that, to obtain

Table 7.3 Proportion of total population required to be trapped in 100 or fewer trappings for specified coefficient of variation of N (after Zippin, 1956).

	Coefficient of variation (%)			
	30	20	10	5
200	0.55	0.60	0.75	0.90
300	0.50	0.60	0.75	0.85
500	0.45	0.55	0.70	0.80
1 000	0.40	0.45	0.60	0.75
10 000	0.20	0.25	0.35	0.50
100 000	0.10	0.15	0.20	0.30

a coefficient of variation (CV = estimate/standard error × 100) of 30%, more than half the animals would have to be removed from a population of less than 200. For this reason, Turner (1962) found it impractical for estimating populations of insects caught in pitfall traps; the proportion of the population caught was too low.

After calculating estimates for N and p, the expected number of captures should be calculated, using equation (7.1), and the goodness of fit to the observed sequence of captures tested, using a χ^2 test. The results of these calculations for the worked example are given in Table 7.3 and they show that a constant-probability model can be accepted (χ^2 test statistic of 1.06, $s - 2$ = three degrees of freedom, probability of a value this large or larger 0.786). If a significant difference is found, then the assumption that p is constant must be rejected (see section 3.1.1.3).

Simplified procedures for when, at most, three sampling occasions have been undertaken are given in Seber (1982). For two samples, a good approximation to the maximum-likelihood estimator is:

$$\hat{N} = \frac{u_1}{1 - (u_2/u_1)}$$

Population estimation based on two samples should only be used for $p > 0.9$. Generally, three or more sampling occasions should be undertaken so that the assumed constant probability of capture can be tested.

Removal-trapping methods require the number captured to fall with increasing sampling. If this does not occur, the method may fail. Seber and Whale (1970) show that the population can be estimated if the following criterion is satisfied:

$$\sum_{j=1}^{s} (s+1-2j)u_j > 0$$

Graphical and regression methods with constant probability of capture. Because graphical and regression methods have been so widely used and are still frequently applied, we briefly introduce them here. These methods are inferior to the maximum-likelihood approach and do not give valid estimates of the standard error of the estimated population size. If the number caught on the ith occasion are plotted against the total catch up to occasion $i - 1$, a straight line may be fitted by eye (Menhinick, 1963) or a linear regression line calculated (Zippin, 1956). The point where this line cuts the x axis gives the estimated population size. Fitting by eye is acceptable only when the points lie fairly close to a straight line. Fitting by regression is not really acceptable, as the values are not independent.

Dealing with variable probabilities of capture and the general maximum-likelihood model. It will often be observed that the probability of capture changes during the course of a study. This may be due to the catch consisting of a number of subclasses with differing p's. For example, larger fish are more easily caught when electric-fishing. Thus, the first sweep along a reach of a stream will tend to have the highest rate of capture. Generally, the best way to deal with this problem is to divide the population into a number of groups, each of which can be independently estimated with a constant-probability model. In an electric fish survey of a stream, it might be appropriate to undertake independent analyses for eel, trout, and minnow and to further subdivide the trout into size (age) classes. However, the ability to avoid capture varies even between individuals of a single age or size class and, if this effect is large, the mean probability of capture will change in what the observer perceives as a homogeneous population. When a constant-probability model has been rejected and it is not possible to subdivide the population, then the generalized removal method of Otis *et al.* (1978) is available.

However, if subdivison is inappropriate a different approach is required. A removal experiment does not generate sufficient information to allow the estimation of separate probabilities of capture for each sampling occasion. Otis *et al.* (1978) noted that, as animals with a higher probability of capture tend to be caught first, the mean probability of capture (p_k) on sampling occasion k tends to decline with successive samples, so that after, say, m samples p can be assumed constant. A family of models can therefore be constructed, in which the first $1, 2, 3 \ldots (m - 2)$ samples are assumed to have different values for p and all later samples a constant p. The approach is to work upwards from $m = 1$ and stop when a χ^2 test shows that the model can no longer be rejected. The complicated procedure for calculating maximum-likelihood estimates of the p_k values is given in outline in Otis *et al.* (1978). The method is available in computer programs such as CAPTURE (see Table 3.1, p. 74).

7.2.4 Collecting

Relative methods may be used simply as collecting methods, e.g. for animals for mark and recapture and for age determination in the construction of time-specific life-tables, provided the chances of all age and sex groups being captured are equal (e.g. Davies *et al.* (1971), who found that parous rates in mosquitoes collected in light traps and at baits were only slightly different). They may also be used to determine the proportion of forms in a population (e.g. Bishop, 1972).

7.3 Relative methods: catch per unit effort

Methods grouped here are those in which the movement or action that results in the capture or observation of the insect is made by the observer.

7.3.1 Observation by radar

The use of radar in bird migration studies started in the 1960s and developed into a powerful tool for the study of insect movement in the upper air during the 1970s (Schaeffer, 1979). It has been found most valuable for studying large-scale night-time migrations in both birds (Bruderer, 1997a, b) and insects (Riley & Reynolds, 1990; Wolf et al., 1990; Smith et al., 1993). Radar techniques, like the analogous sonar systems used for aquatic life, suffer from the general problem that they may not give sufficient taxonomic information to allow specific identification without supporting information. The obvious exceptions to this are large single-species aggregations, such as a desert-locust swarm. However, the periodic change in signal strength of a bird linked to the flapping of wings, the speed, or the height of flight may give information that can lead to the identification of the species.

Detection of a flying object is related to the wavelength used. The upper part of the C band (3.8–7.5 cm) and the lower part of the S band (7.5–15 cm) are optimal for small birds, while the L band (15–30 cm) will not detect small passerines (Bruderer, 1997a). The reflectivity of birds seems to be related to their water content, as the reflectivity of feathers is negligible, whereas for insects reflectivity is related to the amount of chitin present.

Rainey and Joyce (1990) states that a radar with operating parameters of 3.18 cm wavelength, peak power of 2.5 kW, pulse length 0.1 μs, and pulse repetition rate of 1760 Hz could detect a single spruce budworm moth of 1 cm body length and 35 mg in weight at 750 m and a desert locust 5 cm long at a considerably greater distance. Hobbs and Wolf (1989) describe an airborne radar system capable of detecting a moth with a radar cross-section of 1 cm^2 at a range of about 1 km. Individual detection efficiency is altered by the size of the animal, the direction of movement relative to the radar, the position within the radar beam, interference from rain or other air-borne material, the terrain, and the degree of flocking or swarming of the animal. Absolute counts of animals can rarely be obtained; as Bruderer (1997b) states: 'many ornothologists erroneously assume that counting echoes provides direct data on the number and distribution of birds, neglecting the difficulties'. These difficulties are even greater for insects, where radar can only realistically give indices of density at different heights. The abundance of an animal can be expressed in terms of 'movement rates', which is the number per minute, as was done by Reynolds et al. (1997) in a study of Hawaiian sea birds and a bat. Radar has contributed most to the study of the timing, size, and temporal variability of migrations, e.g. bird migration over Israel (Alfiya, 1995; Leshem & Yom Tov, 1996), passerine migration across the Straits of Gibraltar (Hilgerloh, 1989, 1991), and seasonal insect flight activity in Texas (Beerwinkle et al., 1994, 1995).

Radar remains a relatively expensive technique but is likely to become more important and widely used in the future. An example of the use of a low-cost hand-held Doppler-shift radar is in the study of Blake *et al.* (1990) to measure the flying speed of the barn swallow, *Hirundo rustica*. Xu *et al.* (1996) describe the use of ground-penetrating radar to detect termite nests.

7.3.2 Hydroacoustic methods

Sonar systems send out pulses of sound and detect the echoes off underwater objects, such as fish, crustaceans, and cephalopods, and can even detect aggregations of planktonic crustaceans or the boundary between waters of different density. Acoustic methods can only be used for quantitative surveys of species which live in the water column and are only practicable in waters not dominated by aquatic macrophytes, logs, or drifting leaves. When these conditions are met, dual or split-beam systems are able to give quantitative estimates of fish number and size distribution within a number of depth bands. For example, Kubecka *et al.* (1994) estimated brown trout, *Salmo trutta*, populations in freshwater lochs and Duncan and Kubecka (1996) studied fish density and distribution in the River Thames. Acoustic surveys are extensively used for marine surveys of commercial species, where normally the method gives an index of abundance, which can be used to detect temporal and spatial changes. The tendency for fish and other aquatic organisms to form tight shoals can create problems for quantitative estimation (Appenzeller & Leggett, 1992). Absolute density estimates can be obtained for shoaling species using echo integration techniques, although errors of estimation can be large. An example of the use of acoustic surveys for marine crustaceans is in the study of Brierley and Watkins (1996) on Antarctic krill, *Euphausia superba*, and zooplankton. As in the case of radar, a weakness of the method is that it requires supporting information to distinguish between species.

While fish-finding sonars have become cheap and easy to use, quantitative fish-counting systems are expensive and require skilled operators. As in the case of radar, it is anticipated that technological advances in sonar engineering and in signal-processing software will increase both the accuracy and the availability of quantitative fish-counting sonars.

7.3.3 Fish counters

Automatic fish counters are widely used in rivers for counting migratory species, such as salmon, *Salmo salar*. The detectors, which often work by measuring the change in conductivity caused by a passing fish, are usually placed into narrow channels. Fish counters give false counts both by not detecting passing fish, particularly if they pass as a group, and by counting drifting material. Accuracy can be tested by checking the record of fish detections against video recordings. Such counters are useful for studying seasonality of migration and fish movement in relation to factors such as flow or temperature and for producing annual indices of abundance.

7.3.4 Electric fishing

Electric fishing is the method of choice for estimating fish populations in small streams and ponds, where absolute estimates of population number can be made (Kennedy & Stange, 1988). We place it under relative methods because it is never 100% efficient, and population size is best estimated using removal-trapping methods. Electric fishing is usually undertaken with a pulsed direct current (DC), which both draws fish towards the anode and stuns them. The power is supplied either from an electric generator or from batteries in the case of backpack units. Electric fishing is dangerous if undertaken by untrained personnel. An important early book on the method is Vibert (1967) and more recent developments are discussed in Cowx (1990).

The normal procedure for estimating population size in streams is to isolate a reach with stop nets and then work downstream with a hand-held anode, removing any stunned fish. This procedure is repeated three or more times (the exact number depending on sampling efficiency) and the population number estimated by removal methods (p. 255). This is not possible in larger bodies of water, where Persat and Copp (1990) argue that good estimates can be obtained by sampling at a large number of individual points, providing the area sampled at each can be estimated. However, on larger waters where boat-mounted gear is used, it is normal to express the number caught as catch per unit effort, e.g. fish caught per $100\,m^2$ area swept.

Electric fishing efficiency is altered by the following factors.
1 The conductivity of the water. In low-conductivity fresh waters, such as are normal in areas with non-calcareous bedrock, such as the Amazon basin, the efficiency is too low for the method to be used.
2 Turbidity and flow, which affect the ability to spot and collect stunned fish.
3 Variation in water depth can have a large effect and electric fishing is rarely useful in waters >2 m deep.
4 The skill and tiredness of the operator.
5 The size and species of fish. Large fish are more easily stunned than small fish.

7.3.5 Aural detection

Strictly speaking, sounds are 'products of animals' and this section should come in the next chapter; however, this method has mostly been used with birds and the approach is similar to distance-sampling methods, with a 100% detection assumed out to a known distance, beyond which none are heard. A simple estimator, due to W.H. Petrabough, is:

$$\hat{D} = \bar{b} / L\pi r_a^2$$

where b = average number of songs heard per stop, L = the total number of listening stops made, and r_a = the radius of audibility. The estimation of r_a presents a major source of error. A further problem is that birds and insects have a definite periodicity of calling; Gates and Smith (1972) developed a time-dependent model to allow for this. Bird surveys often use the response of birds to tape-recorded calls to detect the presence of species that are difficult to spot. Such a technique is used particularly in forests.

With many insects, one 'caller' stimulates others, so that the probability of hearing a song might be better described by a skewed rather than a Poisson distribution with time. This method might be used with some insects (male cicadas, acridids) as a second or third method to compare estimates obtained with capture–recapture or other techniques (see p. 73).

Indices of abundance of cetaceans and bats can be obtained by recording and quantifying their navigational calls (e.g. Verheggen, 1994).

7.3.6 Exposure by plough

The surveying of large but aggregated soil animals by conventional methods (Chapter 6) is often extremely laborious. Roberts and Smith (1972) found that a rapid survey of scarabaeid larvae could be made by ploughing a transect across grassland and counting those insects exposed in the furrow or on the overturned turf. Such transects would also detect patterns of aggregation; in general, there was a linear relationship between numbers per unit plough transect and absolute population, as determined from core samples. Relatively more, however, are exposed in dense populations, because they cause the soil to break up.

7.3.7 Collecting with a net or similar device

A number of approaches are included under this heading: in aquatic habitats, dipping and netting are widely practised relative methods; insects and other small animals on terrestrial vegetation may be collected with a sweep-net or, for those on trees, with a beating tray. The latter approaches an absolute method and is discussed in Chapter 4 (p. 156).

7.3.7.1 Aquatic habitats

A wide variety of nets, including seine, beam-trawl, and push nets are used to sample fish and crustaceans (Fig. 7.4). Most nets are so inefficient that, even when it is possible to estimate the volume or area swept out, the catch cannot be used to estimate population density. A probable exception is quantitative sampling with a seine net, which is discussed in Chapter 5. Beam trawls are used extensively for sampling benthic fish and crustaceans. Because their efficiency is low and changes with substrate and aquatic conditions, such as turbidity, they are difficult to calibrate and thus can rarely be used to estimate absolute density.

Plankton nets (see Chapter 5) may also produce indices of abundance under conditions where it is not possible to estimate the volume of water filtered, such as is normally the case in highly turbulent water. In fresh waters, drag-nets can be used for sampling aquatic invertebrates in an analogous fashion to a sweep-net.

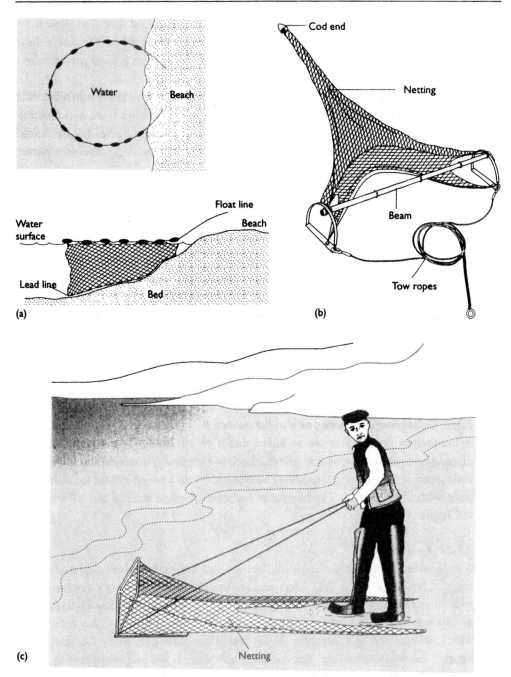

Fig. 7.4 Nets for sampling fish. (a) Seine netting from above and in cross-section. (b) Beam-trawl. (c) Push net.

7.3.7.2 Terrestrial insects

Aerial insects may be collected by random strokes through the air with a light net (Parker, 1949; Linsley *et al.*, 1952; Nielsen, 1963) or with net or gauze cones on a car (Almand *et al.*, 1974; Roberts & Kumar, 1994), a ship (Yoshimoto & Gressitt, 1959; Yoshimoto *et al.*, 1962), or an aeroplane (Glick, 1939; Odintsov, 1960; Gressitt *et al.*,

1961); because of the impedance to air flow due to the gauze, which becomes more severe at higher speeds, it is difficult—but not impossible—to obtain direct measure of aerial density from such sampling cones.

The sweep-net is perhaps the most widely used piece of equipment for sampling insects from vegetation; its advantages are simplicity, speed, and high return for little cost—and it will collect comparatively sparsely dispersed species. However, only those individuals on the top of the vegetation that do not fall off or fly away on the approach of the collector are caught. The influence of these behavioural patterns on the efficiency of the method has been investigated by the comparison of sweep-net samples with those from cylinders or suction apparatus (p. 158) (Beall, 1935; Romney, 1945; Johnson et al., 1957; Race, 1960; Heikinheimo & Raatikainen, 1962; Rudd & Jensen, 1977; Harper et al., 1993), by the comparison of sweep-net samples with capture–recapture estimates (Nakamura et al., 1967), by comparison with pit-falls (Cherrill & Sanderson, 1994), or by a long series of sweeps in the same habitat (Hughes, 1955; Fewkes, 1961; Saugstad et al., 1967). Changes in efficiency may be due to:

1 changes in the habitat;
2 changes in species composition;
3 changes in the vertical distribution of the species being studied;
4 variation in the weather conditions;
5 the influence of the diel cycle on vertical movements.

A sweep-net cannot be used on very short vegetation and once plants become more than about 30 cm tall, further increases in height mean that the net will be sampling progressively smaller proportions of any insect population whose vertical distribution is more or less random.

As Johnson et al. (1957) and Heikinheimo and Raatikainen (1962) show, even related species may differ in their availability for sweeping and therefore the methods are unsuitable for synecological work; these same workers also showed with a nabid bug and a leafhopper that the vertical distribution of the various larval instars and the adults differed (Fig. 7.5). Within the adult stage, the vertical distribution in vegetation may alter with age. Ovipositing females of the grass mirid, *Leptopterna dolabrata*, are relatively unavailable to the sweep-net; the same appears to be true of mature healthy males and females of the leafhopper, *Javesella* (= *Calligypona*) *pellucida*, in cereals, although parasitized individuals do not change their behaviour with age and so the percentage of parasitism would be overestimated by sweeping (Heikinheimo & Raatikainen, 1962; see also p. 359).

Weather factors profoundly affect the vertical distribution and hence the availability of insects to the sweep-net. These have been studied by Romney (1945), Hughes (1955), and Saugstad et al. (1967); Hughes found with the chloropid fly, *Meromyza*, that the regressions of numbers swept against the various factors gave the following coefficients: wind speed −0.1774, time since saturation −0.0815, air temperature +0.0048, radiation intensity +0.0010, and radiation penetration −0.0367. Thus the major influences on the catches of this fly are the first two factors listed, maximum efficiency being achieved at low wind speeds and immediately after a

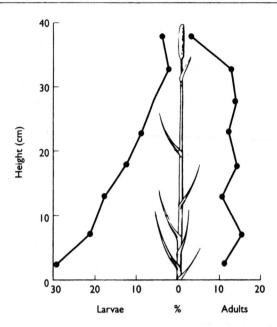

Fig. 7.5 The vertical distribution of larval instars IV and V and adults of *Javesella pellucida* on timothy grass (after Heikinheimo & Raatikainen, 1962).

shower. Many arthropods have a diel periodicity of movement up and down the grass, the maximum numbers being on the upper parts a few hours after sunset (Romney, 1945; Fewkes, 1961; Benedek *et al.*, 1972).

The bags of sweep-nets are usually made of linen, thick cotton, or some synthetic fibre; the mouth is most often round; a square-mouthed net does not give more consistent results (Beall, 1935), but a D-shaped mouth is useful for collecting from short vegetation, especially young crops. There may be considerable variation in the efficiency of different collectors; usually the more rapidly the net is moved through the vegetation, the larger the catch (Balogh & Loksa, 1956).

The number of sweeps necessary to obtain a mean that is within 25% of the true value has been investigated by Gray and Treloar (1933) for collections from lucerne (alfalfa), and was found to vary from taxon to taxon, with an average of 26 units each of 25 sweeps. These authors suggested that this high level of variability was due to heterogeneity in the insect's spatial distribution, and much of this may have been attributable to the diurnal periodicity demonstrated by Fewkes (1961). In contrast, Luczak and Wierzbowska (1959) considered 10 units of 25 sweeps adequate for grassland spiders and Banks and Brown (1962) found that sweep-net catches of the shield bug, *Eurygaster integriceps*, on wheat had sampling errors of only 10% and reflected absolute population differences determined by other methods.

7.3.8 Visual searching and pooting
Methods based on simple counts of the number of animals present are described in Chapter 9. The aspirator or pooter (Fig. 7.6) is convenient for rapidly collecting

small insects in fixed-time estimations (catch/unit effort). When using a pooter, care should be taken to ensure that the operator cannot accidentally inhale damaging particles. Evans (1975) describes two pooter designs that work on the Venturi principle, with the operator blowing. The one shown in Fig. 7.6b is emptied by carefully removing the top cork and the gauze filter; this operation and the realignment of the tubes before reuse must be carried out carefully. While searching a fixed area, if completely efficient, provides an absolute estimate of population size, this is rarely possible. Sampling efficiency tends to vary with the weather and the time of day.

7.4 Relative methods: trapping

The methods described in this section, in contrast to the last, are those in which it is

Fig. 7.6 The aspirator or pooter. (a) A simple model. (b) A blow model (after Evans, 1975). Mouthpiece is 9 mm diameter and the blow tube 11 mm external diameter.

the animal rather than the observer that makes the action that leads to its enumeration. Basically, traps may be divided into those that catch animals randomly and those that attract them in some way. (The word 'attract' is used in its widest sense without any connotation of desire; the studies of Verheijen (1960), for example, have shown that animals are trapped by artificial light through interference with the normal photic orientation and not strictly because of attraction.) It is important to distinguish between these two types, as those that are based on attraction allow the possibility of a further source of error. But a strict division is impossible, as some traps, more particularly water and sticky traps, are intermediate in position.

7.4.1 Interception traps

These are traps that intercept the animals, more or less randomly, as they move through the habitat—air, water, or land. Indices of absolute population may theoretically be obtained more easily from this type of trap than from others, as there is no variation due to attraction.

7.4.1.1 Air—flight traps

Here we are concerned with stationary flight traps that are not believed to attract the insects; those that have some measure of attraction are discussed below (p. 294); others described elsewhere are moving (rotary) nets, which may give absolute samples (p. 145) or, if the quantity of air they filter is uncertain, they are regarded as aerial sweeps (p. 266).

One of the simplest types is the suspended cone net, as used by Johnson (1950); this may have a wind-vane attachment to ensure that it swivels around to face into the wind; Taylor (1962) has shown that its efficiency at different wind speeds may be calculated. Nets are particularly useful for weak flyers, such as aphids (Davis & Landis, 1949), or at heights or in situations where the wind speed is always fairly high (Gressitt et al., 1960), so that the insects cannot crawl out after capture. Similar traps have been used on moving ships.

The Malaise trap is more elaborate; it consists basically of an open-fronted tent of cotton or nylon net, black or green in colour; the 'roof' slopes upwards to the innermost corner at which there is an aperture leading to a trap. It was developed by Malaise (1937) as a collector's tool; modified designs have been described by Gressitt and Gressitt (1962), Townes (1962), and Butler (1965), and a basically similar trap by Leech (1955). Gressitt and Gressitt's (Fig. 7.7) and Butler's models are much simpler to construct and transport than Townes'. The Gressitt model is large (7 m long and 3.6 m high). Butler's is smaller, being made from a bed mosquito-net by cutting out part of one side and a hole in the roof into which the collecting trap—a metal cylinder and a polythene bag—is placed. Townes (1962) gives very full instructions for the construction of his model, which is more durable and traps insects from all directions. When maximum catches are desired (for collecting), Malaise traps should be placed across 'flight paths', such as woodland paths, but in windy situations they cannot be used. The studies of Juillet (1963), who compared a Malaise trap with

Fig. 7.7 Malaise trap. (a) Sketch of the Gressitt type. (b) Plastic collecting tube (after Gressitt & Gressitt, 1962).

others, including the rotary trap, which is believed to give unbiased catches, suggest that, for the larger Hymenoptera and some Diptera, this trap is unbiased, but it is unsatisfactory for Coleoptera and Hemiptera. However, Roberts (1970, 1972) found that, even for Diptera, the form and colour of the trap would greatly influence the catch. Hansen (1988) found black traps to catch more specimens and species in both burnt and unburnt areas of open range, although the differences were not statistically significant.

A very large 'net-tent' ('filet-tente') extending in an arc of 37 m was designed by Aubert (1969) for trapping migratory insects in mountainous areas. An 'electrified net' has been used, sometimes with bait, for tsetse flies (Rogers & Smith, 1977). Walker and Lenczewski (1989) and Walker and Whitesell (1993) describe portable traps for catching migrating butterflies.

Flying Coleoptera and other insects that fall on hitting an obstacle during flight may be sampled with a window trap, which is basically a large sheet of glass or Plexiglas held vertically, with a trough containing water with a wetting agent and a little preservative below it (Chapman & Kinghorn, 1955; Van Huizen, 1977; Fig. 7.8) or with a layer of soy-bean oil (Flamm et al., 1993). Living insects may be collected by replacing the trough with opaque cylinders whose ends have transparent collecting

Grooved wooden
supports

Glass

Collecting
trough

Fig. 7.8 Window trap.

tubes; insects move towards the light and accumulate in the tubes (Nijholt &
Chapman, 1968). A modified flight-interception trap for use in tree canopies is
described by Springate and Basset (1996).

One of the most important uses of stationary flight traps is to determine the
direction of flight (see p. 391). Nielsen (1960) has observed that migratory butterflies will enter stationary nets and remain trapped, whereas those engaging in trivial
movement will fly out again. The direction of migration can be determined by a
number of stationary flight traps arranged so as to sample from four fixed directions.
Stationary nets may be used (Nielsen, 1960) or a Malaise trap modified to collect the
insects from each side separately (Roos, 1957). A more extensive trap is the 'robot
observer', which was used in work on blowflies and consisted of a zigzag wall of
net, with 10 V-shaped bays, each with a $2\,m^2$ aperture and ending in a fly-trap;
the catches from the 10 opposing bays on each side are directly comparable
(J. MacLeod, pers. comm.).

Special traps have been designed to collect and distinguish between the insects
entering and leaving the 'homes' of vertebrates, principally burrows and houses
(Myers, 1956; Muirhead-Thomson, 1963; Service, 1963). Exit traps are usually net
cones that are inverted above the exit hole. Entrance traps have to be inconspicuous
from the outside, and a metal-gauze cylinder with a dark canvas 'skirt', which could
be spread out to prevent passage round it, was used in rabbit burrows by Myers, who
concluded, however, that the entrance trap did not sample as randomly as the exit
trap. In houses, ingress and egress may be measured by identical traps, but Service's
(1963) results show that, in this habitat, the former is measured less efficiently. The
traps consist of a net cone leading into a net or net and Perspex box; Saliternik (1960)
trapped the mosquitoes leaving cesspits with a similar trap. This could be a measure
of emergence (see Chapter 10).

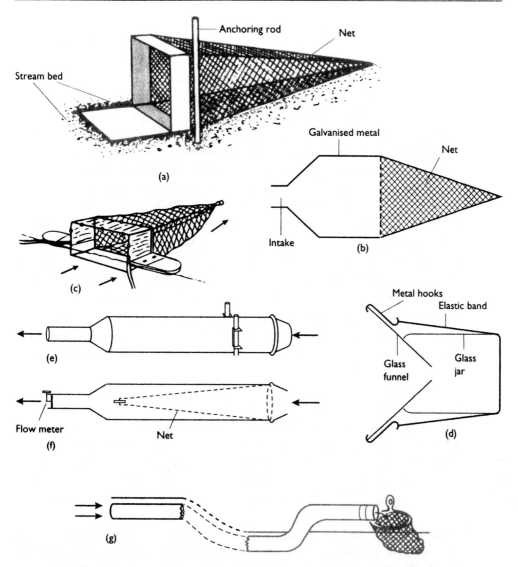

Fig. 7.9 Aquatic interception traps. (a) For drifting invertebrates in streams. (b) Cushing-type trap for drifting invertebrates in streams. (c) Elliott floating trap (after Elliott, 1970). (d) Piezynski trap for water mites. (e) and (f) External and sectional views of Müller-type tubular trap with flowmeter (after Elliott, 1970); (g) a pipe trap latter Eliott, 1970): the dotted portion represents a length of pipe sufficient for a drop in level.

Flying vertebrates may be trapped alive by placing fine nets, mist nets, vertically across flight paths. Cannon nets may be fired over resting flocks of waders and other birds. These techniques should only be carried out by trained (and in many countries licensed) persons. The equipment can be obtained from several suppliers including Avinet at www.qty.com/index.html and the Association of Field Ornithologists at www.afonet.org/mistnets.html.

7.4.1.2 Water—drift samplers and fish traps
Drift samplers. Waters (1969) and Elliott (1970) have reviewed methods of studying

organisms drifting in streams. The simplest type is based on Waters's (1962) design (Fig. 7.9), a 1 m long Nitex net with a square mouth and a board sunk flush with the stream bed in front of the mouth. Nets of this type may become clogged, but, if the mouth is narrowed (Fig. 7.9b), this is alleviated and hence back-flow is prevented and filtration increased (Cushing, 1964; Mundie, 1964; Anderson, 1967). A collecting vessel may be placed at the end of the net.

Elliott (1967) devised a floating trap (Fig. 7.9), similar to Waters's but having the advantage for 'drift sampling' that crawling forms were definitely excluded.

If the volume of water passing through a drift sampler is measured, it then provides an absolute population estimate, analogous to a suction trap. Elliott (1970) was able to measure the water-flow through his floating net using a current meter. The flowmeter may be incorporated into the sampler (Fig. 7.9e, f); originally designed by K. Müller, they have been modified by Elliott (1967, 1970) and Cloud and Stewart (1974). A third approach to the measurement of drift is to pipe a portion over a sufficient distance so that it can be brought above the surface and filtered (Fig. 7.9g; Elliott, 1970). Such methods are also particularly appropriate by weirs and falls. Anderson and Lehmkuhl (1968) funnelled the entire stream flow through their trap. Sampling period is important, because drift often shows a diel periodicity (Elliott, 1969), which is influenced by moonlight and other variations in light intensity (Anderson, 1966).

Pieczynski (1961) has found that what is in effect an unbaited glass lobster-pot (Fig. 7.9d) will trap numbers of water mites and other free-swimming animals. Usinger and Kellen (1955) used a similar trap, constructed of plastic screen rather than glass, and caught large numbers of aquatic insects; their trap is illustrated in Usinger (1963, p. 55). A modification with a collecting jar reached through a narrower entrance on the roof has been found particularly useful for water-beetles (James & Redner, 1965). These traps are the aquatic equivalents of the pitfall trap.

Fish and crustacean traps. Humans have trapped fish for thousands of years and a great many methods are still currently used (see Brant (1984) for a review of methods). Most familiar is the lobster-pot, which is used throughout the world. The scientist new to a region will find it useful to study local methods, which are often designed to perform well under the specific conditions of water and animal behaviour. Most fish traps work on the principle that it is easier for a fish to enter than leave and thus the catch need not be all the fish that have entered. Commonly used types of fish trap, illustrated in Fig. 7.10, include fish-weirs, bag-nets, pot traps, salmon putts, and minnow traps. All fish traps are highly selective. Aquatic traps can be used with an attractant, such as light (see p. 291) or a bait.

The filter screens of industrial cooling water intakes at power stations, oil refineries, and chemical works are widely used as traps in river, estuarine, and coastal studies. These cooling water systems pump huge volumes of water (about $30\,m^3/s$ for a 1000 MW nuclear-powered station) and thus act as powerful samplers. The intakes are often placed on the seabed and any fish or crustacean swimming or crawl-

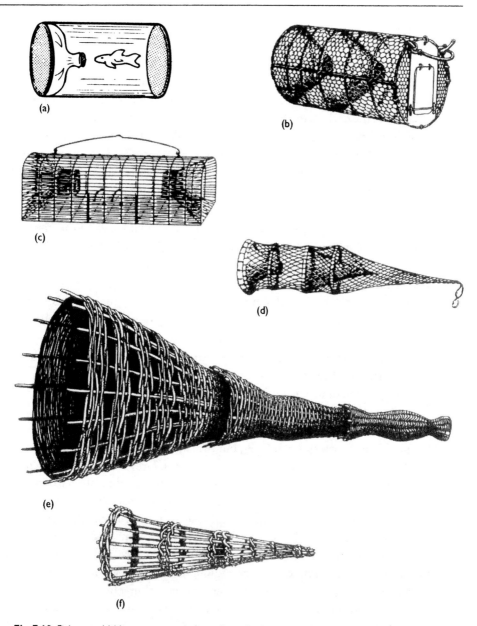

Fig. 7.10 Fish traps. (a) Minnow trap made from plastic drink bottle with the neck cut off and inverted. (b) Eel grig. (c) Wire-mesh double-entry trap. (d) Dutch bag-net. (e) Putt—traditional Severn estuary trap made from woven willow. (f) Putcher—a traditional Severn estuary trap made from woven willow. d, e and f after Jenkins (1974) *Nets and Coracles*, David & Charles, London.

ing close to the intake is sucked in. Examples of studies using power-station intake data are van den Broek (1979), Henderson (1989), and Henderson and Seaby (1994).

Gill-nets. Gill-nets are designed to work by catching swimming fish, which attempt to pass through the net but, when they find the mesh is too small, cannot reverse out, generally because their gill covers catch on the net; however, fish are caught for a variety of reasons, e.g. some catfish entangle their spiny pectoral fins. The size range of fish caught is related to mesh size and it is common practice during general fisheries surveys to lay banks of nets with a range of mesh sizes. The efficiency is related to the ability of fish to detect the net. Fish with good vision are caught more often when light levels are low. However, this does not imply that they are always more effective in the dark, as they will only catch swimming fish and many fish are inactive at night (while other species become active). Monofilament nets are more difficult than multifilament nets for fish to detect and thus tend to be more efficient. However, they are more easily damaged by predators, such as piranhas, crabs, or caymans, which are attracted by the catch, and thus they may be less practical.

Fish abundance is normally expressed as catch per unit effort, where effort is given in units of net length × time. This expression might suggest that the catch in a net increases linearly with time, which is not true (Minns & Hurley, 1988).

7.4.1.3 Land—pitfall and other traps

Like the lobster-pot, the pitfall trap was an adaptation by the ecologist of the technique of the hunter; basically it consists of a smooth-sided hole into which the hunted animal may fall but cannot escape. Often a glass, plastic, or metal container is sunk into the soil so that the mouth is level with the soil surface. Pitfalls can be used to capture arthropods, amphibians, snakes, and mammals. Insects may be emptied with a hand-operated or mechanical suction apparatus (p. 158), thereby avoiding the disturbance to the surroundings that would result from continued removal and resinking. Pitfalls can be arranged as trapping webs or grids for the estimation of population parameters (see p. 379), but the experimental results of Parmenter et al. (1989) on a captive population of darkling beetles, Eleodes sp., at known density give little confidence that trapping webs will reliably estimate population parameters.

While pitfalls are used most frequently for arthropods, they can be effective with vertebrates. When used to sample amphibians and lizards, pitfalls may be used in conjunction with a drift fence (Morton et al., 1988; Friend et al., 1989; Heyer et al., 1994). An advantage of pitfalls over snap or live traps, such as the Longworth, is that they can produce multiple captures, although a captured animal may influence the probability of further captures. A description of an effective pitfall for small mammals is given in Walters (1989) and the relative merits of pitfalls over other types of traps for mammals are considered by Innes and Bendell (1988).

Pitfall traps have been used extensively for studies on surface-dwellers such as spiders (Bultman, 1992; Koponen, 1992; Bauchhenss, 1995), Collembola (Budaeva, 1993), centipedes, millipedes (Kurnik, 1988), ants (Abensperg-Traun & Steven, 1995), beetles, especially Carabidae (e.g. Kowalski, 1976; Uetz, 1977; Cameron & Reeves, 1990; Epstein & Kulman, 1990; Togashi et al., 1990; Braman & Pendley, 1993), and even crabs (Williams et al., 1985). More elaborate traps have been designed to facilitate emptying (Rivard, 1962), for use under snow (Steigen, 1973), with rain-guards (Fichter, 1941; Steiner et al., 1963), or with timing devices that allow the catch from each time period to be segregated (Blumberg & Crossley, 1988). Artificial (Stammer, 1949) or natural (Walsh, 1933) baits can be used in pitfall traps, and baited traps have been found useful for collecting beetles from the burrows of mammals (Welch, 1964). For example, Rieske and Raffa (1993) sampled pales weevil, *Hylobius pales*, pitch-eating weevil, *Pachylobius picivorus*, and pine root collar weevil, *Hylobius radicis*, using pitfalls baited with ethanol and turpentine. However, the effects of baits are variable and in many cases, as their attractiveness will change with age, they may well introduce a further source of error, and simple traps are generally to be preferred in population studies (Greenslade & Greenslade, 1971).

Preservatives, such as formalin or alcohol, have sometimes been used in traps, but, like baits, these may affect the catches of species differentially (Luff, 1968; Greenslade & Greenslade, 1971). While ethylene-glycol antifreeze has been found to work better (Clark & Blom, 1992) for the Carabidae, even this has been found to produce a different catch and sex ratio when compared with traps holding just water (Holopainen, 1992). The most suitable procedure for water-filled traps may be to add a small amount of detergent as a wetting agent.

Pitfall traps have many advantages; they are cheap (empty food or drink containers may be used), they are easy and quick to operate, and a grid of traps can provide an impressive set of data (e.g. Gist & Crossley, 1973). However, many studies have shown that catch size is influenced by a wide range of factors, apart from population size (Petruska, 1969; Hayes, 1970; Greenslade, 1973; Luff, 1975). As pointed out by Topping and Sunderland (1992) in a study of the spider fauna of winter wheat, the catch may not reflect either changes in abundance, relative abundance between species, or the actual sex ratios. Pitfall traps must therefore be used with extreme caution, but, as Gist and Crossley (1973), Uetz and Unzicker (1976), and others have found, they may provide valuable information, provided proper attention is paid to potential sources of variation. Pitfalls are frequently used in studies of community richness and for habitat assessment; but they sample simply the surface-dwelling community and sweep-nets or suction sampling will reveal a different species pattern (Samu & Sarospataki, 1995). Indeed, pitfalls are particularly appropriate for heavy ground-living beetles, which are poorly sampled by suction samplers such as the D-Vac (Mommertz *et al.*, 1996). For carabids, catches can reflect density (Hokkanen & Holopainen, 1986), but the capture efficiency, measured as the proportion of trap encounters resulting in capture, can be low. Halsall and Wratten (1988) found efficiencies ranging from zero to 40%, depending on both species and conditions.

An interesting model of the action of pitfall traps has been developed by Jansen and Metz (1979), who consider that the number of animals trapped depends on: (i) their population density; (ii) their movement, assumed to be Brownian; (iii) the boundary of the pitfall and its 'absorptiveness'; and (iv) the outer boundary of the area and the extent to which the animals penetrate it (the probability of absorption). Movement of cursorial arthropods is affected by temperature, moisture, and other weather conditions (Mitchell, 1963), food supply (Briggs, 1961), the characters of the habitat (e.g. the amount of impedance by the vegetation), and the age, sex, and condition of the individuals (Petruska, 1969; Hayes, 1970). If due account is taken of the thresholds, corrections can be made for weather conditions (p. 249); the other variables need to be kept in view—they may or may not be significant in the particular study. The catches immediately after a pitfall trap is placed in position are commonly found to be higher than those subsequently achieved; Joosse (1975) and Greenslade (1973) termed these 'digging-in effects'. Jansen and Metz (1979) obtained similar effects from their theoretical model, due to depletion of the individuals, moving in a Brownian manner, around the trap.

The boundary of the pitfall trap influences the catch, in respect of both its magnitude and its 'capture efficiency' (the probability of absorption). Traps are generally circular and catch size will theoretically be related to the length of the perimeter (Morrill *et al.*, 1990); if approach, in the region of the trap, is effectively linear, then rectangular or any non-reflexed polygonal trap aperture will also catch in proportion to its diameter (Luff, 1975). However Luff's (1975) studies showed that the capture efficiency of trap boundaries varied with respect to both the nature of the trap (e.g. glass jar or tin) and the species of ground-beetle. The only generalization that was possible was that smaller-aperture traps seemed to have higher efficiencies for the smaller beetles, whilst the larger traps caught a higher proportion of the larger beetles that encountered the trap boundary (aperture). Abensperg-Traun & Steven (1995) concluded that larger traps caught more ant species (especially those with larger individuals); however, if results were expressed as catch per mouth area, then the smallest (18 mm diameter) traps were the most efficient.

Gist and Crossley (1973) used exclusion barriers to isolate an area (provide a 'non-absorptive'—*sensu* Jansen and Metz (1979)—outer boundary) and a grid of pitfalls to 'trap out' the area, so that removal-trapping methods could be used to estimate population size (p. 255). Such studies and the 'digging-in effect' suggest that, within a discrete habitat, a grid of pitfall traps could influence the population density of a highly mobile animal.

Two other factors may influence the efficiency of pitfall traps: the level of the trap lip and the retaining efficiency. Greenslade (1964) found that trap catches were qualitatively or quantitatively affected by the precise placement of the level of lip of the trap container, whether at the surface of the soil or the surface of the litter. Retaining efficiency depends on escape; the smoother the side, the less likely this is. Glass and most plastic containers are therefore to be preferred to metal traps, which quickly corrode, providing footholds. Small winged species may escape from large-aperture traps by flight or they may be eaten by larger species; it is, of course, for these reasons that investigators often use preservative solutions.

Walking animals may also be caught on sticky traps, bands placed round tree trunks (see also p. 352), and glass plates laid on the ground. Mellanby (1962) found that the latter caught numbers of springtails and other animals; he suggested a gauze roof over the trap to prevent flying insects being caught. Details of the design of sticky traps are given below.

Mammal traps. A wide variety of traps for both large and small mammals are described by Bateman (1979). Most ecologists are only concerned with the live trapping of small mammals, which is normally accomplished with a trap that closes when the animal touches a lever or treddle. The most frequently used traps in Britain are Longworth traps (Penlon Ltd., Radley Road, Abingdon, Oxon, OX14 3PH, and Wathins & Doncaster at www.watdon.com) and in the USA the Sherman trap (H.B. Sherman Inc., PO Box 20267, Tallahassee, Florida 32316). Traps are often baited with food, both to attract and later to sustain the captive. References regarding the selectivity of these traps are given in section 7.1.

7.4.1.4 Grain-probe traps

Commercial probe traps are available to detect beetle infestation in grain stores. The trap is a perforated cylinder, one end of which is sealed while the other has a collecting vial. The trap is placed within the grain and works rather like a pitfall, in that it traps beetles that crawl through the perforations. White *et al.* (1990) discuss the design of probe traps and Subramanyam *et al.* (1993) compare the capture efficiencies of the Storgard (manufactured by Trace, Salinas, California, USA) and Grain Guard (manufactured by Grain Guard, Verona, Wisconsin, USA) commercial traps. The Storgard trap comprises a 45 cm long, 3 cm internal diameter, polyethylene tube with 4 mm × 2 mm rectangular perforations, while the Grain Guard version is a 37 cm polycarbonate cylinder of 2 cm internal diameter with 2.5 mm round, downward-pointing perforations.

Probe traps might also find application in studies of the infauna of habitats, such as aquatic leaf-litter banks, where there is a lack of fine particulate matter that could clog or fill the trap.

7.4.2 Flight traps combining interception and attraction

The distinction between water and sticky traps and those flight traps described above (p. 270) should not be regarded as rigorous, for, as Taylor (1962) has shown, a white sticky trap catches insects almost as if they are inert particles, whereas a net may sample selectively. The two approaches may be combined, as in the sticky net described by Provost (1960). However, many aphids and weevils are particularly attracted to yellow (Broadbent, 1948; Riley & Schuster, 1994), ceratopogonid midges to black (Hill, 1947), citrus midge to orange (Pena & Duncan, 1992), thrips to white (Lewis, 1959), blue (Brodsgaard, 1989), or yellow (Cho *et al.*, 1995), frit-flies to blue (Mayer, 1961), and bark-beetles to red (Entwistle, 1963); therefore the quality and magnitude of water- and sticky-trap catches will be greatly influenced by the colour of the trap and this will be an additional source of variation in the efficiency of these traps. It is probable that the attraction of different colours to a given species may vary with both age and sex.

7.4.2.1 Sticky traps

These are an extension of the 'fly-paper'; the animal settles or impacts on the adhesive surface and is retained. A variety of adhesives may be used; those resins and greases developed for trapping moths ascending fruit-trees have proved particularly useful; castor oil may be satisfactory for minute insects (Cameron *et al.*, 1995). One of the most commonly used sticky materials is polyisobutylene, and instructions for the removal of insects from this matrix are given by Murphy (1985). The insects are separated from most fruit-tree banding resins by warming and then scraping the resin plus insects into an organic solvent, such as trichlorethylene or hot paraffin, from which they may be filtered. The separation from the greases is easier; a mixture of benzene and isopropyl alcohol rapidly dissolves the adhesive. Thus, when possible, a grease will be used in preference to a resin, but only weak insects will be trapped by a grease—mosquitoes (Provost, 1960), mites (Staples & Allington, 1959), and aphids (Close, 1959). Indeed, a grease may be more efficient for such insects, as the effective area of the trap is not reduced by the numbers of large insects that are also trapped if a powerful adhesive is used (Close, 1959). Greases may become too fluid at high temperatures.

Sticky traps are of several basic designs. Até strands, an African bird-lime, were used by Golding (1941, 1946) in warehouses, and Provost (1960) used sticky nets, but most traps have been either large screens or small cylinders, boxes, or plates. Cylindrical plastic traps are commonly used for small insects. Large screens, consisting of a wooden lattice or series of boards, have been used to measure movement, at a range of heights, of aphids and beetles (Dudley *et al.*, 1928; Taft & Jernigan, 1964). Individual boards, white and yellow in colour, were used by Roesler (1953) and Wilde (1962) for studies of various flies and psyllids.

Ibbotson (1958) introduced the idea of glass plates for sticky traps. These were 8 in square, the upper surface coated with the adhesive and the under painted yellow (or as desired). Glass has the advantage that the adhesive is easily scraped off with a knife or the whole plate immersed in a solvent. Ibbotson used his trap for the frit-fly; Staples & Allington (1959) used grease-coated microscope slides for trapping mites; Chiang (1973) used an oil-coated strip over a white disc for midges; Wakerley (1963) found Ibbotson's trap suitable for the carrot fly, *Psila rosae*, and Maxwell (1965) a similar trap for the apple maggot fly, *Rhagoletis*. These plates may be exposed either vertically or horizontally; the exposure will affect the catch (Table 7.4; Heathcote, 1957a). The influence of colour is less with vertical plates, where most insects are

Table 7.4 The catches of some species of aphid on three types of trap, expressed as a ratio of the number caught by a suction trap (after Heathcote, 1957a).

	Water	Cylindrical sticky	Flat sticky
Aphis fabae group	2.28	1.92	0.84
Tuberculoides annulatus	3.91	0.45	0.18
Cavariella aegopodii	0.36	2.00	0.27
Myzus persicae	1.34	0.34	0.78
Hyalopterus pruni	0.33	1.50	0.40
Drepanosiphum platanoidis	1.29	0.25	0.01
Brevicoryne brassicae	0.47	0.06	0.07
Sitobion spp.	0.08	0.17	0.06

caught by wind impaction, i.e. the catches are less biased, but horizontal plates catch more aphids in the landing phase (A'Brook, 1973) and gave high catches of *Adelges piceae* (Lambert & Franklin, 1967). However, fixed, vertical, plate traps will sample a different proportion of the passing air, depending upon the wind direction, and eddies will develop at the sides of the plates and around sticky boxes (Fröhlich, 1956). Small glass plates may be mounted at right angles to a wind vane and so present a constant exposure to the wind.

A revolving sticky trap was found most efficient for mosquitoes (Dow & Morris, 1972), but most *Stomoxys* were caught on the leeward side of white sticky traps (Williams, 1973). A good design, sampling at random from the passing air, is the cylindrical sticky trap of Broadbent *et al.* (1948) (Fig. 7.11a), which consists basically of a piece of plastic material covering a length of stove-pipe. Taylor (1962) showed that, if the wind speed was known, the catches of small insects on a white sticky trap of this type could be converted to a measure of aerial density (see Table 7.1, p. 252); however, yellow traps catch more aphids than white ones, and these catch more than black (Broadbent *et al.*, 1948; Taylor & Palmer, 1972; A'Brook, 1973). Care should therefore be exercised in applying these corrections to a species whose reaction to the colour of the trap is unknown. Colour will presumably have the greatest effect at low wind speeds. Cylindrical sticky traps have been utilized in studies on the cacao mealy bug (Cornwell, 1960) and small ones, 2.5 cm diameter, were used by Lewis (1959) for trapping thrips.

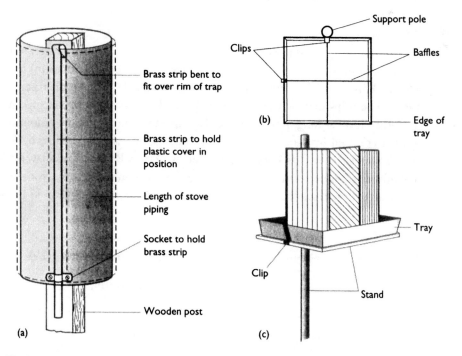

Fig. 7.11 (a) Cylindrical sticky trap (after Broadbent *et al.*, 1948). (b) and (c) Plan and isometric view of a water trap with baffle (Coon and Rinicks's design).

The relation of the size of the sticky trap to the catch has been investigated by Heathcote (1957b) and Staples and Allington (1959), who found that, although catch usually increases with size, it is not proportional to size, so that the smallest trap catches the largest number per unit area (Table 7.5).

Simpson and Berry (1973) exposed successive portions of a Tanglefoot-coated strip of plastic on the leading edge of a trap, with a vane, so that it always faced into the wind; in this way they were able to get a 24 h record of wind-borne aphids. A sticky trap that automatically changed the collecting surface is described by Rohitha & Stevenson (1987).

Mention was made above of the attraction of aphids and other insects in the landing phase to particular-coloured horizontal plates. Prokopy (1968, 1977) and Kring (1970) have shown with fruit flies (Trypetidae = Tephritidae) that yellow panels and red spheres attract flies in different behavioural conditions: the yellow plate is a 'super leaf', the red sphere a fruit. The use of relative methods sensitive to behavioural phase to partition the population in this way is a powerful approach. Sticky-trap catches are frequently biased towards one sex. For example, Kersting and Baspinar (1995), using yellow sticky traps to sample for the leafhopper *Circulifer haematoceps*, found >85% of the catch to be male, while D-Vac collections were 74% female. The sticky-trap catch may also be biased towards females (e.g. Riley & Schuster, 1994), but it is often able to detect population change. For example, Palumbo *et al.* (1995) compared yellow sticky traps, direct counting, and a suction sampler and found all three methods were able to show population trends in sweet-potato whitefly, *Bemisia tabaci*.

7.4.2.2 Water traps

These are simply glass, plastic, or metal bowls or trays filled with water, to which a small quantity of detergent and a preservative (usually a little formalin) have been added. Omission of the detergent will give less than half the catch (Harper & Story, 1962). The traps may be transparent or painted various colours and placed at any height. Yellow has long been recognized as one of the most effective colours. Yellow bowls were used by Moericke (1950) and others for trapping aphids, by Fröhlich (1956) and Fritzsche (1956) for a weevil, and by D'Arcy Burt & Blackshaw (1987) for bibionid flies. Fluorescent yellow trays caught twice as many cabbage-root flies, *Erioischia*, as yellow trays (Finch & Skinner, 1974). White trays were used by South-

Table 7.5 The effect of trap size on the catch of black, cylindrical, sticky traps (after Heathcote, 1957b).

	Trap diameter (cm)		
	3	6	12
Total catch	837	1473	2017
Catch/unit area (3 cm trap = 1)	837	736.5	504.2

wood *et al.* (1961) for the frit-fly and Pollet and Grootaert (1987) found most species of dolichopodid fly to be most attracted to white, although soil-dwelling and arboreal species were most numerous in the red and the blue traps, respectively. A variety of colours were tested by Harper and Story (1962) for the sugar-beet fly, *Tetanops myopaeformis*. The total numbers taken in the different-coloured traps were as follows: yellow 330, white 264, black 202, red 107, blue 64, green 53.

For many purposes, sticky traps would be chosen in preference to water traps, because the relationship between wind speed and the catch of the water trap is likely to prove less simple than that for the sticky trap investigated by Taylor (1962) and because water traps must be frequently attended or they overflow in heavy rain or dry out in the sun: although a model has been designed with a reservoir for automatically maintaining a constant level (Adlerz, 1971). On the other hand, water traps have certain advantages; the insects that are caught are in good condition for identification, as the catch is easily separated by straining or individual insects can be picked out with a pipette or forceps. Furthermore, when the population is sparse, a water trap will, with aphids at least, make catches when a sticky trap of a manageable size would not (Heathcote, 1957a). The 'catching power' of a water trap can be further increased by standing two upright baffles of aluminium sheeting at right angles to each other to form a cross in the tray (Coon & Rinicks, 1962; Fig. 7.11b,c). These divisions enable one to separate the insect according to the quadrant in which it has been captured; this may be related to the direction of flight at the time of capture, although eddy effects could introduce error (see also p. 271). The results of Coon and Rinicks show that the effect of the baffle varies from species to species; in the four cereal aphids they studied, it increased the catch of *Rhopalosiphum maidis* 10-fold and nearly doubled the very low catch of *Schizaphis agrostis* (= *Toxoptera graminum*), but had no effect on those of the two others, *Macrosiphum granarium* and *Rhopalosiphum fitchii*.

A floating water trap, which collected the insects settling on the surface of ponds, was devised by Grigarick (1959), the animals being drowned through the action of the detergent. The tray of his trap was surrounded by a wooden frame; therefore, it would seem that the contrast presented by this surround might bias the catch and a more transparent float (e.g. an air-filled polythene tube) would be preferable.

7.4.3 Light and other visual traps

7.4.3.1 Mode of action and limitations

Light traps are probably the most widely used insect traps and there are several thousand references to them. Originally, paraffin and acetylene lamps were used by collectors (Frost, 1952); later the tungsten-filament electric light and, after its development by Robinson and Robinson (1950), the ultraviolet trap became widely used.

The uses of light traps in ecology are subject to all the general limitations of relative methods outlined above (p. 247); but the variation in efficiency of the trap from insect to insect, from night to night, and from site to site is more serious than in almost any other type of trap, because light traps are entirely artificial, relying on the disturbance of normal behaviour for their functioning. It is not justified to claim on a priori grounds, as Mulhern (1953) has done, that, being mechanical, light traps are more reliable than hand-collection per unit of habitat. Furthermore, in the discussion of the pros and cons of the many models of light trap, it is often implied that the bigger the catch, the 'better' the trap; although it is true that the larger figures are often more acceptable for statistical analysis, it is unwise to assume that they are biologically more unbiased.

The exact mechanisms that lead to an insect's capture by a light trap are far from clear. Mikkola (1972) has shown that electroretinograms evoked by different types of light were at variance with the actual responses of the moths to traps. It seems reasonable to suppose that there is an area, sometimes called the radius of the trap or 'catchment area', within which the insect comes under the influence of the light. The brighter the trap, the greater this area, and hence, when there is bright moonlight, the amount of contrast between the trap and its surroundings will be lessened and the catchment area reduced (e.g. Miller et al., 1970; Taylor, 1986; Nag & Nath, 1991). For example, a study of light-trap catches of rice green leafhoppers (Nephotettix spp.) showed that the effective radius of a 100 W tungsten light varied from 245.20 m during the new moon to 16.72 m at full moon (Mukhopadhyay, 1991). Bowden and Church (1973) found that the extent of the reduction was consistent with the random, rather than fixed, capture of the insect within the catchment area. The behavioural studies of Verheijen (1960) and Mikkola (1972) suggest that the result may appear random because, even if there are zones of attraction, corresponding perhaps to the insect seeking open flight routes through the vegetation, there are probably 'annuli of repulsion' and 'zones of dazzle' when the normal photic orientation is disturbed.

In a series of studies, Bowden has been concerned with the changes in the catchment area ('trap radius') with time — the lunar cycle; he has shown how the precise distribution of moonlight throughout the night needs to be considered in relation to the insect's periodicity of flight (Bowden, 1973; Bowden & Church, 1973). Using this information, standardized catches may be calculated; these eliminate the variations due to the changes in the catchment area brought about by the effect of moonlight (Bowden & Morris, 1975). When this is done, it is found that there are some insects that are more active on nights with a full moon (Anderson & Poorbaugh, 1964; Bidlingmayer, 1964; Bowden & Gibbs, 1973).

Variation in the catchment area in space, in regard to the height of flight of the moths, the position and design of the trap, and the role of wind in moving a column of air across the trap, have been studied by Taylor and Brown (1972) and Taylor and French (1974). They postulate a volume equivalent to the amount of air originally occupied by the catch, i.e. 100% efficient extraction. This will, of course, be less than the catchment area, but one can visualize that the two approaches might be combined to make light traps a measure of absolute population. Catches generally fall with increasing wind speed, e.g. Culicoides (Edwards et al., 1987); turnip moth, A.

segetum (Esbjerg, 1987); migratory moths (Gregg *et al.*, 1994). Other factors known to influence trap efficiency include temperature, relative humidity, and rainfall.

Light traps have been found useful in survey work, both for particular species (Geier, 1960; Otake & Kono, 1970; Grimm *et al.*, 1993; Bishop *et al.*, 1994) and on an extensive scale for a taxonomic group or region (Rings & Metzler, 1990; El Hag & El Meleigi, 1991; Richert & Huelbert, 1991; Worwood & Harrington, 1994).

7.4.3.2 The effect of trap design on catch

From the above, we can conclude that catches in a light trap will be influenced by the following design features.

1 The amount of contrast between the light source and surroundings: the greater the contrast, the greater the 'catchment area'.

2 Most animals will have a tendency to withdraw from the high light intensity immediately adjacent to the lamp.

3 The extent to which an animal may be able to change from approach to avoidance will depend on its flight speed; the faster, heavier flyers are unable to stop and change course quickly. Indeed, Stanley and Dominick (1970) found with a Pennsylvania-type trap with nested funnels that the large moths were trapped in the central smallest funnel, whilst coccinellids were most frequently caught in the widest outer funnel.

The effect of the illumination of the environment of the catch was utilized by Common (1959), who designed a transparent light trap for Lepidoptera. The main advantage of this over the Robinson trap, of which it is a modification, is that the number of Coleoptera (especially Scarabaeidae) caught is greatly reduced; large numbers of beetles will damage many of the moths in a catch.

An increase in the intensity of the lamp, frequently brought about by the substitution of an ultraviolet lamp for an incandescent one, usually leads to an increased catch (e.g. Williams *et al.*, 1955; Barr *et al.*, 1963; Belton & Kempster, 1963). Because of the role of intensity in the mechanism of trapping, the actual size of the light source is of less importance (Belton & Kempster, 1963) and a small trap may often be substituted for a larger one without the catch being reduced in proportion to the change in light output (Smith *et al.*, 1959).

Many workers have considered that the differences in quality of the light, from incandescent lamps, from mercury-quartz lamps emitting ultraviolet (UV) and visible light, and from black (or 'blue') lamps emitting only or mainly UV, affect the catch; but Mikkola's (1972) study shows this to be very complex. Whatever the mechanism, there are considerable differences, e.g. many Diptera and Miridae are taken in the largest numbers in traps with incandescent (tungsten-filament) lamps, Corixidae and many Lepidoptera (especially noctuids) are more abundant in 'UV traps' and are even better represented in 'black-light traps', as are Trichoptera (Frost, 1953; Williams *et al.*, 1955; Southwood, 1960; Tshernyshev, 1961).

Some of these differences are explained by the increased intensity of the UV traps leading to the repulsion of the insect when close to the lamp; slow flyers, such as many Diptera, Nematocera, and geometrid and pyralid moths, stop their approach before they enter the trap, whilst the heavier, faster flyers, like noctuid and sphingid moths, pass straight into the trap. Suction fans have therefore been incorporated in light traps to catch those insects that avoid capture at the 'last moment' (e.g. Glick *et al.*, 1956; Downey, 1962). The same phenomenon was demonstrated in the opposite way by Hollingsworth *et al.* (1961), who found that a suitably placed windbreak could increase the catch of a trap; in the exposed situation, many species would avoid trapping by being blown off course, but, in the lee of the windbreak, this would not occur and the larger moths would be caught particularly efficiently. The addition of baffles to the trap also serves to catch those insects whose repulsion by the high intensity close to the lamp would otherwise cause them to escape; by adding four intersecting baffles to a trap, Frost (1958a) almost doubled the numbers caught.

When differences due to flight speed and momentum have been eliminated, other specific differences in susceptibility to trapping remain, e.g. amongst the mosquitoes, *Aedes* seem especially susceptible to trapping (Love and Smith, 1957; Loomis, 1959b). Males are often taken in much greater numbers than females at light traps; this may be because the males engage in more nocturnal flight activity than the females, e.g. Miridae (Southwood, 1960).

The height above the vegetation at which the trap is exposed will influence the catch. As the aerial density of most insects decreases with height, in general, the higher the trap, the smaller the catch (Frost, 1958c; Taylor & Brown, 1972). There are a few exceptions; Taylor and Carter (1961) have shown that some Lepidoptera may occur at a maximum density at as much as 10 m above the ground and many ornithophilic mosquitoes and blackflies also have maxima some distance above the ground.

7.4.3.3 Techniques and types of trap

Many light traps retain the insects alive; Merzheevskaya and Gerastevich (1962) did this by covering the collecting funnel of a Pennsylvania-type trap with a cloth bag and inserting another shallow cone in the neck. The Robinson trap retains the insects in the drum of the trap, but Belton and Kempster (1963) showed that, unless the opening was under 1 in (2.5 cm) in diameter, many small moths, such as geometrids, would leave the trap in the morning, more especially if the sun shone on it; however, such a small cone opening will restrict the entry of other species.

A large number of killing agents have been used in light traps; some of the chlorinated hydrocarbons, such as tetrachloroethylene, trichloroethylene, and tetrachloroethane, are particularly useful. They may be poured on to a block of plaster of Paris at the bottom of the killing bottle, where their heavy vapours will be slowly evolved and remain relatively concentrated (Williams, 1958; Haddow & Corbet, 1960). However, it must be noted that inhalation of their vapours over a long period may be dangerous to humans. Delicate insects, such as Ephemeroptera or Psocoptera, may be collected straight into alcohol and chafer beetles into kerosene (Frost, 1964). Potassium and sodium cyanides are traditional, but dangerous, poisons; the last-named is more suitable in moist climates (Frost, 1964).

White (1964) has designed a killing tin to preserve the light-trap catch in good condition (Fig. 7.12). It has the further advantage that only specimens too large to pass through the screen across the top of the funnel are retained; the mesh of this screen may be so fine as to retain the whole catch or, for example, it may be of such a size that the unwanted Diptera and small Trichoptera escape (E.G. White, pers. comm.). When only the smaller species are required, the larger ones may be excluded, and so prevented from damaging the catch, by covering the entrance to the trap with a coarse screen (e.g. Downey, 1962). If all specimens are to be captured, a series of graded screens in the killing bottle will tend to separate the insects by size and aid in the subsequent sorting (Frost, 1964). In connection with the latter operation, the sorting tray described by Gray (1955) may be useful.

7.4.3.4 The Rothamsted trap (Fig. 7.13a)

Originally used by C.B. Williams as a research tool for studying the influence of climate on the numbers of flying insects, this trap takes moderate numbers of most groups, including nematocerous flies and the larger moths; it would seem to be less selective than most other light traps and is useful in studies on the diversity of restricted groups. It is described in detail by Williams (1948); it has a roof, which allows its operation in all weather but reduces the catch. The collections from different periods of the night may be separated by an automatic device. It may be purchased from the Burkard Co. (http://www.burkard.co.uk).

7.4.3.5 The Robinson trap (Fig. 7.13b)

This was the first trap using UV light and was designed by Robinson and Robinson

Fig. 7.12 White's light-trap filling tin.

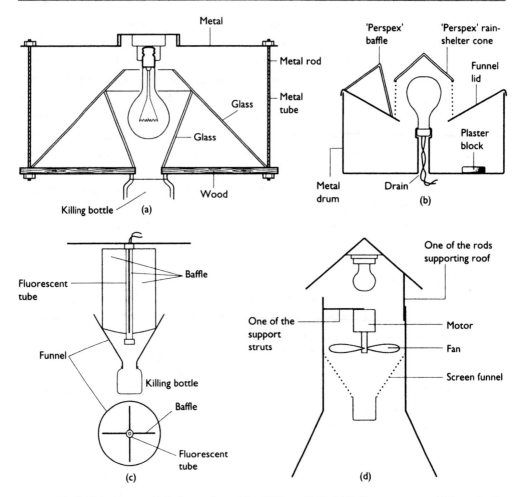

Fig. 7.13 Light traps. (a) Rothamsted trap (after Williams, 1948). (b) Robinson-type trap. (c) Pennsylvanian trap (sectional and plan views). (d) New Jersey trap.

(1950) to make maximum catches of the larger Lepidoptera. It is therefore without a roof and, although the bulb is protected by a celluloid cone and there is a drainage hole in the bottom, it cannot be used in all weathers. The drum is partly filled with egg boxes or similar pieces of cardboard, in which the insects can shelter. The catch may be retained alive or killed by placing blocks of plaster of Paris saturated with a killing agent on the floor. Large numbers of fast-flying nocturnal insects (e.g. Noctuidae, Sphingidae, Corixidae, Scarabaeidae) are trapped (Williams, 1951; Williams *et al.* 1955); but Common's (1959) transparent trap catches fewer beetles. The 'Muguga trap', widely used in eastern Africa, is a modification of this type of trap (Brown *et al.*, 1969; Siddorn & Brown, 1971). It may be purchased from Watkins & Doncaster, Cranbrook, Kent, UK (www.watdon.com).

7.4.3.6 *The Pennsylvanian and Texas traps* (Fig. 7.13c)

These traps are basically similar, consisting of a central fluorescent tube surrounded by four baffles; below the trap is a metal funnel and a collecting jar (Frost, 1957;

Hollingsworth *et al.*, 1963). The Pennsylvanian trap has a circular roof to prevent the entry of rain into the killing bottle and it is presumably a reflection of the climates of the two states that the Texas trap is roofless. A wire-mesh grid directly below the funnel in the Texas trap will allow rain and small insects to escape, whilst the larger specimens are deflected into a collecting container (Dickerson *et al.*, 1970). Frost's (1958b) experiments suggest that the largest catches would be obtained with a 15 W black-light fluorescent tube; Graham *et al.* (1961) used three argon-glow lamps in the Texas trap in survey work on the pink bollworm, *Platyedra gossypiella*, and found that moths were trapped over a radius of 200 ft (61 m) from the lamp. Fluorescent tubes may be run for many hours or even years from batteries (Clark & Curtis, 1973).

The Minnesota and Monks Wood light traps are basically similar to the Pennsylvanian model, but the roof is cone-shaped rather than flat and this appears to reduce the catch (Frost, 1952, 1958b). The effects of many different modifications of this type of trap on the catches of small moths have been reported by Tedders and Edwards (1972).

7.4.3.7 The New Jersey trap (Fig. 7.13d)

Developed by T.J. Headlee, this is primarily a trap for sampling mosquitoes; it combines light and suction — the lamp causes the insects to come into the vicinity of the trap and they are drawn in by the suction of the fan (Mulhern, 1942). The trap is therefore particularly useful for weak flyers, which may fail to be caught by the conventional light trap, as, when close to the lamp, they are repelled by its high intensity. Fitted with a UV lamp, the New Jersey trap has been used to collect other groups of Nematocera, as well as mosquitoes (Zhogolev, 1959). Kovrov and Monchadskii (1963) found that, if the trap was modified to emit polarized light, many insects, but not mosquitoes, were caught in larger numbers. Like the Rothamsted and Pennsylvanian traps, this trap may be fitted with an automatic-interval collector. A tiny battery-operated model, the Centers for Disease Control (CDC) trap, has been developed (Buckley & Stewart, 1970; Vavra *et al.*, 1974; Fontenille *et al.*, 1997). For comparative studies using different traps, it is important to standardize air flow and direction (Loomis, 1959a; Wilton & Fay, 1972), although, as has been pointed out several times, relative trapping methods, and especially light traps, may sample different proportions of the population in different areas, even when the traps are identical. Acuff (1976) found that light traps collected a wider range of mosquito species than bait and Malaise traps. For blood-meal and arbovirus studies, the insects should be collected into saline (7.5 g/L) with some antiseptic (Walker & Boreham, 1976).

7.4.3.8 The Haufe–Burgess visual trap (Fig. 7.14a)

As Verheijen's (1960) studies have shown that the trapping effect of light depends on the degree of contrast between the lamp and the surroundings, it is not surprising that light traps are ineffective in the twilight night of the Arctic, and Haufe and Burgess (1960) designed the present trap for mosquitoes; it has been found to catch other groups as well. Basically, the trap is an exposed-cone type of suction trap, on the outside of which is a revolving cylinder painted with $1\frac{5}{8}$-in-wide (4.1 cm) black

Fig. 7.14 Visual traps. (a) Haufe–Burgess mosquito trap, sectional view (after Haufe & Burgess, 1960) and external view of the attraction cylinder. (b) Manitoba or canopy trap.

and white stripes. A horizontal disc extends out at the top of the cylinder and the insects are drawn through slits at the junction between the cylinder and the disc. Harwood (1961) retained the light-bulb above the trap (as in the New Jersey light trap) and exposed various animals' bait below the trap, but the combination of so many attractants makes the interpretation of the results very difficult.

7.4.3.9 The Manitoba or canopy trap (Fig. 7.14b)

Tabanidae are attracted to the highlights of a black or red sphere (Bracken *et al.*, 1962) and large numbers may be collected by suspending a sphere of these colours (e.g. a balloon) beneath a polythene collecting cone (Bracken & Thorsteinson, 1965; Thorsteinson *et al.*, 1965). The collecting container may be of the type used on 'tents' (see Fig. 7.7; Adkins *et al.*, 1972) and with modifications flies can be marked (see Chapter 3).

7.4.3.10 The pit–light trap

This trap, described by Heap (1988), is designed to capture ground-living insects and combines features of a light trap and a pitfall trap. It is claimed to capture a wide range of insects, with a greater number of specimens and range of soil-dwelling species than can be captured using non-baited pitfall traps.

7.4.3.11 Aquatic light traps (Fig. 7.15)

There are two types: floating and submerged. At present, light traps are little used for sampling either invertebrates or fish but reports of their use are increasing in number. Light-trapping for young fish is particularly appropriate amongst underwater obstructions, such as fallen wood and macrophytes, where netting becomes inefficient or impossible (Conrow *et al.*, 1990; Dewey & Jennings, 1992).

Baylor and Smith (1953) have described an ingenious floating trap, principally for Cladocera, which utilizes their 'attraction' to yellow to get them into the trap and their 'repulsion' to blue to drive them into the collecting net (Fig. 7.15b,c). It is important that the blue light is not visible laterally; thus the level of the blue Cellophane must not extend above the lip of the funnel. Baylor and Smith suggest that the trap could be improved with refinements, such as the polarization of the yellow light, which would increase the 'fishing area' without increasing the intensity (see also Kovrov & Monchadskii, 1963). A floating light trap, especially suitable for emerging insects, may be made out of a plastic bucket with a black light inside the lid and a funnel replacing the base (Carlson, 1971). A model for use in shallow water is described by Apperson and Yows (1976).

Subaquatic light traps can be used in marine and freshwater environments and take a variety of forms. They are most frequently used to sample postlarval and juvenile fish, when common form is a Plexiglas box with slits. Both electric lights and chemical light sticks may be used as a light source. Gehrke (1994) found that, when larval golden perch, *Macquaria ambigua*, and silver perch, *Bidyanus bidyanus*, were exposed to light gradients, silver perch displayed stronger phototactic behaviour than golden perch, and both species were most responsive to light in the 601 nm waveband. The intensity of phototactic responses in both species was greater at higher irradiance levels. Traps fitted with 12 h yellow light sticks attracted more golden perch larvae than traps with blue, green, orange, or red lights. The greater efficacy of yellow light sticks may be due to reduced attenuation; however, yellow light sticks also emit a greater intensity of light over a longer time than other colours tested. Ponton (1994) tested an improved version of the quatrefoil light trap (Fig. 7.15) and concluded that it appeared useful for sampling in particular freshwater microhabitats of several neotropical young fish. Kissick (1993) compared the attraction of larval and juvenile fish to traps lighted by either electric bulbs or photochemical light sticks. No significant differences in collection rates of bluegills, *Lepomis macrochirus*, and brook silversides, *Labidesthes sicculus*, were found. Significantly more threadfin shad, *Dorosoma petenense*, occurred in chemical light traps. Light-stick output will vary with temperature and is of only limited duration, but it has the advantage over electric light that it is less likely to fail. Examples of studies using lights to sample juvenile fish are Gregory and Powles (1985, 1988), Thorrold (1992), Choat *et al.* (1993), Doherty *et al.* (1994), and Zigler and Dewey (1995).

Fig. 7.15 Aquatic light traps. (a) Hungerford's subaquatic light-trap. (b) and (c) Sectional view of Baylor and Smith's floating trap. (c) Detail of test-tube and light source in Baylor and Smith's floating trap. (d) Commercial larval fish light trap from ARI at *www.aquaticresearch.com* (after Ulrich, 1986).

Light traps can also be used to sample squid Thorrold (1992). Larval fish and aquatic invertebrate light traps are available from Aquatic Research Instruments (web address: http://www.aquaticresearch.com).

7.5 Traps that attract the animals by some natural stimulus or a substitute

When these traps are entirely natural, they have the advantage over artificial ones in that the variations in efficiency reflect real changes in the properties of the population; the changes in the numbers of a phytophagous insect colonizing a plant or of a bloodsucking species biting the host may not reflect population changes, but they do accurately reflect changes in the colonizing or feeding rates of the species — that is, with the possible exception of artificial-bait traps, these traps give measures of availability that are biologically meaningful.

7.5.1 Shelter traps

Cubical boxes ($283 \, cm^3$ in volume), with one open side and painted red, have been used in mosquito surveys; they are often termed 'artificial resting units' (Goodwin, 1942; Goodwin & Love, 1957; Burbutis & Jobbins, 1958). As might be expected, only certain mosquitoes settle in these shelters, particularly the species *Culex*, *Culiseta*, and *Anopheles*. Loomis and Sherman (1959) found that visual counts of the resting mosquitoes with a torch were accurate to within 10 of the true value. Dales (1953) found that large numbers of *Tipula* would shelter in and be unable to escape from small metal truncated cones (15 cm aperture at top, 25 cm high, and 28 cm at base), and Vale (1971) tested a range of artificial refuges for tsetse flies.

Ground-dwellers (cryptozoa), such as woodlice, centipedes, and carabids, which shelter beneath logs and stones, may be trapped by placing flat boards in the habitat. The 'cryptozoa boards' can be placed systematically in the habitat and used to study the dispersion of the animals (Cole, 1946); however, the proportion of the population that shelters beneath them will vary with soil moisture, aspect, and other factors (Paris, 1965; Jenson, 1968).

Many insects shelter in cracks or under bark scales on tree trunks, for overwintering or as a mode of life. Artificial shelters, consisting of several grooved boards bolted together, have been placed in trees to trap earwigs (Chant & McLeod, 1952), placed in the field to trap overwintering *Bryobia* mites (Morgan & Anderson, 1958), and screwed on to the tree bark beside a fungus to trap the beetle, *Tetratoma fungorum* (Paviour-Smith, 1964). A variety of shelter traps (cryptic shelters) were used by Dahlsten *et al.* (1992) in a study of the Douglas fir tussock moth, *Orgyia pseudosugata*; the most successful were wood blocks drilled with 2.5 cm holes. Fager (1968) used oak boxes filled with sawdust, with holes drilled in them, as 'synthetic logs' in studying the fauna of decaying wood.

Trap nests, made by drilling holes in woody stems (e.g. sumac) or in pieces of dowelling, which are split and then bound together, will be colonized by solitary

Hymenoptera if the open ends are exposed in a board (Fig. 7.16); the binding may be removed and the nest contents easily exposed (Medler & Fye, 1956; Levin, 1957; Medler, 1964; Fye, 1965; Freeman & Jayasingh, 1975).

7.5.2 Host-plant traps

Insect-free potted host plants exposed in a habitat may be colonized by insects; these can be removed and counted and their numbers will be directly proportional to the colonization potential of the population in that habitat. Host-plant traps are there-fore most useful for measuring emergence and colonization (Fritzsche, 1956; Grigarick, 1959; Smith, 1962; Waloff & Bakker, 1963; Bucher & Cheng, 1970) and susceptible logs will record the same phenomena in bark beetles (Chapman, 1962). The actual adult insects or the resulting eggs or larvae may be counted. Michelbacher and Middlekauff (1954) studied population changes in *Drosophila* in tomato fields by placing as 'traps' ripe tomatoes that had been slit vertically on the sides and squeezed slightly; the number of eggs laid in the flesh on either side of the cuts could be easily ascertained in the laboratory. It is, of course, important to ensure that the plants, fruit, or logs remain in the condition where they are attractive to the insects. Wireworms (Elateridae) are attracted to the carbon dioxide (CO_2) respired by germinating seeds and Williams *et al.* (1992) used this behaviour to sample them by placing germinating seed into the soil for later retrieval.

7.5.3 Baited traps

The study of the behavioural responses of insects to scents from either their own species (pheromones) or their food sources (kairomones) is one of the most recent and most active fields of entomological research. Such scents are used by humans for a variety of purposes: for confusing the pest in its approach to the prey (plant)

Elastic band

Ends sealed
with beeswax

Fig. 7.16 Levin's trap nests for solitary Hymenoptera.

or in traps, either to measure or monitor the population or to reduce it by killing large numbers, directly or by the incorporation of a sterilant into the bait. An enormous variety of traps has been described; the present account must be regarded as merely an outline of some of the types, with an emphasis on those most useful in ecological work. (Traps utilizing vertebrates as bait are described in the following section.)

7.5.3.1 Types of trap

Basically traps can be divided into two categories.

1 'Lobster-pot' traps (Fig. 7.17): these lure in animals through an inverted funnel, when they are generally captured alive. Traps of this type have been used widely, with CO_2 as bait for bloodsucking insects (Bellamy & Reeves, 1952; Wilson & Richardson, 1970), with carrion as bait for flies (MacLeod & Donnelly, 1956; Gillies, 1974), or with various extracts for bark beetles, when they are termed 'field olfactometers' (Chararas, 1977).

2 Impaction traps (Fig. 7.17): in these, the animals is held generally in an adhesive resin or grease or in a water trap in proximity to the lure. Designs of this type have been widely used in studies on chemical lures (Howell et al., 1975; Lewis & Macaulay, 1976; Anderbrant et al., 1989).

Pheromone traps are important monitoring tools for insect pests and many different trap designs have been developed. The relative effectiveness of six commercially available pheromone traps—Multi-pher 1, 2, and 3 non-sticky traps and Scentry wing, Pherocon 1C, and Delta sticky traps—for six orchard-pest moths was compared by Knodel and Agnello (1990). Sticky traps generally captured more moths than non-sticky traps, regardless of species.

7.5.3.2 Baits or lures

The baits used have ranged from the actual biological material (e.g. virgin female insects or rotting fruit) to the extracted chemicals or their 'mimics'. Broadly, a distinction may be made between kairomones and pheromones, although some behavioural processes, such as the attraction of scolytids to a host tree, may involve both. Furthermore, some of the lures have been discovered empirically and their classification is a matter of conjecture.

Kairomones. Carrion is used in trapping flies, especially blowflies (Dodge & Seago, 1954); the freshness of the bait, the precise position of the trap, and the presence of other flies have been found to affect the results (MacLeod & Donnelly, 1956; Fukuda, 1960; Kawai and Suenaga, 1960). Rotting fruit will attract certain butterflies (Sevastpulo, 1963), bananas or vinegar with yeast, *Drosophila* (Mason, 1963), hydrolysed proteins and similar compounds, fruit flies (McPhail, 1939; Neilson, 1960; Bateman, 1972), and proteinaceous materials, eye gnats, *Hippelates* (Mulla et al., 1960). Various chemicals will act as lures; these are probably either the actual kairomone (as with carbon dioxide) or are related to it. Various fruit flies have been attracted by a variety of chemicals (Bateman, 1972), but complex synthetic lures, e.g. Medlure and Cuelure, are particularly effective (Beroza et al., 1961; Fletcher, 1974).

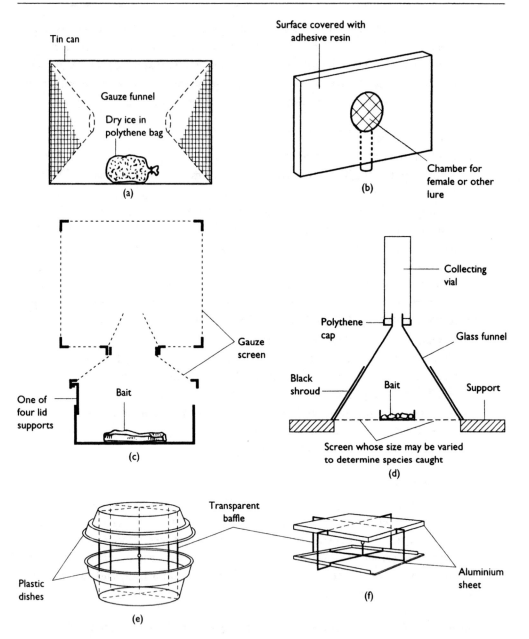

Fig. 7.17 Bait traps. (a) Carbon dioxide trap. (b) Sticky trap (Coppel et al., 1960). (c) MacLeod's carrion trap. (d) Citrus Experimental Station general bait trap. (e) Water trap (after Lewis & Macaulay, 1976). (f) 'Lantern trap' (after Lewis & Macaulay, 1976). Chemical bait is exposed in the centre of the baffles, which are covered with adhesive resin in the water and lantern traps.

Many scolytids are attracted by turpinoids (Chararas, 1977), the cupesid beetle, *Priacma serrata*, by chlorine (Atkins, 1957), the cabbage rootfly, *Erioschia brassicae*, by allylisothiocyanate (Finch & Skinner, 1974), and the pine weevil, *Hylobius abietis*, by the methyl ester of linoleic acid (Hesse et al., 1955). Carbon dioxide, conveniently supplied as 'dry ice', has been used as a lure for many bloodsucking insects: mosquitoes (Bellamy & Reeves, 1952; Gillies, 1974; Service, 1995), ticks (Garcia, 1962), tsetse flies (Rennison & Robertson, 1959), sandflies (Rioux & Golvan, 1969), and blackflies (Fallis & Smith, 1964). Carbon dioxide has been combined with a Malaise trap for tabanids (Blume *et al.*, 1972; Roberts, 1976), although Wilson and Richardson (1970) found *Chrysops* were less attracted to it than were *Tabanus*. Anderson and Yee (1995) trapped simulids on CO_2-baited three-dimensional horse/cow models with sticky ears and bodies. Tsetse fly can be attracted by baiting with natural and synthetic host odour (Hargrove *et al.*, 1995).

Pheromones. These may mostly be categorized as sex attractants and the simplest approach is to expose a virgin female as the lure (Lewis *et al.*, 1971; Goodenough & Snow, 1973). The identification and synthesis of insect pheromones have been an important scientific achievement of the last 30 years. This has allowed the selective trapping of many moths and, more recently, some Hemiptera and Coleoptera. There is an extensive literature on optimum trap design and some recent studies are Anshelevich *et al.* (1993, 1994) on monitoring the honeydew moth, *Cryptoblabes gnidiella*, and European vine moth, *Lobesia botrana*, respectively, Drapek *et al.* (1990) on the corn ear worm, *Heliothis zea*, and Kehat *et al.* (1994) on the peach twig borer, *Anarsia lineatella*. Particular progress has been made with codling moth (Reidl *et al.*, 1986), various tortricoid moths (Anshelevich *et al.*, 1994; Chang, 1995), noctuid moths (Witz *et al.*, 1992), pyralid moths (Anshelevich *et al.*, 1993), and bark beetles (Scolytidae) (Schlyter *et al.*, 1987).

Pheromone traps have been mostly used to detect and delimit outbreaks of insect pests. In some cases, the number of individuals caught can be used to give an index of population size or impact. For example, Gage *et al.* (1990) found that returns from pheromone traps for gypsy moth were correlated with the amount of defoliation.

7.5.4 The use of vertebrate hosts or substitutes as bait for insects

The methods described in this section are used principally to determine the biting rate of intermittent ectoparasites under various conditions, the host range, and the relative importance of different hosts and vectors and other measurements of significance in epidemiological studies. Some give unbiased indices of these behavioural characteristics, but in others the presence of the trap may repel some insects or attract others independently of the presence of the host. Dyce and Lee (1962) found that the presence of the drop-cone trap about 30 cm above the rabbit would deter certain species of mosquito, but not others, from biting it. In contrast, Colless (1959) showed that from one-third to half the catch of a Malayan trap is independent of the presence of the host. Comparisons of the responsiveness of tsetse flies to stationary and mobile baits showed that, whereas visual and olfactory stimuli are important with the former, visual cues predominate with moving bait (Vale, 1974). As there is often a marked diurnal periodicity in biting rate (Haddow, 1954), it is important that assessments should be made throughout the 24 h; indeed, the 'biting cycle' is often longer (McClelland & Conway, 1971), so that if there is an unusual age distribution estimates need to be made over at least one full cycle. Human bait may be moving (see below) or stationary, vector landing rates being described in terms such as 'collector-nights' (Attenborough *et al.*, 1997).

These methods therefore give measures of availability and, when attractiveness is constant, indices of absolute population (i.e. relative abundance). Under certain circumstances, they might be used to measure absolute population density by the removal technique (see p. 255). One must note that the attractiveness of individual baits differs (Saunders, 1964; Khan *et al.*, 1970). Many special traps are reviewed by Service (1995).

Reference should also be made to the Manitoba or canopy trap (p. 290), the CO_2 trap (p. 296), and perhaps even the Haufe–Burgess mosquito trap (p. 289), which could have been included here on the grounds that they are substitutes for vertebrate hosts.

7.5.4.1 Moving baits

As ectoparasite incidence is often low and non-random and as landing rate may vary in relation to the time of exposure in one spot (Fig. 7.18), a moving bait may give more extensive data than one stationary over a long period. The 'fly-round', first developed by W.H. Potts, has been used extensively in work on the ecology of the tsetse fly. In this, the catching party consisting of the human bait and the collectors would walk through the habitat, making a number of stops, at which flies were caught. Ford *et al.* (1959) proposed a transect fly-round that followed arbitrary straight lines with equidistantly placed halting sites. These transect rounds have advantages in the statistical analysis necessary for the determination of dispersion, but the catches are still variable and Glasgow (1961) concluded that a 7500-yard (approx. 6850 m) fly-round done once a week could not detect less than a fivefold change in the mean catch (i.e. in the index of absolute population). More males than females are usually (Glasgow & Duffy, 1961), but not always (Morris, 1960), caught on fly-rounds. This appears to be due to the fact that only teneral females feed on moving hosts, therefore mature females are seldom taken; furthermore, the males follow the moving objects (even vehicles) not so much to feed (except those that are teneral) as to be in the vicinity to pair with the young females (Bursell, 1961). Oxen and other animals may be used instead of humans as the bait on fly-rounds; the species of *Glossina* taken will be influenced by the bait animal (Jordan, 1962). Studying the biting habits of *Chrysops*, Duke (1955) found that biting rate per individual of a group of human baits was greater than that for a single individual working on his/her own. Single human baits were used by Murray (1963) in studies on *Aedes* and found to give a better picture of seasonal variation than light traps.

Ticks and fleas sit on grass and other vegetation and seek to attach themselves to their hosts as they pass by, a behaviour termed questing. Rothschild (1961) and Bates (1962) sampled fleas with what they termed artificial rabbits and birds: a skin with an internal heating device (a hot-water bottle) that was pulled across the ground. Ticks have been sampled by dragging blankets and pieces of cloth over the vegetation;

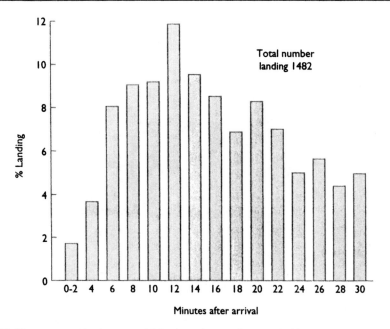

Fig. 7.18 The variation in landing rate of *Culicoides* with time after arrival of human bait (after Jamnback & Watthews, 1963).

Wilkinson (1961) has compared various methods and concludes that the 'hinged-flag sampler', a modification of Blagoveschenskii's (1957) technique, was most efficient and gave a good correlation with counts on cattle in the same enclosure. However, Randolph and Storey (1999) found that questing activity has a diurnal rhythm and is also influenced by the immediate microclimate; thus correction factors may be necessary. The hinged-flag sampler consists of a piece of plywood (50 cm square), covered with cloth and with a handle strongly hinged to the centre of one side. The sampler is held as rigidly as possible, with the lower edge making contact with the grass. In very rough country, this cannot be used and cloth-covered leggings are the only available method. Ticks may be removed from these samplers, retained, and stored on transparent adhesive tape (e.g. Sellotape).

7.5.4.2 Stationary baits

The most natural situation is achieved when the host is freely exposed in the normal habitat; when a human is the host, this is, of course, possible, and it may be feasible for one individual to act as both bait and collector. In order to obtain a complete collection, the bait may need to enclose him/herself periodically in a cage. Klock and Bidlingmayer (1953) devised one that was essentially an umbrella, with blinds that could be released and would rapidly drop to the ground (the 'drop net'). Blagoveschenskii *et al.* (1943) dropped a bell-shaped cover over their human bait at intervals and Myers (1956) extended this principle to the rabbit with the 'drop-cone trap'; a muslin cone is suspended by a rope over a pulley about 30 cm above a tethered bait (Fig. 7.19); it may be quickly lowered by an observer some distance away. But, as Dyce and Lee (1962) showed, the presence of the drop-cone trap just above the bait

Fig. 7.19 (a) The Myers drop-cone trap. (b) Anderson and Defoliart's method for trapping blackflies. (c) Harrison's trap for poultry mites. (d) The Morris trap.

may affect the behaviour of some species. Dyce and Lee (1962) also suggest that the effect of humans handling the traps in which other species are exposed may invalidate comparisons between the biting rate on humans and these other animals.

Blackflies (Simuliidae) are less easily disturbed when feeding on their hosts, and Anderson and Defoliart (1961) investigated their host preferences by exposing various birds and mammals in net or wire cages for 15 min and then covering these with a blackout box with a removable insect trap on top (Fig. 7.19b). The flies would soon be attracted to the light and move into the muslin trap. The blackout box was conveniently made of a strong cardboard box strengthened with wood. Disney (1972) lowered the box on a cantilever; it then becomes a type of 'drop-cone trap' (see above). Apart from any influence due to the trap, the catches at a bait may be influenced by the length of time it has been present (see Fig. 7.19) (Jamnback & Watthews, 1963) and there are considerable local variations due to site (e.g. Saunders, 1964). Selection due to the influence of the trap itself will be greater with those methods where the animal is more or less enclosed.

Although traps of many different designs have been invented for flying blood-sucking insects, and often named after the inventor (Service, 1995), the majority can be classified into one of five types.

1 The Malayan or bed-net trap, originally designed by B.A.R. Garter and, apparently independently, by R.C. Shannon. Basically, this consists of a gauze structure or modified mosquito-net, operating rather like a Malaise trap over the bait, generally human. It is not surprising, therefore, that Colless (1959) found that an unbaited trap only caught up to half as many mosquitoes as a baited one. Initially, the sides or part of the sides is open, but these flaps can be dropped by pulling a string, say, every hour, and the catch collected. Wharton *et al.* (1963) used a similar trap with monkeys, enclosed in a wire-netting cage, as bait, They may even be used with well-tethered cattle (Mpofu *et al.*, 1993).

2 The stable trap, described by E.H. Magoon and E.C. Earle. It consists of a fixed or portable sectional screen and wooden shed, into which mosquitoes and other biting flies can enter by horizontal slits in the lower walls. After feeding, most will fly upwards and are trapped (Bates, 1944). As the name suggests, these traps originally utilized cattle or horses as bait, but they have also been widely used with dogs. A lightweight form, which can be hoisted into trees, uses birds as bait (Flemings, 1959). Trap huts are often of similar construction, but the slits are replaced by entry traps, which protect the bait from being bitten.

3 The Lumsden trap was invented by W.H.R. Lumsden (1958) and combines bait with a weak suction device. It is constructed of a Perspex hood leading to a suction fan, a gauze funnel, and a collecting tube ('an inverted suction trap'). A bait animal is tethered, or placed in a cage, below the hood. The fan is switched on periodically and the resulting air current carries the insects feeding on the bait into the collecting tube. Portable, battery- or generator-operated versions were originally developed by Snow *et al.* (1960) and Minter (1961); these catch large numbers of *Culicoides* and *Phlebotomus*, but the air flow seems insufficient to capture many of the mosquitoes. A model with a more powerful suction device used rabbits as bait and successfully trapped mosquitoes (Service, 1995).

4 Traps for small animals. These are generally open below (Worth & Jonkers, 1962; Turner, 1972) or the design may be like the CO_2 trap (see Fig. 7.18). The No. 10 Trinidad (Worth & Jonkers, 1962) and No. 17 Trinidad (Davies *et al.*, 1971) traps are versatile and easily constructed gauze-covered models (Service, 1995). Another trap of this general type has been constructed by Harrison (1963) for studying the poultry red mite, *Dermanyssus*. It consists of a box, containing the bait chicken, with a number of holes near the base (Fig. 7.19); these may be closed with a time-switch- or hand-operated barrier. The trap could be used for other ectoparasites that leave the host after feeding (e.g. Cimicidae).

5 Harris, Morris, or tent traps. R.H.T.P. Harris devised a trap that apparently attracted and retained tsetse flies, the trap being the bait. His design was considerably modified and improved by Morris and Morris (1949) (Fig. 7.19d). A portable folding version has been described by Morris (1961). The flies enter the traps from below and only a few seem able to escape; however, if the traps were emptied less than once every 24 h, losses of trapped flies occurred due to predation by ants (Smith & Rennison, 1961, part IV). A high proportion of the flies caught in the Morris trap are females and this is their great advantage compared with the fly-round. However, the dispersion pattern suggested by trap catches does not coincide with that found by fly-rounds or searching the resting sites (Glasgow & Duffy, 1961). Thus their value as an index of absolute population must be doubted, and the time of day of maximum catch in the Morris trap seems to differ from that at oxen (Smith & Rennison, 1961). Several species of *Glossina* and many tabanids are caught by Morris traps, but others are not taken even when they attack tethered oxen in the vicinity (Morris, 1960, 1961; Jordan, 1962; Vale 1971). Black-coloured cloth on the trap seemed to attract more tabanids (Morris, 1961) and, in the wet season, more tsetse flies than did brown hessian (Morris, 1960). The variations in catch in these traps remain something of a mystery, in spite of extensive tests (Hargrove, 1977). Fredeen (1961) used basically similar traps (termed silhouette traps) to collect Simuliidae, which, as Wenk and Schlörer (1963) have shown, are attracted to the silhouette of their hosts.

7.6 Using sound

Many insects respond to sounds of their own and other species, so recordings could be used as the lure in traps (Belton, 1962; Cade, 1975). Kanda *et al.* (1988) used dry ice, a hamster, and a speaker emitting sound at 500 Hz, which is the wing-beat frequency to attract female *Mansonia*, while Ikeshoji & Yap (1987) and co-workers have developed a number of devices that attract male mosquitoes and trap them on cylindrical or other shaped sticky traps (Service, 1995).

References

Abensperg-Traun, M. & Steven, D. (1995) The effects of pitfall trap diameter on ant species richness (Hymenoptera: Formicidae) and species composition of the catch in a semi-arid eucalypt woodland. *Aust. J. Ecol.* **20**(2), 282–7.

A'Brook, J. (1973) Observations on different methods of aphid trapping. *Ann. Appl. Biol.* **74**, 263–77.

Acuff, V.R. (1976) Trap biases influencing mosquito collecting. *Mosquito News* **36**, 173–6.

Adkins, T.R., Ezell, W.G., Sheppard, D.C., & Askey, M.M. (1972) A modified canopy trap for collecting Tabanidae (Diptera). *J. Med. Entomol.* **9**, 183–5.

Adler, G.H. & Lambert, T.D. (1997) Ecological correlates of trap response of a neotropical forest rodent, *Proechimys semispinosus. J. Trop. Ecol.* **13**(1), 59–68.

Adlerz, W.C. (1971) A reservoir-equipped Moericke trap for collecting aphids. *J. Econ. Entomol.* **64**, 966–7.

Agnello, A.M., Reissig, W.H., Spangler, S.M., Charlton, R.E., & Kain, D.P. (1996) Trap response and fruit damage by obliquebanded leafroller (Lepidoptera: Tortricidae) in pheromone-treated apple orchards in New York. *Environ. Entomol.* **25**(2), 268–82.

Alfiya, H. (1995) Surveillance radar data on nocturnal bird migration over Israel, 1989–1993. *Israel J. Zool.* **41**(3), 517–22.

Almand, L.K., Sterling, W.L., & Green, C.L. (1974) A collapsible truck-mounted aerial net for insect sampling. *Texas Agric. Exp. Sta. Mise Pub.* **1189**, 1–4.

Anderbrant, O., Lofqvist, J., Jonsson, J., & Marling, E. (1989) Effects of pheromone trap type, position and color on the catch of pine sawfly, *Neodiprion sertifer* (Geoff). (Hymenoptera, Diprionidae). *J. Appl. Entomol.* **107**(4), 365–9.

Anderson, J.R. & Defoliart, G.R. (1961) Feeding behaviour and host preferences, of some black flies (Diptera: Simulidae). *Ann. Entomol. Soc. Am.* **54**, 716–29.

Anderson, J.R. & Poorbaugh, J.H. (1964) Observation on the ethology and ecology of various Diptera associated with northern California poultry ranches. *J. Med. Entomol.* **1**, 131–47.

Anderson, J.R. & Yee, W.L. (1995) Trapping black flies (Diptera: Simuliidae) in northern California: I. Species composition and seasonal abundance on horses, host models, and in insect flight traps. *J. Vector Ecol.* **20**, 7–25.

Anderson, N.H. (1966) Depressant effect of moonlight on activity of aquatic insects. *Nature* **209**, 319–20.

Anderson, N.H. (1967) Biology and downstream drift of some Oregon Trichoptera. *Can. Entomol.* **99**, 507–21.

Anderson, N.H. & Lehmkuhl, D.M. (1968) Catastrophic drift of insects in a woodland stream. *Ecology* **49**, 198–206.

Andreassen, H.P. & Bondrup Nielsen, S. (1991) Home range size and activity of the wood lemming, *Myopus schisticolor. Holarctic Ecol.* **14**(2), 138–41.

Anshelevich, L., Kehat, M., Dunkelblum, E., & Greenberg, S. (1993) Sex pheromone traps for monitoring the honeydew moth, *Cryptoblabes gnidiella*: effect of pheromone components, pheromone dose, field ageing of dispenser, and type of trap on male captures. *Phytoparasitica* **21**(3), 189–98.

Anshelevich, L., Kehat, M., Dunkelblum, E., & Greenberg, S. (1994) Sex pheromone traps for monitoring the European vine moth, *Lobesia botrana*: effect of dispenser type, pheromone dose, field aging of dispenser, and type of trap on male captures. *Phytoparasitica* **22**(4), 281–90.

Appenzeller, A.R. & Leggett, W.C. (1992) Bias in hydroacoustic estimates of fish abundance due to acoustic shadowing: evidence from day–night surveys of vertically migrating fish. *Can J. Fish. Aquat. Sci.* **49**(10), 2179–89.

Apperson, C.S. & Yows, D.G. (1976) A light trap for collecting aquatic organisms. *Mosquito News* **36**, 205–6.

Atkins, M.D. (1957) An interesting attractant for *Priacma serrata* (Lec.) (Cupesidae: Coleoptera). *Can. Entomol.* **89**, 214–19.

Attenborough, R.D., Burkot, T.R., & Gardner, D.S. (1997) Altitude and risk of bites from mosqitoes infected with malaria and filariosis among the Mianmia people of Papua New Guinea. *Trans. Roy. Soc. Trop. Med. Hyg.* **9**, 8–10.

Aubert, J. (1969) Un appareil de capture de grandes dimensions destiné au marquage d'insectes migrateurs. *Mitt. Schweiz. entomol. Gesell.* **42**, 135–9.

Bagenal, T.B. (1972) The variability in the number of perch, *Perca fluviatilis* L., caught in traps. *Freshw. Biol.* **2**, 27–36.

Balogh, J. & Loksa, I. (1956) Untersuchunger über die Zoozönose des Luzernenfeldes. *Acta Zool. Acad. Sci. Hung.* **2**(1–3), 17–114.

Banks, C.J. & Brown, E.S. (1962) A comparison of methods of estimating population density of adult sunn pest *Eurygaster integriceps* Put. (Hemiptera: Scutelleridae) in wheat fields. *Entomol. Exp. Appl.* **5**, 255–60.

Barr, A.R., Smith, T.A., Boreham, M.M., & White, K.E. (1963) Evaluation of some factors affecting the efficiency of light traps in collecting mosquitoes. *J. Econ. Entomol.* **569**, 123–7.

Bateman, J. (1979) *Trapping: A Practical Guide.* David & Charles, Newton Abbot.

Bateman, M.A. (1972) The ecology of fruit flies. *Ann. Rev. Entomol.* **17**, 493–518.

Bates, J.K. (1962) Field studies on the behaviour of bird fleas. 1. Behaviour of the adults of three species of bird flea in the field. *Parasitology* **52**, 113–32.

Bates, M. (1944) Notes on the construction and use of stable traps for mosquito studies. *J. Nat. Malaria Soc.* **3**, 135–45.

Bauchhenss, E. (1995) Ground-living spiders of sandy soil areas in Northern Bavaria (Germany) (Arachnida: Araneae). *Zool. Beitr.* **36**(2), 221–50.

Baylor, E.R. & Smith, F.E. (1953) A physiological light trap. *Ecology* **34**, 223–4.

Beall, G. (1935) Study of arthropod populations by the method of sweeping. *Ecology* **16**, 216–25.

Beerwinkle, K.R., Lopez, J.D., Jr, Witz, J.A., *et al.* (1994) Seasonal radar and meteorological observations associated with nocturnal insect flight at altitudes to 900 meters. *Environ. Entomol.* **23**(3), 676–83.

Beerwinkle, K.R., Lopez, J.D., Jr, Schleider, P.G., & Lingren, P.D. (1995) Annual patterns of aerial insect densities at altitudes from 500 to 2400 meters in east-central Texas indicated by continuously-operating vertically-oriented radar. *Southw. Entomol. Suppl.* **18**, 63–79.

Bellamy, R.W. & Reeves, W.C. (1952) A portable mosquito bait trap. *Mosquito News* **129**, 256–8.

Belton, P. (1962) Effects of sound on insect behaviour. *Proc. Entomol. Soc. Manitoba* **18**, 1–9.

Belton, P. & Kempster, R.H. (1963) Some factors affecting the catches of Lepidoptera in light traps. *Can. Entomol.* **95**, 832–7.

Benedek, P., Erdelyi, C., & Fesus, I. (1972) General aspects of diel vertical movements of some arthropods in flowering lucerne stands and conclusions for the integrated pest control of lucerne. *Acta Phytopathol. Hung.* **7**, 235–49.

Beroza, M., Green, N., Gertler, S.I., Steiner, L.R., & Miyashita, D.H. (1961) New attractants for the Mediterranean fruit fly. *J. Agric. Food Chem.* **9**, 361–5.

Bidlingmayer, W.L. (1964) The effect of moonlight on the flight activity of mosquitoes. *Ecology* **45**, 87–94.

Bishop, A.L., McKenzie, H.J., Spohr, L.J., & Barchia, I.M. (1994) *Culicoides brevitarsis* Kieffer (Diptera: Ceratopogonidae) in different farm habitats. *Aust. J. Zool.* **42**(3), 379–84.

Bishop, J.A. (1972) An experimental study of the cline of industrial melanism in *Biston betularia* (L.) (Lepidoptera) between urban Liverpool and rural North Wales. *J. Anim. Ecol.* **41**, 209–43.

Blagoveschenskii, D.I. (1957) Biological principles of the control of ixodid ticks. [In Russian.] *Entomol. Obozr.* **36**, 125–33.

Blagoveschenskii, D.I., Sregetova, N.G., & Monchadskii, A.S. (1943) Activity in mosquito attacks under natural conditions and its diurnal periodicity. [In Russian.] *Zool. Zhur.* **22**, 138–53.

Blake, R.W., Kolotylo, R., & De La Cueva, H. (1990) Flight speeds of the barn swallow, *Hirundo rustica. Can. J. Zool.* **68**(1), 1–5.

Blumberg, A.Y. & Crossley, D.A.J. (1988) Diurnal activity of soil-surface arthropods in agroecosystems: design for an inexpensive time-sorting pitfall trap. *Agric. Ecosys. Environ.* **20**(3), 159–64.

Blume, R.R., Miller, J.A., Eschle, J.L., Matter, J.I., & Pickens, M.O. (1972) Trapping tabanids with modified Malaise traps baited with CO_2. *Mosquito News* **32**, 90–5.

Bowden, J. (1973) The influence of moonlight on catches of insects in light-traps in Africa. Part 1.

The moon and moonlight. *Bull. Entomol. Res.* **63**, 113–28.

Bowden, J. & Church, B.M. (1973) The influence of moonlight on catches of insects in light traps in Africa. Part II. The effect of moon phase on light trap catches. *Bull. Entomol. Res.* **63**, 129–42.

Bowden, J. & Gibbs, D.G. (1973) Light-trap and suction-trap catches of insects in the northern Gezira, Sudan, in the season of southward movement of the intertropical front. *Bull. Entomol. Res.* **62**, 571–96.

Bowden, J. & Morris, M. (1975) The influence of moonlight on catches of insects in light-traps in Africa. III. The effective radius of a mercury-vapour light trap rid the analysis of catches using effective radius. *Bull. Entomol. Res.* **65**, 303–48.

Bracken, G.K. & Thorsteinson, A.J. (1965) Orientation behaviour of horse flies; influence of some physical modifications of visual decoys on orientation of horse flies. *Entomol. Exp. Appl.* **8**, 314–18.

Bracken, G.K., Hanec, W., & Thorsteinson, A.J. (1962) The orientation of horseflies and deer flies (Tabanidae: Diptera). *Can. J. Zool.* **40**, 685–95.

Braman, S.K. & Pendley, A.F. (1993) Activity patterns of Carabidae and Staphylinidae in centipedegrass in Georgia. *J. Entomol. Sci.* **28**(3), 299–307.

Brant, A.V. (1984) *Fish Catching Methods of the World.* Fishing Newsbooks, Farnham, UK.

Brierley, A.S. & Watkins, J.L. (1996) Acoustic targets at South Georgia and the South Orkney Islands during a season of krill scarcity. *Mar. Ecol. Progr. Ser.* **138**(1–3), 51–61.

Briggs, J.B. (1961). A comparison of pitfall trapping and soil sampling in assessing populations of two species of ground beetles (Col.: Carabidae). *Rep. E. Malling Res. Sta.* **1960**, 108–12.

Broadbent, L. (1948) Aphis migration and the efficiency of the trapping method. *Ann. Appl. Biol.* **35**, 379–94.

Broadbent, L., Doncaster, J., Hull, R., & Watson, M. (1948) Equipment used for trapping and identifying alate aphids. *Proc. Roy. Entomol. Soc. Lond. A* **23**, 57–8.

Brodsgaard, H.F. (1989) Colored sticky traps for *Frankliniella occidentalis* (Pergande) (Thysanoptera, Thripidae) in glasshouses. *J. Appl. Entomol.* **107**(2), 136–40.

Brown, E.S., Betts, E., & Rainey, R.C. (1969) Seasonal changes in the distribution of the African armyworm, *Spodoptera exempta* (Wlk.) (Lep., Noctuidae) with special reference to eastern Africa. *Bull. Entomol. Res.* **58**, 661–728.

Bruderer, B. (1997a) The study of bird migration by radar: Part 1: The technical basis. *Naturwissenschaften* **84**(1), 1–8.

Bruderer, B. (1997b) The study of bird migration by radar. Part 2: Major achievements. *Naturwissenschaften* **84**(2), 45–54.

Bucher, G.E. & Cheng, H.H. (1970) Use of trap plants for attracting cutworm larvae. *Can. Entomol.* **102**, 797–8.

Buckley, D.I. & Stewart, W.W.A. (1970) A light-activated switch for controlling battery operated light traps. *Can. Entomol.* **102**, 911–12.

Budaeva, L.I. (1993) Peculiarities of surface dwelling collembolan communities (Collembola) in Khakasia steppe. *Zool. Zhur.* **72**(3), 45–52.

Bultman, T.L. (1992) Abundance and association of cursorial spiders from calcareous fens in southern Missouri. *J. Arachnol.* **20**(3), 165–72.

Burbutis, P.P. & Jobbins, D.M. (1958) Studies on the use of a diurnal resting box for the collection of *Culiseta melanura* (Coquillet). *Bull. Brooklyn Entomol. Soc. (NS)* **53**, 53–8.

Bursell, E. (1961) The behaviour of tsetse flies (*Glossina swynnesioni* Austen) in relation to problems of sampling. *Proc. Roy. Entomol. Soc. Lond. A* **36**, 9–20.

Butler, G.D. (1965) A modified Malaise insect trap. *Pan. Pacif. Entomol.* **41**(1), 51–3.

Cade, W. (1975) Acoustically orienting parasitoids: fly phototaxis to cricket song. *Science* **190**, 1312–13.

Cameron, E.A. & Reeves, R.M. (1990) Carabidae (Coleoptera) associated with gypsy moth, *Lymantria dispar* (L.) (Lepidoptera: Lymantriidae), populations subjected to *Bacillus thuringiensis* Berliner treatments in Pennsylvania (USA). *Can. Entomol.* **122**(1–2), 123–30.

Cameron, M.M., Pessoa, F.A.C., Vasconcelos, A.W., & Ward, R.D. (1995) Sugar meal sources for the phlebotomine sandfly *Lutzomyia longipalpis* in Ceara State, Brazil. *Med. Vet. Entomol.* **9**(3), 263–72.

Carley, C.J. & Knowlton, F.F. (1968) Trapping wood rats: effectiveness of several techniques and differential catch by sex and age. *Texas J. Sci.* **19**, 248–51.

Carlson, D. (1971) A method for sampling larval and emerging insects using an aquatic black light trap. *Can. Entomol.* **103**, 1365–9.

Chang, V.C.S. (1995) Trapping *Cryptophlebia illepida* and *C. ombrodelta* (Lepidoptera: Tortricidae) in *Macadamia* in Hawaii. *Int. J. Pest Manag.* **41**(2), 104–8.

Chant, D.A. & McLeod, J.H. (1952) Effects of certain climatic factors on the daily abundance of the European earwig, *Forficula auricularia* L.

(Dermaptera: Forficulidae), in Vancouver, British Columbia. *Can. Entomol.* **84**, 174–80.

Chapman, J.A. (1962) Field studies on attack flight and log selection by the Ambrosia beetle, *Trypodendron lineatum* (Oliv.) (Coleoptera: Scolytidae). *Can. Entomol.* **94**, 74–92.

Chapman, J.A. & Kinghorn, J.M. (1995) Window-trap for flying insects. *Can. Entomol.* **82**, 46–7.

Chararas, C. (1977) Attraction chimique exercée sur certains Scolytidae par les pinacées et les cupressaées. *Comp. Insect. Milieu Trop. Coll. Int. CNRS* **265**, 165–86.

Cherrill, A.J. & Sanderson, R.A. (1994) Comparison of sweep-net and pitfall trap samples of moorland Hemiptera: evidence for vertical stratification within vegetation. *Entomologist* **113**(1), 70–81.

Chiang, H.C. (1973) A simple trap for certain minute flying insects. *Ann. Entomol. Soc. Am.* **669**, 704–5.

Cho, K., Eckel, C.S., Walgenbach, J.F., & Kennedy, G.G. (1995) Comparison of colored sticky traps for monitoring thrips populations (Thysanoptera: Thripidae) in staked tomato fields. *J. Entomol. Sci.* **30**(2), 176–90.

Choat, J.H., Doherty, P.J., Kerrigan, B.A., & Leis, J.M. (1993) A comparison of towed nets, purse seine, and light-aggregation devices for sampling larvae and pelagic juveniles of coral reef fishes. *US Nat. Mar. Fish. Serv. Fish. Bull.* **91**(2), 195–209.

Christensen, J.T. (1996) Home range and abundance of *Mastomys natalensis* (Smith, 1834) in habitats affected by cultivation. *Afr. J. Ecol.* **34**(3), 298–311.

Clark, J.D. & Curtis, C.E. (1973) A battery-powered light trap giving two years continuous operation. *J. Econ. Entomol.* **66**, 393–6.

Clark, W.H. & Blom, P.E. (1992) An efficient and inexpensive pitfall trap system. *Entomol. News* **103**(2), 55–9.

Close, R. (1959) Sticky traps for winged aphids. *NZ J. Agric. Res.* **2**, 375–9.

Cloud, T.J. & Stewart, K.W. (1974) Seasonal fluctuations and periodicity in the drift of caddis fly larvae (Trichoptera) in the Brazos River, Texas. *Ann. Entomol. Soc. Am.* **67**, 805–11.

Cole, L.C. (1946) A study of the cryptozoa of an Illinois woodland. *Ecol. Monogr.* **16**, 49–86.

Colless, D.H. (1959) Notes on the culicine mosquitoes of Singapore. VI. Observations on catches made with baited and unbaited trap-nets. *Ann. Trop. Med. Parasitol. Liverpool* **53**, 251–8.

Common, I.F.B. (1959) A transparent light trap for the field collection of Lepidoptera. *J. Lepidop. Soc.* **13**, 57–61.

Conrow, R., Zale, A.V., & Gregory, R.W. (1990) Distributions and abundances of early life stages of fishes in a Florida (USA) lake dominated by aquatic macrophytes. *Trans. Am. Fish. Soc.* **119**(3), 521–8.

Coon, B.F. & Rinicks, H.B. (1962) Cereal aphid capture in yellow baffle trays. *J. Econ. Entomol.* **55**, 407–8.

Coppel, H.C., Casida, J.E., & Dauterman, W.C. (1960) Evidence for a potent sex attractant in the introduced pine sawfly. *Diprion similis* (Hymenoptera: Diprionidae). *Ann. Entomol. Soc. Am.* **53**, 510–12.

Cornwell, P.B. (1960) Movements of the vectors of virus diseases of cacao in Ghana. *Bull. Entomol. Res.* **51**, 175–201.

Cowx, I.G. (ed.) (1990) *Developments in Electric Fishing.* Fishing News Books, Blackwell Scientific Publications, Oxford, UK.

Cushing, C.E. (1964) An apparatus for sampling drifting organisms in streams. *J. Wildl. Manag.* **28**, 592–4.

Dales, R.P. (1953) A simple trap for Tipulids (Dipt.). *Entomol. Monthly Mag.* **89**, 304.

D'Arcy Burt, S. & Blackshaw, R.P. (1987) Effects of trap design on catches of grassland Bibionidae (Diptera: Nematocera). *Bull. Entomol. Res.* **77**(2), 309–16.

Dahlsten, D.L., Rowney, D.L., Copper, W.A., & Wenz, J.M. (1992) Comparison of artificial pupation shelters and other monitoring methods for endemic populations of douglas-fir tussock moth, *Orgyia pseudotsugata* (McDunnough) (Lepidoptera, Lymantriidae). *Can. Ent.* **124**, 259–369.

Davies, J.B., Corbet, P.S., Gillies, M.T., & McCrae, A.W.R. (1971) Parous rates in some Amazonian mosquitoes collected by three different methods. *Bull. Entomol. Res.* **61**, 125–32.

Davis, E.W. & Landis, B.J. (1949) An improved trap for collecting aphids. *USDA Bur. Entomol. Plant Quar. ET* **278**, 3 pp.

Dewey, M.R. & Jennings, C.A. (1992) Habitat use by larval fishes in a backwater lake of the upper Mississippi River. *J. Freshw. Ecol.* **7**(4), 363–72.

Dickerson, W.A., Gentry, C.R., & Mitchell, W.G. (1970) A rainfree collecting container that separates desired Lepidoptera from small undesired insects in light traps. *J. Econ. Entomol.* **63**, 1371.

Disney, R.H.L. (1972) Observations on chicken-biting blackflies in Cameroon with a discussion of parous rates of *Simulium damnosum*. *Ann. Trop. Med. Parasitol.* **66**, 149–58.

Dodge, H.R. & Seago, J.M. (1954) Sarcophagidae and other Diptera taken by trap and net on

Georgia mountain summits in 1952. *Ecology* **35**, 50–9.

Doherty, P.J., Fowler, A.J., Samoilys, M.A., & Harris, D.A. (1994) Monitoring the replenishment of coral trout (Pisces: Serranidae) populations. *Bull. Mar. Sci.* **54**(1), 343–55.

Dow, R.P. & Morris, C.D. (1972) Wind factors in the operation of a cylindrical bait trap for mosquitoes. *J. Med. Entomol.* **9**, 60–6.

Downey, J.E. (1962) Mosquito catches in New Jersey mosquito traps and ultra-violet light traps. *Bull. Brooklyn Entomol. Soc.* **57**, 61–3.

Drapek, R.J., Coop, L.B., Croft, B.A., & Fisher, G.C. (1990) *Heliothis zea* pheromone trapping: studies of trap and lure combinations and field placement in sweet corn. *Southw. Entomol.* **15**(1), 63–70.

Drickamer, L.C., Lenington, S., Erhart, M., & Robinson, A.S. (1995) Trappability of wild house mice (*Mus domesticus*) in large outdoor pens: implication for models of t-complex gene frequency. *Am. Midland Naturalist* **133**(2), 283–9.

Dry, W.W. & Taylor, L.R. (1970) Light and temperature thresholds for take-off by aphids. *J. Anim. Ecol.* **39**, 493–504.

Dudley, J.E., Searles, E.M., & Weed, A. (1928) Pea aphid investigations. *Trans. IV Int. Congr. Entomol.* **2**, 608–21.

Duke, B.O.L. (1955) Studies on the biting habits of *Chrysops*. III. *Ann. Trop. Med. Parasitol. Liverpool* **49**, 362–7.

Duncan, A. & Kubecka, J. (1996) Patchiness of longitudinal fish distributions in a river as revealed by a continuous hydroacoustic survey. *ICES J. Mar. Sci.* **53**(2), 161–5.

Dyce, A.L. & Lee, D.J. (1962) Blood-sucking flies (Diptera) and myxomatosis transmission in a mountain environment in New South Wales. II. Comparison of the use of man and rabbit as bait animals in evaluating vectors of myxomatosis. *Aust. J. Zool.* **10**, 84–94.

Eastop, V. (1995) Selection of aphid species by different kinds of insect traps. *Nature* **1769**, 936.

Edwards, P.B., Kettle, D.S., & Barnes, A. (1987) Factors affecting the numbers of Culicoides (Diptera: Ceratopogonidae) in traps in coastal South-East Queensland (Australia) with particular reference to collections of *Culicoides subimmaculatus* in light traps. *Aust. J. Zool.* **35**(5), 469–86.

El Hag, E.T.A. & El Meleigi, M.A. (1991) Insect pests of spring wheat in central Saudi Arabia. *Crop Prot.* **10**(1), 65–9.

Elliott, J.M. (1967) Invertebrate drift in a Dartmoor stream. *Arch. Hydrobiol.* **63**, 202–37.

Elliott, J.M. (1969) Diel periodicity in invertebrate drift and the effect of different sampling periods. *Oikos* 20, 524–8.

Elliott, J.M. (1970) Methods of sampling invertebrate drift in running water. *Ann. Limnol.* 6, 133–59.

Entwistle, P.F. (1963) Some evidence for a colour sensitive phase in the flight period of Scolytidae and Platypodidae. *Entomol. Exp. Appl.* 6, 143–8.

Epstein, M.E. & Kulman, H.M. (1990) Habitat distribution and seasonal occurrence of carabid beetles in east-central Minnesota (USA). *Am. Midl. Nat.* 123(2), 209–25.

Esbjerg, P. (1987) The influence of diurnal time and weather on sex trap catches of the turnip moth (*Agrotis segetum* Schiff.) (Lepidoptera, Noctuidae). *J. Appl. Entomol.* 103(2), 177–84.

Evans, L.J. (1975) An improved aspirator (pooter) for collecting small insects. *Proc. Br. Entomol. Nat. Hist. Soc.* 8, 8–11.

Fager, E.W. (1968) The community of invertebrates in decaying oak wood. *J. Anim. Ecol.* 37, 121–42.

Fallis, A.M. & Smith, S.M. (1964) Ether extracts from birds and CO_2 as attractants for some ornithophilic simuliids. *Can. J. Zool.* 42, 723–30.

Feldhamer, G.A. & Maycroft, K.A. (1992) Unequal capture response of sympatric golden mice and white-footed mice. *Am. Midl. Nat.* 128(2), 407–10.

Fewkes, D.W. (1961) Diel vertical movements in some grassland (Nabidae Heteroptera). *Entomol. Monthly Mag.* 97, 128–30.

Fichter, E. (1941) Apparatus for the comparison of soil surface arthropod populations. *Ecology* 22, 338–9.

Finch, S. & Skinner, G. (1974) Some factors affecting the efficiency of water-traps for capturing cabbage root flies. *Ann. Appl. Biol.* 77, 213–26.

Flamm, R.O., Pulley, P.E., & Coulson, R.N. (1993) Colonization of disturbed trees by the southern pine bark beetle guild (Coleoptera: Scolytidae). *Environ. Entomol.* 22, 62–70.

Flemings, M.B. (1959) An altitude biting study of *Culex tritaeniorhynchus* (Giles) and other associated mosquitoes in Japan. *J. Econ. Entomol.* 52, 490–2.

Fletcher, B.S. (1974) The ecology of a natural population of the Queensland fruit fly, *Dacus tryoni* IV. *Aust. J. Zool.* 21, 541–65.

Fontenille, D., Diallo, M., Mondo, M., Ndiaye, M., & Thonnon, J. (1997) First evidence of natural vertical transmission of yellow fever virus in *Aedes aegypti* its epidemic vector. *Trans. Roy. Soc. Trop. Med. Hyg.* 91, 533–5.

Ford, J., Glasgow, J.P., Johns, D.L., & Welch, J.R. (1959) Transect fly-rounds in field studies of *Glossina*. *Bull. Entomol. Res.* 50, 275–85.

Fredeen F.J.H. (1961) A trap for studying the attacking behaviour of black flies *Simulium arcticum* Mall. *Can. Entomol.* 93, 73–8.

Freeman, B.E. & Jayasingh, D.B. (1975) Population dynamics of *Pachodynerus nasidens* (Hymenoptera) in Jamaica. *Oikos* 26, 86–91.

Friend, G.R., Smith, G.T., Mitchell, D.S., & Dickman, C.R. (1989) Influence of pitfall and drift fence design on capture rates of small vertebrates in semi-arid habitats of Western Australia. *Aust. Wildl. Res.* 16, 1–10.

Fritzsche, R. (1956) Untersuchungen zur Bekämpfung der Rapsschädlinge. IV. Beiträge zur Ökologie und Bekämpfung des Grossen Rapsstengelrüsslers (*Ceuthorrhynchus napi* Gyll.). *Nachr. Bl. Deutsch Pflschntzdienst Berl.* 10, 97–105.

Fröhlich, G. (1956). Methoden zur Bestimmung der Befalls-bzw. Bekämpfungster-mine verschiedener Rapsschädlinge, insebesondere des Rapsstengeirüsslers (*Ceuthorrhynchus napi* Gyll.). *Nachr. Bl. Deutsch Pfl. Schutzdienst Berl.* 10, 48–53.

Frost, S.W. (1952) Light traps for insect collection, survey and control. *Bull. Pennsylv. Agric. Exp. Sta.* 550, 1–32.

Frost, S.W. (1953) Response of insects to black and white light. *J. Econ. Entomol.* 46, 376–7.

Frost, S.W. (1957) The Pennsylvania insect light trap. *J. Econ. Entomol.* 50, 287–92.

Frost, S.W. (1958a) Insects captured in light traps with and without baffles. *Can. Entomol.* 909, 566–7.

Frost, S.W. (1958b) Traps and lights to catch night-flying insects. *Proc. X Int. Congr. Entomol.* 2, 583–7.

Frost, S.W. (1958c) Insects attracted to light traps placed at different heights. *J. Econ. Entomol.* 51, 550–1.

Frost, S.W. (1964) Killing agents and containers for use with insect light traps. *Entomol. News* 75, 163–6.

Fukuda, M. (1960) On the effect of physical condition of setting place upon the number of flies collected by fish baited traps. *Endemic Dis. Bull. Nagasaki Univ.* 2, 222–8.

Fye, R.E. (1965) The biology of the Vespidae, Pompilidae and Sphecidae (Hymenoptera) from trap nets in Northwestern Ontario. *Can. Entomol.* 97, 716–44.

Gage, S.H., Wirth, T.M., & Simmons, G.A. (1990) Predicting regional gypsy moth (Lymantriidae) population trends in an expanding population using pheromone trap catch and spatial analysis. *Environ. Entomol.* 19, 370–7.

Garcia, R. (1962) Carbon dioxide as an attractant for certain ticks (Acarina: Argasidae and Ixodidae). *Ann. Entomol. Soc. Am.* **55**, 605–6.

Gates, C.E. & Smith, W.B. (1972) Estimation of density of mourning doves from aural information. *Biometrics* **28**, 345–59.

Gehrke, P.C. (1994) Influence of light intensity and wavelength on phototactic behaviour of larval silver perch *Bidyanus bidyanus* and golden perch *Macquaria ambigua* and the effectiveness of light traps. *J. Fish Biol.* **44**(5), 741–51.

Geier, P.W. (1960) Physiological age of codling moth females (*Cydia pomonella* L.) caught in bait and light traps. *Nature* **185**, 709.

Gerard, D., Bauchau, V., & Smets, S. (1994) Reduced trappability in wild mice, *Mus musculus domesticus*, heterozygous for Robertsonian translocations. *Anim. Behav.* **47**(4), 877–83.

Gillies, M.T. (1974) Methods for assessing the density and survival of blood sucking Diptera. *Ann. Rev. Entomol.* **19**, 345–62.

Gist, C.S. & Crossley, D.A. (1973) A method for quantifying pitfall trapping. *Environ. Entomol.* **2**, 951–2.

Glasgow, J.P. (1953) The extermination of animal populations by artificial predation and the estimation of populations. *J. Anim. Ecol.* **22**, 32–46.

Glasgow, J.P. (1961) The variability of fly-round catches in field studies of *Glossina*. *Bull. Entomol. Res.* **51**, 781–8.

Glasgow, J.P. & Duffy, B.J. (1961) Traps in field studies of *Glossina pallidipes* Austen. *Bull. Entomol. Res.* **52**, 795–814.

Glick, P.A. (1939) *The Distribution of Insects, Spiders and Mites in the Air*. Technical Bulletin 673, USDA, 150 pp.

Glick, P.A., Hollingsworth, J.P., & Eitel, W. (1956) Further studies on the attraction of pink bollworm moths to ultra-violet and visible radiation. *J. Econ. Entomol.* **49**, 158–61.

Golding, F.D. (1941) Two new methods of trapping the cacao moth (*Ephestia cautella*). *Bull. Entomol. Res.* **32**, 123–32.

Golding, F.D. (1946) A new method of trapping flies. *Bull. Entomol. Res.* **37**, 143–54.

Goodenough, J.L. & Snow, J.W. (1973) Tobacco budworms: nocturnal activity of adult males as indexed by attraction to live virgin females in electric grid traps. *J. Econ. Entomol.* **66**, 543–4.

Goodwin, M.H. (1942) Studies on artificial resting places of *Anopheles quadrimaculatus* Say. *J. Nat. Malaria Soc.* **1**, 93–9.

Goodwin, M.H. & Love, G.J. (1957) Factors influencing variations in populations of *Anopheles quadrimaculatus* in southwestern Georgia. *Ecology* **38**, 561–70.

Graham, H.M., Glick, P.A., & Hollingsworth, J.P. (1961) Effective range of argon glow lamp survey traps for pink bollworm adults. *J. Econ. Entomol.* **54**, 788–9.

Gray, H. & Treloar, A. (1933) On the enumeration of insect populations by the method of net collection. *Ecology* **14**, 356–67.

Gray, P.H.H. (1955) An apparatus for the rapid sorting of small insects. *Entomologist* **889**, 92–3.

Greenslade, P. (1973) Sampling ants with pitfall traps: digging-in effects. *Insectes Soc.* **20**, 343–53.

Greenslade, P. & Greenslade, P.J.M. (1971) The use of baits and preservatives in pitfall traps. *J. Aust. Entomol. Soc.* **10**, 253–60.

Greenslade, P.J.M. (1964) Pitfall trapping as a method for studying populations of Carabidae (Coleoptera). *J. Anim. Ecol.* **33**, 301–10.

Gregg, P.C., Fitt, G.P., Coombs, M., & Henderson, G.S. (1994) Migrating moths collected in tower-mounted light traps in northern New South Wales, Australia: influence of local and synoptic weather. *Bull. Entomol. Res.* **84**(1), 17–30.

Gregory, R.S. & Powles, P.M. (1985) Chronology, distribution, and sizes of larval fish sampled by light traps in macrophytic Chemung Lake (Ontario, Canada). *Can J. Zool.* **63**(11), 2569–77.

Gregory, R.S. & Powles, P.M. (1988) Relative selectivities of Miller high-speed samplers and light traps for collecting ichthyoplankton. *Can J. Fish. Aquat. Sci.* **45**(6), 993–8.

Gressitt, J.L. & Gressitt, M.K. (1962) An improved Malaise trap. *Pacific Insects* **4**, 87–90.

Gressitt, J.L., Leech, R.E., & O'Brien, C.W. (1960) Trapping of air-borne insects in the Antarctic area. *Pacific Insects* **2**, 245–50.

Gressitt, J.L., Sedlacek, J., Wise, K.A.J., & Yoshimoto, C.M. (1961) A high speed airplane trap for air-borne organisms. *Pacific Insects* **3**, 549–55.

Grigarick, A.A. (1959) Bionomics of the rice leafminer, *Hydrellia griseola* (Fallen.), in California (Diptera: Ephydridae). *Hilgardia* **29**, 1–80.

Grimm, F., Gessler, M., & Jenni, L. (1993) Aspects of sandfly biology in southern Switzerland. *Med. Vet. Entomol.* **7**(2), 170–6.

Haddow, A.J. (1954) Studies on the biting habits of African mosquitoes: an appraisal of methods employed, with special reference to the twenty-four-hour catch. *Bull. Entomol. Res.* **45**, 199–242.

Haddow, A.J. & Corbet, P.S. (1960) Observations on nocturnal activity in some African Tabanidae (Diptera). *Proc. Roy. Entomol. Soc. Lond. A* **35**, 1–5.

Halgren, L.A. & Taylor, L.R. (1968) Factors affecting flight responses of alienicolne of *Aphis fabae* Scop. and *Schizaphis graminum* Rondeni (Homoptera: Aphididae). *J. Anim. Ecol.* **37**, 583–93.

Halsall, N.B. & Wratten, S.D. (1988) The efficiency of pitfall trapping for polyphagous predatory Carabidae. *Ecol. Entomol.* **13**, 293–9.

Hansen, J.D. (1988) Trapping methods for rangeland insects in burned and unburned sites: a comparison. *Great Basin Nat.* **48**(3), 383–7.

Hargrove, J.W. (1977) Some advances in the trapping of tsetse (*Glossina* spp.) and other flies. *Ecol. Entomol.* **2**, 123–37.

Hargrove, J.W., Holloway, M.T.P., Vale, G.A., Gough, A.J.E., & Hall, D.R. (1995) Catches of tsetse (*Glossina* spp.) (Diptera: Glossinidae) from traps and targets baited with large doses of natural and synthetic host odour. *Bull. Entomol. Res.* **85**(2), 215–27.

Harper, A.M. & Story, T.P. (1962) Reliability of trapping in determining the emergence period and sex ratio of the sugar-beet root maggot *Tetanops myopaeformis* (Röder) (Diptera: Otitidae). *Can. Entomol.* **94**, 268–71.

Harper, A.M., Schaber, B.D., Entz, T., & Story, T.P. (1993) Assessment of sweepnet and suction sampling for evaluating pest insect populations in hay alfalfa. *J. Entomol. Soc. Br. Columbia* **90**, 66–76.

Harrison, I.R. (1963) Population studies on the poultry red mite *Dermanyssus gallinae* (Deg.). *Bull. Entomol. Res.* **53**, 657–64.

Harwood, R.F. (1961) A mobile trap for studying the behaviour of flying bloodsucking insects. *Mosquito News* **21**, 35–9.

Haufe, W.O. & Burgess, L. (1960) Design and efficiency of mosquito traps based on visual response to patterns. *Can. Entomol.* **92**, 124–40.

Hayes, W.B. (1970) The accuracy of pitfall trapping for the sand-beach isopod *Tylos punctatus*. *Ecology* **51**, 514–16.

Heap, M.A. (1988) The pit-light, a new trap for soil-dwelling insects. *J. Australian Ent. Soc.* **27**, 239–40.

Heathcote, G.D. (1957a) The comparison of yellow cylindrical, flat and water traps and of Johnson suction traps, for sampling aphids. *Ann. Appl. Biol.* **45**, 133–9.

Heathcote, G.D. (1957b) The optimum size of sticky aphid traps. *Plant Pathol.* **6**, 104–7.

Heikinheimo, O. & Raatikainen, M. (1962) Comparison of suction and netting methods in population investigations concerning the fauna of grass leys and cereal fields, particularly in those concerning the leafhopper, *Calligypona pellucida* (F.). *Valt. Maatalourk. Julk. Helsingfors* **191**, 31 pp.

Henderson, P.A. (1989) On the structure of the inshore fish community of England and Wales. *J. Mar. Biol. Assoc. UK* **69**, 145–63.

Henderson, P.A. & Seaby, R.M.H. (1994) On the factors influencing juvenile flatfish abundance in the lower Severn Estuary. *Netherl. J. Sea Res.* **32**, 321–30.

Henderson, P.A., Irving, P.W., & Magurran, A.E. (1997) Fish pheromones and evolutionary enigmas: a reply to Smith. *Proc. Roy. Soc. Lond. B.* **264**, 451–3.

Hesse, G., Kauth, H., & Wachter, R. (1955) Frasslockstoffe beim Fichtenrüsselkäfer *Hylobius abietis*. *Z. angew. Entomol.* **37**, 239–44.

Heydemann, B. (1961) Untersuchungen über die Aktivitäts und Besiedlungsdichte bei epigaischen spinnen. *Verh. deutsch zool. Ges.* **1961**, 538–56.

Heyer, W.R., Donnelly, M.A., McDiarmid, R.W., Hayek, L.A., & Foster, M.S. (eds) (1994) *Measuring and Monitoring Biological Diversity. Standard Methods for Amphibians.* Smithsonian Institution Press, Washington DC.

Hilgerloh, G. (1989) Autumn migration of trans-Saharan migrating passerines in the Straits of Gibraltar (Mediterranean Sea). *Auk* **106**(2), 233–9.

Hilgerloh, G. (1991) Spring migration of passerine trans-Saharan migrants across the Straits of Gibraltar. *Ardea* **79**(1), 57–62.

Hill, M.A. (1947) The life-cycle and habits of *Culicoides impuctatus* Goet. and *C. obsoletus* Mg., together with some observations on the life-cycle of *Culicoides odibilis* Aust., *Culicoides pallidicornis* Kief. *Culicoides cubitalis* Edw. and *Culicoides chiopterus* Mg. *Ann. Trop. Med. Parasitol.* **41**, 55–115.

Hobbs, S.E. & Wolf, W.W. (1989) An airborne radar technique for studying insect migration. *Bull. Entomol. Res.* **79**(4), 693–704.

Hokkanen, H. & Holopainen, J.K. (1986) Carabid species and activity densities in biologically and conventionally managed cabbage fields. *J. Appl. Entomol.* **36**, 257–61.

Hollingsworth, J.P., Briggs, C.P., Glick, P.A., & Graham, H.M. (1961) Some factors influencing light trap collections. *J. Econ. Entomol.* **54**, 305–8.

Hollingsworth, J.P., Hartsock, J.G., & Stanley, J.M. (1963) *Electrical Insect Traps for Survey Purposes.* 42-3-1, Agricultural Research Service (ARS), USDA, 10 pp.

Holopainen, J.K. (1992) Catch and sex ratio of Carabidae (Coleoptera) in pitfall traps filled with ethylene glycol or water. *Pedobiologia* **36**, 257–61.

Howell, J.F., Cheikh, M., & Harris, E.J. (1975) Comparison of the efficiency of three traps for the Mediterranean fruit fly baited with minimum amounts of trimedlure. *J. Econ. Entomol.* **68**, 277–9.

Hughes, R.D. (1955) The influence of the

prevailing weather conditions on the numbers of *Meromyza variegata* Meigèn (Diptera; Chloropidae) caught with a sweep net. *J. Anim. Ecol.* **24**, 324–35.

Ibbotson, A. (1958) The behaviour of frit fly in Northumberland. *Ann. Appl. Biol.* **46**, 474–9.

Ikeshoji, T. & Yap, H.H. (1987) Monitoring and chemosterilization of a mosquito population, *Culex quinquefasciatus* (Diptera, Culicidae) by sound traps. *Appl. Ent. Zool.* **22**, 474–81.

Innes, D.G.L. & Bendell, J.F. (1988) Sampling of small mammals by different types of traps in northern Ontario, Canada. *Acta Theriol.* **33**(26–43), 443–50.

James, H.G. & Redner, R.L. (1965) An aquatic trap for sampling mosquito predators. *Mosquito News* **25**, 35–7.

Jamnback H. & Watthews, T. (1963) Studies of populations of adult and immature *Culicoides sanguisuga* (Diptera: Ceratopogonidae). *Ann. Entomol. Soc. Am.* **56**, 728–32.

Jansen, M.J.W. & Metz, J.A.J. (1979) How many victims will a pitfall make? *Acta Biotheor.* **28**, 98–122.

Jenson, P. (1968) Changes in cryptozoan numbers due to systematic variation of covering boards. *Ecology* **49**, 409–18.

Johnson, C.G. (1950) The comparison of suction trap, sticky trap and tow-net for the quantitative sampling of small airborne insects. *Ann. Appl. Biol.* **37**, 268–85.

Johnson, C.G. (1952) The role of population level, flight periodicity and climate in the dispersal of aphids. *Trans. IX Int. Congr. Entomol.* **1**, 429–31.

Johnson, C.G. (1960) A basis for a general system of insect migration and dispersal flight. *Nature* **186**, 348–50.

Johnson, C.G. (1963) Physiological factors in insect migration by flight. *Nature* **198**, 423–7.

Johnson, C.G., Southwood, T.R.E., & Entwistle, H.M. (1957) A new method of extracting arthropods and molluscs from grassland and herbage with a suction apparatus. *Bull. Entomol. Res.* **48**, 211–18.

Joosse, E.N.G. (1975) Pitfall-trapping as a method for studying surface dwelling Collembola. *Z. Morph. Oekol. Tiere* **55**, 587–96.

Jordan, A.M. (1962) The ecology of the jiisca group of tsetse flies (*Glossina*) in Southern Nigeria. *Bull. Entomol. Res.* **53**, 355–85.

Juillet, J.A. (1963) A comparison of four types of traps used for capturing flying insects. *Can. J. Zool.* **41**, 219–23.

Kanda, T., Loong, K.P., Chiang, G.L., Cheong, W.H., & Lim, T.W. (1988) Field study on sound trapping and the development of trapping method for both sexes of Mansonia in Malaysia. *Trop. Biomed.* **5**(1), 37–42.

Kawai, S. & Suenaga, O. (1960) Studies of the methods of collecting flies. III. On the effect of putrefaction of baits (fish). [In Japanese, Eng. summary.]. *Endemic Dis. Bull. Nagasaki Univ.* **2**, 61–6.

Kehat, M., Anshelevich, L., Dunkelblum, E., & Greenberg, S. (1994) Sex pheromone traps for monitoring the peach twig borer, *Anarsia lineatella* Zeller: effect of pheromone components, pheromone dose, field aging of dispenser, and type of trap on male captures. *Phytoparasitica* **22**, 291–8.

Kennedy, G.J.A. & Stange, C.D. (1988) Efficiency of electric fishing for salmonids in relation to river width. *Fish. Manag.* **12**, 55–60.

Kennedy, J.S. (1961) A turning point in the study of insect migration. *Nature* **189**, 785–91.

Kersting, U. & Baspinar, H. (1995) Seasonal and diurnal flight activity of *Circulifer haematoceps* (Hom., Cicadellidae), an important leafhopper vector in the Mediterranean area and the Near East. *J. Appl. Entomol.* **119**(8), 533–7.

Khan, A.A., Maibach, H.I., Strauss, W.G., & Fisher, J.L. (1970) Differential attraction of the yellow fever mosquito to vertebrate hosts. *Mosquito News* **30**, 43–7.

Kissick, L.A. (1993) Comparison of traps lighted by photochemicals or electric bulbs for sampling warmwater populations of young fish. *N. Am. J. Fish. Manag.* **13**(4), 864–7.

Klock, J.W. & Bidlingmayer, W.L. (1953) An adult mosquito sampler. *Mosquito News* **13**, 157–9.

Knodel, J.J. & Agnello, A.M. (1990) Field comparison of nonsticky and sticky pheromone traps for monitoring fruit pests in western New York (USA). *J. Econ. Entomol.* **83**(1), 197–204.

Kono, T. (1953) On the estimation of insect population by time unit collecting. [In Japanese.] *Res. Popul. Ecol.* **2**, 85–94.

Koponen, S. (1992) Spider fauna (Araneae) of the low Arctic Belcher Islands, Hudson Bay. *Arctic* **45**(4), 358–62.

Kovrov, B.G. & Monchadskii, A.S. (1963) The possibility of using polarized light to attract insects. [In Russian.] *Entomol. Obozr.* **42**, 49–55. (*Transl. Entomol. Rev.* **42**, 25–8)

Kowalski, R. (1976) Obtaining valid population indices from pitfall traps. *Bull. Acad. Pol. Sci. Ser. Sci. Biol. II* **23**, 799–803.

Krebs, C.J., Singleton, G.R., & Kenney, A.J. (1994) Six reasons why feral house mouse populations might have low recapture rates. *Wildl. Res.* **21**(5), 559–67.

Kring, J.B. (1970) Red spheres and yellow panels

combined to attract apple maggot flies. *J. Econ. Entomol.* **63**, 466–9.

Kubecka, J., Duncan, A., Duncan, W.M., Sinclair, D., & Butterworth, A.J. (1994) Brown trout populations of three Scottish lochs estimated by horizontal sonar and multimesh gill nets. *Fish. Res.* **20**(1), 29–48.

Kurnik, I. (1988) Millipedes from xerothermic and agricultural sites in South Tyrol (Austria). *Berichte Naturwissensch. Med. Vereins Innsbruck* **75**, 109–14.

Lambert, H.L. & Franklin, R.T. (1967) Tanglefoot traps for detection of the balsam woolly aphid. *J. Econ. Entomol.* **60**, 1525–9.

Landin, J. (1968) Weather and diurnal periodicity of flight by *Helophorus bretipalpis* Bedel (Col.: Hydrophilidae). *Opusc. Entomol.* **33**, 28–36.

Landin, J. & Stark, E. (1973) On flight thresholds for temperature and wind velocity, 24-hour flight periodicity and migration of the beetle, *Helophorus brevipalpis* Bedel (Col. Hydrophilidae). *Zoon. Suppl.* **1**, 105–14.

Leech, H.B. (1955) Cheesecloth flight trap for insects. *Can. Entomol.* **85**, 200.

Le Pelley, R.H. (1935) Observations on the control of insects by hand-collection. *Bull. Entomol. Res.* **26**, 533–41.

Leshem, Y. & Yom Tov, Y. (1996) The magnitude and timing of migration by soaring raptors, pelicans and storks over Israel. *Ibis* **138**(2), 188–203.

Levin, M.D. (1957) Artificial nesting burrows for *Osmia lignaria* Say. *J. Econ. Entomol.* **50**, 506–7.

Lewis, T. (1959) A comparison of water traps, cylindrical sticky traps and suction traps for sampling thysanopteran populations at different levels. *Entomol. Exp. Appl.* **2**, 204–15.

Lewis, T. & Macaulay, E.D.M. (1976) Design and elevation of sex-attractant traps for pea moth, *Cydia nigricana* (Steph.) and the effect of plume shape on catches. *Ecol. Entomol.* **1**, 175–87.

Lewis, W.J., Snow, J.W., & Jones, R.L. (1971) A pheromone trap for studying populations of *Cardiochiles nigriceps*, a parasite of *Heliothis virescens*. *J. Econ. Entomol.* **64**, 1417–21.

Linsley, E.G., MacSwain, J.W., & Smith, R.F. (1952) Outline for ecological life histories of solitary and semi-social bees. *Ecology* **33**, 558–67.

Lockart, R.B. & Owings, D.H. (1974) Seasonal changes in the activity pattern of *Dipodomys spectabilis*. *J. Mammal.* **55**, 291–87.

Loomis, E.C. (1959a) A method for more accurate determination of air volume displacement of light traps. *J. Econ. Entomol.* **52**, 343–5.

Loomis, E.C. (1959b) Selective response of *Aedes nigronaculis* (Ludlow) to the Minnesota light trap. *Mosquito News* **19**, 260–3.

Loomis, E.C. & Sherman, E.J. (1959) Comparison of artificial shelter and light traps for measurements of *Culex tarsalis* and *Anopheles freeborni* populations. *Mosquito News* **19**, 232–7.

Love, G.I. & Smith, W.W. (1957) Preliminary observations on the relation of light trap collections to mechanical sweep net collections in sampling mosquito populations. *Mosquito News* **17**, 9–14.

Luczak, J. & Wierzbowska, T. (1959) Analysis of likelihood in relation to the length of a series in the sweep method. *Bull. Acad. Pol. Sci. Ser. Sci. Biol.* **7**, 313–18.

Luff, M.L. (1968) Some effects of formalin on the numbers of Coleoptera caught in pitfall traps. *Entomol. Monthly Mag.* **104**, 115–16.

Luff, M.L. (1975) Some features influencing the efficiency of pitfall traps. *Oecologia* **19**, 345–57.

Lumsden, W.H.R. (1958) A trap for insects biting small vertebrates. *Nature* **181**, 819–20.

McClelland, G.A.H. & Conway, G.R. (1971) Frequency of blood-feeding in the mosquito *Aedes aegypti*. *Nature* **232**, 485–6.

MacLeod, J. & Donnelly, J. (1956) Methods for the study of blowfly popultions. 1. Bait trapping. Significance of limits for comparative sampling. *Ann. Appl. Biol.* **44**, 80–104.

McPhail, M. (1939) Protein lures for fruit flies. *J. Econ. Entomol.* **32**, 758–61.

Magurran, A.E., Irving, P.W., & Henderson, P.A. (1996) Why is fish alarm 'pheromone' ineffective in the wild? *Proc. Roy. Soc. Lond. B.* **263**, 1551–6.

Malaise, R. (1937) A new insect-trap. *Entomol. Tidskr.* **58**, 148–60.

Martin, F.R., McCreadie, J.W., & Colbo, M.H. (1994) Effect of trap site, time of day, and meteorological factors on abundance of host-seeking mammalophilic black flies (Diptera: Simuliidae). *Canadian Entomologist* **126**, 283–9.

Masaki, J. (1959) Studies on rice crane fly (*Tipula aino* Alexander, Tipulidae, Diptera) with special reference to the ecology and its protection. [In Japanese.] *J. Kanto Tosan Agric. Exp. Sta.* **13**, 1–195.

Mason, H.C. (1963) Baited traps for sampling *Drosophila* populations in tomato field plots. *J. Econ. Entomol.* **56**, 897–8.

Mathis, A. & Smith, R.J.F. (1992) Avoidance of areas marked with a chemical alarm substance by fathead minnows (*Pimephales promelas*) in a natural habitat. *Can. J. Zool.* **70**, 1473–6.

Maxwell, C.W. (1965) Tanglefoot traps as indicators of apple maggot fly activities. *Can. Entomol.* **97**, 1–10.

Mayer, K. (1961) Untersuchungen über das Wahlyerhalten der Fritfiiege (*Oscinella frit* L.) beim Anflug von Kulturpflanzen im Feldversuch mit der Fangschalenmethode. *Mitt. biol. Bund. Anst. Ld- Forstw. Berlin* **106**, 1–47.

Medler, J.T. (1964) Biology of *Rygchium foraminatum* in trap-nests in Wisconsin (Hymbroptera: Vespidae). *Ann. Entomol. Soc. Am.* **57**, 56–60.

Medler, J.T. & Fye, R.E. (1956) Biology of *Ancistrocerus antilope* (Panzer) in trap nests in Wisconsin. *Ann. Entomol. Soc. Am.* **49**, 97–102.

Mellanby, K. (1962) Sticky traps for the study of animals inhabiting the soil surface. In: Murphy, P.W. (ed.) *Progress in Soil Zoology*. Butterworths, London, 226–7.

Menhinick, E.F. (1963) Estimation of insect population density in herbaceous vegetation with emphasis on removal sweeping. *Ecology* **44**, 617–21.

Merzheevskaya, O.I. & Gerastevich, E.A. (1962) A method of collecting living insects at light. [In Russian.] *Zool. Zhur.* **41**, 1741–3.

Michelbacher, A.E. & Middlekauff, W.W. (1954) Vinegar fly investigations in Northern California. *J. Econ. Entomol.* **47**, 917–22.

Mikkola, K. (1972) Behavioural and electrophysiological responses of night-flying insects, especially Lepidoptera, to near-ultraviolet visible light. *Ann. Zool. Fenn.* **99**, 225–54.

Miller, T.A., Stryker, R.G., Wilkinson, R.N., & Esah, S. (1970) The influence of moonlight and other environmental factors on the abundance of certain mosquito species in light-trap collections in Thailand. *J. Med. Entomol.* **7**, 555–61.

Minns, C.K. & Hurley, D.A. (1988) Effects of net length and set time on fish catches in gill nets. *N. Am. J. Fish. Manag.* **8**, 216–23.

Minter, D.M. (1961) A modified Lumsden suction-trap for biting insects. *Bull. Entomol. Res.* **529**, 233–8.

Mitchell, B. (1963) Ecology of two carabid beetles, *Bembidion lampros* (Herbst.) and *Trechus quadristriatus* (Schrank). *J. Anim. Ecol.* **32**, 377–92.

Moericke, V. (1950) Über den Farbensinn der Pfirsichblattlaus Wy, odespersicae Sulz. *Z. Tierpsychol.* **7**, 265–74.

Mommertz, S., Schauer, C., Koesters, N., Lang, A., & Filser, J. (1996) A comparison of D-Vac suction, fenced and unfenced pitfall trap sampling of epigeal arthropods in agro-ecosystems. *Ann. Zool. Fenn.* **33**, 117–24.

Moran, P.A.P. (1951) A mathematical theory of animal trapping. *Biometrica* **38**, 307–11.

Morgan, C.V.G. & Anderson, N.H. (1958) Techniques for biological studies of tetranychid mites, especially *Bryobia arborea* M. & A. and *B. praetiosa* Koch. (Acarina: Tetranychidae). *Can. Entomol.* **90**, 212–15.

Morrill, W.L., Lester, D.G., & Wrona, A.E. (1990) Factors affecting efficacy of pitfall traps for beetles (Coleoptera: Carabidae and Tenebrionidae). *J. Entomol. Sci.* **25**(2), 284–93.

Morris, K.R.S. (1960) Trapping as a means of studying the game tsetse, *Glossina pallidipes* Aust. *Bull. Entomol. Res.* **51**, 533–57.

Morris, K.R.S. (1961) Effectiveness of traps in tsetse surveys in the Liberian rain forest. *Am. J. Trop. Med. Hyg.* **10**, 905–13.

Morris, K.R.S. & Morris, M.G. (1949) The use of traps against tsetse in West Africa. *Bull. Entomol. Res.* **39**, 491–523.

Morton, S.R. Gillam, M.W., Jones, K.R., & Fleming, M.R. (1988) Relative efficiency of different pit-trap systems for sampling reptiles in spinifex grasslands. *Aust. Wildl. Res.* **15**, 571–7.

Mpofu, S.M., Masendu, H.T., Kanyimo, K.H., & Mtetwa, C. (1993) Laboratory colonization of *Anopheles quadriannulatus* from sympatry with other sibling species of the *Anopheles gambiae* complex in Zimbabwe. *Medical and Veterinary Entomology* **7**, 122–6.

Muirhead-Thomson, R.C. (ed.) (1963) Practical entomology in malaria eradication, Part 1, Chapter 1. MHO/PA/62.63, WHO, mimeographed.

Mukhopadhyay, S. (1991) Lunation-induced variation in catchment areas of light-traps to monitor rice green leafhoppers (*Nephotettix* spp.) in West Bengal. *Ind. J. Agric. Sci.* **61**(5), 337–40.

Mulhern, T.D. (1942) New Jersey mechanical trap for mosquito surveys. *NJ Agric. Exp. Sta. Circ.* **421**, 1–8.

Mulhern, T.D. (1953) Better results with mosquito light traps through standardizing mechanical performance. *Mosquito News* **13**, 130–3.

Mulla, M.S., Georghiou, G.P., & Dorner, R.W. (1960) Effect of ageing and concentration on the attractancy of proteinaceous materials to *Hippelates* gnats. *Ann. Entomol. Soc. Am.* **53**, 835–41.

Mundie, J.H. (1964) A sampler for catching emerging insects and drifting materials in streams. *Limnol. Oceanogr.* **9**, 456–9.

Murphy, W.L. (1985) Procedure for the removal of insect specimens from sticky-trap material. *Ann. Entomol. Soc. Am.* **78**(6), 881.

Murray, W.D. (1963) Measuring adult populations of the pasture mosquito, *Aedes migromaculis* (Ludlow). In: *Proceedings of the 27th*

Conference of the Californian Mosquito Control Association 1959, 67–71.

Myers, K. (1956) methods of sampling winged insects feeding on the rabbit *Oryctolagus cuniculus* (L.). Aust. *CSIRO Wildl. Res.* 1, 45–58.

Myllymaki, A., Paasikallio, A., Pankakoski, E., & Kanervo, V. (1971) Removal experiments on small quadrats as a means of rapid assessment of abundance of small mammals. *Ann. Zool. Fenn.* 8, 177–85.

Mystowska, E.T. & Sidorowicz, J. (1961) Influence of the weather on captures of micromammalia II. *Insectiv. Acta Theriol.* 5, 263–73.

Nag, A. & Nath, P. (1991) Effect of moon light and lunar periodicity on the light trap catches of cutworm *Agrotis ipsilon* (Hufn.) moths. *J. Appl. Entomol.* 111(4), 358–60.

Nakamura, K., Ito, Y., Miyashita, K., & Takai, A. (1967) The estimation of population density of the green rice leafhopper, *Nephotettix cincticeps* Uhler. in spring field by the capture–recapture method. *Res. Popul. Ecol.* 9, 113–29.

Neilson, W.T.A. (1960) Field tests of some hydrolysed proteins as lures for the apple maggot, *Rhagoletis pomonella* (Walsh). *Can. Entomol.* 92, 464–7.

Nielsen, B.O. (1963) The biting midges of *Lyngby aamose* (Culicoides: Ceratopogonidae). *Nat. Jutl.* 10, 1–48.

Nielsen, E.T. (1960) A note on stationary nets. *Ecology* 41, 375–6.

Nijholt, W.W. & Chapman, J.A. (1968) A flight trap for collecting living insects. *Can. Entomol.* 100, 1151–3.

Odintsov, V.S. (1960) Air catch of insects as a method of study upon entomofauna of vast territories. [In Russian.] *Entomol. Obozr.* 39, 227–30.

Otake, A. & Kono, T. (1970) Regional characteristics in population trends of smaller brown planthopper, *Laodelphax striatellus* (Fallen) (Hemiptera: Delphacidae), a vector of rice stripe disease: an analytical study of light trap records. *Bull. Shikoku Exp. Sta.* 21, 127–47.

Otis, D.L., Burnham, K.P., White, G.C., & Anderson, D.R. (1978) Statistical inference from capture data on closed animal populations. *Wildl. Monogr.* 62, 1–135.

Palumbo, J.C., Tonhasca, A., Jr, & Byrne, D.N. (1995) Evaluation of three sampling methods for estimating adult sweetpotato whitefly (Homoptera: Aleyrodidae) abundance on cantaloupes. *J. Econ. Entomol.* 88(5), 1393–400.

Paris, O.H. (1965) The vagility of P-32-labelled isopods in grassland. *Ecology* 46, 635–48.

Parker, A.H. (1949) Observations on the seasonal and daily incidence of certain biting midges (*Culicoides* Latreille – Diptera,

Ceratopogonidae) in Scotland. *Trans. Roy. Entomol. Soc. Lond.* 100, 179–90.

Parmenter, R.R., Macmahon, J.A., & Anderson, D.R. (1989) Animal density estimation using a trapping web design: field validation experiments. *Ecology* 70(1), 169–79.

Paviour-Smith, K. (1964) The life history of *Tetratoma fungorum* F. (Col., Tetratomidae) in relation to habitat requirements, with an account of eggs and larval stages. *Entomol. Monthly Mag.* 100, 118–34.

Pena, J.E. & Duncan, R. (1992) Sampling methods for *Prodiplosis longifila* (Diptera: Cecidomyidae) in limes. *Environ. Entomol.* 21, 996–1001.

Persat, H. & Copp, G.H. (1990) Electricfishing and point abundance sampling for the ichthyology of large rivers. In: Cowx, I.G. (ed.) *Developments in Electric Fishing*. Fishing News Books, Blackwell Scientific Publications, Oxford UK.

Petruska, F. (1968) Members of the group Silhini as a component part of the insects fauna of sugar beet fields in the Unicov Plain. *Acta Univ. Palackianae Olomucenscs Fac. Rerum Nat.* 28, 189–200.

Petruska, F. (1969) On the possibility of escape of various components of the epigeic fauna of the fields from the pitfall traps containing formalin (Coleoptera). [In Czech, English summary.] *Acta Univ. Palackianae Olomucensis Fac. Rerum Nat.* 31, 99–124.

Plesner Jensen, S. & Honess, P. (1995) The influence of moonlight on vegetation height preference and trappability of small mammals. *Mammalia* 59(1), 35–42.

Pollet, M. & Grootaert, P. (1987) Ecological data on Dolichopodidae (Diptera) from a woodland ecosystem: I. Color preference, detailed distribution and comparison of different sampling techniques. *Bull. Inst. Roy. Sci. Nat. Belg. Entomol.* 57, 173–87.

Pollock, K.H. & Otto, M.C. (1983) Robust estimation of population size in closed animal populations from capture–recapture experiments. *Biometrics* 39, 1035–49.

Ponton, D. (1994) Sampling neotropical young and small fishes in their microhabitats: an improvement of the quatrefoil light-trap. *Arch. Hydrobiol.* 131(4), 495–502.

Prokopy, R.J. (1968) Sticky spheres for estimating apple maggot adult abundance. *J. Econ. Entomol.* 61, 1082–5.

Prokopy, R.J. (1977) Host plant influences on the reproductive biology of Tephritidae. In: Labeyrie, V. (ed.) Comportement des insectes et milieu trophique. Coll. *Int. C.N.R.S.* 265, 305–34.

Provost, M.W. (1960) The dispersal *Aedes*

taeniorhynchus. III. Study methods for migratory exodus. *Mosquito News* **20**, 148–61.

Race, S.R. (1960) A comparison of two sampling techniques for lygus bugs and stink bugs on cotton. *J. Econ. Entomol.* **53**, 689–90.

Rainey, R.C. & Joyce, R.J.V. (1990) An airborne radar system for desert locust control. *Phil. Tran. Roy. Soc. B* **328**, 585–606.

Randolph, S.E. & Storey, K. (1999) Impact of microclimate on immature tick–rodent host interactions (Acari: Ixodidae): implications for parasite transmission. *J. Med. Entomol.* **36**, 741–8.

Reidl, H., Howell, J.F., McNally, P.S., & Westigard, P.H. (1986) Codling moth management: use and standardisation of pheromone trapping systems. *Univ. Calif. Agric. Exp. Sta. Bull.* **1918**.

Rennison, B.D. & Robertson, D.H.H. (1959) The use of carbon dioxide as an attractant for catching tsetse. *Rep. E. Afr. Trypan. Res. Organ.* **1958**, 26.

Reynolds, M.H., Cooper, B.A., & Day, R.H. (1997) Radar study of seabirds and bats on Windward Hawaii. *Pacific Sci.* **51**(1), 97–106.

Richert, A. & Huelbert, D. (1991) Evaluation of catches made by light traps over five years (1984 to 1988) on the phenological basis of Hohenfinow (district Eberswalde-Finow) (Germany) for the faunistics of Lepidoptera. *Beitr. Entomol.* **41**(1), 251–64.

Rickart, E.A., Heaney, L.R., & Utzurrum, R.C.B. (1991) Distribution and ecology of small mammals along an elevational transect in southeastern Luzon, Philippines. *J. Mammal.* **72**(3), 458–69.

Ricker, W.E. (1975) *Computation and Interpretation of Biological Statistics of Fish Populations*. Bull. Fish Res. Bd. Can. 191, Ottawa, 382 pp.

Rieske, L.K. & Raffa, K.F. (1993) Potential use of baited pitfall traps in monitoring pine root weevil, *Hylobius pales*, *Pachylobius picivorus*, and *Hylobius radicis* (Coleoptera: Curculionidae) populations and infestation levels. *J. Econ. Entomol.* **86**(2), 475–85.

Riley, D.G. & Schuster, D.J. (1994) Pepper weevil adult response to colored sticky traps in pepper fields. *Southw. Entomol.* **19**(2), 93–107.

Riley, J.R. & Reynolds, D.R. (1990) Nocturnal grasshopper migration in West Africa: transport and concentration by the wind, and the implications for air-to-air control. *Phil. Trans. Roy. Soc. Lond. B* **328**(1251), 655–72.

Rings, R.W. & Metzler, E.H. (1990) The Lepidoptera of Fowler Woods State Nature Preserve, Richland County, Ohio (USA). *Great Lakes Entomol.* **23**(1), 43–56.

Rioux, J.A. & Golvan, Y.J. (1969) Epidemiologie des leishmanioses dans le sud de la France. *Monogr. Inst. Nat. Santé Rech. Med.* **37**, 1–220.

Rivard, I. (1962) Un piège à fosse amélioré pour la capture d'insectes actifs à la surface du sol. *Can. Entomol.* **94**, 270–1.

Roberts, D. & Kumar, S. (1994) Using vehicle-mounted nets for studying activity of Arabian sand flies (Diptera: Psychodidae). *J. Med. Entomol.* **31**(3), 388–93.

Roberts, R.H. (1970) Colour of Malaise trap and collection of Tabanidae. *Mosquito News* **30**, 567–71.

Roberts, R.H. (1972) The effectiveness of several types of Malaise traps for the collection of Tabanidae and Culicidae. *Mosquito News* **32**, 542–7.

Roberts, R.H. (1976) The comparative efficiency of six trap types for the collection of Tabanidae (Diptera). *Mosquito News* **36**, 530–7.

Roberts, R.J. & Smith, T.J.R. (1972) A plough technique for sampling soil insects. *J. Appl. Ecol.* **9**, 427–30.

Robinson, H.S. & Robinson, P.J.M. (1950) Some notes on the observed behaviour of Lepidoptera in flight in the vicinity of light-sources together with a description of a light-trap designed to take entomological samples. *Entomol. Gaz.* **1**, 3–15.

Roesler, R. (1953) Über eine Methode zur Feststellung der Flugzeit schädlicher Fliegenarten (Kirschfliege, Kohlfliege, Zwiebelfliege). *Mitt. Biol. Reichsant. (Zent Anst.) Ld- Forstw. Berlin* **75**, 97–9.

Rogers, D.J. & Smith, D.T. (1977) A new electric trap for tsetse flies. *Bull. Entomol. Res.* **67**, 153–9.

Rohitha, B.H. & Stevenson, B.E. (1987) An automatic sticky trap for aphids (Hemiptera: Aphididae) that segregates the catch daily. *Bull. Entomol. Res.* **77**, 67–72.

Romney, V.E. (1945) The effect of physical factors upon catch of the beet leafhopper (*Eutettix tenellus* (Bak.)) by a cylinder and two sweep-net methods. *Ecology* **26**, 135–48.

Roos, T. (1957) Studies on upstream migration in adult stream-dwelling insects. *Inst. Freshw. Res. Drottringholm Rep.* **38**, 167–93.

Rothschild, M. (1961) Observations and speculations concerning the flea vector of myxomatosis in Britain. *Entomol. Monthly Mag.* **96**, 106–9.

Rudd, W.G. & Jensen, R.L. (1977) Sweep net and ground cloth sampling for insects in soybeans. *J. Econ. Entomol.* **70**, 301–4.

Saliternik, Z. (1960) A mosquito light trap for use on cesspits. *Mosquito News* **20**, 295–6.

Samu, F. & Sarospataki, M. (1995) Design and use of a hand-hold suction sampler, and its comparison with sweep net and pitfall trap sampling. *Fol. Entomol. Hung.* **56**, 195–203.

Saugstad, E.S., Bram, R.A., & Nyquist, W.E. (1967) Factors influencing sweep-net sampling of alfalfa. *J. Econ. Entomol.* **60**, 421–6.

Saunders, D.S. (1964) The effect of site and sampling and method on the size and composition of catches of tsetse flies (Glossina) and Tabanidae (Diptera). *Bull. Entomol. Res.* **55**, 483.

Schaeffer, G.W. (1979) An airborne radar technique for investigation and control of migrating insects. *Phil. Trans. Roy. Soc. Lond.* **287**, 459–65.

Schlyter, F., Lofqvist, J., & Byers, J.A. (1987) Behavioral sequence in the attraction of the bark beetle *Ips typographus* to pheromone sources. *Physiol. Entomol.* **12**(2), 185–96.

Seber, G.A.F. (1982) *Estimation of Animal Abundance and Related Paramters*. Griffin, London.

Seber, G.A.F. (1986) A review of estimating animal abundance. *Biometrics* **42**, 267–92.

Seber, G.A.F. & Whale, J.F. (1970) The removal method for two and three samples. *Biometrics* **26**, 393–400.

Service, M.W. (1963) The ecology of the mosquitoes of the northern Guinea savannah of Nigeria. *Bull. Entomol. Res.* **54**, 601–32.

Service, M.W. (1995) *Mosquito Ecology Field Sampling Methods*, 2nd edn. Chapman & Hall, 988 pp.

Sevastpulo, D.G. (1963) Field notes from East Africa — Part XI. *Entomologist* **96**, 162–5.

Siddorn, J.W. & Brown, E.S. (1971) A Robinson light trap modified for segregating samples at predetermined time intervals, with notes on the effect of moonlight on the periodicity of catches of insects. *J. Appl. Ecol.* **8**, 69–75.

Simpson, R.G. & Berry, R.E. (1973) A twenty-four hour directional aphid trap. *J. Econ. Entomol.* **66**, 291–2.

Smith, A.D., Riley, J.R., & Gregory, R.D. (1993) A method for routine monitoring of the aerial migration of insects by using a vertical-looking radar. *Phil. Trans. Roy. Soc. Lond. B* **340**(1294), 393–404.

Smith, B.D. (1962) The behaviour and control of the blackcurrant gall mite *Phytoptus ribis* (Nal.). *Ann. Appl. Biol.* **50**, 327–34.

Smith, I.M. & Rennison, B.D. (1961) Studies of the sampling of *Glossina pallidipes* Aust. I, II. *Bull. Entomol. Res.* **52**, 165–89.

Smith, P.W., Taylor, J.C., & Apple, J.W. (1959) A comparison of insect traps equipped with 6- and 15-watt black light lamps. *J. Econ. Entomol.* **52**, 1212–14.

Snow, W.E., Pickard, E., & Sparkman, R.E.

(1960) A fan trap for collecting biting insects attacking avian hosts. *Mosquito News* **20**, 315–16.

Southwood, T.R.E. (1960) The flight activity of Heteroptera. *Trans. Roy. Entomol. Soc. Lond.* **112**, 173–220.

Southwood, T.R.E. (1962) Migration of terrestrial arthropods in relation to habitat. *Biol. Rev.* **37**, 171–214.

Southwood, T.R.E., Jepson, W.F., & Van Emden, H.F. (1961) Studies on the behaviour of *Oscinella frit* L. (Diptera) adults of the panicle generation. *Entomol. Exp. Appl.* **4**, 196–210.

Spradbery, J.P. & Vogt, W.G. (1993) Mean life expectancy of Old World screwworm fly, *Chrysomya bezziana*, inferred from the reproductive age-structure of native females caught on wormlure-baited sticky traps. *Med. Vet. Entomol.* **7**(2), 147–54.

Springate, N.D. & Basset, Y. (1996) Diel activity of arboreal arthropods associated with Papua New Guinean trees. *J. Nat. Hist.* **30**(1), 101–12.

Stahl, C. (1954) Trapping hornworm moths. *J. Econ. Entomol.* **47**, 879–82.

Stammer, H.J. (1949) Die Bedeutung der Äthylenglycolfallen für tierökologische und phänologische Untersuchungen. *Verh. Deutsch Zool. Kiel* **1948**, 387–91.

Stanley, J.M. & Dominick, C.B. (1970) Funnel size and lamp wattage influence on light-trap performance. *J. Econ. Entomol.* **63**, 1423–6.

Staples, R. & Allington, W.B. (1959) The efficiency of sticky traps in sampling epidemic populations of the eriophyid mite *Aceria tulipae* (K.), vector of wheat streak mosaic virus. *Ann. Entomol. Soc. Am.* **52**, 159–64.

Starr, D.F. & Shaw, J.C. (1944) Pyridine as an attractant for the Mexican fruit fly. *J. Econ. Entomol.* **37**, 760–3.

Steigen, A.L. (1973) Sampling invertebrates active below a snow cover. *Oikos* **24**, 373–6.

Steiner, P., Wenzel, F., & Baumert, D. (1963) Zur Beeinflussung der Arthropodenfauna nordwestdeutscher Kartoffelfelder durch die Anwendung synthetischer Kontaktinsektizide. *Mitt. Biol. Bund Anst. Ld- Forstw. Berlin* **109**, 1–38.

Subramanyam, B., Hagstrum, D.W., & Schenk, T.C. (1993) Sampling adult beetles (Coleoptera) associated with stored grain: comparing detection and mean trap catch efficiency of two types of probe traps. *Environ. Entomol.* **22**, 33–42.

Taft, H.M. & Jernigan, C.E. (1964) Elevated screens for collecting boll weevils flying between hibernation sites and cottonfields. *J. Econ. Entomol.* **57**, 773–5.

Taylor, L.R. (1962) The efficiency of cylindrical

sticky insect traps and suspended nets. *Ann. Appl. Biol.* **50**, 681–5.

Taylor, L.R. (1963) Analysis of the effect of temperature on insects in flight. *J. Anim. Ecol.* **32**, 99–112.

Taylor, L.R. & Brown, E.S. (1972) Effects of light-trap design and illumination on samples of moths in the Kenya highlands. *Bull. Entomol. Res.* **62**, 91–112.

Taylor, L.R. & Carter, C.I. (1961) The analysis of numbers and distribution in an aerial population of Macrolepidoptera. *Trans. Roy. Entomol. Soc. Lond.* **113**, 369–86.

Taylor, L.R. & French, R.A. (1974) Effects of light-trap design and illumination on samples of moths in an English woodland. *Bull. Entomol. Res.* **63**, 583–94.

Taylor, L.R. & Palmer, J.M.P. (1972) Aerial sampling. In: van Emden, H.F. (ed.) *Aphid Technology*. Academic Press, London, 189–234.

Taylor, R.A.J. (1986) Time series analysis of numbers of Lepidoptera caught at light traps in East Africa, and the effect of moonlight on trap efficiency. *Bull. Entomol. Res.* **76**(4), 593–606.

Tedders, W.L. & Edwards, G.W. (1972) Effects of black light trap design and placement on catch of adult hickory shuckworms. *J. Econ. Entomol.* **65**, 1624–7.

Thorrold, S.R. (1992) Evaluating the performance of light traps for sampling small fish and squid in open waters of the central Great Barrier Reef lagoon. *Mar. Ecol. Progr. Ser.* **89**(2–3), 277–85.

Thorsteinson, A.J., Bracken, G.K., & Hanec, W. (1965) The orientation behaviour of horse flies and deer flies (Tabanidae, Diptera). III. The use of traps in the study of orientation of Tabanids in the field. *Entomol. Exp. Appl.* **8**, 189–92.

Togashi, I., Nakata, K., Sugie, Y., Hashimoto, M., & Tanaka, A. (1990) Beetles belonging to the subtribe Carabina captured by pitfall traps in Ishikawa Prefecture (Japan). *Bull. Biogeogr. Soc. Japan* **45**(1–22), 103–6.

Topping, C.J. & Sunderland, K.D. (1992) Limitations to the use of pitfall traps in ecological studies exemplified by a study of spiders in a field of winter wheat. *J. Appl. Ecol.* **29**(2), 485–91.

Townes, H. (1962) Design for a Malaise trap. *Proc. Entomol. Soc. Wash.* **64**, 253–62.

Tshernyshev, W.B. (1961) Comparison of field responses of insects to the light of a mercury-quartz lamp and clear ultra-violet radiation of the same lamp. [In Russian.] *Entomol. Obozr.* **40**, 568–70. (Trans. *Entomol. Rev.* **40**, 308–9).

Turner, E.C. (1972) An animal-baited trap for the collection of *Culicoides* spp. (Diptera: Ceratopogonidae). *Mosquito News* **32**, 527–30.

Turner, F.B. (1962) Some sampling characteristics of plants and arthropods of the Arizona desert. *Ecology* **43**, 567–71.

Uetz, G.W. (1977) Co-existence in a guild of wandering spiders. *J. Anim. Ecol.* **46**, 531–41.

Uetz, G.W. & Unzicker, J.D. (1976) Pit fall trapping in ecological studies of wandering spiders. *J. Arachnol.* **3**, 101–11.

Ulrich, G.W. (1986) Construction of a compact submersible aquatic light trap. *Pan-Pac. Entomol.* **52**, 144–9.

Usinger, R.L. (ed.) (1963) *Aquatic Insects of California*. University of California Press, Berkeley, Los Angeles, 508 pp.

Usinger, R.L. & Kellen, W.R. (1955) The role of insects in sewage disposal beds. *Hilgardia* **23**, 263–321.

Vale, G.A. (1971) Artificial refuges for tsetse flies (*Glossina* spp.). *Bull. Entomol. Res.* **61**, 331–50.

Vale, G.A. (1974) The responses of tsetse flies (Diptera, Glossinidae) to mobile and stationary baits. *Bull. Entomol. Res.* **64**, 545–88.

van den Broek, W.L.F. (1979) A seasonal survey of fish populations in the Lower Medway Estuary, Kent, based on power station screen sampling. *Estuar. Coast. Mar. Sci.* **9**, 1–15.

Van Huizen, T.H.P. (1977) The significance of flight activity in the life cycle of *Amara plebeja* Gyll. (Coleoptera, Carabidae). *Oecologia* **29**, 27–41.

Vavra, J.R., Carestia, R.R., Frommer, R.L., & Gerberg, E.J. (1974) Field evaluation of alternative light sources as mosquito attractants in the Panama Canal Zone. *Mosquito News* **34**, 382–4.

Verheggen, L. (1994) Monitoring van territoriaal vleermuizen in de paartijd. *Nieusbr. Vleermuiswerkgr. Nederl.* **17**, 5–8.

Verheijen, F.J. (1960) The mechanisms of the trapping effect of artificial light sources upon animals. *Arch. Néerl. Zool.* **13**, 1–107.

Vibert, R. (ed.) (1967) *Fishing with Electricity. Its Application to Biology and Management*. Fishing News (Books) Ltd., Farnham, UK.

Vickery, W.L. & Bider, J.R. (1978) The effect of weather on *Sorex cinereus* activity. *Can. J. Zool.* **56**, 291–7.

Wakerley, S.B. (1963) Weather and behaviour in carrot fly (*Psila rosae* Fab. Dipt. Psilidae) with particular reference to oviposition. *Entomol. Exp. Appl.* **6**, 268–78.

Walker, A.R. & Boreham, P.F.L. (1976) Saline, as a collecting medium for Culicoides (Diptera: Ceratopogonidae) in blood-feeding and other studies. *Mosquito News* **36**, 18–20.

Walker, T.J. & Lenczewski, B. (1989) An inexpensive portable trap for monitoring butterfly migration. *J. Lepidop. Soc.* **43**(4), 289–98.

Walker, T.J. & Whitesell, J.J. (1993) A superior trap for migrating butterflies. *J. Lepidop. Soc.* **47**(2), 140–9.

Waloff, N. & Bakker, K. (1963) The flight activity of Miridae (Heteroptera) living on broom, *Sarothamnus scoparius* (L.). *J. Anim. Ecol.* **32**, 461–80.

Walsh, G.B. (1933) Studies in the British necrophagous Coleoptera. II. The attractive powers of various natural baits. *Entomol. Monthly Mag.* **69**, 28–32.

Walters, B.B. (1989) Differential capture of deer mice with pitfalls and live traps. *Acta Theriol.* **34**(29–43), 643–7.

Waters, T.F. (1962) Diurnal periodicity in the drift of stream invertebrates. *Ecology* **43**, 316–20.

Waters, T.F. (1969) Invertebrate drift—ecology and significance to stream fishes. In: *Symposium on Salmon and Trout in Streams*. University of British Columbia, Vancouver, pp. 121–34.

Webster, J.P., Brunton, C.F.A., & MacDonald, D.W. (1994) Effect of *Toxoplasma gondii* upon neophobic behaviour in wild brown rats, *Rattus norvegicus*. *Parasitology* **109**(1), 37–43.

Welch, R.C. (1964) A simple method of collecting insects from rabbit burrows. *Entomol. Monthly Mag.* **100**, 99–100.

Wenk, P. & Schlörer, G. (1963) Wirtsorientierung und Kopulation bei blutsaugenden Simuliiden (Diptera). *Zeit. Trop. Med. Parasitol.* **14**, 177–91.

Wharton, R.H., Eyles, D.E., & Warren, M.C.W. (1963) The development of methods for trapping the vectors of monkey malaria. *Ann. Trop. Med. Parasitol. Liverpool* **57**, 32–46.

White, E.G. (1964) A design for the effective killing of insects caught in light traps. *NZ Entomol.* **3**, 25–7.

White, G.C., Anderson, D.R., Burnham, K.P., & Otis, D.L. (1982) *Capture–Recapture and Removal Methods for Sampling Closed Populations.* Los Alamos National Laboratory, Los Alamos.

White, N.D.E., Arbogast, R.T., Fields, P.G., *et al.* (1990) The development and use of pitfall and probe traps for capturing insects in stored grain. *J. Kansas Entomol. Soc.* **63**, 506–25.

Wilde, W.H.A. (1962) Bionomics of the pear psylla, *Psylla pyricola* Foerster in pear orchards of the Kootenay valley of British Columbia, 1960. *Can. Entomol.* **94**, 845–9.

Wilkinson, P.R. (1961) The use of sampling methods in studies of the distribution of larvae of *Boophilus microplus* on pastures. *Aust. J. Zool.* **9**, 752–83.

Williams, B.G., Naylor, E., & Chatterton, T.D. (1985) The activity patterns of New Zealand

mud crabs under field and laboratory conditions. *J. Exp. Mar. Biol. Ecol.* **89**(2–3), 269–82.

Williams, C.B. (1939) An analysis of four years captures of insects in a light trap. Part I. General survey; sex proportion; phonology and time of flight. *Trans. Roy. Entomol. Soc. Lond.* **89**, 79–132.

Williams, C.B. (1940) An analysis of four years captures of insects in a light trap. Part II. The effect of weather conditions on insect activity; and the estimation and forecasting of changes in the insect population. *Trans. Roy. Entomol. Soc. Lond.* **90**, 228–306.

Williams, C.B. (1948) The Rothamsted light trap. *Proc. Roy. Entomol. Soc. Lond. A* **23**, 80–5.

Williams, C.B. (1951) Comparing the efficiency of insect traps. *Bull. Entomol. Res.* **42**, 513–17.

Williams, C.B., French, R.A., & Hosni, M.M. (1955) A second experiment on testing the relative efficiency of insect traps. *Bull. Entomol. Res.* **46**, 193–204.

Williams, C.B., Singh, B.P., & El Ziady, S. (1956) An investigation into the possible effects of moonlight on the activity of insects in the field. *Proc. Roy. Entomol. Soc. Lond. A* **319**, 135–44.

Williams, D.F. (1973) Sticky traps for sampling populations of *Stomoxys calcitrans*. *J. Econ. Entomol.* **66**, 1279–80.

Williams, G. (1958) Mechanical time-sorting of pitfall captures. *J. Anim. Ecol.* **27**, 27–35.

Williams, L., Schotzko, D.J., & McCaffrey, J.P. (1992) Geostatistical description of the spatial distribution of *Limonius californicus* (Coleoptera: Elateridae) wireworms in the Northwestern United States with comments on sampling. *Environ. Entomol.* **21**, 983–95.

Wilson, B.H. & Richardson, C.C. (1970) Attraction of deer flies (*Chrysops*) (Diptera: Tabanidae) to traps baited with dry ice under field conditions in Louisiana. *J. Med. Entomol.* **7**, 625.

Wilton, D.P. & Fay, R.W. (1972) Air-flow direction and velocity in light trap design. *Entomol. Exp. Appl.* **15**, 377–86.

Witz, J.A., Lopez, J.D.J., & Latheef, M.A. (1992) Field density estimates of *Heliothis virescens* (Lepidoptera: Noctuidae) from catches in sex pheromone-baited traps. *Bull. Entomol. Res.* **82**(2), 281–6.

Woiwood, I.P. & Harrington, R. (1994) Flying in the face of change: the Rothamsted Insect Survey. In: Leigh, R.A. & Johnston, A.E. (eds) *Long-term Experiments in Agricultural and Ecological Sciences*. CABI, Wallingford, UK, pp. 321–42.

Wolf, W.W., Westbrook, J.K., Raulston, J., Pair, S.D., & Hobbs, S.E. (1990) Recent airborne radar observation of migrant pests in the USA. *Phil. Trans. Roy. Soc. Lond. B* **328**(1251), 619–30.

Woodman, N., Slade, N.A., Timm, R.M., & Schmidt, C.A. (1995) Mammalian community structure in lowland, tropical Peru, as determined by removal trapping. *Zool. J. Linn. Soc.* **113**(1), 1–20.

Worth, C.B. & Jonkers, A.H. (1962) Two traps for mosquitoes attracted to small vertebrate animals. *Mosquito News* **22**, 18–21.

Xu, X., Wu, J., Wu, X., Li, D., & Rao, Q. (1996) A study on ground penetrating radar exploration of subterranean termites nests in dykes and dams. *Acta Entomol. Sin.* **39**(1), 46–52.

Yoshimoto, C.M. & Gressitt, J.L. (1959) Trapping of airborne insects on ships on the Pacific (Part II). *Proc. Hawaiian Entomol. Soc.* **17**(1), 150–5.

Yoshimoto, C.M., Gressitt, J.L., & Wolff, T. (1962) Airborne insects from the Galathea expedition. *Pacific Insects* **4**, 269–91.

Zhogolev, D.T. (1959) Light-traps as a method for collecting and studying the insect vectors of disease organisms. [In Russian.] *Entomol. Obozr.* **38**, 766–73.

Zigler, S.J. & Dewey, M.R. (1995) Phototaxis of larval and juvenile northern pike. *N. Am. J. Fish. Manag.* **15**(3), 651–3.

Zippin, C. (1956) An evaluation of the removal method of estimating animal populations. *Biometrics* **12**, 163–89.

Zippin, C. (1958) The removal method of population estimation. *J. Wildl. Manag.* **22**, 82–90.

Zoebisch, T.G., Stimac, J.L., & Schuster, D.J. (1993) Methods for estimating adult densities of *Liriomyza trifolii* (Diptera: Agromyzidae) in staked tomato fields. *J. Econ. Entomol.* **86**(2), 523–8.

8 Indices: the Use of Signs, Products, and Effects

Comparative surveys of species richness for some animal groups can be undertaken by surveying signs or products, such as footprints, faeces, nests, burrows, or cast skins. Measures of the size of populations based on the magnitude of their products or effects are often referred to as population indices. The relationship of these indices to the absolute population varies from equivalence, when, for example, the number of exuviae are counted, to no more than an approximate correlation, when the index is obtained from general measures of damage.

8.1 Arthropods

8.1.1 Exuviae

The larval or pupal exuviae of insects with aquatic larval stages are often left in conspicuous positions around the edges of bodies of water and, where it is possible to gather these, they will provide a measure of range, species diversity, emergence rate, and the absolute population of newly emerged adults. The method is most easily applied to large insects, such as dragonflies (Corbet, 1957; Wiesmath, 1989; Hartnoll & Bryant, 1990; Pollard & Berrill, 1992; Kielb *et al.*, 1996). Surveys of chironomid communities in steams and rivers have frequently been undertaken by collecting floating pupal exuviae (Fend & Carter, 1995; Kownacki, 1995; Franquet & Pont, 1996; Cranston *et al.*, 1997). The exuviae of the last immature, subterranean stage of some insects are also often conspicuous and, with cicadas, Strandine (1940) and Dybas and Lloyd (1962, 1974) found that the number of larval cases per unit area was correlated with the number of emergence holes and provided a useful index of population; different species could be recognized from the exuviae.

Paramonov (1959) drew attention to the possibility of obtaining an index of the population of arboreal insects from their exuviae, more particularly from the head capsules of lepidopterous larvae, which may be collected in the same way as frass (see below). He found that the head capsules could be separated from the other debris by flotation. In most species, the stage of the larva can be determined from the size of the capsule and, if all the capsules could be collected, an elegant measurement of absolute population would be obtained. However, a significant, and probably fluctuating, number fail to be recovered and this variable needs to be carefully investigated in a given situation before the method is employed to measure either absolute population or even an index of it. Higashiura (1987) estimated absolute larval density of each stage of a larval population of the gypsy moth, *Lymantria dispar*, using head-capsule collections sorted by hand, and produced an age-specific life-table. He concluded that the results were similar to those based on frass collection. Pupal exuviae on the branches and trunks of peach trees were used to provide relative estimates of the moth *Synanthedon*, which accorded with estimates made with pheromone traps (Yonce *et al.*, 1977) and Floater (1996) used exuviae and frass of processionary caterpillars, *Ochrogaster lunifer*, left at the foot of trees, to estimate the number of individuals.

8.1.2 Frass

The faeces of insects are generally referred to as frass; as this is based on the incorrect use of a German word, the term feculae has been suggested (Frost, 1928) but is seldom used. (The German word frass = insect damage or food, whilst *kot* = faeces.) The frass drop—the number of frass pellets falling to the ground—was first used as an index of both population and insect damage by a number of forest entomologists in Germany (Rhumbler, 1929; Gösswald, 1935). The falling frass is collected in cloth or wooden trays or funnels under the trees (Fig. 8.1). In order for such collections to be of maximum value for population estimation, one should be able to identify the species and the developmental stage; information is also required on the quantity of frass produced per individual per unit of time and the proportion of this that falls to the ground and is collected.

8.1.2.1 Identification

Although the frass of early instars of several species may be confused, in the later instars it is generally distinctive and keys for the identification of frass pellets have

50 cm

Weight to prevent
fluttering in wind

Fig. 8.1 Frass collector, constructed of cloth (after Tinbergen, 1960).

been prepared (Morris, 1942; Weiss & Boyd, 1950, 1952; Hodson & Brooks, 1956; Solomon, 1977). With the nun moth, *Lymantria monacha*, Eckstein (1938) concluded that frass size was a more reliable indicator of larval instar than the width of the head capsule; in the sawfly *Diprion hercyniae* (Morris, 1949), the moth *Hypsipyla robusta* (Mo & Tanton, 1995) and the army worm *Pseudaletia unipuncta* (Pond, 1961), instar can easily be determined from pellet size. Water-soluble constituents are washed out when the pellets are exposed to rain and so pellet weight may fall by as much as 30%; however, in these species the volume is not affected. Bean (1959) considered that in the spruce budworm, *Choristoneura fumiferana*, the width of the pellet rather than its volume was closely correlated with larval instar; furthermore, this author also found that larvae feeding on pollen ejected frass pellets that were considerably larger than those produced by the same-instar larvae feeding on foliage.

8.1.2.2 The rate of frass production

This can be measured in terms of dry weight or number of pellets, but the two may not vary in direct proportion (Fig. 8.2a). Whatever units are used, the rate of frass production is affected by many factors: the phase and generation (Fridén, 1958; Iwao, 1962), the temperature and humidity (Morris, 1949; Green & de Freitas, 1955; Pond, 1961; Liebhold & Elkinton, 1988; Kamata & Yanbe, 1994), the available food, photoperiod (Kamata & Yanbe, 1994), both the plant species and its condition (Morris, 1949; Green & de Freitas, 1955; Dadd, 1960; Pond, 1961; Waldbauer, 1964), the diel periodicity of larval feeding (Liebhold & Elkinton, 1988), the developmental rate, as determined by the phenology of the season (Tinbergen, 1960), and the presence of adult parasites (Green & de Freitas, 1955) (Fig. 8.2). Therefore it is not possible to assume that the number of pellets produced in different areas or different seasons reflects very precisely changes in the actual populations. Direct comparisons should be made with other independent measurements of the population. Tinbergen (1960) found that, in 6 out of 7 years, frass drop provided a reliable index of the absolute population. Zandt (1994) compared frass production with direct counting from sampled branches and water-basin traps as techniques for measuring population size of sawfly larvae. He concluded that, while frass production could successfully estimate mean caterpillar intensity per tree, the sampling error was too great to estimate the population intensity in a particular tree. When methods are found to give different estimates, this must be investigated further to determine whether the absolute method has become inaccurate or if the frass-drop rate per individual has changed. The latter can be measured by the use of a 'coprometer', which collects the frass voided each hour into a separate compartment (Green & Henson, 1953; Green & de Freitas, 1955).

8.1.2.3 Efficiency of collection or sighting

Another variable is the proportion of the frass that actually reaches the ground and is collected. As Morris (1949) points out, a greater proportion of the frass will be

(a) Food plant

(b)

(c)

(d)

(e)

retained on the foliage in calm weather than under more windy conditions and, where the young larvae produce webbing, the frass will tend to be caught up with this and its fall delayed (Bean, 1959).

Most of the above work has referred to Lepidoptera and sawflies; the frass of Coleoptera may also be identified and Eckstein (1939) and Campbell (1960) used frass drop as an index of population in studies of stick insects, Phasmida.

It can be concluded that frass-drop measurements may, with caution, be used as a second method to check trends established by another technique (see p. 4). If comparisons are limited to a certain area where the insect has but a single host plant, the chances that a good relative estimate will be obtained are maximal.

If bird droppings can be reliably sighted, then these may be used to make estimates of population, using the principles of line transects (Bibby *et al.*, 1992).

8.1.3 Other products

Populations of web-building spiders can be estimated by counting the number of webs; the visibility of the webs can be increased by dusting with lycopodium powder (Cherrett, 1964) or by spraying with a fine mist of water from a knapsack sprayer. Webs over a large area can be rapidly assessed using the latter method.

Indices of the populations of colonial nest-building caterpillars have been obtained by counting the number of nests, rather than the caterpillars; the great advantage of this approach lies in extensive work, as it enables the population level to be measured over many hectares (Tothill, 1922; Legner & Oatman, 1962; Morris & Bennett, 1967; Ito *et al.*, 1970). Some mites also produce webbing and this may be used as an index (Newcomer, 1943). Cercopid larvae may be estimated by counting the spittle masses, but this is less accurate for the first instars (Whittaker, 1971).

8.2 Vertebrate products

Bird, reptile, and mammal abundance can be estimated from counting nests or burrows. The main problem is to ensure that the structure is occupied. Nest counting is particularly appropriate for sea birds and other colonial breeders that synchronize their reproduction. The census techniques to be used will vary with the habitat (Bibby *et al.*, 1992) and, for sea birds, are discussed by Walsh *et al.* (1995).

Bird and mammal numbers may be estimated by counting faeces. For many mammals, their dung is much more conspicuous than the animal. An interesting example is the Amazonian manatee, which is extremely shy and difficult to spot at

---

Fig. 8.2 (*Opposite.*) Variations in frass-drop rate with different factors: (a) the influence of food plant on the weight and number of pellets per instar from *Protoparce sexta* (Sphingidae) (data from Waldbauer, 1964); (b) the influence of phase and foodplant on the number of pellets per day from *Leucania separata* (Noctuidae) (data from Iwao, 1962); (c) the influence of temperature and humidity on the number of pellets per day from *Dendrolimus pini* (Lasiocampidae) (from Gösswald, 1935); (d) the influence of temperature and condition of the foliage on the number of pellets per hour from 20 larvae of the sawfly, *Neodiprion lecontei* (Diprionidae) (data from Green & de Freitas, 1955); (e) the influence of the presence of a dipterous parasite on the number of pellets per hour from 20 larvae of the sawfly, *N. lecontei* (modified from Green & de Freitas, 1955).

the water surface. However, it produces floating dung that is characteristic and easily spotted. Generally, a simple count of the droppings in a known area will give an index of abundance over the indeterminate period over which the droppings have remained intact. Rainfall and other physical variables will greatly change this time period, as will biotic factors, such as the abundance of dung-beetles. There are two ways to overcome this problem: either remove all the droppings from the study area prior to starting the census or mark some fresh dung and subsequently never count anything which looks older. Animal droppings are often highly clumped in their distribution and may conform to a negative binomial distribution (p. 31); statistical methods to compare densities are discussed by White and Eberhardt (1980).

The presence of mammals, such as squirrels, mice, and birds, can often be detected by the characteristic tooth or beak marks they leave on the remains of their food, such as nuts or pine cones. A wide variety of feeding signs have been used in mammalian studies. An unusual example is the use by Turner (1975) of bite marks on cattle as an index of vampire bat, *Desmodus rotundus*, abundance.

Mammal density has also been monitored using hair samplers, which snag hair from a passing animal. These can work as a barbed-wire fence does, catching the wool of sheep, or on sticky tape, maybe inside a tube placed so that the animals will use it as a runway (Suckling, 1978). These devices are probably only useful for demonstrating the presence of a species at a locality, for the quantity gathered will reflect many variables other than population size. However, DNA identification techniques may soon make it practicable to identify an individual from a single hair.

Footprints and runways may also be used. Mooty and Karns (1984) showed that the density of white-tailed deer inferred from tracks was correlated with the density inferred from pellet counts. Whatever type of sign is used, the data can be analysed in two principal ways. The density can be expressed as counts per area of search or alternatively signs can be used, instead of direct animal observations, in distance-sampling methods (Chapter 9).

8.3 Effects

8.3.1 Effects due to an individual insect

These are of great value to the ecologist, as they are immediately convertible into an absolute estimate. When, at a given stage of the life cycle, each animal has some unique effect (for example, the commencement of a leaf mine), a count of these, after all the animals have passed this stage, will provide a precise measure of the total number passing through that stage. Counts of the actual animals would need to be integrated through time to provide such a total (see p. 405). These effects are therefore measures of the number of animals entering a stage, the estimation of which is discussed in Chapter 10.

Cicada larvae construct conspicuous and distinctive turreted emergence holes, which may be counted to measure the total emerging population (Dybas & Davis, 1962). Populations of solitary Hymenoptera can be assessed from the number of nest holes (Bohart & Lieberman, 1949); this will give the number of reproducing females, but in some species a female will construct more than one nest.

Measures of absolute population can be obtained for plant feeders when they enter the plant. Leaf-miners allow particularly elegant studies, as, at the end of a generation, it is possible to determine the number of larvae that commenced mining, the number that completed development, and often some indication of the age and cause of death in those that failed (Askew, 1968; Pottinger & Le Roux, 1971; Leroi, 1972, 1974; Payne *et al.*, 1972). When the larva cuts characteristic cases at different ages, even more information may be gathered (De Gryse, 1934) and, because of the 'historical' record of spatial position, precise studies of interspecific competition are possible (Murai, 1974). Galls are easily counted, but many harbour a variable number of insects and it is usually impossible to determine externally whether these have been parasitized or not (Bess & Haramoto, 1959; Nakamura *et al.*, 1964; Howse & Dimond, 1965; Redfern, 1968; Redfern & Cameron, 1993).

Some stem-borers cause the growing shoot to die; when multiple invasion is sufficiently rare to be overlooked, estimates of these 'dead hearts' may be taken as equivalent to the total number of larvae invading, and the same approach can be applied to insects in grains or seeds (Jepson & Southwood, 1958; Gomez & Bernardo, 1974). Oviposition punctures have been found to provide an accurate index of the number of eggs per plant unit for the weevil *Hypera postica* (Harcourt *et al.*, 1974).

8.3.2 General effects—damage

The effect of an insect population on a plant stand is the product of two opposing processes: the eating rate of the insects and the growing rate of the plant. The eating rate of the insects will depend on their age, the temperature conditions, and the other factors that affect frass production (see above), as well as on population size. An example of the complex interaction of these processes for various cruciferous plants and two different pests is given by Taylor and Bardner, 1968a,b). Therefore measures of damage are only approximate indices of population size, although they may have advantages over other methods (e.g. Johnson & Burge, 1971). Damage *per se* is, of course, an important parameter for many entomologists; an outline of its assessment will be given.

8.3.2.1 Criteria

Economic damage. This measure is the one least related to the population size and describes the effects of the animals in economic terms; it is of considerable importance to the agriculturist and forester, as it indicates the need for control measures, which become desirable once the economic threshold is reached. The economic threshold is defined by Stern *et al.* (1959) as 'the density at which control measures should be determined to prevent an increasing pest population from reaching the economic injury level'. Changing economic conditions will alter the economic injury level from season to season and its relation to the size of the insect population will vary greatly from species to species. Pests that cause blemishes to fruits or destroy the leading shoots of growing trees cause damage whose economic level is very high relative to the amount of plant material destroyed (Smith, 1969; Southwood & Norton, 1973; Stern, 1973; Mumford & Knight, 1997).

Loss of yield. This is a measure of the extent to which the weight, volume, or number of the marketable parts of the plant is changed—normally reduced, but occasionally increased (Harris, 1974). Although more of a biological measure than economic damage, variations in yield may be caused by many factors other than pest numbers (Möllerström, 1963). Furthermore, the same number of insects attacking a plant can have very different effects, depending on the timing of the attack in relation to the growth stages of the plant. For example, the larva of the frit-fly invades and kills a young shoot of the oat plant. Early in the season the plant will have time to develop another tiller, which will contribute towards the yield of grain, and so the latter is not affected. Later in the season, any tillers destroyed would have contributed towards the crop, but there is insufficient growing time for the tillers that replace them to ripen grain; this is the time of the maximum effect of attack on yield. At a still later time, any tillers that are young enough to be susceptible to the frit-fly are so far behind the bulk of the crop that their contribution to the eventual yield would be insignificant. If the frit-fly attacks such tillers, yield will not be reduced; it might even be increased. Prasad (1961) studied the reduction in weight of cabbage heads following the artificial colonization by newly emerged larvae of the small white cabbage butterfly, *Pieris rapae*, at various dates after transplantation. It was found (Fig. 8.3) that the greatest reduction in yield occurred when the larvae were introduced 7 weeks

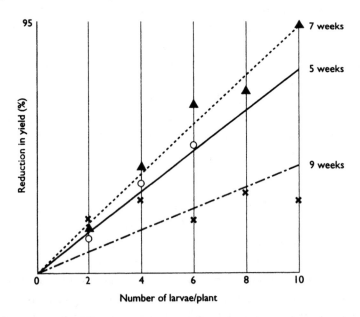

Fig. 8.3 The relationship of damage to the age of the plant: the reduction in the yield of cabbage plants when infested with different numbers of *Pieris rapae* larvae 5, 7, and 9 weeks after transplantation (after Prasad, 1961).

after transplantation, as the leaves damaged at this time contribute to the head. On the other hand, attacks soon after transplanting affect the growth of the plant.

The amount of plant consumed. This is perhaps the most meaningful biological measure and could be expressed precisely in kilojoules as the amount of the primary production consumed by the insect (see Chapter 14). The actual amount of damage may be assessed in terms of dry-weight reduction (e.g. Ortman & Painter, 1960; Buntin, 1991) or leaf area destroyed (e.g. Abbott *et al.*, 1993). The area of the leaf may be measured electronically (Hatfield *et al.*, 1976) or by using a flatbed scanner with a computer, while photocopying provides a rapid permanent record. The area of leaf blotching due to the removal of cell contents (by leafhoppers and mirids) may be estimated by photography with diazo-process film (Fry & Henneberry, 1977) or by colour scanning, followed by image analysis, in which the area occupied by different colours is calculated by a computer. For example, Baur *et al.* (1990) quantified leaf-area reduction in the grey alder, *Alnus incana*, by initially photographing leaves in the field against a white background. Subsequently, on the photographic prints, the missing outlines of the leaves were reconstructed and these images entered into a computer, using a video camera for analysis by an image-processing program. They estimated measurement error as <2% of the total leaf area and considered this method suitable for quantifying leaf-mine size and necrotic tissue area.

It is often easier to measure some index of the amount of plant destroyed rather than the actual quantity. For example, Coaker (1957) used various degrees of battering of the leaf margin of cotton as an index of mirid damage; Dimond and Allen (1974) found that the length of pine needles closely correlated with the numbers of first-instar adelgid, *Pineus* larvae; and Coombs (1963) assessed the level of damage in stored grain from the weight of fine dust present in a given volume of grain. Nuckols and Connor (1995) visually examined each sampled leaf for damage caused by chewing herbivores, skeletonizers, sap-feeders, leaf-miners, and gall-formers and scored the effect in terms of the percentage of the leaf area affected. The degree of defoliation can also be measured by the amount of light passing through the canopy (Higley, 1992).

Estimates of the quantity of leaf consumed may be confused by the subsequent enlargement of the hole by the growth of the leaf, Reichle *et al.* (1973) allowed for this by studying the expansion of standard holes punched in young leaves and Lowman (1987) showed for a range of tropical forest trees that leaf holes expand proportionately with leaf growth.

Remote-sensing techniques (Chapter 15), using optical and thermal imaging from aircraft, provides a method of surveying insect damage over large areas, generally in forests (Wear *et al.*, 1964; Klein, 1973; Riley, 1989), but also in orchards and crops (Stern, 1973; Harris *et al.*, 1976; Wallen *et al.*, 1976; Riley, 1989). Infestations are often detected by their secondary effects, e.g. corn-leaf aphids, *Rhopalosiphum maidis*, are detected photographically by the sooty mould which grows on the honeydew (Wallen *et al.*, 1976). As yet, satellite images have found little application for the surveying of plant damage by insects, though the availability of higher-resolution images may make them important in the future.

Insect defoliation of trees affects the growth rate and is reflected in the annual growth rings (Lessard & Buffam, 1976; Creber, 1977), whilst aphids may also affect the time of leaf fall and other characters (Dixon, 1971). Considering more particularly the effects of defoliates on trees, Henson and Stark (1959) proposed that insect populations could be defined as follows.

1 Tolerable: populations that do not utilize the entire excess biological productivity of the host, i.e. the insect and the host-plant populations could continue at this level indefinitely.

2 Critical: populations that utilise more than the excess biological productivity of the hosts, but less than the total productivity. Such population levels cannot be continued indefinitely, and Henson and Stark (1959) (Fig. 8.4) and Churchill *et al.* (1964) show the long-term effects of insect attack on perennial hosts.

3 Intolerable: populations that are depleting the host at a rate greater than the current rate of production.

The condition of the host tree (expressed as the number of leaf needles per branch tip) and the history of attack in previous years will interact with the intensity of larval population (larvae per tip) in the determination of the appropriate description of population in these terms (Fig. 8.4).

8.3.2.2 Determining the relationship between damage and insect populations
As has been indicated above, the level of damage a plant suffers is influenced by

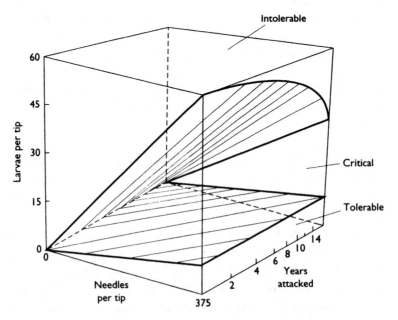

Fig. 8.4 The relationship of tree condition, insect population intensity, and previous history to intolerable, critical, and tolerable populations (after Henson & Stark, 1959).

many factors, such as soil, climate, and age and health of the plant, as well as the size of the insect population. Therefore attempts to correlate insect numbers to damage when these variables are measured in different areas and seasons are seldom satisfactory. However, it may be possible to separate the roles of the different factors by a carefully planned sampling programme, emphasizing the comparison of nearby fields (Sen & Chakrabarty, 1964). The comparisons will be even more exact if they are made in the same field, as then climatic and soil effects will be identical. Theoretically there are four techniques for making such comparisons.

1 The introduction of a known number of animals on to pest-free plants (e.g. Andrzejewska & Wojcik, 1970; Ogunlana & Pedigo, 1974; Harrison & Maron, 1995).

2 The exclusion of the animals from certain plots by mechanical barriers.

3 The exclusion of the animals or the reduction of their populations in some plots by the use of pesticides.

4 A comparison of the growth of individual plants attacked by the pest compared with unattached plants growing in the same crop (Bardner, 1968).

Methods 1 and 2 both suffer from the fact that the barriers erected to exclude or retain the animals will severely modify the climate and growth conditions of the plant; the exclusion technique is, of course, widely used in studies on the effects of grazing by vertebrates.

The third method has been used successfully by Dahms and Wood (1957) and Jepson (1959). Working on the frit-fly, *Oscinella frit*, the last-named author found that a detailed knowledge of the periods of oviposition of the pest allowed frequent 'blanket spraying' to be replaced by a limited number of timed sprays or granular applications (Jepson & Mathias, 1960).

References

Abbott, I., Van Heurck, P., Burbidge, T., & Williams, M. (1993) Damage caused by insects and fungi to eucalypt foliage: spatial and temporal patterns in Mediterranean forest of Western Australia. *For. Ecol. Manag.* **58**(1–2), 85–110.

Andrzejewska, L. & Wojcik, Z. (1970) The influence of Acridoidea on the primary production of a meadow (field experiment). *Ekol. Polska* **18**, 89–109.

Askew, R.R. (1968) A survey of leaf-miners and their parasites on laburnum. *Trans. Roy. Entomol. Soc. Lond.* **120**, 1–37.

Bardner, R. (1968) Wheat bulb fly, *Leptohylemyia coarctata* Fall. and its effect on the growth and yield of wheat. *Ann. Appl. Biol.* **61**, 1–11.

Baur, R., Fritsche, A., Camenzind, R., & Benz, G. (1990) Quantitative analysis of leaf area loss caused by insects: a method combining photography and electronic image processing. *J. Appl. Entomol.* **109**(2), 182–8.

Bean, J.L. (1959) Frass size as an indicator of spruce budworm larval instars. *Ann. Entomol. Soc. Am.* **52**, 605–8.

Bess, H.A. & Haramoto, F.H. (1959) Biological control of Pamakani, *Eupatorium adenophorum*, in Hawaii by a tephritid gall fly, *Procecidochares utilis*. 2. Population studies of the weed, the fly and the parasites of the fly. *Ecology* **40**, 244–9.

Bibby, C.J., Burgess, N.D., & Hill, D.A. (1992) *Bird Census Techniques*. Acadmic Press, San Diego, 257 pp.

Bohart, G.E. & Lieberman, F.V. (1949) Effect of an experimental field application of DDT dust on *Nomia melanderi*. *J. Econ. Entomol.* **42**, 519–22.

Buntin, G.D. (1991) Effect of insect damage on

the growth, yield, and quality of *Sericea lespedeza* forage. *J. Econ. Entomol.* **84**(1), 277–84.

Campbell, K.G. (1960) Preliminary studies on population estimation of two species of stick insects (Phasmatidae) occurring in plague numbers in Highland Forest areas of southeastern Australia. *Proc. Linn. Soc. NSW* **85**, 121–37.

Cherrett, J.M. (1964) The distribution of spiders on the Moor House National Nature Reserve, Westmorland. *J. Anim. Ecol.* **33**, 27–48.

Churchill, G.B., John, H.H., Duncan, D.P., & Hodson, A.C. (1964) Long-term effects of defoliation of aspen by the forest tent caterpillar. *Ecology* **45**, 630–3.

Coaker, T.H. (1957) Studies of crop loss following insect attack on cotton in East Africa. II. Further experiments in Uganda. *Bull. Entomol. Res.* **48**, 851–66.

Coombs, C.W. (1963) A method of assessing the physical condition of insect-damaged grain and its application to a faunistic survey. *Bull. Entomol. Res.* **54**, 23–35.

Corbet, P.S. (1957) The life-history of the Emperor dragonfly, *Anax imperator* Leach (Odonata: Aeshnidae). *J. Anim. Ecol.* **26**, 1–69.

Cranston, P.S., Cooper, P.D., Hardwick, R.A., Humphrey, C.L., & Dostine, P.L. (1997) Tropical acid streams: the chironomid (Diptera) response in northern Australia. *Freshw. Biol.* **37**(2), 473–83.

Creber, G.T. (1977) Tree rings: a natural data storage system. *Biol. Rev.* **52**, 349–82.

Dadd, R.H. (1960) Observations on the palatability and utilisation of food by locusts, with particular reference to the interpretation of performances in growth trials using synthetic diets. *Entomol. Exp. Appl.* **3**, 283–304.

Dahms, R.G. & Wood, E.D. (1957) Evaluation of green bug damage to small grains. *J. Econ. Entomol.* **50**, 443–6.

De Gryse, J.J. (1934) Quantitative methods in the study of forest insects. *Sci. Agric.* **14**, 477–95.

Dimond, J.B. & Allen, D.C. (1974) Sampling populations of pine leaf chermid *Pineus pinifoliae* (Homoptera: Chermidae). III. Neosisterites on white pine. *Can. Entomol.* **106**, 509–18.

Dixon, A.F.G. (1971) The role of aphids in wood formation II. The effect of the lime aphid, *Eucallipterus tiliae* L. (Aphididae), on the growth of lime, *Tilia* × *vulgaris* Hayne. *J. Appl. Ecol.* **8**, 393–9.

Dybas, H.S. & Davis, D.D. (1962) A population census of seventeen-year periodical cicadas (Homoptera: Cicadidae: Magicicada). *Ecology* **43**, 432–59.

Dybas, H.S. & Lloyd, M. (1962) Isolation by

habitat in two synchronized species of periodical cicadas (Homoptera: Cicadidae: Magicicada). *Ecology* **43**, 444–59.

Dybas, H.S. & Lloyd, M. (1974) The habitats of 17-year periodical cicadas (Homoptera: Cicadidae: Magicicada spp.). *Ecol. Monogr.* **44**, 279–324.

Eckstein, K. (1938) Die Bewertung des Kotes der Nonnenraupe, *Psilura monacha* L., als Grundlage für die Festellung ihres Auftretens und zu ergreifenden Massregeln. *Allgem. Forst. Jagdztg.* **114**, 132–48.

Eckstein, K. (1939) Das Bohrmehl des Waldgärtners, *Myelophilus pimperda* L., nebst Bemerkungen über den 'Frass' der Borkenkäfer und anderen Insekten. *Arb. physiol. angew. Ent.* **6**, 32–41.

Fend, S.V. & Carter, J.L. (1995) The relationship of habitat characteristics to the distribution of Chironomidae (Diptera) as measured by pupal exuviae collections in a large river system. *J. Freshw. Ecol.* **10**(4), 343–59.

Floater, G.J. (1996) Estimating movement of the processionary caterpillar *Ochrogaster lunifer* Herrich-Schaffer (Lepidoptera: Thaumetopoeidae) between discrete resource patches. *Aust. J. Entomol.* **35**(3), 279–83.

Franquet, E. & Pont, D. (1996) Pupal exuviae as descriptors of the chironomid (Diptera: Nematocera) communities of large rivers. *Arch. Hydrobiol.* **138**(1), 77–98.

Fridén, F. (1958) *Frass-drop Frequency in Lepidoptera*. Almqvist & Wiksells Boktryckeri, Uppsala.

Frost, S.-W. (1928) Insect scatology. *Ann. Entomol. Soc. Am.* **21**, 35–46.

Fry, K.E. & Henneberry, T.J. (1977) Measuring leaf damage by the cotton leaf perforator. *J. Econ. Entomol.* **709**, 141–2.

Gomez, K.-A. & Bernardo, R.C. (1974) Estimation of stem borer damage in rice fields. *J. Econ. Entomol.* **67**, 509–16.

Gösswald, K. (1935) Über die Frasstätigkeit von Forstschädlingen unter verschiedener Temperatur und Luftfeuchtigkeit und ihre praktische und physiologische Bedeutung. I. *Z. angew. Entomol.* **21**, 183–7.

Green, G.W. & de Freitas, A.S. (1955) Frass drop studies of larvae of *Neodiprion americanus banksianae*, Rob. and *Neodiprion lecontii*, Fitch. *Can. Entomol.* **87**, 427–40.

Green, G.W. & Henson, W.R. (1953) A new type of coprometer for laboratory and field use. *Can. Entomol.* **85**, 227–30.

Harcourt, D.G., Mukerji, M.K., & Guppy, J.C. (1974) Estimation of egg populations of the alfalfa weevil, *Hypera postica* (Colcoptera: Curculionidae). *Can. Entomol.* **106**, 337–47.

Harris, M.K., Hart, W.G., Davis, M.R., Ingle,

S.I., & Van Cleave, H.W. (1976) Aerial photography shows caterpillar infestation. *Pecan Quart.* **10**, 12–18.

Harris, P. (1974) A possible explanation of plant yield increases following insect damage. *Agro-ecosystems* **1**, 219–25.

Harrison, S. & Maron, J.L. (1995) Impacts of defoliation by tussock moths (*Orgyia vetusta*) on the growth and reproduction of bush lupine (*Lapinus arboreus*). *Ecol. Entomol.* **20**, 223–9.

Hartnoll, R.G. & Bryant, A.D. (1990) Size-frequency distributions in decapod Crustacea: the quick, the dead, and the cast-offs. *J. Crustac. Biol.* **10**, 14–19.

Hatfield, J.L., Stanley, C.D., & Carlton, R.E. (1976) Evaluation of an electronic foliometer to measure leaf area in corn and soybeans. *Agron. J.* **68**, 434–6.

Henson, W.R. & Stark, R.W. (1959) The description of insect numbers. *J. Econ. Entomol.* **529**, 847–50.

Higashiura, Y. (1987) Larval densities and a life-table for the gypsy moth, *Lymantria dispar*, estimated using the head-capsule collection method. *Ecol. Entomol.* **12**(1), 25–30.

Higley, L.G. (1992) New understandings of soybean defoliation and their implications for pest management. In: Copping, L.G., Green, M.B., & Rees, T.R. (eds) *Pest Management of Soybean*. Elsevier, Amsterdam.

Hodson, A.C. & Brooks, M.A. (1956) The frass of certain defoliators of forest trees in the north central United States and Canada. *Can. Entomol.* **88**, 62–8.

Howse, G.M. & Dimond, J.B. (1965) Sampling populations of pine leaf adelgid, *Pineus pinifoliae* (Fitch). I. The gall and associated insects. *Can. Entomol.* **97**, 952–61.

Ito, Y., Shibazaki, A., & Iwahashi, O. (1970) Biology of *Hyphantria cunea* Drury (Lepidoptera: Arctiidae) in Japan. XI. Results of road-survey. *Appl. Entomol. Zool.* **5**, 133–44.

Iwao, S. (1962) Studies on the phase variation and related phenomena in some lepidopterous insects. *Mem. Coll. Agric. Kyoto Univ. (Entomol. no. 12)* **84**, 1–80.

Jepson, W.F. (1959) The effects of spray treatments on the infestation of the oat crop by the frit fly (*Oscinella frit* L.). *Ann. Appl. Biol.* **47**, 463–74.

Jepson, W.F. & Mathias, P. (1960) The control of frit fly, *Oscinella frit* (L.) in sweet corn (*Zea mays*) by Thirpet (*O,O*-diethyl *S*-ethyethio-methyl phosphorodithioate). *Bull. Entomol. Res.* **51**, 427–33.

Jepson, W.F. & Southwood, T.R.E. (1958) Population studies on *Oscinella frit* L. *Ann. Appl. Biol.* **46**, 465–74.

Johnson, C.G. & Burge, G.A. (1971) Field trials of anti-capsid insecticides on farmers cocoa in Ghana, 1956–60. 2. Effects of different insecticides compared by counting capsids, and capsid counting compared with counting the percentage of newly damaged trees. *Ghana. J. Agric. Sci.* **4**, 33–8.

Kamata, N. & Yanbe, T. (1994) Frass production of last instar larvae of the beech caterpillar, *Quadricalcarifera punctatella* (Motschulsky) (Lep., Notodontidae), and method of estimating their density. *J. Appl. Entomol.* **117**(1), 84–91.

Kielb, M.A., Bright, E., & O'Brien, M.F. (1996) Range extension of *Stylogomphus ablistylus* (Odonata: Gomphidae) for the Upper Peninsula of Michigan. *Great Lakes Entomol.* **29**(2), 87–8.

Klein, W.H. (1973) Beetle killed pine estimates. *Photogramm. Eng.* **39**, 385–8.

Kownacki, A. (1995) The use of chironomid pupal exuviae for ecological characterization of the upper Vistula (southern Poland). *Acta Hydrobiol.* **37**(1), 41–50.

Legner, E.F. & Oatman, E.R. (1962) Sampling and distribution of summer eyespotted bud moth *Spilonota ocellana* (D. & S.) larvae and nests on apple trees. *Can. Entomol.* **94**, 1187–9.

Leroi, B. (1972) A study of natural populations of the celery leaf-miner *Philophylla heraclei* L. (Diptera Tephritidae). 1. Methods of counting of larval populations. *Res. Popul. Ecol.* **13**, 201–15.

Leroi, B. (1974) A study of natural populations of the celery leaf-miner, *Philophylla heraclei* 1. (Diptera, Tephritidae). 11. Importance of changes of mines for larval populations. *Res. Popul. Ecol.* **15**, 163–82.

Lessard, G. & Buffam, P.E. (1976) Effects of *Rhyacionia neomexicana* on height and radial growth in Ponderose Pine reproduction. *J. Econ. Entomol.* **69**, 755–60.

Liebhold, A.M. & Elkinton, J.S. (1988) Estimating the density of larval gypsy moth, *Lymantria dispar* (Lepidoptera: Lymantriidae), using frass drop and frass production measurements: sources of variation and sample size. *Environ. Entomol.* **17**(2), 385–90.

Lowman, M.D. (1987) Relationships between leaf growth and holes caused by herbivores. *Aust. J. Ecol.* **12**, 189–91.

Mo, J. & Tanton, M.T. (1995) Estimation of larval instars of *Hypsipyla robusta* Moore (Lepidoptera: Pyralidae) by larval frass widths. *Aust. Entomol.* **22**(2), 59–62.

Möllerström, G. (1963) Different kinds of injury to leaves of the sugar beets and their effect on yield. *Medd. Växtskyddsanst.* **12**, 299–309.

Mooty, J.J. & Karns, P.D. (1984) The relationship between white-tailed deer track counts and pellet-group surveys. *J. Wildl. Manag.* **48**, 275–9.

Morris, R.F. (1942) The use of frass in the identification of forest insect damage. *Can. Entomol.* **74**, 164–7.

Morris, R.F. (1949) Frass-drop measurement in studies of the European spruce sawfly. *Univ. Michigan Sch. For. Conserv. Bull.* **12**, 1–58.

Morris, R.F. & Bennett, C.W. (1967) Seasonal population trends and census methods for *Hyphantria cunea. Can. Entomol.* **99**, 9–17.

Mumford, J.D. & Knight, J.D. (1997) Injury, damage and threshold concepts. In: Dent, D.R. & Walton, M.P. (eds) *Methods in Ecological and Agricultural Entomology.* CAB International, Wallingford, UK, pp. 202–20.

Murai, M. (1974) Studies on the interference among larvae of the citrus leaf miner, *Phyllocristis citrella* Stainton (Lepidoptera: Phyllocristidae). *Res. Popul. Ecol.* **16**, 80–111.

Nakamura, M., Kondo, M., Ito, Y., Miyashita, K., & Nakamura, K. (1964) Population dynamics of the chestnut gall-wasp, *Dryocosmus buriphilus.* 1. Description of the survey station and the life histories of the gall wasp and its parasites. *Japan. J. Appl. Entomol. Zool.* **8**, 149–58.

Newcomer, E.J. (1943) Apparent control of the Pacific mite with xanthone. *J. Econ. Entomology* **369**, 344–5.

Nuckols, M.S. & Connor, E.F. (1995) Do trees in urban or ornimental plantings receive more damage by insects than trees in natural forests? *Ecol. Entomol.* **20**, 253–60.

Ogunlana, M.O. & Pedigo, P. (1974) Economic injury levels of the potato leafhopper on soybeans in Iowa. *J. Econ. Entomol.* **67**, 29–32.

Ortman, E.E. & Painter, R.H. (1960) Quantitative measurements of damage by the greenbug, *Toxoptera graminum* to four wheat varieties. *J. Econ. Entomol.* **53**, 798–802.

Paramonov, A. (1959) A possible method of estimating larval numbers in tree crowns. *Entomol. Manag. Mag.* **95**, 82–3.

Payne, J.A., Tedders, W.L., Cosgrove, G.E., & Foard, D. (1972) Larval mine characteristics of four species of leaf-mining Lepidoptera in Pecan. *Ann. Entomol. Soc. Am.* **65**, 74–81.

Pollard, J.B. & Berrill, M. (1992) The distribution of dragonfly nymphs across a pH gradient in south-central Ontario lakes. *Can. J. Zool.* **70**(5), 878–85.

Pond, D.D. (1961) Frass studies of the armyworm, *Pseudaletia unipuncta. Ann. Entomol. Soc. Am.* **54**, 133–40.

Pottinger, R.P. & Le Roux, E.J. (1971) The biology and dynamics of *Lithocolletis blancardella*

(Lepidoptera: Gracillariidae) on apple in Quebec. *Mem. Entomol. Soc. Can.* **77**, 1–437.

Prasad, S.K. (1961) Quantitative estimation of damage to cabbage by cabbage worm, *Pieris rapae* (Linn.). *Ind. J. Entomol.* **23**, 54–61.

Redfern, M. (1968) The natural history of spear thistle-heads. *Field St.* **2**, 669–717.

Redfern, M. & Cameron, R.A.D. (1993) Population dynamics of the yew gall *Taxomyia taxi* and its chalcid parasitoids: a 24 year study. *Ecol. Entomol.* **18**, 365–78.

Reichle, D.E., Goldstein, R.A., Van Hook, R.I., & Dobson, G.J. (1973) Analysis of insect consumption in a forest canopy. *Ecology* **54**, 1076–84.

Rhumbler, L. (1929) Zur Begiftung des Kiefernspanners (*Bupalus piniarius* L.) in der Oberförsterei Hersfeld-Ost 1926. *Z. angew. Entomol.* **15**, 137–58.

Riley, J.R. (1989) Remote sensing in entomology. *Ann. Rev. Entomol.* **34**, 1–247.

Sen, A.R. & Chakrabarty, R.P. (1964) Estimation of loss of crop from pests and diseases of tea from sample surveys. *Biometrics* **20**, 492–504.

Smith, R.F. (1969) The importance of economic injury levels in the development of integrated pest control programs. *Qual. Plant. Mater. Veg.* **17**, 81–92.

Solomon, J.D. (1977) Frass characteristics for identifying insect borers (Lepidoptera: Cossidae and Sesiidae; Coleoptera: Cerambyciidae) in living hardwoods. *Can. Entomol.* **109**, 295–303.

Southwood, T.R.E. & Norton, C.A. (1973) Economic aspects of pest management strategies and decisions. In: Geier, P.W., Clark, L.R., Anderson, D.J., & Nix, H.A. (eds) Insects: studies in population management. *Mem. Ecol. Soc. Aust.* **1**, 168–95.

Stern, V.M. (1973) Economic thresholds. *Ann. Rev. Entomol.* **18**, 259–80.

Stern, V.M., Smith, R.F., Van Den Bosch, R., & Hagen, K.S. (1959) The integrated control concept. *Hilgardia* **29**, 81–101.

Strandine, E.J. (1940) A quantitative study of the periodical cicada with respect to soil of three forests. *Am. Midl. Nat.* **24**, 177–83.

Suckling, G.C. (1978) A hair sampling tube for the detection of small mammals in trees. *Aust. Wildl. Res.* **5**, 249–52.

Taylor, W.E. & Bardner, R. (1968a) Effects of feeding by larvae of *Phaedon cochleariae* (F.) and *Plutella maulipennis* (Curt.) on the yield of radish and turnip plants. *Ann. Appl. Biol.* **62**, 249–54.

Taylor, W.E. & Bardner, R. (1968b) Leaf injury and food consumption by larvae of *Phaedon cochleariae* (Col. Chrysomelidae) and *Plutella maculipennis* (Lep. Plutellidae) feeding on turnip and radish. *Entomol. Exp. Appl.* **11**, 177–84.

Tinbergen, L. (1960) The natural control of insects in pinewoods. 1. Factors influencing the intensity of predation by song birds. *Arch. Néerl. Zool.* **13**, 266–343.

Tothill, J.D. (1922) The natural control of the fall webworm (*Hyphentia cunea* Drury) in Canada. *Bull. Can. Depart. Agric. (n.s.)* **3** (*Entomol. Bull.* **19**), 1–107.

Turner, D.C. (1975) *The Vampire Bat*. Johns Hopkins Press, Baltimore, Maryland.

Waldbauer, G.P. (1964) Quantitative relationships between the numbers of fecal pellets, fecal weights and the weight of food eaten by tobacco hornworms, *Protoparce sexta* (Johan.) (Lepidoptera: Sphingidae). *Entomol. Exp. Appl.* **7**, 310–14.

Wallen, V.R., Jackson, H.R., & MacDiarmid, S.W. (1976) Remote sensing of corn aphid infestation, 1974 (Hemiptera: Aphididae). *Can. Entomol.* **108**, 751–4.

Walsh, P.M., Halley, D.J., Harris, M.P., *et al.* (1995) *Seabird Monitoring Handbook for Britain and Ireland*. JNCC, Peterborough.

Wear, J.F., Pope, R.B., & Lauterbach, P.G. (1964) Estimating beetle-killed Douglas fir by photo and field plots. *J. For.* **62**, 309–15.

Weiss, H.B. & Boyd, W.M. (1950) Insect feculae I. *J. NY Entomol. Soc.* **58**, 154–68.

Weiss, H.B. & Boyd, W.M. (1952) Insect feculae II. *J. NY Entomol. Soc.* **60**, 25–30.

White, G.C. & Eberhardt, L.E. (1980) Statistical analysis of deer and elk pellet group data. *J. Wildl. Manag.* **44**, 121–31.

Whittaker, J.B. (1971) Population changes in *Neophilaenus lineatus* (L.) (Homoptera: Cercopidae) in different parts of its range. *J. Anim. Ecol.* **40**, 425–43.

Wiesmath, I. (1989) Faunistic and ecological studies of damselflies and dragonflies (Odonata: Zygoptera and Anisoptera) on bodies of water in the Tuebingen area (West Germany). *Jahreshefte Gesellsch. Naturkunde Württemberg* **144**(0), 297–314.

Yonce, C.E., Gentry, C.R., Tumlinson, J.H. *et al.* (1977) Seasonal distribution of the lesser peach tree borer in Central Georgia as monitored by pupal skin counts and pheromone trapping techniques. *Environ. Entomol.* **6**, 203–6.

Zandt, H.S. (1994) A comparison of three sampling techniques to estimate the population size of caterpillars in trees. *Oecologia* **97**(3), 399–406.

9 Wildlife Population Estimates by Census and Distance-measuring Techniques

Counting the number of sightings forms the basis for estimating density for many animal groups. This is particularly the case for large or easily seen animals, such as birds, large grassland mammals, whales, crocodilians, and large, active insects, such as butterflies. While it may be possible to count animals from a suitable vantage point or while moving along a transect, the count can only be converted to a density estimate if the area scanned can be estimated. This simple approach (see Chapter 4) is often difficult to undertake, for two reasons: first, it may not be possible to estimate the area scanned sufficiently accurately and, secondly, not all of the animals present may have been spotted. Distance sampling methods have been developed to allow for these problems by assuming that the likelihood that an individual will be observed will decline in a mathematically definable way with distance. The methods discussed here are also useful for small, sessile, or slow-moving organisms, such as barnacles, corals, and molluscs, and some can also be applied to data collected non-visually, e.g. a bird census may be based on birdsong or an electric fish survey on the detection of the distinctive electrical signals. One problem, as can be the case with a visual survey, is to ensure that the same individual is only counted once. Distance sampling and nearest-neighbour methods give absolute estimates of animal density. These represent the fourth and final approach to absolute population estimation, the others being mark–recapture (Chapter 3), removal trapping (Chapter 7), and direct counting of a unit volume or area of habitat (Chapters 4 to 6). We briefly introduce wildlife census methods here, rather than with other direct counting methods, because they are the natural starting-point from which to introduce distance sampling methods.

By convention, a census is defined as the counting of all the individuals belonging to the group of interest within a defined area and survey when only a proportion are counted. When the study area can be divided into quadrats which can vary in size or shape, e.g. a rectangular 10 ha area of forest or a 0.1 m^2 core or grab sample, then a census can be made of a randomly selected set and standard statistical theory used to estimate average density and its variance for the entire area. Survey methods, such as distance and nearest-neighbour sampling, aim to estimate density using observations on the distance between animals or from a selected line or point to the animals and thus do not require the worker accurately to map out or define the sampling area. They are thus particularly appropriate for the estimation of population density for large animals living at low density in a habitat that is difficult to traverse. For example, distance sampling using a Fourier series is frequently the method of choice for estimating population size of primates in neotropical forest. The animals are counted while walking along specially cut forest trails. The high density of the forest and the low density and mobility of the apes would make it futile to try to census a number of predefined quadrats.

9.1 Census methods

If it proves possible to count all of the individuals, n, within a known area, a, then this is termed a census and the estimated density, D, is simply:

$$\hat{D} = \frac{n}{a} \qquad (9.1)$$

Counting often requires the observer to move over the census area and thus favours the use of strip transects (long, thin quadrats) of length L and width $2w$, along which the observer moves in a straight line. Strips have been surveyed by foot, car, boat, aircraft, remote-operated vehicle, or submarine (e.g. Starr et al., 1996). The strip width must be determined at the outset; if this cannot be done, or if animals may be missed, then line-transect methods (p. 336) should be used.

Aerial census methods, described in detail by Norton-Griffiths (1978), are frequently used for wildlife surveys (e.g. Pinder, 1996). Remote-sensing methods in general are discussed further in Chapter 15. The counts obtained suffer from error, which varies with the counting rate, and bias, because of the tendency of observers to undercount. From a study of bias during aerial surveys of moose, Alces alces, LeResche and Rausch (1974) concluded that this bias was sufficient to invalidate the method as a means of absolute population estimation. A survey of 17 studies on large mammals by Caughly (1974) found that the proportion of the population counted varied from 23 to 89%. One possible means of reducing or assessing the bias is to undertake counts from photographs or video film. However, as Harris and Lloyd (1977) show for a study of sea-bird colony size by aerial photography, estimates may still vary substantially between observers and a tendency to undercount remains, because not all nests are visible. Ways of handling the bias inherent in aerial surveys are discussed by Caughly (1974). A comparison of census methods for bird-nest estimation was made by Dodd and Murphy (1995) (see also Chapter 15).

Some animals concentrate at particular seasons or times of the day, when a census becomes possible. For example, Allsteadt and Vaughan (1992) were able to census caymans during the dry season by counting the reflective eyes by torchlight at night. Technological solutions to improve detection rates can also make census methods feasible; Naugle et al. (1996) were able to detect white-tailed deer using an aerial infrared imaging system. Local concentrations of conspicuous insects can also allow a direct census (Chapter 4).

9.2 Point- and line-survey methods

9.2.1 Indices of abundance using transects

These methods are frequently used in conservation studies, where minimizing cost is important and when the objective is to detect any change in abundance rather than absolute magnitude. A typical example applied to insects is the study by Thomas (1983) on a number of species of British butterflies. The protocol was as follows.

1 Undertake a preliminary inspection to define the extent of the butterfly colony: this is termed the flight area, A.

2 Mark out a zigzag trail of length L m, which should either exceed 1000 m in length or be long enough to record 40 individuals.

3 Walk the trail and count individuals, N, seen within an imaginary box from 0 to 5 m ahead, with a fixed width, which varies with species but is typically 4–6 m.

4 Calculate the index as:

$$p = \frac{NA}{L} \tag{9.2}$$

Thomas (1983) showed that this abundance index was correlated to the absolute population density, as determined by mark–recapture, and could be converted to an estimate of absolute abundance by regression analysis.

9.2.2 Methods based on flushing

The simplest approach is to estimate the number disturbed from a known width of habitat. Under certain circumstances, e.g. sea-bird colonies, all individuals may be flushed and the problem is simply obtaining an accurate count (Bibby et al., 1992), but usually only a proportion is disturbed. If this is constant, then the numbers themselves give an index of absolute population and, incorporating the proportion and the area covered, an estimate of absolute population density. If the proportion flushed is unknown or varies, only a relative measure of availability is given.

This method has also been used for the assessment of populations of the red locust (*Nomadacris septemfasciata*) in its outbreak areas. The observer moved across the area in a motor vehicle (Land-rover) and counted the insects disturbed ahead (Scheepers & Dunn, 1958). Although the response of a locust to flushing depends on its condition (Nickerson, 1963), this method appears to have a fairly constant efficiency — about 75% of the locusts within the path of the vehicle rise (Symmons *et al.*, 1963). In an attempt to cover a larger area more readily, flushing by a low-flying aircraft was tried, but the efficiency of flushing, even when increased by spraying a noxious chemical, was low in sparse populations. It was suggested that a siren might be used to disturb a higher proportion of locusts (Symmons *et al.*, 1963), and such an approach has been found to be effective with nesting sea birds (Bullock & Gomersall, 1981; Bibby *et al.*, 1992).

Hayne (1949) developed the first estimator of density for a flushing experiment that could claim to be robust, for studies on bird populations. This simple method was designed to estimate the numbers of grouse, which it was assumed would flush as the observer came within a certain radius, r, of the bird. The observer moves along a transect through the habitat noting the radial distance at which each bird is flushed. The density estimate, $\hat{D}_H$, is then given by:

$$\hat{D}_H = \frac{n}{2L}\left(\frac{1}{n}\sum\frac{1}{r_i}\right) \tag{9.3}$$

where n is the total number of animals counted, L the transect length, and r the sighting distance to the ith animal. The approximate variance of this estimate is:

$$\text{var}\left(\hat{D}_H\right) \approx D_H^2\left[\frac{\text{var}(n)}{n^2} + \frac{\sum\left(\frac{1}{r_i}-R\right)^2}{R^2 n(n-1)}\right] \tag{9.4}$$

where R is the mean of the reciprocals of the sighting distances.

Burnham and Anderson (1976) point out that an underlying assumption is that the average sighting angle, θ, is 32.7°. To test this assumption, calculate:

$$Z = \frac{\sqrt{n}(\theta - 32.7)}{21.56} \tag{9.5}$$

and reject the assumption if the absolute value of Z is greater than 1.96.

If this assumption is rejected, then the following adjustment can be applied (Burnham & Anderson, 1976):

$$\hat{D}_{MH} = \hat{c}\,\frac{n}{2L}\left[\frac{1}{n}\sum_{i=1}^{n}\frac{1}{r_i}\right] \tag{9.6}$$

where:

$$\hat{c} = 1.9661 - 0.0295(\theta)$$

More than six other estimators have been developed (Gates et al., 1968; Gates, 1969); an unbiased estimator with minimal variance is considered to be that of Gates (1969) (Kovner & Patil, 1974), although some others may be more robust. This estimator is:

$$\hat{D} = [2n - 12L\bar{r}] \tag{9.7}$$

where n = number of animals sighted, L = length of the line transect, and $\bar{r}$ = average radial distance over which the observer encountered animals. The assumptions are made that encounter distances (r) follow a negative exponential distribution and that the flushing of one animal will not affect another. The effect of varying the two latter assumptions has been explored by Sen et al. (1974) with several sets of data for birds, and they concluded that, although r often followed a negative exponential function, a Pearson type III distribution was more general. They also concluded that the flushing of one animal was not independent of the flushing of another, and the best fit was obtained if cluster size was considered to follow a Poisson distribution (i.e. the distribution of 'flushings' would fit the negative binomial).

The estimation of each r can, of course, be a major source of error. The observer should seek to fix the position where the animal was seen and then measure the distance; a rangefinder may be helpful (Bibby *et al.*, 1992). If, because of the risk of disturbing other animals, r has to be estimated, the observer should regularly check the accuracy of the estimation, at another site if necessary.

The transect route may simply follow an existing path and this may be satisfactory for relative estimates. A more acceptable basis for absolute population estimates will be provided if the single route is replaced with a number of shorter straight, transepts, each with starting-point and direction determined randomly. If the habitat has different zones, each should be traversed separately—a stratified sampling system (Chapter 2) (Bibby *et al.*, 1992).

Line-transect theory normally assumes that the animals (objects under observation) are immobile prior to detection. This assumption will not cause serious error provided the movement relative to that of the observer is random and slow. For cases where there is substantial movement, a different estimator was developed by Yapp (1956):

$$\hat{D} = \frac{E}{2rV} \tag{9.8}$$

where E = number of encounters per unit time, i.e. n/t, V = average velocity of the organism relative to the observer, which is given by:

$$V^2 = \bar{u}^2 + \bar{w}^2 \tag{9.9}$$

where $\bar{u}$ = average velocity of the observer and $\bar{w}$ = average velocity of the animal. This estimator is based on the kinetic theory of gases and assumes that the observer walks in a straight line at a constant speed, the animals move at random at a constant speed, and there is a fixed and known radius r within which the observer can recognize the animals.

The applicability of this approach has been investigated by Skellam (1958), who concluded that equation (9.8) was valid, but that the derivation of V contained an approximation, the effect of which would be most serious when the speed of the observer and the animal were equal and almost negligible when the two were very dissimilar. Skellam also showed that equation (9.9) could be regarded as a Poisson variable, provided the animals did not move back on their tracks or move in groups; under these conditions, the variance would equal the mean. Aggregation will increase the variance. The major practical difficulty in the application of this method—for the estimation of populations of butterflies, dragonflies, birds, and other animals for which it would seem to be appropriate—is the determination of the average speed of the animal. Care must of course be taken with the units: if E is expressed as the number of encounters per hour and r in meters, the speed must be expressed in m/h. Burnham *et al.* (1980) concluded that the inherent assumptions greatly restricted the valid use of this method.

9.2.3 Line-transect methods: the Fourier-series estimator

Line-transect methods have been developed for situations when it is not possible to count all the animals within a strip transect. The methods are based on the idea that only animals lying on the centre line of the strip transect along which the observer moves will be certain to be detected and that the probability of detection will fall with perpendicular distance from this line. This was also the case with the Haynes and Gates estimators discussed above, but they made strong assumptions about the shape of the detection function (rectangular and negative exponential, respectively), which will often not hold. The techniques presented here have been reviewed in detail by Buckland *et al.* (1993). For many animals, the observer may walk the transect, but counts can also be undertaken from motor vehicles, ships, and remote-operated vehicles.

For these methods, the following assumptions are made.
1 Objects on the line are always detected.
2 The observer does not influence the recorded positions. For mobile animals, the position must be that prior to any response to the presence of the observer. The theory has been developed under the assumption that the objects are immobile, but slow movement in relation to the observer creates little inaccuracy.
3 Distances and/or angles are measured accurately.
4 The objects are correctly identified.

The basic field procedure is for the transect route to be a straight line of length L, randomly placed with respect to the animals or objects to be counted, and the perpendicular distance to each detected object of interest, x, is recorded. In practice, a number of lines, arranged as a regular grid and randomly placed in the study, may be used.

Further, it is usually easier to record the sighting distance, r, which is the distance from the observer to the object, and the sighting angle, θ, which is the angle of the object from the transect line, and to calculate x, rather than measuring x directly (see Fig. 9.1). Methods of population estimation based on r and θ were reviewed by Hayes and Buckland (1983), but they are not presented here, as they are considered inferior to those based on the perpendicular distance (Buckland *et al.*, 1993).

If all n objects in a strip of length L and width $2w$ are counted, then the estimated density is:

$$\hat{D} = \frac{n}{2wL} \tag{9.10}$$

If a proportion, P, of the animals present is detected, then the equation becomes:

$$\hat{D} = \frac{n}{2wLP} \tag{9.11}$$

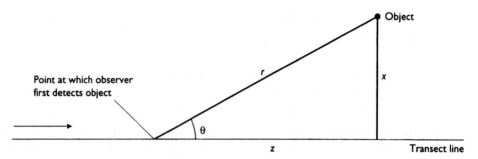

Fig. 9.1 The relationship between sighting distance, r, sighting angle, θ, and the perpendicular distance, x, from the transect line.

Line-transect methods use the distribution of the perpendicular detection distances to estimate P. Figure 9.2 shows a typical histogram of the decline in the number of observations with perpendicular distance. A detection function, $g(x)$, defined as the probability of detecting an object at distance x, can be fitted to these data. Given assumption 1 above, $g(0)$, the probability of detecting an object lying on the line, is assumed to equal 1.

The probability of detecting an object within a strip of area $2wL$, P, is:

$$P = \frac{\int_0^w g(x)\,dx}{w} \tag{9.12}$$

which, on substitution into equation (9.10), gives:

$$\hat{D} = \frac{n}{2L\int_0^w g(x)\,dx} \tag{9.13}$$

As it is assumed that $g(0) = 1$, the probability density function evaluated at $x = 0$ is:

$$f(0) = \frac{1}{\int_0^w g(x)\,dx} \tag{9.14}$$

and thus the general estimator of density is often expressed as:

$$\hat{D} = \frac{nf(0)}{2L} \tag{9.15}$$

Density is expressed as numbers per square unit of length, where length of the transect and distance to the objects are expressed in the same length units, such as metres or kilometres. A flexible function with the desired properties to use as a detection function is a cosine series. This is the Fourier model of Crain *et al.* (1979), a general model that has been shown to give good results for a wide variety of data.

When perpendicular distance has been measured as a continuous variable and not grouped into size classes, the Fourier series estimate $\hat{f}(0)$ is given by:

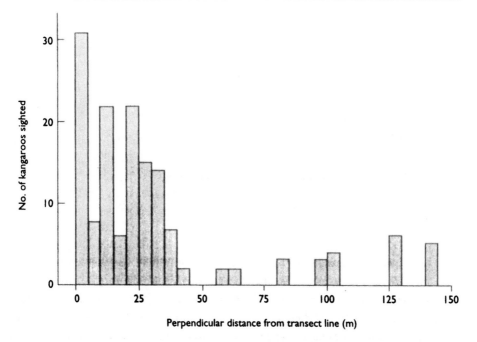

Fig. 9.2 A typical histogram of the decline in the number of observations with perpendicular distance from the transect line (after Coulson & Raines, 1985).

$$\hat{f}(0) = \frac{1}{w^*} + \sum_{k=1}^{m} \hat{a}_k \tag{9.16}$$

where:

$$\hat{a}_k = \frac{2}{nw^*} \left[\sum_{i=1}^{n} \cos\left(\frac{k\pi x_i}{w^*}\right) \right] \tag{9.17}$$

and w^* is the transect half-width, which, without truncation, is the largest perpendicular distance observed, x_i the perpendicular distance of the ith animal, n the number of observations, and m the number of cosine terms determined by the stopping rule. You choose the first value of m such that

$$\frac{1}{w^*} \left(\frac{2}{n+1}\right)^{\frac{1}{2}} \geq |\hat{a}_{m+1}| \tag{9.18}$$

Experience suggests that m should be <7.

Having calculated $\hat{f}(0)$, equation 9.15 is then used to estimate $\hat{D}$. The sampling variance of $\hat{D}$ requires considerable computation. First calculate the variance and covariance of the parameters a_k:

$$\text{var}(\hat{a}_k) = \frac{1}{n-1} \left[\frac{1}{w^*} \left(a_{2k} + \frac{2}{w^*} \right) - a_k^2 \right] \tag{9.19}$$

$$\mathrm{cov}(\hat{a}_k, a_j) = \frac{1}{n-1}\left[\frac{1}{w^*}\left(a_{k+j} + a_{k-j}\right) - a_k a_j\right]$$ (9.20)

Then sum the variance–covariance matrix to give:

$$\mathrm{var}\left(\frac{1}{\hat{a}}\right) = \sum_{j=1}^{m}\sum_{k=1}^{m}\mathrm{cov}(\hat{a}_j, \hat{a}_k)$$ (9.21)

Finally, the estimated variance is calculated using:

$$\mathrm{var}(\hat{D}) = \hat{D}^2\left[\frac{\mathrm{var}(n)}{n} + \frac{\mathrm{var}(1/\hat{a})}{(1/\hat{a})^2}\right]$$ (9.22)

It is normally assumed that the variance of the number of animals counted, $\mathrm{var}(n)$ is a Poisson variable, so that it is equal to n. As this need not be so, Burnham *et al.* (1980) suggest that a sounder approach is to undertake a number of replicate transects, calculate D for each, and then find the variance of these replicates. A sophisticated software package to calculate density from line transect data called DISTANCE or a later version called TRANSECT is available from www.ruwpa.st-and.ac.uk/distancebook/ and a simpler program that includes a variety of methods for plotless density estimation from Pisces Conservation Ltd at www.irchouse.demon.co.uk.

Buckland *et al.* (1993) argue that extreme observations of perpendicular distance should be removed from the data set prior to calculating population density. As a rough rule, they suggest that 5% of the data should be truncated.

The Fourier-series method can also be applied to grouped data. Such data are generated when it is impossible to assign observations to accurately measured distances, so they are allocated to distance bands, such as 0–5, 5–10, 10–15, and 15–20 m. It may also be advantageous to group data when bias in favour of certain distances is suspected. A description of these methods is beyond the scope of the present text, but they are offered in specialized computer programs, such as DISTANCE. If possible, grouping data should be avoided, as it reduces accuracy (Southwell & Weaver, 1993).

From the above Fourier model, it can be seen that the key computational issue is the estimation of $g(x)$ or $f(0)$, and a variety of other functions could have been used to relate the number of detections to perpendicular distance. The approach of Buckland *et al.* (1993) is to estimate $g(x)$ from a small group of models chosen for their robustness, a shape criterion, and computational efficiency. They argue that the formulation of the model can be a two stage-process. First, a key function is selected — two initial candidates are a uniform or a half-normal function, because each requires just zero and one estimated parameter respectively. Secondly, a series expansion is added in order to improve the fit, so that the function is represented as:

$$g(x) = \mathrm{key}(x)[1 + \mathrm{series}(x)]$$ (9.23)

Table 9.1 Key functions and series expansions useful for fitting the detection function for distance sampling (Buckland *et al.*, 1993).

Key function	Series expansion
Uniform, $1/w$	Cosine, $\sum\limits_{i=1}^{m} a_j \cos\left(\dfrac{j\pi y}{w}\right)$
Uniform, $1/w$	Simple polynomial, $\sum\limits_{j=1}^{m} a_j \left(\dfrac{y}{w}\right)^{2j}$
Half-normal, $\exp\left(\dfrac{-y^2}{2\sigma^2}\right)$	Cosine, $\sum\limits_{i=1}^{m} a_j \cos\left(\dfrac{j\pi y}{w}\right)$
Half-normal, $\exp\left(\dfrac{-y^2}{2\sigma^2}\right)$	Hermite polynomial, $\sum\limits_{j=2}^{m} a_j H_{2j}(y_s)$
Hazard-rate, $1 - \exp\left[-\left(\dfrac{y}{\sigma}\right)^{-b}\right]$	Cosine, $\sum\limits_{i=1}^{m} a_j \cos\left(\dfrac{j\pi y}{w}\right)$

Buckland *et al.* (1993) consider the key functions and series expansions listed in Table 9.1 to be useful.

Examples of the application of distance-sampling methods are given in Buckland *et al.* (1993) and recent applications are in Ensign *et al.* (1995) for benthic fish in streams and Kelley (1996) for wood duck.

9.2.4 Point transects

Instead of traversing a transect, the observer may move to a number of fixed points and record the distance, r, to individual animals. These methods are almost only used for bird surveys, where the patchy suitability of the habitat to the birds may make transects inappropriate because they cut across a number of habitat types. Point transects are often easier to undertake because the observer need only estimate distance and markers may be placed in advance to aid the estimation of distance. Population density is given by:

$$\hat{D} = \frac{nh(0)}{2\pi k} \tag{9.24}$$

where n is the number of animals observed, k the number of point transects undertaken, and $h(0)$ the slope of the probability density function of detection distances evaluated at zero distance. In similar fashion to what was described for the line transect method, the central problem is to estimate $h(0)$. This is normally accomplished using programs such as DISTANCE.

For the case of a half-normal detection function, the maximum-likelihood estimator for density has the particularly simple form (Buckland *et al.*, 1993):

$$\hat{D} = \frac{n^2}{2\pi k \sum\limits_{i=1}^{n} r_i^{\,2}}$$
(9.25)

9.3 Spatial distribution and plotless density estimators

In their simplest form, these methods are based on the premise that, if the individuals of a population are randomly distributed, the population density can be estimated either from the distance between individuals (nearest-neighbour methods) or from randomly chosen points to the nearest individual (closest-individual methods). If these 'basic distance' methods are to be applied in their simpler forms, tests of randomness must be carried out. For populations sampled in quadrats, with a mean greater than one, the Poisson index of dispersion (Chapter 2) is a satisfactory test of randomness. Seber (1982) gives other test procedures.

If the density is known from some other method, these methods may themselves be used as a test of randomness (Waloff & Blackith, 1962). Pielou's (1969) index (I_α) is particularly useful for this purpose:

$$I_\alpha = \bar{a}_i^2 m \pi$$
(9.26)

where m = density per unit area calculated from another method and $\bar{a}$ = the mean distance from a random point to the closest individual (see below); then $I_\alpha = 1$ for random, >1 for contagious, and <1 for regular distributions, suggesting competition. Underwood (1976) has used this approach in a study of intertidal molluscs.

Although widely used with vegetation, especially trees, these approaches have not been much used by zoologists. In the main, this is undoubtedly due to the difficulty, and high cost in time, of finding the exact positions of animals. Furthermore, habitat heterogeneity is often fine-grained, so that interpretations of pattern in terms of intraspecific interactions (e.g. by using Pielou's index) may be invalid, more especially if the second- or third-closest neighbour is used (Waloff & Blackith, 1962). They are, however, particularly appropriate for animals living or nesting in fairly uniform, two-dimensional habitats, such as muddy or sandy littoral zones, or on the soil surface of arable crops. The practical difficulties are reduced if the positions of the animals, nests, or other artefacts can be recorded photographically.

Whilst the non-randomness of many populations may preclude using the basic equations for exact population estimation, they do have the potential to provide an order-of-magnitude check (p. 4) on estimates obtained by other methods. Like mark and recapture, they are independent of sample size.

A complete review of plotless density estimators is provided by Engeman *et al.* (1994). They used simulated data to assess the utility and robustness of the various estimators with different spatial patterns and different sample sizes. Their major conclusions are incorporated in the accounts below.

9.3.1 Closest-individual or distance method[1]

A point is selected at random and then, searching round it in concentric rings, the closest individual is found and its distance (X) from the point measured. The process should be continued to find the second-, third-, . . . ith-closest individual. Referred to by Engeman *et al.* (1994) as an 'ordered distance estimator', they recommend using the third as being the most practical, as did Keuls *et al.* (1963), who used it for estimating the density of the snail *Limnaea truncatula*. The estimate of density, $\hat{D}$, is (Morisita, 1957):

$$\hat{D} = \frac{sr - 1}{\pi \sum_{i=1}^{s} X_i} \tag{9.27}$$

where r = the rank of the individual in distance from the randomly selected point, e.g. for the nearest neighbour $r = 1$, for the second-nearest $2 \ldots$; X = the distance between the randomly selected point and the individual; and s = the number of random sampling points. The variance of D is given by Seber (1982) as:

$$\mathrm{var}\left[\hat{D}\right] = \frac{\hat{D}^2}{(sr - 2)} \tag{9.28}$$

A somewhat different approach, the variable-area transect, was developed by Parker (1979) and considered by Engerman *et al.* (1994) to be the simplest and most practical. The search area is a fixed-width strip of width w. Commencing from a random point, the area is searched until the ith individual is found. When i is the number of individuals searched for from each random point and l is the length searched from the random point to the ith individual, then:

$$\hat{D} = \frac{(si - 1)}{w \sum l} \tag{9.29}$$

9.3.2 Nearest-neighbour methods[1]

In these methods, an animal is selected at random and searching is continued until another, the nearest neighbour, is encountered. The distance between the two (r) is measured. As with the closest-individual method, there are variants that use the second- . . . ith-nearest neighbour. Strictly speaking, this is not a random method (Pielou, 1969); a truly random method would involve numbering all the animals in advance and following random numbers for the selection of the first animal. Also, in its simplest form, the method is not robust if the distribution is non-random (Engeman *et al.*, 1994). The basic expression is due to Clark and Evans (1954):

$$\hat{D} = \frac{1}{4\bar{r}^2} \tag{9.30}$$

---

[1] The use of these methods to determine dispersion is discussed in section 2.5.3 (pp. 50–53).

where $\hat{D}$ = density per unit area and $\bar{r}$ = mean distance between nearest neighbours.

A number of other methods have been devised that are more robust if the distribution is non-random. These are based on the separate consideration of each quadrat in the circle surrounding the original animal (angle-order estimators) or on a combination of information on the closest individual and nearest neighbour (methods of M.G. Kendal and P.A.P. Moran); however, the algorithms for their calculation are extremely complex (Engerman *et al.*, 1994) and they have not been much used by zoologists.

References

Allsteadt, J. & Vaughan, C. (1992) Population status of *Caiman crocodilus* (Crocodylia: Alligatoridae) in Cano Negro, Costa Rica. *Brenesia* 38, 57–64.

Bibby, C.J., Burgess, N.D., & Hill, D.A. (1992) *Bird Census Techniques*. Academic Press, San Diego, 257 pp.

Buckland, S.T., Anderson, D.R., Burnham, K.P., & Laake, J.L. (1993) *Distance Sampling: Estimating Abundance of Biological Populations*. Chapman & Hall, London.

Bullock, I.D. & Gomersall, C.H. (1981) The breeding populations of terns in Orkney and Shetland in 1980. *Bird Study* 28, 187–200.

Burnham, K.P. & Anderson, D.R. (1976) Mathematical models for non-parametric inferences from line transect data. *Biometrics* 32, 325–36.

Burnham, K.P., Anderson, D.R., & Laake, J.L. (1980) *Estimation of Density from Line Transect Data*. Wildlife Monograph 72.

Caughly, G. (1974) Bias in aerial survey. *J. Wildlife Management* 38, 921–33.

Clark, P.I. & Evans, F.C. (1954) Distance to nearest neighbor as a measure of spatial relationships in populations. *Ecology* 35, 445–53.

Coulson, G.M. & Raines, J.A. (1985) Methods for small-scale surveys of grey kangaroo populations. *Australian Wildlife Research* 12, 119–25.

Crain, B.R., Burnham, K.P., Anderson, D.R., & Laake, J.L. (1979) Nonparametric estimation of population density for line transect sampling using Fourier series. *Biometr. J.* 21, 731–48.

Dodd, M.G. & Murphy, T.M. (1995) Accuracy and precision of techniques for counting great blue heron nests. *J. Wildl. Manag.* 59(4), 667–73.

Engeman, R.M., Sugihara, R.T., Pank, L.F., & Dusenberry, W.E. (1994) A comparison of plotless density estimators using Monte Carlo simulation. *Ecology* 75(6), 1769–79.

Ensign, W.E., Angermeier, P.L., & Dolloff, C.A. (1995) Use of line transect methods to estimate abundance of benthic stream fishes. *Can. J. Fish. Aquat. Sci.* 52(1), 213–22.

Gates, C.E. (1969) Simulation study of estimators for the line transect sampling method. *Biometrics* 25, 317–28.

Gates, C.E., Marshall, W.H., & Olson, D.P. (1968) Line transect method of estimating grouse population densities. *Biometrics* 24, 135–45.

Harris, M.P. & Lloyd, C.S. (1977) Variations in counts of seabirds from photographs. *Bird Study* 34, 187–90.

Hayes, R.J. & Buckland, S.T. (1983) Radial-distance models for the line-transect method. *Biometrics* 39, 29–42.

Hayne, D.W. (1949) Two methods of estimating populations from trapping records. *J. Mammal.* 30, 399–411.

Kelley, J.R., Jr (1996) Line-transect sampling for estimating breeding wood duck density in forested wetlands. *Wildl. Soc. Bull.* 24(1), 32–6.

Keuls, M., Over, H.I., & De Wit, C.T. (1963) The distance method for estimating densities. *Stat. Neerl.* 17, 71–91.

Kovner, J.L. & Patil, S.A. (1974) Properties of estimators of wildlife population density for the line transect method. *Biometrics* 30, 225–30.

LeResche, R.E. & Rausch, R.A. (1974) Accuracy and precision of aerial moose censusing. *J. Wildl. Manag.* 38, 175–82.

Morisita, M. (1957) A new method for the estimation of density by spacing method applicable to nonrandomly distributed populations. *Physiology and Ecology* 7, 134–44. (In Japanese: English translation from USDA Forest Service, Washington DC, translation number 11116.)

Naugle, D.E., Jenks, J.A., & Kernohan, B.J. (1996) Use of thermal infrared sensing to estimate density of white-tailed deer. *Wildl. Soc. Bull.* 24(1), 37–43.

Nickerson, B. (1963) An experimental study of the

effect of changing light intensity on the activity of adult locusts. *Entomol. Monthly Mag.* **99**, 139–40.

Norton-Griffiths, M. (1978) *Counting Animals.* African Wildlife Leadership Foundation, Nairobi.

Parker, K.R. (1979) Density estimation by variable area transect. *J. Wildl. Manag.* **43**, 484–92.

Pielou, E.C. (1969) *An Introduction to Mathematical Ecology.* Wiley Interscience, New York, London.

Pinder, L. (1996) Marsh deer *Blastocerus dichotomus* population estimate in the Parana River, Brazil. *Biol. Conserv.* **75**(1), 87–91.

Scheepers, C. & Dunn, D.L. (1958) Enumerating populations of adults of the red locust, *Nomadacris septemjasciata* (Serville), in its outbreak areas in East and Central Africa. *Bull. Entomol. Res.* **49**, 273–85.

Seber, G.A.F. (1982) *The Estimation of Animal Abundance and Related Parameters.* Griffin, London.

Sen, A.R., Tourigny, I., & Smith, G.E.J. (1974) On the line transect sampling method. *Biometrics* **30**, 329–40.

Skellam, J.G. (1958) The mathematical foundations underlying the use of line transects in animal ecology. *Biometrics* **14**, 385–400.

Southwell, C. & Weaver, K. (1993) Evaluation of analytical procedures for density estimation from line-transect data: data grouping, data truncation and the unit of analysis. *Wildl. Res.* **20**(4), 433–44.

Starr, R.M., Fox, D.S., & Hixon, M.A., *et al.* (1996) Comparison of submersible-survey and hydroacoustic-survey estimates of fish density on a rocky bank. *US Nat. Mar. Fish. Serv. Fish. Bull.* **94**(1), 113–23.

Symmons, P.M., Dean, G.J.W., & Stortenbere, R.C.W. (1963) The assessment of the size of populations of adults of the red locust, *Nomadacris septemfasciata* (Serville), in an outbreak area. *Bull. Entomol. Res.* **54**, 549–69.

Thomas, J.A. (1983) A quick method for estimating butterfly numbers during surveys. *Biol. Conserv.* **27**, 195–211.

Underwood, A.J. (1976) Nearest neighbour analysis of spatial dispersion of intertidal prosobranch gastropods within two substrata. *Oecologia* **26**, 257–66.

Waloff, N. & Blackith, R.E. (1962) The growth and distribution of the mounds of *Lasius flavus* (Fabricius) (Hym.: Formicidae) in Silwood Park, Berkshire. *J. Anim. Ecol.* **31**, 421–37.

Yapp, W.B. (1956) The theory of line transects. *Bird St.* **3**, 93–104.

10 Observational and Experimental Methods for the Estimation of Natality, Mortality, and Dispersal

In this chapter, we discuss the methods of directly quantifying the various processes that produce population change. The magnitude of these different 'pathways' may also be obtained by the subtraction or integration of census figures in a budget; methods of calculation and of analysis of budgets are discussed in the next chapter, but there is no hard and fast distinction between the contents of the two chapters. As is indicated below, in the appropriate sections, the terms 'natality' and 'dispersal' are used in their widest sense.

10.1 Natality

Natality is the number of births. For an egg-layer, such as birds, most insects, or fish, this is, strictly speaking, the number of living eggs laid; however, from the practical point of view of constructing a population budget, the number of individuals entering any post-ovarian stage, i.e. a larval instar, pupa, fledgling, or adult, can be considered as the 'natality' of that stage.

10.1.1 Fertility

Fertility is the number of viable eggs laid by a female and fecundity is a measure of the total egg production; the latter is often easier to measure. In those animals where all the eggs are mature on emergence, the total potential fecundity may be estimated by examining the ovaries, as Davidson (1956) did with subimagines of a mayfly, *Cloeon*. The ovaries were removed and lightly stained in methylene blue and the eggs were separated by sieving through bolting silk and then counted in a Sedwick–Rafter plankton-counting cell. For fish and other animals that produce many eggs, fecundity is frequently estimated by a gravimetric method (Bagenal & Braum, 1978). A sample of eggs is counted and the total number of eggs estimated from the weight of the ovaries. Some fish are batch spawners, so that the ovary comprises a population of eggs at a number of stages of development. For these species, the fecundity of a single batch can be estimated by only counting the eggs in the largest size group, but total annual fecundity is difficult to estimate, as there is no direct method for estimating the number of batches produced during the spawning season. Many crustaceans, including copepods, prawns, and crabs, and certain spiders and cockroaches carry the eggs externally, where the eggs can be counted, as may eggs in birds' nests. Crustaceans may remove dead eggs from the egg mass; therefore the developmental stage of the eggs should be noted, as the number of eggs tends to decline with age. Other crustaceans, such as mysids, darwinulid ostracods, and cladocerans, carry their eggs and early stages internally, where they can often be counted under a microscope without dissection.

When eggs are matured throughout much of adult life, fecundity can be measured directly by keeping females captive under as natural conditions as possible and recording the total number of eggs laid (e.g. Huffaker & Spitzer, 1950; Spiller, 1964). If viable eggs can be distinguished from non-viable ones—usually by the onset of development—then fertility may be measured (e.g. Fewkes, 1964). Egg cannibalism can occur with many animals, particularly following disturbance; the influence of this behaviour may be calculated from the loss of a proportion of a number of marked eggs (Rich, 1956).

It is often found that fecundity is proportional to weight. For insects, this relationship has been used to estimate female fecundity (e.g. Prebble, 1941; Richards & Waloff, 1954; Waloff & Richards, 1958; Colless & Chellapah, 1960; Lozinsky, 1961; Murdie, 1969; Taylor, 1975). As female weight is directly related to size, measures of size, such as wing length (Gregor, 1960) or pupal length (Miller, 1957) or both (Hard, 1976), may be substituted for weight. Once the relation between size and fecundity has been established (regression analysis is a convenient method), it may be applied to estimate the natality in field populations by measuring wild females or female pupae—the length of the empty pupal cases is a particularly suitable criterion (Miller, 1957). The same approach can be used for fish and crustaceans. The size (age)–fecundity relationships is frequently required for fisheries population models. Following a comparison of a range of possible relationships, Zivkov and Petrova (1993) considered a logistic curve to be the best model of the length–fecundity relationship over the full adult size range, although other models were equally appropriate over reduced size ranges.

Within a given population of insects, the rate of oviposition will be influenced by temperature, and the extent to which the potential fecundity is realized is influenced by the longevity of the females. The influence of temperature can be studied in the laboratory and incorporated into the regression equation; the observations of Richards and Waloff (1954) show that it may be justified to apply data from the laboratory to populations in the field. Information on longevity may be obtained from wild populations by marking and recapture. An equation incorporating all these variables will be in the form (Richards & Waloff, 1954):

No. of eggs = regression coeff. × weight ± reg. coeff. × temp.
          + reg. coeff. × longevity + constant

Such equations cannot, however, be applied outside the populations for which they were derived. In fish, for example, there is often considerable zoogeographical variation in size-specific fecundity, as shown by Henderson (1998) for the dab, *Limanda*

limanda, in the north-east Atlantic. Further, females that have developed under different regimes of nutrition and crowding will have different potential fecundities (Blais, 1953; Clark, 1963). Therefore, for example, it is necessary to have different equations for different stages in an insect-pest outbreak (Miller, 1957). Larval density, of course, affects size and longevity directly (Miller & Thomas, 1958), but the fact that new equations are necessary shows that the effects on fecundity are not just reflections of changes in these properties. Evidence is accumulating that changes may occur in the genetic constitution of fluctuating populations and the fecundity of different forms of a species may be quite different. Therefore, laboratory measurements of fertility should be continually checked, for it is important in any population study to establish not only the potential fertility, but any variation in it, as this may be the essence of a population regulation mechanism (p. 426).

When large numbers of eggs are laid together in a group, the weight of the individual female will fall sharply after each oviposition and slowly rise again until the next group is deposited. Where individual animals can be marked, it may be possible, by frequently recapturing and weighing, to establish the time of oviposition of actual egg batches; Richards and Waloff (1954) were able to do this with a field population of grasshoppers. It is sometimes possible to mark some insects with a radionuclide (e.g. zinc-65 (^{65}Zn)) and then estimate egg production by measuring the loss of isotope (Mason & McGraw, 1973). The actual number of eggs per egg mass may vary with size or age of the female (Richards & Waloff, 1954) or from generation to generation (Iwao, 1956).

10.1.2 Numbers entering a stage

When an animal can be trapped as it passes from one part of its habitat to another part at a specific stage of its life history, the summation of the catches provides a convenient measure of the total population entering the next stage. Such traps are commonly referred to in entomological research as emergence traps. The particular design will depend on the insect and its habitat; a number are described by Peterson (1934), Nicholls (1963), and Ives *et al.* (1968). Basically, the traps consist of a metal or cloth box, which covers a known area of soil, and glass collecting vials with 'lobster-pot' type baffles are inserted in some of the corners (Fig. 10.1). Being positively phototactic, a newly emerged insect will make its way into the collection vials, which are emptied regularly; Turnock (1957) used an adhesive resin in the collecting containers of his trap, so that they could be emptied at less frequent intervals. Releasing a known number of newly emerged individuals into it can test the efficiency of a trap for a particular insect. The construction of the trap will influence its effect on the microclimate; all traps tend to reduce the daily temperature fluctuations (Fig. 10.2) and the deeper they are (the greater the insulating layer of air), the smaller the fluctuations for the same covering. Cloth-covered traps lose the heat less quickly than metal ones, and wood and screen traps do not heat the soil to the same degree as metal ones (Higley & Pedigo, 1985). Although the daytime deficit may be approximately balanced by the greater warmth at night (Fig. 10.2), over a

Fig. 10.1 Emergence traps for terrestrial habitats. (a) and (b) Metal box and tub, respectively. (c) Cloth-covered, Calnaido type. (d) Celluloid baffle of collecting tube. (e) Sectional view of collecting tube with pair of baffles designed to separate, partially at least, large and small insects. (f) Cloth trap for twigs. (a, b, and c from Southwood & Siddorn, 1965; f from Glen, 1976.)

period of several weeks these small daily excesses or deficits will accumulate to levels where they might influence the development rate of pupae (Southwood & Siddorn, 1965). For this reason, caution should be exercised in using emergence-trap data for phenology. Rice and Reynolds (1971) collected twice as many pink bollworms (*Platyedra*) from metal-screened emergence cages as from plastic-screened ones. Special traps have been described by many workers, including one for fleas (Bates, 1962) and others for ceratopogonid midges (Campbell & Pelham-Clinton, 1960; Neilson, 1963; Davies, 1966; Braverman, 1970), for anthomyiid flies (van Dinther, 1953), for cecidomyiid midges (Speyer & Waede, 1956; Nijveldt, 1959; Guennelon & Audemard, 1963), for beetles (Richards & Waloff, 1961; Polles & Payne, 1972;

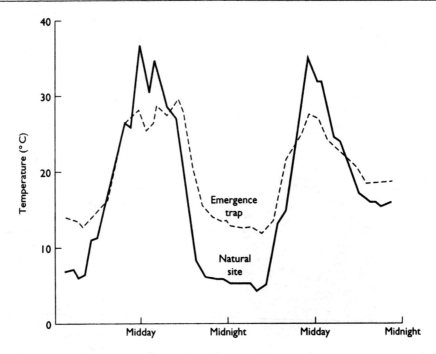

Fig. 10.2 The soil surface temperature over 48 h in a Calnaido cloth-covered emergence trap compared with that from a natural site (after Southwood & Siddorn, 1965).

Boethel *et al.*, 1976), and for lepidopterous larvae emerging from maize 'ears' (Straub *et al.*, 1973).

A number of special traps have been designed for bark-dwelling insects. Moeck (1988) describes a trap constructed from a plastic pail and funnel that was found to be 97.5% efficient at collecting insects from lodgepole pine blots. For bark-beetles and other insects occurring at high densities, it may even be useful to subdivide the trap to record the dispersion pattern (Reid, 1963). Simpler traps have been developed by Nord and Lewis (1970) and Glen (1976), essentially consisting of a wire and wood frame, covered with black cloth (Fig. 10.1f); Glen found that mean temperature was hardly affected.

Emergence traps (Fig. 10.3) have been used extensively in aquatic environments for measuring the quality, quantity, and biomass of insects emerging from the habitat and have been reviewed by Davies (1984). Recent examples of their application are in Dewalt *et al.* (1994), Boettger & Rudow (1995), Speth (1995), and Aagaard *et al.* (1997). In shallow water, amongst emergent vegetation, and in sheltered situations, floating box traps have been utilized (Adamstone & Harkness, 1923; Macan, 1949; Vallentyne, 1952; Sommerman *et al.*, 1955; Judge, 1957; Morgan & Waddell, 1961). In exposed situations, submerged funnel traps have been used, as these avoid damage from wind and rain (Grandilewskaja-Decksbach, 1935; Brundin, 1949; Jónasson, 1954; Palmén, 1955, 1962; Mundie, 1956; Darby, 1962; Mulla *et al.*, 1974; McCauley, 1976). A careful comparison of these two types of trap

(a)

(b)

(c)

(d)

Fig. 10.3 Emergence traps for aquatic habitats. (a) Floating box trap for use under exposed situations; the screen may be omitted in sheltered situations (after Morgan et al., 1963). (b) Submerged funnel trap, with details of collecting jar (c) (after Borutsky, 1955). (d) Submerged trap for chironomids (after Welch et al., 1988).

for emerging insects have been made by Morgan *et al.* (1963) and Kimerle and Anderson (1967), from which the following conclusions may be drawn. Funnel traps are generally much less efficient than floating box traps. A number of factors probably contribute to this; one is the decomposition of the catch in the small airspace of the collecting jar of the funnel trap. Such losses will be proportionally greater the denser the population of emerging insects and the longer the intervals between emptying the trap, which should therefore be done daily. The ascending pupae and larvae are strongly phototactic and the slight shade produced by the gauze of the funnel trap will cause it to be avoided; furthermore, some insects, e.g. Odonata, will crawl out again. It is impossible to exclude small predatory insects from the traps and they may consume much of the catch. Lastly, the effective trapping area is reduced if the trap becomes tipped. However, in very exposed situations, these traps must be used, and Morgan *et al.* (1963) recommend a design, similar to that in Fig. 10.3a, to overcome as many of these faults as possible. Borutsky (1955) originally devised the collecting jar; it is very important that the Perspex be kept clean, as algal growth will soon render it opaque. Figure 10.3d shows the preserving-funnel trap described by Welch *et al.* (1988), which is a modification of the basic design of Davies (1984). The sampling efficiency of these traps was tested with and without preservative, with transparent and opaque collecting tubes, and with and without the funnel, which stops the insects falling back into the water. Traps holding preservative (normally phenoxyethanol, but diesel fuel, salt water with formalin, and Mistovan were also tried) caught as many chironomids as those without. However these chironomids were rarely in the adult form, as they died before eclosion. It was found that the preservatives were unable to stop decay in traps left in place for 12 or 18 days. Contrary to the general finding that opaque traps catch fewer animals (Davies, 1984), no significant difference was found by Welch *et al.* (1988), although they did find that most chironomids remained as pupae in the opaque traps. Funnels are clearly important, as those without them achieved only 17% of the pupal and 58% of the adult catch of chironomids obtained by those with funnels. Further information on the choice of aquatic emergence trap and their relative efficiencies are given in Morgan (1971), Mundie (1971), LeSage and Harrison (1979), and Davies (1984).

The most important causes of loss of efficiency of floating box traps are waves, which will swamp the catch, and the shading effect of the trap itself on the ascending larvae and pupae. Morgan *et al.* (1963) point out that, as it is probably only those animals that ascend near the edges of the trap that can take successful avoiding reaction, the larger the area of the trap the smaller this edge effect relative to the size of the catch. The trap designed by Morgan *et al.* (Fig. 10.3a) is constructed mainly from clear Perspex; the apron projecting into the water retains floating exuviae within the trap and helps to reduce wave damage; the latter function is also served by the lateral screens. The wooden frame should be as narrow as possible and painted white to minimize shadow; wire netting stretched under the frame reduces predation by fish. The trap needs to be emptied every 2 days and this is a difficult process.

A floating tray is inserted under the trap and the whole towed to a boat-house or similar site, where the unidirectional light source ensures that the edge on the darkest side may be lifted to allow the entrance of an entomological pooter (= aspirator (see Fig. 7.6)) without any of the insects escaping. On dull days or at night, the trap may be tilted towards a lamp during emptying. In view of the time-consuming nature of this operation, it would seem worthwhile to determine whether shrouding the traps for a short while in the day would drive a large propor-tion of the catch into a collecting tube inserted in the roof, which is the method of collection from tents (p. 163).

Kimerle and Anderson (1967) had a sliding base in their trap, which greatly facilitated the retrieval of the whole catch, and Corbet (1966) arranged for the traps to be closed and then emptied in the laboratory. Carlson (1971) increased catch size (and difficulties of interpretation regarding the catchment area) by incorporating a battery-operated light.

Langford and Daffern (1975) designed traps with floats for work in large, fast-flowing rivers; they found these robust, but there was some evidence that the numbers of Trichoptera, but not Ephemeroptera, caught were affected by the presence of floats. Wrubleski and Rosenberg (1984) noted that polystyrene floats, as fitted to the 'week'-model tent trap of LeSage and Harrison (1979), tended to become colonized by chironomids; a polyvinyl chloride (PVC) pipe, which is easily cleaned, would be preferable.

In very shallow, fast-running streams, it may be necessary to use a tent trap (Ide, 1940; Anderson & Wold, 1972); basically, it resembles a gauze tent that is attached to the substrate. A rather unsatisfactory feature of the design is that the observer has to enter the trap to remove the insects; not only, as Ide mentions, is he exposed to noxious insects (e.g. Simuliidae), but the resultant trampling of the substrate must affect future emergence. Possibly a smaller tent should be used—one that can be emptied from the exterior.

In flowing waters, an emergence trap receives input from the insects which were living within the substrate below the trap, plus drifting animals brought in from upstream. To quantify the drifting component, Steffan (1997) describes an emergence trap fitted with an underwater sieve-plate, so that any drifting fauna enters the trap while those of benthic origin are excluded.

Using a trap similar in appearance to the submerged-funnel type, Lindeberg (1958) sampled chironomids from very shallow rock pools. The funnel was made of glass and the whole upper part of the trap was airtight, so that it could be kept full of water even though it projected well above the level of water in the pool. Cook and Horn (1968) developed a sturdy trap for damselflies and similar insects emerging from rushes in shallow water in areas frequented by cattle. It consisted of fencing wire, covered with fibreglass window screening.

Material containing the resting stage of the animal may be collected and exposed under field conditions or in the laboratory. Bark-beetle emergence from logs may be measured under natural conditions by placing the logs in screened cages (McMullen

Fig. 10.4 (a) Emergence tin to determine in the laboratory the emergence of insects from samples (after van Emden, 1962). (b) Funnel trap for the collection of descending arboreal larvae.

& Atkins, 1959; Clark & Osgood, 1964) and emergence of flies from heads of grass and cereals determined by placing these in muslin bags in the field (Southwood *et al.*, 1961). The emergence of mites from overwintered eggs on twigs or bark was measured by Morgan and Anderson (1958) by attaching the substrate to the centre of a white card and surrounding it with a circle of a fruit-tree banding resin. As the mites crawled out they were caught in the resin and, being red, were easily counted.

Unseasonable emergence may be forced in the laboratory (Wilbur & Fritz, 1939; Terrell, 1959; van Emden, 1962; McKnight, 1969; Gruber & Prieto, 1976). Commercial containers, such as shoeboxes, ice-cream cartons, and biscuit tins, have been found useful in such studies (Fig. 10.4a).

Insects that are arboreal for part of their life may often be trapped on their upward or downward journeys. Wingless female moths and others moving up tree trunks may be trapped in sackcloth and other trap bands (DeBach, 1949; Reiff, 1955; Otvos, 1974) or in inverted funnel traps (Varley & Gradwell, 1963; Ives *et al.*, 1968; Agassiz, 1977). Larvae descending to pupate may be caught in funnel traps (Fig. 10.4b) (Ohnesorge, 1957; Pilon *et al.*, 1964).

Emergence traps of the types used for aquatic insects can also be placed over gravel spawning substrate to quantify trout or salmon fry emergence (e.g. Weaver & Fraley, 1993). An assessment of the vertical movement of benthic marine copepods upwards from sea-grass beds and their subsequent resettlement has been made by Bell *et al.* (1989) and Kurdziel and Bell (1992), using emergence-type traps.

Freshwater animals that migrate to different habitats at set developmental stages, such as salmon, *Salmo salar*, can be counted by placing automatic counters in weirs (see p. 263).

10.1.3 The birth-rate from mark-and-recapture data

The details of marking animals and of the equations for estimating the birth-rate are given in Chapter 3. This birth-rate will include any immigration; it is thus more correctly referred to as the dilution rate.

10.2 Mortality

10.2.1 Total

Measurements of total mortality are also commonly obtained by the subtraction of population estimates for successive stages (e.g. Miller, 1958; Cook & Kettlewell, 1960; see Chapter 11).

10.2.1.1 Successive observations on the same cohort

When it is possible to make these, mortality may be measured directly. Completely natural cohorts can, in general, only be followed in sedentary or relatively immobile animals (e.g. the egg and pupal stages of insects, scale insects, some aphids, birds' eggs, benthic worms, barnacles, and other attached epibenthic animals). Colonies present in the field in accessible positions may be repeatedly examined, each colony being identified by labelling (e.g. MacLellan, 1962; van Emden, 1963), and, with an animal as large as the hornworm (Sphingidae), Lawson (1959) was able to follow the larvae, as well as the eggs. In other situations, it may be necessary to delimit the sample population or even to 'plant' it. Examples of the former are Ives and Prentice's (1959) and Turnock and Ives's (1962) studies, where they placed pupa-free moss-filled trays or blocks of peat with screen bottoms and sides under the host trees of certain sawflies at pupation. Later in the season, these natural traps may be removed and the pupae classified. More mobile animals will need to be retained by a more extensive enclosure (Vlijm *et al.*, 1968; Ashby, 1974) or 'field cage' (Dobson *et al.*, 1958; see p. 25). It is important to remember, however, that the screens and cages are bound to alter, albeit only slightly, the situation one is trying to assess. The planting of known populations in the field and their subsequent re-examination gives a measure of the level of mortality. Sometimes, it is possible to recover the remains of the dead individuals in such experiments, and then further information may be gained about the cause of death (Graham, 1928; Morris, 1949; Buckner, 1959; Pavlov, 1961).

When an insect leaves a mark of its presence, as many that bore into plant tissue do, a single count at, for example, the pupal stage enables the still living and the dead to be distinguished (see also p. 325). Furthermore, the latter may be divided into those whose burrows have been opened by predators and those killed by parasites and disease. Such a study, which of course is, in effect, population estimates of two successive stages (i.e. total young larvae = total burrows, and pupae), was made by Gibb (1958) for the eucosmid moth, *Ernarmonia conicolana*, in pine cones.

10.2.1.2 The recovery of dead or unhealthy individuals

This gives another measure of mortality. Although Gary (1960) has devised a trap for collecting dead and unhealthy honey-bees, it is unusual to be able to recover the non-survivors of a mobile stage unless they are marked with a tracking device (p. 127). However, unhatched eggs can often be examined to ascertain the cause of their death (e.g. Bess, 1961; Way & Banks, 1964).

10.2.2 The death-rate from mark-and-recapture data

Details of these models and the calculation of the death-rate, which includes any emigration, are given in Chapter 3 (see also p. 385).

10.2.3 Climatic factors

Apart from direct observations on a known cohort, the main method of establishing the role of climate in the total mortality has been experimental. A known number of individuals may be exposed to field conditions (e.g. Lejeune *et al.*, 1955) or predictions may be made from laboratory experiments (MacPhee, 1961, 1964; Green, 1962; Sullivan & Green, 1964; Kensler, 1967; Carter, 1972; Abdel Rahman, 1974; Neuenschwander, 1975); in the latter case, it is important to allow for the effect of acclimatization. Cold-hardiness may be influenced by the presence of free water in the surroundings and of food in the gut (Eguagie, 1974).

10.2.4 Biotic factors

As with climatic factors, knowledge of the role of various biotic ones is often obtained from successive observations on the same cohort (see above); this section is restricted to techniques for the recognition of the role of individual factors. It is important to remember that the effect of a parasite will vary with the host, as has been demonstrated by Loan and Holdaway (1961) with the braconid parasite of a weevil, which has no effect on the adult male but causes the egg production of the female to fall off quickly. The quantitative evaluation of natural enemies has been reviewed by Kiritani and Dempster (1973) and Sunderland (1988); there are four methods of direct assessment, together with the indirect approach by laboratory experiment.

10.2.4.1 Direct observation of predation

Sometimes it is possible to estimate predation on the basis of sightings of predators consuming prey (e.g. Kiritani *et al.*, 1972; Kiritani & Dempster, 1973). The observed frequency of feeding (F) is determined by routine counts, together with the length of time captured prey are retained (R) and the diurnal rhythm of feeding to give the proportion of feeding (C) that occurs during the time interval when the observations are made. Now, if the probability of observing feeding (P_f) is retention time in hours (R) divided by 24, then an estimate of the total number taken is:

$$n = \frac{FC}{P_f}$$ (10.1)

A series of values of n may be plotted against time and the area under the graph gives the total prey killed in that habitat. This 'sight-count' method was devised and used by Kiritani et al. (1972) in studying spider predation of leafhoppers. It depends on a high accuracy in observing all instances of predation at a given time and on the values of C and R being fairly constant.

As the young of many birds remain in the nest, it is possible to record their food. The simplest method is to observe feeding through the glass side of a nesting box (Tinbergen, 1960). However, identification and counting of the prey are often difficult, and Promptow and Lukina (1938) and Betts (1954, 1958) have shown that parent birds may be induced to put the whole, or at least part, of the food for the young into an artificial gape, from which it may be removed. Photography and video can also be used to record the parental visits, when it is often possible to identify the food carried or to identify nest predators (Major & Gowing, 1994).

When a single parasite larva emerges from each individual of an arboreal host and drops to the ground to pupate, the larvae may be collected in trays or cone traps (see p. 320) to give an absolute measure of their own population, which is equivalent to the number of hosts they have killed. Such conditions are fulfilled, for example, in many tachinid parasites of Lepidoptera. When the percentage parasitism of the host is known accurately, such figures may also be used to calculate the actual host population (Dowden et al., 1953; Bean, 1958).

10.2.4.2 Examination of the prey (host)

The examination of vertebrates for ectoparasites has been described on p. 167. However, a description of methods for the detection of internal vertebrate parasites is beyond the scope of this book. These methods include dissection, histological examination of tissues, forced removal from the gut, and examination of faeces.

Parasitic insects may often be detected in their hosts by dissection (e.g. Miller, 1955; Hafez, 1961; Evenhuis, 1962) or by breeding out the parasites from a sample of hosts (e.g. Richards, 1940; Sasaba & Kiritani, 1972). As the rate of parasitism varies throughout a generation of the host, a single assessment will not give an adequate degree of precision unless the hosts are closely synchronized. Parasites may sometimes be detected within their hosts without dissection by the use of soft X-rays (wavelength >0.25 Å) and fine-grain film (Holling, 1958). It is possible that elutriation (p. 226) could also be used to diagnose parasitism. The proportion of hosts attacked by a parasite is referred to as the 'apparent parasitism'. The problems involved in the combination of a series of apparent mortalities and its interpretation are discussed in the next chapter.

Sometimes, it is possible to find a good proportion of the corpses of insects killed by a predator, generally when the predator collects the food together in one place, e.g. spiders' webs (Turnbull, 1960), hunting wasps' nests (Rau & Rau, 1916; Richards & Hamm, 1939; Evans & Yoshimoto, 1962), thrushes' anvils (Goodhart, 1958), shrikes' larders (Lefranc & Worfolk, 1997). Ants provide unusual opportunities for this type of assessment, because in many species the foragers return to the nest with virtually intact prey, along trails; special apparatus has been designed to sample returning foragers and their prey (Chauvin, 1966; Finnegan, 1969). The frequency of beak marks on butterflies, arising from unsuccessful attacks by birds, has often been recorded; its interpretation is difficult, but it may be used as evidence for seasonal variation in predation intensity (Shapiro, 1974). As mentioned above, the food brought by their parents to nestling birds may be recorded photographically and subsequently identified to broad categories.

A semi-natural assessment of the role of predators may be made by 'planting' a known number of prey in natural situations; Buckner (1958) describes how sawfly cocoons were exposed in this way and the type of predator determined by the markings left on the opened cocoons.

The diagnosis and determination of the extent of infection by pathogens in an insect population are a complex and specialized problem outside the scope of this book; reference should be made to works such as Steinhaus (1963), Wittig (1963), Smirnoff (1967), Weiser (1969), Weiser and Briggs (1971), Thomas (1974), Tanada and Kaya (1993) and Lacey (1997). Good examples of the application of these methods to insect populations in the field are provided by Tanada (1961), Neilson (1963), Neilson and Morris (1964), Young (1974), and Entwistle (1976). It should be noted that some methods of collection (e.g. sweep-net) may lead to an overestimate of the proportion of diseased insects in the population (Newman & Carner, 1975).

10.2.4.3 Examination of the predator
Analysis of faeces and pellets. For some, particularly vertebrate, predators, the faeces can be examined for the remains of prey. If the number of prey individuals can be counted, then, given the gut evacuation rate, the number of prey consumed can be estimated. Unfortunately, for many poikilotherms, the evacuation rate is affected by the size of the meal (Sunderland, 1988), although this may not be the case with fish (Elliott & Persson, 1978). For specialist predators, quantification may even be possible, if assimilation data are available, by using the weight of faeces produced to estimate the number of prey consumed. Laboratory studies, such as that of Hewitt and Robbins (1996) on the grizzly bear, *Ursus arctos*, need to be undertaken to relate the volume of food eaten to the faecal volume produced, if the relative importance of different foods is to be estimated from their presence and quantity in the faeces. Captive studies were also undertaken by Cottrell *et al.* (1996) on harbour seals, *Phoca vitulina*, to assess the survival in recognizable form of fish remains, such as eye lenses, scales, vertebrae, and otoliths, which suggested that no one single structure, such as an otolith, should be used. Generally, care must be taken, because the species

list from stomach contents (or pellets) and the actual food eaten will often not correspond (Hartley, 1948; Custer & Pitelka, 1975). Cavallini and Volpi (1995) compared gut and faecal-content analysis for the red fox, *Vulpes vulpes*, and concluded that there were significant differences, with birds in particular declining in dominance from stomach to faeces. Many predation studies have used faecal analysis, some examples being: Cavallini and Serafini (1995), the small Indian mongoose *Herpestes auropunctatus*; Kohira and Rexstad (1997), wolves, *Canis lupus*; Le Jacques & Lode (1994), European genets, *Genetta genetta*; Mukherjee *et al.* (1994), the Asian lion, *Panthera leo persica*; and Santori *et al.* (1995), the marsupials *Metachirus nudicaudatus* and *Didelphis aurita*.

Some predators regurgitate pellets, comprising the indigestible parts of their meals, from which their prey can be identified. For example, the remains of insects and other prey are found in the pellets of owls (Hartley, 1948; Miles, 1952; Southern, 1954; Jedrzejewski *et al.*, 1996), cormorants, *Phalacrocorax auritus* (Johnson *et al.*, 1997), brown skua, *Catharacta skua* (Moncorps *et al.*, 1998), and white stork, *Ciconia ciconia* (Barbraud & Barbraud, 1997). Jedrzejewski *et al.* (1996) quantified the numbers of prey of each species taken from 1 ha of woodland by collecting the pellets under roost trees and identifying the prey eaten and their number in each pellet from skeletal remains and the presence of hair and feathers. This quantification required estimates of owl density, a coefficient to correct for the fact that some of the prey disappears during digestion, and the number of pellets produced per day. When both pellets and faeces are available for analysis, the best results are likely to come from pellet analysis, as the remains are more easily identified. Johnson and Ross (1996) examined fish remains in the pellets and faeces of the double-crested cormorant (*P. auritus*) and found that about 90% of all the diagnostic fish remains found were from pellets.

Gut-content analysis. One of the most direct and frequently used methods for the identification of predator–prey linkages is the examination of gut contents. For small animals, this normally requires the animals to be killed, so that the gut can be dissected. It is generally undesirable in a study (quite apart from conservational and legal considerations) to kill large numbers of vertebrates, although this is considered acceptable in fisheries studies. For certain larger animals, such as fish or cayman, it is often possible to wash out the stomach contents without causing long-term harm to the animal. Similarly, the young of some birds will regurgitate the food if their necks are manipulated shortly after feeding (Errington, 1932; Lack & Owen, 1955); with others, a neck ring needs to be used to prevent swallowing. Emetics may sometimes be administered to recover stomach contents (Radke & Frydendall, 1974).

Even when the predator eats large prey, macroscopic examination alone may be insufficient to identify all the prey consumed: Douglas (1992) found that microscopic examination of the gut contents of the grass snake, *Psammophylax rhombeatus*, improved both the detection and the identification of both mammalian and lizard remains. The ability to detect the remains of different animals in invertebrates often requires skill and experience, because the meal is often broken into small pieces. However, given experience, the gut contents of even microscopic animals, such as copepods and ostracods, can be identified. By measuring the size of remains in the gut, it may be possible to infer the size range of the prey. For example, using samples taken from the guts of large predatory fish, Smale (1986) identified and estimated the size of fish and squid from otoliths and 'beaks', respectively.

While many predatory insects are fluid feeders, skeletal remains can be found and identified from the guts of Dermaptera, coccinellids, and carabids (Sunderland, 1988). In general, more refined methods of detection are needed for insects and these are discussed below.

Differential digestion is one of the problems in interpreting data from stomach and faecal contents (Buckner, 1966).

Chromatography. This is generally of little application, but Knutsen and Vogt (1985) used gas chromatography to identify dietary components in the guts of the lobster, *Homarus gammarus.*

Electrophoresis. Electrophoresis can be used to separate prey enzymes, such as esterases, to produce banding patterns that are species-specific. The different enzymes (and other proteins) are separated because they migrate through a gel at different rates when an electrical field is applied. Specific stains detect the individual enzymes. For the method to be applied, reference gels must be constructed for potential prey, starved predators, and predator guts holding known prey. Because of their diversity and high activity and thus ease of staining, esterase banding patterns are frequently used, although other enzymes are also potentially available. Giller (1986), for example, used esterase banding to detect the prey of notonectids. Sunderland (1988) suggests that this method is best used for predator–prey systems with few species, as this enhances the likelihood that banding patterns for the various species will be unique. The method suffers from numerous pitfalls, perhaps one of the most important being that, if not all the prey is consumed, then the prey banding pattern from the gut may not contain all the bands seen on the reference gels. The method may also be unable to distinguish between some species. Diallo *et al.* (1997) describes a technique based on the electrophoresis of superoxide dismutase to distinguish between tsetse blood meals of human and non-human origin.

Serological methods. Originally developed by vertebrate immunologists, serological techniques have been used extensively in entomology for gut analysis of species that suck fluids, particularly for the identification of blood meals of biting flies. Sunderland (1988) reviews the use of these techniques for the quantification of predation. Relevant studies include those of Brooke and Proske (1946), Dempster *et al.* (1959), Loughton *et al.* (1963), Dempster (1964), Rothschild (1966, 1970, 1971), Frank (1967), Davies (1969), Pickavance (1970, 1971), Service (1973, 1976), Healey and Cross (1975), and Vickermann and Sunderland (1975). For the population ecologist, the value of serological tests may be limited by the fact that, unless certain assumptions are made, they cannot be quantified. The other important feature is that the specificity of the antisera must be well known so that the correct interpretation of the results may be made; this requires knowledge of the species present in the study area.

Before describing the range of techniques that have been applied, we first summarize the steps in the preparation of the antisera, the methodology for which has been reviewed by Davies (1969), Boreham and Ohiagu (1978), and Washino and Tempelis (1983).

1 Preparation of antigen. Numbers of prey are starved to empty the gut, killed with cyanide or by freezing, and, with large arthropods, the legs, wings, etc. removed. They are then crushed in a pestle and mortar with buffered normal saline (0.85% sodium chloride (NaCl), pH 7.0) and maintained at 4°C overnight, with constant stirring. Particulate matter is removed by centrifugation and a second extraction may be carried out. It is preferable to extract insects with a large volume of buffered saline and concentrate the solution later in order to obtain the maximum amount of antigenic material. The pooled extracts are freeze-dried, reconstituted in a small volume of buffered saline, and dialysed for 24h at 4°C against buffered saline filtered through a Seitz EKS pad. The antigen may be stored at −20°C or freeze-dried in small aliquots. Working with the beetle *Phytodecta olivacea*, Dempster (1960) found it necessary to add M/1000 potassium cyanide to prevent the deposition of melanin.

2 Production of antiserum. The important feature of serological tests to identify prey–predator interactions is the preparation of high-titre specific antisera. A large number of different regimens have been used to prepare antisera. The two most suitable methods are injection of alum-precipitated antigen intramuscularly into rabbits (Weitz, 1952) and injection of an emulsion of antigen and Freund's complete adjuvant into the lymph nodes of rabbits (Boreham & Gill, 1973). In the former method, 0.1 ml of 10% potash alum is added to 2.5 ml of antigen and, after adjusting the pH to 6.8 with 1 N sodium hydroxide, washing the precipitate three times with buffered saline; the resulting suspension is injected intramuscularly into the hind limbs of rabbits. In the second method, equal parts of Freund's complete adjuvant and the antigen are mixed to form an emulsion by drawing up the mixture into a syringe and expelling it through a fine-gauge needle. The emulsion is divided into four aliquots and injected into the two axillary and two inguinal lymph nodes of rabbits. With both methods, a second injection is given 7–10 days after the first and, approximately 10 days later, a blood sample is taken to test the antibody titre. If the titre is satisfactory (the antiserum reacts with a dilution of the antigen at a dilution of 1 in 5000 or more), 50 ml of blood can be taken from the rabbit and the serum separated. If the titre is not satisfactory, a third or even fourth injection may be given to boost the titre. The advantage of the lymph-node method is that the amount of antigen required is only about

10% of that for the alum-precipitate method. In addition, the antisera tend to be more specific.

3 Collection of predator meals. The predator may be killed with cyanide and either the whole animal (small species) or the gut (larger ones) is smeared on to Whatman no. 1 filter-paper, labelled, dried, and stored in a desiccator, or the whole predator may be stored at below −20°C.

4 Predator testing. The filter-paper containing a predator is soaked in a small volume of saline (0.2–0.5 ml, depending on the size) overnight, or the whole predator is crushed in a similar volume of saline. After centrifugation, the extract is ready for testing. The basic test is to allowing the antigen and antibody to interact and the four main approaches are precipitation methods, agglutination methods, complement fixation, and immunoassays.

Precipitation methods. There are a number of methodologies.

1 The tube precipitin test. The test antigen and antiserum are mixed in a tube and the amount of precipitate formed can be measured turbidimetrically (Buffone, 1980). For a constant concentration of antiserum, the curve relating the amount of precipitation to the amount of antigen present rises to a maximum and then falls. Thus, for quantitative studies, series dilutions of the antigen are required to determine to which side of the maximum the sample belongs.

2 The capillary ring test (Weitz, 1956; Dempster, 1960; Boreham, 1975). The predator extract is overlaid on the antiserum in a glass capillary tube and a positive result is obtained by the formation of a visible ring of precipitate at the interface of the two liquids. The main advantages are that it is adaptable to large-scale use and gives a rapid result (2 h). However, both liquids must be completely clear and this can present difficulties. While principally used to detect mosquito blood meals (Washino & Tempelis, 1983), it has also been used to identify the predators of Lepidoptera (Gardner *et al.*, 1981) and gall-wasps (Nakamura & Nakamura, 1977) and the prey of beetles (Denison & Hodkinson, 1983).

3 Single radial diffusion. Both antigen and antiserum are placed in the same well in a gel. The antigen diffuses outwards and a ring of precipitate is formed around the well. The diameter of this ring is proportional to the amount of antigen present. While used in the laboratory, it has not been used in field studies (Sunderland, 1988).

4 Double diffusion (Ouchterlony, 1948; Pickavance, 1970). Antigen and antisera are allowed to diffuse towards each other through a gel (e.g. 0.6% agar) and a line of precipitation is formed where the reactants meet at equivalent proportions. This technique has the advantage that a single predator can be tested against several antisera simultaneously and that cloudy predator extracts can be used. It may be further improved by staining with a protein stain, such as thiazine red (Pickavance, 1970). Davies (1969) has used a modification of this basic test. This method has been used to detect the prey of carabids (Lesiewicz *et al.*, 1982).

5 Immunoelectrophoresis (also termed immunoelectro-osmophoresis and electroimmunodiffusion). This is a class of methods basically similar to the Ouchterlony technique, except that either one reactant is initially separated into its constituent parts or the two reactants are brought together rapidly by electrophoresis. The standard technique is to electrophorese antigens along an agar film, remove a longitudinal strip of the agar, fill the resulting well with antiserum, and wait until precipitin arcs are formed. This technique can be used to distinguish between species of prey that cross-react (Hance & Rossignol, 1983). Rocket immunoelectrophoresis has been used to detect the protein of sand eel, *Ammodytes marinus*, in the faeces of the common cormorant, *P. vitulina*, and grey seals, *Halichoerus grypus* (Pierce *et al.*, 1990). The most sensitive and widely used electrophoretic technique is crossover immunoelectrophoresis. If electrophoresis is undertaken in a medium at pH 8.2, most antibodies migrate to the cathode and most proteins to the anode. Therefore, the antiserum is placed near the anode and the antigen near the cathode and the two are pulled towards each other until a reaction and precipitation occurs. The entire procedure, which is much more sensitive than simple double-diffusion, is completed in a maximum of a few hours. Examples include the identification of predators of the sugar-cane borer, *Eldana saccharina* (Leslie & Boreham, 1981), and Australian soldier fly, *Inopus rubriceps* (Doane *et al.*, 1985).

Agglutination methods. There are a number of methods.

1 Latex agglutination. First, the immunoglobulin G (IgG) is separated from the antisera by caprylic acid and then attached to polystyrene latex particles (Steinbuch & Audran, 1969). When this sensitized latex is mixed with the homologous predator meal extract, agglutination will occur. The great advantage of this test is that it can be done in the field, the only equipment required being a microscope slide and pipettes. The reagents are known to be stable for at least 3 months at 4°C and the test is at least as sensitive as the precipitin test. Boorman *et al.* (1977) used this method to identify the blood meals of ceratopongonid midges.

2 Passive haemagglutination inhibition. This technique allows discrimination between prey in related genera (Greenstone, 1977). It depends on attaching the antigen to red blood cells (commonly from sheep); placed with the appropriate antibody, these red blood cells will agglutinate. The procedure involves the preparation of both the specific antibody and sensitized red blood cells (i.e. with the antigen attached). The extracts from the gut of the predator are added to antibody and the mixture is incubated before adding the sensitized red blood cells. When the specific prey is present in the predator's gut, the antibody will combine with it, leaving no free antibodies to combine with the sensitized red cells, so these will settle to form a small spot at the bottom of the tube. When the specific prey is absent and in the controls (with the antibody alone), the sensitized red blood cells will agglutinate, i.e. the prey antigen is not present to inhibit the reaction. The method involves a significant measure of skill and experience and a ready supply of antigens; each assay takes about $2\frac{1}{2}$ days, although some 150 tests may be handled in the same batch. Full details of this sophisticated method are given in Greenstone (1977). It has been used

to identify the blood meals of mosquitoes, ceratopogonids, and sandflies (Washino & Tempelis, 1983).

3 Complement fixation. This methodology is described by Palmer (1980). The complement group of proteins in mammals can produce cytolysis of antibody-sensitized bacteria and red blood cells. The sample to be tested is added to a suspension of antibodies and red blood cells and any antigen present blocks the activity of the antibody and thus reduces cytolysis.

4 Fluorescence immunoassay. McKiney *et al.* (1972) used this method, in which the globulin part of the antiserum is conjugated with a fluorescent compound, to identify mosquito blood meals. The test sample is placed on a cellulose actetate disc, incubated in the conjugated antiserum for about 30 min, and then washed and read either visually or by a fluorometer under ultraviolet (UV) light.

5 Enzyme-linked immunosorbent assay (ELISA). Commercial kits for automated application of this important technique are available. The basic idea is to link an enzyme to the antibody, using glutaraldehyde or periodate to form a conjugate, which can then react with any antigen present in the test sample to form a complex. This complex is attached to a solid, so that non-reactants can be washed away, and is detected by using the linked enzyme (e.g. alkaline phosphatase or horseradish peroxidase) to catalyse a reaction with a coloured end-product. ELISA is a highly sensitive method, suitable for the large-scale testing of field samples (Sunderland, 1988). The method has been used for blood-meal identification (Washino & Tempelis, 1983) and Chow *et al.* (1993) compare a number of different ELISA protocols for identifying the meals of the mosquito *Aedes aegypti*.

Potential problems with serological methods. The main potential pitfalls are itemized below.

1 Cross-reaction. The precipitin test cannot be satisfactorily used to analyse prey species that are taxonomically closely related, since the antisera will tend to cross-react. Cross-reactions can often be removed by absorption (Weitz, 1952; Dempster, 1960; Service, 1976), but for closely related species, if this is attempted, the titre is reduced. However, if the fauna of the study area is accurately known, this problem may not be important. For example, the antisera prepared by Dempster (1960) for *Phytodecta* cross-reacted with other Chrysomelidae and with Cassididae and Coccinellidae. Chrysomelidae and Cassididae were considered unimportant in the study area and reaction with ladybirds was removed by absorption. In his study of predators of rice borers, Rothschild (1971) was unable to distinguish between predation on *Chilo suppressalis* and *Tryporyza incertulas* because of cross-reactions.

2 Specificity of the antiserum. The second limitation is that the antiserum may not react equally well with all stages of the animal used as antigen, e.g. S.R. Spencer and P.F.L. Boreham (unpublished) found that an antiserum prepared against adult *Locusta migratoria* reacted poorly with eggs and first and second instars. Their table also shows cross-reactions occurring with closely related Orthoptera species.

Similar results were obtained with the spruce budworm *Choristoneura* (Loughton *et al.*, 1963). Working with *Pieris rapae*, Ashby (1974) found that antiserum prepared from pupae was more sensitive than that prepared from larvae. It is therefore preferable to use all stages of the insect as antigen when preparing an antiserum, or at least to test the completed product carefully.

3 These methods give no indication as to whether the prey was taken alive, scavenged, or actually eaten by another prey now itself in the gut (Sutton, 1970; Tod, 1973). This criticism will also apply to microscopical examination of the predator meal.

4 The rate of digestion of the predator depends upon intrinsic factors as well as extrinsic ones, such as temperature. Measurement of the rate of digestion gives an indication of the length of time a meal can be detected. This is partly a matter of the relative size of the predator and the meal, but the data of Hall *et al.* (1953), Dempster (1960), and Loughton *et al.* (1963) for a number of predators (ranging in size from mites and mirids to earwigs and reduviid bugs) with lepidopterous and coleopterous prey are reasonably consistent: eggs were detectable for 18–48h after feeding and large larvae for about 5 days. For precise population studies, it is desirable to know not only that a given predator has eaten the prey within the last xh but also whether it has eaten one or more individuals. By studying the dispersion of the prey in the field and the rate of movement of the predator, Dempster (1960) was able to argue, supported by laboratory feeding trials, that the chance of a positive representing more than one feed was very small. Therefore it was possible to estimate the number of prey consumed by the predator in the total population during the digestion period. For example, if, in a sample of 200 predators, six were shown to contain meals of the prey, which remain detectable for 1 day, we may say that the total number of prey destroyed by that predator during the previous 24h = 6 × total population of predator ÷ 200 (Dempster, 1960).

Labelled prey. The prey may be labelled with a dye, a rare element, or a radioactive isotope, but it is important that the label does not modify the behaviour of the predator. The egg predators of a moth were detected by spraying the eggs with an alcoholic suspension of a powdered fluorescent dye. The dye could be detected, under UV light, in the gut of the predators; this was particularly easy if the gut was macerated and dried (Hawkes, 1972).

Feeding can also easily be established if the prey is marked with a rare element (see p. 119) or a radioactive isotope and the label is subsequently detected in the predator. Ito *et al.* (1972) used europium for this purpose, but found it to be rapidly excreted by insects. However, Hamann and Iwannek (1981) found europium to remain detectable for 3 weeks in tsetse flies, while lanthanum was detectable for only a few days. If each prey carried a similar burden of rare element or radionuclide, the level of these labels in the predator would be a measure of the prey consumed.

Theoretically, therefore, this method could be of value with voracious predators that consume prey at frequent intervals, a situation in which the precipitin test is unsuitable for quantitative studies. However, in practice, it is difficult to arrange for all the prey to carry an equal burden of label; van Dinther and Mensink (1971) found this impossible to achieve with fly eggs, whether they were labelled directly or via the female.

Most studies using this technique have involved prey tagged with phosphorus-32 (^{32}P) (see Chapter 3 for marking methods) and have had as their objective the identification of the predators (Jenkins & Hassett, 1950; James, 1961, 1966; Jenkins, 1963; Walker et al., 1992). An approximate measure of the number of prey taken may be obtained from the level of activity in the predator. However, Baldwin et al. (1955) showed, in laboratory tests, that the same level of radioactivity (in counts per minute) could result from the consumption of different numbers of prey. This may be due to the variable radionuclide burdens of the prey, but could arise from different assimilation rates on the part of the predator (p. 367). If a large number of results were obtained and the radioactivity of the prey were normally distributed, it would be justifiable to calculate the number of prey eaten from the mean counts for 1, 2, ... x prey. In essence, this was the approach of Pendleton and Grundmann (1954), who made a thistle plant radioactive by placing 1 μC ^{32}P in 3 ml of water in a hollow in the pith; the aphids feeding on the plant gave an average of 250 ct/min and the predators that ate whole aphids gave counts roughly in multiples of 250 ct/min; those, such as spiders, that consume only part of the aphid gave lower counts and so the lowest group of counts (120–180) was taken to represent the killing of a single aphid and approximate multiples of this the killing of two or more.

Greater precision for more continuous studies might be introduced by using a modification of Crossley's (1963) approach, which was developed for measuring the consumption of vegetation by insects. He points out that, if the biological half-life of the nuclide is long relative to the lifespan of the insect, the amount in the insect (Q_t) may be given by:

$$Q_t = \frac{ra}{K}\left(1 - e^{-Kt}\right) \tag{10.2}$$

where r = rate of ingestion, a = proportion of ingested nuclide assimilated, K = average elimination constant, and t = time. (This approach could not be used in a situation when the radionuclide in the predator reaches equilibrium with that of its food, when the relationship $Q_e = ra/K$ holds; this is discussed on p. 517.) Now, if the isotope used had a long biological half-life (e.g. ^{45}Ca, ^{65}Zn) for any predator, Q_t could be measured and, if a and K were known from laboratory studies, it should be possible to calculate r which would give a measure of the total number of isotope-marked prey eaten. If the proportion of these in the total population were known (it would probably have to be fairly high), it could be assumed that the marked and unmarked prey were taken equally and the total prey consumption easily estimated. Gamma emitters and the use of a scintillation counter would be particularly suitable for measuring radioactivity in these studies, since it would eliminate problems of counting geometry associated with whole-body counts.

Moulder and Reichle (1972) were able to assess the role of spider predation in a forest at Oak Ridge which had been tagged with caesium-137; they found the caesium concentrations of potential prey reasonably similar, although there were seasonal variations. Allowances had to be made for this, for the portion of the prey discarded by the spider, and for the different size categories of spiders and prey. This was, of course, a rather special situation where the whole environment was labelled (including every prey) with an isotope with a long half-life.

Extreme caution should be exercised in all work of this type to ensure that the predator's intake of radionuclide came only from the prey. Baldwin *et al.* (1955) point out that, if mosquito larvae are returned to a pond directly after marking by immersion in a solution of ^{32}P, their radioactive excreta contaminates the whole pond. They overcome this by washing the larvae twice and keeping them for 2 days in fresh water before release. If the prey is marked through the host plant, it must be remembered that some predators also feed, to a greater or lesser extent, on the plant. Van der Meijden (1973) determined the fate of lepidopterous larvae by labelling them with paint containing an isotope (see p. 122), the labels could be recovered and it could be determined whether they were close to a pupa or isolated, when the larvae were assumed to have been killed by ants (or other causes).

10.2.4.4 Exclusion techniques

These methods seek to demonstrate the effects of predators, parasites, or competitors by artificially excluding them and measuring the increase in the prey's population under the new conditions; this increase represents the action of the predator (or parasite) in the exposed situation. Conceptually, the method is very attractive but its weakness is that the techniques of excluding the predators often affect the microclimate and other aspects of the habitat, so that it is difficult to be certain that the observed effects are entirely due to the predator (Fleschner, 1958).

Mechanical or other barriers. Mammals and birds are relatively easily excluded by wire netting or nets, and it is unlikely that such methods have marked side-effects on the prey species, unless, of course, they also exclude herbivorous mammals, whose activities have an important role in regard to the vegetation. Buckner (1959) used mammal exclusion cages in his study of the predation of larch sawfly, *Pristiphora*, cocoons; in this case, both the protected and exposed cocoons were 'planted' and therefore there was the risk that the resulting artificially high density might have acted as a bait to the mammals.

Predatory insects need to be excluded by smaller muslin cages of the sleeve-cage type (Smith & DeBach, 1942), but, as Fleschner (1958) and Hunter and West (1990) have shown, such cages may affect the population directly and therefore, even if some cages are left open for predator access, as the cages affect the prey's rate of increase, the natural situation cannot always be measured (DeBach & Huffaker, 1971). Weseloh (1990) tethered gypsy-moth, *Lymantira*, larvae in cages with

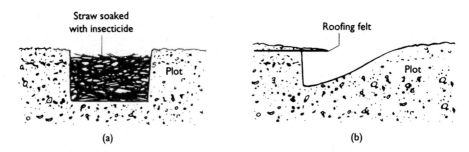

Fig. 10.5 Barriers to carabid beetles. (a) Exclusion barrier of straw soaked with insecticide. (b) Emigration barrier that allows beetles to enter plot but not to leave (based on Wright *et al.*, 1960).

various-sized meshes to determine the role of predators of different sizes in distinct microhabitats.

Non-flying insects can be more easily excluded by mechanical or insecticidal barriers. Wright *et al.* (1960) and Coaker (1965) demonstrated the effects of carabid beetles on populations of the cabbage-root fly (*Erioischia brassicae*) by means of trenches round the plots, filled with straw soaked with an insecticide (Fig. 10.5). Spiders (Clarke & Grant, 1968) and ants can often be eliminated by similar barriers or bands on trees (p. 279). Various combinations of exclusion cages may be used to separate the roles of vertebrates (wire cages), ground-dwelling invertebrates (mechanical barriers), and flying invertebrates (net cages—variations in net size can produce further separation) (Ashby, 1974; Eickwort, 1977).

Elimination of predator or parasite. Vertebrates may be shot (Dowden *et al.*, 1953) or trapped (Buckner, 1959), until their numbers are extremely low. As this must be done over a fairly large area to be effective, it will lead to major disturbance in the habitat; this is undesirable in ecological studies and often on conservation grounds. Fleschner (1958) found that hand-picking of invertebrate predators from a part of the tree was both a feasible and reliable method.

As some predatory insects are very susceptible to certain insecticides to which the prey are resistant, it is possible to assess their effects by an insecticidal check method (DeBach, 1946; DeBach & Huffaker, 1971). It is very important to ensure that the pesticide has no side-effects on the prey species. Although this method was found reliable for studies on predators of *Aonidiella* scale insects (DeBach, 1955), it was not satisfactory for various plant-feeding tetranycid mites (Fleschner, 1958).

A biological check method has also been described, in which large populations of ants are built up; these attack and often prevent predator and parasite action (Fleschner, 1958). Although such observations do give a measure of predator importance, there are so many other factors involved that they cannot be considered as equivalent to the quantitative estimation of predation.

10.2.4.5 Experimental assessment of natural enemies

There is now an extensive theoretical background, supported by field and laboratory studies, on the interactions of prey with their natural enemies; the subject is reviewed by Murdoch and Oaten (1975), Beddington *et al.* (1976), Hassell (1976a, b, 1978, 2000), Hassell *et al.* (1976), and May (1976). The interactions may be viewed in relation to two population variables (Beddington *et al.*, 1976; Hassell, 1976a; Hassell *et al.*, 1976):

1 the death-rate of the prey;
2 the rate of increase of the predator.

The two are, of course, interrelated; the prey death-rate depends on the numbers of predators and on their searching efficiency. To some extent, this classification parallels that between functional and numerical responses (Solomon, 1949; Holling, 1959a, b, 1961, 1965, 1966), but is somewhat broader, in that prey density is now only one of the independent variables against which predator efficiency or reproductive rate is viewed. The major relationships are summarized in Table 10.1. The following sections indicate some approaches to the measurement of these relationships. Once this has been done the best procedure, as almost invariably in ecology, is to plot the data; when this has been completed, it will, in many cases, be possible to select the most appropriate model and determine precisely the value of the major

Table 10.1 Some relationships whose measurement are important in assessing the role of natural enemies (other variables: A = age of predator (developmental response), B = age of other predators of same species, C = climate and other externals, D = other prey (switching), E = other animals (competition and mutualism)).

Component	In relation to	Other variables	Affecting
Attack rate a Handling time T_h	prey density 'functional response'	A,C,D,E.	
Interference constant m Quest constant q or Time lost through encounters bt_w	predator density	A,B,C,D,E.	death rate of prey
Predator dispersion P_i (number/ith area) T_i (time/ith area)	prey dispersion N_i 'aggregative component of numerical response'	A,C,D,E.	
Predator fecundity Survival rate of predator of different stages Duration of different stages of predator	prey density 'breeding component of numerical response'	B,C,D,E.	rate of increase of predator

component parameters. For some other relationships, clear models have yet to be developed.

Death rate of prey/prey density. This is most elegantly measured in the laboratory, with different prey densities per predator and the numbers of prey killed or hosts parasitized (for insect parasitoids) recorded.

The numbers of prey eaten per predator at different prey densities can also be obtained from field data, although it is rare that an adequate estimate of the density of searching predators is available. Such results, however, should only be interpreted as a functional response *sensu stricto* if the density is relatively constant. Wide fluctuations in predator density (and food choice, if polyphagous) may confuse the relationship by different levels of predator interference occurring (see below).

There are two essential components to a functional response: an instantaneous attack rate (*a*) and a saturation term, often called the 'handling time' (T_b) but also including any effects of satiation. Additional variables are the speed of movement of the predator (Glen, 1975), its reactive distance (both of which may be influenced by habitat characteristics), and the proportion of attacks that are successful (Nelmes, 1974) and of prey that are fully consumed. The number of prey dying and the number eaten may not be the same, either because the predator kills more than it consumes (Buckner, 1966; Kruuk, 1972; Toth & Chew, 1972) or because it disturbs the prey, driving it from the host plant, etc., and exposing it to other mortalities. Nakasuji *et al.* (1973) showed this for spiders with lepidopterous larvae and W. Milne (unpublished) for birds with aphids. The type of habitat will influence predator–prey interactions through its effect on speed of movement and reactive distance, as well as through dispersion (Martin, 1969).

Many functional responses have been found to be of the Holling type II form (Fig. 10.6). This has been traditionally described by Holling's disc equation:

$$N_a = \frac{aNTP}{1 + aT_b N} \tag{10.3}$$

where N_a = total number of prey attacked, a = attack rate, N = total number of prey, P = predator density, usually 1 in experimental systems, T_b = handling time, and T = total time (searching time + $T_b N_a$, but excluding any regular sleeping time, i.e. sleeping not related to the amount of feeding).

The method of estimating a and T_b from functional-response data using the above equation is described in Holling (1959b). This technique, however, does not allow for the removal of prey during the experiment (i.e. it assumes prey replaced as eaten or a systematically searching predator) and will lead to incorrect estimates if prey depletion is significant (Royama, 1971; Rogers, 1972) (cf. removal trapping, p. 255). Rogers (1972) discusses a more satisfactory method for attaining a and T_b, which assumes random exploitation of the available prey. Using this technique, it

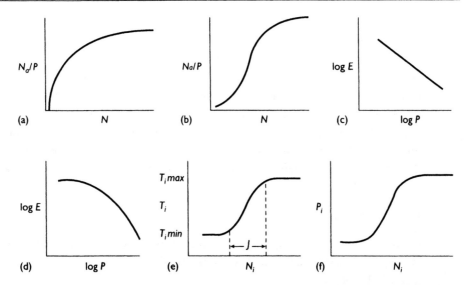

Fig. 10.6 Some characteristic forms of predator–prey relationships. (a) Holling type II functional response. (b) Holling type III functional response. (c) Interference—Hassel and Varley model. (d) Interference—Rogers and Hassel or Beddington model. (e) and (f). Aggregative response measured in time and numbers.

is necessary to distinguish between predators (which remove prey as eaten) and parasitoids (which can re-encounter hosts, thus involving more handling time). Random predator:

$$N_a = N[1 - \exp\{-a(PT - N_a T_b)\}] \tag{10.4}$$

Random parasitoid:

$$N_a = N\left[1 - \exp\left\{-\frac{TaP}{(1 + aT_b N)}\right\}\right] \tag{10.5}$$

where P = number of predators (parasitoids).

Although the Holling type II response is usually regarded as the form typical of invertebrate predators and parasitoids, there is increasing evidence that sigmoid type III responses are also widespread, at least amongst insects (Murdoch & Oaten, 1975; Hassell, 1978). Experimental evidence suggests that these responses result from some component of a, or the time initially available for search (T), now being a rising function of prey density. The increased number of parameters involved in describing such responses makes their adequate estimation less straightforward than for type II responses and requires a non-linear, least-squares technique.

Death rate of prey/predator density. The presence of other predators may enhance (cooperation) or reduce (interference) the prey death-rate due to a single predator. Instances of cooperation are particularly common amongst vertebrates and range from the group-hunting tactics of carnivorous mammals to the mere confusion of prey by several birds fishing in the same region. The defence mechanisms of, for example, a colony of aphids would probably be more easily disrupted by a number of coccinellid larvae than by a singleton, but this does not seem to have been quantitatively demonstrated; the more general result with invertebrates is interference (Hassell *et al.*, 1976). It is important to remember that, in laboratory experiments, predators or parasitoids are often prevented from dispersing when crowded; therefore unreal levels of interference and superparasitism may be observed (Hassell *et al.*, 1976). Interference effects, when plotted in a log scale, are commonly linear (Fig. 10.6c) or curvilinear (Fig. 10.6d). When the relationship is linear, then the component coefficients may be extracted, using the model of Hassell and Varley (1969):

$$\log E = \log Q - m \log P \qquad (10.6)$$

where P = predator density, m = interference constant, Q = quest constant (= E when $P = 1$), and E = searching efficiency of the predator. In general, E may be defined (Hassell, 1976b) as:

$$E = \frac{1}{P} \log_e \left[\frac{N}{N - N_{ba}} \right] \qquad (10.7)$$

where N_{ba} = prey actually eaten or parasitized. Under the special circumstances when every prey encountered is actually eaten and the number attacked is small compared with the total prey population, then:

$$E = \frac{N_a}{NP} \qquad (10.8)$$

The curvilinear relationships were discussed and modelled by Rogers and Hassell (1974) and also by Beddington (1975), who provided an expression for searching efficiency (E):

$$E = \frac{aT}{[1 + bt_w(P - 1)]} \qquad (10.9)$$

where a = attack rate = N_a/NT_sP' (where T_s = total search time and P' = predator number), P = predator density, b = rate of encounter between parasites, and t_w = time wasted per encounter.

Predator dispersion/prey dispersion. Predators may aggregate in a region of high prey density. This aggregative or behavioural component of the numerical response has considerable significance in the stability of predator–prey systems (Hassell & May, 1974; Murdoch & Oaten, 1975). Observations may be made in the laboratory or in the field, measuring the number of predators found in regions of different prey density. When this can be done sequentially (e.g. by time-lapse photography), the series of numbers obtained may be used to calculate the proportion of time spent in each area. More rarely, continuous observations can be made to give data on actual searching time per area (Murdie & Hassell, 1973).

The results may be plotted as numbers of predators per unit area (P_i) or the time spent by predators per unit area (T_i) against prey density per unit area (N_i) (Fig. 10.6e, f). In their analysis of the significance of the form of the aggregative response for stability, Hassell and May (1974) showed that the greater the difference between the maximum and minimum times per unit area and the more closely the region J (Fig. 10.6e) corresponded to the average range of the prey's density, the more stabilizing the aggregative response.

Fecundity, developmental, and survival rates of predators in relation to prey density. The importance of these relationships is reviewed by Beddington *et al.* (1976), who suggest certain expressions for them but stress that the lack of documentation for all the components for one species 'represents a major gap in experimental ecological work'. The methodology for such studies is, of course, basically similar to that used for any animal (e.g. Turnbull, 1962).

The role of other prey. 'Other prey' is often important in allowing the survival of natural enemies over a period when the prey species being considered is sparse. However, the few quantitative studies in this area have been concerned with 'switching', i.e. a sudden change in the predator's preference between various prey: switching is reviewed by Murdoch and Oaten (1975). The idea arose from L. Tinbergen's theory of a 'search image'; some supporting evidence is given by Murton (1971). Royama (1970) suggested, however, that the predator maximizes the 'profitability' of its searching; the profitability of feeding on a particular prey is compounded of size, population density, and ease of capture. Royama's own work, together with that of Bryant (1973), in which the prey of insectivorous birds was compared with the available prey, tends to support his concept, larger prey items tending to be relatively favoured — a finding that also accords with Hespenheide's (1975). However, even in these studies, the prey spectrum was not markedly different from that available. No significant preference or switching has been discerned for trout (*Salmo trutta*) (Elliott, 1970), coccinellid larvae (Murdoch & Marks, 1973), mites (Santos, 1976), or generally for predatory littoral snails (Murdoch, 1969).

A number of different indices have been used to detect preference, of which V.S. Ivlev's 'elective index' and various forage ratios are examples (Jacobs, 1974; Lawton *et al.*, 1974; Mustafa, 1976). These were reviewed by Cock (1978), who follows Lawton *et al.* (1974) in using the random equations (10.4) and (10.5) to predict the numbers of each prey type (subscripts 1 and 2 in the notation) eaten:

$$N_{a1} = N_1[1 - \exp\{-a_1 P(T - T_{h2}N_{a1} - T_{h2}N_{a2})\}] \qquad (10.10)$$

$$N_{a2} = N_2[1 - \exp\{-a_2 P(T - T_{h2}N_{a2} - T_{h1}N_{a1})\}] \qquad (10.11)$$

when the ratio of the two prey types becomes:

$$\frac{N_{a1}}{N_{a2}} = \frac{N_1[1 - \exp(-a_1 PT_s)]}{N_2[1 - \exp(-a_2 PT_s)]} \qquad (10.12)$$

where $T_s = (T - T_h N_{a1} - T_{h2} N_{a2})$, T = total time available for searching and feeding, and T_h = handling time. A similar procedure based on the random parasitoid equation (10.5) should be used with parasitoids.

Cock (1978) (see also Hassell, 1978) recommends the following procedure for detecting preference.

1 Carry out function-response experiments, using each prey separately.

2 Estimate a_1, a_2, T_{h1}, and T_{h2} from the random equations (10.10) and (10.11) as appropriate (i.e. predator or parasitoid).

3 Any preference resulting from differences in the functional-response parameters (i.e. $a_1 \neq a_2$ and/or $T_{h1} \neq T_{h2}$) can now be conveniently displayed in terms of N_{a1}/N_{a2} (calculated from equation (10.11)) plotted against N_1/N_2 or, alternatively, as the proportion of one of the species in the total diet against the proportion available, e.g. $N_{a1}/(N_{a1} + N_{a2})$ against $N_1/(N_1 + N_2)$. Such innate preference will then be detected as a deviation from a slope of unity passing through the origin.

4 Carry out a further experiment in which various ratios of the two prey types are presented together, and so contrast predicted and observed results. Ideally, this procedure should then be repeated for a range of total prey densities that encompass those used in the functional-response experiments (step 1). Any difference between the predicted preference from step 3 and that observed will now be due either to an active rejection of one of the prey or some change in a_1, a_2, T_{h1}, or T_{h2} as a result of the predator experiencing the two prey types together.

In the field situation the position may be further complicated, because polyphagous predators will often have wider trivial ranges than their prey, whose microhabitat preferences may be more restricted. Thus, if a predator modifies its hunting pattern from one microhabitat to another in response to relative differences in prey availability, then the switching effect might well be greater than that estimated from preference alone when prey is randomly mixed. Jolicoeur and Brunel's (1966) observations on the cod, *Gadus morhua*, may be an illustration of this.

Changes during the development of the predator. This concept was formalized by Murdoch (1971) and termed the developmental response. Clearly, many components (T_h, a, m, etc.) will change with the age of the predator; Hassell *et al.* (1976) show how age structure of predator and prey may alter the parameter values of the functional response. Such data are most easily gained from laboratory experiments (Turnbull, 1962; Thompson, 1975; Evans, 1976; Wratten, 1976).

Changes due to other animals. Competition between predators is often recorded (e.g. between parasitoids, leading to multiparasitism). This has yet to be considered in detail *vis-à-vis* the general theory of competition between two populations (May, 1976). Mutualism between two predators or between a non-predator (buffalo) and a predator (egret) has not been related to the theoretical framework for this relationship (May, 1973, 1976), although this has been done for the interactions between bacterial mutualists and their leguminous host plants (Parker, 1999).

The important variables for measurement would be prey killed, prey density, and density and survival of both types of predator.

Climatic and similar effects. The appropriate experimental methods are referred to on p. 358.

10.3 Dispersal

The term dispersal covers any movement away from the initial locality. It is useful to define neighbourhood dispersal for the process by which individuals migrate into and via adjacent areas and jump dispersal for longer-distance dispersal, where individuals are transported or purposively move quickly to a new area. Neighbourhood dispersal may be described by random-walk or diffusion equations and an example would be the movement of newly hatched larvae away from their egg mass. Given the wide range of processes that are encompassed by jump dispersal—the transport of ostracods in the feet and feathers of ducks or the displacement of a fish by a flash flood are two examples—there cannot be a single mathematical approach. In expanding populations, neighbourhood dispersal progresses on a continuous front, whereas jump dispersal creates isolated local populations. Natural populations will often exhibit both types of dispersal or even exhibit a continuum running between these two extremes (Plank, 1967).

10.3.1 Detecting and quantifying jump dispersal
Often, the only practical methods for studying long-distance movements involve the use of marked animals, which are released in a known locality and subsequently caught or observed elsewhere. Methods for marking are given in Chapter 3. Such an approach is most applicable to large, long-lived animals, such as fish, reptiles, birds, or mammals, although they are applicable to some insects, such as butterflies (e.g. Urquhart, 1960). For some insects, long-distance aerial movements can be quantified using suction traps (Chapter 4) and fish or crustacean movements via estuaries can be detected in a similar fashion using cooling-water intakes (p. 274). Migratory birds can be counted by observers placed on known routes or counted using radar (p. 262). It may be possible to determine if the rate of long-distance immigration between populations is of any significance either by a comparison of gene frequencies between populations or from the frequency of parasites on hosts. However, such methods cannot estimate the magnitude of the movements.

10.3.2 Quantifying neighbourhood dispersal
When neighbourhood dispersal is modelled as a diffusion process, then, as Fagan (1997) argues, there are three approaches that can be taken. The traditional, concentration-curve, approach aims to study the spatial distribution of a large number of individuals away from the release point at known instants in time, then estimating dispersal parameters by fitting an appropriate model to the observed

change of numbers with distance (Fig. 10.7a). A second approach is to plot the movements of individuals and calculate step lengths and turning angles for a random-walk model, the parameters of which are then used to predict the population dispersal rate (Fig. 10.7b). The third method, termed by Fagan (1997) the boundary-flux approach, is to count through time the cumulative number of individuals that have moved out from a release point and then to use these data to estimate the dispersal coefficient (Fig. 10.7c). The approaches presented below assume that dispersal can be successfully modelled as a diffusion process, as has been found often to be so by Okubo (1980, 1989). All of the methods presented assume that the

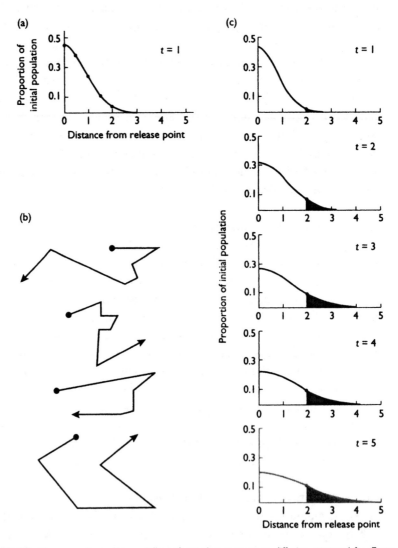

Fig. 10.7 Possible approaches to the modelling of animal movement as a diffusion process (after Fagan, 1997). (a) The traditional, concentration-curve, approach. (b) Plotting the movements of individuals and calculating step lengths and turning angles for a random-walk model, the parameters of which are then used to predict the population dispersal rate. (c) The boundary-flux approach, in which counts through time of the cumulative number of individuals which have moved out from a release point are made.

dispersal (diffusion) coefficient is constant over the period of observation and each animal behaves in a similar way. Neither of these assumptions can be strictly true, although they often lead to reasonable assumptions. When a model gives a poor fit, it is generally because one or both of these assumptions are violated. Diffusion models tend to be most in error close to the release point shortly after release and at later times close to the leading edge of the dispersing population. A number of issues are raised by the requirement of these methods to release and/or recognize the animal in the field, which will be discussed first.

10.3.2.1 The use of marked or introduced animals

When marked animals are released where they were found or directly marked in the field (Greenberg & Bornstein, 1964), their subsequent movements may be altered by the experience. An alternative approach is to release unmarked animals in an 'empty' habitat. Some workers have used laboratory-bred material for these releases; although this may be acceptable (MacLeod & Donnelly, 1956; Fletcher & Economopoulos, 1976), it should not be adopted without a check that the behaviour is normal. The use of animals captured in the wild also poses problems; the level of migratory activity in insects is commonly related to age (e.g. Johnson, 1969) and the process of marking may so disturb them that dispersal is exceptional during the first few days after release (Clark, 1962; Greenslade, 1964). Dean (1973) studied the movements of aphids in an uninfested crop by placing pots of cereals, with dense colonies of aphids, at various points.

10.3.2.2 Methods based on a radial solution of the diffusion equation

The basic approach is for animals to be released at a known point and subsequently recaptured in traps placed at increasing distances (in annuli or in the form of a cross) from the release point (e.g. Doane, 1963). The influence of irregularities in environmental favourableness around different traps may be reduced by using the ratio of unmarked to marked animals in each trap (Gilmour et al., 1946; MacLeod & Donnelly, 1963). As it is assumed that the animals can disperse equally in all directions, a one-dimensional model can be used where distance is expressed as the radial distance from the point of release.

First, it should be determined if there is drift or non-randomness in the direction of dispersal. One approach to this is to superimpose upon the map of the release and recapture sites a horizontal and vertical grid, and the mean and the variance are calculated for each day in terms of these grids (Clark, 1962). If the means differ significantly, then there was markedly more dispersal in one direction than another — drift. If the means are similar but the variances differ, then movement was non-random. Clark (1962) describes a method for testing whether the resulting spatial distribution is circular or elliptical.

Another approach is that of Paris (1965), who, in a series of experiments, determined the number of woodlice in different radii, and compared the results for eight radii, using Friedman's (1940) analysis of variance by ranks test, which is non-parametric.

Both the above methods are limited to situations when a group of marked animals is released from a central point. When the members of a natural population have been marked individually, it may be possible to use the approach of Frank (1964), who demonstrated that individual limpets, *Acmaea digitalis*, were not moving randomly and hence concluded that they had home ranges. Let the habitat be regarded as a grid of identical squares, and assume that the probability of movement from one square to the next over each time step is constant. Then the movement can be modelled as a Markov process, for which the transition probability matrix at the end of a given number of time periods for the probabilities can be easily calculated. The actual movements may then be compared with the expected, using the χ^2-test. Work on bird navigation has revealed some ability to orientate in a time-compensated manner to the sun, moon, certain constellations, and perhaps the magnetic field; attention has therefore been directed at analysing the effect of a small bias on an otherwise random movement (Matthews, 1974; Kendall, 1974).

Provided the drift in one direction is not excessive, the dispersal of a marked population from a point source may be represented by Fig. 10.8, the shaded panels representing the successive density/distance profiles (Inoue *et al.*, 1973). Five interrelated variables may be calculated to determine the properties of a dispersing population as revealed by such an experiment.

1 The extent to which the population is heterogeneous with some dispersing while others remain almost sedentary.

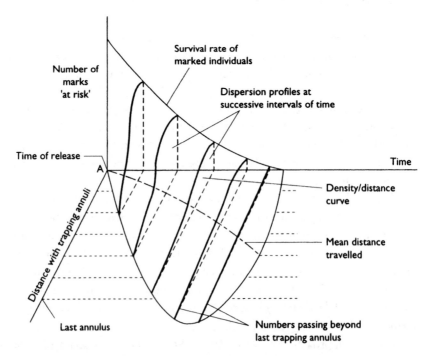

Fig. 10.8 A schematic representation of the dispersion of marked individuals released at a fixed point A, showing some of the variables that may be estimated (modified from Inoue et al., 1973).

2 The numbers or relative densities at successive distances from the release point.

3 The fall-off of density with distance, i.e. the form of the leading edge of the shaded segments in Fig. 10.8.

4 The mean distance travelled (by the emigrating component if some are stationary) (Fig. 10.8).

5 The proportion, or number, of marked individuals that have passed beyond the area.

Account must also be taken of the density at the start, because these variables may well be density-modified (Taylor & Taylor, 1976). Furthermore, the tail of the distribution, representing the most mobile individuals, may not be found in small samples.

The detection of heterogeneity with respect to the rate of dispersal for the individuals of a population. In an experiment on the dispersal of marked *Drosophila* from a central point of a cross of traps, Dobzhansky and Wright (1943) demonstrated that the flies were heterogeneous with respect to the distance they travelled. The departure from normality of the curve of the frequency curve of numbers with distance from the point of release on a given day was detected by calculating the kurtosis by the formula:

$$K = \frac{R\sum_{i=1}^{y} x_i^4 n_i}{\left(\sum_{i=1}^{y} x_i^2 n_i\right)^2} \tag{10.13}$$

where R = total animals caught (recaptured) in all traps, x_i = distance of the recapture point (i) from the point of release, n_i = total number of animals caught in traps at the same distance (x_i) from the release point, and y = the last equidistant set of recapture points (the first point will be at the centre, i.e. $x_i = 0$).

Heterogeneity of dispersal is an expected property of populations; Shaw (1970) has shown how even alate aphids may differ greatly in their potential for dispersal.

The numbers at various distances from the release point. The approach with the fewest assumptions is that of Fletcher (1974), who studied the Queensland fruit fly, *Dacus tryoni*, and calculated the proportion of the migrant flies that were in a particular annulus (F_i). Let n_i/g_i represent the total number of marked flies trapped (recaptured) divided by the number of traps at annulus i. Then:

$$\hat{F}_i = \frac{n_i}{g_i}\left(x_{i+1}^2 - x_i^2\right) \bigg/ \sum_{i-1}^{y} \frac{n_i}{g_i}\left(x_{i+1}^2 - x_i^2\right) \tag{10.14}$$

where x_i = the distance the inner radius of the ith annulus is from the central release point and therefore x_{i+1} = the distance of the outer radius of the same annulus from the central point, and y = total number of annuli. This method cannot be used when significant numbers have migrated beyond the outer annulus; proportions are used, because the catchment area of the traps cannot be ascertained.

In their study of *Drosophila*, Crumpacker and Williams (1973) assumed that the zone of attraction of one trap would overlap the next, but the number of marked flies (M) in the annulus could be estimated for a particular day as:

$$\hat{M}_i = r_i \left(\frac{U_i}{u_i} \right) \left(\frac{A_i}{\alpha_i} \right) \tag{10.15}$$

where r_i = marked flies recaptured on the particular day in the ith annulus, u_i = the unmarked flies captured, U_i = the total unmarked flies, A_i = the total area of the annulus, and α_i = the attractive area of the annulus. On their assumption, $A_i/\alpha_i = 1$; clearly M_i and U_i are both unknown, but these could easily be computed from a series of equations.

The probability that a marked fly reached the ith annulus is estimated by:

$$\hat{q}_i = \frac{M_i}{\displaystyle\sum_{i=1}^{y} M_i} \tag{10.16}$$

Given the assumption that $A_i = \alpha_i$, this may be expressed as:

$$\hat{q} = \frac{A_i r_i / u_i}{\displaystyle\sum_{i=1}^{y} \frac{A_i r_i}{u_i}} \tag{10.17}$$

where the first annulus is numbered 1 and the last y.

Of course, these probabilities will be influenced by heterogeneity in the population, discussed above (see also p. 385 for a method assuming random movement.)

The fall-off of density with distance. This topic was comprehensively reviewed by Freeman (1977) and Taylor (1978); the latter shows that the equations belong to two 'families' (Table 10.2). The shapes of the curves generated vary significantly (Fig. 10.9), and both Taylor (1978) and Freeman (1977) found that the equations in family II give the best fits to most field data for insects; this implies that the lengths of individual movements are not random. Random dispersal would be better fitted by equations in family I. The general equation for family II is:

$$N = \exp(\eta + bX^c). \tag{10.18}$$

Taylor (1978) suggests that the parameter c is a measure of non-randomness — $c < 2$ indicates a tendency to aggregation, $c \approx 2$ random, and $c > 2$ repulsion, leading to regularity (see Fig. 2.4 and p. 51). Taylor and Taylor (1976, 1978) have taken this concept of density dependence further in their general model (see below).

A simple test for random movement — a random-walk, Brownian motion or diffusion model — is to compare the fall-off of numbers with distance with half a normal distribution (Fletcher, 1974). Considered as a diffusion phenomenon, the

Table 10.2 Equations for the dispersal of organisms from a release point following the classification of Taylor (1978), where N = number dispersing to distance X and η, b and c are constants.

General form for family	Equations	Author
I. $N = \eta + b\, f(X)$	$N = \eta + c/X$	Paris, 1965
	$N = \eta + b.\log_e X$	Wolfenbarger, 1946
	$N = \eta + b.\log_e X + c/X$	Wolfenbarger, 1946
II. $N = \exp(\eta + bX^c)$	$N = \exp(\eta + c/X)$	Taylor, 1978
	$N = \exp(\eta + b.\log_e X)$	MacLeod & Donnelly, 1963
	$N = \exp(\eta + b\sqrt{X})$	Hawkes, C., 1972
	$N = \exp(\eta + bX)$	Gregory & Read, 1949
	$N = \exp(\eta + bX^2)$	Dobzhansky & Wright, 1943

Fig. 10.9 Various forms of the dispersal–distance curves. (a) and (b) Half-distributions for various equations (Table 10.2) (after Taylor, 1978) based on Dobzhansky and Wright's (1943) data. (a) Family I type curves. (b) Family II type curves, which imply a density relationship. (c) Representations of the whole distributions for family II type curves with different values of c in the general equation.

density of marked insects per unit area (m') at a particular point is given by (Scotter *et al.*, 1971; Aikman & Hewitt, 1972):

$$m' = \frac{M}{4\pi Dt} \exp\left(\frac{-x^2}{4Dt}\right) \tag{10.19}$$

where M = effective (i.e. surviving and moving) number of marked animals released, x = the radial distance from the point of release, t = time, and D = dispersal rate coefficient. This equation may be solved by iteration, but D will be found to have two values, one as the density at a particular distance builds up and one as it gradually falls as they disperse even further (as shown by Clark (1962) in his study of grasshoppers).

It should be noted that the magnitude of the maximum x in relation to the scale of the animal's habitat will strongly influence the type of frequency distribution found with distance; many species will show a different pattern within the population habitat and outside it.

The mean distance travelled and rate of dispersal. This is most easily calculated from data on the proportional frequency (F_i) (Fletcher, 1974) or actual numbers (P_i) (Crumpacker & Williams, 1973) of marked individuals estimated for successive annuli (see above). The estimate is:

$$\hat{\bar{d}} = \sum_{i=0}^{y} F_i \frac{1}{2}(x_{i+1} + x_i) \tag{10.20}$$

where x_i is the distance the inner radius of the ith annulus is from the central point, x_{i+1} is the same for the outer radius, and y is the outermost annulus. If numbers have been estimated, M_i is substituted for F_i in the above equation. These estimates are reliable only so long as no animals pass beyond the sampling annuli.

If a random walk can be assumed, then a measure of the rate of dispersal is given by the dispersal coefficient, D. In the experiment with *Drosophila* referred to above, Dobzhansky and Wright (1943, 1947) determined the speed of dispersal by comparing the change in variance on successive days. If the speed of dispersal is constant, the variance should change by a constant amount from day to day. They suggested that, as the curve of numbers on distance was not normal, variance should be estimated:

$$s^2 = \frac{\pi \sum_{i=2}^{y} x_i^3 F_i}{\left(\sum_{i=2}^{y} x_i \bar{r}_i\right) + r_1} \tag{10.21}$$

where x_i = the distance of the recapture point from the central release point, $\bar{r}_i$ = the mean number of animals captured per trap at a distance (x_i) from the release point, F_i = the mean number of animals captured per site at a given distance (x_i) from the release point, y = the equidistant set of traps or capture sites furthest from the release point, and r_i = the number of recaptures in the central trap (i.e. where $x = 0$). If animals are released at a central point, it is possible, making the assumptions of random flight movements and constant speed, to calculate the theoretical curve for the fall-off in numbers of marked individuals with distance. By relating this curve to actual data of the ratio of marked to unmarked flies, Gilmour *et al.* (1946) were able to calculate the number of marked blowflies beyond the last ring of traps.

The number of marked animals that have left the area. Using a standard diffusion model, with the dispersal rate (diffusion) coefficient (D) calculated as above, assuming trapping or searching methods 100% efficient (i.e. $M_i = r_i$), Scotter *et al.* (1971) give the equation as:

$$\hat{M}_i = a\left[\exp\left(\frac{-x_i^2}{4Dt}\right) - \exp\left(\frac{-x_{i+1}^2}{4Dt}\right)\right]$$

(10.22)

where a = number of marked flies released, t = time, x_i = distance of the inner radius of the ith annulus from the central point, and x_{i+1} = distance of the outer radius from the ith annulus from the point. Where $x_i = 0$, i.e. the innermost radius, there is a unique solution to D. Then the probability (P) that a particular insect travels more than a certain distance beyond the radius of the outer annulus (y) is:

$$P = \exp\left(\frac{-y^2}{4Dt}\right)$$

(10.23)

The actual number of animals migrating into or out of a population may also be determined from mark-and-recapture analysis (see Chapter 3). If there are no births or deaths, then the loss and dilution rates may be taken as equivalent to migration rates; alternatively, the two components may be separated by measuring natality and mortality separately (see pp. 348–77).

The simple Lincoln index may be used to calculate the proportion of a population that has migrated if the total population is known from some other method and there are no births or deaths. A number of marked individuals are released; knowing this number, the total population, and the size of subsequent samples, expected recapture values may be calculated. The proportion of the marked insects that have emigrated (a_e) is estimated by the ratio of actual to expected recaptures:

$$a_e = \frac{r}{r_e}$$

(10.24)

where r = actual recaptures and r_e = expected recaptures = an/N (where a = number of marked animals released, n = total size of sample, and N = total population known from some other estimate). Such an approach is of particular value for a highly mobile animal.

10.3.2.3 Methods based on a two-dimensional solution of the diffusion equation

The use of quadrat counts of unmarked individuals. Where the habitat is uniform and it is possible to assume random diffusion, movement may be separated from mortality by using the method of Dempster (1957), who solved the diffusion equation using a standard numerical approximation. The changes in the numbers of animals (f) with time are given by (Skellam, 1951):

$$\frac{df}{dt} = D\left(\frac{d^2f}{dx^2} + \frac{d^2f}{dy^2}\right) - \mu f$$

(10.25)

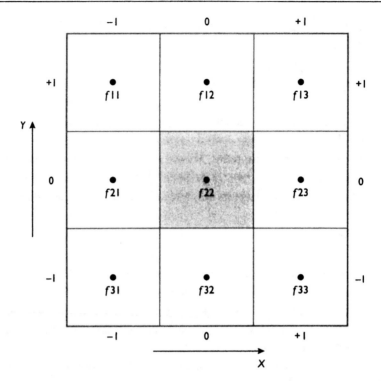

Fig. 10.10 The populations (f_{11}, f_{12}, . . .) along the two axes (x and y) across a central quadrat surrounded by quadrats of equal area (after Dempster, 1957).

where D = the dispersal coefficient, x and y are the two dimensions in which dispersal occurs, and μ = mortality rate.

Now, if we consider a block of nine quadrats (Fig. 10.10), for the central quadrat, f_{22}, the second derivatives can be approximated by:

$$\frac{d^2f}{dx^2} + \frac{d^2f}{dy^2} = \frac{1}{3}(2f_{11} - f_{12} + 2f_{13} - f_{21} - 4f_{22} - f_{23} + 2f_{31} - f_{32} + 2f_{33}) \qquad (10.26)$$

This expression can be substituted into equation (10.25) and, if the number of animals in each quadrat is known at two points in time, an expression with only two unknowns, D and μ, is formed. Now imagine a situation with a number of central squares so that a series of equations can be produced. While the absolute minimum number of equal-sized quadrats needed to solve for D and μ would be 15, arranged in a block of 3×5, a larger number would be preferable—Dempster (1957) used 18. Equations for the four middle squares can be calculated, the df/dt value being the changes in numbers—normally a decrease—between time periods. The equations are then solved by normal mathematical procedures.

For example (from Dempster, 1957): consider square B_2 in Fig. 10.11. The number of insects in this square has changed from 50 500 to 26 475.

Therefore $df/dt = -24 025$

and

	A	B	C
1	44 230 14 600	45 650 16 145	6478 5840
2	47 370 22 580	50 500 26 475	16 650 6623
3	50 720 30 015	103 900 52 680	89 160 57 860

Fig. 10.11 The estimated number of locusts entering the first (upper figure) and second instar (lower figure) in a number of quadrats (from Dempster, 1957).

$$\left(\frac{d^2f}{dx^2}+\frac{d^2f}{dy^2}\right)=\frac{1}{3}[(2\times44\,230)-45\,650+(2\times6478)-47\,370-(4\times50\,500)$$
$$-16\,650+(2\times50\,720)-103\,900+(2\times89\,160)]$$
$$=-11\,465$$

Substituting in equation 10.25 we get

$$-24\,205=(-11\,465D)-50\,500\mu$$

At least 3, preferably more, similar equations are obtained. Then the normal equation for μ is found by multiplying each equation by the coefficient of μ in it and adding all equations. The normal equation for α is found in the same way. These two equations are then solved.

The boundary-flux approach. This method solves the two-dimensional diffusion equation for the dispersal coefficient, D, and the mortality rate, μ, by observing the cumulative capture of animals dispersing outwards from a release point. Fagan (1997) illustrated the approach with a study of dispersal in two mantids. The experiments were undertaken in quadrats delimited by tanglefoot bands (p. 279) that would capture any mantid passing over them. The mantids were initially released in the centre of the quadrat, the tanglefoot bands at the periphery of the area inspected daily for 3 weeks, and the number of mantids captured counted. As the mantids could not escape from the tanglefoot, the bands acted as an absorbing boundary, and the count is the flux of mantids at a known time and distance from the release point.

For a square plot, Fagan (1997) gives the cumulative number caught at the boundary as:

$$C(t) = \sum_{j=0}^{\infty} \sum_{k=0}^{\infty} \left\{ \frac{\left[\left(16(-1)^{(1+j+k)} D\left[(1+2j)^2 + (1+2k)^2\right] \left[\exp\left(\frac{-t\Omega_{jk}}{L^2}\right) - 1\right]\right) \right]}{\left[(1+2j)(1+2k)\Omega_{jk}\right]} \right\} \tag{10.27}$$

$$\Omega_{jk} = D\pi^2\left[(1+2j)^2 + (1+2k)^2\right] + \mu L^2 \tag{10.28}$$

where t is time and L is the length of one side of the square. This large and ugly expression was solved numerically for D, with the summations approximated as a double-Fourier series. The mortality rate was estimated from other data.

10.3.2.4 The rate of population interchange between two areas

Richards and Waloff (1954) described a method for studying the movement between two grasshopper colonies; their basic assumption was that the survival rates in the two colonies were similar. Iwao (1963), who has derived equations that are applicable to populations where both the survival rates and sampling ratios differ, further developed their method. A series of three sets of observations on days 1 (t_1), 2 (t_2), and 3 (t_3) is necessary, as with Bailey's triple catch (p. 93). A number (a_1) of animals are marked and released in both areas on day 1. On day 2 a sample (n_2) is taken and the number of already marked individuals recorded; all the individuals (a_2) are then given a distinctive mark and released. On the third day, samples are again taken in both areas and the number of already marked individuals recorded, together with the details of their marks.

Thus the estimate of the emigration rate from area x to area y during the time interval from day 1 to day 2 is given by:

$$_{xy}\hat{e} = \frac{\left(\dfrac{_{yy}r_{31}\,_y a_2 + _{yy}r_{21}\,_{yy}r_{32}}{_{yy}r_{32}\,_y a_1}\right)_y a_{1xy}r_{21}}{_x a_{1yy}r_{21}} \tag{10.29}$$

where the notation has been adapted to conform with that in Chapter 3, the anterior subscripts being added to denote the areas; with both the anterior and posterior subscripts, the symbol nearest the character represents the actual condition and that furthest away its previous history. Thus $_{xy}r_{21}$ represents the recaptures in area y on day 2 that were marked in area x on day 1. To recapitulate the notation for the above equation:

$_x a_1$ = no. of marked individuals released in area x on day 1
$_y a_1$ = no. of marked individuals released in area y no day 1
$_y a_2$ = no. of marked individuals released in area y no day 2

$_{yy}r_{21}$ = recaptures in area y on day 2 marked in area y on day 1

$_{yy}a_{31}$ = recaptures in area y on day 3 marked in area y on day 1

$_{yy}r_{32}$ = recaptures in area y on day 3 marked in area y on day 2

The equivalent equation for the estimation of the emigration rate from y to x is:

$$_{yx}\hat{e} = \frac{\left(\dfrac{_{xx}r_{31\,y}a_2 + _{xx}r_{21\,xx}r_{32}}{_{xx}r_{32\,x}a_1}\right)_{x}a_{1\,yx}r_{21}}{_{y}a_{1\,xx}r_{21}} \tag{10.30}$$

If the total population (N) has been estimated by capture–recapture or in some other way, then the actual numbers that are estimated to have emigrated (E) are:

$$_{xy}\hat{E} = _{x}N_{1\,xy}\hat{e}_1 \tag{10.31}$$

and:

$$_{yx}\hat{E} = _{y}N_{1\,yx}\hat{e}_1 \tag{10.32}$$

Survival rates (ϕ) can also be calculated:

$$_{x}\phi_1 = _{xx}\mu_1 + _{xy}\hat{e}_1 \tag{10.33}$$

where

$$_{xx}\mu_1 = \frac{_{yy}r_{31\,y}a_2 + _{yy}r_{21\,yy}r_{32}}{_{yy}r_{32\,y}a_1} \tag{10.34}$$

This expression is the bracketed term in equation (10.29).

10.3.2.5 The description of population displacement in relation to its dispersion

Migration has long been recognized as one of the three pathways of population change; however, its formal incorporation into population mechanisms is more recent. Many models now envisage organisms as forming metapopulations composed of semi-isolated subpopulations, each of which may exchange members (Hanski, 1991; Hassell et al., 1991; Hastings & Harrison, 1994). Taylor and Taylor (1976, 1978) showed, on the basis of the extensive monitoring of adult Lepidoptera in Britain, that populations must be regarded as spatially fluid. Emigration and immigration are linked as two opposing density-related behaviour sets such that the coefficient of displacement is:

$$\Delta = G\rho^p - H\rho^q \tag{10.35}$$

where ρ = population density; p = the density-related moderator in relation to emigration, and q is that for congregation. This may be written:

$$\Delta = \Gamma\left[\left(\frac{\rho}{\rho_0}\right)^p - \left(\frac{\rho}{\rho_0}\right)^q\right] \tag{10.36}$$

where ρ_0 is the density at which emigration balances immigration and:

$$\Gamma = {}^{(p-q)}\!\!\sqrt{H^p g^{-q}} \tag{10.37}$$

· is a scale factor. The net result of displacement is therefore congregation when $\rho < \rho_0$ and emigration when $\rho > \rho_0$. Taylor and Taylor fitted the above expression to various sets of density/distance frequency data for animals and calculated values for the parameters E, ρ_0, p, and q. The significance of ρ_0, the equilibrium density, is most easily grasped, although of course habitat, as well as behavioural characters, will influence its magnitude. Taylor and Taylor's model provides a description of displacement with its link to spatial pattern (dispersion) and so provides a dynamic description of the changes of the population of an animal with time.

10.3.2.6 The measurement and description of home range and territory

The determination of the home range or territory of an individual or, for social animals, a colony is of value in the analysis of competition and density effects, the assessment of resources, and similar problems (Brown & Orians, 1970). Many studies have been made on the territories of vertebrates. There has been little work on insects, except for crickets (Alexander, 1961), dragonflies (Borror, 1934; Moore, 1952, 1957), some grasshoppers (Clark, 1962), and ants (Elton, 1932). Some insects may be marked and recaptured on a number of occasions (e.g. Borror, 1934; Green & Pointing, 1962; Greenslade, 1964); this type of experiment gives data that would be suitable for the computation of the home range, using the methods of vertebrate ecologists. Ten recaptures are usually taken as sufficient to calculate the home range, but considerably more data would be required to use the method of Frank (1964) (p. 380) and to ensure that one was not measuring an artefact.

The term territory implies the exclusion of at least certain other individuals While this is frequently observed for vertebrates, it is rarely observed for invertebrates, although clear examples are crickets and dragonflies (Alexander, 1961; Moore, 1952, 1957). Nevertheless, most invertebrates will have a home range, over which they forage and search for mates and where they rest; that is the area over which they engage in trivial movement (*sensu* Southwood, 1962). As Moore (1957) points out, the home range, especially for insects, should not be regarded as a hard and fast geographical area, but, provided it is measured in a strictly comparative way, it should be possible to detect changes due to inter- or intraspecific competition, to variation in the available resources, or to changes in the behaviour of the animals themselves (e.g. Wellington, 1964). There are a number of methods of calculating the home range of vertebrates; these have been reviewed by Sanderson (1966) and Jennrich and Turner (1969). The conclusion reached by the latter authors was that, in a homogeneous environment, the most statistically stable graphical technique is the minimum-area method, but the calculation of the determinant of the covariance matrix is to be preferred. They provide a table that allows the comparison of home ranges compared by different methods.

Minimum-area method. The points of recapture are mapped and the outermost points joined up to enclose the area, which may be measured (Mohr, 1947; Odum & Kuenzler, 1955). This method has the advantage that no assumption is made about the shape of the range. However, with such a method, the worker may feel it necessary to exclude 'incidental forays outside the area' (Jorgensen & Tanner, 1963). All re-entrant angles must be avoided, for the depth to which they are drawn will depend on individual judgement and thus different results will be obtained from the same data.

The matrix index. Developed by Jennrich and Turner (1969), this index, I_m, is:

$$I_m = 6\pi|S|^{\frac{1}{2}} \qquad (10.38)$$

Here $|S|$ is the determinant of the capture-point covariance matrix:

$$S = \begin{pmatrix} s_{xx} & s_{xy} \\ s_{yx} & s_{yy} \end{pmatrix} \qquad (10.39)$$

defined by the equations:

$$s_{xx} = \frac{1}{n-2}\sum_{i=1}^{n}(x_i - \bar{x})^2, s_{yy} = \frac{1}{n-2}\sum_{i=1}^{n}(y_i - \bar{y})^2 \qquad (10.40)$$

$$s_{xy} = s_{yx} = \frac{1}{n-2}\sum(x_i - \bar{x})(y_i - \bar{y}) \qquad (10.41)$$

$$\bar{x} = \frac{1}{n}\sum_{i=1}^{n}x_i, \bar{y} = \frac{1}{n}\sum_{i=1}^{n}y_i \qquad (10.42)$$

where (x_i, y_i) is the ith ordered capture point (defined on two coordinates, x and y) from a total of n capture points.

Where an animal's habitat is not homogeneous or where it spans two types of ecosystem, the above methods are not appropriate for the calculation of home range. Van Winkle *et al.* (1973) developed a method for frogs living on the edge of a pond, which could be applied to other animals in similar ecotones.

10.3.2.7 The rate of colonization of a new habitat
This property, which is related to migration, may be measured by planting virgin artificial habitats (e.g. new plants in a field or stones in a stream) and determining the rate of colonization (Breymeyer & Pieczynski, 1963; see also pp. 194 and 294).

10.3.2.8 The direction of migration
There are a number of traps (p. 272) that enable one to determine the direction an insect was flying in at the time of capture. It is possible that the proportionality values obtained from a series of such traps round a habitat (Sylven, 1970) might be used in conjunction with either measures of the net change in population or a

measurement of migration, to determine values for immigration and emigration. Furthermore, it has been pointed out how the dilution and 'loss' rates obtained from multiple capture–recapture analysis are compounded of birth and immigration and death and emigration, respectively. Sometimes, it is possible to separate the two components of the dilution rate from a knowledge of natality or the numbers entering the stage. When 'births' are known, the proportion of emigration to immigration, determined from a 'directional trap', might be used to give an indication of the amount of immigration and hence allow the partitioning of the death-rate into mortality and immigration. Caution would need to be used in such an approach, particularly to ensure an adequate distribution of traps around the habitat, and difficulty would be experienced in determining the numbers of insects leaving and arriving in vertical air currents, i.e. only crossing the habitat boundaries at a considerable height.

References

Aagaard, K., Solem, J.O., Nost, T., & Hanssen, O. (1997) The macrobenthos of the pristine stream, Skiftesaa, Hoylandet, Norway. *Hydrobiologia* **348**, 81–94.

Abdel Rahman, I. (1974) The effect of extreme temperatures on California red scale, *Aonidiella aurantii* (Mask.) (Hemiptera: Diaspididae) and its natural enemies. *Aust. J. Zool.* **22**, 203–12.

Adamstone, F.B. & Harkness, W.J.K. (1923) The bottom organisms of Lake Nipigon. Univ. *Toronto St. Biol.* **22**, 121–70.

Agassiz, D. (1977) A trap for wingless female moths. *Proc. Br. Entomol. Nat. Hist. Soc.* **10**, 69–70.

Aikman, D. & Hewitt, G. (1972) An experimental investigation of the rate and form of dispersal in grasshoppers. *J. Appl. Ecol.* **9**, 807–17.

Alexander, R.D. (1961) Aggressiveness, territoriality and sexual behaviour in field crickets (Orthoptera: Gryllidae). *Behaviour* **17**, 130–223.

Anderson, N.H. & Wold, J.L. (1972) Emergence trap collections of Trichoptera from an Oregon stream. *Can. Entomol.* **104**, 189–201.

Ashby, J.W. (1974) A study of arthropod predation of *Pieris rapae* L. using serological and exclusion techniques. *J. Appl. Ecol.* **11**, 419–25.

Bagenal, T.B. & Braum, E. (1978) Eggs and early life history. In: Bagenal, T. (ed.) *Methods for Assessment of Fish Production in Fresh Waters*, 3rd edn. IBP Handbook No. 3, Blackwell Scientific Publications, Oxford, pp. 165–201.

Baldwin, W.F., James, H.G., & Welch, H.E. (1955) A study of predators of mosquito larvae and pupae with a radio-active tracer. *Can. Entomol.* **87**, 350–6.

Barbraud, C. & Barbraud, J.C. (1997) Diet of white stork *Ciconia ciconia* chicks in 'Charente-Maritime': importance of insects. *Alauda* **65**, 259–62.

Bates, J.K. (1962) Field studies on the behaviour of bird fleas. 1. Behaviour of the adults of three species of bird flea in the field. *Parasitology* **52**, 113–32.

Bean, J.L. (1958) The use of larvaevorid maggot drop in measuring trends in spruce budworm populations. *Ann. Entomol. Soc. Am.* **51**, 400–3.

Beddington, J.R. (1975) Mutual interference between parasites or predators and its effect on searching efficiency. *J. Anim. Ecol.* **44**, 331–40.

Beddington, J.R., Hassell, M.P., & Lawton, J.H. (1976) The components of arthropod predation II. The predator rate of increase. *J. Anim. Ecol.* **45**, 165–85.

Bell, S.S., Hicks, G.R.F., & Walters, K. (1989) Experimental investigations of benthic reentry by migrating meiobenthic copepods. *J. Exp. Mar. Biol. Ecol.* **130**, 291–304.

Bess, H.A. (1961) Population ecology of the gipsy moth, *Porthetria dispar* L. (Lepidoptera: Lymantridae). *Bull. Conn. Agric. Exp. Sta.* **646**, 43 pp.

Betts, M.M. (1954) Experiments with an artificial nestling. *Br. Birds* **47**, 229–31.

Betts, M.M. (1958) Further experiments with an artificial nestling gape. *Br. Birds* **49**, 213–15.

Blais, J.R. (1953) The effects of the destruction of the current year's foliage of balsam fir on the fecundity and habits of flight of the spruce budworm. *Can. Entomol.* **85**, 446–8.

Boethel, D.J., Morrison, R.D., & Eikenbary, R.D. (1976) Pecan weevil *Curculio caryae*

(Coleoptera: Curculionidae). 2. Estimation of adult populations. *Can. Entomol.* **1089**, 19–22.

Boettger, K. & Rudow, A. (1995) The Chironomidae (Diptera, Nematocera) of the Kossau a North German lowland stream using the emergence trap method: limnological studies in the nature reserve Kossautal (Schleswig-Holstein): IV. *Limnologica* **25**, 49–60.

Boorman, J., Mellor, P.S., Boreham, P.F.L., & Hewett, R.S. (1977) A latex agglutination test for the identification of blood-meals of *Culicoides* (Diptera: Ceratopogonidae). *Bull. Entomol. Res.* **67**, 305–11.

Boreham, P.F.L. (1975) Some applications of bloodmeal identifications in relation to the epidemiology of vector-borne tropical disease. *J. Trop. Med. Hyg.* **78**, 83–91.

Boreham, P.F.L. & Gill, G.S. (1973) Serological identification of reptile feeds of *Glossina*. *Acta Trop.* **30**, 356–65.

Boreham, P.F.L. & Ohiagu, C.E. (1978) The use of serology in evaluating invertebrate predator–prey relationships: a review. *Bull. Entomol. Res.* **68**, 171–94.

Borror, D.J. (1934) Ecological studies of *Agria moesta* Hogen (Odonata: Coenagrionidae) by means of marking. *Ohio J. Sci.* **34**, 97–108.

Borutsky, E.V. (1955) A new trap for the quantitative estimation of emerging chironomids. [In Russian.] *Trudy Vses. Gidrobiol. Obshch.* **6**, 223–6.

Braverman, Y. (1970) An improved emergence trap for *Culicoides*. *J. Econ. Entomol.* **63**, 1674–5.

Breymeyer, A. & Pieczynski, E. (1963) Review of methods used in the Institute of Ecology. Polish Academy of Sciences, for investigating migration. [In Polish.] *Ekol. Polska B* **9**, 129–44.

Brooke, M.M. & Proske, H.O. (1946) Precipitin test for determining natural insect predators of immature mosquitoes. *J. Nat. Malaria Soc.* **5**, 45–56.

Brown, J.L. & Orians, G.H. (1970) Spacing patterns in mobile animals. *Ann. Rev. Ecol. Syst.* **1**, 239–62.

Brundin, L. (1949) Chironomiden und andere Bodentiere der Südschwedischen Urgebirgsseen. *Rep. Inst. Freshw. Res. Drottringholm* **20**, 915 pp.

Bryant, D.M. (1973) The factors influencing the selection of food by the house martin (*Delichon urbica* (L.). *J. Anim. Ecol.* **42**, 539–64.

Buckner, C.H. (1958) Mammalian predators of the larch sawfly in eastern Manitoba. *Proc. X Int. Congr Entomol.* **49**, 353–61.

Buckner, C.H. (1959) The assessment of larch sawfly cocoon predation by small mammals. *Can. Entomol.* **91**, 275–82.

Buckner, C.H. (1966) The role of vertebrate predators in the biological control of forest insects. *Ann. Rev. Entomol.* **11**, 449–70.

Buffone, G.J. (1980) Immunonephelometric and turbometric measurement of specific plasma proteins. In: Rose, N.R. & Friedman, H. (eds) *Manual of Clinical Immunology*. American Society for Microbiology, Washington, DC.

Campbell, J.A. & Pelham-Clinton, E.C. (1960) A taxonomic review of the British species of '*Culicoides*' Latreille (Diptera, Ceratopogonidae). *Proc. Roy. Soc. Edin. B* **67**, 181–302.

Carlson, D. (1971) A method for sampling larval and emerging insects using an aquatic black light trap. *Can. Entomol.* **103**, 1365–9.

Carter, C.L. (1972) *Winter Temperature and Survival of the Green Spruce Aphid* Elatobium abietinum. Forest Record No. 84, Forestry Commission, London, 10 pp.

Cavallini, P. & Serafini, P. (1995) Winter diet of the small Indian mongoose, *Herpestes auropunctatus*, on an Adriatic Island, J. *Mammal.* **76**, 569–74.

Cavallini, P. & Volpi, T. (1995) Biases in the analysis of the diet of the red fox *Vulpes vulpes*. *Wildl. Biol.* **1**, 243–8.

Chauvin, R. (1966) Un procédé pour récolter automatiquement les proies que les *Formica polyctena* rapportent au rid. *Insectes Soc.* **13**, 59–67.

Chow, E., Wirtz, R.A., & Scott, T.W. (1993) Identification of blood meals in *Aedes aegypti* by antibody sandwich enzyme-linked immunosorbent assay. *J. Am. Mosquito Contr. Assoc.* **9**(2), 196–205.

Clark, D.P. (1962) An analysis of dispersal and movement in *Phaulacridium vittatum* (Sjöst.) (Acrididae). *Aust. J. Zool.* **10**, 382–99.

Clark, E.W. & Osgood, E.A. (1964) An emergence container for recovering southern pine beetles from infested bolts. *J. Econ. Entomol.* **57**, 783–4.

Clark, L.P. (1963) The influence of population density on the number of eggs laid by females of *Cardiaspina albitextura* (Psyllidae). *Aust. J. Zool.* **11**, 190–201.

Clarke, R.D. & Grant, P.R. (1968) An experimental study of the role of spiders as predators in a forest litter community. Part 1. *Ecology* **49**(6), 1152–4.

Coaker, T.H. (1965) Further experiments on the effect of beetle predators on the numbers of the cabbage root fly, *Erioischia brassicae* (Bouché), attacking brassica crops. *Ann. Appl. Biol.* **56**, 7–20.

Cock, M.I.W. (1978) The assessment of preference. *J. Anim. Ecol.* **47**, 805–16.

Colless, D.H. & Chellapah, W.T. (1960) Effects of body weight and size of bloodmeal upon egg production in *Aedes aegypti* (Linnaeus) (Diptera, Culicidae). *Ann. Trop. Med. Parasitol.* **54**, 475–82.

Cook, L.M. & Kettlewell, H.B.D. (1960) Radioactive labelling of lepidopterous larvae: a method of estimating late larval and pupal mortality in the wild. *Nature* **1879**, 301–2.

Cook, P.P. & Horn, H.S. (1968) A sturdy trap for sampling emergent Odonata. *Ann. Entomol. Soc. Am.* **61**, 1506–7.

Corbet, P.S. (1966) Diel periodicities of emergence and oviposition in riverine Trichoptera. *Can. Entomol.* **98**, 1025–34.

Cottrell, P.E., Trites, A.W., & Miller, E.H. (1996) Assessing the use of hard parts in faeces to identify harbour seal prey: results of captive-feeding trials. *Can. J. Zool.* **74**(5), 875–80.

Crossley, D.A. (1963) Consumption of vegetation by insects. In: Schultz, V. & Klement, A.W. (eds) *Radioecology.* Rheinhold, New York, pp. 431–40.

Crumpacker, D.W. & Williams, J.S. (1973) Density, dispersion and population structure in *Drosophila pseudoobscura. Ecol. Monogr.* **43**(4), 499–538.

Custer, T.W. & Pitelka, F.A. (1975) Correction factors for digestion rates for prey taken by snow buntings (*Plectrophenax nivalis*). *Condor* **77**, 210–12.

Darby, R.E. (1962) Midges associated with California rice fields, with special reference to their ecology (Diptera: Chironomidae). *Hilgardia* **32**, 1–206.

Davidson, A. (1956) A method of counting ephemeropteran eggs. *Entomol. monthly Mag.* **92**, 109.

Davies, I.J. (1984) Sampling aquatic insect emergence. In: Downing, J.A. & Rigler, F.H. (eds) *A Manual of Methods for the Assessment of Secondary Productivity in Fresh Waters.* Blackwell Scientific Publications, Oxford.

Davies, J.B. (1966) An evaluation of the emergence or box trap for estimating sandfly (*Culicoides* spp. Heleidae) populations. *Mosquito News* **26**, 170–2.

Davies, R.W. (1969) The production of antisera for detecting specific triclad antigens in the gut contents of predators. *Oikos* **20**, 248–60.

Dean, G.T. (1973) Aphid colonization of spring cereals. *Ann. Appl. Biol.* **75**, 183–93.

DeBach, P. (1946) An insecticidal check method for measuring the efficacy of entomophagous parasites. *J. Econ. Entomol.* **39**, 695–7.

DeBach, P. (1949) Population studies of the long-tailed mealy bug and its natural enemies on citrus trees in Southern California, 1946. *Ecology* **30**, 14–25.

DeBach, P. (1955) Validity of the insecticidal

check method as a measure of the effectiveness of natural enemies of diaspine scale insects. *J. Econ. Entomol.* **48**, 584–8.

DeBach, P. & Huffaker, C.B. (1971) Experimental techniques for evaluation of the effectiveness of natural enemies. In: Huffaker, C.B. (ed.) *Biological Control.* Plenum Press, New York, pp. 113–40.

Dempster, J.P. (1957) The population dynamics of the Moroccan locust (*Dociostaurus maroccanus* Thunberg) in Cyprus. *Anti-Locust Bull.* **27**, 1–60.

Dempster, J.P. (1960) A quantitative study of the predators on the eggs and larvae of the broom beetle, *Phytodecta olivacea* Forster, using the precipitin test. *J. Anim. Ecol.* **29**, 149–67.

Dempster, J.P. (1964) The feeding habits of the Miridae (Hetcroptera) living on broom (*Sarothamnus scoparius* (L.) Wimm.). *Entomol. Exp. Appl.* **7**, 149–54.

Dempster, J.P., Richards, O.W., & Waloff, N. (1959) Carabidae as predators on the pupal stage of the chrysomelid beetle, *Phytodecta olivacea* (Forster). *Oikos* **10**, 65–70.

Denison, D.F. & Hodkinson, I.D. (1983) Structure of the predatory beetle community in a woodland soil ecosystem. I. Prey selection. *Pedobiologia* **25**, 109–15.

Dewalt, R.E., Stewart, K.W., Moulton, S.R., & Kennedy, J.H. (1994) Summer emergence of mayflies, stoneflies and caddisflies from a Colorado mountain stream. *Southw. Nat.* **39**(3), 249–56.

Diallo, B.P., Truc, P., & Laveissiere, C. (1997) A new method for identifying blood meals of human origin in tsetse flies. *Acta Trop.* **63**(1), 61–4.

van Dinther, J.B.M. (1953) Details about some flytraps and their application to biological research. *Entomol. Ber.* **14**, 201–4.

van Dinther, J.B.M. & Mensink, F.T. (1971) Use of radioactive phosphorus in studying egg predation by carabids in cauliflower fields. *Meded. Fak. Landb. Welensch. Gent.* **36**, 283–93.

Doane, J.F. (1963) Dispersion on the soil surface of marked adult *Ctenicera destructor* and *Hypolithus bicolor* (Coleoptera: Elateridae), with notes on fight. *Ann. Entomol. Soc. Am.* **56**, 340–5.

Doane, J.F., Scotti, P.D., Sutherland, O.R.W., & Pottinger, R.P. (1985) Serological identification of wireworm and staphylinid predators of the Australian soldier fly (*Inopus rubriceps*) and wireworm feeding on plant and animal food. *Entomol. Exp. Appl.* **38**, 65–72.

Dobson, R.M., Stephenson, J.W., & Lofty, J.R. (1958) A quantitative study of a population of wheat bulb fly, *Leptohylemyia coarctala* (Fall.), in the field. *Bull. Entomol. Res.* **49**, 95–111.

Dobzhansky, T. & Wright, S. (1943) Genetics of natural populations: X. Dispersion rates in *Drosophila pseudoobscura*. *Genetics* **28**, 304–40.

Dobzhansky, T. & Wright, S. (1947) Genetics of natural populations. XV. Rate of diffusion of a mutant gene through a population of *Drosophila pseudoobscura*. *Genetics* **32**, 303–24.

Douglas, R.M. (1992) Microscopic investigation of the digestive tract contents of spotted grass snakes, *Psammophylax rhombeatus rhombeatus* (Reptilia: Colubridae). *J. Afr. Zool.* **106**(5), 401–11.

Dowden, P.B., Jaynes, H.A., & Carolin, V.M. (1953) The role of birds in a spruce budworm outbreak in Maine. *J. Econ. Entomol.* **46**, 307–12.

Eguagie, W.E. (1974) Cold hardiness of *Tingis ampliata* (Heteroptera: Tingidae). *Entomol. Exp. Appl.* **17**, 204–14.

Eickwort, K.R. (1977) Population dynamics of a relatively rare species of milkweed beetle (*Labidomera*). *Ecology* **58**, 527–38.

Elliott, J.M. (1970) Diel changes in invertebrate drift and the food of trout (*Salmo trutta* L.) *J. Fish Biol.* **2**, 161–5.

Elliott, J.M. & Persson, L. (1978) The estimation of daily rates of food consumption for fish. *J. Anim. Ecol.* **47**, 977–91.

Elton, C. (1932) Territory among wood ants (*Formica rufa* L.) at Picket Hill. *J. Anim. Ecol.* **1**, 69–76.

van Emden, H.F. (1962) A preliminary study of insect numbers in field and hedgerow. *Entomol. Monthly Mag.* **98**, 255–9.

van Emden, H.F. (1963) A field technique for comparing the intensity of mortality factors acting on the cabbage aphid, *Brevicoryne brassicae* (L) (Hem., Aphididae) in different areas of a crop. *Entomol. Exp. Appl.* **6**, 53–62.

Entwistle, P.F. (1977) The development of an epizootic of a nuclear polyhedrosis virus disease in European spruce sawfly, *Gilpinia hercyaniae*. In: *Proc. 1ˢᵗ. Int. Colloquium Invert. Pathology*. Printing Dept. Queens University, Kingston, Canada, pp. 184–8.

Errington, P. (1932) Technique of raptor food habits study. *Condor* **34**, 75–86.

Evans, H.E. & Yoshimoto, C.M. (1962) the ecology and nesting behaviour of the Pompilidae (Hymenoptera) of the Northeastern United States. *Misc. Pub. Entomol. Soc. Am.* **3**(3), 65–119.

Evans, H.F. (1976) The role of predator–prey size ratio in determining the efficiency of capture by *Anthrocoris nemorum* and the escape reactions of its prey, *Acyrthosiphon pisum*. *Ecol. Entomol.* **1**, 85–90.

Evenhuis, H.H. (1962) Methods to investigate the population dynamics of aphids and aphid parasites in orchards. *Entomophaga* **7**, 213–20.

Fagan, W.F. (1997) Introducing a 'boundary-flux' approach to quantifying insect diffusion rates. *Ecology* **78**(2), 579–87.

Fewkes, D.W. (1964) The fecundity and fertility of the Trinidad sugar-cane froghopper, *Aeneolamia varia saccharine* (Homoptera, Cercopidae). *Trop. Agric. Trin.* **419**, 165–8.

Finnegan, R.J. (1969) Assessing predation by ants on insects. *Insectes Soc.* **16**, 61–5.

Fleschner, C.A. (1958) Field approach to population studies of Tetranychid mites on citrus and Avocado in California. *Proc. X Int. Congr. Ent.* **2**, 669–76.

Fletcher, B.S. (1974) The ecology of a natural population of the Queensland fruit fly, *Dacus tryoni* V. The dispersal of adults. *Aust. J. Zool.* **22**, 189–202.

Fletcher, B.S. & Economopoulos, A.P. (1976) Dispersal of normal and irradiated laboratory strains and wild strains of the olive fly *Dacus oleae* in an olive grove. *Entomol. Exp. Appl.* **20**, 183–9.

Frank, J.H. (1967) The insect predators of the pupal stage of the winter moth, *Operophtera boumata* (L.) (Lepidoptera: Hydriomenidae). *J. Anim. Ecol.* **36**, 375.

Frank, P.W. (1964) On home range of limpets. *Am. Nat.* **98**, 99–100.

Freeman, G.H. (1977) A model relating numbers of dispersing insects to distance and time. *J. Appl. Ecol.* **14**, 477–87.

Friedman, M. (1940) A comparison of alternative tests of significance for the problem of *m* rankings. *Ann. Math. Stat.* **11**, 86–92.

Gardner, W.A., Shepard, M., & Noblet, T.R. (1981) Precipitin test for examining predator–prey interactions in soybean fields. *Can. Entomol.* **113**, 365–9.

Gary, N.E. (1960) A trap to quantitatively recover dead and abnormal honey bees from the hive. *J. Econ. Entomol.* **53**, 782–5.

Gibb, J.A. (1958) Predation by tits and squirrels on the eucosmid *Ernarmonia conicolana* (Heyl.). *J. Anim. Ecol.* **27**, 375–96.

Giller, P.S. (1986) The natural diet of the Notonectidae: field trials using electrophoresis. *Ecol. Entomol.* **11**(2), 163–72.

Gilmour, D., Waterhouse, D.F., & McIntyre, G.A. (1946) An account of experiments undertaken to determine the natural population density of the sheep blowfly, *Lucilia cuprina* Wied. *Bull. Coun. Sci. Indust. Res. Aust.* **195**, 1–39.

Glen, D.M. (1975) Searching behaviour and prey-density requirements of *Blephandopterus angulatus* (Fall.) (Heteroptera: Miridae) as a predator of the Lime Aphid, *Eucallipterus tiliae* (L.) and Leafhopper, *Ainetoidea aineti* (Dahlbom). *J. Anim. Ecol.* **44**, 116–34.

Glen, D.M. (1976) An emergence trap for barkdwelling insects, its efficiency and effects on temperature. *Ecol. Entomol.* **1**, 91–4.

Goodhart, C.B. (1958) Thrush predation on the snail *Cepaea hortensis. J. Anim.* Ecol. **279**, 47–57.

Graham, S.A. (1928) The influence of small mammals and other factors upon larch sawfly survival. *J. Econ. Entomol.* **21**, 301–10.

Grandilewskaja-Decksbach, M.L. (1935) Materialien zur Chironomidenbiologie verschiedener Becken. Zur Frage über die Schwankungen der Anzahl und der Biomasse der Chironomidenlarven. *Trudŷ Limnol. Sta. Kosine* **19**, 145–82.

Green, G.W. (1962) Low winter temperatures and the European pine shoot moth, *Rhyacionia buoiiana* (Schiff.) in Ontario. *Can. Entomol.* **94**, 314–36.

Green, G.W. & Pointing, P.J. (1962) Flight and dispersal of the european pine shoot moth, *Rhyacionia buoliana* (Schiff.). II. Natural dispersal of egg-laden females. *Can. Entomol.* **94**, 299–314.

Greenberg, B. & Bornstein, A.A. (1964) Fly dispersion from a rural Mexican slaughter house. *Am. J. Trop. Med. Hyg.* **13**(6), 881–6.

Greenslade, P.I.M. (1964) The distribution, dispersal and size of a population of *Nebria brevicollis* (F.) with comparative studies on three other carabidae. *J. Anim. Ecol.* **33**, 311–33.

Greenstone, M.H. (1977) A passive haemagglutination inhibition assay for the identification of stomach contents of invertebrate predators, *J. Appl. Ecol.* **14**, 457–64.

Gregor, F. (1960) Zur Eiproduktion des Eichenwicklers (*Tortrix viridana* L.) *Zool. Listy.* **9**, 11–18.

Gruber, F. & Prieto, C.A. (1976) Collecting chamber suitable for recovery of insects from large quantities of host plant material. *Environ. Entomol.* **5**, 343–4.

Guennelon, G. & Audemard, M.H. (1963) Enseignements écologiques donnés par la méthode de captures par cuisses-éclosion de la cécidomyie des lavandes (*Thomasmiana lavandulae* Barnes). Critique de la méthode. Conclusions pratiques. *Ann. Epiphytol.* C **149**, 35–48.

Hafez, M. (1961) Seasonal fluctuations of population density of the cabbage aphid, *Brevicoryne brassicae* (L.) in the Netherlands, and the role of its parasite, *Aphidius* (*Diaeretiella*) *rapae* (Curtis). *Tijdschr. Pl. Ziekt.* **67**, 445–548.

Hall, R.R., Downe, A.E.R., MaClellan, C.R., & West, A.S. (1953) Evaluation of insect predator–prey relationships by precipitin test studies. *Mosquito News* **13**, 199–204.

Hamann, H. & Iwannek, K.H. (1981) The use of the indicator activation method for the labelling of tsetse flies. *Zeitschr. angew. Entomol.* **91**, 206–12.

Hance, T. & Rossignol, P. (1983) Essai de quantification de la predation des Carabidae par le test ELISA. *Meded. Fak. Landbouww. Rijksuniv. Gent* **49**, 849–55.

Hanski, I. (1991) Single-species metapopulation dynamics: concepts, models and observations. *Biol. J. Linn. Soc.* **42**, 17–38.

Hard, J.S. (1976) Estimation of hemlock sawfly (Hymenoptera: Diprionidae) fecundity. *Can. Entomol.* **108**, 961–6.

Hartley, P.H.T. (1948) The assessment of the food of birds. *Ibis* **90**, 361–81.

Hassell, M.P. (1976a) Arthropod predator–prey systems. In: May, R.M. (ed.) *Theoretical Ecology, Principles and Applications*. Blackwell Scientific Publications, Oxford, pp. 71–93.

Hassell, M.P. (1976b) *The Dynamics of Competition and Predation*. Edward Arnold, London.

Hassell, M.P. (1978) *The Dynamics of Arthropod Predator–Prey Relationships*. Princeton Monograph on Population Biology 13, Princeton, New Jersey, 245 pp.

Hassell, M.P. (2000) *The Spatial and Temporal Dynamics of Insect Host–Parasitoid Interaction*. Oxford University Press, Oxford (in press).

Hassell, M.P. & May, R.M. (1974) Aggregation of predators and insect parasites and its effect on stability. *J. Anim. Ecol.* **43**, 567–94.

Hassell, M.P. & Varley, C.C. (1969) A new inductive population model for insect parasites and its bearing on biological control. *Nature* **223**, 1133–7.

Hassell, M.P., Comins, H.N., & May, R.M. (1991) Spatial structure and chaos in insect population dynamics. *Nature* **353**, 255–8.

Hassell, M.P., Lawton, J.H., & Beddington, J.R. (1976) The components of arthropod predation. 1. The prey death rate. *J. Anim. Ecol.* **45**, 135–64.

Hastings, A. & Harrison, S. (1994) metapopulation dynamics and genetics. *Ann. Rev. Ecol. Syst.* **25**, 167–88.

Hawkes, R.B. (1972) A fluorescent dye technique for marking insect eggs in predation studies. *J. Econ. Entomol.* **65**, 1477–8.

Healey, J.A. & Cross, T.F. (1975) Immunoelectroosmophoresis for aerological identification of predators of the sheep-tick *Ixodes ricinus. Oikos* **26**, 97–101.

Henderson, P.A. (1998) On the variation in dab *Limanda limanda* recruitment: a zoogeographic study. *J. Sea Research* **40**, 131–42.

Hespenheide, H.A. (1975) Prey characteristics and predator niche width. In: Cody, M.L.

& Diamond, J.M. (eds) *Ecology and Evolution of Communities*. Harvard University Press, Cambridge, Massachusetts, pp. 158–80.

Hewitt, D.G. & Robbins, C.T. (1996) Estimating grizzly bear food habits from fecal analysis. *Wildl. Soc. Bull.* **24**(3), 547–50.

Higley, L.G. & Pedigo, L.P. (1985) Examination of some adult sampling techniques for the seedcorn maggot (*Delia platura*). *J. Agric. Entomol.* **2**(1), 52–60.

Holling, C.S. (1958) A radiographic technique to identify healthy, parasitised and diseased sawfly prepupae within cocoons. *Can. Entomol.* **90**, 59–61.

Holling, C.S. (1959a) The components of predation as revealed by a study of small mammal predation of the European pine sawfly. *Can. Entomol.* **91**, 293–320.

Holling, C.S. (1959b) Some characteristies of simple types of predation and parasitism. *Can. Entomol.* **91**, 385–98.

Holling, C.S. (1961) Principles of insect predation. *Ann. Rev. Entomol.* **6**, 163–82.

Holling, C.S. (1965) The functional response of predators to prey density and its role in mimicry and population regulation. *Mem. Entomol. Soc. Can.* No. 45, 60 pp.

Holling, C.S. (1966) The functional response of invertebrate predators to prey density. *Mem. Entomol. Soc. Can.* No. 48, 86 pp.

Huffaker, C.B. & Spitzer, C.H. (1950) Some factors affecting red mite populations on pears in California. *J. Econ. Entomol.* **43**, 819–31.

Hunter, M.D. & West, C. (1990) Variation in the effects of spring defoliation on the late season phytophagous insects of *Quercus robur*. In: Watt, A.D., Leather, S.R., Hunter, M.D., & Kidd, N.A.C. (eds) *Population Dynamics of Forest Insects*. Intercept, Andover, UK, pp. 123–35.

Ide, F.P. (1940) Quantitative determination of the insect fauna of rapid water. *Univ. Toronto Stud. Biol. Ser.* 47 (Publ. Ontario Fish. Res. Lab. 59), 20 pp.

Inoue, T., Kamimmura, K., & Watanabe, M. (1973) A quantitative analysis of dispersal in a horse-fly *Tabanus iyoensis* Shiraki and its application to estimate the population size. *Res. Popul. Ecol.* **14**, 209–33.

Ito, Y., Yamanaka, H., Nakasuji, F., & Kiritani, K. (1972) Determination of predator–prey relationship with an activable tracer, Europium-151. *Kontyu* **40**, 278–83.

Ives, W.G.H. & Prentice, R.M. (1959) Estimation of parasitism of larch sawfly cocoons by *Bessa harveyi* Tnsd. in survey collections. *Can. Entomol.* **91**, 496–500.

Ives, W.G.H., Turnock, W.J., Buckner, J.H., Heron, R.I., & Muldrew, I.A. (1968) Larch

sawfly population dynamics: techniques. *Manitoba Entomol.* **2**, 5–36.

Iwao, S. (1956) On the number of eggs per egg-mass of the paddy rice borer, *Schoenobius incertellus* Walker and the percentage of their parasitization. [In Japanese.] *Gensei (Kochi Konchu Dokokai)* **5**, 45–9.

Iwao, S. (1963) On a method for estimating the rate of population interchange between two areas. *Res. Popul. Ecol.* **5**, 44–50.

Jacobs, J. (1974) Quantitative measurement of food selection: a modification of the forage ratio and Iviev's selectivity index. *Oecologia* **14**, 413–17.

James, H.G. (1961) Some predators of *Aedes stimulans* (Walk) and *Aedes trichurus* (Dyar) (Diptera: Culicidae) in woodland pools. *Can. J. Zool.* **39**, 533–40.

James, H.G. (1966) Location of univoltine *Aedes* eggs in woodland pool areas and experimental exposure to predators. *Can. Entomol.* **98**, 550–5.

Jedrzejewski, W., Jedrzejewska, B., Szymura, B., & Zub, K. (1996) Tawny owl (*Strix aluco*) predation in a pristine decidous forest (Bialowieza National Park, Poland). *J. Anim. Ecol.* **65**, 105–20.

Jenkins, D.W. (1963) Use of radionuclides in ecological studies of insects. In: Schultz, V. & Klement. A.W. (eds) *Radioecology*. Rheinhold, New York, pp. 431–40.

Jenkins, D.W. & Hassett, C.C. (1950) Radioisotopes in entomology. *Nucleonics* **6**(3), 5–14.

Jennrich, R.I. & Turner, F.B.(1969) Measurement of non-circular home range. *J. Theor. Biol.* **22**, 227–37.

Johnson, C.G. (1969) *Migration and Dispersal of Insects by Flight*. Methuen, London, 763 pp.

Johnson, J.H. & Ross, R.M. (1996) Pellets versus faeces: their relative importance in describing the food habits of double-crested cormorants. *J. Great Lakes Res.* **22**(3), 795–8.

Johnson, J.H., Ross, R.M., & Smith, D.R. (1997) Evidence of secondary consumption of invertebrate prey by double-crested cormorants. *Colonial Waterbirds* **20**(3), 547–51.

Jolicoeur, P. & Brunel, D. (1966) Application du diagramme hexagonal à l'étude de la sélection de ses proies par la morue. *Vie Milieu* **17**, 419–33.

Jónasson, P.M. (1954) An improved funnel trap for capturing emerging aquatic insects, with some preliminary results. *Oikos* **5**, 179–88.

Jorgensen, C.D. & Tanner, W.W. (1963) The application of the density probability function to determine the home ranges of *Ula stansburiana stansburiana* and *Cnemidophorus tigris tigris*. *Herpetologica* **19**, 105–15.

Judge, W.W. (1957) A study of the population of emerging and littoral insects trapped as adults from tributary waters of the Thames River at London, Ontario. *Am. Midl. Nat.* **58**, 394–412.

Kendall, D.G. (1974) Pole seeking, Brownian motion and bird navigation. *J. Roy. Stat. Soc. (B)* **36**, 365–417.

Kensler, C.B. (1967) Desiccation resistance of intertidal crevice species as a factor in their zonation. *J. Anim. Ecol.* **36**, 391–406.

Kimerle, R.A. & Anderson, N.H. (1967) Evaluation of aquatic insect emergence traps. *J. Econ. Entomol.* **60**, 1255–9.

Kiritani, K. & Dempster, J.P. (1973) Different approaches to the quantitative evaluation of natural enemies. *J. Appl. Ecol.* **10**, 323–30.

Kiritani, K., Kawahara, S., Sasaba, T., & Nakasuji, F. (1972) Quantitative evaluation of predation by spiders on the green rice leafhopper, *Nephotettix cincticeps* Uhler, by a sight-count method. *Res. Popul. Ecol.* **13**, 187–200.

Knutsen, H. & Vogt, N.B. (1985) An approach to identifying the feeding patterns of lobsters using chemical analysis and pattern recognition by the method of SIMCA. II Attempts at assigning stomach contents of lobsters, *Homarus gammarus* (L) to infauna and detritus. *J. Exp. Mar. Biol. Ecol.* **89**, 121–34.

Kohira, M. & Rexstad, E.A. (1997) Diets of wolves, *Canis lupus*, in logged and unlogged forests of southeastern Alaska. *Can. Field Nat.* **111**(3), 429–35.

Kruuk, H. (1972) Surplus killing by carnivores. *J. Zool.* **166**, 233–44.

Kurdziel, J.P. & Bell, S.S. (1992) Emergence and dispersal of phytal-dwelling meiobenthic copepods. *J. Exp. Mar. Biol. Ecol.* **163**(1), 43–64.

Lacey, L. (ed.) (1997) *Manual of Techniques in Insect Pathology.* Academic Press, San Diego, 409 pp.

Lack, D. & Owen, D.F. (1955) The food of the swift. *J. Anim. Ecol.* **24**, 120–36.

Langford, T.E. & Daffern, J.R. (1975) The emergence of insects from a British river warmed by power station cooling-water. Part I. *Hydrobiologia* **46**, 71–114.

Lawson, F.R. (1959) The natural enemies of the hornworms on tobacco (Lepidoptera: Sphingidae). *Ann. Entomol. Soc. Am.* **52**, 741–55.

Lawton, J.H., Beddington, J.R., & Bonser, R. (1974) Switching in invertebrate predators. In: Usher, M.B. & Williamson, M.H. (eds) *Ecological Stability.* Chapman & Hall, London, pp. 141–58.

Lefranc, N. & Worfolk, T. (1997) *Shrikes.* Pica Press, Robertsbridge, Sussex, UK.

Le Jacques, D. & Lode, T. (1994) Feeding habits of European genets (*Genetta genetta* L., 1758)

from a grove in western France. *Mammalia* **58**(3), 383–9.

Lejeune, R.R., Fell, W.H., & Burbidge, D.P. (1955) The effect of flooding on development and survival of the larch sawfly *Pristiphora erichsonii* (Tenthredinidae). *Ecology* **36**, 63–70.

LeSage, L. & Harrison, A.D. (1979) Improved traps and techniques for the study of emerging aquatic insects. *Entomol. News* **90**(2), 65–78.

Lesiewicz, D.S., Lesiewicz, J.L., Bradley, J.R., & Van Duyn, I.W. (1982) Seriological determination of carabid (Coleoptera, Adephaga) predation of corn earworm (Lepidoptera, Noctuidae) in field corn. *Environ. Entomol.* **11**, 1183–6.

Leslie, G.W. & Boreham, P.F.L. (1981) Identification of arthropod predators of *Elana saccharina* Walker (Leidoptera: Pyralidae) by cross-over electrophoresis. *J. Entomol. Soc. South Afr.* **44**, 381–8.

Lindeberg, B. (1958) A new trap for collecting emerging insects from small rockpools, with some examples of the results obtained. Suom. hyönt. *Aikak. (Ann. Entomol. Fenn.)* **24**, 186–91.

Loan, C. & Holdaway, F.G. (1961) *Microctonus aethiops* (Nees) auctt. and *Perilitus rutilus* (Nees) (Hymenoptera: Braconidae): European parasites of *Sitona* weevils (Coleoptera: Curculionidae). *Can. Entomol.* **93**, 1057–78.

Loughton, B.G., Derry, C., & West, A.S. (1963) Spiders and the spruce budworm. *Mem. Entomol. Soc. Can.* **31**, 249–68.

Lozinsky, V.A. (1961) On the correlation existing between the weight of pupae and the number and weight of eggs of *Lymantria dispar* L. [In Russian.] *Zool. Zh.* **40**, 1571–3.

Macan, T.T. (1949) Survey of a moorland fishpond. *J. Anim. Ecol.* **18**, 160–86.

McCauley, V.J.E. (1976) Efficiency of a trap for catching and retaining insects emerging from standing water. *Oikos* **27**, 339–45.

McKiney, R.M., Spillane, J.T., & Holden, P. (1972) Mosquito blood meals: identification by fluorescent antibody method. *Am J. Trop. Med. Hyg.* **21**, 999–1003.

McKnight, M.E. (1969) Distribution of hibernating larvae of the Western Budworm, *Choristoneura occidentalis* on Douglas Fir in Colorado. *J. Econ. Entomol.* **62**, 139–42.

MacLellan, C.R. (1962) Mortality of codling moth eggs and young larvae in an integrated control orchard. *Can. Entomol.* **94**, 655–66.

MacLeod, I. & Donnelly, J. (1956) Methods for the study of blowfly populations. II. The use of laboratory-bred material. *Ann. Appl. Biol.* **44**, 643–8.

MacLeod, I. & Donnelly, J. (1963) Dispersal and interspersal of blowfly populations. *J. Anim. Ecol.* **32**, 1–32.

McMullen, L.H. & Atkins, M.D. (1959) A portable tent-cage for entomological field studies. *Proc. Entomol. Soc. BC* **56**, 67–8.

MacPhee, A.W. (1961) Mortality of winter eggs of the european red mite *Panonychus ulmi* (Koch), at low temperatures, and its ecological significance. *Can. J. Zool.* **39**, 229–43.

MacPhee, A.W. (1964) Cold-hardiness, habitat and winter survival of some orchard Arthropods in Nova Scotia. *Can. Entomol.* **96**, 617–36.

Major, R.E. & Gowing, G. (1994) An inexpensive photographic technique for identifying nest predators at active nests of birds. *Wildlife Research* **21**, 657–66.

Martin, F.J. (1969) Searching success of predators in artificial leaflitter. *Am. Midl. Nat.* **819**, 218–27.

Mason, W.H. & McGraw, K.A. (1973) Relationship of ^{65}Zn excretion and egg production in *Trichoplusia ni* (Hubner). *Ecology* **54**, 214–16.

Matthews, G.V.T. (1974) On bird navigation with some statistical undertones. *J. Roy. Stat. Soc. (B)* **36**, 349–64.

May, R.M. (1973) Qualitative stability in model ecosystems. *Ecology* **54**, 638–41.

May, R.M. (1976) Models for two interacting populations. In: May, R.M. (ed.) *Theoretical Ecology, Principles and Applications*. Blackwell Scientific Publications, Oxford, pp. 49–70.

van der Meijden, E. (1973) Experiments on dispersal, late-larval predation, and pupation in the cinnabar moth (*Tyria jacobaeae* L.) with a radioactive label (^{192}Ir.). *Netherl. J. Zool.* **23**, 430–45.

Miles, P.M. (1952) Entomology of bird pellets. *Amat. Entomol. Soc. Leaflet* **24**. 8 pp.

Miller, C.A. (1955) A technique for assessing larval mortality caused by parasites. *Can. J. Zool.* **33**, 5–17.

Miller, C.A. (1957) A technique for estimating the fecundity of natural populations of the spruce budworm. *Can. J. Zool.* **35**, 1–13.

Miller, C.A. (1958) The measurement of spruce budworm populations and mortality during the first and second larval instars. *Can. J. Zool.* **36**, 409–22.

Miller, R.S. & Thomas, J.L. (1958) The effects of larval crowding and body size on the longevity of adult *Drosophila melanogaster*. *Ecology* **39**, 118–25.

Moeck, H.A. (1988) A bucket emergence trap for corticolous insects. *Can. Entomol.* **120**(10), 855–8.

Mohr, C.O. (1947) Table of equivalent populations of North American small mammals. *Am. Midl. Nat.* **37**, 223–49.

Moncorps, S., Chapuis, J.L., Haubreux, D., &

Bretagnolle, V. (1998) Diet of the brown skua *Catharacta skua loennbergi* on the Kerguelen archipelago: comparisons between techniques and between islands. *Polar Biol.* **19**(1), 9–16.

Moore, N.W. (1952) On the so-called 'territories' of dragonflies (Odonata – Anisoptera). *Behaviour* **4**, 85–100.

Moore, N.W. (1957) Territory in dragonflies and birds. *Bird St.* **4**, 125–30.

Morgan, C.V.G. & Anderson, N.H. (1958) Techniques for biological studies of tetranychid mites, especially *Bryobia arborea* M. and A. and *B. praetiosa* Koch (Acarina: Tetranychidae). *Can. Entomol.* **90**, 212–15.

Morgan, N.C. (1971) Factors in the design and selection of insect emergence traps. In: Edmondson, W.T. & Winberg, G.G. (eds) *A Manual on Methods for the Assessment of Secondary Productivity in Fresh Waters*. Blackwell Scientific Publications, Oxford, pp. 93–108.

Morgan, N.C. & Waddell, A.B. (1961) Insect emergence from a small trout loch, and its bearing on the food supply of fish. *Sci. Invest. Freshw. Fish. Scot.* **25**, 1–39.

Morgan, N.C., Waddell, A.B., & Hall, W.B. (1963) A comparison of the catches of emerging aquatic insects in floating box and submerged funnel traps. *J. Anim. Ecol.* **32**, 203–19.

Morris, R.F. (1949) Differentiation by small mammals between sound and empty cocoons of the European spruce sawfly. *Can. Entomol.* **81**, 114–20.

Moulder, B.C. & Reichle, D.E. (1972) Significance of spider predation in the energy dynamics of forest-floor arthropod communities. *Ecol. Monogr.* **42**(4), 473–98.

Mukherjee, S., Goyal, S.P., & Chellam, R. (1944) Refined techniques for the analysis of Asiatic lion *Panthera leo persica* Scats. *Acta Theriol.* **39**(4), 425–30.

Mulla, M.S., Norland, R., Ikeshoji, T., & Kramer, W.L. (1974) Insect growth regulators for the control of aquatic midges. *J. Econ. Entomol.* **67**(2), 165–70.

Mundie, J.H. (1956) Emergence traps for aquatic insects. *Mitt. int. Verein. theor. angew. Limnol.* **7**, 1–13.

Mundie, J.H. (1971) Techniques for sampling emerging aquatic insects. In: Edmondson, W.T. & Winberg, G.G. (eds) *A Manual on Methods for the Assessment of Secondary Productivity in Fresh Waters*. Blackwell Scientific Publications, Oxford, pp. 80–108.

Murdie, G. (1969) The biological consequences of decreased size caused by crowding or rearing temperatures in apterae of the pea aphid, *Acyrthosiphon pisum* Harris. *Trans. Roy. Entomol. Soc. Lond.* **121**, 443–55.

Murdie, G. & Hassell, M.P. (1973) Food distribution, searching success and predator-prey models. In: Hioms, R.W. (ed.) *The Mathematical Theory of the Dynamics of Biological Populations.* Academic Press. London, pp. 87–101.

Murdoch, W.W. (1969) Switching in general predators: experiments on predators specificity and stability of prey populations. *Ecol. Monogr.* 39, 335–54.

Murdoch, W.W. (1971) The developmental response of predators to changes in prey density. *Ecology* 52, 132–7.

Murdoch, W.W. & Marks, J.P. (1973) Predation by coccinellid beetles: experiments on switching. *Ecology* 54, 160–7.

Murdoch, W.W. & Oaten, A. (1975) Predation and population stability. *Adv. Ecol. Res.* 9, 1–131.

Murton, R.K. (1971) The significance of a specific search image in the feeding behaviour of the wood pigeon. *Behaviour* 40, 10–42.

Mustafa, S. (1976) Selective feeding behaviour of the common carp, *Esomus danricus* (Ham.) in its natural habitat. *Biol. J. Linn. Soc.* 8, 279–84.

Nakamura, M. & Nakamura, K. (1977) Population dynamics of the chestnut gall wasp (Cynipidae, Hym.). 5. Estimation of the effect of predation by spiders on the mortality of imaginal wasps based on the precipitin test. *Oecologia* 27, 97–116.

Nakasuji, R., Yamanaka, H., & Kiritani, K. (1973) The disturbing effect of microphantid spiders on the larval aggregation of the tobacco cutworm, *Spodoptera litura* (Lepidoptera: Noctuidae). *Kontyu* 41, 220–7.

Neilson, M.M. (1963) Disease and the spruce budworrn. *Mem. Entomol. Soc. Can.* 31, 272–88.

Neilson, M.M. & Morris, R.F. (1964) The regulation of European spruce sawfly numbers in the Maritime Provinces of Canada from 1937–1963. *Can. Entomol.* 96, 773–84.

Nelmes, A.J. (1974) Evaluation of the feeding behaviour of *Prionchulus punctatus* (Corb.) a nematode predator. *J. Anim. Ecol.* 43, 553–65.

Neuenschwander, P. (1975) Influence of temperature and humidity on the immature stages of *Hemerobius pacificus*. *Environ. Entomol.* 4(2), 215–20.

Newman, G.G. & Carner, G.R. (1975) Disease incidence in soy bean loopers collected by two sampling methods. *Environ. Entomol.* 4, 231–2.

Nicholls, C.F. (1963) Some entomological equipment. *Res. Inst. Can. Dep. Agric. Belleville Inf. Bull.* 2, 85 pp.

Nijveldt, W. (1959) Overhet gebruik van vangekegels bij het galmugonderzoek. *Tijdschr. Pl. Zieki.* 65, 56–9.

Nord, J.C. & Lewis, W.C. (1970) Two emergence traps for wood-boring insects. *J. Georgia Entomol. Soc.* 5, 155–7.

Odum, E.P. & Kuenzler, E.J. (1955) Measurement of territory and home range size in birds. *Auk* 72, 128–37.

Ohnesorge, B. (1957) Untersuchungen über die Populationsdynamik der kleinen Fichtenblattwespe, *Pristiphora abietina* (Christ) (Hym. Tenthr.). I. Teil. Fertilitdt und Mortalitdt. *Z. angew. Entomol.* 40, 443–93.

Okubo, A. (1980) *Diffusion and Ecological Problems: Mathematical Models.* Springer-Verlag, Berlin.

Okubo, A. (1989) Dynamical aspects of animal grouping: swarms, schools, flocks, and herds, *Adv. Biophys.* 22, 1–94.

Otvos, I.S. (1974) A collecting method for pupae of *Lambderia fiscellaria fiscellaria* (Lepidoptera: Geometridea). *Can. Entomol.* 106, 329–31.

Ouchterlony, O. (1948) *In vitro* method for testing the toxin producing capacity of diphtheria bacteria. *Acta Pathol. Microbiol. Scand.* 25, 186–91.

Palmén, E. (1955) Diel periodicity of pupal emergence in natural populations of some chironomids (Diptera). *Ann. Zool. Soc. Vanamo* 17(3), 1–30.

Palmén, E. (1962) Studies on the ecology and phonology of the chironomids (Dipt.) of the Northern Baltic. *Ann. Entomol. Fenn.* 28(4), 137–68.

Palmer, D.F. (1980) Complement fixation test. In: Rose, N.R. & Friedman, H. (eds) *Manual of Clinical Immunology.* American Society for Microbiology, Washington, DC.

Paris, O.H. (1965) The vagility of P^{32}-labelled isopods in grassland. *Ecology* 46, 635–48.

Parker, M.A. (1999) Mutualism in metapopulations of legumes and rhizobia. *Amer. Nat.* 153 (suppl.), 48–60.

Pavlov, I.F. (1961) Ecology of the stem moth *Ochsenheimeria vaculella* F.-R. (Lepidoptera Tineoidea). [In Russian.] *Ent. Obozr.* 40, 818–27. (Transl. *Entomol. Rev.* 40, 461–6.)

Pendleton, R.C. & Grundmann, A.W. (1954) Use of P^{32} in tracing some insect–plant relationships of the thistle, *Cirsium undulatum. Ecology* 35, 187–91.

Peterson, A. (1934) *A Manual of Entomological Equipment and Methods,* Part 1. Edwards Bros., Ann Arbor.

Pickavance, J.R. (1970) A new approach to the immunological analysis of invertebrate diets. *J. Anim. Ecol.* 39, 715–24.

Pickavance, J.R. (1971) The diet of the immigrant planarian *Dugesia tigrina* (Girard). II. Food in the wild and comparison with some British species. *J. Anim. Ecol.* 40, 637–50.

Pierce, G.J., Boyle, P.R., Diack, J.S.W., & Clark, I. (1990) Sandeels in the diets of seals: application of novel and conventional methods of analysis to feces from seals in the Moray Firth area of Scotland (UK). *J. Mar. Biol. Assoc. UK* **70**(4), 829–40.

Pilon, J.G., Tripp, H.A., McLeod, J.M., & Ilnitzkey, S.L. (1964) Influence of temperature on prespinning eonymphs of the Swainejack-pine sawfly, *Neodiprion swainei* Midd. (Hymenoptera: Diprionidae). *Can. Entomol.* **96**, 1450–7.

Plank, V.D. (1967) Spread of plant pathogens in space and time. In: Gregory, P.H. & Monteith, J.L. (eds) *Airborne Microbes*. Cambridge University Press, Cambridge.

Polles, S.G. & Payne, J.A. (1972) An improved emergence trap for adult pecan weevils. *J. Econ. Entomol.* **65**, 1529.

Prebble, M.L. (1941) The diapause and related phenomena in *Gilpinia polytoma* (Hartig). IV. Influence of food and diapause on reproductive capacity. *Can. J. Res. D* **19**, 417–36.

Promptow, A.N. & Lukina, E.W. (1938) Die Experimente beim biologischen Studium und die Erndhrung der Kohlmeise (*Parus major* L.) in der Brutperiode. [In Polish.] *Zool. Zh.* **17**, 777–82.

Radke, W.I. & Frydendall, M.J. (1974) A survey of emetics for use in stomach contents recovery in the house sparrow. *Am. Midl. Nat.* **92**, 164–72.

Rau, P. & Rau, N. (1916) The biology of the mud-daubing wasps as revealed by the contents of their nests. *J. Anim. Behav.* **6**, 27–63.

Reid, R.W. (1963) Biology of the mountain pine beetle, *Dendrocionus monticoiae* Hopkins, in the East Kootenay Region of British Columbia. III. Interaction between the beetle and its host, with emphasis on brood mortality and survival. *Can. Entomol.* **95**, 225–38.

Reiff, M. (1955) Untersuchungen zum Lebenszyklus der Frostspanner *Cheimatobia* (*Operophthera*) *brumata* L. und *Hibernia defoliaria* Ch. *Mitt. schweiz. Entomol. Ges.* **269**, 129–44.

Rice, R.E. & Reynolds, H.T. (1971) Seasonal emergence and population development of the pink bollworm in Southern California. *J. Econ. Entomol.* **64**, 1429–32.

Rich, E.R. (1956) Egg cannibalism and fecundity in *Tribolium*. *Ecology* **37**, 109–20.

Richards, O.W. (1940) The biology of the small white butterfly (*Pieris rapae*), with special reference to the factors controlling its abundance. *J. Anim. Ecol.* **9**, 243–88.

Richards, O.W. & Hamm, A.H. (1939) The biology of the British Pompilidae

(Hymenoptera). *Trans. Soc. Br. Entomol.* **6**, 51–114.

Richards, O.W. & Waloff, N. (1954) Studies on the biology and population dynamics of British grasshoppers, *Anti-locust Bull.* **17**, 184 pp.

Richards. O.W. & Waloff, N. (1961) A study of a natural population of *Phytodecta olivacea* (Forster) (Coleoptera: Chrysomelidae). *Phil. Trans. B* **244**, 205–57.

Rogers, D.J. (1972) Random search and insect population models. *J. Anim. Ecol.* **41**, 369–83.

Rogers, D.J. & Hassell, M.P. (1974) General models for insect parasite and predator searching behaviour: interference. *J. Anim. Ecol.* **43**, 239–53.

Rothschild, G.H.L. (1966) A study of a natural population of *Conomelus anceps* (Germar) (Homoptera: Delphacidae) including observations on predation using the precipitin test. *J. Anim. Ecol.* **35**, 413–34.

Rothschild, G.H.L. (1970) Observations on the ecology of the rice ear bug, *Leptocoris oratorius* (F.) (Hemiptera: Alydidae) in Sarawak (Malaysian Borneo). *J. Appl. Ecol.* **7**, 147–67.

Rothschild, G.H.L. (1971) The biology and ecology of rice-stem borers in Sarawak (Malaysian Borneo). *J. Appl. Ecol.* **8**, 287–322.

Royama, T. (1970) Factors governing the hunting behaviour and selection of food by the great tit (*Parus major* L.). *J. Anim. Ecol.* **39**, 619–68.

Royama, T. (1971) A comparative study of models for predation and parasitism. *Res. Popul. Ecol. Suppl.* **1**, 90 pp.

Sanderson, G.C. (1966) The study of mammal movements – a review. *J. Wildl. Manag.* **309**, 215–35.

Santori, R.T., Moraes, D.A.D., & Cerqueira, R. (1995) Diet composition of *Metachirus nudicaudatus* and *Didelphis aurita* (Marsupialia, Didelphoidea) in southern Brazil. *Mammalia* **59**(4), 511–16.

Santos, M.A. (1976) Prey selectivity and switching response of *Zetzellia maki*. *Ecology* **579**, 390–4.

Sasaba, T. & Kiritani, K. (1972) Evaluation of mortality factors with special reference to parasitism of the green rice leafhopper, *Nephotettix cinticeps* Uhler (Hemiptera: Deltocephalidae). *Appl. Entomol. Zool.* **7**, 83–93.

Scotter, D.R., Lamb, K.P., & Hassan, E. (1971) An insect dispersal parameter. *Ecology* **52**, 174–7.

Service, M.W. (1973) Study of the natural predators of *Aedes cantaris* (Meigen) using the precipitin test. *J. Med. Entomol.* **10**, 503–10.

Service, M. (1976) *Mosquito Ecology, Field Sampling Techniques*. Applied Science Publishers, London.

Shapiro, A.M. (1974) Beak-mark frequency as an index of seasonal predation intensity on common butterflies. *Am. Nat.* **108**, 229–32.

Shaw, M.J.P. (1970) Effects of population density on alienicolae of *Aphis fabae* Scop. *Ann. Appl. Biol.* **65**, 191–4, 197–203, 205–12.

Skellam, J.G. (1951) Random dispersal in a theoretical population. *Biometrika* **38**, 196–218.

Smale, M.J. (1986) The feeding habits of six pelagic and predatory teleosts in eastern Cape coastal waters (South Africa). *J. Zool. Ser. B* **1**, 357–410.

Smirnoff, W.A. (1967) A method for detecting viral infection in populations of *Neodiprion swainei* by examination of pupae and adults. *Can. Entomol.* **99**, 214–16.

Smith, H.S. & DeBach, P. (1942) The measurement of the effect of entomophagous insects on population densities of the host. *J. Econ. Entomol.* **35**, 845–9.

Solomon, M.E. (1949) The natural control of animal population. *J. Anim. Ecol.* **18**, 1–35.

Sommerman, K.M., Sailer, R.I., & Esselbaugh, C.O. (1955) Biology of Alaskan black flies (Simuliidae, Diptera). *Ecol. Monogr.* **25**, 345–85.

Southern, H.N. (1954) Tawny owls and their prey. *Ibis* **96**, 384–410.

Southwood, T.R.E. (1962) Migration of terrestrial arthropods in relation to habitat. *Biol. Rev.* **37**, 171–214.

Southwood, T.R.E. & Siddorn, J.W. (1965) The temperature beneath insect emergence traps of various types. *J. Anim. Ecol.* **34**, 581–5.

Southwood, T.R.E., Jepson, W.F., & van Emden, H.F. (1961) Studies on the behaviour of *Oscinella frit* L. (Diptera) adults of the panicle generation. *Entomol. Exp. Appl.* **4**, 196–210.

Speth, S. (1995) The emergence of Ephemeroptera, Plecoptera and Trichoptera (Insecta) from two streams of the north German Lowland (Osterau and Rodenbek, Schleswig-Holstein). *Limnologica* **25**(3–4), 237–50.

Speyer, W. & Waede, M. (1956) Eine Methodezur Vorhersagedes Weizengallmiicken-fluges. *Nachr. bl. deutsch. Pfl. Sch. Dienst. Stuttg.* **8**, 113–21.

Spiller, D. (1964) Numbers of eggs laid by *Anobium punctatum* (Degeer). *Bull. Entomol. Res.* **559**, 305–11.

Steffan, A.W. (1997) Drift emergence-traps for the recording of hatching running water-insects (Ephemeroptera, Plecoptera, Trichoptera, Diptera). *Entomol. Gen.* **21**(4), 293–306.

Steinbuch, M. & Audran, R. (1969) The isolation of Ig Cr from mammalian sera with the aid of caprylic acid. *Arch. Biochem. Biophys.* **134**, 279–84.

Steinhaus, E.A. (1963) Background for the diagnosis of insect diseases. In: Steinhaus, E.A. (ed.) *Insect Pathology, an Advanced Treatise*, vol. 2.

AcademicPress, New York, London, pp. 549–89.

Straub, R.W., Fairchild, L.M., & Keaster, A.J. (1973) Corn earworm: use of larval traps on corn ears as a method of evaluating corn lines of resistance. *J. Econ. Entomol.* **66**(4), 989–90.

Sullivan, C.R. & Green, G.W. (1964) Freezing point determination in immature stages of insects. *Can. Entomol.* **96**, 158.

Sunderland, K.D. (1988) Quantitative methods for detecting invertebrate predation occurring in the field. *Ann. Appl. Biol.* **112**, 201–24.

Sutton, S.L. (1970) Predation on woodlice: an investigation using the precipitin test. *Entomol. Exp. Appl.* **13**, 279–85.

Sylven, E. (1970) Field movement of radioactively labelled adults of *Dasyneura brassicae* Winn. (Dipt., Cecidomyiidae). *Entomol. Scand.* **1**, 161–87.

Tanada, Y. (1961) The epizootiology of virus disease in field populations of the armyworm, *Pseudaletia unipuncta* (Hamorth). *J. Insect Pathol.* **3**, 310–23.

Tanada, Y. & Kaya, H.K. (1993) *Insect Pathology.* Academic Press, San Diego, 666 pp.

Taylor, L.R. (1975) Longevity, fecundity and size: control of reproductive potential in a polymorphic migrant, *Aphis fabae* Scop. *J. Anim. Ecol.* **44**, 135–63.

Taylor, L.R. & Taylor, R.A.J. (1976) Aggregation, migration and population mechanics. *Nature* **265**, 415–21.

Taylor, L.R. & Taylor, R.A.J. (1978) Dynamics of spatial behaviour. In: *Population Control by Social Behaviour.* Institute of Biology Symposium, London, pp. 181–212.

Taylor, R.A.J. (1978) The relation between density and distance of dispersing insects. *Ecol. Entomol.* **3**, 63–70.

Terrell, T.T. (1959) Sampling populations of overwintering spruce budworm in the Northern Rocky Mountain region. *Res. Note Intermountain For. Range Exp. Sta. Ogden Utah* **61**, 8 pp.

Thomas, G.M. (1974) Diagnostic techniques. In: Cantwell, G.E. (ed.) *Insect Diseases*, vol. 1. Marcel Dekker, New York.

Thompson, D.J. (1975) Towards a predator–prey model incorporating age structure. *J. Anim. Ecol.* **44**, 907–16.

Tinbergen, L. (1960) The natural control of insects in pinewoods. 1. Factors influencing the intensity of predation by songbirds. *Arch. Néerl. Zool.* **13**, 266–343.

Tod, M.E. (1973) Notes on beetle predators of molluscs. *Entomologist* **106**, 196–201.

Toth, R.S. & Chew, R.M. (1972) Development and energetics of *Notonecta undulata* during predation on *Culex tarsalis. Ann. Entomol. Soc. Am.* **5**, 1270–9.

Turnbull, A.L. (1960) The prey of the spider *Linyphia triangularis* (Clerck) (Araneac: Linyphiidae). *Can. J. Zool.* **38**, 859–73.

Turnbull, A.L. (1962) Quantitative studies of the food of *Linyphia triangularis* Clerck (Araneac: Linyphiidae). *Can. Entomol.* **94**, 1233–49.

Turnock, W.J. (1957) A trap for insects emerging from the soil. *Can. Entomol.* **89**, 455–6.

Turnock, W.J. & Ives, W.G.H. (1962) Evaluation of mortality during cocoon stage of the larch sawfly, *Pristiphora erichsonii* (Htg.). *Can. Entomol.* **94**, 897–902.

Urquhart, F.A. (1960) *The Monarch Butterfly.* University of Toronto Press, Toronto, 361 pp.

Vallentyne, J.R. (1952) Insect removal of nitrogen and phosphorus compounds from lakes. *Ecology* **33**, 573–7.

Varley, G.C. & Gradwell, G.R. (1963) The interpretation of insect population change. *Proc. Ceylon Assoc. Adv. Sci. (D)* **18**, 142–56.

Vickerman, G.P. & Sunderland, K.D. (1975) Arthropods in cereal crops: nocturnal activity, vertical distribution and aphid predation. *J. Appl. Ecol.* **12**, 755–66.

Vlijm. L., Van Dijck, T.S., & Wijmans, S.Y. (1968) Ecological studies on carabid beetles. III. Winter mortality in adult *Calathus melanocephalus* (Linn.) egg production and locomotory activity of the population which has hibernated. *Oecologia* **1**, 304–14.

Walker, I., Henderson, P.A., & Sterry, P. (1991) On the patterns of biomass transfer in the benthic fauna of an Amazonian black-water river, as evidenced by ^{32}P label experiment. *Hydrobiologia* **215**, 153–62.

Waloff, N. & Richards, O.W. (1958) The biology of the chrysomelid beetle, *Phytodecta olivacea* (Forster) (Coleoptera: Chrysomelidae). *Trans. Roy. Entomol. Soc. Lond.* **110**, 99–116.

Washino, R.K. & Tempelis, C.H. (1983) Mosquito host bloodmeal identification: methodology and data analysis. *Ann. Rev. Entomol.* **28**, 179–201.

Way, M.I. & Banks, C.J. (1964) Natural mortality of eggs of the black bean aphid, *Aphis fabae* (Scop.), on the spindle tree, *Euonymus europaeus. Ann. Appl. Biol.* **54**, 255–67.

Weaver, T.M. & Fraley, J.F. (1993) A method to measure emergence success of westslope cutthroat trout fry from varying substrate compositions in a natural stream channel. *N. Am. J. Fish. Manag.* **13**(4), 817–22.

Weiser, J. (1969) *An Atlas of Insect Diseases.* Irish University Press, Shannon, and Academia, Prague, 292 pp.

Weiser, J. & Briggs, J.D. (1971) Identification of pathogens. In: Burges, H.D. & Hussey, N.W. (eds) *Microbial Control of Insects and Mites.* Academic Press, London, New York, pp. 13–66.

Weitz, B. (1952) The antigenicity of sera of man and animals in relation to the preparation of specific precipitating antisera. *J. Hyg. Camb.* **50**, 275–94.

Weitz, B. (1956) Identification of blood-meals of blood-sucking arthropods. *Bull. World Health Org.* **15**, 473–90.

Welch, H.E., Jorgenson, J.K. & Curits, M.F. (1988) Measuring abundance of emerging Chironomidae (Diptera): experiments on trap size and design, set duration, and transparency, *Can. J. Fish. Aquat. Sci.* **45**(4), 738–41.

Wellington, W.G. (1964) Qualitative changes in populations in unstable evironments. *Can. Entomol.* **96**, 436–51.

Weseloh, R.M. (1990) Gypsy moth predators: an example of generalist and specialist natural enemies. In: Watt, A.D., Leather, S.R., Hunter, M.D., & Kidd, N.A.C. (eds) *Population Dynamics of Forest Insects.* Intercept, Andover, UK, pp. 233–43.

Wilbur, D.A. & Fritz, R. (1939) Use of shoebox emergence cages in the collection of insects inhabiting grasses. *J. Econ. Entomol.* **32**, 571–3.

van Winkle, W., Martin, D.C., & Sebetich, M.J. (1973) A home-range model for animals inhabiting an ecotone. *Ecology* **54**, 205–9.

Wittig, G. (1963) Techniques in insect pathology. In: Steinhaus, E.A. (ed.) *Insect Pathology, an Advanced Treatise*, vol. 2. Academic Press, London, New York, pp. 591–636.

Wratten, S.D. (1976) Searching by *Adatea bipunctata* (L.) (Coleoptera: Coccinellidae) and escape behaviour of its aphid and cicadellid prey on lime (*Tilia × vulgaris* Hayne). *Ecol. Entomol.* **1**, 139–42.

Wright, D.W., Hughes, R.D., & Worrall, J. (1960) The effect of certain predators on the numbers of cabbage root fly *Erioischia brassicae* (Bouché) and on the subsequent damage caused by the pest. *Ann. Appl. Biol.* **48**, 756–63.

Wrubleski, D.A. & Rosenberg, D.M. (1984) Overestimates of Chironomidae (Diptera) abundance from emergence traps with polystyrene floats. *Am. Midl. Nat.* **111**(1), 195–7.

Young, E.C. (1974) The epizootiology of two pathogens of the coconut palm rhinoceros beetle. *J. Invert. Pathol.* **24**, 82–92.

Zivkov, M. & Petrova, G. (1993) On the pattern of correlation between the fecundity, length, weight and age of pikeperch *Stizostedion luciopera. J. Fish Biol.* **43**, 173–82.

11 The Construction, Description, and Analysis of Age-specific Life-tables

11.1 Types of life-table and the budget

The construction of a number of life-tables is an important component in the understanding of the population dynamics of a species (Carey, 1993). Although some animal ecologists, such as Richards (1940), had expressed their results showing the successive reductions in the population of an insect throughout a single generation, Deevey (1947) was really the first to focus attention on the importance of this approach. Life-tables have long been used by actuaries for determining the expectation of life of an applicant for insurance and thus the column indicating the expectation of life at a given age (the e_x column) is an essential feature of human life-tables. However, the fundamental interests of the ecologist and, even more, of the applied biologist are essentially different from those of the actuary and it is a mistake to believe that the approaches and parameters of primary interest in the study of human populations are also those of greatest significance to the animal ecologist. Because many insects have discrete generations and their populations are not stationary, the age-specific life-table is more widely applicable than the time-specific life-table. The differences between these two types are as follows.

An age-specific (or horizontal, or cohort) life-table is based on the fate of a real cohort; conveniently the members of a population belonging to a single generation. The population may be stationary or fluctuating.

A time-specific (or vertical, or current) life-table is based on the fate of an imaginary cohort found by determining the age structure, at one instant in time, of a sample of individuals from what is assumed to be a stationary population with considerable overlapping of generations, i.e. a multi-stage population. Age determination is a prerequisite for time-specific life-tables (Chapter 12). A modification of this approach is the variable life-table of Gilbert *et al.* (1976), which is an inductive, strategic, computer model of the population; this is varied until it provides a reasonable description of the population (see Chapter 12).

The data in age-specific life-tables may be corrected so as to start with a fixed number of individuals, e.g. 1000; however, this practice, which simplifies the calculation of life expectancy, causes the very important information on actual population size to be lost. It will be seen in this chapter that the variations in population size, from generation to generation, provide the frame of reference against which the roles of the various factors are analysed; therefore, in much work on insect populations, the type of table required lists the actual absolute populations at different stages and records the action of mortality factors, where these are known (see Table 11.4). Such a table, giving just the observed data, is well described by the term budget, proposed by Richards (1961), which also has the advantage that it emphasizes the distinction between this approach and that of the actuary.

11.2 The construction of a budget

The ideal budget will contain absolute estimates of the total population of as many stages as possible. At some points in the generation, it may be possible to determine the total number entering a stage directly, as described in Chapter 9 (see also p. 272). For other stages, there will be a series of estimates, made on successive sampling dates, using methods based on the numbers per unit area (Chapters 4–6) or from mark-and-recapture (Chapter 3), nearest-neighbour (p. 345), or removal-trapping and distance sampling (Chapter 7) techniques. The problem now arises as to how to determine from these estimates the total number of individuals that pass through a particular stage in one generation.

The degree of synchronization in the life cycle is an important factor affecting the ease or difficulty of this step. The ideal situation is when there is a point of time when all the individuals of a generation are in a given stage; a census at this time will provide a 'peak estimate', which may be used in a life-table. As the overlap in time of successive stages increases, it will be necessary to integrate a number of estimates to obtain the total population; special techniques for doing this are described below. When there is complete overlap of all stages, methods based on the age structure of the population are most appropriate (see Chapter 12). Once the series of estimates of total population at each stage have been developed, it can be assumed that the differences between these represent mortality and/or dispersal. It may be possible, and is an advantage, to check this assumption by direct measurement of these factors (Chapter 9).

A budget is, to some extent, self-checking; erroneous population measurements may be exposed by increases in numbers at a stage when natality or immigration is impossible or by other inconsistencies. Thus, the more terms in a budget, the greater the confidence that can be placed in it (Richards, 1959); this confidence is additional to that obtained from the statistical confidence limits of the individual estimates. The latter are based solely on the information gathered by the given method for that particular stage and indicate that the true value is likely (to the extent of the chosen probability level) to lie within them. It is reasonable to claim that when there is agreement with other estimates, by other methods made simultaneously (p. 4) or sequentially, this substantially increases the probability that the true value lies close to the estimate.

Manly (1990) tabulates 23 methods, which use stage-frequency data, for estimating some combination of the numbers entering a stage, stage duration, and stage-specific survival rates. These methods vary in their assumptions and requirements and, in determining which to use, the first consideration must be the validity of the assumptions, followed secondly by the availability of computing facilities. There has been little published on the relative merits of these methods, although Manly and Seyb (1989), Munholland *et al.* (1989), and Manly (1990) do compare some approaches. Of the early and computationally less demanding methods, the graphical method is probably the simplest and most robust (Ruesink, 1975); that of Kiritani and Nakasuji (1967), as modified by Manly (1976), is also relatively simple and reliable (Manly, 1974c). From the beginning of the 1970s, increased computer availability and power allowed the development of stochastic models, for which the parameters can be estimated by maximum likelihood. The initial work of Ashford *et al.* (1970), who described a mathematically complex method, based on the general theory of stochastic branching processes and using maximum-likelihood estimates, had little direct impact, as it was difficult for biologists to understand and implement. Birley's (1977) method, which was further developed by Bellows and Birley (1981), is flexible because survival can vary between the different stages; however, this can result in the desire to estimate more parameters than the observations will allow. This is termed overparameterization. A theoretical framework for a class of widely applicable models was presented by Kempton (1979) and models of this type were compared by Manly and Seyb (1989). Manly (1990) considered that the Kempton model offered the most efficient means of estimating the parameters by maximum liklihood. Finally, there have been methods developed for populations followed over a number of generations or which have overlapping generations, which will not be discussed further here. When the population is sampled over a number of generations, parameters for the different stages may be estimated by fitting age-structured simulation models. Schneider and Ferris (1986) describe an algorithm for estimating the stage-specific duration and survival parameters for such a model. An age-structured population can also be envisaged as a surface obtained from a plot of age by time by population size. Wood (1994) shows how age- and time-specific mortality and birth-rates can be estimated by fitting a smoothed population surface to age-structured data, using spline interpolation.

11.2.1 Southwood's graphical method

This is the crudest and simplest of the methods of integration. It gives reasonable estimates of the numbers entering a stage if mortality, which may be heavy, occurs entirely at the end of the stage. If a constant mortality rate applies throughout the stage, then an overestimate of the numbers as the median age of the stage will result. Successive estimates are plotted on graph paper, most conveniently allowing one square per individual and per day (Fig. 11.1). The points are joined up and the number of squares under the line counted; this total is then divided by the mean developmental time under field conditions to give the estimate of the numbers reaching the median age for the stage. This technique has been found to be robust

Total number of 'egg–days' = 106
As development period = 5 days
∴ total individual eggs reaching
age of 2.5 days ≃ 106 ÷ 5 = 21.2

Fig. 11.1 The determination of the total number of individuals in a stage from a series of estimates by graphical summation (hypothetical example with unrealistically small numbers for simplicity).

(Ruesink, 1975) and has been quite widely used (e.g. Southwood & Jepson, 1962; Helgeson & Haynes, 1972).

11.2.2 Richards and Waloff's first method
This method (Richards & Waloff, 1954) assumes a single impulse of recruitment and a constant mortality rate; it is therefore only applicable to a stage with a well-marked peak in abundance. By plotting the regression of the fall-off of numbers with time after the peak and then extending this back to the time when the stage was first found, the total number entering the stage can be estimated (Fig. 11.2).

A population exposed to a steady mortality may be expressed as:

$$Y_t = N_0 \phi^t \tag{11.1}$$

where Y_t = population on day t, N_0 = the peak population (ideally the number hatched), and ϕ = the fraction of the population which survives to the end of a unit of time (e.g. a day), the survival rate per unit time. Hence:

$$\log Y_t = \log N_0 + t \log \phi \tag{11.2}$$

and, after the peak, this will describe a straight line, which is conveniently obtained from the regression of $\log Y_t$ on t; the regression coefficient will be $\log \phi$, the logarithm of the average, and supposedly constant, survival rate. The number entering the stage will then be given by the value of Y_t, found by inserting into the equation a value of t corresponding to the start of the stage. Alternatively, if t is numbered from the start of the stage in question, $\log N_0$ will equal the number entering the stage.

More reliable estimates are obtained if Y_t is taken as the accumulated population,

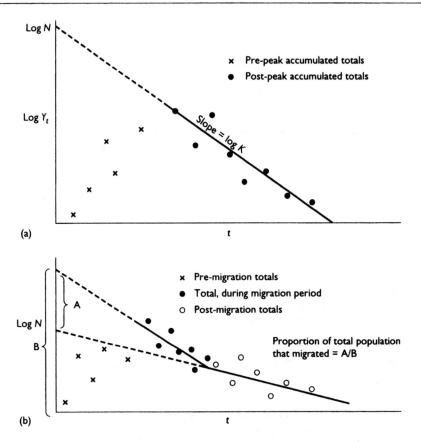

Fig. 11.2 The estimation of the number of individuals entering a stage by Richards and Waloff's first method. (a) Simple case without migration. (b) When the animal may migrate.

i.e. after the peak of a particular instar, the values of Y_t are the total of that instar, plus any individuals that have passed through that instar and are in subsequent ones. For example, Y_t for the second instar would consist of the populations of second instar, third instar, and fourth instar, etc. estimated for day t, but Y_t for the third instar would consist only of the populations of the third and subsequent instars. Dempster (1956) found that, if Y_t was based solely on the numbers of the particular instar, the estimates of population were too high, the slope of the line being too steep due to the confusion of moulting with mortality; the use of accumulated totals avoids this.

Waloff and Bakker (1963) used a modification of this method to determine the total change in population due to migration. Basically, separate regressions were calculated for the migration and post-migration period and the slopes (log ϕ) compared. The value of log ϕ for the migration period represents migration and mortality, for the post-migration period just mortality. Therefore, if the difference between these is expressed as a percentage of the larger, this will be the percentage of the initial population that was lost during the migratory period.

11.2.3 Manly's method

Manly's (1974a, b) method is essentially a modification of Richards and Waloff's first method, but it avoids the need for a distinct peak. Recruitment is assumed to be normal, but Manly considered that, so long as the entry was unimodal, reasonable estimates would be provided. Simulation studies suggest that significant biases will arise if recruitment to the first stage is 'far from normal' (Manly, 1974c). The basic equation is:

$$Y_t = N_0 \int_{-\infty}^{t} \exp[-\theta(t-x)] f(x)(dx) \tag{11.3}$$

where Y_t = the number of animals in a particular stage and later stages at time t, N_0 = total number entering the stage, $\exp(-\theta(t-x))$ = the probability that an animal that enters the population at time x is still alive at time t (θ being the constant age-specific death-rate), and $f(x)$ is the frequency of the stage at time x. Because of Manly's assumption of normality, the frequency function for this distribution can be inserted and equation (11.3) becomes:

$$Y_t = N_0^* \exp(-\theta t) \int_{-\infty}^{t-\mu^*/\sigma} (2\pi)^{-\frac{1}{2}} \exp\left(-\frac{1}{2}x^2\right) dx \tag{11.4}$$

where:

$$N_0^* = \exp\left[\theta\left(\mu + \frac{1}{2}\theta\sigma^2\right)\right] N_0$$

and:

$$\mu^* = \mu + \theta\sigma^2$$

where μ is mean time of entry to the stage and σ is its standard deviation.

The four unknowns, N_0, μ, σ, and θ, may be estimated for a series of at least four values of Y_t by using a computer to find the best fit by non-linear regression. Manly (1974a) also describes how the values can be estimated from graphical methods.

11.2.4 Ruesink's method

Ruesink's (1975) method, like that of Birley, allows the calculation of the stage survival rate. It does this by a comparison of the rate of recruitment into the stage in question with the rate of recruitment to the next stage. The calculations are most easily performed by computer and the basic equation is:

$$\phi_{jj+1} = \frac{C_{j+1}(t+n) - C_{j+1}(t-n)}{D_j(t+n) - D_j(t-n)} \tag{11.5}$$

where ϕ_{jj+1} = the survival between the jth stage and the next stage, C_{j+1} = the total number of individuals that have entered stage $(j+1)$ in the periods $(t+n)$ and $(t-n)$;

likewise D_j = the total number of individuals that have left stage j; n = an arbitary time period. Ruesink (1975) points out that, if there are large sampling errors, n must be large to provide meaningful estimates. The values of C_{j+1} and D_j are obtained either graphically or by computer. In the former method, a continuous line is drawn through the points from field data for the numbers of that stage; this is, in essence, the approach of Richards and Waloff's first method.

11.2.5 Dempster's method

This method requires an independent estimate of the total natality of the first stage, together with a series of population estimates – at least two more than the number of stages (Dempster, 1961). If the mortality rate (μ) is assumed to be a constant, then:

$$Y_0 - Y_t = N\alpha_{(0 \to t)} - \frac{(I_0 + I_t)}{2}t\mu_1 - \frac{(II_0 + II_t)}{2}t\mu_2 \cdots \frac{(Ad_0 + Ad_t)}{2}t\mu_a \qquad (11.6)$$

where Y_0 and Y_t = total population at days 0 and t (successive sampling dates), N = the total number hatching (or emerging), which is found independently; $\alpha_{(0 \to t)}$ = the proportion of the total hatch that occurs between days 0 and t; I_0 and I_t = the total numbers of first-instar larvae on the first and second sampling dates, respectively, II_0 and II_t are the numbers of the second-instar larvae; further terms are inserted, as appropriate, until the adult (Ad_0 and Ad_t) stage is reached; t is the time interval between day 0 and day t (the two sampling occasions), and $\mu_1, \mu_2 \dots \mu_a$ are the average daily mortality of first- and second-stage larvae and adults, respectively.

The unknowns, $\mu_1, \mu_2 \dots \mu_a$ may be found by solving a series of simultaneous equations, provided there is one more equation than there are unknowns. Alternatively, if sufficient observations have been taken, the unknowns can be found by multiple regression.

Where there is migration, a further term needs to be inserted into the equation; let M = the net result of migration in terms of increase or decrease of population. A measure of the amount of migration may be obtained, e.g. from one or more suction, rotary, or interception traps placed close to the population, and from these data the statistic x may be calculated:

$$x = \frac{\text{number trapped between days 0 and } t}{\text{total number trapped}}$$

Thus, x is the migratory analogue of α. The following term is then inserted in the equation: $Mx_{(0 \to t)}$, the value of M and its sign being unknown; if M is positive, migration has resulted in a gain (immigration), if negative in a loss (emigration). Estimates obtained in this way should, where possible, be compared with others obtained from directional interception traps and other approaches (pp. 271 and 391).

This is a method capable of wide application; however, it is not robust. In order to obtain reliable estimates, a large number of accurate population samples are required (Manly, 1974c).

11.2.6 Richards and Waloff's second method

This method may be used when recruitment and mortality overlap widely. The total number (N_i) of the ith stage taken in all the daily samples will be given by:

$$N_i = N_0 \int_0^a \phi^t dt = \frac{N_0(\phi^a - 1)}{\log_e \phi}$$ (11.7)

where N_0 = the total number entering the stage, ϕ = the survival rate (for a unit of time), and a = the duration of the stage (Richards, 1959; Richards et al., 1960; Richards & Waloff, 1961). Now, if the number of eggs laid (N_0) and duration of the egg stage (a) are known and N_i is obtained from samples, then the equation can be solved for ϕ. The percentage mortality is $100(1 - \phi^a)$, and this may be used to calculate the number surviving the egg stage and entering the first larval instar. Thus, N_0, a, and N_i will again be known values in the equation for the first instar, so that the percentage mortality for this instar and the numbers entering the next may be found. The process can be repeated to the end of the life cycle.

In order to apply this method, one must have an accurate estimate for the initial N_0; Richards and Waloff (1961) obtained this from the population of adult females, assessed independently, and the oviposition rate, measured under the field conditions. It is also necessary to determine the duration of the stage (a) experimentally and, as Richards et al. (1960) pointed out, differences of as little as half a day in the estimate of the duration of an instar can make a large difference in the estimated mortality. It is clear, therefore, that inaccuracies in these independent estimates of a and N_0 or discrepancies between the actual and assumed field conditions, which will influence the values assigned to a and N_0, are potential sources of error with this method.

11.2.7 Kiritani, Nakasuji and Manly's method

As originally developed (Kiritani & Nakasuji, 1967), this method required that samples be taken at regular intervals throughout the generation and, unless the number entering stage i are known independently, only survival rates can be estimated. Manly (1976) has modified the method to remove these restrictions. Samples are taken at irregular intervals $(b_1, b_2 \ldots)$ and the area under the frequency-trend curve is estimated by the trapezoidal rule:

$$\hat{A}_i = \frac{1}{2} \sum_{l=1}^{n} (b_l + b_{l+1}) \hat{f}_{il}$$ (11.8)

where f_{il} = the number of the ith instar estimated from the samples taken on the lth occasion, which is at the end of the sampling intervals b_1; there are $(n + 1)$ sampling intervals, the last interval extending from the last occasion when the stage was present to the next sampling occasion (when it was found to be absent). It will be noted that these f_{il} values correspond to $I, II \ldots Ad$ of Dempster's method but, whereas in his method two numbers are averaged over one sampling interval, in this method the numbers are 'spread' into the sampling intervals on either side and then the sum divided by two.

Then the survival rate of the ith stage is estimated:

$$\hat{\phi}_i = \left| -A_i \right| \sum_{j=1}^{q} A_j \tag{11.9}$$

where j = the next stage and q = the last stage.

Manly (1976) shows how further estimates can be obtained if the median stage (instar value, which will not normally, of course, be a whole number) is plotted against time, so that the time for a 'median' animal to pass from one stage to another can be read off. Now, the daily survival rate raised to the power of the duration of the stage should equal the stage survival rate, as determined above:

$$\hat{\phi}_i = \hat{\phi}_d^{\hat{a}_i} \tag{11.10}$$

where $\hat{\phi}_d$ = daily survival rate and $\hat{a}_i$ = the duration of the ith stage, as determined above. (This calculation can be done for two or more instars at once, which may give a better estimate of duration.) Thus:

$$\log \hat{\phi}_d = \log \frac{\hat{\phi}_i}{\hat{a}_i} \tag{11.11}$$

If, on the other hand, time is taken over several instars (i.e. the divisor is, say, $a_1 + a_2 + a_3$), then, having found ϕ_i and ϕ_d, the appropriate stage durations (a_i values) can be found. Then the numbers entering each stage may be found:

$$\hat{N}_{0i} = -\ln \phi_d \left[\sum_{j=i}^{q} A_j \right] \tag{11.12}$$

where $\hat{N}_{0i}$ = the number entering the ith stage and the others are as above.

It will be apparent that this method depends on the assumption of a constant daily survival rate for all stages. This method is, like the two following methods, able to model multi-cohort data.

11.2.8 Kempton's method

Kempton (1979) gave the probability that an individual will be in stage j at time t as:

$$p_j(t) = w(t) \int_0^t \left\{ g_j(y) - g_{j+1}(y) \right\} dy, \quad j < q$$

and: $\tag{11.13}$

$$p_q(t) = w(t) \int_0^t g_q(y) dy,$$

where $w(t)$ is the probability of surviving to time t and $g_j(t)$ is the probability-density function of the time of entry to stage j. The integral term in eqn 11.13 calculates the probability that an individual alive at time t is at stage j. For example, the integral from 0 to t of $g_j(y)$ gives the probability that an individual has entered stage j by time t.

Kempton (1979) noted that a stage-structured model required three components.

1 A description of stage survival. When constant survival rates cannot be assumed, a gamma, Weibull, or log-normal distribution in life expectancy might be reasonable assumptions.

2 A model for the time of entry into stage 1. Conceptually, each individual can be considered to start in stage 0 and to enter stage 1 at a time that is either normally or gamma-distributed.

3 A model for the distribution of the duration of each stage. It is advantageous to assume that this is either constant or has a normal, gamma, or inverse normal distribution.

The particular distributions chosen for 1, 2, and 3 above will determine the functions $w(t)$ and $g_j(t)$ and the difficulty in estimating $p_j(t)$. In a worked example, Kempton (1979) assumed survival per unit time to be constant and stage duration and time of entry to stage 1 to be gamma-distributed, with a common scale parameter. He further assumed that the number of individuals counted in stage j at time t was Poison-distributed, with mean $N_0 p_j(t_i)$, where N_0 is the total number of individuals at time 0. He was then able to give a likelihood function, which he was able to maximize. While considered by Manly and Seyb (1989) to be one of the most efficient methods for modelling multi-cohort data, access to a computer program is generally required to use this method.

11.2.9 The Bellows and Birley method
Bellows and Birley (1981) refined the model originally developed by Birley (1977). The method allows each stage to have a different survival rate.

The number of individuals in stage j at time t, Y_t, is given by:

$$Y_j(t) = N_1 \sum_{i=0}^{t} P_j(i)\phi_j^{t-i}\{1 - H_j(t-i)\} \tag{11.14}$$

where N_1 is the total number of individuals entering stage 1 and $P_j(i)$ is the proportion of these individuals which have entered stage j at time t. The survival rate of stage j is ϕ_j and $H_j(t)$ is the probability that an individual remains for less than or equal to t time units in stage j.

The proportion of the population entering stage j at time t is a function of the numbers that have arrived at other times:

$$P_{j+1}(t) = \sum_{i=0}^{t} P_j(i)\phi_j^{t-i}b_j(t-i) \tag{11.15}$$

where $b_j(t)$ denotes a duration of t time units in stage j.

The model requires assumptions about the distribution of stage duration times and the distribution of entry to stage 1. Bellows and Birley (1981) suggest that stage duration distribution is inverse normal, because development time is normally distributed. Manly (1990), however, considered the Weibull distribution as advantageous because of its simple form and he also used it to model the time of entry to stage 1, so that:

$$H_j(t) = 1 - \exp\left\{-\left(\frac{t}{Q_j}\right)^{\alpha_j}\right\}$$

(11.16)

and:

$$P_1(t) = \exp\left[-\{(t-1)/Q_0\}^{\alpha_0}\right] - \exp\left[-(t/Q_0)^{\alpha_0}\right]$$

(11.17)

where α and Q_j are parameters to be estimated so that $H_j(t)$ and $P_1(t)$ may be found.

While Bellows and Birley (1981) suggested using non-linear regression to estimate the parameters of the model, Manly (1990) noted that, if stage-frequency data could be assumed to be Poisson variates, then the general maximum-likelihood algorithm MAXLIK can be used.

As in the case of Kempton's method, a computer program is required to undertake the calculations.

11.3 The description of budgets and life-tables

11.3.1 Survivorship curves

The simplest description of a budget is the graphical representation of the fall-off of numbers with time—the survivorship curve. The number living at a given age (l_x) are plotted against the age (x); the shape of curve will describe the distribution of mortality with age. Slobodkin (1962) recognized four basic types of curve (Fig. 11.3); in type I, mortality acts most heavily on the old individuals; in type II (a

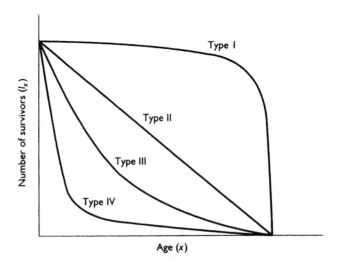

Fig. 11.3 Types of survivorship curves (after Slobodkin, 1962).

straight line when the l_x scale is arithmetic), a constant number die per unit of time; in type III (a straight line when the l_x scale is logarithmic), the mortality rate is constant; and, in type IV, mortality acts most heavily on the young stages. Deevey (1947) also drew attention to these different types, but only recognized three and, in order to avoid confusion between his classification and Slobodkin's, it should be noted that Deevey plotted survivors (l_x) on a log scale and thus type II of Slobodkin was not recognized by him; type II of Deevey = type III of Slobodkin and type III of Deevey = type IV of Slobodkin.

Fish and many marine organisms with high fecundities show type IV curves. In insects, mortality often occurs in distinct stages, so that survivorship curves show a number of distinct steps (Ito, 1961).

11.3.2 Stock–recruitment (Moran–Ricker) curves

The density-dependent relationship between the number of adults in a population and recruitment can be expressed by the stock–recruitment curve. This approach is particularly important in fisheries research, where independent estimates of the number of adult and larval fish are available. The management objective in fisheries is to maintain the adult population at a level that will produce sufficient recruits to replace the adults killed by both natural causes and fishing. If recruitment is density-dependent, then, at least in theory, it is possible to exploit a population without appreciably reducing recruitment. Ricker (1954) and Beverton and Holt (1957) originally derived the two basic stock–recruitment relationships that are commonly used.

The curve developed by W.E. Ricker is expressed by the equation:

$$R = CNe^{-DN} \tag{11.18}$$

where N is the size of the population, R is the number of recruits, C and D are constants, and e is the base of the natural logarithm.

The parameter D measures the tendency of the population to adjust to changes in population number and C measures the number of individuals that would survive to adulthood if only density-independent mortality was operating. This curve reaches a maximum when $N = 1/D$ and $R = C/D_e$ and has been found to describe the stock relationships of some fish, including salmon (Gulland, 1983). Given estimates of N and R for a number of years, the parameters C and D are estimated by regression. In practice, it is difficult to obtain reliable estimates of C and D.

The Beverton–Holt curve is given by the equation:

$$R = \frac{1}{A + \dfrac{B}{N}} \tag{11.19}$$

where N and R are as above and A and B are constants to be estimated.

These stock–recruitment curves, examples of which are shown in Fig. 11.4, are particularly appropriate for the description of density-dependent relationships in

Fig. 11.4 Examples of stock–recruitment relationships for fish. (a) Plaice, *Pleuronectes platessa*, in the western English Channel (after Iles, 1994). The fitted model is the Saila–Lorda equation with Scheffé confidence bands. (b) English sole, *Pleuronectes vetulus* (after Iles & Beverton, 1998); o, original data, +, adjusted for Ekman transport; Beverton–Holt curves were fitted to the original data (solid line) and the adjusted data (dotted line).

fisheries studies, because of the biological characteristics of fish populations. First, the extremely high fecundity of fish makes a Ricker-type relationship feasible. Secondly, adults and young fish often differ greatly in lifestyle and distribution, so that different methods are used to estimate their numbers, making it natural to divide the adult stock from the recruits. Both because of the difficulty of fitting appropriate curves to noisy data and because of their lack of biological realism, the utility of stock–recruitment relationships has frequently been questioned.

An alternative use of this approach was developed by Rogers (1979) and depends on plotting the data on log–log scales. Then the finite capacity for increase (λ) describes a straight line, parallel to the 45° line, until density-dependent factors begin to operate (Fig. 11.5). He termed this the Moran curve (after P.A.P. Moran, who put forward such methods of expressing population data). A long series of successive populations are plotted; the upper limit will be likely to delineate the point at which density dependence starts to operate — where the curve inflects. The vertical departures from the line before the point of inflection will represent the action of density-dependent factors. Departures after this point (the right-hand side of Fig. 11.5) represent the action of both density-dependent and independent factors. The Moran curve may be fitted by eye to a long series of data (as in Fig. 11.5) when $\lambda = e^{rm}$. Alternatively, the straight portion may be calculated from a knowledge of maximal potential fertility (as in Fig. 11.5a) when $\lambda = e^{rp}$. Even when the Moran curve cannot be determined with certainty, the plotting of successive popu lations on a log–log scale allows the comparison of the population dynamics of different species (Southwood, 1981). An example is provided by the tight cycling of *Bupalus* (Fig. 11.5) compared with that of another moth, *Tyria* (Fig. 11.5), where density-dependent factors seem to be absent between the second and third generations.

Further approaches to the analysis of time-series are given in Chapter 15 (p. 545),

Fig. 11.5 Examples of the Moran curve. (a) Moran plot for the tsetse fly, *Glossina morsitans submorsitans*, populations in the Yankori Game Reserve, Nigeria (after Rogers, 1979); the numbers represent the month and the subscripts the year. (b) Sequential Moran plot for the pine moth, *Bupalus piniarius*, populations. (c) Sequential Moran plot for the cinnabar moth, *Tyria jacobaeae* (after Southwood, 1981).

and a detailed account of various tests for density dependence in such series is provided by Rothery (1998).

11.3.3 The life-table and life expectancy

For the further description of the data collected in the form of a budget or any type of life-table (Chapter 12), it is often most convenient if these are corrected so as to commence with a fixed number, usually 1000. A table may then be constructed with the following columns (Deevey, 1947):

x the pivotal age for the age class in units of time (days, weeks, etc.)

l_x the number surviving at the beginning of age class x (out of 1000 originally born)

d_x the number dying during the age interval x

e_x the expectation of life remaining for individuals of age x

Table 11.1 A hypothetical life-table.

x	l_x	d_x	L_x	T_x	e_x	$1000\,q_x$
1	1000	300	850	2180	2.18	300
2	700	200	600	1330	1.90	286
3	500	200	400	730	1.46	400
4	300	200	200	330	1.10	667
5	100	50	75	130	1.30	500
6	50	30	35	55	1.10	600
7	20	10	15	20	1.00	500
8	10	10	5	5	0.50	1000

In practice, the table may have two further columns (Table 11.1) to facilitate the calculation of the expectation of life, as follows.

1 The number of animals alive between age x and $x + 1$ is found. This is calculated using:

$$L_x = \int_x^{x+1} l_x d_x \approx \frac{l_x + l_{x+1}}{2} \tag{11.20}$$

if the age intervals are reasonably small.

2 The total number of animals × age units beyond the age x, which is given by:

$$T_x = L_x + L_{x+1} + L_{x+2} \ldots L_w \tag{11.21}$$

where w = the last age. In practice, it is found by summing the L_x column from the bottom upwards.

The expectation of life is theoretically:

$$e_x = \frac{\int_x^w l_x d_x}{l_x} \tag{11.22}$$

and is therefore given by:

$$e_x = \frac{T_x}{l_x} \tag{11.23}$$

When the survivorship curve is of type I (Fig. 11.3), e_x will decrease with age; it will be constant for type II and it will increase for types III and IV.

A further column is sometimes added to life-tables: the mortality rate per age interval (q_x), usually expressed as the rate per 1000 alive at the start of the interval:

$$1000\,q_x = 1000\frac{d_x}{l_x} \tag{11.24}$$

11.3.4 Life and fertility tables and the net reproductive rate

Hitherto, we have been concerned solely with the description of one of the pathways

Table 11.2 Life and fertility table for the beetle, *Phyllopertha horticola* (modified from Laughlin, 1965).

x (in weeks)	l_x	m_x	$l_x m_x (V_x)$
0	1.00	—	Immature
49	0.46	—	Immature
50	0.45	—	Immature
51	0.42	1.0	0.42
52	0.31	6.9	2.13
53	0.05	7.5	0.38
54	0.01	0.9	0.01

of population change — mortality. In this section and the next, methods of describing natality and its interaction with mortality in the population will be discussed.

A life and fertility table (or fecundity schedule) may be constructed by preparing a life-table with x and l_x columns, as before, except that the l_x column refers entirely to females and should represent the number of females alive, during a given age interval, as a fraction of an initial population of one, or, expressed another way, the life expectancy at birth to age x as a fraction of one (Birch, 1948) (Table 11.2).

A new column is then added on the basis of observations; this is the m_x or age-specific fertility column (often termed the fecundity column, but, as it refers to live births, fertility is more appropriate), which records the number of living females born per female in each age interval. In practice, it is frequently necessary to assume a 50:50 sex ratio, when $m_x = N_x/2$, N_x being the total natality per female of age x.

Columns l_x and m_x are then multiplied together to give the total number of female births (female eggs laid) in each age interval (the pivotal age being x); this is the $l_x m_x$ column.

The number of times a population will multiply per generation is described by the net reproductive rate, R_0, which is:

$$R_0 = \int_0^\infty l_x m_x d_x = \sum l_x m_x \qquad (11.25)$$

Thus, from Table 11.2, $R_0 = 2.94$; this net reproductive rate may be expressed in another way, as the ratio of individuals in a population at the start of one generation to the numbers at the beginning of the previous generation. Thus:

$$R_0 = \frac{N_{t+\tau}}{N_t} \qquad (11.26)$$

where τ = generation time.

Clearly, values of R_0 in excess of 1 imply an increasing population and of less than 1 a decreasing population; when $R_0 = 1$, the population will be stationary. Where it is difficult to define the generation time, the value of R_0 as a description is limited, and this led A.J. Lotka, a student of human demography, to propose the consideration of the growth rates of populations.

11.3.5 Population growth rates

As A.J. Lotka pointed out, the growth rate of a population is r in the equation:

$$\frac{dN}{dt} = rN \qquad (11.27)$$

where N is the number of individuals at any given time (t). On integration, this gives:

$$N_t = N_0 e^{rt} \qquad (11.28)$$

The parameter r in this equation describes population growth. Under conditions of an unlimited environment and with a stable age distribution, this parameter becomes a constant.

The maximum value of the parameter r that is possible for the species under the given physical and biotic environment is denoted as r_m and is variously termed the intrinsic rate of natural increase, the Malthusian parameter, the innate capacity for increase, or various combinations of these (Leslie & Ranson, 1940; Birch, 1948; Caughley & Birch, 1971), whilst $e^r = \lambda$ is the finite capacity for increase. It must be stressed that this value is the maximum under various natural conditions, i.e. it allows for some mortality; the theoretical potential, r_p, with maximal fertility and zero mortality will be higher (Southwood, 1969). Under optimal physical conditions, r_m will approach r_p; at the limits of a species' environmental range, r_m will approach zero; and, beyond the boundary of the range, r_m will be negative. If the population is constrained by its environment (perhaps through intraspecific effects), another parameter, r_s, may be determined, which is the rate at which the population would change (it may be positive or negative) if the age distribution were stable (Caughley & Birch, 1971). As environmental constraints (due to population density) approach zero, so r_s comes to equal r_m. The parameter r_s is of particular interest to students of vertebrate populations.

The intrinsic rate of natural increase (r_m) is of value as a means of describing the growth potential of a population under given climatic and food conditions (Messenger, 1964; Watson, 1964). It is an important parameter in inductive strategic and management models for insect-pest populations and in fisheries work (Jensen, 1975).

When a stable age distribution has been achieved, but the population is still growing in an unlimited environment, and given $l_x m_x$ values from a life-table, r_m may be calculated from the expression:

$$\sum e^{-r_m x} l_x m_x = 1 \qquad (11.29)$$

However, ecologists are interested in r_m or the 'time r' in a broader context — its relationship to life-history parameters, such as generation size and fecundity (Smith,

1954; Fenchel, 1974; Southwood *et al.*, 1974)—whilst the establishment of a stable age distribution in an unlimited environment may be virtually impossible. Sometimes, the instantaneous birth-rate per individual ($\sim r_p$) has been used instead of the instantaneous rate of natural increase per individual, thereby eliminating the mortality pathway.

An approximation for r_m, the capacity for increase (r_c), has long been used by insect ecologists, and Laughlin (1965) presented a graph that showed its error over r_m in relation to the relative length of the reproductive period and the net reproductive rate. It is defined:

$$r_c = \frac{\ln R_0}{T_c} \tag{11.30}$$

where R_0 is as defined in equations (11.25) and (11.26) and T_c = cohort generation time—the mean age of the females in the cohort at the birth of female offspring or the pivotal age where $l_x m_x = 0.5\, R_0$ (Bengstron, 1969).

The simple and general relationship between r_m and r_c (and between τ and T_c) has been elucidated by May (1976):

$$r_c = r_m \left(1 - \frac{r\sigma^2}{2T_c} + \cdots \right) \tag{11.31}$$

where σ^2 the variance of the $l_x m_x$ distribution—is defined as:

$$\sigma^2 = \left(\frac{\sum x^2 l_x m_x}{R_0} \right) - T_c^2 \tag{11.32}$$

The further correction terms, omitted from equation (11.31), involve higher-order moments (skewness, etc.); if we can assume that $l_x m_x$ is normal (Gaussian), then the equation is exact. May (1976) rearranged the equation for r_c to give an expression of the relative error involved in calculating r_c rather than r_m:

$$\frac{r_m - r_c}{r_c} \approx \frac{1}{2} (\ln R_0)(CV)^2 \tag{11.33}$$

where $CV = \sigma/T_c$, the variation in the distribution of the age of reproduction within a cohort. A consideration of equation (11.33) immediately shows that, provided the $l_x m_x$ distribution is approximately normal, the relative error of r_c compared with r_m will be small, provided that:

1 $R_0 \approx 1$, i.e. the population essentially just replaces itself, e.g. some human populations, other vertebrates with low fecundities (e.g. albatross, whales).

2 CV is small, i.e. reproduction tends to occur in all individuals at about the same age and is not spread out over a long reproductive life. This is true of many arthropods, but not of long-lived, virtually iteroparous species, such as the tropical rain-forest butterflies, *Heliconius*.

CV is clearly related to Laughlin's (1965) comparison of reproductive period to total period (to oldest reproductive age).

Another parameter that is involved in expression of life-table data is generation time (τ), which is defined:

$$\tau = \frac{\ln R_0}{r_m} \tag{11.34}$$

and has been considered to have little biological significance; however, May (1976) shows that:

$$\tau = T_c\left[1 - \frac{r_m \sigma^2}{2T_c} + \cdots\right] \tag{11.35}$$

Thus the difference is significant only if the $l_x m_x$ distribution has a high variance, particularly if this implies (as it generally will) a significant departure from the normal (i.e. additional terms are required in the above). Lefkovitch (1963) suggests that variation in the stored-products beetle, *Lasioderma serricorne*, is not normally distributed; Messenger (1964) also obtained various anomalous results, possibly for the same reason. The subject would repay further investigation in the light of May's (1976) paper, but such errors are not among the most substantial facing ecologists!

11.3.6 The calculation of r

Where a life-table can be constructed for a stable age distribution in an unlimited environment, then r_m may be calculated by numerical methods on a computer, using equation (11.29). For many insect species, where the conditions listed above are met, r_c is adequate and may be calculated from equation (11.30).

Several species of aphid and spider mites have been found by Wyatt and White (1977) to have steep cumulative $l_x m_x$ curves early in adult life, which flatten out at a time period that happens to equal the period of pre-reproductive life, i.e. the development time (ξ) from birth to reproduction (Fig. 11.6). The value on the $\Sigma l_x m_x$ curve at a particular time represents the total births until that time per number of original females. Wyatt and White (1977) point out that it is much easier to determine the pre-reproductive development period (ξ) and the total reproduction/original female (R_ξ) in the first part of the reproductive period of length = ξ than the $l_x m_x$ values. The slope given by these values will be slightly steeper than r_c (Fig. 11.6), so that:

$$r = \frac{\varepsilon\left(\log_e R_\xi\right)}{\xi} \tag{11.36}$$

where ε is a constant that corrects the slope of this line to correspond with r_c; Wyatt and White (1977) found $a = 0.74$ to hold for many aphids and mites, but, when a

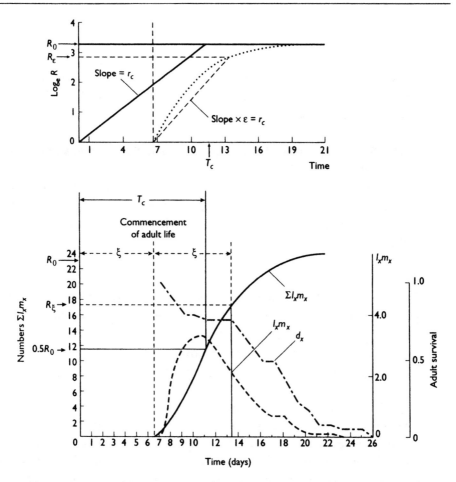

Fig. 11.6 The intrinsic rate of natural population increase. Population growth data for the mite, *Tetranychus urticae* (from Bengstron, 1979), displayed to illustrate the values used to make approximate estimates following the methods of Laughlin (1965) and Bengstron (1969) (r_l) and Wyatt and White (1977) (r_c/ε) (see text for further explanation).

whole series of determinations are to be made under different climatic or biotic conditions, the value of the constant should be confirmed, because it is very dependent on the form of the $\Sigma l_x m_x$ curve. This method additionally assumes, of course, that CV, as defined for equation (11.33) is small.

11.4 The analysis of life-table data

In the previous section, we were concerned with the description of a life-table by a single parameter, r_m. The present section will discuss the methods of 'taking the life-table apart' so as to determine the role of each factor. Several life-tables are necessary; ideally these will be a series for a number of generations of the same population, but some information can be obtained from life-tables from different populations or by measuring density, mortality, and the associated factors in different parts of the habitat. Such analysis is not only of considerable theoretical interest, but will also eventually provide a rational basis for prediction of animal populations: enabling, for example, the forecasting of climatically induced pest outbreaks and the making of prognoses of the effects of changes in cultural or control practices. The analysis of population on multiple sites has been developed by Dennis and Taper (1994) and Dennis *et al.* (1998) (see Chapter 15).

11.4.1 The comparison of mortality factors within a generation (Table 11.3)

11.4.1.1 Apparent mortality

This is the measured mortality, the numbers dying as a percentage of the numbers entering that stage, i.e. d_x as a percentage of l_x (see also p. 406). Its main value is for simultaneous comparison either with independent factors or with the same factor in different parts of the habitat.

11.4.1.2 Real mortality

This is calculated on the basis of the population density at the beginning of the generation, i.e. $100 \times d_i/l_c$ = the deaths in the ith age interval, where is l_c the size of the cohort at the commencement of the generation. The real-mortality row in Table 11.3 is the only % row that is additive and it is useful for comparing the role of population factors within the same generation.

11.4.1.3 Indispensable (or irreplaceable) mortality

This is that part of the generation mortality that would not occur should the mortality factor in question be removed from the life system, after allowance is made for the action of subsequent mortality factors. It is often assumed that these will still destroy the same percentage, independent of the change in prey density; clearly this assumption will not always be justified. To take an example of the calculation of the indispensable mortality from Table 11.3, consider the egg-stage mortality; if there is no egg mortality, 1000 individuals enter the larval stage, where a 40% mortality leaves 600 survivors to pupate; in the pupal stage, a 90% mortality leaves 60 survivors, i.e. 30 more than when egg mortality occurs, and thus its indispensable mortality = $30/1000 \times 100 = 3\%$.

When it is known that the subsequent mortalities are unrelated to density, the indispensable component of a factor may be used for assessing its value in control programmes (e.g. Huffaker & Kennet, 1965). If the exact density relationship of subsequent factors is known, then a corrected, and more realistic, indispensable mortality can be calculated.

11.4.1.4 Mortality–survivor ratio

Introduced by Bess (1945), this measure represents the increase in population that would have occurred if the factor in question had been absent. If the final population is multiplied by this ratio, then the resulting value represents, in individuals, the indispensable mortality due to that factor.

Table 11.3 Various measures for the comparison of mortality factors.

Measure	Eggs		Larvae		Pupae		Adults
			Stage				
l_x	1000		500		300		30
d_x		500		200		270	
% apparent mortality	50		40		90		
% real mortality	50		20		27		
% indispensible mortality	3		2		27		
Mortality/ survivor ratio	1.00		0.66		9.00		
log population	3.00		2.70		2.48		1.48
k-values		0.30		0.22		1.00	

11.4.2 The simple statistical relationship of population size to a factor

Originally, such relationships were investigated by straightforward regression analysis. However, such an approach ignores autocorrelation, which is almost inevitable in time-series data. Given time series of sufficient length, more sophisticated approaches are available (see Chapter 15).

11.4.3 Survival and life-budget analysis

Very significant advances in the analysis of population-census data were made by Watt (1961, 1963, 1964), Morris (1959, 1963a, b), and their co-workers. Watt developed the mixed deductive–inductive model, which blended field data and mathematical analysis in a way that allowed 'feedback' and correction in the model's development (Conway, 1973). Morris introduced the concept of key-factor analysis, which aimed to determine the factor (or factors) that were of the greatest predictive value in forecasting future population trends; he also sought to estimate the magnitude of density dependence in the system. Subsequent studies (Southwood, 1967; Hassell & Huffaker, 1969; Maelzer, 1970; St Amant, 1970; Luck, 1971; Brockelman & Fagen, 1972; Benson, 1973) have shown that many difficulties arise in the interpretation of results from the Morris method. At this time, the method of Varley and Gradwell (1960), the development of which was stimulated by Morris' original paper, was to be preferred. However, Varley and Gradwell's method still has serious conceptual and statistical problems (Manly, 1990; Royama, 1992; Sibley & Smith, 1998). These are addressed by Brown et al. (1993) with a method they called structured demographic accounting, by Vickery (1991) who tested for bias and density dependence by randomization and simulation, and subsequently by Sibley and Smith (1998) with the development of λ contribution analysis. Below, we first describe the traditional (Varley & Gradwell's) method, because it has been widely used and is simpler to understand. The calculations required for λ contribution analysis are then described; once these have been undertaken, the graphical presentation and identification of key factors is the same as for the traditional method.

Sibley and Smith (1998) have summarized the technical problems suffered by traditional key-factor analysis. First, it does not correctly account for births, as the k (mortality) values only measure changes in population caused by death and emigration. A way of including variation in births is by calculating birth-rate mortality, defined as the difference between the maximum potential birth-rate and the observed birth-rate. Secondly, the number of individuals in each final age class declines with age, so that the estimation of k values becomes progressively more unreliable with age. However, all k values are given equal weight in the calculation of K, which can lead to serious errors. When applied to populations with overlapping generations, no account is taken of the different contributions made by the age classes present to population growth. Thus, the mortality experienced by older individuals with low reproductive potential is given equal weight with that of young adults (Sibley & Smith, 1998).

Royama (1996) points out that, in any key-factor analysis, the potential importance of a factor that varies little through time is overlooked and that a key factor with a large effect may be arbitrarily created by the choice of stage division in a life-table.

11.4.3.1 Varley and Gradwell's method

This method, developed by Varley and Gradwell (1960, 1963a, b, 1965), Varley (1963), and Varley *et al.* (1973), demands the data from a series of successive life-tables. The whole generation is considered; thus it is immediately apparent in which age interval the density-dependent and key factors lie, and it provides a direct method of testing the role of changes in natality from generation to generation.

Varley and Gradwell's method may be outlined as follows:

1 The maximum potential natality is found by multiplying the number of females of reproductive age by the maximum mean fecundity and this figure is entered in the budget (Table 11.4).

2 The values in the budget are converted to logarithms.

3 A convenient generation basis in this method is adult to adult; the base chosen will affect the recognition of key factors (Varley & Gradwell, 1965). The total generation 'mortality' is given by subtracting the log of the population of adults entering the reproductive stage from the log-maximum potential natality of the previous generation — this value is referred to as K^* (Table 11.4).

4 The series of age-specific mortalities is calculated by subtracting each log population from the previous one (Table 11.4); these are referred to as k values, so that:

$$K = k_0 + k_1 + k_2 + k_3 + \ldots k_i \tag{11.37}$$

---

* This K should not be confused with the more widespread use of the symbol for the carrying capacity of the habitat as defined by the logistic equation.

Table 11.4 A budget prepared for Varley & Gradwell's analysis.

	Numbers/10 m² (observed, unless marked*)	Log numbers/10 m²	k's
Maximum potential natality	× 30 × 100	3.176	
(no. reprod. ♀ × maximum natality	= 1500		
k^0 (variation in natality)			0.076
Eggs laid	1260	3.100	
k^1 (egg loss)			0.171
Eggs hatching	850	2.929	
k^2 (predation, etc.)			0.498
3rd stage larvae	270	2.431	
k^3 (apparent parasitism of larvae)	100		0.201
[Larvae surviving parasitism]	170*	2.230	
k^4 (predation and other larval mortality)			0.276
Pupae	90	1.954	
k^5 (overwintering loss)			0.301
Adults emerging	45	1.653	
k^6 (dispersal, etc.)			0.352
Adults reproducing	20	1.301	
		K	= 1.875

Where precise estimates of mortality and migration (see Chapter 10) are available, these are also incorporated into the equation. These series of k values—one series for each generation—provide a complete picture of population changes. In the subsequent steps of this analysis, the role of each k factor is examined separately, but it must be remembered that sampling errors are 'hidden' in each k and may be responsible for spurious results (Kuno, 1971). The identification of the k value with a specific factor presents no problem when it is based on an apparent mortality, but, when it is the difference of successive estimates, it should, strictly, be referred to as 'overwintering loss' or 'loss of young adults'. It may be possible from other knowledge to indicate the major components of these losses; such assumptions could be checked by testing the correlation between the k value and an independent measure of the loss factor, e.g. the abundance of a predator, over a number of generations. The difference between the log-maximum potential natality and log-actual natality, k_0, has a special significance, for it does not represent mortality in the strict sense, but, rather, the variation in natality. This is compounded of two separate causes: death of the reproducing females before the end of the reproductive life and variations in the fertility of the females. The two could be separated by determining the form of survival curve of the reproducing female (Chapter 12).

5 The next step involves the recognition of the key factor for the index of population trend from adult to adult. An assessment may be made by visual correlation: K and k_0 to k_i are plotted against generation and it may easily be seen which k is most closely correlated with K (Fig. 11.7). A quantitative evaluation of the roles of each k is provided by Podoler and Rogers's (1975) method, in which the individual k

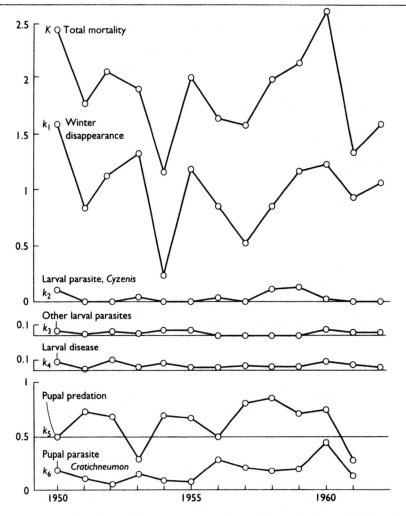

Fig. 11.7 The recognition of the key factor by visual correlation of various k values with K (after Varley & Gradwell, 1960).

values are plotted (on the y axis) against total mortality (K, on the x axis) and the regression is calculated. The relative importance of each factor will be proportional to its regression coefficient. Because sampling errors incorporated in k will also appear in K, this is not a precise statistical test for the importance of each mortality, but for the role of the estimates of each mortality in contributing to changes in the value of K.

6 The various k values are then tested for direct density dependence. Each k is plotted against the number (N_t) entering the stage (age interval) on which it acts (Fig. 11.8) and, if the regression is significant, then density dependence may be suspected. But the principal difficulty now arises: the variables are not independent (Varley & Gradwell, 1960; Bulmer, 1975). The regression is:

$$k = \log \alpha + b \log N_t \tag{11.38}$$

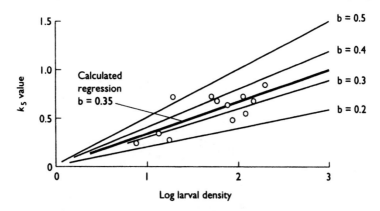

Fig. 11.8 The recognition of density dependence of a k factor by plotting the value of this factor against the log density at the start of the stage on which it acts (after Varley & Gradwell, 1963a).

where b = the slope of the density dependence (see also May *et al.*, 1974) and α is a constant. But:

$$k = \log N_t - \log N_{t+1} \tag{11.39}$$

Thus the regression could be spurious, due to sampling errors. The second step, therefore, is to plot the log numbers entering the stage ($\log N_t$) against the log number of survivors ($\log N_{t+1}$). The regressions of $\log N_t$ on $\log N_{t+1}$ and of $\log N_{t+1}$ on $\log N_t$ should be calculated and, if both the regression coefficients depart significantly from 1.0, then the density dependence may be taken as real. Bulmer (1975) has devised a more precise test, based on the ratio:

$$R = \frac{\sum_{t=1}^{n}(\log N_t - \log \overline{N})^2}{\sum_{t=1}^{n-1}k^2} \tag{11.40}$$

where N = the mean population level at time t over the years studied; R is the reciprocal of von Neumann's ratio, which is known to have a certain distribution depending on the value of N. Bulmer (1975) shows by simulation that a value of R smaller than the lower limit of R (due to random variation, as predicted from tables of von Neumann's ratio) is an indication of density dependence. However, as he points out, this test will eliminate any trend (i.e. delayed density dependence) and it is not strictly valid if the estimates of N_t and N_{t+1} contain sampling errors (as they normally do). However, Bulmer concludes that 'a series of 62 observations is only likely to reveal the existence of rather strong density-dependence and its existence may be masked by temporal trends (i.e. delayed effects) in the data'; so the practical utility of this test is unfortunately minimal (unless the organisms have very short generation times, so that a series of about 100 generations may be accumulated).

Fig. 11.9 Time-sequence plots showing how the different density relationships may be recognized from the patterns produced.

7 If density dependence is believed to be real, attention may now be refocused on the plot of k, against the numbers entering the stage (Fig. 11.9). The slope of the line, the regression coefficient, should be determined as this will give a measure of how the factor will act; the closer the regression coefficient is to 1.0, the greater the stabilizing effect of that regulatory factor. If the coefficient is exactly 1.0, the factor will compensate completely for any changes in density; if the coefficient is less than 1.0, the factor will be unable to compensate completely for the changes in density caused by other disturbing factors, whilst a coefficient of more than 1.0 implies overcompensation.

Using equation (11.38), Stubbs (1977) was able to detect density dependence in 46 sets of life-budget data and to draw significant conclusions. Hassell (1975) developed a more general model for density dependence, for single-species populations:

$$k = b \log(1 + aN_t) \tag{11.41}$$

where k is as defined in equation (11.38), b is a measure of density dependence, $a = 1/N_c$ and N_c = a critical density for the organism at which scramble competition starts to operate. Hassell (1975) was able to estimate the level of density dependence in populations of several species by fitting his model to published population budgets.

8 The detection of temporal trends (*sensu* Bulmer, 1975) or of delay in density dependence may be undertaken by plotting the k value against log initial density (as in Fig. 11.9) and then joining the points up in a time-sequence plot (Varley, 1947, 1953; Morris, 1959; Varley & Gradwell, 1965). The different types of factor will trace different patterns (Fig. 11.9); direct density-dependent factors will trace a more or less straight line or narrow band of points, delayed density-dependent factors circles or spirals, and density-independent factors irregular or zigzag plots, whose amplitude reflects the extent to which they fluctuate. Solomon (1964) discusses how the order in which various factors act will influence the magnitude of their effect. This type of plot, the linking of consecutive points serially, is of particular value in investigating what Hassell (1966) has appropriately termed 'intergeneration relationships', and it is the only method by which a delayed relationship (a delayed density-dependent factor) can be recognized.

In conclusion, Varley and Gradwell's method allows the following.

1 The recognition of the key factor (or factors) or the period in which it acts.

2 The investigation of the role of natality in population dynamics.

3 The consideration of the role of mortality factors at every stage of the life cycle, the recognition of the different density relationships of these factors, and an indication of their mode of operation: direct density-dependent factors tend to stabilize, delayed density-dependent factors lead to oscillations, density-independent factors lead to fluctuations, and inverse density-dependent factors will tend to accentuate the fluctuations.

Many life-tables have now been analysed by these methods and comparative evaluations of the role of density-dependent and density-independent factors have been made (e.g. Stiling, 1987, 1988) and Roland (1998) used them to compare the population dynamics of winter moth (*Operophtera brumata*) in different parts of the world. Density dependence, which may be hidden by stochastic variation (Hassell, 1987), can be studied on various spatial scales. The survival of whitefly, *Aleurotrachelus jelinekii*, was measured on individual leaves on two branches of *Viburnum tinus* (Southwood & Reader, 1976; Hassell *et al.*, 1987; Southwood *et al.*, 1989; Hassell, 1998). It was found that the dynamics could be understood only with a long (16 generations) series of observations, which allowed the recognition of different density-dependent relationships in K_0 at the low and higher population densities, as measured on the bush, and of additional density dependence in the survival from egg to adult on a leaf-to-leaf basis. From this detailed, but complex, picture, it was possible to develop a model with multiple equilibria, as predicted by other studies (Southwood & Comins, 1979).

Working on data from individual farms, Randolph (1994, 1997) and Randolph and Rogers (1997) have constructed life budgets for the African tick, *Rhipicephalus appendiculatus*. This was done from a series of measurements of the population of different stages by subtracting the monthly $\log_{10}$ number + 1 of one stage of the tick from the $\log_{10}$ number + 1 of the preceding stage with the time interval equivalent to the development period. These k values were then plotted against the density of the earlier stage and against climatic variables. Clear examples were found of density-dependent mortality and climate-related density-independent mortality in different stages.

Using several thousand time series of moths and aphids sampled in Britain by light and suction traps, Woiwod and Hanski (1992) found significant evidence of density dependence by Bulmer's methods in 79% of the moth and 88% of the aphid series greater than 20 years in length.

11.4.3.2 Sibley's λ contribution analysis

Whereas traditional key-factor analysis seeks to determine the influences on mortality, this method aims to identify those elements in the life-table that most influence the population growth rate, measured as either λ or r ($\lambda = e^r$) (Sibley & Smith, 1998). Given an age-specific birth rate, b_x, the rate of change of λ with respect to b_x is given by:

$$\frac{\partial \lambda}{\partial b_x} = \frac{\lambda^{-x} l_x}{\sum_{i=1}^{M} i \lambda^{-i-1} b_i l_i}$$

(11.42)

where M is the age of last reproduction.

Similarly, the rate of change of λ with respect to k_x is given by:

$$\frac{\partial \lambda}{\partial k_x} = \frac{\sum_{i=x+1}^{M} \lambda^{-i} b_i l_i}{\sum_{i=1}^{M} i \lambda^{-i-1} b_i l_i}$$

(11.43)

and the total population growth rate is the summation:

$$l_{total} = \sum \frac{\partial \lambda}{\partial p_i} p_i$$

(11.44)

where p_i is a life-history parameter $-b_x$ or k_x.

Note that equation (11.44) is analogous to the traditional key-factor equation:

$$K = k_0 + k_1 + k_2 + k_3 + \ldots k_i$$

The difference is that, in equation (11.44), each term is weighted by its contribution to the population growth rate and birth-rates as well as mortality rates are included.

Having calculated the various terms:

$$\frac{\partial \lambda}{\partial b_x} b_x$$

and

$$\frac{\partial \lambda}{\partial k_x} k_x$$

for each year they are plotted over the years on a graph with λ_{total} and a visual examination and regression analysis are undertaken, as for traditional key-factor analysis, to identify the key factors.

References

Ashford, J.R., Read, K.L.Q., & Vickers, G.G. (1970) A system of stochastic models applicable to studies of animal population dynamics. *J. Anim. Ecol.* **39**, 29–50.

Bellows, T.S. & Birley, M.H. (1981) Estimating developmental and mortality rates and stage recruitment from insect stage-frequency data. *Res. Popul. Ecol.* **23**, 232–44.

Bengstron, M. (1969) Estimating provisional values for intrinsic rate of natural increase in population growth studies. *Aust. J. Sci.* **32**, 24.

Benson, J.F. (1973) Some problems of testing for density-dependence in animal populations. *Oecologia* **13**, 183–90.

Bess, H.A. (1945) A measure of the influence of natural mortality factors on insect survival. *Ann. Entomol. Soc. Am.* **38**, 472–82.

Beverton, R.J.H. & Holt, S.J. (1957) On the dynamics of exploited fish populations. *Fisheries Investigations, Ser. 2* **19**, 533 pp. Min Agric. Fish. Food. HMSO, London.

Birch, L.C. (1948) The intrinsic rate of natural increase of an insect population. *J. Anim. Ecol.* **17**, 15–26.

Birley, M.H. (1977) The estimation of insect density and instar survivorship functions from census data. *J. Anim. Ecol.* **46**, 497–510.

Brockelman, W.Y. & Fagen, R.M. (1972) On modelling density-independent population change. *Ecology* **53**, 944–8.

Brown, D., Alexander, N.D.E., Marrs, R.W., & Albon, S. (1993) Structured accounting of variance of demographic change. *J. Anim. Ecol.* **62**, 490–502.

Bulmer, M.G. (1975) The statistical analysis of density-dependence. *Biometrics* **31**, 901–211.

Carey, J.R. (1993) *Applied Demography for Biologists.* Oxford University Press, New York, 206 pp.

Caughley, G. & Birch, L.C. (1971) Rate of increase. *J. Wildl. Manag.* **35**, 658–63.

Conway, G.R. (1973) Experience in insect pest modeling: a review of models, uses and future directions. *Mem. Aust. Ecol. Soc.* **1**, 103–30.

Deevey, E.S. (1947) Life tables for natural populations of animals. *Quart. Rev. Biol.* **22**, 283–314.

Dempster, J.P. (1956) The estimation of the numbers of individuals entering each stage during the development of one generation of an insect population. *J. Anim. Ecol.* **259**, 1–5.

Dempster, J.P. (1961) The analysis of data obtained by regular sampling of an insect population. *J. Anim. Ecol.* **30**, 429–32.

Dennis, B. & Taper, M.L. (1994) Density dependence in time series observations: estimation and testing. *Ecol. Monogr.* **64**, 205–24.

Dennis, B., Kemp, W.P., & Taper, M.L. (1998) Joint density dependence. *Ecology* **79**, 426–41.

Fenchel, T. (1974) Intrinsic rate of natural increase: the relationaship with body size. *Oecologia* **14**, 317–26.

Gilbert, N., Gutierrez, A.R., Frazer, B.D., & Jones, R.E. (1976) *Ecological Relationships*. Addison-Wesley, Reading, San Francisco, 157 pp.

Gulland, J.A. (1983) *Fish Stock Assessment*. John Wiley & Sons, New York. 223 pp.

Hassell, M.P. (1966) Evaluation of parasite or predator responses. *J. Anim. Ecol.* **35**, 65–75.

Hassell, M.P. (1975) Density-dependence in single-species populations. *J. Anim. Ecol.* **44**, 283–95.

Hassell, M.P. (1987) Detecting regulation in patchily distributed animal populations. *J. Anim. Ecol.* **56**, 705–13.

Hassell, M.P. (1998) The regulation of populations by density-dependent processes. In: Dempster, J.P. & McLean, I.F.G. (eds) *Insect Populations in Theory and in Practice*. Kluwer Academic Publishers, Dordrecht, pp. 29–51.

Hassell, M.P. & Huffaker, C.B. (1969) The appraisal of delayed and direct density-dependence. *Can. Entomol.* **101**, 353–61.

Hassell, M.P., Southwood, T.R.E., & Reader, P.M. (1987) The dynamics of the viburnum whitefly (*Aleurotrachelus jelinkii*): a case study of population regulation. *J. Anim. Ecol.* **56**, 283–300.

Helgeson, R.G. & Haynes, D.L. (1972) Population dynamics of the cereal leaf beetle, *Aulema melanopus* (Coieoptera: Chrysomelidae): model for age-specific mortality. *Can. Entomol.* **104**, 797–814.

Huffaker, C.B. & Kennet, C.E. (1965) Ecological aspects of control of olive scale *Parlatonia oleae* (Colvee) by natural enemies in California. *Proc. XII Int. Congr. Entomol.* 585–6.

Iles, T.C. (1994) A review of stock-recruitment relationships with reference to flatfish populations. *Netherlands J. Sea Research* **32**, 399–420.

Iles, T.C. & Beverton, R.J.H. (1998) Stock, recruitment and moderating processes in flatfish. *J. Sea Research* **39**, 41–55.

Ito, Y. (1961) Factors that affect the fluctuations of animal numbers, with special reference to insect outbreaks. *Bull. Nat. Inst. Agric. Sci. C* **13**, 57–89.

Jensen, A.L. (1975) Comparison of logistic equations for population growth. *Biometrics* **31**, 853–62.

Kempton, R.A. (1979) Statistical analysis of frequency data obtained from sampling an insect population grouped by stages. In: Ord, J.K., Patil, G.P., & Taille, C. (eds) *Statistical Distributions in Scientific Work*. International Cooperative Publishing House, Maryland.

Kiritani, K. & Nakasuji, F. (1967) Estimation of the stage-specific survival rate in the insect population with overlapping stages. *Res. Popul. Ecol.* **9**, 143–52.

Kuno, E. (1971) Sampling error as a misleading artifact in key factor analysis. *Res. Popul. Ecol.* **13**, 28–45.

Laughlin, R. (1965) Capacity for increase: a useful population statistic. *J. Anim. Ecol.* **349**, 77–91.

Lefkovitch, L.P. (1963) Census studies on unrestricted populations of *Lasioderma serricorne* (F.) (Coleoptera: Anobiidae). *J. Anim. Ecol.* **32**, 221–31.

Leslie, P.H. & Ranson, R.M. (1940) The mortality, fertility and rate of natural increase of the vole (*Microtus agrestis*) as observed in the laboratory. *J. Anim. Ecol.* **9**, 27–52.

Luck, R.F. (1971) An appraisal of two methods of analysing insect life tables. *Can. Entomol.* **1039**, 1261–71.

Maelzer, D.A. (1970) The regression of $\log N_{n+1}$ on $\log N_n$ as a test of density dependence: an exercise with computer-constructed density-independent populations. *Ecology* **51**, 810–22.

Manly, B.F.J. (1974a) Estimation of stage-specific survival rates and other parameters for insect populations developing through several stages. *Oecologia* **15**, 277–85.

Manly, B.F.J. (1974b) A note on the Richards, Waloff & Spradbery method for estimating stage-specific mortality rates in insect populations. *Biom. Z.* **17**, 77–83.

Manly, B.F.J. (1974c) A comparison of methods for the analysis of insect stage-frequency data. *Oecologia* **17**, 335–48.

Manly, B.F.J. (1976) Extensions to Kiritani and Nakasuji's method for analysing insect stage-frequency data. *Res. Popul. Ecol.* **17**, 191–9.

Manly, B.F.J. (1990) *Stage-structured Populations, Sampling, Analysis and Simulation*. Chapman & Hall, London.

Manly, B.F.J. & Seyb, A. (1989) A comparison of three maximum likelihood models for stage-frequency data. *Res. Popul. Ecol.* **31**, 367–80.

May, R.M. (1976) Estimating r: a pedagogical note. *Am. Nat.* **110**, 496–9.

May, R.M., Conway, G.R., Hassell, M.P., & Southwood, T.R.E. (1974) Time delays, density-dependence and single-species oscillations. *J. Anim. Ecol.* **43**, 747–70.

Messenger, P.S. (1964) Use of life tables in a bioclimatic study of an experimental aphid–braconid wasp host–parasite system. *Ecology* **45**, 119–31.

Morris, R.F. (1959) Single-factor analysis in population dynamics. *Ecology* **40**, 580–8.

Morris, R.F. (ed.) (1963a) The dynamics of epidemic spruce budworm populations. *Mem. Entomol. Soc. Can.* **31**, 1–332.

Morris, R.F. (1963b) Chapters 6, 7, and 18 in 'The dynamics of epidemic spruce budworm populations'. *Mem. Entomol. Soc. Can.* **31**, 30–7, 116–29.

Munholland, P.L., Kalbfleisch, J.D., & Dennis, B. (1989) A stochastic model for insect life history data. In: McDonald, L.L., Manly, B.F.J., Lockwood, J.A., & Logan, J.A. (eds) *Estimation and Analysis of Insect Populations*. Springer-Verlag, Berlin.

Podoler, H. & Rogers, D. (1975) A new method for the identification of key factors from life-table data. *J. Anim. Ecol.* **44**, 85–114.

Randolph, S.E. (1994) Population dynamics and density-dependent seasonal mortality indices of the tick *Rhipicephalus appendiculatus* in eastern and southern Africa. *Medical and Veterinary Entomology* **8**, 351–68.

Randolph, S.E. (1997) Abiotic and biotic determinants of the seasonal dynamics of the tick *Rhipicephalus appendiculatus* in South Africa. *Medical and Veterinary Entomology* **11**, 25–37.

Randolph, S.E. & Rogers, D.J. (1997) A generic population model for the African tick *Rhipicephalus appendiculatus*. *Parasitology* **115**, 265–79.

Richards, O.W. (1940) The biology of the small white butterfly (*Pieris rapae*), with special reference to the factors controlling abundance. *J. Anim. Ecol.* **9**, 243–88.

Richards, O.W. (1959) The study of natural populations of insects. *Proc. Roy. Entomol. Soc. Lond. C* **23**, 75–9.

Richards, O.W. (1961) The theoretical and practical study of natural insect populations. *Ann. Rev. Entomol.* **6**, 147–62.

Richards, O.W. & Waloff, N. (1954) Studies on the biology and population dynamics of British grasshoppers. *Anti-Locust Bull.* **17**, 182 pp.

Richards, O.W. & Waloff, N. (1961) A study of a natural population of *Phytodecta olivacea* (Forster) (Coleoptera: Chrysomeloidea). *Phil. Trans. B* **244**, 205–57.

Richards, O.W., Waloff, N., & Spradbery, J.P. (1960) The measurement of mortality in an insect population in which recruitment and mortality widely overlap. *Oikos* **11**, 306–10.

Ricker, W.E. (1954) Stock and recruitment. *J. Fisheries Res. Bd. Canada* **11**, 559–623.

Rogers, D.J. (1979) Tsetse population dynamics and distribution: a new analytical approach. *J. Anim. Ecol.* **48**, 825–49.

Roland, J. (1998) The population dynamics of *Operophtera brumata* (Lepidoptera: Geometridae). In: Dempster, J.P. & McLean, I.F.G. (eds) *Insect Populations in Theory and in Practice*. Kluwer Academic Publishers, Dordrecht, pp. 309–21.

Rothery, P. (1998) The problems associated with the identification of density dependence in population data. In: Dempster, J.P. & McLean, I.F.G. (eds) *Insect Populations in Theory and in Practice*. Kluwer Academic Publishers, Dordrecht, pp. 97–133.

Royama, T. (1992) *Analytical Population Dynamics*. Chapman & Hall, London.

Royama, T. (1996) A fundamental problem in key factor analysis. *Ecology* **77**, 87–93.

Ruesink, W.G. (1975) Estimating time-varying survival of arthropod life stages from population density *Ecology* **56**, 244–7.

St Amant, J.L.S. (1970) The detection of regulation in animal populations. *Ecology* **51**, 823–8.

Schneider, S.M. & Ferris, H. (1986) Estimation of stage-specific developmental times and survivorship from stage-frequency data. *Res. Popul. Ecol.* **28**, 267–80.

Sibley, R.M. & Smith, R.H. (1998) Identifying key factors using lambda contribution analysis. *J. Anim. Ecol.* **67**, 17–24.

Slobodkin, L.B. (1962) *Growth and Regulation of Animal Populations*. Holt, Rinehart & Winston, New York, 184 pp.

Smith, F.E. (1954) Quantitative aspects of population growth. In: Boell, E. (ed.) *Dynamic of Growth Processes*. Princeton University Press, Princeton, New Jersey.

Solomon, M.E. (1964) Analysis of processes involved in the natural control of insects. *Adv. Ecol. Res.* **2**, 1–58.

Southwood, T.R.E. (1967) The interpretation of population change. *J. Anim. Ecol.* **36**, 519–29.

Southwood, T.R.E. (1969) Population studies of insects attacking sugar cane. In: Williams, J.R. et al. (eds) *Pests of Sugar Cane*. Elsevier, Amsterdam, pp. 427–59.

Southwood, T.R.E. (1981) Stability in field population of insects. In: Hiorns, R.W. (ed) *Mathematical Theory of the Dynamics of Biological Populations II*. London Mathematical Society, London, pp. 31–46.

Southwood, T.R.E. & Comins, H.N. (1979) A synoptic population model. *J. Anim. Ecol.* **45**, 949–65.

Southwood, T.R.E. & Jepson, W.F. (1962) Studies on the populations of *Oscinella frit* L. (Dipt.: Chloropidae) in the oat crop. *J. Anim. Ecol.* **31**, 481–95.

Southwood, T.R.E. & Reader, P.M. (1976) Population census data and key factor analysis for the *Viburnum* whitefly, *Aleurotrachelus jelinekii* (Frauenf.) on three bushes. *J. Anim. Ecol.* **45**, 313–25.

Southwood, T.R.E., Hassel, M.P., Reader, P.M., & Rogers, D.J. (1989) Population dynamics of the viburnum whitefly (*Aleurotrachelus jelinekii*). *J. Anim. Ecol.* **58**, 921–42.

Southwood, T.R.E., May, R.M., Hassell, M.P., & Conway, G.R. (1974) Ecological strategies and population parameters. *Am. Nat.* **108**, 791–804.

Stiling, P.D. (1987) The frequency of density dependence in insect host-parasitoid systems. *Ecology* **68**, 844–56.

Stiling, P. (1988) Density-dependent processes and key factors in insect populations. *J. Anim. Ecol.* **57**, 581–93.

Stubbs, M. (1977) Density-dependence in the life-cycles of animals and its importance in *K* and *r*-strategies. *J. Anim. Ecol.* **46**, 677–88.

Varley, G.C. (1947) The natural control of population balance in the knapweed gallfly (*Urophora jaceana*). *J. Anim. Ecol.* **16**, 139–87.

Varley, G.C. (1953) Ecological aspects of population regulation. *Trans. IX Int. Congr. Entomol.* **2**, 210–14.

Varley, G.C. (1963) The interpretation of change and stability in insect populations. *Proc. Roy. Entomol. Soc. Lond. C* **27**, 52–7.

Varley, G.C. & Gradwell, G.R. (1960) Key factors in population studies. *J. Anim. Ecol.* **29**, 399–401.

Varley, G.C. & Gradwell, G.R. (1963a) The interpretation of insect population changes. *Proc. Ceylon Assoc. Adv. Sci. (D)* **18**, 142–56.

Varley, G.C. & Gradwell, G.R. (1963b) Predatory insects as density dependent mortality factors. *Proc. XVI Int. Zool. Congr.* **1**, 240.

Varley, G.C. & Gradwell, G.R. (1965)

Interpreting winter moth population changes. *Proc. XII Int. Congr. Entomol.* 377–8.

Varley, G.C., Gradwell, G.R., & Hassell, M.P. (1973) *Insect Population Ecology, an Analytical Approach*. Blackwell Scientific Publications, Oxford, 212 pp.

Vickery, W.L. (1991) An evaluation of bias in *k*-factor analysis. *Oecologia* **85**, 413–19.

Waloff, N. & Bakker, K. (1963) The flight activity of Miridae (Heteroptera) living on broom, *Sarothamnus scoparius* (L.) Wimn. *J. Anim. Ecol.* **32**, 461–80.

Watson, T.F. (1964) Influence of host plant condition on population increase of *Tetranychus telarius* (Linnaeus) (Acarina: Tetranychidae). *Hilgardia* **35**(11), 273–322.

Watt, K.E.F. (1961) Mathematical models for use in insect pest control. *Can. Entomol.* **93** (suppl. 19), 62 pp.

Watt, K.E.F. (1963) Mathematical population models for five agricultural crop pests. *Mem. Entomol. Soc. Can.* **32**, 83–91.

Watt, K.E.F. (1964) Density dependence in population fluctuations. *Can. Entomol.* **96**, 1147–8.

Woiwod, I.P. & Hanski, I. (1992) Patterns of density dependence in moths and aphids. *J. Anim. Ecol.* **61**, 619–29.

Wood, S.N. (1994) Obtaining birth and mortality patterns from structured population trajectories. *Ecol. Monogr.* **64**, 23–44.

Wyatt, I.I. & White, P.F. (1977) Simple estimation of intrinsic increase rates for aphids and tetranychid mites. *J. Appl. Ecol.* **14**, 757–66.

12 Age-grouping, Time-specific Life-tables, and Predictive Population Models

This chapter is concerned with techniques for animals whose generations overlap widely; age-grouping is a prerequisite for these methods, which have been most widely applied to vertebrate populations. Analysis is easiest with two extreme types of population: the stationary and the exponentially expanding. If the population can be assumed to be stationary, then the fall-off in numbers in successive age-groups will reflect the survivorship curve and thus a time-specific (or vertical) life-table can be constructed on this basis (p. 447). If the population is expanding and unconstrained by its environment, the age structure may become stabilized and then mortality can be estimated from the difference between expected and actual growth rates (p. 420). Age-grouping may also provide useful information on the fertility or potential fertility of the population (p. 348).

12.1 Age-grouping

Considerable research has been undertaken on estimating age and thus this is only a selective introduction to the literature. In the following, taxonomically arranged sections, the methods available for age-grouping (or age-grading) both vertebrates and invertebrates are outlined. While a method may be generally applicable to the members of a taxon, the ease and accuracy with which it may be applied can vary greatly between species, so, whenever possible, the advice of someone experienced with the group should be sought. There are still many animals, such as leeches, flatworms, and sea anemones, that cannot be aged.

It is common practice, particularly in animals which reproduce in synchrony once a year, such as some fish, to age-group animals using a size-frequency histogram to identify the different cohorts. There are two major problems with this method: first, individual variation in growth can be considerable both within and between cohorts, resulting in erroneous age assignment, and, secondly, growth usually decelerates with age, making it difficult or impossible to discriminate between the older age classes. An extreme example of these pitfalls is shown by the study of Galinou Mitsoudi and Sinis (1995) on the long-lived bivalve *Lithophaga lithophaga*, where it was found that individuals of 5.0 ± 0.2 cm in length ranged in age from 18 to 36 years. Size-frequency distributions can be used if they are supported by prior knowledge on the age-length relationship and individual variation (Goodyear, 1997).

12.1.1 Ageing young by developmental stage

Age cannot be inferred from developmental stage without reference to the environment. The speed of development may be temperature-dependent or influenced by factors such as oxygen and food availability. Rearing experiments may be required to understand the influence of environmental conditions on development. It is sometimes possible to relate development to the degree-days (the product of temperature and time) needed to reach a certain stage. However, it has been found that this measure can also vary with temperature. If allowance is made for the environment, developmental stage can be a useful method for ageing juveniles.

The immature stages of most insects and crustaceans are easily distinguished, larval instars being recognized by the diameter of the head or carapace, the length or presence of the appendages, and other structural features, although even so basic a feature as the width of the larval head capsule may be misleading (Kishi, 1971). For the lobster, *Homarus gammarus*, where the number of moults is not fixed, Henocque (1987) has suggested that the number of segments in the antennules increases with each successive moult. It is even possible to age-group within the instars of some heteropterons larvae, as the number of eye facets increases for some time after moulting; this was noted in lace-bugs (Tingidae) (Southwood & Scudder, 1956) and in Corixidae, where it is even more marked; the interocular distance decreases throughout the instar (E.C. Young, unpublished). In the last larval instar of Exopterygota, the wing pads frequently darken shortly before the final moult.

Different ages of the insect pupal stage may often be determined by dissection, particularly in the Diptera, where various categories based on pigmentation can be recognized, e.g. unpigmented, eyes of pharate adult pigmented, head pigmented, pharate adult fully pigmented (Schneider & Vogel, 1950; van Emden *et al.*, 1961). Using the falling phenolic content, measured by the reduction of a dye, Chai and Dixon (1971) were able to assess the age of sawfly cocoons.

In fisheries research, the egg and the larval and post-larval development of fish are divided into a number of distinct developmental stages for ageing. Typical features used for the egg include the presence of the embryonic shield, pigment, and eye development, the number of myomeres, and the length of the embryo. Most fish hatch with a yolk-sac and the larvae are termed post-larvae after yolk-sac resorption. Features used to stage larval and post-larval fish include fin rays, scale formation, and pigmentation. Daily growth bands are laid down by fish on scales and otoliths, which can be used to age larval and juvenile stages (Sepulveda, 1994; Antunes & Tesch, 1997; Narimatsu & Munehara, 1997). Amphibian tadpoles also can be easily assigned to developmental stage (Brown, 1990).

For birds, it is often easy to distinguish juveniles or young adults from older individuals because of their plumage (e.g. Hohman & Cypher, 1986; Graves,

1997) and they are frequently divided into three age classes: unfledged, juvenile, and adult.

For mammals, one of the most useful features for ageing juveniles is the pattern of tooth eruption (e.g. Anderson, 1986; Taylor, 1988; Bianchini & Delupi, 1990; Ancrenaz & Delhomme, 1997), although a wide range of other skeletal and developmental features have also been used.

12.1.2 Ageing by using structures

12.1.2.1 Annelids

In temperate regions, annual growth rings are formed in the jaws of the polychaetes *Harmothoe imbricata, Harmothoe derjugini, Halosydna nebulosa, Hermilepidonotus robustus, Lepidonotus squamatus*, and an *Eunoe* sp. (Britaev & Belov, 1993). An interesting example of how habitat clues can aid ageing is the study by Nishi and Nishihira (1996), in which they used the growth rings of coral to age the polychaete Christmas-tree worm, *Spirobranchus giganteus*, which lives embedded within the skeleton of the coral.

12.1.2.2 Crustaceans

Crustaceans, particularly the long-lived Malacostraca, are difficult to age. Sheehy *et al.* (1994, 1996) have shown that the amount of lipofuscin in the left olfactory-lobe cell mass of the brain of Australian red-claw crayfish, *Cherax quadricarinatus*, or in the eye-stalk ganglia of the European lobster, *H. gammarus*, is positively correlated with age. The lipofuscin concentration was measured using confocal fluorescence microscopy and image analysis. Krill can also be aged by the amount of fluorescent pigment (Berman *et al.*, 1989).

The time since the last moult has been determined for the spider crab, *Maja squinado*, and European lobster, *H. gammarus*, by measuring the natural radionuclide activity ratio, thorium-228/radium-228 (^{228}Th/^{228}Ra), in the exoskeleton (Le Foll *et al.*, 1989).

12.1.2.3 Insects

Some of the types of criteria used are indicated below. Early methods for age-grouping of mosquitoes are reviewed by Hamon *et al.* (1961) and Muirhead-Thomson (1963) and for arthropods of medical importance as a whole by Detinova (1962, 1968); Russian workers have contributed greatly to this subject. Most methods rely on criteria that are indications of some physiological process, particularly reproduction and excretion, or on general wear and tear; therefore the precise chronological equivalent of a given condition may vary from habitat to habitat or even between individuals.

Biochemical markers. Thomas and Chen (1989, 1990) reported that fluorescent pteridines in the compound eyes of adult screw-worms, *Cochliomyia hominivorax*, increased in concentration through time and could be used for age determination.

Age was calculated using a regression model that also took account of the size of the head capsule. Similarly, Camin *et al.* (1991) reported that the homogenized head capsules of Mediterranean fruit flies, *Ceratitis capitata*, measured spectrofluorometrically showed significant differences in fluorescence from 0 to 28 days after emergence. This pteridine-accumulation method has also been shown to work for melon fly, *Bactrocera cucurbitae* (Mochizuki *et al.*, 1993). For tsetse flies, it is not possible to accurately age recently emerged flies using this method, but Msangi and Lehane (1991) showed that compounds in the abdomens of young *Glossina morsitans morsitans* displayed high levels of fluorescence, which decreased linearly with time for the first 5 days post-emergence. In a detailed study of the head capsules of adults of the fly *Anastrepha ludens*, Tomic Carruthers *et al.* (1996) identified a range of pteridines and concluded that deoxysepiapterin, a pteridine with yellow fluorescence, accumulated sufficiently and over a sufficiently long period to be useful for ageing. Fluorescence has been found not to work for ageing the fly *Culicoides variipennis sonorensis* (Mullens & Lehane, 1995).

Cuticular bands. The demonstration by Neville (1963) that there were apparently daily growth layers present in some areas of the cuticle of certain insects opened up the possibility of an ageing criterion for insects as precise as that available from fish scales or otoliths.

Further studies have shown that such cuticular banding is widespread, but care is needed in its interpretation. In some Heteroptera, food availability affects band deposition (Neville, 1970), but not in the blowfly, *Lucilia* (Tyndale-Biscoe & Kitching, 1974). The banding was thought to depend on a rhythm that was merely influenced by temperature, although in Coleoptera the number of bands may vary in different parts of the same animal (Neville, 1970); however, Tyndale-Biscoe and Kitching (1974) showed that in *Lucilia* they were dependent on temperature changes—precisely a rise of 3.5°C or more over a threshold of 15.5°C. With the normal daily temperature rhythm, one band (a dark layer + a light one, seen most clearly under polarized light) is laid down each day; but this commences at apolysis, so the number of bands formed in the pharate adult must be determined before they can be used to estimate post-ecdysal adult age. In the Exopterygota, where there seems to be more rhythmicity in their deposition, bands may be seen by sectioning the hind tibia and have been demonstrated in most orders (Neville, 1970). Similar rings have been noted in honey-bees and beetles. In the Diptera, Schlein and Gratz (1972, 1973) showed that the skeletal apodemes exhibited more bands than other regions of the cuticle; they describe a staining technique, but Tyndale-Biscoe and Kitching (1974) could count banding in *Lucilia* without polarized light or staining. Band formation appears to be limited to the early part of adult life, up to 10–13 days in *Anopheles* mosquitoes (Schlein & Gratz, 1973), and as many as 63 bands have been observed in a large grasshopper (Neville, 1970). Zuk (1987) found that the field crickets *Gryllus pennsylvanicus* and *Gryllus veletis* added one growth ring per day until about day 25–30, when cuticle growth is completed and rings are no longer added.

In practice, this was not a serious limitation to the method, as few animals were found in the wild with more than 18 rings.

Sclerotization and colour changes in the cuticle and wings. In all insects, the newly emerged adult is pale and the cuticle untanned — in most species, the major part of tanning is completed during a short teneral period.

However, complete sclerotization may not occur for a considerable time, especially if a period of diapause or aestivation intervenes; e.g. Lagace and van den Bosch (1964) found in the weevil, *Hypera*, that the elytra remained thin and almost teneral for much of the summer diapause. During teneral development in the Corixidae and Notonectidae, the cuticle is pigmented at different rates; the sequence of pigmentation of the mesotergum is remarkably constant, proceeding forwards, in a number of stages, from the posterior region (Young, 1965). The wings of young dragonflies are milky, becoming clear when mature (Corbet, 1962a, b).

Various excretory pigments are accumulated throughout adult life and especially during diapause; some of these occur in the cuticle and may cause progressive colour changes. In other instances, cuticular colour change is a product of the sclerotization process. Dunn (1951) showed that the region of the discal cell and the associated veins of the hind-wing of the Colorado beetle (*Leptinotarsa*) change from yellowish in the newly emerged individual to reddish after hibernation, and the green shield-bug, *Palomena prasina*, becomes redder, often a dark bronze, during hibernation (Schiemenz, 1953); in contrast, the pink colouring of another pentatomid, *Piezodorus lituratus*, disappears while overwintering. Many Miridae change colour during or after hibernation, and yellows on the forewings of various Heteroptera often darken with age, becoming orange or even reddish (Southwood & Leston, 1959). The coloration of the bodies of adult dragonflies changes with age (Corbet, 1962a, b).

Developmental changes in the male genitalia. The males of many insects have a period of immaturity during which they may be recognized by the condition of the male genitalia (e.g. in *Dacus tryoni*, as described by Drew (1969)). In most Nematocera, the male hypogydium rotates soon after emergence; Rosay (1961) found in mosquitoes that the time for rotation was similar in the species of *Aedes* and *Culex* studied, but was strongly influenced by temperature (at 17°C–58h, at 28°C–12h). The aedeagus of the weevil, *Hypera*, remains relatively unsclerotized throughout the period of aestivation, lasting several months (Lagace & van den Bosch, 1964). In carabid beetles, comparable changes and the colour of the accessory glands may be used to determine age class (van Dijk, 1972).

Changes in the internal non-reproductive organs. The fat-body is perhaps the most useful of these; in some insects, it is large at the start of adult life, gradually waning (e.g. in the moth *Argyroploce* (Waloff, 1958)), and, in others, it is slowly built up prior to diapause. The availability and quality of food affect the condition of the fat-body and other organs (Fedetov, 1947, 1955, 1960; Haydak, 1957), as does parasitism.

The deposition of coloured excretory products in various internal organs, especially the malpighian tubules, provides another measure of age. Haydak (1957) found in worker honey-bees that the Malpighian tubules are clear for the first 3 days of life, milky from 3 to 10 days, and in older bees generally (not always) yellowish-green. Female *Culicoides* acquire a burgundy-red pigment in the abdominal wall during the development of the ovarian follicles and, as this remains, parous midges can easily be recognized without dissection (Dyce, 1969).

In newly emerged mosquitoes, bees, moths, and other endopterygote insects, part of the gut is filled with the brightly coloured meconium; this is usually voided within a day or two, at the most (Haydak, 1957; Rosay, 1961).

The detailed studies of Haydak (1957) and others on the honey-bee have shown that ranges of organs, especially various glands, change their appearance with age.

The condition of the ovaries and associated structures. Criteria associated with changes in the female reproductive system have been used widely in ageing studies; they are valuable, for, even when they do not provide a precise chronological age, they do give information on the extent of egg-laying, which is for some purposes more useful. The principal characters that have been used may be summarized as follows.

1 Egg rudiments. In those insects that do not develop additional egg follicles during adult life, a count of the number of egg rudiments will indicate the potential fecundity; at the start of adult life, the count will be high, gradually falling off, though a large number of rudiments may still remain at the end of life (Waloff, 1958; Corbet, 1962b).

2 Follicular relics. After an egg has been laid, the relics of the follicle, which are often pigmented, will remain in the bases of the ovariole pedicels or in the oviducts for a variable period, enabling a parous female to be recognized (Corbet, 1960, 1961; Hamon *et al.*, 1961; Saunders, 1962, 1964; Anderson, 1964; Vlijm & van Dijk, 1967; Lewis *et al.*, 1970; van Dijk, 1972; Hoc & Charlwood, 1990; Hoc & Wilkes, 1995). In some insects, the total amount of relics accumulates with an increasing number of cycles (Anderson, 1964); in others, they are eventually lost. V.P. Polovodova demonstrated in *Anopheles* that each time an egg develops it causes local stretching of the stalk of the oviduct. These dilatations contain the remains of the follicles and, if the ovaries are carefully stretched and examined (see Giglioli, 1963), the number of dilatations may be counted. In certain mosquitoes, the maximum number of bead-like dilatations in any ovariole can be taken as equal to the number of ovarian cycles; with other species, difficulties have been encountered (Muirhead-Thomson, 1963).

3 Granulation of the basal body. In the blackfly *Simulium woodi*, the basal body of each ovariole is a group of six to eight cells in the calyx wall, enclosed by the end of the ovariolar sheath. The granulation of these cells was reported by Hoc and Wilkes (1995) to progressively increase in intensity with each ovulation.

4 Ovarian tracheoles. The tracheoles supplying the ovaries of nulliparous females are tightly coiled ('tracheal skeins'); as the eggs mature, these become stretched, so that they do not resume their previous form, even in interovular periods. These changes have been observed and used for ageing in mosquitoes (Hamon *et al.*, 1961; Detinova, 1962; Kardos & Bellamy, 1962), in dragonflies (Corbet, 1961), and in calypterate flies (Anderson, 1964). A standard method of dissection for age-grading mosquitoes by this method is given by Davies *et al.* (1971).

5 Ovariole cycles and combined evidence. In many insects, the ovaries pass through a series of cycles; the stage in any given cycle may be easily recognized by the size of the most mature egg rudiments. Kunitskaya (1960) recognized six stages in fleas, but five stages seem normal within the Diptera, as initially recognized in mosquitoes by Christophers (1960) (e.g. ceratopogonids (Linley, 1965), simulilds (Le Berre, 1966), anthomyiids (Jones, 1971)); by combining these with information on follicular relics, Tyndale-Biscoe and Hughes (1969) were able to recognize 17 different stages in *Musca* and Vogt *et al.* (1974) 16 in *Lucilia*. Taken in conjunction with other evidence that indicates whether the female is nulliparous or parous, the age may be determined over two ovarian cycles. Vogt and Morton (1991) assigned sheep blowfly, *Lucilia cuprina*, to age class on the basis of parity and stage of ovarian development. In viviparous insects and others (e.g. Schizopteridae) where one large egg is laid at a time, the ovaries develop alternately and this may allow a larger number of reproductive cycles to be recognized (Saunders, 1962, 1964). In some insects, the number of functional ovarioles decreases after the first two cycles and this may be used as an index of age.

Indices of copulation. Although frequency of copulation is not a reliable index of chronological age, it may be 'calibrated' for a particular population and gives *per se* information of ecological significance. The presence of sperms in the spermatheca will provide evidence of pairing; in some mosquitoes, a gelatinous mating plug remains in the oviduct for a short period (Muirhead-Thomson, 1963). In insects where the sperms are deposited in a spermatophore, the remains of these in the spermatheca will give a cumulative measure of copulation (Waloff, 1958).

In some, insects, the appendages of the male will leave characteristic 'copulation marks'. In dragonflies, these are on the compound eyes or occipital triangle of the female; in Zygoptera, some sticky secretion may remain on the sides of the female's thorax (Corbet, 1962a); in tsetse flies, they are on the sixth sternum (Squire, 1952). In Cimicidae with a broad spermalege, every pairing will leave a characteristic groove (Usinger, 1966). The mated status of male calliphorine flies can be determined, as traces of the accessory-gland secretion remain in the lateral penis ducts for up to a week after mating (Pollock, 1969).

Changes in weight. Insects that feed little or not at all during adult life will become progressively lighter; Waloff (1958) showed that individual variation due to size differences could be reduced if a weight/length ratio was used. With Lepidoptera, weight/wing length is a convenient measure and Waloff found in a number of species that this falls off strikingly with age, e.g. male *Argyroploce* with an expectation of life of 10 days had a ratio of 1.93, but when the expectation of life was 2.5 days the ratio was 1.13.

In other insects, the adult weight will fluctuate, reflecting ovarian cycles (Waloff, 1958; see also p. 348) or changes associated with diapause.

'Wear and tear'. As an adult insect becomes older, its cuticle and appendages become damaged by contact with the environment and this 'wear and tear' may be used as an index of age. Kosminskii (1960) has found that fleas may be age-graded by the extent to which the ctenidal (genal comb) bristles are broken, whilst Michener *et al.* (1955) and Daly (1961) used the wear on the mandibles of wild bees. However, the most widely applicable index of this type is provided by the tearing or battering of the wings, e.g. in Lepidoptera, Diptera (Saunders, 1962), and Hymenoptera (Michener *et al.*, 1955). Although apparently crude, the number of tears ('nicks') in the wings has been found to correlate well with other indices of ageing (Michener & Lange, 1959; Saunders, 1962). Phoretic water mites die or fall off mosquitoes, and possibly other insects that emerge from water, with time; under certain conditions (Corbet, 1970), they may be used to indicate age (Gillett, 1957; Corbet, 1963).

12.1.2.4 Molluscs

Bivalve molluscs gradually increase the thickness of the shell, so that growth checks can show as a series of bands in suitably prepared cross-sections. In temperate regions, the horse mussel, *Modiolus modiolus*, has alternating patterns of light (summer) and dark (winter) growth lines in the middle nacreous layer of the shell (Anwar *et al.*, 1990), while *Mytilus galloprovincialis* in the Black Sea produces dark summer and light winter bands (Shurova & Zolotarev, 1988). By counting such annual bands, Galinou Mitsoudi & Sinis (1995) were able to show that piddock, *L. lithophaga*, could have a longevity of over 54 years. For some species, it may be possible to count annuli in thin cross-sections using transmitted light; alternatively, the cross-section can be polished and etched to emphasize the bands and an acetate-peel impression of the surface taken for examination (Richardson *et al.*, 1993). For species such as the European flat oyster, *Ostrea edulis*, acetate peels are taken from a cross-section of the umbo (Richardson *et al.*, 1993) or chondrophore (Tian & Shimizu, 1997). While clear annual rings on bivalve shells are often present, in some highly variable habitats they may be unclear and confused by other growth checks; when this is the case, Garcia (1993) found detailed examination using scanning electron microscopy useful.

Pulmonates may be aged by growth bands. Raboud (1986) aged *Arianta arbustorum* from the Swiss Alps, using thin sections of the shell margins. In some habitats, harsh winter conditions can give the shells of gastropods, such as *Hydrobia ulva*, clear winter-growth checks. Terrestrial helicid snails are known generally to show bands caused by winter-growth checks in northern temperate habitats and summer-growth checks in Australia; these can be used for ageing (Baker & Vogelzang, 1988).

Cross-sections of squid statoliths show daily growth rings (Rodhouse & Hatfield, 1990) and Raya *et al.* (1994) have used these to age the cuttlefish, *Sepia hierredda*. Larval gastropods have been aged by this method (Grana Raffucci & Appeldoorn, 1997). Daily growth rings have also been observed in young giant scallop, *Placopecten magellanicus* (Parson *et al.*, 1993). But shell growth patterns and checks may be related to the tidal cycle (Richardson, 1987; Richardson *et al.*, 1990).

12.1.2.5 Fish

The allocation of fish to year class is normally accomplished by the examination of structures such as scales (Moreira *et al.*, 1991; Loubens & Panfili, 1992; Almeida *et al.*, 1995; Singh & Sharma, 1995; Vidalis & Tsimenidis, 1996), otoliths (Berg, 1985; Hassager, 1991; Hales & Belk, 1992; Oxenford *et al.*, 1994; Worthington *et al.*, 1995; Gordo, 1996), spines (Tucker, 1985; Esteves *et al.*, 1995) opercular bones (Achionye Nzeh, 1994), fin rays (Lai *et al.*, 1987), and vertebrae (Parsons, 1993; Officer *et al.*, 1996), which are marked with bands because of seasonal growth checks. In temperate regions, rapid summer and slow winter growth result in an annual pattern, comprising a summer and a winter zone, which differ in appearance. The annulus is usually defined as the winter zone. These methods may be inappropriate for tropical fish living in environments with little seasonality. Bands will be formed whenever the growth of a fish is checked, as could be caused by drought, famine, periods of low oxygen, reproduction, or disease. For example, the scales of the small goby, *Pomatoschistus microps*, show one growth check per year in the Tagus estuary, Portugal, but two in the British Isles. This may be because growth checks are caused by both reproduction and winter and in Portugal the two checks coincide (Moreira *et al.*, 1991).

Computer-aided systems have been developed for the identification and counting of annuli (Cailliet *et al.*, 1996).

Scales must be taken from part of the fish known to acquire scales early in life, where scale replacement is unlikely and where they exhibit complete and clear growth patterns. For example, bluefish, *Pomatomus saltatrix*, or black sea-bass, *Centropristis striata*, scales should be removed from the area behind the pectoral fin. A number of scales should be removed and examined from each fish. Age determinations may be made from direct observation, scale impressions, or photographs. A microfiche viewer is convenient for scale examination. For thick scales which do not show a clear structure with transmitted light, impressions in laminated plastic film can be taken for examination (Dery, 1983). Frequently encountered problems include difficulty in identifying the first annulus (Mann & Steinmetz, 1985) and indistinct patterns towards the edge of the scale (Hesthagen, 1985).

Otoliths are removed by dissection from the head of the fish and may be stored dry prior to examination. Examination for annuli may be undertaken using whole, baked, broken, or crossed-sectioned otoliths, depending on the species. Visibility of the annuli is enhanced by baking or burning, because the hyaline zones turn brown in contrast to the white opaque zones (e.g. Aprahamian, 1987). Richter and McDermott (1990) tested the use of histological stains to enhance otolith annulus visibility and obtained best results with acidified neutral red for sole, turbot, brill, and scad, with aqueous aniline blue for cod, hake, whiting, plaice, and grey mullet, and with eosin Y for pelagic species with small, translucent otoliths. Worthington *et al.* (1995) suggested that the relationship between otolith weight and fish age may be used to age tropical-reef fish and found the method satisfactory provided the weight–age relationship was frequently recalibrated.

Transverse sections of a dorsal spine show annuli in species such as the sailfish, *Istiophorus platypterus* (Alvarado Castillo & Felix Uraga, 1995). For sharks and other elasmobranches, growth rings in the vertebral centra and dorsal spines can be used for ageing (Tucker, 1985; Parsons, 1993; Officer *et al.*, 1996). Age determination of elasmobranches can be problematic; Polat and Gumus (1995), for example, could not age spiny dogfish, *Squalus acanthias*, using vertebrae, but were successful using the annuli on the surface of the mantle and the basal diameter of the second dorsal spine. The visibility of growth bands in the vertebrae of black-mouth cat-sharks, *Galeus melastomus*, was enhanced by decalcification with 5% nitric acid solution (Correia & Figueiredo, 1997). Tropical catfish can be difficult to age, and Al Hassan and Al Sayab (1994) show that eye-lens weight, which is often used for small mammals, can indicate age. The pectoral spines of catfish can also be used (Blouin & Hall, 1990). Brennan & Cailliet (1989) concluded that, for white sturgeon, pectoral-fin sections were the most practical ageing structure in terms of ease of collection, processing, legibility, and precision of interpretation.

Beamish and Medland (1988) estimated lamprey age using statolith banding. However, annuli were not recognizable in the statoliths of the southern brook lamprey, *Ichthyomyzon gagei*, because growth was even throughout the year, so that length–frequency-distribution analysis is the only method available.

12.1.2.6. Reptiles and amphibians

Examination of skeletal structures is commonly used for reptile and amphibian ageing (Francillon *et al.*, 1984; Castanet, 1986; Smirina, 1994). Francillon Viellot *et al.* (1990) aged the newts *Triturus cristatus* and *Triturus marmoratus* by counting annual growth rings in cross-sections of femurs and phalanges, staining with Ehrlich haemtoxylin. Hutton (1986) found that male and non-breeding female Nile crocodiles, *Crocodilus niloticus*, could be aged from laminae in osteoderms, and Tucker (1997) in a study of the freshwater crocodile, *Crocodilus johnstoni*, found that these growth lines could be reliably counted in unstained 60–80 μm sections viewed by Nomarski interference microscopy.

12.1.2.7 Birds

Reliable methods for ageing adult birds are few, probably because of the widespread use of rings. Klomp and Furness (1992) found that the count of endosteal layers in the tibia could accurately determine age in fulmar, *Fulmarus glacialis*, shag, *Phalacro-corax aristotelis*, redshank, *Tringa totanus*, and great skua, *Catharacta skua*. Chae and Fujimaki (1996) suggested that the extent of skull pneumatization could be used to age hazel grouse, *Bonasa bonasia*. Fornasari *et al.* (1994) considered it possible to age hawfinches, *Coccothraustes coccothraustes*, using a set of criteria that scored rectrix abrasion, width of the outer web of the sixth primary feather, and sharpness of the greater alula feather.

12.1.2.8 Mammals

While skeletal development assessed from radiography can be used to assign indi-viduals to age classes (Nassar Montoya *et al.*, 1992), the most widely applicable methods use the teeth. One of the most accurate estimates of the age at death is the count of incremental lines in the cementum of teeth (Martin, 1996), and particular applications of this method are to: red foxes, *Vulpes vulpes* (Goddard & Reynolds, 1993; Cavallini & Santini, 1995); coyotes, *Canis latrans* (Jean *et al.*, 1985); marmots (Mashkin & Kolesnikov, 1990); and beaver, *Castor fiber* (Stiefel & Piechock, 1986). This method is not always reliable; in wart-hog, the incremental lines are often indistinct (Mason, 1984) and, in tropical ungulates, more than one band per annum may be laid down in the cementum (Kay & Cant, 1988). Other tooth characteristics used include: incisor height in deer (Moore *et al.*, 1995); degree of root closure and dentine layers, as observed using radiography in beaver, *C. fiber*, and pine martens, *Martes martes* (Hartman, 1992; Helldin, 1997); and tooth wear (Ohtaishi *et al.*, 1990; Hartman, 1995). Tooth wear will vary with the amount of abrasive material in the diet (Fandos *et al.*, 1993).

Growth rings can sometimes be observed in bone. Iason (1988) found that pol-ished cross-sections of the mandibles of mountain hares, *Lepus timidus*, from Scot-land showed annual growth rings, which are laid down in the autumn. Growth lines in bone have also been used to age Florida manatees, *Trichechus manatus latirostris* (Marmontel *et al.*, 1996).

Eye-lens weight, which increases linearly in weight with age, is considered the best character with which to age small mammals from genera such as *Microtus*, *Clethrionomys*, *Peromyscus*, *Mus*, *Apodemus*, *Pitymys*, and *Rattus* (Rowe *et al.*, 1985; Pascal *et al.*, 1988; Quere & Vincent, 1989; Heise *et al.*, 1991). It has also been found useful for larger species, such as the Japanese hare, *Lepus brachyurus brachyurus* (Ando *et al.*, 1992).

12.2 Time-specific life-tables and survival rates

This method of building a demographic picture of a population may be used if it is justified to assume that, over the period of life of the individuals in the table, the recruitment rate has been constant and the mortality rate within each age class has been steady (although, of course, it may change with age — the table will reveal this). Time-specific life-tables are also termed static, stationary, or vertical life-tables.

In contrast to age-specific life-tables (p. 418), where successive observations of a cohort are needed, the time-specific life-table is based solely on the age-grouping of the individuals collected at a single instant of time. Thus, a relative method (Chapter 7) may be used for sampling the population, provided that it samples at random with respect to the different age-groups. As we are studying relative abundance, the youngest age-group may be equated with a convenient number (e.g. 1, 100, 1000), and the other values corrected accordingly to give the l_x column. The remaining columns of the table can then be calculated as described in Chapter 11 (p. 417). An example of this type of life-table and its calculation is that given by Paris and Pitelka (1962) for the woodlouse, *Armadilidium*.

A number of time-specific life-tables were constructed for the mosquito *Aedes aegypti* by Southwood *et al.* (1972); they were for limited parts of the yearly cycle, when the assumption of steady recruitment appeared justified. Modified key-factor analysis may be applied to these life-tables; it will indicate differences between populations under the different conditions (with *Aedes* in different breeding sites) at the period of steady recruitment, but it will not disclose anything about the periods of population change. A further analysis of this data was undertaken by Rogers (1983).

An extensive collection of age-grouping data is provided by the 'catch curves' obtained from commercial fishing, and their analysis has been described by Ricker (1944, 1948), Wohlschlag (1954), Beverton and Holt (1957), Ebert (1973), and others. Another approach to the analysis is provided, when log numbers are plotted against age, by the comparison of the actual shape of the plot and the straight line that would result from a Slobodkin type III constant-survival rate (Fig. 12.1). Temporal variations in the rate of recruitment and in both the age- and time-specific survival rates will affect the shape of the curve. Age-specific survival rates may be obtained from mark-and-recapture data (Chapter 3) and a measure of recruitment can sometimes be obtained from another source, e.g. with insects from emergence traps. By obtaining information from such other sources, the departures of the time-specific survivorship curve (Fig. 12.1) from linearity can often be interpreted; if recruitment and time-specific survival rates are constant, then the fall-off with age in Fig. 12.1 implies increasing mortality with increasing age (Wohlschlag, 1954).

The hunting records of terrestrial vertebrates provide further extensive age-grouped data. Surveys of the proportions of various ages and sexes are made before and after hunting and the numbers of the different groups killed are also known; from this information, it is possible to calculate natality (referred to as production), survival rates, and absolute population. The methods of computation have been reviewed by Hanson (1963) and Seber (1982), and some of these might be combined with the 'removal-sampling' approach and used with insects. Some of the methods do not require hunting returns; of these, the equation for the calculation of differential survival of age classes is of particular potential interest to insect ecologists (Hanson, 1963; Kelker & Hanson, 1964). It could be applied to a situation where generations overlap, but each was comparatively well synchronized; from biological

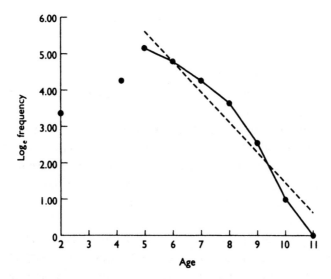

Fig. 12.1 A time-specific survivorship curve. The logarithms of the frequencies plotted against age groups in a season's sample of the fish, *Leucichthys sardinella* (after Wohlschlag, 1954).

knowledge, we must select three times: t_1 shortly before generation II becomes adult, t_2 just after generation II has become adult, and t_3 a subsequent occasion. Assuming that the mortality of both sexes is equal, then the ratio of the survival of generation I to that of generation II during the period t_1 to t_2 is given by:

$$\frac{S_{II}}{S_I} = \frac{(\delta t_2 - \female t_2)(\delta t_3 \female t_1 - \delta t_1 \female t_3)}{(\delta t_3 - \female t_3)(\delta t_2 \female t_1 - \delta t_1 \female t_2)}$$

where the symbols represent the number of males and females in the samples on dates t_1, t_2, and t_3, as defined above (see also p. 88).

The use, in entomology, of age-grouping, time-specific life-tables, and survivorship-curve analysis of this type will probably prove most important in attempts to investigate the dynamics of the adult population. Detailed studies of this type will be necessary if the full meaning of k_0 (p. 428) is to be investigated.

Key-factor analysis of age-specific life-tables or budgets (Chapter 11) and survival analysis of time-specific life-tables provide a quantitative description of the population processes. From this, predictions may be made of future population trends or of the response of the population to some change: a control measure or harvesting. A simple management model could be constructed to utilize the information obtained in this way. However, both types of life-table are inappropriate tools for the understanding of the population dynamics of the many populations in which there is continuous breeding and overlapping generations. As will be discussed below, it is possible to tackle the problem 'the other way round', i.e. to predict population growth under 'Malthusian conditions'—'unlimited' resources—when $r = dN/dt$ (see p. 420) and to compare the actual field results with those from the predictions. In most cases, the same model may be used to predict the impact of change on the population processes.

12.2.1 Physiological time

These models predict the rate of population growth with time; population growth is dependent on generation time, which depends on development rate. Development rate is influenced by temperature and therefore, in strongly seasonal regions, where the same stage may be exposed to significantly different temperatures, either a correction must be made for the different development rates or 'physiological time' may be used (Hughes, 1962, 1963; Gilbert et al., 1976; Hardman, 1976; Atkinson, 1977). Theoretically there is no reason why this should not always be used; however, it seems advisable to restrict it to models where it is really necessary. As most biologists can only easily visualize their populations against calendar time, the use of physiological time makes more difficult an intuitive check on the output of the model. Physiological time is commonly expressed as day-degrees (D°) (Hughes, 1962; Hardman, 1976) or hour-degrees (h°) (Atkinson, 1977), being the cumulative product of total time × temperature (above the threshold). Allen (1976) describes a sine-wave method for computing day-degrees, and Scopes and Biggerstaff (1977) show how a temperature integrator may be usefully used in the field as an effective measure of accumulated physiological temperature. Jones et al. (1997) produced a life-table for koa seed-worm based on degree-day demography and pointed out that this considerably reduced the variability within the data set. For example, at different temperatures, the adult period varied from 72 to 32 days but, in contrast, on a physiological scale the observed range was only 2.7 to 6.2 days. Atkinson's (1977) studies suggest that the common practice of determining the physiological time for development at constant temperatures may be misleading when applied to fluctuating field conditions.

In most population models, it is necessary to divide time, physiological or calendar, into intervals. The length of the interval is critical; if it is too large, the changes between intervals are too sudden and the heterogeneity within the time-interval age class too great. Too small an interval will add to the complexities and costs of the work and the field data may be inadequate in quantity. Gilbert et al. (1976) tested a number of different time units and concluded that, for their aphid, a quarter-instar period (a 'quip') was suitable; they make no claims for the generality of this unit, but it or a half-instar period seems to be of a reasonable order of magnitude to test initially. In their case, instar length was similar (in physiological time) for the first three instars; with other insects, the instar period might have to be based on the average or perhaps the shortest instar (excluding very short first instars) development time.

12.2.2 Life-table parameters

Predictive models require input information on the magnitude and variation of the development times of the different stages; a complete example of this type of data is given by Gomez et al. (1977) for Culex. When key-factor or similar analysis is to be undertaken, it is debatable whether the mortalities of the different stages detected under apparently 'Malthusian conditions' (e.g. Hardman, 1976) should be included in the prediction or whether, as seems safer, the prediction should assume 100% survival. Complete models, including the adult stage, will require age-specific fecundity schedules.

12.2.3 Recruitment in the field
A reliable estimate of recruitment is essential; depending on the form of the model, this could be a measure of adult emergence, daily or annual egg production, or some combination of the two (see also Chapters 10 and 11). These models generally assume that age distribution within a stage, at any one time, is effectively uniform.

12.2.4 Empirical models
The egg-ratio method was developed by Edmondson (1968) for the estimation of birth-rate in zooplankton communities. If F_e is the number of eggs observed per female and D is the mean development time of the eggs in days, then the number of offspring produced per female per day, B, is:

$$B = \frac{F_e}{D}$$

If the number of females alive on day 1 is given a value of 1, then the number on day 2 will be $1 + B$. The instantaneous birth-rate, b, is then given by the equation:

$$b = \ln(1 + B)$$

Thus the basic inputs for this model are an observed recruitment value (or in this case its equivalent) and an experimentally determined life-table parameter. The instantaneous rate of population change, r, is:

$$r = b - m$$

where m is the instantaneous mortality rate, and the size of the population is given by:

$$N_t = N_0 e^{rt}$$

where N_0 and N_t are the size of the population at times 0 and t, respectively.

Population growth in the absence of death ($r = b$) can be predicted and the real growth rate, as expressed by the field population, can be compared with this. See Caswell (1972) for details of the calculation of b if mortality exists.

The model of Southwood *et al.* (1972) allowed variable hatching and developmental times to be incorporated into the prediction. The size of the egg cohort of a particular time period is known from field observation. It is also necessary to know the developmental spectra of the various stages (e.g. the shape of the hatching curve for eggs of the same age). If physiological time is not used, but developmental periods differ at different seasons, this variation may also be incorporated. The model calculates the contribution of each egg cohort to the numbers of subsequent stages and then sums the contributions from each cohort to predict the total population of that stage in that time interval. These predictions may be compared with actual data from field sampling to estimate mortality levels; it is assumed that at a particular time the members of an age-group suffer mortality at random, i.e. irrespective of their cohort origin. Vandermeer (1975) has queried the validity of this assumption, which is also basic for the Lewis–Leslie matrix model if the age classes are too wide.

Shiyomi (1967) developed a model for the reproduction of aphids, recognizing that, as aphids are not randomly distributed in space, reproduction per parent will be variable and will conform to the negative binomial. It enables a prediction to be made of the probability that a certain number of offspring will arise on a plant from a particular density of females. This model is noteworthy as an early example of a model that takes into account a spatial factor; subsequent research has demonstrated the importance of spatial variation in ecological models.

12.2.5 Intrinsic-rate models and variable life-tables

Hughes (1962, 1963) laid the foundations of this approach when he showed that stable age distributions often occurred in expanding aphid populations; under these conditions, the potential rate of population increase could be predicted by a model based on the intrinsic rate of increase in Lotka's population-growth equation. The models have been extensively developed in collaboration with N. Gilbert (Hughes & Gilbert, 1968; Gilbert & Hughes, 1971) and colleagues (Gilbert & Gutierrez, 1973; Gilbert et al., 1976). They have been termed 'variable life-tables' by Gilbert et al. (1976), who give a detailed account of the basic philosophy and the construction of the models, including the facsimile reproduction of the printout of the computer programs; reference should be made to this work for details. The variable life-table also owes much of its methodology to C.S. Holling's component analysis, for the approach is to identify the major components of population change, insert these in the model, test the model against the field data, examine the places where it does not fit (as Gilbert et al. (1976) stress, these are the most interesting results), and then modify that part of the model in a biologically meaningful way and rerun, comparing the prediction against field data. The initial components of a variable life-table might be expected to include the age–fecundity schedule, the fecundity–density relationship, the development schedule (normally on a physiological time-scale), the impact of density on development, and mortality schedules in relation to various natural enemies.

12.2.6 Lewis–Leslie matrices

During the 1940s, E.G. Lewis and P.H. Leslie independently developed a deterministic model in matrix form for the age structure of a reproducing population. The general form of these models, in matrix notation, is:

$$a_t M = a_{t+1}$$

where M is a square matrix that describes the transition of the population over one time period and thus contains terms describing age-specific fecundities and survival rates, a_t is a column vector of the age structure of the population at time t (the start) and a_{t+1} is the column vector at time $t+1$. Set out in full, it is:

$$
\begin{bmatrix}
f_0 & f_1 & f_2 & f_3 \\
p_0 & 0 & 0 & 0 \\
0 & p_1 & 0 & 0 \\
0 & 0 & p_2 & 0
\end{bmatrix}
\times
\begin{bmatrix}
a_0 \\
a_1 \\
a_2 \\
a_3
\end{bmatrix}
=
\begin{bmatrix}
f_0 a_0 + f_1 a_1 + f_2 a_2 + f_3 a_3 \\
p_0 a_0 + 0 + 0 + 0 \\
0 + p_1 a_1 + 0 + 0 \\
0 + 0 + p_2 a_2 + 0
\end{bmatrix}
$$

where the f values represent the age-specific fecundity of each age class (0, 1, 2, and 3), as expressed in daughters born between t and $t+1$, and $p_0 \ldots p_2$ represents the probability of a female (of the age indicated by the subscript) surviving into the next age class. It will be noted that these probabilities will always form a diagonal that commences at the left-hand end of the second row and ends (reaches the last row) in the penultimate column. In practice, f_0 is normally zero. The column vector $(a_0 \ldots a_3)$ represents the number of animals at each age-group $t=0$ to $t=3$ at time t.

From any starting vector a_t, the future population can be calculated by repeated multiplication, so that after k time periods:

$$a_{t+k} = M^k a_t$$

As k increases, the age distribution gradually becomes constant. It can be shown that the long-term rate of growth of the population, λ (the per capita rate of increase, $r = \ln \lambda$ (see p. 432), and the stationary (stable) age distribution are given by the dominant eigenvector and eigenvalue of matrix M. The population will increase in size when $\lambda > 1$. Eigenvalues and vectors are obtained by iteration, using standard computer packages. Lucid accounts of the Lewis–Leslie matrix and its development are given by Williamson (1967, 1972) and Usher (1972).

Population growth from a starting vector, considerably different from the stationary age distribution, can show short-term oscillations as the population age structure changes. The magnitude of these oscillations is determined by the relative magnitudes of the eigenvalues of matrix M, which has led to the suggestion that they may be compared to assess the relative stability of different populations. In practice, because of natural variation in age-specific fecundities and survivals and, in particular, the introduction of non-linear features, such as density dependence, at high population values, these approaches are probably of little value.

The male sex was introduced into the model by Williamson (1959) by doubling the number of rows and columns, whilst Pennycuick et al. (1968) produced a computer model incorporating density dependence. They made the elements in the matrix M dependent on the size of the population (the sum of the appropriate vector column) by multiplying the values in the top row by a function F and the p values (in the line below the diagonal) by another function S. The functions F and S are expressions that modify fecundity and survival to give sigmoid curves; they combine the present size of the population, the equilibrium population, and the degree of density dependence, in a rather complex manner (May *et al.*, 1974). The problem with such density-dependent models is that the eigenvalues and vectors can no longer be used to predict long-term population behaviour, because the schedule of births and deaths is no longer constant.

A requirement of the Lewis–Leslie matrix is that the time intervals are exactly equal, and yet insects are generally age-grouped by stages that are of unequal length. Lefkovitch (1965) considered this problem and describes a method of stage grouping taking account of development times and the influence of one stage on another. Many elements in the matrix will contain terms for more than one stage, and it is clear that, unless a stable age distribution is reached, the survival probabilities will change as the proportion of the stages in the elements changes. Vandermeer (1975) emphasizes the need to make age categories as brief as possible, so that the assumption that all individuals within each age category have the same ages (and survival probabilities) can be used without introducing a major source of error.

Longstaff (1977) has utilized these techniques to study the population dynamics of Collembola in culture. Besides deterministic models, he developed a stochastic model. The assumption was made that the parameter values were normally distributed around the mean and the actual value was selected by a random-number generator. The accuracy of the deterministic and stochastic models varied, but the predictors were sufficiently close to the form of the actual populations for it to seem likely that the population parameters derived from the models (the density-dependent function and the equilibrium population) are meaningful.

The Lewis–Leslie matrix approach has often been used in applied research on fish and mammals to gain insight into the response of a population to changes in age (size) selective mortality, e.g. Smith and Trout (1994) used this approach to model the effect of different rabbit-control policies.

References

Achionye Nzeh, C.G. (1994) Age and growth determination in *Sarotherodon galilaeus* by use of scales and opercular bones (Pisces: Cichlidae). *Rev. Biol. Trop.* **42**(1–2), 371–4.

Al Hassan, L.A.J. & Al Sayab, A.A. (1994) Eye lens diameter as an age indicator in the catfish, *Silurus triostegus. Pakistan J. Zool.* **26**(1), 81–2.

Allen, J.C. (1976) A modified sine wave method for calculating degree days. *Environ. Entomol.* **5**, 388–96.

Almeida, P.R., Moreira, F.M., Dimingos, I.M., *et al.* (1995) Age and growth of *Liza ramada* (Risso, 1826) in the River Tagus, Portugal. *Sci. Mar.* **59**(2), 143–7.

Alvarado Castillo, R.M. & Felix Uraga, R. (1995) Age determination in *Istiophorus platypterus* (Pisces: Istiophoridae) in the south of the Gulf

of California, Mexico. *Rev. Biol. Trop.* **44**(1), 233–9.

Ancrenaz, M. & Delhomme, A. (1997) Teeth eruption as a means of age determination in captive Arabian oryx, *Oryx leucoryx* (Bovidae, Hippotraginae). *Mammalia* **61**(1), 135–8.

Anderson, J.L. (1986) Age determination of the nyala *Tragelaphus angasi. South Afr. J. Wildl. Res.* **16**(3), 82–90.

Anderson, J.R. (1964) Methods for distinguishing nulliparous from parous flies and for estimating the ages of *Fannia canicularis* and some other cyclorrophous Diptera. *Ann. Entomol. Soc. Am.* **57**, 226–36.

Ando, A., Yamada, F., Taniguchi, A., & Shiraishi, S. (1992) Age determination by the eye lens weight in the Japanese hare, *Lepus brachyurus*

brachyurus and its application to two local populations. *Sci. Bull. Fac. Agric. Kyushu Univ.* **46**(3–4), 169–75.

Antunes, C. & Tesch, F.W. (1997) A critical consideration of the metamorphosis zone when identifying daily rings in otoliths of European eel, *Anguilla anguilla* (L.). *Ecol. Freshw. Fish* **6**(2), 102–7.

Anwar, N.A., Richardson, C.A., & Seed, R. (1990) Age determination, growth rate and population structure of the horse mussel *Modiolus modiolus*. *J. Mar. Biol. Assoc. UK* **70**(2), 441–58.

Aprahamian, M.W. (1987) Use of the burning technique for age determination in eels (*Anguilla anguilla* (L.)) derived from the stockings of elvers. *Fish. Res.* **6**(1), 93–6.

Atkinson, P.R. (1977) Preliminary analysis of a field population of citrus red scale, *Aonidiella aurontii* (Maskell) and the measurement and expression of stage duration and reproduction for life tables. *Bull. Entomol. Res.* **67**, 65–87.

Baker, G.H. & Vogelzang, B.K. (1988) Life history, population dynamics and polymorphism of *Theba pisana* (Mollusca: Helicidae) in Australia. *J. Appl. Ecol.* **25**, 867–87.

Beamish, F.W.H. & Medland, T.E. (1988) Age determination for lampreys. *Trans. Am. Fish. Soc.* **117**(1), 63–71.

Berg, R. (1985) Age determination of eels, *Anguilla anguilla*: comparison of field data with otolith ring patterns. *J. Fish Biol.* **26**(5), 537–44.

Berman, M.S., McVey, A.L., & Ettershank, G. (1989) Age determination of Antarctic krill using fluorescence and image analysis of size. *Polar Biol.* **9**(4), 267–72.

Beverton, R.J.H. & Holt, S.J. (1957) *On the Dynamics of Exploited Fish Populations*. Ministry of Agriculture, Fisheries and Food HMSO, London.

Bianchini, J.J. & Delupi, L.H. (1990) Age determination of pampas deer (*Odocoileus bezoarticus*) by comparative study of tooth development and wear. *Physis Secc. C Contin. Organ. Terr.* **48**(114–15), 27–40.

Blouin, M.A. & Hall, G.R. (1990) Improved method for sectioning pectoral spines of catfish for age determination. *J. Freshw. Ecol.* **5**(4), 489–90.

Brennan, J.S. & Cailliet, G.M. (1989) Comparative age-determination techniques for white sturgeon in California (USA). *Trans. Am. Fish. Soc.* **118**(3), 296–310.

Britaev, T.A. & Belov, V.V. (1993) Age determination in polynoid polychaetes, using growth rings on their jaws. *Zool. Zh.* **72**(11), 15–21.

Brown, H.A. (1990) Morphological variation and age-class determination in overwintering tadpoles of the tailed frog, *Ascaphus truei*. *J. Zool.* **220**(2), 171–84.

Cailliet, G.M., Botsford, L.W., Brittnacher, J.G., *et al.* (1996) Development of a computer-aided age determination system: evaluation based on otoliths of bank rockfish off California. *Trans. Am. Fish. Soc.* **125**(6), 874–88.

Camin, V., Baker, P., Carey, J., Valenzuela, J., & Arredondo Peter, R. (1991) Biochemical age determination for adult Mediterranean fruit flies (Diptera: Tephritidae). *J. Econ. Entomol.* **84**(4), 1283–8.

Castanet, J. (1986) Skeletochronology of reptiles: III. Application. *Ann. Sci. Nat. Zool. Biol. Anim.* **8**(3), 157–72.

Caswell, H. (1972) On instantaneous and finite birth rates. *Limnol. Oceanogr.* **17**, 787–91.

Cavallini, P. & Santini, S. (1995) Age determination in the red fox in a Mediterranean habitat. *Zeitschr. Saeugetierk.* **60**(3), 136–42.

Chae, H.Y. & Fujimaki, Y. (1996) Age determination of the hazel grouse based on skull pneumatization. *Japan. J. Ornithol.* **45**(1), 17–22.

Chai, F.-C. & Dixon, S.E. (1971) A technique for ageing cocoons of the sawfly *Neodiprion serlifer* (Hymenoptera: Diprionidae). *Can. Entomol.* **103**, 80–3.

Christophers, S.R. (1960) Aedes aegypti *(L.) the Yellow Fever Mosquito*. Cambridge University Press, Cambridge.

Corbet, P.S. (1960) The recognition of nulliparous mosquitoes without dissection. *Nature* **187**, 525–6.

Corbet, P.S. (1961) The recognition of parous dragonflies (Odonata) by the presence of follicular relics. *Entomologist* **94**, 35–7.

Corbet, P.S. (1962a) Age-determination of adult dragonflies (Odonata). *Proc. XI Int. Congr. Entomol.* **3**, 287–9.

Corbet, P.S. (1962b) *A Biology of Dragonflies*. Witherby, London.

Corbet, P.S. (1963) Reliability of parasitic water mites (*Hydracarina*) as indicators of physiological age in mosquitoes (Diptera: Culicidae). *Entomol. Exp. Appl.* **6**, 215–33.

Corbet, P.S. (1970) The use of parasitic water-mites for age-grading female mosquitoes. *Mosquito News* **30**, 436–8.

Correia, J.P. & Figueiredo, I.M. (1997) A modified decalcification technique for enhancing growth bands in deep-coned vertebrae of elasmobranchs. *Environ. Biol. Fish.* **50**(2), 225–30.

Daly, H.V. (1961) Biological observations on *Hemihalictus lustrans*, with a description of the larva (Hymenoptera: Halictidae). *J. Kansas Entomol. Soc.* **34**, 134–41.

Davies, J.B., Corbet, P.S., Gillies, M.T., & McCrae, A.W.R. (1971) Parous rates in some Amazonian mosquitoes collected by three different methods. *Bull. Entomol. Res.* **61**, 125–32.

Dery, L.M. (1983) Use of laminated plastic to impress fish scales. *Progr. Fish Cult.* **45**, 88–9.

Detinova, T.S. (1962) Age-grouping methods in Diptera of medical importance with special reference to some vectors of malaria. *Monogr. Ser. World Health Org.* **47**, 216 pp.

Detinova, T.S. (1968) Age structure of insect populations of medical importance. *Ann. Rev. Entomol.* **13**, 427–50.

Drew, R.A.L. (1969) Morphology of the reproductive system of *Strumeta tryoni* (Froggatt) (Diptera: Trypetidae) with a method of distinguishing sexually mature males. *J. Aust. Entomol. Soc.* **8**, 21–32.

Dunn, E. (1951) Wing coloration as a means of determining the age of the Colorado beetle (*Leptinotarsa decemlineta* Say). *Ann. Appl. Biol.* **38**, 433–4.

Dyce, A.L. (1969) The recognition of multiparous and parous *Culicoides* without dissection. *J. Aust. Entomol. Soc.* **8**, 11–15.

Ebert, T.A. (1973) Estimating growth and mortality rates from size data. *Oecologia (Berlin)* **11**, 281–98.

Edmondson, W.T. (1968) A graphical model for evaluating the use of the egg ratio for measuring birth and death rates. *Oecologia* **1**, 1–37.

Esteves, E., Simoes, P., Da Silva, H.M., & Andrade, J.P. (1995) Ageing of swordfish, *Xiphias gladius* Linnaeus, 1758, from the Azores, using sagittae, anal-fin spines and vertebrae. *Arquipel. Bol. Univ. Acores Cienc. Biol. Mar.* **13A**, 39–51.

Fandos, P., Orueta, J.F., & Aranda, Y. (1993) Tooth wear and its relation to kind of food: the repercussion on age criteria in *Capra pyrenaica*. *Acta Theriol.* **38**(1), 93–102.

Fedetov, D.M. (1947, 1955, 1960) The noxious little tortoise, *Eurygaster integriceps* Put. [In Russian.] **1** (272 pp.) and **2** (1947); **3** (278 pp.) (1955); **4** (239 pp.) (1960).

Fornasari, L., Pianezza, F., & Carabella, M. (1994) Criteria for the age determination of the hawfinch *Coccothraustes coccothraustes*. *Ring. Migr.* **15**(1), 50–5.

Francillon, H., Barbautl, R., Castanet, J., & De Ricqles, A. (1984) The biology of the desert toad *Bufo pentoni*: age determination by the skeletochronological method, population structure and dynamics. *Rev. Ecol. Terre Vie* **39**(2), 209–24.

Francillon Viellot, H., Arntzen, J.W., & Geraudie, J. (1990) Age, growth and longevity of sympatric *Triturus cristatus*, *Triturus marmoratus* and their hybrids (Amphibia, Urodela): a skeletochronological comparison. *J. Herpetol.* **24**(1), 13–22.

Galinou Mitsoudi, S. & Sinis, A.I. (1995) Age and growth of *Lithophaga lithophaga* (Linnaeus, 1758) (Bivalvia: Mytilidae), based on annual growth lines in the shell. *J. Mollusc. St.* **61**(4), 435–53.

Garcia, F. (1993) Interpretation of shell marks for the estimation of growth of the European carpet clam *Ruditapes decussatus* L. of the Bay of Fos (Mediterranean Sea). *Oceanol. Acta* **16**(2), 199–203.

Giglioli, M.E.C. (1963) Aids to ovarian dissection for age determination in mosquitoes. *Mosquito News* **23**, 156–9.

Gilbert, N. & Gutierrez, A.P. (1973) A plant–aphid–parasite relationship. *J. Anim. Ecol.* **42**, 323–40.

Gilbert, N. & Hughes, R.D. (1971) A model of an aphid population-three adventures. *J. Anim. Ecol.* **40**, 525–34.

Gilbert, N., Gutierrez, A.P., Frazer, B.D., & Jones, R.E. (1976) *Ecological Relationships*. Addison-Wesley, Reading, San Francisco.

Gillett, J.D. (1957) Age analysis of the biting cycle of the mosquito *Taeniorhynchus (Mansonioides) africana* Theobald, based on the presence of parasitic mites. *Ann. Trop. Med. Hyg.* **51**, 151–8.

Goddard, H.N. & Reynolds, J.C. (1993) Age determination in the red fox (*Vulpes vulpes* L.) from tooth cementum lines. *Gibier Faune Sauvage* **10**, 173–87.

Gomez, C., Rabinovich, J.E., & Machado-Allison, C.E. (1977) Population analysis of *Culex pipiensfatigans* Wielt. (Diptera: Culicidae) under laboratory conditions. *J. Med. Entomol.* **13**, 453–63.

Goodyear, C.P. (1997) Fish age determined from length: an evaluation of three methods using simulated red snapper data. *Fish. Bull.* **95**(1), 39–46.

Gordo, L.S. (1996) On the age and growth of bogue, *Boops boops* (L.), from the Portuguese coast. *Fish. Manag. Ecol.* **3**(2), 157–64.

Grana Raffucci, F.A. & Appeldoorn, R.S. (1997) Age determination of larval strombid gastropods by means of growth increment counts in statoliths. *Fish. Bull.* **95**(4), 857–62.

Graves, G.R. (1997) Age determination of free-living male black-throated blue warblers during the breeding season. *J. Field Ornithol.* **68**(3), 443–9.

Hales, L.S.J. & Belk, M.C. (1992) Validation of otolith annuli of bluegills in a southeastern thermal reservoir. *Trans. Am. Fish. Soc.* **121**(6), 823–30.

Hamon, J., Grjedine, A., Adam, J.P., *et al.* (1961) Les méthodes de valuation de l'age physiologique des moustiques. *Bull. Soc. Entomol. Fr.* **669**, 137–61.

Hanson, W.R. (1963) Calculation of productivity, survival, and abundance of selected vertebrates from sex and age ratios. *Wild. Monogr.* **9**, 1–60.

Hardman, J.M. (1976) Life table data for use in deterministic and stochastic simulation models predicting the growth of insect populations under Malthusian conditions. *Can. Entomol.* **108**, 897–906.

Hartman, G. (1992) Age determination of live beaver by dental X-ray. *Wildl. Soc. Bull.* **20**(2), 216–20.

Hartman, G.D. (1995) Age determination, age structure, and longevity in the mole, *Scalopus aquaticus* (Mammalia: Insectivora). *J. Zool.* **237**(1), 107–22.

Hassager, T.K. (1991) Comparison of three different otolith-based methods for age determination of turbot (*Scophthalmus maximus*). *Dana* **9**, 39–43.

Haydak, M.H. (1957) Changes with age in the appearance of some internal organs of the honeybee. *Bee World* **38**, 197–207.

Heise, S., Stubbe, M., Wieland, H., & Nessen, B. (1991) The analysis of the population structure of the common vole (*Microtus arvalis*, Pallas, 1779). *Zeitschr. angew. Zool.* **78**(1), 19–30.

Helldin, J.O. (1997) Age determination of Eurasian pine martens by radiographs of teeth *in situ*. *Wildl. Soc. Bull.* **25**(1), 83–8.

Henocque, Y. (1987) Age determination of the European lobster, *Homarus gammarus*, from its antennules. *Mer* **25**(1), 1–12.

Hesthagen, T. (1985) Validity of the age determination from scales of brown trout (*Salmo trutta* L.). *Inst. Freshw. Res. Drottningholm Rep.* **62**, 65–70.

Hoc, T.Q. & Charlwood, J.D. (1990) Age determination of *Aedes cantans* using the ovarian oil injection technique. *Med. Vet. Entomol.* **4**(2), 227–33.

Hoc, T.Q. & Wilkes, T.J. (1995) Age determination in the blackfly *Simulium woodi*, a vector onchocerciasis in Tanzania. *Med. Vet. Entomol.* **9**(1), 16–24.

Hohman, W.L. & Cypher, B.L. (1986) Age-class determination of ring-necked ducks (*Aythya collaris*). *J. Wildl. Manag.* **50**(3), 442–5.

Hughes, R.D. (1962) A method for estimating the effects of mortality on aphid populations. *J. Anim. Ecol.* **31**, 389–96.

Hughes, R.D. (1963) Population dynamics of the cabbage aphid, *Brevicoryne brassicae* (L.). *J. Anim. Ecol.* **32**, 393–424.

Hughes, R.D. & Gilbert, N. (1968) A model of an aphid population—a general statement. *J. Anim. Ecol.* **37**, 553–63.

Hutton, J.M. (1986) Age determination of living Nile crocodiles (*Crocodilus niloticus*) from the cortical stratification of bone. *Copeia* **2**, 332–41.

Iason, G.R. (1988) Age determination of mountain hares (*Lepus timidus*): a rapid method and when to use it. *J. Appl. Ecol.* **25**(2), 389–96.

Jean, Y., Bergeron, J.M., Bisson, S., & Larocque, B. (1985) Relative age determination of coyotes, *Canis latrans*, from southern Quebec (Canada). *Can. Field Nat.* **100**(4), 483–7.

Jones, M.G. (1971) Observations on changes in the female reproductive system of the wheat bulb fly, *Leptohylemyia coarctata* (Fall.). *Bull. Entomol. Res.* **61**, 55–68.

Jones, V.P., Tome, C.H.M., & Caprio, L.C. (1997) Life tables for the koa seedworm (Lipidoptera: Tortricidae) based on degree-day demography. *Econ. Entomol.* **26**, 1291–8.

Kardos, E.H. & Bellamy, R.E. (1962) Distinguishing nulliparous from parous female *Culex larsalis* by examination of the ovarian tracheation. *Ann. Entomol. Soc. Am.* **54**, 448–51.

Kay, R.F. & Cant, J.G.H. (1988) Age assessment using cementum annulus counts and tooth wear in a free-ranging population of *Macaca mulatta*. *Am. J. Primatol.* **15**(1), 1–16.

Kelker, G.H. & Hanson, W.R. (1964) Simplifying the calculation of differential survival of age-classes. *J. Wildl. Manag.* **28**, 411.

Kishi, Y. (1971) Reconsideration of the method to measure the larval instars by use of the frequency distribution of head-capsule widths or lengths. *Can. Entomol.* **103**, 101–15.

Klomp, N.I. & Furness, R.W. (1992) A technique which may allow accurate determination of the age of adult birds. *Ibis* **134**(3), 245–9.

Kosminskii, R.B. (1960) The method of determining the age of the fleas *Leptopsylla segnis* Schönh. 1811 and *L. taschenbergi* Wagn. 1898 (Suctoria—Aphaniptera) and an experiment on the age analysis of a population of *L. segnis*. [In Russian.] *Med. Parazitol.* **29**, 590–4.

Kunitskaya, N.T. (1960) On the reproductive organs of female fleas and determination of their physiological age. [In Russian]. *Med. Parazitol.* **29**, 688–701.

Lagace, C.F. & van den Bosch, R. (1964) Progressive sclerotization and melanization of certain structures in males of a field population of *Hypera brunneipennis* (Coleoptera: Curculionidae). *Ann. Entomol. Soc. Am.* **57**, 247–52.

Lai, H.L., Gunderson, D.R., & Low, L.L. (1987) Age determination of Pacific cod, *Gadus macrocephalus*, using five ageing methods. *US Nat. Mar. Fish. Serv. Fish. Bull.* **85**(4), 713–24.

Le Berre, R. (1966) *Contribution à l'Étude Biologique et Écologique de* Simulium damnosum (*Diptera, Simuliidae*). Off. Rech. Sci. Tech. Outre-Mer, 204 pp.

Lefkovitch, L.P. (1965) The study of population growth in organisms grouped by stages. *Biometrics* **21**, 1–18.

Le Foll, D., Brichet, E., Reyss, J.L., Lalou, C., & Latrouite, D. (1989) Age determination of the spider crab *Maja squinado* and the European lobster *Homarus gammarus* by thorium-228 radium-228 chronology: possible extension to other crustaceans. *Can. J. Fish. Aquat. Sci.* **46**(4), 720–4.

Lewis, D.J., Lainson, R., & Shaw, J.J (1970) Determination of parous rates in phlebotomine sandflies with special reference to Amazonian species. *Bull. Entomol. Res.* **60**, 209–19.

Linley, J.R. (1965) The ovarian cycle and egg stage in *Leptoconops* (*Holoconops*) *becquaerti* (Kiefr.) (Diptera, Ceratopogaoidae). *Bull. Entomol. Res.* **56**, 37–56.

Longstaff, B.C. (1977) The dynamics of collembolan populations: a matrix model of single species population growth. *Can. J. Zool.* **55**, 314–24.

Loubens, G. & Panfili, J. (1992) Age determination of *Prochilodus nigricans* (Teleostei, Prochilodidae) in Beni (Bolivia): setting of a procedure and application. *Aquat. Living Resources* **5**(1), 41–56.

Mann, R.H.K. & Steinmetz, B. (1985) On the accuracy of age-determination using scales from rudd, *Scardinius erythrophthalmus* of known age. *J. Fish Biol.* **26**(5), 621–8.

Marmontel, M., O. Shea, T.J., Kochman, H.I., & Humphrey, S.R. (1996) Age determination in manatees using growth-layer-group counts in bone. *Mar. Mammal Sci.* **12**(1), 54–88.

Martin, H. (1996) Age determination in adults by analysis of dental cementum: methods and limits. *Bull. Mém. Soc. Anthropol. Paris* **8**(3–4), 433–40.

Mashkin, V.I. & Kolesnikov, V.V. (1990) Age determination of marmots (Marmota, Sciuridae) by the pattern of the wearing down of the chewing surface of teeth. *Zool. Zh.* **69**(6), 124–31.

Mason, D.R. (1984) Dentition and age determination of the warthog, *Phacochoerus aethiopicus*, in Zululand, South Africa. *Koedoe* **27**, 79–120.

May, R.M., Conway, G.R., Hassell, M.P., & Southwood, T.R.E. (1974) Time delays, density dependence and single-species oscillations. *J. Anim. Ecol.* **43**, 747–70.

Michener, C.D. & Lange, R.B. (1959) Observations on the behaviour of Brazilian Halictid bees (Hymenoptera, Apoidea). IV. *Augochloropsis*, with notes on extralimital forms. *Am. Mus. Noritates* **1924**, 1–41.

Michener, C.D., Cross, E.A., Daly, H.V., Rettenmeyer, C.W., & Wille, A. (1955) Additional techniques for studying the behaviour of wild bees. *Insectes Soc.* **2**, 237–46.

Mochizuki, A., Shiga, M., & Imura, O. (1993) Pteridine accumulation for age determination in the melon fly, *Bactrocera* (*Zeugodacus*) *cucurbitae* (Coquillett) (Diptera: Tephritidae). *Appl. Entomol. Zool.* **28**(4), 584–6.

Moore, N.P., Cahill, J.P., Kelly, P.F., & Hayden, T.J. (1995) An assessment of five methods of age determination in an enclosed population of fallow deer (*Dama dama*). *Biol. Environ.* **95B**(1), 27–34.

Moreira, F., Costa, J.L., Almeida, P.R., Assis, C., & Costa, M.J. (1991) Age determination in *Pomatoschistus minutus* (Pallas) and *Pomatoschistus microps* (Kroyer) (Pisces: Gobiidae) from the upper Tagus estuary, Portugal. *J. Fish Biol.* **39**(3), 433–40.

Msangi, A. & Lehane, M.J. (1991) A method for determining the age of very young tsetse flies (Diptera: Glossinidae) and an investigation of the factors determining head fluorescent levels in newly emerged adults. *Bull. Entomol. Res.* **81**(2), 185–8.

Muirhead-Thomson, R.C. (1963) *Practical Entomology in Malaria Eradication*, Part 1. MHO/PA/62.63, WHO, Geneva, pp. 64–71.

Mullens, B.A. & Lehane, M.J. (1995) Fluorescence as a tool for age determination in *Culicoides variipennis sonorensis* (Diptera: Ceratopogonidae). *J. Med. Entomol.* **32**(4), 569–71.

Narimatsu, Y. & Munehara, H. (1997) Age determination and growth from otolith daily growth increments of *Hypoptychus dybowskii* (Gasterosteiformes). *Fish. Sci.* **63**(4), 503–8.

Nassar Montoya, F., Sainsbury, A.W., Kirkwood, J.K., & Du Boulay, G.H. (1992) Age determination of common marmosets (*Callithrix jacchus*) by radiographic examination of skeletal development. *J. Med. Primatol.* **21**(5), 259–64.

Neville, A.C. (1963) Daily growth layers in locust rubber-like cuticle, influenced by an external rhythm. *J. Insect Physiol.* **9**, 177–86.

Neville, A.C. (1970) Cuticle ultrastructure in relation to the whole insect. In: Neville, A.C.

(ed.) Insect ultrastructure. *Symp. Roy. Entomol. Soc. Lond.* **5**, 17–39.

Nishi, E. & Nishihira, M. (1996) Age-estimation of the Christmas tree worm *Spirobranchus giganteus* (Polychaeta, Serpulidae) living buried in the coral skeleton from the coral-growth band of the host coral. *Fish. Sci.* **62**(3), 400–3.

Officer, R.A., Gason, A.S., Walker, T.I., & Clement, J.G. (1996) Sources of variation in counts of growth increments in vertebrae from gummy shark, *Mustelus antarcticus*, and school shark, *Galeorhinus galeus*: implications for age determination. *Can. J. Fish. Aquat. Sci.* **53**(8), 1765–77.

Ohtaishi, N., Kaji, K., Miura, S., & Wu, J. (1990) Age determination of the white-lipped deer *Cervus albirostris* by dental cementum and molar wear. *J. Mammal. Soc. Japan* **15**(1), 25–9.

Oxenford, H.A., Hunte, W., Deane, R., & Campana, S.E. (1994) Otolith age validation and growth-rate variation in flyingfish (*Hirundichthys affinis*) from the eastern Caribbean. *Mar. Biol.* **118**(4), 585–92.

Paris, O.H. & Pitelka, A. (1962) Population characteristics of the terrestrial isopod *Armadillidium vulgare* in California grassland. *Ecology* **43**, 229–48.

Parson, G.J., Robinson, S.M.C., Roff, J.C., & Dadswell, M.J. (1993) Daily growth rates as indicated by valve ridges in postlarval giant scallop (*Placopecten magellanicus*) (Bivalvia: Pectinidae). *Can. J. Fish. Aquat. Sci.* **50**(3), 456–64.

Parsons, G.R. (1993) Age determination and growth of the bonnethead shark *Sphyrna tiburo*: a comparison of two populations. *Mar. Biol.* **117**(1), 23–31.

Pascal, M., Damange, J.P., Douville, P., & Guedon, G. (1988) Age determination in pine vole *Pitymys duodecimcostatus* (De Selys-Longchamps, 1839). *Mammalia* **52**(1), 85–92.

Pennycuick, C.J., Compton, R.M., & Beckingham, L. (1968) A computer model for simulating the growth of a population, or of two interacting populations. *J. Theor. Biol.* **18**, 316–29.

Polat, N. & Gumus, A. (1995) Age determination of spiny dogfish (*Squalus acanthias* L. 1758) in Black Sea waters. *Israeli J. Aquacult. Bamidgeh* **47**(1), 17–24.

Pollock, J. (1969) Test for the mated status of male sheep blowflies. *Nature* **223**, 1287–8.

Quere, J.P. & Vincent, J.P. (1989) Age determination in wood mice (*Apodemus sylvaticus* L., 1758) by dry weight of the lens. *Mammalia* **53**(2), 287–94.

Raboud, C. (1986) Age determination of *Arianta arbustorum* (L.) (Pulmonata) based on growth

breaks and inner layers. *J. Mollusc. St.* **52**(3), 243–7.

Raya, C.P., Fernandez Nunez, M., Balguerias, E., & Hernandez Gonzalez, C.L. (1994) Progress towards ageing cuttlefish *Sepia hierredda* from the northwestern African coast using statoliths. *Mar. Ecol. Progr. Ser.* **114**(1–2), 139–47.

Richardson, C.A. (1987) Microgrowth patterns in the shell of the Malaysian cockle *Anadara granosa* (L.) and their use in age determination. *J. Exp. Mar. Biol. Ecol.* **111**(1), 77–98.

Richardson, C.A., Seed, R., & Naylor, E. (1990) Use of internal growth bands for measuring individual and population growth rates in *Mytilus edulis* from offshore production platforms. *Mar. Ecol. Progr. Ser.* **66**(3), 259–65.

Richardson, C.A., Collis, S.A., Ekaratne, K., Dare, P., & Key, D. (1993) The age determination and growth rate of the European flat oyster, *Ostrea edulis*, in British waters determined from acetate peels of umbo growth lines. *Ices* **50**(4), 493–500.

Richter, H. & McDermott, J.G. (1990) The staining of fish otoliths for age determination. *J. Fish Biol.* **36**(5), 773–80.

Ricker, W.E. (1944) Further notes on fishing mortality and effort. *Copeia* **1944**, 23–44.

Ricker, W.E. (1948) Methods of estimating vital statistics of fish populations. *Indiana Univ. Publ. Sci. Ser.* **15**, 101.

Rodhouse, P.G. & Hatfield, E.M.C. (1990) Age determination in squid using statolith growth increments. *Fish. Res.* **8**(4), 323–34.

Rogers, D.J. (1983) Interpretation of sample data. In: Youdeowei, A. & Service, M.W. (eds) *Pest and Vector Management in the Tropics*. Longman, London, pp. 139–60.

Rosay, B. (1961) Anatomical indicators for assesing the age of mosquitoes: the teneral adult (Diptera: Culicidae). *Ann. Entomol. Soc. Am.* **54**, 526–9.

Rowe, F.P., Bradfield, A., Quy, R.J., & Swinney, J. (1985) The relationship between lens weight and age in the wild house mouse (*Mus musculus*). *J. Appl. Ecol.* **22**, 55–61.

Saunders, D.S. (1962) Age determiantion for female tsetse flies and the age composition of samples of *Glossina pallidipes* Aust., *G. palpalisfuscipes* Newst. and *G. brevipalpis* Newst. *Bull. Entomol. Res.* **53**, 579–95.

Saunders, D.S. (1964) Age-changes in the ovaries of the sheep ked, *Melophagus ovinus* (L.) (Diptera: Hippoboscidae). *Proc. Roy. Entomol. Soc. Lond. A* **39**, 68–72.

Schiemenz, H. (1953) Zum Farbwechsel bei heimischen Heteropterenunter besonderer Berücksichtigung von *Palomena* Muls. & Rey. *Beitr. Entomol.* **3**, 359–71.

Schlein, I. & Gratz, N.G. (1972) Age determination of some flies and mosquitoes by daily growth layers of skeletal apodemes. *Bull. World Health Org.* **47**, 71–6.

Schlein, I. & Gratz, N.G. (1973) Determination of the age of some anopheline mosquitoes by daily growth layers of skeletal apodemes. *Bull. World Health Org.* **49**, 371–5.

Schneider, F. & Vogel, W. (1950) Neuere Erfahrungen in der Bekdmpfung der Kirschenfliege (*Rhagoletis cerasi*) Schweiz. *Z. Obst-weinbau* **59**, 37–47.

Scopes, N.E.A. & Biggerstaff, S.H. (1977) The use of a temperature integrator to predict the developmental period of the parasite *Aphidius matricariae* (Hal.). *J. Appl. Ecol.* **14**, 1–4.

Seber, G.A.F. (1982) *The Estimation of Animal Abundance and Related Parameters*. Griffin, London.

Sepulveda, A. (1994) Daily growth increments in the otoliths of European smelt *Osmerus eperlanus* larvae. *Mar. Ecol. Progr. Ser.* **108**(1–2), 33–42.

Sheehy, M.R.J., Greenwood, J.G., & Fielder, D.R. (1994) More accurate chronological age determination of crustaceans from field situations using the physiological age marker, lipofuscin. *Mar. Biol.* **121**(2), 237–45.

Sheehy, M.R.J., Shelton, P.M.J., Wickins, J.F., Belchier, M., & Gaten, E. (1996) Ageing the European lobster *Homarus gammarus* by the lipofuscin in its eyestalk ganglia. *Mar. Ecol. Progr. Ser.* **143**(1–3), 99–111.

Shiyomi, M. (1967) A statistical model of the reproduction of aphids. *Res. Popul. Ecol.* **9**, 167–76.

Shurova, N.M. & Zolotarev, V.N. (1988) Seasonal growth layers in the shells of the mussel in the Black Sea (USSR). *Biol. Morya* **1**, 18–22.

Singh, D. & Sharma, R.C. (1995) Age and growth of a Himalayan teleost *Schizothorax richardsonii* (Gray) from the Garhwal Hills (India). *Fish. Res.* **24**(4), 321–9.

Smirina, E.M. (1994) Age determination and longevity in amphibians. *Gerontology* **40**(2–4), 133–46.

Smith, G.C. & Trout, R.C. (1994) Using Leslie matrices to determine wild rabbit population growth and the potential for control. *J. Appl. Ecol.* **31**, 223–30.

Southwood, T.R.E. & Leston, D. (1959) *Land and Water Bugs of the British Isles*. Warne & Sons, London.

Southwood, T.R.E. & Scudder, G.G.E. (1956) The bionomics and immature stages of the thistle lace bugs (*Tingis ampliata* H-S. and *T. Cardui* L.; Hem., Tingidae). *Trans. Soc. Br. Entomol.* **12**, 93–112.

Southwood, T.R.E., Murdie, G., Yasuno, M., Tonn, R.J., & Reader, P.M. (1972) Studies on the life budget of *Aedes aegypti* in Wat Samphaya, Bangkok, Thailand. *Bull. World Health Org.* **46**, 211–26.

Squire, F.A. (1952) Observations on mating scars in *Glossina palpalis* (R.-D.). *Bull. Entomol. Res.* **42**, 601–4.

Stiefel, A. & Piechock, R. (1986) Circannual layers in the cementum of molar teeth in the beaver (*Castor fiber*) as an aid to exact age determination. *Zool. Abhandl.* **41**(14), 165–75.

Taylor, R.D. (1988) Age determination of the African buffalo, *Syncerus caffer* (Sparrman) in Zimbabwe. *Afr. J. Ecol.* **26**(3), 207–20.

Thomas, D.B. & Chen, A.C. (1989) Age determination in the adult screwworm (Diptera: Calliphoridae) by pteridine levels. *J. Econ. Entomol.* **82**(4), 1140–4.

Thomas, D.B. & Chen, A.C. (1990) Age distribution of adult female screwworms (Diptera: Calliphoridae) captured on sentinal animals in the coastal lowlands of Guatemala. *J. Econ. Entomol.* **83**(4), 1422–9.

Tian, Y. & Shimizu, M. (1997) Growth increment patterns in the shell of the cockle *Fulvia mutica* (Reeve) and their use in age determination. *Nippon Suisan Gakkaishi* **63**(4), 585–93.

Tomic Carruthers, N., Robacker, D.C., & Mangan, R.L. (1996) Identification and age-dependence of pteridines in the head and adult Mexican fruit fly, *Anastrepha ludens*. *J. Insect Physiol.* **42**(4), 359–66.

Tucker, A.D. (1997) Validation of skeletochronology to determine age of freshwater crocodiles (*Crocodilus johnstoni*). *Mar. Freshw. Res.* **48**(4), 343–51.

Tucker, R. (1985) Age validation studies on the spines of the spurdog (*Squalus acanthias*) using tetracycline. *J. Mar. Biol. Assoc. UK* **65**(3), 641–52.

Tyndale-Biscoe, M. & Hughes, R.D. (1969) Changes in the female reproductive system as age indicators in the bushfly *Musca vetustissima* Wlk. *Bull. Entomol. Res.* **59**, 129–41.

Tyndale-Biscoe, M. & Kitching, R.L. (1974) Cuticular bands as age criteria in the sheep blowfly, *Lucillia cuprina* (Wied.) (Diptera, Calliphoridae). *Bull. Entomol. Res.* **64**, 161–74.

Usher, M.B. (1972) Developments in the Leslie matrix model. In: Jeffers, J.N.R. (ed.) Mathematical models in ecology. *Symp. Br. Ecol. Soc.* **12**, 29–60.

Usinger, R.L. (1966) *Monograph of Cimicidae*. Thomas Say Foundation, Entomological Society of America, College Park, Maryland.

Vandermeer, J.H. (1975) On the construction of

the population projection matrix for a population grouped in unequal stages. *Biometrics* **31**, 239–42.

van Dijk, T.S. (1972) The significance of the diversity in age composition of *Calathus melanocephalus* L. (Col. Carabidae) in space and time at Schiermannikoog. *Oecologia (Berlin)* **10**, 111–36.

van Emden, H.F., Jepson, W.F., & Southwood, T.R.E. (1961) The occurrence of a partial fourth generation of *Oscinella frit* L. (Diptera: Chloropidae) in southern England. *Entomol. Exp. Appl.* **4**, 220–5.

Vidalis, K. & Tsimenidis, N. (1996) Age determination and growth of picarel (*Spicara smaris*) from the Cretan continental shelf (Greece). *Fish. Res.* **28**(4), 395–421.

Vlijm, L. & van Dijk, T.S. (1967) Ecological studies on carabid beetles. II. General pattern of population structure in *Calathus melanocephalus* at Schiermonnikoog. *J. Morphol. Okol. Tiere* **58**, 396–404.

Vogt, W.G. & Morton, R. (1991) Estimation of population size and survival of sheep blowfly, *Lucilia cuprina*, in the field from serial recoveries of marked flies affected by weather dispersal and age-dependent trappability. *Res. Popul. Biol.* **33**, 141–63.

Vogt, W.G., Woodburn, T.L., & Tyndale-Biscoe, M. (1974) A method of age determination in *Lucilia cuprina* (Wied) (Diptera, Calliphoridae) using cyclic changes in the female reproductive system. *Bull. Entomol. Res.* **64**, 365–70.

Waloff, N. (1958) Some methods of interpreting trends in field populations. *Proc. X Int. Congr. Entomol.* **2**, 675–6.

Williamson, M.H. (1959) Some extensions of the use of matrices in population theory. *Bull. Math. Biophys.* **21**, 13–17.

Williamson, M.H. (1967) Introducing students to the concepts of population dynamics. In: Lambert, J. (ed.) The teaching of ecology. *Symp. Br. Ecol. Soc.* **7**, 169–76.

Williamson, M.H. (1972) *The Analysis of Biological Populations*. Arnold, London.

Wohlschlag, D.E. (1954) Mortality rates of whitefish in an arctic lake. *Ecology* **35**, 388–96.

Worthington, D.G., Doherty, P.J., & Fowler, A.J. (1995) Variation in the relationship between otolith weight and age: implications for the estimation of age of two tropical damselfish (*Pomacentrus moluccensis* and *P. wardi*). *Can. J. Fish. Aquat. Sci.* **52**(2), 233–42.

Young, E.C. (1965) Teneral development in British Corixidae. *Proc. Roy. Entomol. Soc. Lond. A* **40**, 159–68.

Zuk, M. (1987) Age determination of adult field crickets: methodology and field applications. *Can. J. Zool.* **65**(6), 1564–6.

13 Species Richness, Diversity, and Packing

A great deal of time and expertise has been expended on the compilation of faunal lists for particular habitats, but the consequent increase in our understanding of the structure and functioning of animal communities or of the impact upon them of natural or human-induced change is still meagre. The body of quantitative theory on communities has continued to develop, especially following the lead given by the 'Hutchinson–MacArthur school'. Theory, while directing the way we organize and analyse our data, also poses hypotheses and suggests new analytical approaches at a rate that almost defeats our ability to test their validity.

Linked to increased interest in conservation, the term biological diversity or bio-diversity (Wilson, 1988; Reaka-Kudla *et al.*, 1997)—a measure of the total genetic and ecological diversity—has entered our vocabulary. A task for ecologists is to give a quantitative measure to this rather spiritual concept. Only by creating and applying reliable measures of diversity can we measure how it varies both spatially and temporally and thus recognize the influences that create and destroy it. There are numerous measures of diversity or ways of estimating species richness; which method is to be preferred will depend on the quality of the data, the sampling effort, community properties, such as species number and community stability, and the objectives of the study.

In this chapter, we shall focus primarily on species diversity. It is axiomatic that the objectives and methods of analysis should be fully and carefully considered before the field programme is undertaken. The methods of sampling are those outlined in Chapters 2 and 4 to 8, although the differential response of species to trapping is an indication for caution in respect of many of the methods in Chapter 7. The present chapter aims to describe the methods of handling and analysing data on animal communities (from guilds to continental faunas). Details of the theoretical studies and examples of fieldwork are given in MacArthur and Wilson (1967), MacArthur (1972), May (1973, 1976a, b), Cody and Diamond (1975), McClure and Price (1976), Pianka (1976a, b), and Pimm (1991).

13.1 Diversity

Diversity, even restricted as it is (for the purposes of this chapter) to the variety of animal species, is one of those common-sense ideas that prove elusive and multifaceted when precise quantification is sought (Peet, 1974). A useful classification, due to Whittaker (1972), is:

α diversity: the diversity of species within a community or habitat

β diversity: a measure of the rate and extent of change in species along a gradient, from one habitat to others

γ diversity: the richness in species of a range of habitats in a geographical area (e.g. island), which is a consequence of the α diversity of the habitats, together with the extent of the β diversity between them

Thus α and γ diversity are qualities that simply have magnitude and could, theoretically, be described entirely by a single number (a scalar). In contrast, β diversity is analogous to a vector, as it has magnitude and direction. Their descriptions therefore require different approaches. For example in a study of secondary succession Southwood *et al.* (1979) recognized four communities for which values of α could be calculated, but these would not reflect the changes in composition as succession proceeded (a direction — on the time axis). To do this, β-diversity needed to be calculated for each pair of communities and each value described in relation to the differences in successional age of the two communities being compared. In fact they found that the magnitudes of the values of β were approximately related to the differences, on a log scale, in successional age.

The restriction of the measurement of diversity to the consideration of the numbers of species, and the failure to take account of the variety, or otherwise, of form represented by these species has been criticized by Van Valen (1965), Hendrickson and Ehrlich (1971), and Findley (1973). The latter has suggested that taxonomic distance between the components of a fauna should be used to give a measure of its 'phenetic packing'.

13.1.1 Description of α and γ diversity

When a fauna (or a flora) is sampled, it is found that a few species are represented by a lot of individuals and a large number of species by few individuals. These relative abundances must be considered to represent the basic pattern of niche utilization in the community (or area), and one approach to diversity has been to seek to expose the distributions that underlie the species patterns; others have been more pragmatic and sought the best description (e.g. Williams, 1964). The result has been an 'explosive speciation' of diversity indices (perhaps only rivalled in ecological methodology by 'new designs' for light traps), which initially brought confusion to the subject; in addition, the ubiquitousness of some relationships and the apparent constancy of certain numerical values have added a measure of mystique. The penetrating analysis by May (1975a) of the underlying mathematical similarity of the assemblage of indices and the development of the diversity-ordering approach (Bela, 1995) have now allowed the subject to develop more rationally. This process has also been aided by systematic reviews, such as that by Magurran (1988).

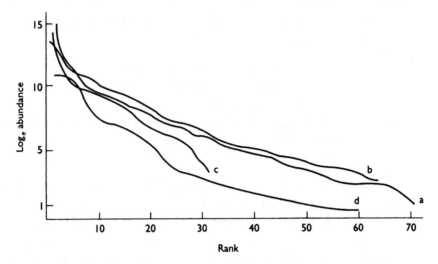

Fig. 13.1 Rank order abundance plots for British inshore marine fish communities. (a) Sizewell. (b) Dungeness. (c) Kingsnorth. (d) West Thurrock (after Henderson, 1989).

The estimate of total species richness, S_{max} (see below) is a straightforward measure of species diversity but does require high sampling effort, as, for small numbers of samples, the estimate is proportional to the number of samples collected. Such a measure only expresses one aspect of the data set. If the log number of individuals (or abundance) of each species is plotted against the rank, the plot will often approximate to a straight line (Fig. 13.1) (see also below and Fig. 13.4). Such plots emphasize that the species number (S): abundance of individuals (N) relationship has two features.

1 Species richness.

2 Equitability or evenness—the pattern of distribution of the individuals between the species (a faunal sample of 100 individuals representing 10 species could consist of 10 individuals of each (extreme equitability) or, at the other extreme 91 individuals of one species (the dominant) and one each of the other nine). The more horizontal the line when plotted as in Fig. 13.1 the more equitable.

Both attributes of the data may be expressed using a model that describes the species-abundance relationship, if a suitable one can be identified. The popular alternative (which can easily be expressed to non-biologists, such as planners) is to summarize both attributes with a single number—a diversity index.

13.1.2 Species richness

Species inventories for particular habitats or localities are frequently required for purposes such as conservation management. Because a complete census is rarely feasible, the community must be sampled. An important problem that then arises is to estimate via sampling the total species (or other taxon) number, S_{max}, for the locality. This will both give a measure of the completeness of the inventory and also allow comparison with the species richness of other localities. An estimate of the maximum species number is also useful when assessing whether the further informa-

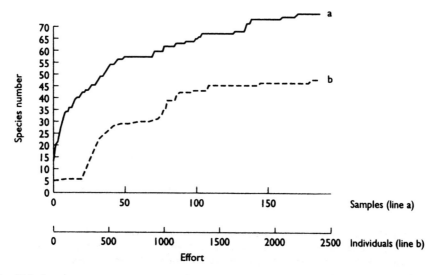

Fig. 13.2 Cumulative species number of: (a) fish recorded from monthly samples collected between 1981 and 1995 at Hinkley Point in the lower Severn Estuary (after Henderson, 1989, with more recent data from Pisces Conservation Ltd.); (b) Heteroptra from oak trees (after Southwood, 1996).

tion to be gained from continued sampling justifies the cost. In the following sections, the different approaches that can be taken to estimate S_{max} are described. At present, no clear consensus as to the best approach is available, although recent studies by Walther and Morand (1998) on parasite and P.A. Henderson and A.E. Magurran (pers. comm.) on fish communities show the Chao estimator (p. 468) used with presence–absence data to perform best.

13.1.2.1 Extrapolating the species accumulation curve: rarefaction

The plot of the cumulative number of species collected, $S(n)$, against a measure of the sampling effort (samples taken or animals captured) is termed the species accumulation curve. The sampling effort can be measured in many different ways; some examples are the number of quadrats taken, the total number of animals handled, the hours of observation, or the volume of water filtered. The species accumulation curve will only prove useful if applied to a defined habitat or area that is reasonably homogeneous.

Two illustrations of this curve (Fig. 13.2) show the cumulative species number of fish recorded from monthly samples collected in 1981 and 1995 at Hinkley Point in the lower Severn Estuary and the cumulative number of species of Heteroptera against the number of individuals collected over several years from oak trees. The sampling method used for fish, described by Henderson and Holmes (1991), was to use the power-station cooling-water pumps as a sampler and thus each month the volume of water filtered was a constant $3.24 \times 10^5\,\mathrm{m}^3$, giving a total volume sampled to the end of 1995 of about $5.18 \times 10^7\,\mathrm{m}^3$. As is normal when sufficient sampling effort is expended, the curve is asymptotic. Yet, even given this exceptionally high sampling effort, new species are still occasionally caught. The inshore fish fauna of the north-eastern Atlantic is well known and Henderson (1988) argues that the total

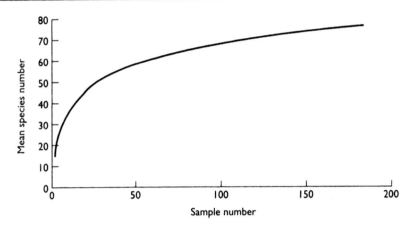

Fig. 13.3 The average species accumulation curve from 20 randomizations of the Hinkley Point species accumulation curve shown in Fig. 13.2. The calculations were performed using species diversity and richness II software from Pisces Conservation Ltd., IRC House, Lymington SO41 8GN, England.

species number at this latitude in fully marine waters is about 80, although offshore or deep-water species may occasionally appear. Likewise, with the oak Heteroptera, when a limited number of trees in two nearby woods were sampled by knock-down (p. 156), yet more new species continued to be found.

The order in which samples are taken alters the shape of a species accumulation curve because of both random error and sample heterogeneity. In the case of the Hinkley data, which are a time series, more fish species were caught during the autumn than at other times of the year, so an accumulation curve commencing in September will rise faster than one starting in January. Further, examination of Fig. 13.2 shows no increase in species number between samples 44 and 76. These samples were collected during an exceptionally cold period, when the inshore fish community was much reduced. To eliminate features caused by random or periodic temporal variation, the sample order can be randomized. A useful procedure is to randomize the sample order r times and calculate the mean and standard deviation of $S(n)$ over the r runs (Colwell & Coddington, 1994). This approach can also be applied to samples collected along transects, provided they are from a single community. The average species accumulation curve from 100 randomizations of the Hinkley Point data is shown in Fig. 13.3.

Extrapolation to estimate the total species complement (species richness) for the habitat, S_{max}, is only possible if the species accumulation curve is derived from a reasonably homogeneous (stable) community. So the first task is to look for heterogeneity. If species abundance, rather than presence–absence data, is available, Colwell and Coddington (1994) suggest comparing the mean randomized species accumulation curve with the curve expected if all the individuals caught over all the samples were randomly assigned to the samples. If the expected curve rises more steeply from the origin, then heterogeneity is greater than could be explained by random sampling error alone. The expected curve can be computed in the following two ways. Both methods start by calculating the mean number of individuals per sample, n. The rarefaction approach generates the curve for x samples by selecting, without replacement, x random draws of n individuals from the pooled sum of the samples (Tipper, 1979; James & Rathbun, 1981). The expected number of species, $E(S_n)$, in a sample of n individuals selected at random from a population of N individuals and S species is given by the hypergeometric distribution:

$$E(S_n) = S - \left(\frac{N}{n}\right)^{-1} \sum_{i=1}^{S}\left(\frac{N-N_i}{n}\right) \qquad (13.1)$$

where N_i is the number of individuals of species i in the population.

The alternative, termed random placement, is easier to compute. With this method xn individuals are assumed to be randomly assigned to each of x collections (Coleman, 1981; Coleman *et al.*, 1982). The increase in species number over a series of samples, $S(n)$, is calculated as:

$$S(n) = S_{TOT} - \sum_{i=1}^{i=S_{TOT}} (1-\alpha)^{n_i} \qquad (13.2)$$

where S_{TOT} is the total species number recorded, n_i the total number of individuals belonging to species i, n the sample number, and α the fraction of the total sampling effort undertaken by sample n.

While non-asymptotic models for the species accumulation curve, based on either a log–linear or log–log relationship between $S(n)$ and n, have been applied (Palmer, 1991), asymptotic models seem more appropriate. If it is assumed that the probability that the next individual captured will be a new species declines linearly with species number, then the species accumulation curve is the negative exponential function (Holdridge *et al.*, 1971; Miller & Wiegert, 1989; Soberon & Llorente, 1993):

$$S(n) = S_{\max}\left(1 - e^{-kn}\right) \qquad (13.3)$$

where k is a fitted constant, $S_{\max}$ the asymptotic species number and n the number of samples.

The asymptotic behaviour of the accumulation curve can also be modelled as the hyperbola:

$$S(n) = \frac{S_{\max}n}{B+n} \qquad (13.4)$$

where $S_{\max}$ and B are fitted constants. This is the Michaelis–Menten equation used in enzyme kinetics and thus there is an extensive literature discussing the estimation of its parameters, which unfortunately presents considerable statistical difficulties. One approach, favoured by Raaijmakers (1987), is to calculate $S_{\max}$ and B, using their maximum-likelihood estimators as follows:

If:

$$X_i = \frac{S(n)}{n}$$

and:

$$Y_i = S(n)$$

then:

$$\hat{B} = \frac{\overline{X}S_{yy} - \overline{Y}S_{xy}}{\overline{Y}S_{xx} - \overline{X}S_{xy}}$$

and:

$$\hat{S}_{max} = Y + B\overline{X} \tag{13.5}$$

where S_{yy}, S_{xx}, and S_{xy} are the sums of squares and cross-products of the deviations $Y_i - \overline{Y}$ and $X_i - \overline{X}$. While this method of estimating S_{max} has been criticized by Lamas *et al.* (1991), if it is decided to use the Michaelis–Menten equation it seems the best approach available.

13.1.2.2 Using parametric models of relative abundance

Given quantitative data for species, abundance models describing the relative abundance of species within the community, such as the log-normal (see section 13.1.3.3), can be used to estimate the total species complement, S_{max}. If the data fit a log-normal distribution, then the number of unsampled species is given by the missing part of the distribution to the left of the veil line. Alternative distributions which have been suggested include the Poisson–log-normal (Bulmer, 1974) and the log series (Williams, 1964), amongst others. It is unlikely that parametric methods yield reliable estimates for S_{max} and they should probably be avoided.

13.1.2.3 Non-parametric estimates

A variety of non-parametric estimators have been suggested. These have been reviewed by Colwell and Coddington (1994), who conclude that we have as yet insufficient experience to judge their relative merits, although recent studies using field data on parasites (Walther & Morand, 1998) and fish (P.A. Henderson and A.E. Magurran, pers. comm.) have found the Chao estimator applied to presence–absence data to be the most reliable and cost-effective. Hellmann and Fowler (1999) compared the jackknife and bootstrap methods and determined that where sample size was small (less than 25% of the community) the second-order jackknife behaved best, though in their evaluation Walther and Morand (1998) found the first-order performed better.

Using the observed number of species represented by one, *a*, or two, *b*, individuals in the sample, Chao (1984) derived the simple estimator:

$$\hat{S}_{max} = S_{obs} + \left(a^2/2b\right) \tag{13.6}$$

where S_{obs} is the actual number of species in the sample.

Chao (1987) gives the variance of this estimate as:

$$\text{var}(\hat{S}_{max}) = b\left[\left(\frac{a/b}{4}\right)^4 + \left(\frac{a}{b}\right)^3 + \left(\frac{a/b}{2}\right)^2\right] \tag{13.7}$$

Note that, when all the species have been observed more than twice, $\hat{S}_{max} = S_{obs}$, the census is considered complete. Equations (13.6) and (13.7) can be applied to presence–absence data when a is the number of species only found in one sample and b the number of species only found in two samples.

Two further estimators were suggested by Chao and Lee (1992). These are a modification of an early estimator calculated as observed species number/coverage. Coverage is simply 1 – (number of species represented by only one individual/the total number of individuals).

Independently, Burnham and Overton (1978) and Heltshe and Forrester (1983) developed the first-order jackknife estimator:

$$\hat{S}_{max} = S_{obs} + a(n - 1/n) \tag{13.8}$$

where n is the number of samples and a the number of species only found in one sample.

Heltshe and Forrester (1983) give the variance of this estimate as:

$$\text{var}(\hat{S}_{max}) = \frac{n-1}{n}\left(\sum_0^{S_{obs}} j^2 f_j - \frac{L^2}{n}\right) \tag{13.9}$$

where f_j is the number of samples holding j of the L species only found in one sample.

Further jackknife estimators are discussed in Burnham and Overton (1979) and Smith and van Belle (1984).

A bootstrap estimate (see p. 27) of species richness can be calculated as follows (Smith and van Belle, 1984).

1 Randomly select, with replacement, n samples from the total available and calculate:

$$S_{max} = S_{obs} + \sum_{i=1}^{S_{obs}} (1 - p_i)^n \tag{13.10}$$

where p_i is the proportion of the n that has species i present.

2 Repeat step 1 a large number of times—say, 50 to 200—and calculate the mean estimate of S_{max} and the variance as follows:

$$\text{var}(S_{max}) = \sum (1 - p_i)^n [1 - (1 - p_i)^n] + \sum \sum \left\{ q_{ij}^n - [(1 - p_i)^n (1 - p_j)^n] \right\} \tag{13.11}$$

where q_{ij} is the proportion of the n bootstraps which hold both species i and j.

Commercially available software to calculate these estimators is listed in Table 13.1.

Table 13.1 Computer packages for the analysis of community structure, including cluster and ordination techniques.

Name	Publisher/Supplier	Methods
Species diversity and richness	Pisces Conservation Ltd, The Square, Pennington, Lymington, Hants SO45 1ZU, UK Web site: http://www. irchouse.demon.co.uk	IBM personal computer (PC) package that calculates a range of species richness estimators and diversity indices and undertakes diversity ordering
CAP	Pisces Conservation Ltd, The Square, Pennington, Lymington, Hants SO45 1ZU, UK Web site: http://www.irchouse.demon.co.uk	Suite of programs for the IBM PC, including similarity analysis, cluster analysis, dendrogram production, principal-component analysis, TWINSPAN, and DECORAMA
PC-ORD	MjM Software, Gleneden Beach, Oregon, USA	Suite of programs for the IBM PC, including principal component analysis, TWINSPAN, and DECORAMA
CLUSTAN	Clustan Ltd, 16 Kingsbridge Rd., Edinburgh EH12 6DZ	Non-hierarchical clustering, cluster analysis, similarity analysis, association analysis
DECORANA/TWINSPAN	Cornell Ecology Programs, Cornell University, Ithaca, New York, USA, or pc versions from Biological Software, 23 Darwin Close, Farnborough, Orpington, Kent BR6 7EP, or Microcomputer Power, 111 Clover Lane, Ithaca, New York, NY 14850, USA	DECORANA undertakes detrended correspondence analysis, with simple reciprocal averaging as an option TWINSPAN undertakes two-way indicator-species analysis
KYST	Bell Labs, Computing Information Library, 600 Mountain Ave, Murrey Hill, New Jersey 07974, USA	Specialized cluster programs
BMDP	BMPD Statistical Software, 1964 Westwood Boulevard, Suite 202, Los Angeles, California 90025, USA	General statistical package, including cluster methods
Genstat	Numerical Algorithms Group, 256 Banbury Rd, Oxford OX2 7DE.	General statistical package, which can be programmed to undertake most forms of classification and ordination
SPSS	SPSS Inc., 444 N. Michigan Avenue, Chicago, IL 60611, USA	General statistical package, including cluster methods
SAS	SAS Institute Inc., Cary, North Carolina, USA	General statistical package, including cluster methods

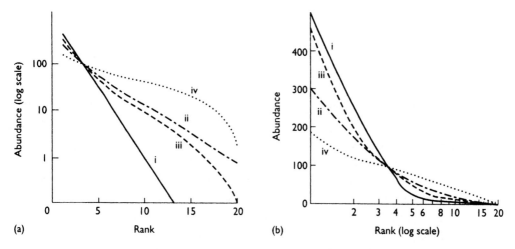

Fig. 13.4 The rank abundance (or dominance–diversity) curves for different underlying distributions when plotted with the log scale on different axes: (a) log abundance (individuals): rank (species sequence); (b) abundance: log rank. i = geometric series, ii = log series, iii = log normal, iv = MacArthur broken stick (after Whittaker, 1972). Note: the exact relative positions of the curves depend on the assumptions about the parameters. Whilst the broken stick and canonical log normal are defined by one parameter (S_T, here taken as 20), geometric and log series and the general log-normal are defined by two parameters: S_T and J, where $J = N_T/m$, m being the number of individuals of the rarest species, often in practice one.

13.1.3 Models for the $S:N$ relationship

The equitability of the species:abundance relationship will be a reflection of the underlying distributions. Four main groups can be recognized: the geometric series, the logarithmic series, the log-normal, and MacArthur's 'broken stick' model. The two former will give approximately straight lines when log abundance is plotted against rank (Figs 13.1 and 13.4a), whilst the latter shows up as a straight plot when abundance is plotted against log rank (Fig. 13.4b) (Whittaker, 1972; May, 1975a).

13.1.3.1 Geometric series

If a species occupies a fraction (k) of a habitat, another species the same fraction of the remainder, a third species k of what now remains and so on, the resulting rank–abundance distribution will, as originally shown by I. Motomura, be a geometric series (May, 1975a). Termed the 'niche-pre-emption hypothesis', this distribution has often been used to describe floral diversity in temperate regions (McNaughton & Wolf, 1970; Whittaker, 1970, 1972). May (1975a) shows that the Odum et al. (1960) formulation (which is expressed in the terms of species and numbers (Fig. 13.1a), rather than rank:abundance (Fig. 13.2)) is the equivalent of this series.

13.1.3.2 Logarithmic (or log) series

Originally suggested as a suitable description of species abundance data by R.A. Fisher (Fisher et al., 1943), this distribution has been criticized as lacking a theoretical justification at the level of species interaction. Kempton and Taylor (1974) and

May (1975a, 1976b) have provided this. It has suggested as a model because the abundance of animals when ranged in rank order is often approximated by the series:

$$\alpha x, \frac{\alpha x^2}{2}, \frac{\alpha x^3}{3} \ldots \tag{13.12}$$

The log series may be regarded as the convenient approximation to the gamma model (of which the negative binomial is a well-known form—see p. 29) with maximal variance (Kempton & Taylor, 1974). Gamma models often arise as a result of a two-stage (compound) process (see p. 37); as May (1975a) points out, if we envisage a geometric-series type of niche pre-emption, where the fractions of niche pre-empted have arisen from the arrival of successive species at uniform intervals of time, the arrival of the species at random time intervals would lead to the log series (see also Boswell & Patil, 1971). This distribution approximates to a straight line on a rank : log abundance plot (Fig. 13.4a) and the relationship between the number of species, S, and individuals N, is given by:

$$S = \alpha \log_e\left(1 + \frac{N}{\alpha}\right), \tag{13.13}$$

where α is an index of diversity.

Alternatively, in terms of a sampling parameter X, for the expected number of species in a total sample of N individuals, S_N is given by:

$$S_N = \frac{\alpha X^N}{N} \tag{13.14}$$

Williams (1964) gives a table and graph for estimating α, given species number and total number of individuals (see Fig. 13.9).

13.1.3.3 Log-normal distribution

Preston (1948) suggested that the log-normal distribution would give the best description of species-abundance patterns, the assumption being that individuals are distributed between species in accordance with the normal or Gaussian distribution and that population growth is geometric (Williams, 1964). May (1975a) enlarges on the biological reasons why the log-normal might be considered to apply to both opportunistic and equilibrium communities. Preston (1948, 1960, 1962) considered the plot of the frequency of species against abundance classes on a log scale (Fig. 13.5). He used logarithms to the base 2, so that each class, or octave, involved a doubling of the size of the population. (The abundance classes could be on any logarithmic scale.) The abundance of the species at the peak (mode) of the distribution is N_0 and the abundance octaves (R) are calculated:

$$R_i = \log_2\left(\frac{N_i}{N_0}\right) \tag{13.15}$$

Preston suggested that sample size is usually too small to obtain the species in the lower (rarer) octaves; these species were, he said, hidden behind the 'veil line'. Field

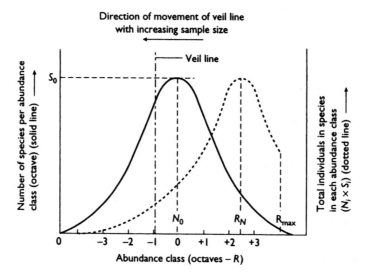

Fig. 13.5 The log-normal representation of abundance and species relationships (modified from May, 1975a).

data will normally only represent those species to the right of the veil line and therefore, from Fig. 13.5, it will be seen that, if the veil line is at N_0 or in any of the positive octaves, the plot will approximate a straight line and be of the form shown in Fig. 13.4b (with axes reversed). However, following very intensive sampling of the Heteroptera in oak trees, Southwood (1996) found a bell-shaped curve, the over-representation of the lowest abundance class (singletons) being due to vagrants. Preston recognized that the plot of individuals per class against abundance class (dotted line in Fig. 13.5) would not describe the complete 'bell', but be truncated at its crest (the 'canonical condition'), so that $R_N = R_{max}$ (Fig. 13.5). This truncated log-normal, ending at the crest (unlike Fig. 13.5), is difficult to fit to data, although Bullock (1971a) and Gage and Tett (1973) did so. Kempton and Taylor (1974) and Bulmer (1974) showed that it is more tractable if regarded as a Poisson log-normal, described by the parameters S_{obs} and σ^2 (variance). As Preston (1948) noted, the dispersion constant:

$$a = \left(2\sigma^2\right)^{-\frac{1}{2}}$$

is independent of sample size.

A useful ratio is:

$$\gamma = \frac{R_N}{R_{max}} = \frac{\log_e 2}{2a(\log_e S_{obs})^{\frac{1}{2}}} \tag{13.16}$$

Preston (1962) analysed a considerable body of field data and found that $\gamma \sim 1$ was general; likewise, Hutchinson (1953) found that $a \sim 0.2$. A rational explanation for both has been given by May (1975a); both arise from the mathematics inherent in the log-normal, given a large value of S_{obs}.

13.1.3.4 MacArthur's broken-stick model

In a theoretical consideration of how niche hypervolume might be divided, MacArthur (1957) postulated three models. One, termed the 'broken stick', assumes that niche boundaries are drawn at random and the abundance of the ith most abundant species is given by:

$$N_i = \frac{N_T}{S_{obs}} \sum_{n=i}^{S_{obs}} \frac{1}{n} \qquad (13.17)$$

The mathematics of this model have been investigated by Webb (1974); it is more even than the log-normal and is characterized simply by the parameter S_{obs}. A ratio commonly calculated in faunistic studies is J, defined as the total number of individuals (N_T) divided by the abundance of the rarest species. From equation (13.17), it is clear that, for the MacArthur model, this will be N_T/N_{obs} and, as May (1975a) points out, then $J = S_{obs}^2$.

13.1.4 Non-parametric indices of diversity

The description of the $S:N$ relationship in terms of the parameters of a model implies that the model is at least approximately applicable. It may be argued that non-parametric indices have the advantage that they make no assumptions of this type. There are also relative indices (e.g. Lloyd & Ghelardi, 1964); these compare the equitability of the distribution with that which would arise if the same sample were distributed according to a certain model — in this case, the MacArthur model. Peet (1975) shows that such relative indices possess mathematically undesirable properties. Comparative reviews of relative merits of the indices described below are provided by Taylor (1978), Kempton (1979), and Magurran (1988). Commercially available software to calculate the diversity indices described below is listed in Table 13.1.

13.1.4.1 Shannon–Wiener function (H)

The Shannon–Wiener function (H) is one devised to determine the amount of information in a code, and is defined:

$$H = -\sum_{i=1}^{S_{obs}} p_i \log_e p_i \qquad (13.18)$$

where p_i = the proportion of individuals in the ith species; or, in terms of species abundance:

$$H = \log_e N - \frac{1}{N} \sum_{i=1}^{\infty} (p_i \log_e p_i) n_i \qquad (13.19)$$

where n_i = the number of species with i individuals. The information measure is nits for base e and bits per individual for base 2 logarithms.

Bulmer (1974) and May (1975a) have shown that, for a large sample, with the log-series model (section 13.1.3.2):

Fig. 13.6 The performance of various diversity indices in relation to the total number of species (S_T) and the underlying model: log-normal (solid lines) ($\gamma = 1.0$ unless labelled otherwise), log series (. . . . .) $\alpha = 5$, geometric series (– – – – –) $k = 0.4$, and MacArthur model (– · – · –). (a) The Shannon–Wiener function. (b) The Simpson–Yule index. (c) The dominance index (after May, 1975a).

$$H \simeq 0.577 + \log_e \alpha \tag{13.20}$$

where 0.577 is Euler's constant. Considering the behaviour of H with changing magnitude of S_{obs}, May (1975a) concluded that it was an insensitive measure of the character of the $S:N$ relationship and is dominated by the abundant species (Fig. 13.6a).

13.1.4.2 Simpson–Yule index (D)

This is a diversity index, proposed by Simpson (1949), to describe the probability that a second individual drawn from a population should be of the same species as the first.* The statistic C (or Y) is given by:

$$C = \sum_i^{S_{obs}} p_i^2 \tag{13.21}$$

where, strictly:

$$p_i^2 = \frac{N_i(N_i - 1)}{N_T(N_T - 1)}$$

but this is usually approximated as:

$$p_i^2 = \left(\frac{N_i}{N_T} \right)^2$$

where N_i is the number of individuals in the ith species and N_T the total individuals in the sample. The index is:

* A few years earlier, a similar type of index had been proposed by G. Udney Yule to compare an author's characteristic vocabulary (frequency of different words in his writings).

$$D = \frac{1}{C}$$

and the larger its value, the greater the equitability (range $1 \rightarrow S_{obs}$). The statistic $1 -$ C gives a measure of the probability of the next encounter (by the collector or any animal moving at random) being with another species (Hurlbert, 1971). May (1975a) (Fig. 13.6b) shows that this index is strongly influenced, for values of $S_{obs} >$ 10, by the underlying distribution.

13.1.4.3 Berger–Parker dominance index

This index is simple, both mathematically and conceptually; it expresses the proportion of the total catch, N_T, that is due to the dominant species, N_{max}:

$$d = \frac{N_{max}}{N_T} \tag{13.22}$$

For reasonable values of S_{obs}, the index is not influenced by this parameter and is relatively independent of the underlying model (Fig. 13.6c). May (1975a) concludes that it seems 'to characterise the distribution as well as any [index], and better than most'.

13.1.4.4 Evenness (equitability)

As diversity is at a maximum when all species within a community are equally abundant, a measure of evenness is the ratio of the observed diversity to the maximum possible for the observed species number. Hurlbert (1971) proposed a slightly more complex definition of evenness, which considered both the maximum, D_{max}, and the minimum, D_{min}, diversity:

$$E_H = \frac{D - D_{min}}{D_{max} - D_{min}} \tag{13.23}$$

Using the Shannon–Weiner index as an example, $D_{min} = 0$, $D_{max} = \log(S)$, where S is the total number of species, so:

$$E_H = \frac{H}{\log S}$$

13.1.4.5 McIntosh diversity measure

McIntosh (1967) suggested the dominance index:

$$D = \frac{N - U}{N - \sqrt{N}} \tag{13.24}$$

where N is the total number of individuals in the sample; U is calculated as:

$$U = \sqrt{\sum n_i^2}$$

where n_i is the number of individuals belonging to the ith species.

13.1.5 Which model or index?

The evaluation of the most appropriate model or index may be based on the consideration of their theoretical properties as against our knowledge of ecology or by testing them with field data for either their fit or their value in discrimination. Valuable studies in these respects are those by Bullock (1971a), Hurlbert (1971), Kempton and Taylor (1974), May (1975a), Taylor *et al.* (1976), Routledge (1979), and Magurran (1988). There can be no universal 'best buy', although there are rich opportunities for inappropriate usage! The objective of the analysis must be clearly conceived before the procedure is selected. It is certainly important to distinguish between data on organisms that are part of the trophic structure of the area (e.g. growing plants or birds seen foraging) and samples that may contain a number of animals that are outside their trivial ranges (e.g. suction-trap catches at several metres above the vegetation). This is illustrated by the study of Kricher (1972) on the diversity of bird populations at different seasons and that of Southwood (1996) on insects from different species of oak tree.

The use of models (e.g. MacArthur's) to gain some insight into resource apportionment may be justified in real samples of 'trophically related' organisms.

From theoretical studies, especially May (1975a), it can be concluded that the models can be arranged in a series corresponding to maximal niche pre-emption or unevenness and moving through to a more uniform resource apportionment, thus:

(uneven) geometric series, log series, log-normal, MacArthur (even)

This appears contrary to Kempton and Taylor's (1974) findings that, whereas most populations of macro-Lepidoptera were well described by the log series, those in unstable habitats with 'rarer' species were better described by the log-normal; it may be that the inclusion of 'additional species' has increased the evenness. May (1975a) also concludes that the log-normal pattern may be expected with 'large or heterogeneous assemblies of species'.

Where comparative studies have included the log series, the conclusion has generally been that this provides the most 'comprehensible result' (e.g. Bullock, 1971a; Hurlbert, 1971; Kempton & Taylor, 1974; May, 1975a; Taylor *et al.*, 1976; Magurran, 1988). Kempton and Taylor (1974) and Taylor *et al.* (1976) suggest that the 'basic environmental structure' of an area will determine the range of populations it can support and, on this assumption, they are able to show that the log series is the best model. They point out that the rank:log abundance plot of the log series is almost linear over the mid-region (see Fig. 13.4a), so α may be interpreted as the slope of this region, 'where the moderately common species reflect most closely the nature of the environment' and fluctuate less violently, from year to year, than the more abundant species. However, there is no theoretical or empirical evidence to support this view; indeed, the only generalization that can be made is that the more abundant species are often those with the largest geographical ranges and thus possibly depend less on local conditions to determine abundance. Taylor and colleagues found that changes in α enable discrimination between diversities in areas where the environment was known to have changed. For the ecologist, the ability of a parameter or index to discriminate between changed conditions may be more relevant than, and is not identical to, the precision of the 'fit' of the underlying model. The richness and evenness parameters of the log-normal (S_{max} and σ^2) and the Shannon and Simpson indices were all found less useful as indications of environmental change and, even when the fit of the log series was not close, its robustness provided a meaningful α.

A novel concept of Taylor *et al.* (1976) is that, viewing diversity as a reflection of basic environmental structure, the two meaningful characteristics are not species richness and evenness, but the following.

1 Diversity, as represented by the 'common α', the slope of the line as dominated by the moderately common species.

2 The fluctuations in numbers from occasion to occasion (e.g. year to year), as shown by X of the log series (see equation (13.14)). This hypothesis has not, however, been further explored.

Perhaps because of the small sample size and the low quality of the data, many published accounts do not take the parametric approach but simply present the Shannon–Wiener index, often to little purpose. Comparative studies have not given much support to the value of the non-parametric indices (Whittaker, 1972; May, 1975a). The Shannon–Wiener index has often been found unsatisfactory (Bullock, 1971a; Hurlbert, 1971; Taylor *et al.*, 1976; Routledge, 1979) and, in spite of some apparently successful uses, should in general be regarded as a distraction, rather than an asset, in ecological analysis. It is strongly influenced by species number and by the underlying model (Fig. 13.6a). The Simpson–Yule index is strongly influenced by the few dominant species, although, bearing this limitation in mind, it could be of value as an indicator of interspecific encounters. As shown in Fig. 13.7 for the Hinkley Point data, all of the non-parametric indices showed the same temporal pattern as species number and gave no additional insight. Thus species number is often a straightforward measure for comparing diversity between samples collected in similar fashion. If the comparison is to be made between samples that differ in sampling effort, then estimates can be made of total species richness, S_{max}, and these can be compared.

Non-parametric diversity indices become useful when used for diversity ordering (Bela, 1995). By generating a family of diversity indices, it is possible to recognize non-comparable communities. If the objective of a study is to compare communities using a single non-parametric diversity measure, then diversity ordering must be undertaken.

13.1.6 Comparing communities—diversity ordering

Different diversity indices may differ in the ranking they give to communities

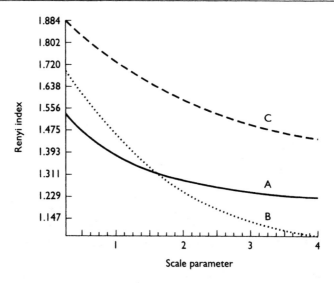

Fig. 13.7 The diversity ordering of three artificial communities, which shows that C is the most diverse and that it is not possible to state which of A or B is the most diverse, because the relative magnitude of the index depends on the magnitude of the scale parameter.

(Hurlbert, 1971; Tóthméresz, 1995). An example from Tóthméresz (1995) illustrates the point. Consider three imaginary communities with the following sets of species abundances, for each of which diversity has been calculated, using both Shannon–Wiener (H) and Simpson's (D):

Community A: {33, 29, 28, 5, 5}, H = 1.3808, D = 3.716
Community B: {42, 30, 10, 8, 5, 5}, H = 1.4574, D = 3.564
Community C: {32, 21, 16, 12, 9, 6, 4}, H = 1.754, D = 5.22

Because H(B) > H(A), it could be argued that B is the most diverse; however, as D(A) > D(B), the opposite conclusion could also be entertained. Communities, such as A and B, which cannot be ordered are termed non-comparable. Such inconsistencies are an inevitable result of summarizing both relative abundance and species number using a single number (Patil & Taillie, 1979). Diversity profiles offer a solution to this problem by identifying those communities that are consistent in their relative diversity. To identify these communities, it is necessary to produce an expression that can generate the various indices by changing the value of preferably a single parameter.

Perhaps the most generally useful expression of this type is that due to Renyi (1961), which is based on the concept of entropy and defined as:

$$H_\beta = \frac{\log \sum_{i=1}^{s} p_i^\beta}{1 - \beta}$$

(13.25)

where β is the order ($\beta \geq 0$, $\beta \neq 0$), p_i the proportional abundance of the ith species, and log the logarithm to a base of choice, often e.

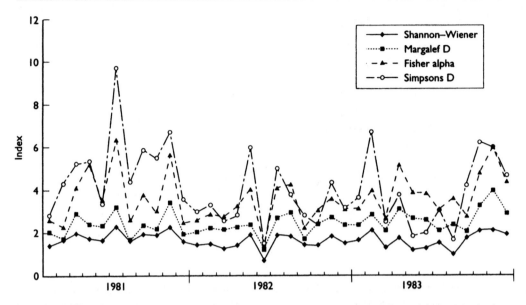

Fig. 13.8 The changing pattern of diversity of the fish community in the Bristol Channel. In this case the data can be considered to be well ordered in that all the indices plotted show the same temporal pattern. Note that the Shannon–Wiener index is the least sensitive to change in the community.

Hill (1973) used an almost identical index, N_a, which is related to H_β by the equality:

$$H_\beta = \log(N_a)$$

He demonstrated that N_a for $a = 0, 1, 2$ gives the total species number, the Shannon–Wiener H, and Simpson's D, respectively. Thus, by using different values of β or a, a range of diversity measures can be generated from the Renyi or Hill equation. To test for non-comparability of communities, H_β is calculated for a range of β values and the results presented graphically. Figure 13.7 shows the diversity ordering of the three artificial communities A, B, and C. This figure shows that C is always greater than A or B and thus can be considered to be more diverse. As A and B cross over, one cannot be considered more diverse than the other by all reasonable measures.

As described by Tóthméresz (1995), numerous other diversity index families have been suggested.

13.1.7 Procedure to determine α diversity
In the light of present knowledge, the following may be taken as a guide to the analysis of diversity data for a single locality.

1 Plot graph(s) of log abundance on rank. It has always been important to examine the form of your data and is easily done on a computer. Do the data form a straight line? Which are the species that depart most from it? Is there anything unusual in their biology (e.g. they could be vagrants)? The consideration of these graphs in the

light of the earlier discussion in this chapter should indicate whether, exceptionally (Taylor *et al.*, 1976), other models or indices should be used in addition to those outlined below. These graphs may be an excellent manner of presenting the data for publication (e.g. Sanders, 1969; Bazzaz, 1975; Taylor *et al.*, 1976; Southwood *et al.*, 1979; Henderson, 1988).

2 Plot the species accumulation curve(s) and calculate total species richness, S_{max}. The species accumulation curve gives insight into the sufficiency of the sampling effort. As there is no best method to calculate S_{max}, it is recommended that most of the methods given in Section 13.1.2 are explored.

3 Determine α of the log series. This may be done rigorously by solving the maximum-likelihood equation for α by numerical methods, using a computer program (see Table 13.1). For most purposes, the value may be read off Williams's (1947) nomograph (Fig. 13.9). Taylor *et al.* (1976) point out that the true large-sample variance is:

$$\text{var}(\hat{a}) = \frac{\alpha}{-\log(1-X)}$$

The standard error contours on Fig. 13.9 exclude variation due to population fluctuations. Which variance is appropriate will clearly depend on the type of comparisons. If there are a series of samples it may be decided to:

(a) calculate the 'common α' (Taylor *et al.*, 1976) by solving numerically:

$$\sum_{i=1}^{z} S_i = \sum_{i=1}^{z} \hat{\alpha} \log\left[1 + \frac{N_i}{\hat{\alpha}}\right]$$

where z = total number of samples and S_i and N_i are the total species and individuals in the ith sample.

(b) Calculate the individual sampling factors (X), which indicate the magnitude of population fluctuation:

$$\hat{X}_i = 1 - e^{-\frac{S_i}{\hat{\alpha}}}$$

4 Calculate the Berger–Parker dominance index (equation (13.22).

5 Calculate other parameters or statistics. This will be done where the graphs (1 above) reveal a special need; the appropriate equations are given earlier in the chapter.

6 If sites are to be compared, undertake diversity ordering to ensure that they can be compared, using a diversity index.

13.1.8 Determining β diversity

β diversity measures the increase in species diversity along transects and is particularly applicable to the study of environmental gradients. It measures two attributes: the number of distinct habitats within a region and the replacement of species by

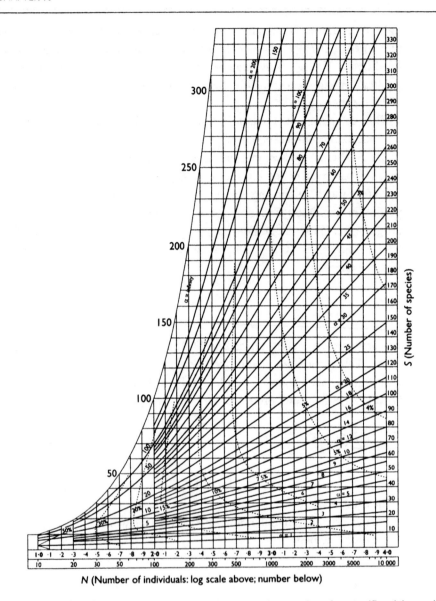

Fig. 13.9 Nomograph for determining the index of diversity (α) for a number of species (S) and the number of individuals (N) in a random sample of a fauna (after Williams, 1947). Note that the standard errors contain only that component of variation due to sampling.

another between disjoint parts of the same habitat. Used together, α and β diversity can give an assessment of the diversity of an area (Routledge, 1977; Southwood *et al.*, 1979). Many of the methods were developed and extensively used in work on plant and bird ecology and details will be found in Bullock (1971b), Whittaker (1972), Goodall (1973), Cody (1974, 1975), Mueller-Dombois and Ellenberg (1974), Cody and Diamond (1975), Pielou (1975), Goldsmith and Harrison (1976), and Greig-Smith (1978). The general message from the review of α-diversity measurements was that the ecological insights gained were by no means proportional to the mathematical sophistication and complexity of the particular method; this is also applicable to β indices.

Wilson and Shmida (1984) considered the following six indices, calculated using presence–absence data collected from a series of stations along a transect.

1 Whittaker's β_w:

$$\beta_w = \frac{S}{\alpha} - 1 \qquad (13.26)$$

where S = the total number of species and α the average species richness of the samples. All samples must have the same size (or sampling effort).

2 Cody's β_c

$$\beta_c = \frac{g(H) + l(H)}{2} \qquad (13.27)$$

where $g(H)$ is the number of species gained and $l(H)$ the number lost moving along the transect.

3 Routledge's β_R, β_I, and β_E:

$$\beta_R = \frac{S^2}{2r + S} - 1 \qquad (13.28)$$

where S is the total species number for the transect and r the number of species pairs with overlapping distributions.

Assuming equal sample sizes:

$$\beta_I = \log(T) - \left[\left(\frac{1}{T}\right)\sum e_i \log(e_i)\right] - \left[\left(\frac{1}{T}\right)\sum \alpha_i \log(\alpha_i)\right] \qquad (13.29)$$

where e_i is the number of samples along the transect in which species i is present, α_i is the species richness of sample i, and T is $\sum e_i$.

The third of Routledge's indices is simply:

$$\beta_E = \exp(\beta_I) - 1 \qquad (13.30)$$

4 Wilson and Schmida's β_T:

$$\beta_T = \frac{[g(H) + l(H)]}{2\alpha} \qquad (13.31)$$

where the parameters are defined as for β_c and β_w.

Based on an assessment of the essential properties of a useful index—ability to detect change, additivity, independence of α, and independence of sample size— Wilson and Schmida (1984) concluded that β_w was best. Harrison *et al.* (1992) apply this measure in modified form to the study of latitudinal gradients in a variety of aquatic and terrestrial groups.

A different approach to β diversity is to measure diversity along a recognized gradient or at least a linear transect. The slope of the line will measure β diversity and sudden changes will reflect community or higher-order boundaries (Odum *et al.*, 1960). Pattern analysis, much used in plant ecology (Kershaw, 1973; Goldsmith & Harrison, 1976; Greig-Smith, 1978), adopts this approach on a small scale.

Whittaker (1972) summarized many studies on diversity gradients along 'coenoclines' by plotting the mean similarity coefficients (see p. 486) for each transect interval; the first part of the resulting curve is extended to zero distance and this allows the 'threshold diversity' (C_0) to be estimated.

Wilson and Mohler (1983) discuss the calculation of β diversity for quantitative data. A direct approach to β diversity would be to compare the separate diversity indices, and the Shannon–Wiener index (or developments of it) has sometimes been used (e.g. Kikkawa, 1968; Hummon, 1974), but, in the light of the shortcomings of this index as a measure of α diversity (see above and Hurlbert, 1971; May, 1975a; Taylor *et al.*, 1976), it seems inappropriate as a general technique. Mountford's (1962) index of similarity is based on the comparison of samples by the log-series distribution, but, as Bullock (1971a) points out, its assumption that the difference in distributions between different sites will be the converse of the similarities between samples from the same site (and some common α) may not be sound biologically. Bullock's (1971b) tests of the index on actual data tended to support these reservations. The most effective approach is based on the use of similarity indices, which are described below. Any of these similarity indices can be used to compare samples along a transect. Generally, the best index for presence–absence data is the Jaccard and for quantitative data the Morisita–Horn (Wolda, 1981, 1983; Magurran, 1998). Magurran (1988) gives a simple method for comparing the distribution of similarity coefficients between transects. Let:

$$\beta = (a+b)(1-\mathcal{J})$$

where a is the number of species in sample A, b the number in sample B, and $\mathcal{J}$ the Jaccard similarity index (p. 486). This index can be calculated for sample pairs along each transect. The distribution of β for each transect can then be compared.

13.2 Similarity and the comparison and classification of samples

When we compare the flora or fauna sampled at different localities, we can approach the task by considering either the similarity or the distinctness of their species assemblages. The conventional approach has been to measure similarity and numerous methods have been suggested, the most successful of which are described below. Legendre and Legendre (1983) give more complete accounts of similarity and distance measures. More recently, the term complementarity has been introduced by Vane-Wright *et al.* (1991) to measure the difference in the biota between potential reserves. Conservationists often wish to maximize the amount of biodiversity protected for a given number and size of reserve. Thus they seek to maximize the complementarity. Colwell and Coddington (1994) have argued for a broadening of the concept to measure distinctness at all spatial scales.

Ecological data sets can often be organized as a two-dimensional matrix, consisting of n samples (or stations) by S species. If the objective is to search for species interrelationships, then an $S \times S$ matrix of species correlations or similarities may be formed and what is termed an R analysis undertaken. If the objective is to identify samples with similar communities, then the $n \times n$ matrix of sample correlations or similarities may be used in a Q analysis.

13.2.1 Measures of complementarity

Colwell and Coddington (1994) argue that the most appropriate measure is the Marczewski–Steinhaus distance:

$$
C = \frac{\sum_{i=1}^{S_{jk}} |X_{ij} - X_{ik}|}{\sum_{i=1}^{S_{jk}} \max(X_{ij} \cdots X_{ik})}
\tag{13.32}
$$

where X_{ij} and X_{ik} are the presence–absence values (either 1 or 0) for species in species lists j and k. This is simply the complement of the familiar Jaccard index (see p. 486).

As a total measure for n lists, Colwell and Coddington (1994) quote Pielou as suggesting:

$$
C_T = \frac{\sum U_{jk}}{n}
\tag{13.33}
$$

where U_{jk} is the number of species only found in one of the two lists j and k and the summation is over all pairs of lists; C_T approaches $nS_T/4$ for large n, where S_T is the total species number in the combined species list.

13.2.2 Similarity indices

These are simple measures of either the extent to which two habitats have species in common (Q analysis) or species have habitats in common (R analysis). Binary similarity coefficients use presence–absence data; following the introduction of computers, more complex quantitative coefficients became practicable. Both groups of indices can be further divided to those which take account of the absence from both communities (double-zero methods) and those which do not. In most ecological applications, it is unwise to use double-zero methods, as they assign a high level of similarity to localities which both lack many species, a problem that becomes particularly acute in species-rich habitats, such as the marine benthos. We therefore only describe binary indices that exclude double-zeros. A good account of similarity and distance measures is given in Legendre and Legendre (1983).

13.2.2.1 Binary coefficients

These have been formulated in a number of slightly different ways and associated with even more originators. When comparing two sites, let a be the number of species held in common and b and c the number of species found at only one of the sites. When comparing two species, the terms are similar, e.g. a is the number of sites where they were both caught. The three simplest coefficients are:

Jaccard (1908):

$$C_j = \frac{a}{a+b+c} \qquad (13.34)$$

Sørensen (1948):

$$C_s = \frac{2a}{2a+b+c} \qquad (13.35)$$

Mountford (1962):

$$C_M = \frac{2a}{2bc-(b+c)a} \qquad (13.36)$$

C_M was designed to be less sensitive to sample size than C_j or C_s; however, it assumes that species abundance fits a log-series model, which may be inappropriate.

A more complex coefficient, proposed by Fager (1957) for the study of zooplankton and used in an approach termed recurrent group analysis, is:

$$C_F = \frac{a}{\sqrt{(a+b)(a+c)}} - \frac{1}{2\sqrt{a+c}} \qquad (13.37)$$

Following evaluation of similarity measures using Rothamsted insect data, Smith (1986) concluded that no index based on presence–absence data was entirely satisfactory, but the Sørensen was the best of those considered.

13.2.2.2 Quantitative coefficients

Based purely on species number, binary coefficients give equal weight to all species and hence tend to place too much significance on the rare species, whose capture will depend heavily on chance. Bray and Curtis (1957) brought abundance into consideration in a modified Sørensen coefficient and this approach, although criticized (Austin & Orloci, 1966), is widely used in plant ecology (Goldsmith & Harrison, 1976; Kent & Coker, 1992); the coefficient as modified essentially reflects the similarity in individuals between the habitats:

$$C_N = \frac{2jN}{(aN+bN)} \qquad (13.38)$$

where aN = the total individuals sampled (NT) in habitat a, bN = the same in habitat b, and jN = the sum of the lesser values for the species common to both habitats (often termed W).

However, for quantitative data, Wolda (1981, 1983) found that the only index not strongly influenced by sample size and species richness was the Morista–Horn index:

$$C_{MH} = \frac{2\sum(an_i \cdot bn_i)}{(da + db)aN \cdot bN} \tag{13.39}$$

where aN = the total number of individuals in site a, an_i = the number of individuals in the ith species in sample a, and

$$da = \frac{\sum an_i^2}{aN^2} \tag{13.40}$$

Smith (1986) also concluded that, for quantitative data, the Morista–Horn index was one of the most satisfactory.

The ready availability of computers now allows easy computation of Morisita's original index of similarity:

$$C_m = \frac{2\sum_{}^{n} X_{ij} X_{ik}}{(\lambda_1 + \lambda_2)N_j N_k} \tag{13.41}$$

where:

$$N_j = \sum X_{ij}$$

$$N_k = \sum X_{ik}$$

$$\lambda_1 = \frac{\sum_{}^{n}[X_{ij}(X_{ij} - 1)]}{N_j(N_i - 1)}$$

and:

$$\lambda_{12} = \frac{\sum_{}^{n}[X_{ik}(X_{ik} - 1)]}{N_k(N_k - 1)}$$

where X_{ij} and X_{ik} are the number of individuals of species i in habitats j and k, respectively; λ is the Simpson (1949) diversity index.

A considerably simpler, and frequently used, index is percentage similarity (Whittaker, 1952), calculated using:

$$P = 100 - 0.5\sum_{i=1}^{S}|P_{a,i} - P_{b,i}| \tag{13.42}$$

where $P_{a,i}$ and $P_{b,i}$ are the percentage abundances of species i in samples a and b, respectively, and S is total species number. This index takes little account of rare species.

13.2.3 Multivariate analysis

Multivariate analysis is used when the objective is to search for relationships between objects, or to classify objects (sites or species) that are defined by a number of attributes. Data sets can be large, e.g. marine benthic or forest beetle faunal studies can easily require analysis of a matrix of 100 samples (stations) by 350 species, and thus multivariate analysis requires a computer. Curiously, most methods were developed before widespread availability of the computer and therefore did not find immediate general application. If the objective is to assign objects into a number of discrete groups, then cluster analysis should be considered. If there is no a priori reason to believe the objects will or could naturally fall into groups, then an ordination technique may be more suitable. Ordination assumes that the objects form a continuum of variation and the objective is often to generate hypotheses about the environmental factor or factors that mould community structure.

There is a considerable literature on multivariate techniques; useful texts for ecologists are Seal (1964), Cooley and Lohnes (1971), Legendre and Legendre (1983), Digby and Kempton (1987), and Kent and Coker (1992).

13.2.3.1 Cluster analysis

When a number of sites or habitats are to be compared, the similarity measures of section 13.2.2 can form the basis of cluster analysis, which seeks to identify groups of sites or stations that are similar in their species composition. Alternatively, distance measures can be used to measure between object similarity (Digby & Kempton, 1987). Accounts of the many methods that have been developed are given in Legendre and Legendre (1983) and Digby and Kempton (1987). These methods require a computer, and many general statistical and specialized packages offer cluster analysis. Some of the most frequently used are given in Table 13.1.

Classification methods comprise two principal types: hierarchial, where objects are assigned to groups, which are themselves arranged into groups, as in a dendrogram, and non-hierarchial, where the objects are simply assigned to groups. The methods are further classified as either agglomerative, where the analysis proceeds from the objects by sequentially uniting them, or divisive, where all the objects start as members of a single group, which is repeatedly divided. For computational and presentational reasons, hierarchical–agglomerative methods are the most popular. However, divisive methods are frequently recommended, because they initially divide the data into broad groups, which can be used in classificatory schemes in the field.

The basic computational scheme used in cluster analysis can be illustrated using single-linkage cluster analysis as an example. This is the simplest procedure and consists of the following steps.

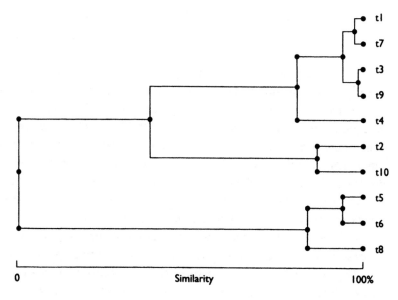

Fig. 13.10 An example of a dendrogram used for the classification of sampling sites.

1 Start with n groups each containing a single object (sites or species).
2 Using the similarity measure of choice, calculate the array of between-object similarities.
3 Find the two objects with the greatest similarity and group them into a single object.
4 Assign similarities between this group and each of the other objects, using the rule that the new similarity will be the greater of the two similarities prior to the join.
5 Continue steps 3 and 4 until one object is formed.

The results from a cluster analysis are usually presented in the form of a dendrogram (Fig. 13.10). The problem with all classification methods is that there can be no objective criteria of the best classification; indeed, even randomly generated data can produce a pleasing dendrogram. Always consider carefully if the groupings identified seem to make sense and reflect some feature of the natural world.

The two-way indicator species-analysis procedure (Hill *et al.*, 1975) also produces dendrograms of the relationship between species and samples but uses the reciprocal-averaging ordination method to order the species and samples. This method is particularly attractive in studies where the objective is to classify communities so that fieldworkers can quickly assign an area to a community type. This is because it identifies indicator species characteristic of each community.

13.2.3.2 Ordination

A number of ordination techniques are commonly used by ecologists and it is not possible to give clear guidance as to the method that should be used. In community studies, the aim of the analysis is to summarize the relationship between the samples (sites) and thus the best method is the one that gives the clearest and most inter-pretable picture. It may thus be appropriate to try a number of methods and compare the results.

Principal-component analysis (PCA) is the oldest and still one of the most fre-quently used ordination techniques in community ecology. General descriptions of the procedure for biologists are given by Legendre and Legendre (1983), Digby and Kempton (1987), and Kent and Coker (1992). The basic concept is to consider the community at each site as a point in n-dimensional space, where each of the axes represents a single species. The points for each site (sample) thus form an n-dimensional ellipsoidal cloud. PCA aims to express the relationship between the sites in a reduced number of dimensions, which can be presented graphically. The first principal axis is the straight line that passes along the greatest dimension of this ellipsoid, the second principal axis is an orthogonal line that passes along the second greatest dimension, and so on. PCA calculates these principal axes and the corre-sponding principal components, which are the coordinates of the sites on the princi-pal axes. The analysis is undertaken on either the between-site variance–covariance or the correlation matrix. If the attributes of these sites are correlated, then much of the relationship between sites will be expressed in only a few principal axes. When successful, PCA will present major features of a complex ecological community in only two or three dimensions, and the ordination of sites along these new axes can be related to the underlying environmental factors that are moulding community structure.

PCA and other ordination methods are available in all common statistical pack-ages for the IBM personal computer (PC) and Apple and mainframe computers (Table 13.1).

Other ordination methods include reciprocal averaging (also termed corre-spondence analysis), detrended correspondence analysis, and non-metric multidi-mensional scaling. These methods, which are described by Legendre and Legendre (1983) and Kent and Coker (1992), may be applied using widely available computer programs (Table 13.1).

13.3 Species packing

The coexistence of organisms, especially of apparent relatedness, has fascinated naturalists since the time of Alfred Russel Wallace, and the extent of interaction between them—the interspecific competition matrix—is the meeting ground of community and population ecology. Field observations on the distribution of organ-isms against a resource gradient will give a measure of the species packing in relation to this particular component of the species' niche; it is also relatively easy to deter-mine the extent to which species overlap on a particular resource gradient. Thus, measures of 'niche breadth' (or width) and 'niche overlap' on this resource gradient can be obtained. It is tempting (and, indeed, desirable) to go beyond this and, having determined similar values on other resource gradients, compound the overlaps in a hopefully appropriate manner to get a true measure of niche overlap. With this 'traveller's cheque' in hand, the jump is made across the chasm from community to population ecology; on arrival, the niche overlap measurements are converted to competition coefficients in classical Lotka–Volterra (or Gause) competition equations. (A safer route to these coefficients is by measuring the actual impact of one species on another (e.g. Vandermeer, 1969).) This analogy is not made to cast disparagement on those who have attempted the 'jump', for nothing is achieved (especially in ecology) by pessimistic contemplation of the difficulties, but rather to caution intending travellers in what is still an exciting area of ecology. Functionally, the analogy is intended to serve as an explanation as to why the explicit treatment in this section makes but a short part of the 'journey'. At present, one can, as Green (1971) pointed out, easily demonstrate from field data that two species do not occupy the same niche, but not, without experiments, that they occupy the same niche.

In this section, the methods of assessing the extent to which species occur together, the usage of a resource spectrum by one species, the comparison of resource utilization between species, and niche overlap will be described.

13.3.1 Measurement of interspecific association

These measurements may be based either on presence–absence data or on abundance figures; as Hurlbert (1971) points out, presence–absence data are preferable if it is desired to measure the extent to which two species' requirements are similar. Interspecific competition (and other factors) may lead to a 'misleading' lack of association if the measure is based on abundance. However, it would seem that, whenever possible, both types of analysis should be undertaken, for a positive association on presence–absence data and a much weaker or negative one on abundance data would suggest (not prove) interspecific competition worthy of further analysis.

13.3.1.1 The departure of the distribution of presence or absence from independence
These methods measure the departure from independence of the distribution of the two species and assume that the probability of occurrence of the species is constant for all samples. Thus, if the distribution of two aphid predators were being compared, it would only be legitimate to include samples that also contained the prey. A good example of the use of these indices is given by Evans and Freeman (1950), who measured the interspecific association of two species of flea on two different rodents (*Apodemus* and *Clethrionomys*); they found that there was a strong negative association on *Apodemus* and a moderate positive one on *Clethrionomys* and suggested that the coarse and longer fur of *Clethrionomys* allowed the two fleas to avoid competition and exist together on that host. However, when applied to habitats whose uniformity is doubtful, these methods may give results of uncertain value, as was found by Macan (1954) in a study of the associations of various species of corixid bug in different ponds. In such a situation, some samples may be from habitats that are outside the environmental ranges of the species; this will have the effect of inflating the value for *d* in the table below and so lead to too many positive associations.

Fager (1957) has demonstrated, by examples, that, if two species are rare and therefore both are absent from most of the samples (this could, as just indicated, be due to unsuitable samples being included), a high level of association will be found. Conversely, if two species occur in most of the samples and so are nearly always found together, no association will be shown with these methods. (Although a biological association is obvious, it is equally correct to say that this does not depart significantly from an association that is due to chance and therefore does not necessarily imply any interspecific relationship.)

13.3.1.2 The two-dimensional Kolmogorov–Smirnov test

Garvey *et al.* (1998) proposed that a statistical technique used by astronomers could be used to detect the association between two species. The observations for each species are plotted as a scatter plot, each pair of coordinates (X, Y) is taken in turn, and the number of points in each of the four surrounding quadrats (with (X, Y) as the origin) is counted. The counts for the four quadrats are expressed as proportions. These proportions are compared with those that would be expected if the two species were independent. The expected values are calculated by multiplying the counts of the proportions of observations $<X$ and $\geq X$ and $<Y$ and $\geq Y$. Within each quadrat, the difference between the observed and expected proportion of points is determined and the maximum difference between the observed and expected over all the quadrats and points found. The original pairs of data points are then randomized 5000 times, using a computer, and the test calculations performed on each randomization. The statistical significance is then determined by comparing the observed maximum difference with the randomly created distribution of maximum differences.

13.3.1.3 The contingency table

The basis of these methods is the 2×2 contingency table (Table 13.2). Such a table should always be drawn up so that A is more abundant than B, i.e. $(a + b) < (a + c)$. A number of statistics are available for analysis of such a table, but the corrected chi-square (χ^2) test makes fewest assumptions about the type of distribution, and the significance of the value obtained can be determined from tables available in standard statistical textbooks. It is calculated:

Table 13.2 The 2×2 contingency table.

		Species A		
		Present	Absent	Total
Species B	Present	a	b	$a + b$
	Absent	c	d	$c + d$
	Total	$a + c$	$b + d$	$n = a + b + c + d$

$$\chi^2 = \frac{n\big[|ad-bc|-(n/2)\big]^2}{(a+c)(b+d)(a+b)(c+d)} \tag{13.43}$$

The test in this form is only valid if the expected numbers (if the distribution was random) are not less than 5, there is only one degree of freedom and so the 5% point is 3.84. Therefore, if a χ^2 of less than this is obtained, any apparent association could well be due to chance and further analysis should be abandoned. If the smallest expected number is less than 5, the exact test should be used.

13.3.1.4 Coefficients of association
If χ^2 is significant, then one of the coefficients of association may be used to given an actual quantitative value for comparison with other species. They are designed so that the coefficient has the same range as the correlation coefficient (r), i.e. $+1 =$ complete positive association, $-1 =$ complete negative association, and $0 =$ no association. Cole (1949) reviews a number of coefficients and points out how, if the above interpretations of values from $+1$ to -1 are to hold for comparative purposes, the plot of the value of the coefficient against the possible number of joint occurrences should be linear; for several of the coefficients, it is not and, for those in which it is, the plot does not pass through the zero. Two coefficients are given below.

Coefficient of mean square contingency. This coefficient makes no assumption about distribution, but it cannot give a value of $+1$ unless $a = d$ and b and $c = 0$. For less extreme forms of association, it is useful and easily calculated:

$$C_{AB} = \sqrt{\frac{\chi^2}{n+\chi^2}} \tag{13.44}$$

where C_{AB} = coefficient of association between A and B, n = total number of occurrences, and the χ^2 value is obtained as above.

This coefficient is recommended by Debauche (1962).

Coefficient of interspecific association. The coefficients originally designed by Cole (1949) have been shown by Hurlbert (1969) to be biased by the species frequencies. He showed that this bias was considerably diminished if the coefficient is defined:

$$C_{AB}^1 = \frac{ad-bc}{|ad-bc|}\left|\left(\frac{\text{Obs}\,\chi^2 - \text{Min}\,\chi^2}{\text{Max}\,\chi^2 - \text{Min}\,\chi^2}\right)^{\frac{1}{2}}\right| \tag{13.45}$$

Min χ^2 is the value of χ^2 when the observed a differs from its expected value ($\hat{a}$) by less than 1.0 (except that when $a - \hat{a} = 0$ or $= 0.5$, the value of Min χ^2 depends on whether ($ad - bc$) is positive or negative), formulated:

$$\text{Min}\,\chi^2 = \frac{n^3(\hat{a}-g[\hat{a}])^2}{(a+b)(a+c)(c+d)(b+d)} \tag{13.46}$$

where $g(\hat{a}) = \hat{a}$, rounded to the next lowest integer when $ad < bc$ or rounded to the next highest integer when $ad < bc$; if $\hat{a}$ is an integer, then $g(\hat{a}) = \hat{a}$.

Max χ^2 is the value of χ^2 when a is as large or small as the marginal totals of the 2×2 table will permit, formulated under specified conditions as follows:

$ad \geq bc$:

$$\chi^2 = \frac{(a+c)(b+d)n}{(a+c)(c+d)}$$

$ad < bc$, $a \leq d$:

$$\chi^2 = \frac{(a+b)(a+c)n}{(a+b)(c+d)}$$

$ad < bc$, $a > d$:

$$\chi^2 = \frac{(b+d)(c+d)n}{(a+b)(a+c)}$$

Obs χ^2 is calculated in the normal manner (see above). Pielou and Pielou (1968) describe a method for species of infrequent occurrence.

13.3.1.5 Proportion of individuals occurring together
Correlation may be utilized if the data can be normalized; otherwise some of the methods of comparing similarity between habitats may be used. For example, Sørensen's coefficient, as used by Whittaker and Fairbanks (1958), may be modified to give the normal range of -1 (no association) to $+1$ (complete association):

$$I_{ai} = 2\left[\frac{J}{A+B} - 0.5\right] \tag{13.47}$$

where J = number of individuals of A and B in samples where both species are present and A and B = total of individuals of A and B in all samples.

13.3.2 Measurement of resource utilization
Among the meaningful resources that may be partitioned by apparently coexisting species are food (defined by quality (type) and size), space (defined in Euclidean terms (e.g. height in vegetation) and physicochemical characters (e.g. climate, substrate type)), and time (seasonal and diel). This by no means exhaustive list gives an idea of the number of 'resource spectra' that contribute to a species' niche (Pianka, 1976b).

In considering the most appropriate way to measure resource utilization, it is useful to start from a theoretical basis (May & MacArthur, 1972). The extent of the separation in the resource utilization between potentially competing species is well expressed in terms of the separation measure (d) between the means of their curves

Fig. 13.11 Theoretical resource-utilization relationships. (a) The 'simple case' of three species with similar (and normal) resource-utilization curves; d, distance apart of means; w, standard deviation of utilization, d/w, resource separation ratio = 2.5 (in these cases) (modified from May & MacArthur, 1972). (b) The more typical case with varying resource-utilization curves: broadest in the region of fewer resources and less interspecific competition (after Pianka, 1976b).

and the widths (1 standard deviation in the normal Gaussian curve) (w) (Fig. 13.11). In this theoretical situation (where resource-utilization functions are Gaussian and equal), then, if the resource separation ratio (d/w) is less than 3, there will be some interaction between the species. Theoretically, the ratio will have a minimum value below which the competitive exclusion principle operates (the two species cannot coexist). May's (1973, 1974) studies suggest that one is the minimum value, but the conversion of this ratio to a competition coefficient is, as already indicated, fraught with biological difficulties (Abrams, 1975; May, 1975b; Heck, 1976).

Resource-utilization curves are seldom normal; indeed, field data suggest that they vary greatly, being broad where resources are scarce and narrow where they are more plentiful (Fig. 13.11b). The shape of the curve influences the extent of packing, as Roughgarden (1974) pointed out on mathematical grounds. This point is illustrated simply in Fig. 13.12, which represents two species pairs, the ones above having leptokurtic resource-utilization curves and the pair below platykurtic ones. It should be noted that each species takes an identical quantity of resource (the areas under the curves are equal) and the ranges ($x^1 \rightarrow x^{11}$) are equal, but, when the lines for w (the distance from the mean of 68% of the resource utilization) are inserted, their ratio (leptokurtic : platykurtic) is 3 : 5. The impact of this on the resource separation ratios (d/w) is shown. The difference implies that, if this were the only resource spectrum, four closely packed ($d/w \sim 1$) 'platykurtic species' could be replaced by seven closely packed 'leptokurtic species'. (These numerical values are, of course, precisely relevant only to Fig. 13.12.)

It will be noted that, whereas the platykurtic species are not as closely packed as leptokurtic ones, the latter, for the same resource utilization, have higher $u(x)_{max}$. If we imagine the supply of the resource $K(x)$ to be far above the utilization (demand) level $u(x)$ in Fig. 13.12b, which could be due to predator pressure, then there will be no restrictions on the shapes of the curves that can replace the platykuritc ones, and

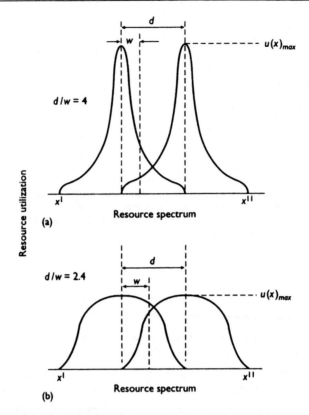

Fig. 13.12 The effect of the shape of the utilization curve on the separation measure (d) and hence species packing. Note that the two species with leptokurtic curves (a) and the two with platykurtic curves (b) both utilize the same quantity of resource (equal areas under curves) and the same range of resources (x^I–x^{II}), but the extent of utilization, of which $u(x)_{max}$ is one measure, differs.

tighter packing, as indicated above, is possible. There are two possible procedures for assessing species packing in respect of a resource dimension.

1 The d/w approach, in which the utilization by each species is considered in turn and the mean and the width ($w = 68\%$ of the utilization on one side of the mean) are determined, the differences between the means giving d.

2 Proportional utilization of the 'Simpson–Yule' index approach, in which the proportion of the species utilizing a particular segment of the resource gradient is determined (p_i), the breadth of utilization ('niche breadth' of Levins (1968)) being equivalent to the Simpson–Yule index.

The former is conceptually simpler, as the investigator can visualize the 'areas' of resource occupied by the species, but it assumes a utilization spectrum that approximates to the normal; thus the resource dimension must be approximately ordered. It has seldom been used in field studies. The proportional utilization approach makes no demands on the ordering on the resource dimension and has been widely used. Of course, p_i is simply the height of the utilization curve at i, the curve whose width (in terms of one standard deviation) is given by w.

13.3.2.1 Species packing in terms of mean and width of resource-utilization spectrum ('d/w method')

This method can easily be used when the animal's use of the resource is for a discrete period and the resource spectrum is continuous; this applies especially to seasonal occurrence, diel occurrence, and food items or other resources, defined by size. Data on seasonal occurrence are widely available and often show striking patterns of species packing (e.g. Greenslade, 1965; Shapiro, 1975; Southwood, 1978). The first steps are to find the mean and the total range of time of occurrence, which can usually be taken as twice the interval between the mean and the earliest occurrence. This method eliminates the extreme, end-of-season, 'tail'; these may be post-reproductives or late developers, which will not survive winter (this may be checked) and will not affect the fitness of that species (they may, however, affect the species that occur with them, and then a new range will need to be found and used when assessing the impact of this putative competitor on those species). If the frequency of occurrence is approximately normally distributed (see p. 10), then the total range will be equivalent to $6w$ (three standard deviations on each side of the mean). If the frequency of occurrence shows kurtosis or other irregularity, w may be calculated by a method analogous to the graphical method for estimating the numbers in a stage (see Fig. 11.1). The mean is drawn in and the number of animal days (or other time period, e.g. 5 days) summed visually. Then, starting from the mean and working in vertical columns towards an end of the range, one sums the 'animal days' until 68% of the total that side of the mean has been accumulated. At this point, a vertical line can be drawn and the distance of this to the mean approximates w. Experience with the graphical method for determining the numbers entering a stage (p. 406) has shown its robustness and, as a method of estimating resource utilization, it is simple and free of assumptions. Large sets of data could easily be handled on a computer, but it is often necessary to make decisions about stragglers, which should be based on biological knowledge about the specific and not on a general rule. This technique allows for different w values either side of the mean and for adjustments, as in the case of post-reproductives mentioned above.

When the values of w are different for the ith and jth species, a 'common w' (w_{ij}) must be calculated:

$$w_{ij} = \left[w_i^2 + \frac{w_j^2}{2} \right]^{\frac{1}{2}} \tag{13.48}$$

The resource separation ratio is:

$$p_{ij} = \frac{d_{ij}}{w_{ij}}$$

where d_{ij} is the difference between the means of resource utilization by the ith and jth species.

Species packing with respect to periodic variation at annual, tidal, or diurnal time-scales must take account of the circularity of the resource, and the ratio for a pair of species may be in the form $d/2w$, because both 'tails' can overlap. Greenslade (1963) treated data on the diurnal/nocturnal partitioning of carabids in this manner.

Size may easily be arranged on a linear scale and it governs many aspects of resource utilization. The size (and hence type) of food may show a reasonable relationship to body size (Hespenheide, 1975; Wilson, 1975; Davidson, 1977) or body part (e.g. bees' tongues (Heinrich, 1976) or birds' bills (Karr & James, 1975)); the trivial range is also approximately related to size when account is taken of feeding habits (predator vs. herbivore) and habitat type (e.g. desert vs. temperate grassland); in such an analysis for ambulatory animals, the span between the front and back legs is a better measure of size than body length.

Other resources, such as plant species or types for herbivores, could be arranged as the resource spectrum, but the ordering would require judgement analogous to that of Cody (1975) in the ordering of bird species (see above). Under such circumstances, the more versatile proportional utilization method is more objective.

13.3.2.2 Species packing in terms of proportional utilization of different resource states (p_i method)

The breadth of utilization ('niche breadth' (Levins, 1968)) is given by the Simpson–Yule index applied to the distribution of the individuals between the resource states:

$$B = \frac{1}{\displaystyle\sum_i^m p_i^2} = \frac{1}{\displaystyle\sum_i^m \frac{N_i^2}{N_T^2}} \qquad (13.49)$$

where p_i = proportion of individuals found in or using resource i, N_i = the number of individuals of the species in question in the ith resource state, N_T = the total number of individuals in all (m) the resource states. In this context, the tendency of this index to 'undervalue' rare events is probably an advantage.

This method is versatile in that the 'resource states' do not have to be ordered along a continuum; they may, for example, be samples from m different categories of food type, found by stomach-content analysis or some other method. In comparative studies, it may be useful to calculate the standardized niche breadth:

$$B_s = \frac{B-1}{m-1} \qquad (13.50)$$

The calculation of B from a series of random samples from a heterogeneous environment is, however, open to criticism, for the degree of uniformity in distribution could well be influenced by the natural clumping (or otherwise) of the organism (see

Chapter 2); one may claim that this is part of the resource, but this is not very illuminating. There is a considerable interpretative advantage if distinct resource states are sampled separately, e.g. parts of the habitat that from other knowledge are known to be distinguished by ecologically significant factors. However, samples often reveal vagrants or transitory members of the fauna; insects that engage in upper-air migration (e.g. aphids, many flies, coccinellids, etc.) regularly occur in places that are not part of their niche, as is also found with fish displaced by torrent, storm, or current (P.A. Henderson and A.E. Magurran, pers. comm.). Studies of colonies of Lepidoptera record the movement of individuals across terrain that is, again, not part of their niche. If these vagrants are relatively few, the Simpson–Yule index's character of emphasizing the high-probability occurrences will minimize their significance. A more general criticism (for the purposes of determining competition) is that 'handfuls of animals' taken from their habitat may not represent usage of the same resource; one may live on the surface of the leaf blade, one on the stem, one may be active by day, another by night, and so on. Thus, although this approach is valid for the assessment of the breadth of resources used, its utilization to calculate 'niche overlap' must be handled with caution.

There is a further problem if the different resource states are present (in the samples) at very different frequencies. Colwell and Futuyma (1971) used as a measure of resource breadth the Shannon–Wiener index (see p. 474):

$$H = \sum p_i \log p_i$$

where p_i is the proportion of individuals using resource i. They proposed weighting resource states so that each is equally represented. The weightings were defined, not in terms of physicochemical or other measurements, but in terms of the rarity of the fauna. Sabath and Jones (1973) found that the weightings did not add to the information gained from unweighted matrices and that the expansion of the matrix eventually prevented discrimination between species distributions.

Another detailed study using the Colwell–Futuyma weightings was made by Hanski and Koskela (1977) on beetles in dung pats. This would seem to be an ideal 'test' of the concept, for the habitat is discrete, its state may also be described by certain physical characters, and the beetles collected are a real assemblage, which cannot necessarily be claimed for trap catches, a difficulty of which Sabath and Jones (1973) were well aware. Hanski and Koskela (1977) conclude that weighting by unusual species is unsatisfactory; but they did weight 'resource states' in relation to the total number of individuals in each. However, once again, they found all the measurements of 'niche breadth' highly correlated and concluded that the main results were fairly insensitive to weighting. It seems, therefore, that, in practice, the simple B has much to recommend it as a measure of the width of resource utilization by a species.

Hurlbert (1978) argues that usage should be scaled by availability, so that equation (13.49) is reformulated as:

$$B' = \frac{1}{\sum \dfrac{p_j^2}{a_j}} \tag{13.51}$$

where a_j is the proportion of the total available resources consisting of resource j.

The variance of both B and B' is given by Smith (1982) as:

$$\text{var}(B) = \frac{4B^4\left[\displaystyle\sum_{j=1}^{n}\dfrac{p_j^3}{a_j^3} - \left(\dfrac{1}{B}\right)^2\right]}{\sum N_j} \tag{13.52}$$

An alternative measure of niche breadth proposed by Smith (1982) is:

$$FT = \sum_{j=1}^{n}\sqrt{p_j a_j} \tag{13.53}$$

which, for large samples, has estimated upper and lower 95% confidence:

$$\sin\left(x - \frac{1.96}{2\sqrt{\sum N_j}}\right)$$

and:

$$\sin\left(x + \frac{1.96}{2\sqrt{\sum N_j}}\right)$$

respectively. The angle is expressed in radians and $x = \text{arc sin}(FT)$.

The proportional overlap in resource utilization between two species (i and j) is (Schoener, 1968; Colwell & Futuyma, 1971):

$$\Theta_{ij} = 1 - 0.5\sum_{h=1}^{m} p_{ih} - p_{jh} \tag{13.54}$$

where p_{ih} = the proportion of species i in resource state h, p_{jh} = the proportion of species j in the same resource state, and m = the total number of resource states. (Note that Θ measures 'overlap', whilst p measures separation.) This method has been used with, for example, leafhoppers (McClure & Price, 1976) and wolf-spiders (Uetz, 1977).

Hurlbert (1978) defined niche overlap as 'the degree to which the frequency of encounter between two species is higher or lower than it would be if each species utilised each resource state in proportion to the abundance of that resource state' and so, L, niche overlap, is given by:

$$L = \sum\left(\frac{p_{ij} p_{ik}}{a_i}\right) \tag{13.55}$$

where p_{ij} and p_{ik} are the proportion of resource i used by species j and k, respectively. Using computer simulation, Ricklefs and Lau (1980) showed niche overlap to be strongly influenced by sample size and Smith and Zaret (1982) showed that the bias in its estimation is minimized when the numbers of the two species are equal. Niche overlap can also be measured using Morisita–Horn similarity indices (p. 487).

13.3.3 Niche size and competition coefficients

All the methods discussed above provide information on species packing and their resource utilization. These measures are themselves intrinsically interesting and may be used in studies of community ecology, island biogeography, etc. (e.g. Price, 1971; McClure and Price, 1976; Mühlenberg et al., 1977). The difficulty of equating resource-utilization measures with niche breadth has already been outlined. The utilization indices (w and B) pose two problems: how may they be combined and, less easily answered, have all the components of the niche been assessed?

If species packing indices (p_{ij} and Θ_{ij}) are to be equated with niche overlap and hence competition, attention also has to be paid to the renewability of resources, interactions at other trophic levels, and the relative availability of other resources (May, 1973, 1974; Schoener, 1974; Levine, 1976; Pianka, 1976b).

Competition coefficients between the ith and jth species for one resource axis may be calculated (May, 1973, 1975b):

1 from resource separation ratio:

$$\alpha_{ij} = \exp\left[\frac{-d_{ij}^2}{4w_{ij}^2}\right]$$

2 from proportional utilization functions:

$$\alpha_{ij} = \frac{\sum p_i p_j}{\sqrt{\left(\sum p_i\right)^2 \left(\sum p_j\right)^2}}$$

where each term is summed for the array of resource states along this dimension. This expression is due to Pianka (1973); May (1975b) compares it with alternatives.

Weighted competition coefficients for food resources may be calculated (Schoener, 1974):

$$\alpha_{ij} = \left(\frac{T_j}{T_i}\right)\left[\frac{\displaystyle\sum_{k=1}^{m}\left(\frac{p_{ik}}{f_k}\right)\left(\frac{p_{jk}}{f_k}\right)b_{ik}}{\sqrt{\displaystyle\sum_{k=1}^{m}\left(\frac{p_{ik}}{f_k}\right)^2\left(\frac{p_{jk}}{f_k}\right)^2 b_{ik}}}\right]$$

where T_j/T_i = the ratio of the total number of food items consumed by an individual of the jth to that consumed by an individual of the ith species, measured over an interval of time that includes all regular fluctuations in consumption for both species; p_{ik} and p_{jk} are the frequencies of food type k in the diets of i and j, respectively, f_k the frequency of food item k in the environment, b_{ik} = the net calories gained by an individual of i from one item of k, or (more approximately) the calories in an item of k, or (still more approximately) the biomass of an item of k (see Chapter 14). The summations are taken over all (m) the food items eaten by one (i) of the two species. The equation is modified from Schoener's so as to put it in the same form as the basic equation (13.59) (Pianka's expression). The above expression may also be used to obtain a coefficient in relation to foraging times in the same habitats; the T values are the ratios of total time, b_{ik} = the calories obtained from an average item of food in habitat k, and the f values are omitted. As T_j/T_i will often approximate to unity, unless the average value of the food items in the different habitats are remarkably different, b_{ik} will not be significant; this will often reduce to the basic (Pianka's) expression. Thus, as indicated above (p. 499), weighting may not in practice always be necessary.

Competition coefficients between species may be expressed as a matrix — the diagonal values representing intraspecific interactions are unity (May, 1973, 1976a).

Species generally interact in more than one resource dimension; as Schoener (1974) points out, the number of dimensions, their type, and the comparative extent of the interactions, are of intrinsic interest and already certain patterns are emerging, e.g. predators partition a habitat through the diel cycle more often than other types.

The combination of competition coefficients from different resource axes is not straightforward, except, as May (1975b) has clearly shown, in the two limiting cases. When the two axes are totally independent, then the coefficients (a values) should be multiplied; when the two axes are completely correlated, then the coefficients should be summed. A method for use with field data, allowing for various degrees of independence between resource axes, has not yet been devised.

Nevertheless, as there are so many difficulties over the assumptions of elegant theoretical models (May, 1973, 1974, 1975b; Abrams, 1975; Heck, 1976; Armstrong, 1977), it seems important to determine from field observations which assumptions are justified and which refinements are, in practice, trivial. Field studies should be made over as many resource dimensions as possible, the d/w ratio used to determine the shape and separation of resource utilization, and, where possible, the impact of perturbations studied (although here too there are difficulties in interpretation (Schoener, 1974)).

References

Abrams, P. (1975) Limiting similarity and the form of the competition coefficient. *Theor. Popul. Biol.* **8**, 356–75.

Armstrong, R.A. (1977) Weighting factors and scale effects in the calculation of competition coefficients. *Am. Nat.* **111**, 810–12.

Austin, M.P. & Orloci, L. (1966) Geometric models in ecology. II. An evaluation of some ordination techniques. *J. Ecol.* **54**, 217–22.

Bazzaz, F.A. (1975) Plant species diversity in old field successional ecosystems in Southern Illinois. *Ecology* **56**, 485–8.

Boswell, M.T. & Patil, G.P. (1971) Chance mechanisms generating the logarithmetic series distribution used in the analysis of number of species and individuals. In: Patil, G.P., Pielou, E.C., & Waters, W.E. (eds) *Statistical Ecology*, Vol. 3. Penn State University Press, Philadelphia, pp. 99–130.

Bray, J.R. & Curtis, C.T. (1957) An ordination of the upland forest communities of southern Wisconsin. *Ecol. Monogr.* **27**, 325–49.

Bullock, J.A. (1971a) The investigations of samples containing many species. 1. Sample description. *Biol. J. Linn. Soc.* **3**, 1–21.

Bullock, J.A. (1971b) The investigations of samples containing many species. II. Sample comparison. *Biol. J. Linn. Soc.* **3**, 23–56.

Bulmer, M.G. (1974) On fitting the Poisson lognormal distribution to species abundance data. *Biometrics* **30**, 101–10.

Burnham, K.P. & Overton, W.S. (1978) Estimation of the size of a closed population when capture probabilities vary among animals. *Biometrika* **65**, 623–33.

Burnham, K.P. & Overton, W.S. (1979) Robust estimation of population size when capture probabilities vary among animals. *Ecology* **60**, 927–36.

Chao, A. (1984) Non-parametric estimation of the number of classes in a population. *Scand. J. Stat.* **11**, 265–70.

Chao, A. (1987) Estimating the population size for capture-recapture data with unequal matchability. *Biometrics* **43**, 783–91.

Chao, A. & Lee, S.-M. (1992) Estimating the number of classes via sample coverage. *J. Am. Stat. Assoc.* **87**, 210–17.

Cody, M.L. (1974) *Competition and the Structure of Bird Communities*. Princeton University Press, Princeton, New Jersey.

Cody, M.L. (1975) Towards a theory of continental species diversity: bird distributions over Mediterranean habitat gradients. In: Cody, M.L. & Diamond, J.M. (eds) *Ecology and Evolution of Communities*. Harvard University Press, Cambridge, Massachusetts, pp. 215–57.

Cody, M.L. & Diamond, J.M. (eds) (1975) *Ecology and Evolution of Communities*. Harvard University Press, Cambridge, Massachusetts.

Cole, L.C. (1949) The measurement of interspecific association. *Ecology* **30**, 411–24.

Coleman, B.D. (1981) On random placement and species-area relations. *Math. Biosci.* **54**, 191–215.

Coleman, M.D., Mares, M.D., Willig, M.R., & Hsieh, Y.-H. (1982) Randomness, area, and species richness. *Ecology* **63**, 1121–33.

Colwell, R.K. & Coddington, J.A. (1994) Estimating terrestrial biodiversity through extrapolation. *Phil. Trans. Roy. Soc. B* **345**, 101–18.

Colwell, R.K. & Futuyma, D.J. (1971) On the measurement of niche breadth and overlap. *Ecology* **52**, 567–76.

Cooley, W.W. & Lohnes, P.R. (1971) *Multivariate Data Analysis*. John Wiley & Sons, New York, London.

Davidson, D.W. (1977) Species diversity and community organization in desert seed-eating ants. *Ecology* **58**, 711–24.

Debauche, H.R. (1962) The structural analysis of animal communities of the soil. In: Murphy, P.W. (ed.) *Progress in Soil Zoology*. Butterworths, London, pp. 10–25.

Digby, P.G.N. & Kempton, R.A. (1987) *Multivariate Analysis of Ecological Communities*. Chapman & Hall, London.

Evans, F.C. & Freeman, R.B. (1950) On the relationship of some mammal fleas to their hosts. *Ann. Entomol. Soc. Am.* **43**, 320–33.

Fager, E.W. (1957) Determination and analysis of recurrent groups. *Ecology* **38**, 586–95.

Findley, J.S. (1973) Phenetic packing as a measure of faunal diversity. *Am. Nat.* **107**, 580–4.

Fisher, R.A., Corbet, A.S., & Williams, C.B. (1943) The relation between the number of species and the number of individuals in a random sample of an animal population. *J. Anim. Ecol.* **12**, 42–58.

Gage, I. & Tett, P.B. (1973) The use of log-normal statistics to describe the benthos of Lochs Etive and Creran. *J. Anim. Ecol.* **42**, 373–82.

Garvey, J.E., Marschall, E.A., & Wright, R.A. (1998) From star charts to stoneflies: detecting relationships in continuous bivariate data. *Ecology* **79**, 442–7.

Goldsmith, F.B. & Harrison, C.M. (1976) Description and analysis of vegetation. In: Chapman, S.B. (ed.) *Methods in Plant Ecology*. Blackwell Scientific Publications, Oxford, London, 85–151.

Goodall, D.W. (1973) Sample similarity and species correlation. In: Whittaker, R.H. (ed.) *Handbook of Vegetation Science Part V.* Junk, The Hague, pp. 107–56.

Green, R.H. (1971) A multivariate statistical approach to the Hutchinson niche: bivalve molluscs of central Canada. *Ecology* **52**, 543–56.

Greenslade, P.J.M. (1963) Daily rhythms of locomotor activity in some Carabidae (Coleoptera). *Entomol. Exp. Appl.* **6**, 171–80.

Greenslade, P.J.M. (1965) On the ecology of some British carabid beetles, with special reference to life histories. *Trans. Soc. Br. Entomol.* **16**, 149–79.

Greig-Smith, P. (1978) *Quantitative Plant Ecology*. Butterworths, London.

Hanski, I. & Koskela, H. (1977) Niche relations among dung-inhabiting beetles. *Oecologia* **28**, 203–31.

Harrison, S., Ross, S.J., & Lawton, J.H. (1992)

Beta-diversity on geographic gradients. *J. Anim. Ecol.* **61**, 151–8.

Heck, K.L. (1976) Some critical considerations of the theory of species packing. *Evol. Theory* **1**, 247–58.

Heinrich, B. (1976) Resource partitioning among some unsocial insects: bumblebees. *Ecology* **57**, 874–89.

Hellmann, J.J. & Fowler, G.W. (1999) Bias, precision and accuracy of four measures of species richness. *Ecological Applications* **9**, 824–34.

Heltshe, J. & Forrester, N.E. (1983) Estimating species richness using the jackknife procedure. *Biometrics* **39**, 1–11.

Henderson, P.A. (1989) On the structure of the inshore fish community of England and Wales. *J. Mar. Biol. Assoc. UK* **69**, 145–63.

Henderson, P.A. & Holmes, R.H.A. (1991) On the population dynamics of dab, sole and flounder within Bridgwater Bay in the lower Severn Estuary, England. *Netherl. J. Sea Res.* **27**(3–4), 337–44.

Hendrickson, J.A. & Ehrlich, P.R. (1971) An expanded concept of 'species diversity'. *Notul. Nat. Acad. Nat. Sci. Philadelphia* **439**, 1–6.

Hespenheide, H.A. (1975) Prey characteristics and predator niche width. In: Cody, M.L. & Diamond, J.M. (eds) *Ecology and Evolution of Communities*. Harvard University Press, Cambridge, Massachusetts, pp. 158–80.

Hill, M.O. (1973) Diversity and evenness: a unifying notation and its consequences. *Ecology* **54**, 427–32.

Hill, M.O., Bunce, R.G.H., & Shaw, M.W. (1975) Indicator species analysis, a divisive polthetic method of classification and its application to a survey of native pinewoods in Scotland. *J. Ecol.* **63**, 597–613.

Holdridge, L.R., Grenke, W.C., Hatheway, W.H., Liang, T., & Tosi, J.A. (1971) *Forest Environments in Tropical Life Zones*. Pergamon Press, Oxford.

Hummon, W.D. (1974) SH: a similarity index based on shared species diversity used to assess temporal and spatial relations among inter-tidal marine Gastrotricha. *Oecologia* **17**, 203–20.

Hurlbert, S.H. (1969) A coefficient of interspecific association. *Ecology* **50**, 1–9.

Hurlbert, S.H. (1971) The non-concept of species diversity: a critique and alternative parameters. *Ecology* **52**(4), 577–86.

Hurlbert, S.H. (1978) The measurement of niche overlap and some derivatives. *Ecology* **59**, 67–77.

Hutchinson, G.E. (1953) The concept of pattern in ecology. *Proc. Acad. Nat. Sci. Philadelphia* **105**, 1–12.

Jaccard, P. (1908) Nouvelles recherches sur la distribution florale. *Bull. Soc. Vandoise Sci. Nat.* **44**, 223–70.

James, F.C. & Rathbun, S. (1981) Rarefaction, relative abundance, and diversity of avian communities. *Auk* **98**, 785–800.

Karr, J.R. & James, F.C. (1975) Eco-morphological configurations and convergent evolution in species and communities. In: Cody, M.L. & Diamond, J.M. (eds) *Ecology and Evolution of Communities*. Harvard University Press, Cambridge, Massachusetts, pp. 258–91.

Kempton, R.A. (1979) Structure of species abundance and measurement of diversity. *Biometrics* **35**, 307–22.

Kempton, R.A. & Taylor, L.R. (1974) Log-series and log-normal parameters as diversity discriminants for the Lepidoptera. *J. Anim. Ecol.* **43**, 381–99.

Kent, M. & Coker, P. (1992) *Vegetation Description and Analysis*. John Wiley & Sons, Chichester, New York.

Kershaw, K.A. (1973) *Quantitative and Dynamic Plant Ecology*. Griffin, London.

Kikkawa, J. (1968) Ecological association of bird species and habitats in eastern Australia: similarity analysis. *J. Anim. Ecol.* **37**, 143–65.

Kricher, J.C. (1972) Bird species diversity: the effect of species richness and equitability on the diversity index. *Ecology* **53**, 278–82.

Lamas, G., Robbins, R.K., & Harvey, D.J. (1991) A preliminary survey of the butterfly fauna of Pakitza, Parque Nacional del Manu, Peru, with an estimate of its species richness. *Publ. Mus. Hist. Nat. UNMSM* **40**, 1–19.

Legendre, L. & Legendre, P. (1983) *Numerical Ecology*. Elsevier Scientific Publishing Company. Amsterdam, Oxford, New York.

Levine, S.H. (1976) Competitive interactions in ecosystems. *Am. Nat.* **110**, 903–10.

Levins, R. (1968) *Evolution in Changing Environments*. Princeton University Press, Princeton, New Jersey.

Lloyd, M. & Ghelardi, R.J. (1964) A table for calculating the 'equitability' component of species diversity. *J. Anim. Ecol.* **33**, 217–25.

Macan, T.T. (1954) A contribution to the study of the ecology of the Corixidae (Hemipt.). *J. Anim. Ecol.* **23**, 115–41.

MacArthur, R.H. (1957) On the relative abundance of bird species. *Proc. Nat. Acad. Sci.* **43**, 293–5.

MacArthur, R.H. (1972) *Geographical Ecology. Patterns in the Distribution of Species*. Harper & Row, New York, London.

MacArthur, R.H. & Wilson, E.O. (1967) *The Theory of Island Biogeography*. Princeton University Press, Princeton, New Jersey.

McClure, M.S. & Price, P.W. (1976) Ecotype

characteristics of coexisting *Erythoneura* leafhoppers (Homoptera: Cicadellidae) on sycamore. *Ecology* 57, 928–40.

McIntosh, R.P. (1967) An index of diversity and the relation of certain concepts to diversity. *Ecology* 48, 1115–26.

McNaughton, S.I. & Wolf, L.L. (1970) Dominance and the niche in ecological systems. *Science* 167, 131–6.

Magurran, A.E. (1988) *Ecological Diversity and its Measurement*. Chapman & Hall, London.

May, R.M. (1973) *Stability and Complexity in Model Ecosystems*. Princeton University Press, Princeton, New Jersey.

May, R.M. (1974) On the theory of niche overlap. *Theor. Popul. Biol.* 5, 297–332.

May, R.M. (1975a) Patterns of species abundance and diversity. In: Cody, M.L. & Diamond, J.M. (eds) *Ecology and Evolution of Communities*. Harvard University Press, Cambridge, Massachusetts, pp. 81–120.

May, R.M. (1975b) Some notes on estimating the competition matrix α. *Ecology* 56, 737–41.

May, R.M. (1976a) Models for two interacting populations. In: May, R.M. (ed.) *Theoretical Ecology*. Blackwell Scientific Publications, Oxford, pp. 78–104.

May, R.M. (1976b) Patterns in multi-species communities. In: May, R.M. (ed.) *Theoretical Ecology*. Blackwell, Oxford, pp. 142–62.

May, R.M. & MacArthur, R.H. (1972) Niche overlap as a function of environmental variability. *Proc. Nat. Acad. Sci.* 69, 1109–13.

Miller, R.J. & Wiegert, R.G. (1989) Documenting completeness, species–area relations, and the species-abundance distribution of a regional flora. *Ecology* 70, 16–22.

Mountford, M.D. (1962) An index of similarity and its application to classificatory problems. In: Murphy, P.W. (ed.) *Progress in Soil Zoology*. Butterworths, London, pp. 43–50.

Mueller-Dombois, D. & Ellenberg, H. (1974) *Aims and Methods of Vegetation Ecology*. John Wiley & Sons, New York, London, Sydney, Toronto.

Mühlenberg, M., Leipold, D., Mader, H.J., & Steinhauer, B. (1977) Island ecology of arthropods I & II. *Oecologia* 29, 117–44.

Odum, H.T., Cantfon, J.E., & Kornicker, L.S. (1960) An organizational hierarchy postulate for the interpretation of species-individual distributions, species entropy, ecosystem evolution and the meaning of a species-variety index. *Ecology* 41, 395–9.

Palmer, M.W. (1991) Estimating species richness: the second-order jackknife reconsidered. *Ecology* 72, 1512–13.

Patil, G.P. & Taillie, C. (1979) An overview of diversity. In: Grassle, J.F., Patil, G.P., Smith, W., & Taillie, C. (eds) *Ecological Diversity in Theory and Practice*. International Cooperative Publishing House, Fairland, Maryland.

Peet, R.K. (1974) The measurement of species diversity. *Ann. Rev. Ecol. Syst.* 5, 285–307.

Peet, R.K. (1975) Relative diversity indices. *Ecology* 56, 496–8.

Pianka, E.R. (1973) The structure of ligard communities. *Rev. Ecol. Syst.* 4, 53–74.

Pianka, E.R. (1976a) *Evolutionary Ecology*. Harper & Row, New York.

Pianka, E.R. (1976b) Competition and niche theory. In: May, R.M. (ed.) *Theoretical Ecology*. Blackwell Scientific Publications, Oxford, pp. 167–96.

Pielou, D.P. & Pielou, E.C. (1968) Association among species of infrequent occurrence: the insect and spider fauna of *Polyporus betulinus* (Builiard) Fries. *J. Theor. Biol.* 21, 202–16.

Pielou, E.C. (1975) *Ecological Diversity*. Wiley, New York.

Pimm, S.L. (1991) *The Balance of Nature?* University of Chicago Press, Chicago.

Preston, F.W. (1948) The commonness and rarity of species. *Ecology* 29, 254–83.

Preston, F.W. (1960) Time and space and the variation of species. *Ecology* 41, 611–27.

Preston, F.W. (1962) The canonical distribution of commonness and rarity. Part 1. *Ecology* 39, 185–215. Part II. *Ecology* 39, 410–32.

Price, P.W. (1971) Niche breadth and dominance of parasitic insects sharing the same host species. *Ecology* 52, 587–96.

Raaijmakers, J.G.W. (1987) Statistical analysis of the Michaelis–Menten equation. *Biometrics* 43, 793–803.

Reaka-Kudla, M.L., Wilson, D.E., & Wilson, E.O. (eds) (1997) *Biodiversity II*. Joseph Henry Press (US National Academy of Science), Washington, DC.

Renyi, A. (1961) On measures of entropy and information. In: Neyman, J. (ed.) *Proceedings of the 4th Berkeley Symposium on Mathematical Statistics and Probability*. University of California Press, Berkeley, California, 547–61.

Riklefs, R.E. & Lau, M. (1980) Bias and dispersion of overlap indexes—results of some Monte-Carlo simulations. *Ecology* 61, 1019–24.

Roughgarden, J. (1974) Species packing and the competition function with illustrations from coral reef fish. *Theor. Popul. Biol.* 5, 163–86.

Routledge, R.D. (1977) On Whittaker's components of diversity. *Ecology* 58, 1120–7.

Routledge, R.D. (1979) Diversity indices: which ones are admissible? *J. Theor. Biol.* 76, 503–15.

Sabath, M.D. & Jones, J.M. (1973) Measurement

of niche breadth and overlap: the Colwell–Futuyma method. *Ecology* **54**, 1143–7.

Sanders, B.L. (1969) Benthic marine diversity and the stability-time hypothesis. In: Diversity and stability in ecological systems. *Brookhaven Symp. Biol.* **22**, 71–81.

Schoener, T.W. (1968) The Anolis lizards of Bimini: resource partitioning in a complex fauna. *Ecology* **49**, 704–26.

Schoener, T.W. (1974) Resource partitioning in ecological communities. *Science* **185**, 27–39.

Seal, H. (1964) *Multivariate Statistical Analysis for Biologists*. Methuen, London.

Shapiro, A.M. (1975) The temporal component of butterfly species diversity. In: Cody, M.L. & Diamond, J.M. (eds) *Ecology and Evolution of Communities*. Harvard University Press, Cambridge, Massachusetts, pp. 181–95.

Simpson, E.H. (1949) Measurement of diversity. *Nature* **163**, 688.

Smith, B. (1986) *Evaluation of Different Similarity Indices Applied to Data from the Rothamsted Insect Survey*. University of York, York.

Smith, E.P. (1982) Niche breadth, resource availability and inference. *Ecolgy* **63**, 1675–81.

Smith, E.P. & van Belle, G. (1984) Non parametric estimation of species richness. *Biometrics* **40**, 119–29.

Smith, E.P. & Zaret, T.M. (1982) Bias in estimating niche overlap. *Ecology* **63**, 1248–53.

Soberon, M.J. & Llorente, B.J. (1993) The use of species accumulation functions for the prediction of species richness. *Conserv. Biol.* **7**, 480–8.

Sørensen, T. (1948) A method of establishing groups of equal amplitude in plant sociology based on similarity of species content and its application to analyses of the vegetation on Danish commons. *Biol. Skr. (K. Danske Vidensk. Selsk. NS)* **5**, 1–34.

Southwood, T.R.E. (1978) The components of diversity. In: Mound, L.A. & Waloff, N. (eds) *The Diversity of Insect Faunas: 9th Symposium of the Royal Entomological Society*. Blackwell Scientific Publications, Oxford, pp. 19–40.

Southwood, T.R.E. (1996) Natural communities: structure and dynamics. *Phil. Trans. R. Soc. Lond. B* **351**, 1113–29.

Southwood, T.R.E., Brown, V.K., & Reader, P.M. (1979) The relationship of insect and plant diversities in succession. *Biol. J. Linn. Soc.* **12**, 327–48.

Taylor, L.R. (1978) Bates, Williams, Hutchinson — a variety of diversities. In: Mound, L.A. & Waloff, N. (eds) *Diversity in Insect Faunas: 9th Symposium of the Royal Entomological Society*. Blackwell Scientific Publications, Oxford, pp. 1–18.

Taylor, L.R., Kempton, R.A., & Woiwood, I.P. (1976) Diversity statistics and the log-series model. *J. Anim. Ecol.* **45**, 255–72.

Tipper, J.C. (1979) Rarefaction and rarefiction — the use and abuse of a method in paleoecology. *Paleobiology* **5**, 423–34.

Tóthméresz, B. (1995) Comparison of different methods for diversity ordering. *J. Veget. Sci.* **6**, 283–90.

Uetz, G.W. (1977) Coexistence in a guild of wandering spiders. *J. Anim. Ecol.* **46**, 531–41.

Vandermeer, J.H. (1969) The community matrix and the number of species in a community. *Ecology* **50**, 362–71.

Vane-Wright, R.I., Humphries, C.J., & Williams, P.H. (1991) What to protect? Systematics and the agony of choice. *Biol. Conserv.* **55**, 235–54.

Van Valen, L. (1965) Morphological variation and width of ecological niche. *Am. Nat.* **999**, 377–90.

Walther, B.A. & Morand, S. (1998) Comparative performance of species richness estimation methods. *Parasitology* **116**, 395–405.

Webb, D.J. (1974) The statistics of relative abundance and diversity. *J. Theor. Biol.* **43**, 277–92.

Whittaker, R.H. (1952) A study of summer foliage insect communities in Great Smoky Mountains. *Ecol. Monogr.* **22**, 1–44.

Whittaker, R.H. (1970) *Communities and Ecosystems*. Macmillan, London.

Whittaker, R.H. (1972) Evolution and measurement of species diversity. *Taxon* **21**, 213–51.

Whittaker, R.H. & Fairbanks, C.W. (1958) A study of plankton copepod communities in the Columbia basin, south eastern Washington. *Ecology* **39**, 46–65.

Williams, C.B. (1947) The logarithmic series and the comparison of island floras. *Proc. Linn. Soc. Lond.* **158**, 104–8.

Williams, C.B. (1964) *Patterns in the Balance of Nature and Related Problems in Quantitative Ecology*. Academic Press, London, New York.

Wilson, D.S. (1975) The adequacy of body size as a niche difference. *Am. Nat.* **109**, 769–84.

Wilson, E.O. (ed.) (1988) *Biodiversity*. National Academy Press, Washington, DC.

Wilson, M.V. & Mohler, C.L. (1983) Measuring compositional change along gradients. *Vegetatio* **54**, 129–41.

Wilson, M.V. & Shmida, A. (1984) Measuring beta diversity with presence–absence data. *J. Ecol.* **72**, 1055–64.

Wolda, H. (1981) Similarity indices, sample size and diversity. *Oecologia* **50**, 296–302.

Wolda, H. (1983) Diversity, diversity indices and tropical cockroaches. *Oecologia* **58**, 290–8.

14 The Estimation of Productivity and the Construction of Energy Budgets

The size of a population and the interactions between populations within an ecosystem may be expressed in terms of biomass (weight of living material) or energy content, as well as in numbers. Biomass and energy are useful to ecologists in that they provide common units for the description of populations of animals and plants of different sizes. By the comparison of energy budgets, we may, for example, compare the strategies of warm-blooded mammals and cold-blooded reptiles or even whales and copepods. Such measures are particularly important in food-web studies, where the focus is on the quantification of the flux of energy or carbon between different trophic levels. A particularly good example of this type of study has been undertaken on the tiny Ythan estuary in Scotland (Baird & Milne, 1981; Raffaelli & Hall, 1996). Great effort is required to quantify each pathway within an ecosystem so that a whole community study with the species coverage achieved in the Ytham cannot at present be produced for a large temperate estuary and is almost inconceivable for a tropical flood-plain or estuary. Comparing the energy flux of different pathways is one way in which we can assess the relative importance of the various interactions within a web (Paine, 1980, 1992). In studies of general predators, e.g. insectivorous birds, predation activity is often best expressed as biomass or energy. Conversely, if the energy requirements are known from metabolic measurements, they may be used to predict food requirements in the field (Stiven, 1961). It must be remembered that the full richness of an organism's requirements and behaviour cannot be expressed in terms of energy. For example, food is more than simply an energy source; its quality in terms of specific amino acids, vitamins, and other constituents will also be important (Boyd & Goodyear, 1971; Iversen, 1974; Onuf *et al.*, 1977; Schroeder, 1977).

While energetic considerations are generally considered to place constraints on the length of food-chains and thus on the structure of food webs, energetic arguments have fulfilled their early promise only in respect of describing the trophic structure of ecosystems. They make only slight contributions to the development of plans for reserve management and the maintenance of biodiversity. Nor have they proved useful to our understanding of population dynamics and the prediction of population collapse or the explosive irruption of a pest or pathogen. It is probably because of these deficiencies that there appears to be less interest in framing population models in terms of energy now than there was during the 1960s and 1970s. The work undertaken during this period defined many present concepts and remains an important source of primary information.

The concept of energy is fundamental in the consideration of the functioning of the ecosystem; Ivlev (1939, 1945) and Lindeman (1942) — now classic papers — pointed the way to this approach, reviewed by Gallucci (1973) and Wiegert (1976). Various trophic levels can be distinguished in the ecosystem. First, there are the primary producers (i.e. plants), which build up complex substances from simple inorganic substances utilizing the energy from sunlight. The total energy retained by the plant as fixed carbon is usually referred to as the gross production and the change in energy level (standing crop or stock) plus any losses due to grazing or mortality during the year, but less imports (e.g. migrants), is termed the net production. The amount of living material present at any given time is the standing crop or biomass. Unfortunately, the use of these terms has been confused, but Macfadyen (1963) has given a useful table of synonymy and Petrusewicz & Macfadyen (1970) have established standard definitions and notation. Information on primary productivity is given in Vollenweider (1969) and Chapman (1976).

The energy equations for an individual may be expressed:

Gross energy = Digestible energy = Metabolizable energy = Resting energy

or	+	+	+
energy intake	Faecal waste↓	Urinary waste↓	Activity
or			+
ingestion			Growth
or			+
consumption			Reproduction

(Note that only the upper term partakes in the equation to the right.)

The energy budget of a population or a trophic level can be expressed in several equations (Petrusewicz & Macfadyen, 1970). These may be related to the equations above, noting that the energy used up in resting activity and the 'work' of growth is equated with respiration, so that:

$$C = D + F = A + U + F = R + P + (F + U) \tag{14.1}$$

where C = consumption (= ingestion), D = digestible energy, F = faecal waste, A = assimilation (metabolizable energy), U = urinary waste, R = respiration, and P = productivity (= growth + reproduction).

This basic equation may be divided into a number of simpler equations for parts of the process, which are of value in the construction of energy budgets, for, if only one term remains known, then this may be found. Thus:

$$P = A - R \tag{14.2}$$

$$A = C - (F + U) \tag{14.3}$$

Productivity is the increase (growth) (Δ) in the biomass of the standing crop (B), plus production eliminated (E), through death, emigration, exuviae, or other products (e.g. feathers of birds at moult, spiders' silk, byssus of mussels), so that:

$$P = \Delta B + E \tag{14.4}$$

One will also note from the composite equation that:

$$P = Pg + Pr + E \tag{14.5}$$

where Pg = productivity as growth and Pr = productivity through reproduction (the biomass of offspring).

These equations demonstrate the usefulness of energy, as opposed to simply biomass, in the description of this aspect of ecosystem or population dynamics. The early studies of Odum and Odum (1955) on a coral reef, of Odum (1957) on a hot spring, of Teal (1957) on a cold spring, and of Whittaker et al. (1974) on a deciduous forest exemplify the value of measurements of biomass and energy in describing and analysing the trophic structure of ecosystems. The quantification of the energy moving between ecological compartments or species, allowing the identification of pathways deemed important in terms of a greater energy flux was an objective in the 1970s International Biological Programme (IBP). This produced a series of handbooks (e.g. Golley & Buechner, 1968; Ricker, 1968; Petrusewicz & Macfadyen, 1970; Edmondson & Winberg, 1971; Grodzinski et al., 1975), which should still be consulted for further information on the measurement of productivity and energy flow. Other useful sources are Paine's (1971) early review, Brafield and Llewellyn's (1982) introductory book, and the review of laboratory methods by Brafield (1985), which, while focused on fish, gives much general information.

14.1 Estimation of standing crop

If a detailed life-table or budget has been constructed (see Chapter 11), then the standing crop of the population at any given time can be determined by converting numbers to biomass (dry weight is usually the most appropriate measure). Given a known, calculated, or experimentally obtained energy density for the organism, biomass can be further converted into energy (Brey et al., 1988).

14.1.1 Measurement of biomass

Biomass can be expressed as wet weight, dry weight (DW), shell-free dry weight (SFDW), ash-free dry weight, or the amount of organic carbon present.

Wet weight for a terrestrial animal or plant is simply the living or freshly killed or harvested weight. For aquatic organisms, such as fish, it should be obtained after blotting to remove surface water. As wet weight can change with the state of dehydration of the organism and changes rapidly after death or following preservation, it is an unsuitable measure of biomass for comparative studies.

Biomass is generally expressed as dry weight, obtained by drying a known wet weight of animals (or plants) until a constant weight is achieved. Unfortunately, no standard protocol can be given. There is agreement that material should be dried at a low temperature to avoid the loss of volatile—especially lipoid—constituents. However, there is no consensus as to what this temperature should be. For insects and other small organisms, freeze-drying or heating to a maximum of 60°C is often recommended. Lovegrove (1966) suggested that plankton and other small organisms can be dried in a desiccator. Paine (1964) originally argued that, for marine organisms with water of hydration in their skeletons, they cannot be dried below 100°C; however, for organisms such as barnacles, he concluded that drying could best be achieved at 80°C (Paine, 1971). Following an extensive series of experiments, Atkinson and Wacasey (1983) showed that a maximum energy density was estimated for benthic marine organisms when dried at 100°C.

Shell-free dry weight is used for molluscs and many marine organisms for which a high proportion of their weight comprises calcareous shell. Shells can be removed either pre- or post-drying by mechanical means (Atkinson & Wacasey, 1983) or by acid digestion in 1 mol/L hydrochloric acid (HCl). Acid decalcification probably results in the loss of some organic carbon. For all organisms, the ash-free dry weight will give a measure of the amount of organic material present and is determined by heating a sample of known dry weight in a muffle furnace at 500°C for 4 h and subtracting the weight of the ash obtained from the original dry weight (Atkinson & Wacasey, 1983). However, at this temperature, some inorganic material may also be lost; calcium carbonate, for example, is broken down at temperatures above 400°C.

With small animals, especially plankton, it may be more convenient to work with the volume of the animals rather than the actual numbers (e.g. in Neess and Dugdale's (1959) method (see below)). A number of devices for measuring the volume of small animals have been developed (e.g. Frolander, 1957; Stanford, 1973).

14.1.2 Determination of energy density

The energy content of a material may be determined directly by oxidation, either by potassium dichromate in sulphuric acid (Teal, 1957) or by burning in oxygen and determining the amount of heat liberated. The latter method—bomb calorimetry—is more convenient and is widely used in ecology; however, it involves drying the material, and volatile substances can be lost. Hartman and Brandt (1995) give a good account of the method as applied to fish.

Phillipson (1964) described a miniature non-adiabatic bomb (or ballistic) calorimeter, which was simple and suitable for handling materials in the order of 5–100 mg dry weight. It is operated by placing the material in a platinum pan enclosed in the 'bomb'. The bomb is filled with oxygen until a pressure of 30 atmospheres is reached, cooled until the temperature, as indicated by a thermocouple, is steady, and then enclosed in a polystyrene insulating jacket. Passing an electric current through a thin wire that presses against it ignites the sample. The temperature rises rapidly, and this is recorded by the deflection of a potentiometer attached to the thermocouple. The maximum deflection in a firing is a measure of the caloric content of the sample. The instrument is calibrated by burning given weights of benzoic acid of known caloric value. This basic principle is still used by the computer-controlled adiobatic calorimeters manufactured by the Parr Instrument Company 211 Fifty-third Street, Moline, Illinois 61265, USA. To obtain an accurate estimate of the energy density, adjustments may be necessary to allow for the amount of fuse wire lost, acid production by sulphur and nitrogen, and the presence of magnesium and calcium carbonates, which decompose endothermically.

Naturally not all the energy in an animal or plant is available to an animal that eats it, and quite frequently only a proportion of the potentially digestible energy is assimilated. Verdun (1972) presents an extreme and iconoclastic view on availability. The assimilation rates for various types and quantities of food need to be determined. The total energy is, however, available to the ecosystem, as micro-organisms can break down almost all organic compounds.

Values for the energy content of various plants have been published by Ovington and Heitkamp (1960) and of animals by Golley (1960), Slobodkin and Richman (1961), and Comita and Schindler (1963). There is, however, considerable variation in caloric value, which, if unrecognized, could be a major source of error in energy budgets. Both plants (Boyd, 1969; Singh & Yadava, 1973; Caspers, 1977) and animals (Wiegert, 1965; Wiegert & Coleman, 1970; Hinton, 1971; Wissing & Hasler, 1971) vary from season to season. These differences in calorific value are due to variations in the levels of fat, which accounts for much of the interspecific variation in energy density between animals (Slobodkin & Richman, 1961; Griffiths, 1977). Particularly large seasonal changes in stored fat levels are observed in birds; small passerines may double their body weight through fat accumulation prior to migration and even show marked diurnal variation (Gosler, 1996). It is also changed by mineral composition, which varies markedly, in plants (Boyd & Blackburn, 1970; Caspers, 1977), insects (Wiegert, 1965), and crustaceans (Griffiths, 1977). Different stages of the same insect also vary markedly (Wiegert, 1965; Hinton, 1971; McNeill, 1971) (Table 14.1). In the froghoppers, *Philaenus* and *Neoplilaenus*, the immature stages contain more ash than the adults; the same appears to be true, to a lesser extent, of the willow-leaf beetle, *Melosoma*, which shows a fall in caloric value throughout development (Hagvar, 1975). In contrast, in the dung-beetle, *Aphodius*, it rises markedly during the last larval instar (Holter, 1975).

14.2 Estimation of energy flow

Energy flow in a population is usually estimated using one or more of the equations listed in the introduction to this chapter, especially equations (14.2) and (14.3):

A(ssimilation) = R(espiration) + P(roduction)

A(ssimilation) = C(onsumption) − F(aecal) + U(rinary waste)*

---

* Because faecal and urinary waste are mixed in insects, digestible energy (D) cannot be obtained (Reichle, 1969).

Table 14.1 Caloric values (Kcal/g dry wt or ash free dry wt*) of different stages of the same insect.

Insect	Data source	Egg	1	2	3	4	5	Pupa	Adult	♂	♀
Homoptera											
*Neophilaenus**	Hinton, 1971		5.44	5.47	5.79	5.80	5.91			5.46–6.00	5.56–
Philaenus	Wiegert, 1965	6.31	4.58	5.27	5.45	5.53	5.71			5.49–5.95	5.16–6.08
*Philaenus**	Wiegert, 1965	6.50	4.98	5.67	5.80	5.78	5.90			5.58–5.95	5.50–6.08
Heteroptera											
Leptopterna	McNeill, 1971	6.30		5.81			5.86			6.05	6.06
Coleoptera											
Melosoma	Hagvar, 1975	6.16	5.70	5.28	5.24			5.71	5.68		
Aphodius	Holter, 1975			4.67	4.66–5.80						
Orthoptera											
*Encoptolopius**	Bailey & Riegert, 1972		5.01	5.23	5.30	5.40	5.16			5.25	5.34
*Chorthippus**	Gyllenberg, 1969			5.05	5.46	5.08				5.61	5.85

Frequently two of the terms in the equations will be found directly and the third by calculation. For example, for a particular animal, it may be difficult to determine the respiratory rate in the field, but ingestion, consumption, and waste excretion (egestion + excretion) per individual can be measured and these values can be converted to energy equivalents (kJ). Then, knowing the numbers of individuals in the field, the total energy assimilation (the energy income) of the population can be calculated. Assuming the numbers of individuals of different ages throughout the season and their energy densities are known, the (net) production can be calculated. The difference between assimilation and production will be a measure of respiration.

The construction of an energy budget requires the initial development of a numerical budget. Normally, it is necessary to calculate the standing crop and its energy equivalent (described above) and from this the production, together with the independent measurement of either consumption and waste excretion rates or respiration. Seasonal variations will affect these values (e.g. Jonasson & Kristiansen, 1967); the energy budget will, perforce, contain a number of approximations and the more variables that can be determined, the more reliance can be placed upon it. Clearly, if all the terms in any of the equations are determined and their magnitudes prove compatible, when they are substituted in the equations, this provides valuable evidence as to their reliability. Thus, as with a numerical budget (p. 405), a detailed energy budget will expose its own errors. Typical examples of single-species energy budgets are those by Plaut *et al.* (1996) for the sea hare, *Aplysia oculifera*, Chaabane *et al.* (1996) for the ground-beetle, *Abax ater*, Henen (1997) for the tortoise. *Gopherus agassizii*, Richman (1958) for crustacea, Marchant and Nicholas (1974) for nematodes, Doohan (1973) for a rotifer, and Edmunds and Davies (1989) for the coral *Porites porites*. The studies of Wiegert (1964a), Lawton (1971), McNeill (1971), Bailey and Riegert (1972), and Holter (1974, 1975) may be taken as model examples of energy-budget construction for insects. An exceptionally long-term study is that of Jenderedjian (1996), who uses data collected over a 60-year period to relate the energy budgets of Oligochaeta to primary production.

14.2.1 The measurement of production

Production consists of the total amount of material synthesized by the animal and is obtained by summing the following.

1 The increase in the standing crop during the season (or other time unit).
2 The biomass of all individuals that died or were eaten during the season.
3 The biomass of the total number of exuviae shed by all the individuals (those alive and those now dead) and of any other product, e.g. the spittle of cercopids (Wiegert, 1964a), the byssus of mussels (Kuenzler, 1961), and the silk of many insects, mites, and spiders.
4 The biomass of any reproductive products (sperms and seminal fluid, eggs), young individuals, or adults that have left the area (i.e. emigration).
Strictly, the amount of nitrogenous waste excreted should also be measured. The

biomass of any individuals that migrated into the population must be subtracted from the total obtained above.

The calculation of production from a budget, as was done, for example, by Golley and Gentry (1964), Wiegert (1964a), McNeill (1971), and Karo (1973), demands that the form of the mortality curve, the actual population at various stages, and immigration and emigration are accurately known. The weights of the various products can be determined in the laboratory, but caution should be exercised in assuming that laboratory determinations hold under field conditions.

A somewhat different approach was developed by Allen (1951) and Neess and Dugdale (1959) for the study of the production of chironomid larvae in a lake. Instead of separating the numbers into different age categories (as in a budget) and then multiplying the age categories by their respective biomasses, Neess and Dugdale use the product of these variables as expressed in the field in terms of weight.

The net production of a developing larval population is the outcome of two opposing processes:
Increase in weight of the individual:

$$k_g Q_t = dQ_t / dt \qquad (14.6)$$

and:

Fall in numbers:

$$-k_m N_t = dN_t / dt \qquad (14.7)$$

where Q_t = weight of the individual at time t, N_t = number of individuals at time t, k_g = growth rate, and k_m = mortality rate. These equations (14.6 and 14.7) can be combined by the elimination of t and integrated to give:

$$\log_e Q_t^{1/k_g} = \log_e C N_t^{1/k_m}$$

or:

$$Q_t^{-k_m} = C N_t^{k_g} \qquad (14.8)$$

The constant C may be substituted by an expression using Q_t or N_t by noting, at time $t = 0$, when Q_0 = weight of a newly hatched larva and N_0 = the total hatch, that:

$$Q_t = \frac{Q_0}{(N_0/N_t)^{-k_g/k_m}} \qquad (14.9)$$

For given values of N_0 and Q_0, the shape of the curves describing the relationship between Q and N is determined solely by the ratio k_g/k_m; if the ratio remains constant, the growth—survivorship curve will trace out the same path. Neess and Dugdale noted that, within the range of values of the ratio up to 2, large changes occur in the area under the curve for small variations in the actual ratio. Now, the area under the curve corresponds to the actual net production. Therefore, if it can be shown that k_g/k_m is a constant, a smooth curve may be drawn of numbers on mean weight and the area beneath it found. As equation (14.9) can be rearranged to:

$$\log Q_t = \log Q_0 + \frac{k_g}{k_m} \log \frac{N_0}{N_t}$$

(14.10)

the constancy of the ratio may be tested by plotting $\log N_t$ against $\log Q_t$. A straight line will indicate that the ratio is constant.

The numbers per unit area can then be plotted against the mean weight, a smooth curve (an 'Allen curve') drawn, and the area below it (ACDE in Fig. 14.1) found. This is the actual net production in biomass per unit area (the units corresponding, of course, to those of Q_t and N_t).

A number of terms may be defined by reference to Fig. 14.1:

ACDE   Actual net production
BCED   Actual net recruitment or increase in standing crop due to this cohort
         $= Q_p N_p$
ABE   Directly recycled production
ACDF   Potential net production $= N_0 Q_p$
AEF   Lost potential net production
ABEF   Lost potential recruitment

This method is most appropriate when the development of the individuals of a generation is closely synchronized or where a particular cohort can be recognized and distinguished from others, as in dragonflies (Benke, 1976).

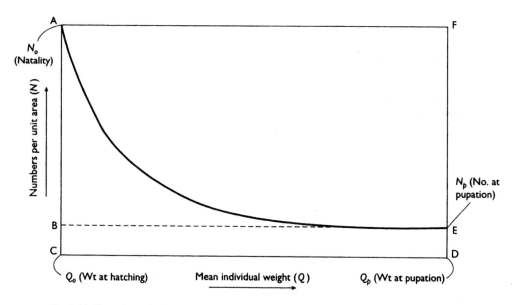

Fig. 14.1 The relationship between numbers and mean individual weight in an insect population during larval development; the area below the curve may be measured to give actual net production. The letters indicate areas representing other productivity terms (see text) (modified from Neess & Dugdale, 1959).

14.2.2 The measurement of feeding and assimilation

14.2.2.1 The quality of the food eaten

In some instances, such as with many arboreal lepidopterous larvae, the nature of an animal's food may be determined by observation in the field; observations in the laboratory may also be used, but with caution, for many animals will eat unnatural foods under artificial conditions. Another method, frequently used by ornithologists and ichthyologists, is to examine the crop or stomach contents (p. 360). Such methods may be used with chewing insects and crustaceans and the remains of different types of plant or animal identified under the microscope (e.g. Hanna, 1957; Mulkern and Anderson, 1959; Henderson and Walker, 1986). This approach can be quantified using a sedimentation technique (Brown, 1961). In insects which such plants, the actual tissue—phloem, xylem, mesophyll cells—from which the nutriment is obtained should be determined. The salivary sheath (Miles, 1972) or the position of the stylets (Wiegert, 1964b; Pollard, 1973) may sometimes be kept in place by a high-voltage shock (Ledbetter & Flemion, 1954) and are precise indications, but examination of a freshly damaged leaf may sometimes be satisfactory (McNeill, 1971).

The identification of the meals of predators by serological techniques is discussed in Chapter 10 (p. 362), where reference is also made to the use of radioisotope-tagged prey. Following a pioneer study by Pendleton and Grundmann (1954), radioisotopes have often been used to trace out food webs (e.g. Odum & Kuenzler, 1963; Paris & Sikora, 1965; Coleman, 1968; Van Hook, 1971; Shure, 1973; Walker et al., 1991). The food plant or primary food source is tagged and subsequent arrival of radioactivity (assessed quantitatively by counting) in the animals demonstrates a trophic link. The rate of build-up of radioactivity in the animal has been taken as proportional to the trophic distance, e.g. a peak in activity soon after tagging would indicate a herbivore, a later and lower peak a predator or saphrophagous species. However, as Shure (1970) points out, such interpretations should be made with care because herbivores may change their host plant and those moving to the labelled plant during the experiment would show accumulation curves resembling a predator's. Appropriate field observation can obviate such misinterpretations. Trophic links in soil invertebrates may be investigated using pesticides for the selective removal of animal groups (Edwards et al., 1969), a method analogous to the 'insecticidal check method' (p. 370). The forage ratio and methods of assessing food preferences are discussed in Chapter 10 (p. 375).

14.2.2.2 Feeding and assimilation rates

There are four basic approaches to the measurement of these rates and related parameters.

1 Radiotracer—measuring the passage of radioisotopes from the food to the animal and their subsequent loss.

2 Gravimetric—by the direct weighing of the food, the faeces, and the animal.

3 Indicator—by marking the food with an inert non-absorbed indicator, the

increase in its concentration as it passes through the alimentary canal measures the amount of matter absorbed.

4 Faecal — by measuring faecal output and relating this to consumption.

Radiotracer techniques. Radiotracers may be used singly or as a double-marking method. When used singly, there are two approaches. The amount of radioisotope used may be small relative to the equilibrium body burden of the animal, so that the rate of increase of assimilation of the isotope is linearly related to the rate of assimilation of the food. Examples are the studies of Engelmann (1961), Strong and Landes (1965), Hubbell *et al.* (1966), Malone and Nelson (1969), and Moore *et al.* (1974). Alternatively, the body burden may be increased so as to reach equilibrium when the following relationship holds:

$$r = \frac{KQ_e}{a} \tag{14.11}$$

where r = feeding rate (measured in bequerels/day), a = proportion of ingested nuclide assimilated, K = the elimination constant* = 0.693 – biological half-life (in days), and Q_e = the whole-body radioactivity (in bequerels) in a steady-state equilibrium. Therefore, if the whole-body radioactivity is measured and the assimilation rate and the biological half-life are known, the feeding rate may be calculated. This approach has been developed by Crossley (1963a, b, 1966), who worked in an environment heavily contaminated with caesium-137. This isotope is almost completely assimilated, so that the term a can be dropped from equation (14.11). Crossley (1963b, c) found that the biological half-life appeared to be linearly related to body weight, except in the pupal stage, of course, when there was no elimination. However, as Odum and Golley (1963) point out and Hubbell *et al.* (1966) confirm, the biological half-life of an isotope in a given animal is variable and will be influenced by temperature, activity, food, and other factors, so that the half-life in the field is likely to be different from that determined in the laboratory.

Working on the isopod, *Armadillidium*, Hubbell *et al.* (1966) also show that the percentage of the isotope ingested that is actually assimilated will vary with the feeding rate; they used strontium-85 (^{85}Sr) and, with this biologically significant element (it is utilized in the exoskeleton), its assimilation rate paralleled the actual assimilation of nutriments from the food. The biological half-life of an isotope may be different in the two sexes, as Williams and Reichle (1968) found in a chrysomelid beetle.

In conclusion, therefore, before radioisotopes can be used to estimate feeding and assimilation rates in field populations, it is necessary to determine the following.
1 The variability of the rate of assimilation of the isotope under field conditions.
2 The variability of the biological half-life of the isotope under field conditions.

---

* This formula only applies to isotopes with long half-lives, so that the effective half-life = the biological half-life. Where radioactive decay is significant, then this loss rate needs to be added to the biological half-life.

3 Whether the level of isotope in the organism is low relative to the equilibrium level or whether it is high, a steady state having been reached. This is done from the equation already given, but some measure of r and a must be obtained experimentally, so that the expected equilibrium body burden of radio nuclide (Q_e) can be compared with that actually observed.

4 That, if the isotope is likely to be generally present in the media, as in studies with aquatic predators and labelled prey, it is only taken up by feeding. Dragonfly larvae appear to absorb zinc-65 (^{65}Zn) on their surface (Kormondy, 1965).

For biologically indeterminate (i.e. with no specific metabolic function) trace substances, whether radioactive or not, Fagerström (1977) considers that the biological half-life will be proportional to body weight raised to $(1-b)$, where b is the exponent relating body weight to metabolic rate, normally taken as 0.8.

Feeding (or ingestion) and assimilation can usually be separated, for, when an animal is fed on radioisotope-tagged food and the fall-off of radioactivity with time is measured, it will be found that, at first, the curve is steep, reflecting the loss of the ingested but non-assimilated component. Later, the curve becomes shallow, and this represents the elimination rate of the assimilated isotope (Odum & Golley, 1963; Fig. 14.2). Furthermore, Odum and Golley point out that the assimilated isotope is eliminated in different ways depending on the element, e.g. iodine mainly through the exuviae, zinc mainly by excretion, especially after moulting, or, with the female, in the eggs.

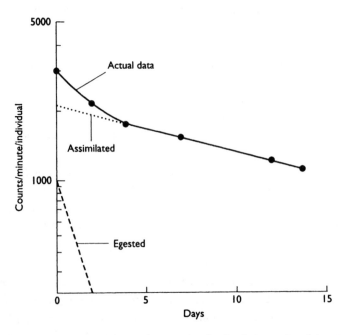

Fig. 14.2 The fall-off of radioactivity with time after ingestion showing the separation of the component that is assimilation from the component that is egested (Odum & Golley, 1963; data based on marine isopod, *Idothea*, ingesting seaweed tagged with zinc-65).

Calow and Fletcher (1972) developed a double-tracer method, in which the isotopes are used as a modification of the indicator technique (see below). They point out that the indicator and gravimetric methods assume that all material in the faeces is derived from the food, although this is not strictly true. By using two isotopes — one, carbon (^{14}C), that is biologically active and the other, chromium (^{51}Cr), that is not absorbed from the gut — they were able to allow for errors arising from this assumption. These two isotopes may be distinguished by their different radiations — β from ^{14}C (detected by liquid scintillation counting) and α from ^{51}Cr (detected by crystal scintillation counting). If their levels (counts/min (ct/min)) are determined in the food and faeces, then:

$$\%A = 1 - \frac{ct/min\ ^{51}Cr(food)}{ct/min\ ^{14}C(food)} \times \frac{ct/min\ ^{14}C(faeces)}{ct/min\ ^{51}Cr(faeces)} \times 100 \qquad (14.12)$$

^{14}C could be replaced by another absorbed isotope (e.g. hydrogen-3 (^{3}H)) and ^{51}Cr by another non-absorbed isotope.

The proper homogenization of material for accurate scintillation counting is sometimes difficult; with ^{14}C-labelled material, this may be overcome by combustion. Burnison and Perez (1974) describe a simple apparatus for this.

Gravimetric techniques. The simplest approach to the measurement of ingestion and one that is frequently applicable is to weigh the food before and after the animal has fed (e.g. Phillipson, 1960; Strong and Landes, 1965; Hubbell *et al.*, 1966; Lawton, 1970). If all values are reduced to dry weight, errors due to the loss of moisture during feeding will be eliminated; otherwise this must be determined from controls.

Aquatic predators may 'lose' part of the body contents of their prey during feeding. This would be most significant when the prey is chewed and is large relative to the predator; Dagg (1974) found that an amphipod could lose nearly 40% of its 'apparent' meal in this way. The quantity of leaf litter destroyed by soil arthropods may be determined by exposing a known quantity in mesh (e.g. nylon) bags (Crossley & Hoglund, 1962; Wiegert, 1974).

In the laboratory, the faeces may usually be collected and their dry weight and energy density found. Those of the faeces will, of course, differ from those of the food (Hubbell *et al.*, 1965) and the difference in energy content between ingestion and egestion may be more striking than the actual weight differences. The collection in the field of the total amount of faeces produced by a population is impossible (p. 320), but a measure could be obtained for a small number of arboreal larvae under semi-experimental conditions.

Hubbell *et al.* (1965) have compared the results of gravimetric and radiotracer methods using ^{85}Sr for determining the feeding and assimilation rates of the woodlouse, *Armadillidium*, and have found close agreement.

The only other weight that can be conveniently measured is the increase in weight of the animal (Fewkes, 1960; Evans, 1962); this is a measure of the amount of growth. When converted to dry weights the two equations hold:

Assimilation = Production + Respiration

Production = Growth + Exuviae and other products

Clearly, then, if respiration can be measured, growth + the weight of the products may be used to give an estimate of assimilation. If wet weights are used, allowance must also be made for the loss of weight due to the evaporation of water from the insect (Strong & Landes, 1965).

Just as the rate of assimilation of radioisotopes has been found to vary under different conditions (p. 517), so does the nutritional assimilation rate and the relation between assimilation and production. For example, Sushchenya (1962) found with the brine shrimp (*Artemia salina*) that the greater the intake of food, the smaller the percentage of it that is assimilated, but a slightly higher percentage of the assimilated energy was used for growth at the higher levels of food intake. Likewise, in the grasshopper, *Poecelocerus*, when food availability was lowered, a lower proportion of the assimilated energy was used for growth (Muthukrishnan & Delvi, 1974). Assimilation efficiencies are also influenced by temperature conditions (in poikilotherms) and even relatively small differences in food, e.g. slugs on lettuces and potatoes (Davidson, D.H. 1977). Thus caution needs to be exercised in the extrapolation of laboratory-determined values to field conditions.

Indicator methods. Assimilation can also be measured if the food can be marked with an indicator that is easily measured quantitatively, is non-toxic at the concentrations used, and is not absorbed by the gut. The percentage assimilation of the food is given by:

$$\%A = 1 - \left(\frac{\text{conc. indicator in food/unit dry wt.}}{\text{conc. indicator in faeces/unit dry wt.}} \right) \times 100 \qquad (14.13)$$

Chromic oxide has been widely used as an indicator in studies on vertebrate nutrition. Corbett *et al.* (1960) and McGinnis and Kasting (1964a, b) have shown how the chemical or, more conveniently, chromic oxide paper may be used with insects as an indicator to measure assimilation from finely divided food. As the indicator must be mixed homogeneously throughout the food, this method will be inappropriate in many instances, especially with natural diets, but Holter (1973) found it very suitable for measuring food consumption in the larvae of the dung beetle, *Aphodius*.

Measurement of faecal output. Provided the assimilation rate is not too sensitive to variations in food quantity, for which there is some evidence (Muthukrishnan & Delvi, 1974), faecal output should be related to consumption and, once standardized for the particular food, could be used for this purpose. Mathavan and Pandian's (1974) studies with various lepidopterous larvae suggest it is reliable in this group (see also p. 321). Lawton (1971) used gut-clearance times to give an independent estimate of consumption in the larva of a damsel fly and Humphreys (1975) obtained a measure of the food consumed in the field by wolf-spiders in the 14 days prior to capture from the guanine content of the excreta during the following 7 days.

14.2.3 The measurement of the energy loss due to respiration and metabolic process
14.2.3.1 Calorimetric
The amount of energy lost through respiration may be determined directly by the measurement of the heat given out by the animal but, as some energy will have been used to vaporize water from the animal, this will be less than the total energy loss and a correction must be applied. It is difficult to work with small animals; few studies have been made with insects using this approach (Heinrich, 1972).

14.2.3.2 The exchange of respiratory gases
The energy equivalents of oxygen and carbon dioxide. During metabolic processes, in which energy is liberated, oxygen is utilized and carbon dioxide produced; the proportion of carbon dioxide evolved to oxygen used depends on the actual metabolic process and is referred to as the respiratory quotient (RQ), which is expressed as:

$$RQ = \frac{\text{Carbon dioxide produced}}{\text{Oxygen utilized}}$$

Thus, if two of the three values are known, the third may be found; the gases may be measured by a number of different techniques and the respiratory quotient by noting the change in volume at constant pressure and temperature when the animal respires in an enclosed space.

The amount of energy liberated in the process can be determined from the quantity of oxygen (or carbon dioxide) involved, if the respiratory quotient is known and certain assumptions are made as to the actual metabolic processes. If carbohydrates alone are being utilized for energy RQ = 1, $0.198\,cm^3$ (at standard temperature and pressure (STP)) of oxygen will be consumed, and the same volume of carbon dioxide will be produced during the liberation of one calorie of energy (1 kcal = 4.18 kJ); with fats alone, RQ = 0.707 and then the consumption of $0.218\,cm^3$ of oxygen and the production of $0.154\,cm^3$ of carbon dioxide will result from the liberation of one calorie. Tables are widely available of the energetic equivalents of unit volumes of oxygen and carbon dioxide for different RQs, assuming that carbohydrate and fat utilization is the only metabolic process involved (Table 14.2). Ecologists have frequently approximated by using the RQ obtained from experiments as equivalent to a 'nitrogen (N) free RQ', disregarding the effect of protein and other metabolisms.

If greater accuracy is to be obtained, the quantity of nitrogen excreted must be measured (Shaw & Beadle, 1949), enabling the component of the gases due to protein catabolism to be established and the 'N-free RQ' determined. The respiratory exchange involved in the production of 1 g of nitrogen from the catabolism of a protein will depend on the proportions by weight of the various elements in its molecule; Kleiber (1961) explains the necessary calculations.

Table 14.2 The caloric equivalents of oxygen and carbon dioxide for various values of RQ, due to utilization of different proportions of carbohydrates and fats (modified from an extensive table in Brody, 1945; original data from Zuntz & Schumberg).

RQ	Oxygen kcal*/litre	Carbon dioxide kcal*/litre	% oxygen consumed by carbohydrate component
0.70	4.686	6.694	0.0
0.75	4.729	6.319	14.7
0.80	4.801	6.001	31.7
0.85	4.683	5.721	48.8
0.90	4.924	5.471	65.9
0.95	4.985	5.247	82.9
1.00	5.047	5.047	100.0

* 1 kcal = 4.18 kJ

However, even this approach, which accounts for fat, carbohydrate, and protein catabolism, does not include all the processes contributing to the RQ. Many others may theoretically influence the RQ and, as these have RQ values outside the 0.7–1.0 range, small amounts may modify the composite RQ disproportionately; e.g. alcohol breakdown has an RQ of 0.67, the synthesis of fat from carbohydrate an RQ of 8.0 (Kleiber, 1961).

It is discouraging for the ecologist to note that Cahn (1956) and Kleiber (1961) report studies that show that the estimate of energy expended obtained indirectly (from the respiratory quotient) may be found to deviate by as much as 25% from the true value calculated by direct calorimetry. As Macfadyen (1963) has pointed out, when doubt exists as to the exact catabolic process, but, assuming that these involve only fats, carbohydrates, and proteins, errors will be reduced if calculations are based on the oxygen uptake rather than on the carbon dioxide produced. This is because the minimum volume of oxygen that is required to produce 1 cal is 90.8% of the maximum volume, but the minimum volume of carbon dioxide that results from the production of 1 cal is as little as 77.7% of the maximum.

The respiratory rate. Starved warm-blooded animals exhibit a level of respiration referred to as the basic metabolism, but there is no comparable standard available for poikilothermic animals. For fish, which have been the subject of much energetics research, energy respired is divided between resting metabolism (energy consumed on tissue maintenance), activity metabolism (energy consumed in movement), and feeding metabolism (energy used digesting food). For cod, *Gadus morhua*, 10–17% of the ingested energy was used for digestion, with the cost increasing with tempera-ture (Soofiani & Hawkins, 1982). Fry (1947) defined the difference between basal and active metabolic rates as the 'scope for activity'. It is important to remember that, although respiration rate and body weight show in general a linear relationship

(Brody, 1945; Engelmann, 1961), with poikilotherms the rate may be varied by many factors: temperature, age, season, oxygen concentration, and even the type of respirometer used for the measurement (Allen, 1959; Golley & Gentry, 1964; Cossins & Bowler, 1987; Hack, 1997; Lyytikainen & Jobling, 1998). When respiration rate is being related to the weight of an invertebrate, it is necessary to eliminate individual variations in weight due to the contents of the gut (Allen, 1959).

The measurement of the energy expenditure or respiration of animals while living naturally has proved particularly difficult. For terrestrial animals, the doubly labelled water technique has been successful (Nagy, 1994). This technique cannot be applied to aquatic organisms such as fish, where one potential approach for the quantification of field metabolic activity is to monitor the physiology of a free animal using physiological telemetry. Information about the physiological activity of the fish is sent using either radio or ultrasonic signals. Metabolic activity has been estimated using heart rate (Lucas *et al.*, 1993), tail beat (Ross *et al.*, 1981), opercular muscle electromyograms (Rogers & Weatherley, 1983), and axial muscle electromyograms (Briggs & Post, 1997). All of these measurements give information on the activity of the animal, which can be converted to energy expenditure using relationships obtained from laboratory experiments. However, the trauma of capture and attachment of transmitting packages to animals may disturb their natural metabolism.

Gas analysis. The precise analysis of the respiratory gas to determine the proportions of oxygen and carbon dioxide is described in detail in textbooks such as Kleiber (1961) and Kay (1964). The methods are extensively reviewed in Petrusewicz and MacFadyen (1970), Edmondson and Winberg (1971), and Grodzinski *et al.* (1975); only a general outline will be given here; normally it is necessary to modify the precise form of the instrument to suit the particular animal and investigation. The main types of respirometer, for both the gaseous phase and dissolved gases, are illustrated in Fig. 14.3; some comparisons of their performance are given in Table 14.3; detailed comparative data are given by Lawton and Richards (1970).

Isotopes. McClintock and Lifson (1958) showed that the total carbon dioxide production of an animal could be determined by using double-labelled water. The heavy isotope of hydrogen is eliminated only as water, but that of oxygen (^{18}O) is eliminated both as carbon dioxide and as water; the difference between the two elimination rates will give a measure of the carbon dioxide evolved. This method is of great value to the ecologist for the determination of respiratory rates under natural conditions; the animals can be fed with double-labelled water, released in the field, recaptured at some later time, and the levels of the isotopes determined by mass spectrometry (neutron activation (p. 119)) and scintillation counting. This technique has been used, for example, with birds (Le Febvre, 1964; Utter & Le Febvre, 1973) and lizards (Nagy, 1975; Bennett & Nagy, 1977). The experimental approach is illustrated by a study on three desert rodents, *Acomys russatus*, *Acomys cahirinus*, and

Sekeetamys calurus, undertaken by Degan *et al.* (1986). The study animals were injected with water containing 97 atoms % oxygen-18 (^{18}O) and $50\,\mu Ci$ tritium/ml. They were retained in captivity for 2 h to allow this water to mix with their body fluids, a blood sample was taken, and they were returned to the wild. Three days later they were recaptured and a second blood sample collected. The water was extracted from the blood samples by microdistillation, and then the tritium and ^{18}O levels were determined by liquid scintillation and gamma-Romatic counting, respectively. Total water volume of each animal was calculated by the dilution of the ^{18}O, water flux by the decline in tritium specific activity, and carbon dioxide (CO_2) production by the declines through time of specific activities of tritium and ^{18}O (Lifson & McClintock, 1966).

Kam and Degen (1997) present a modelling approach that allows the estimation of all energy-budget components in free-living animals using data supplied from a doubly labelled water experiment.

Analysis in the gaseous phase for air-breathing animals. The animal is enclosed in a chamber and the changes caused by its respiration in the composition of the air in the chamber determined in some way. Several techniques are based on the absorption of the evolved carbon dioxide by an alkali solution (e.g. sodium hydroxide), so that the pressure or volume of the air will change proportionally to the consumed oxygen. In constant-volume respirometers, e.g. the Warburg (Dixon, 1951), the reduction in pressure is measured with a manometer.

Alternatively, the pressure can be kept constant and the volume allowed to change. This change can be measured, for example, by the movement of an oil droplet in a capillary tube; Smith and Douglas (1949) developed an apparatus of this type that was subsequently modified by Engelmann (1961), Wiegert (1964a), and Davies (1966). The change in volume can also be utilized in ultramicrorespirometers, which operate on the principle of the Cartesian diver (Zeuthen, 1950; Kay, 1964; Wood & Lawton, 1973).

The amount of carbon dioxide evolved may be determined by alkali titration (Itô, 1964). Arlian (1973) describes methods of construction. An experiment using these types of respirometer cannot be continued for long, as the respiration may become abnormal as an increasing proportion of the available oxygen is utilized.

If the respiratory rate is to be recorded over a period of time, the oxygen utilized by the animal must be replaced. A number of electrolytic respirometers have been developed in which the change in volume resulting from the consumption of oxygen switches on an electrolysis apparatus that generates oxygen; when the previous

Fig. 14.3 (*Opposite.*) Summary sketches of the main types of respirometer (other than calorimeters) (after Petruzewicz & Macfadyen, 1970). (a) Constant-volume Warburg—open. (b) Compensating with mechanical restoration. (c) Simple electrolytic—open. (d) Compensating automatic. (e) Cartesian diver. (f) Gradient diver. (g) Kopf's method of CO_2 determination. (h) Conway microdiffusion method. (i) Constant pressure.

O_2

(a)

(b)

Hg

(c)

O_2 H_2

H_2SO_4

C   X

O_2^+

CuSO$_4$ Cu

(d)

(e)

C

Density gradient e.g. sucrose solution

(f)

NaCl → NaOH

Ag   Pt

mA

(g)

HCL

(h)

X

(i)

C

General key:

X = animal
C = compensating chamber

= Scale which is read

= alkali

= Manometer or indicating fluid

= other fluids

Table 14.3 Comparison of the performance of respirometers (from Petrusewicz & Macfadyen, 1970).

(1) Respirometer	(2) Measures O$_2$, CO$_2$, cals	(3) Container volume (ml)	(4) Normal rates (µl/h)	(5) Normal accuracy	(6) Automation	(7) Sensitivity: smallest detectable change (µl)
WARBURG Normal	O$_2$ and CO$_2$	25	10–500	2 µl/hr	N	0.5
Small vessels	O$_2$ and CO$_2$	5	2–50	1 µl/hr	N	0.5
GILSON Normal	O$_2$ and CO$_2$	16	5–500	2 µl/hr	P	0.5
DIVER Holter	O$_2$ and CO$_2$	0.005–0.05	0.01–1.0	5%	P	0.0001
Gregg and Lints	O$_2$ and CO$_2$	0.005–0.05	0.01–1.0	5%	Y	0.005
Gradient	O$_2$ and CO$_2$	0.001–0.05	0.01–1.0	3%	Y	0.0001
PHILLIPSON automatic	O$_2$ and CO$_2$	50	50–300	8%	Y	10
MACFADYEN electrolytic	O$_2$ and CO$_2$	7	1–100	1 µl/h	Y	0.01
Capillary const. vol.	O$_2$	1.0	2.0 ca	5%	P	0.05
Const. pressure	O$_2$	1–10	0.1–5	2%	N	0.05
CONWAY microdiffusion	CO$_2$	20	50–1000+	20 µl/h	N	20
JENSEN CO$_2$ diffusion	CO$_2$	10	1.0 ca	0.05 µl/h	Y	—
CALVET microcalorimeter	heat	15–100	20–1000+	1%	Y	0.2
WINKLER (dissolved oxygen)	O$_2$	10	100 ca	2%	N	30

Notes: All quantities are converted to equivalents of oxygen volumes, regardless of the parameter actually measured.

(5) Accuracy is expressed as the expected standard error of a series of readings over the normal range or, where accuracy declines at low ranges, as the probable minimum detectable rate.

(6) The column 'Automation' indicates whether application of automatic recording is possible (Y), inherently impossible (N), or potentially possible (P).

volume has been restored, the current is switched off. The activity of the compensating oxygen generator is recorded and, after suitable calibration, this may be converted to give the volume of oxygen utilized (MacFadyen, 1961; Phillipson, 1962; Annis & Nicol, 1975; Arnold & Keith, 1976). In all these volumetric or manometric methods, constant temperature must be maintained throughout the experiment, usually by using a water-bath. Under certain circumstances, the production of ozone is a hazard.

The quantities of carbon dioxide and oxygen in the air may also be assessed by methods utilizing differences in the thermal conductivity, viscosity, magnetic susceptibility, and other properties of the gases (Kleiber, 1961; Kay, 1964).

Analysis of dissolved gases for completely aquatic species. The animal must be enclosed in a limited volume of water out of contact with the atmospheric air. Studies can be undertaken using closed systems in which the study organism is sealed into a chamber in which the gradual decline in oxygen concentration is measured (e.g. Paul *et al.*, 1988). Care must be taken with this approach to ensure that the oxygen level does not drop to levels that suppress respiration. The alternative approach is to use a continuous-flow respirometer, in which the oxygen concentration of the water flowing into and out of the chamber holding the animal is measured (e.g. Yammamoto *et al.*, 1998). Figure 14.4 gives diagrams of both types of experimental apparatus.

The gases dissolved in the water can be determined by a number of methods. Chemical methods, especially the Winkler method, have been widely used (e.g. Teal, 1957; Richman, 1958; Beyers & Smith, 1971; Yammamoto *et al.*, 1998); the procedures of the Winkler and another titration technique, the phenosafranine, are described by Dowdeswell (1959). The accuracy of these can, however, be impaired by the presence of ferrous iron, nitrites, and other 'impurities' in the water (Allee & Oesting, 1934).

Polarometry or polarography, in which the concentration of dissolved oxygen is measured electrolytically, has also been used in many ecological studies and is described by Kay (1964). Oxygen electrodes that are particularly suitable for ecological work were developed by Carey and Teal (1965) and are now widely available from commercial manufacturers. They may also be used to measure the oxygen concentration in aquatic environments.

A tonometric method has been developed by Jones (1959) for use with very small volumes of fluid, and is thus more applicable to samples of fluid from within animals than from without! A small bubble of air is enclosed with the fluid and the carbon dioxide, oxygen, and nitrogen in it allowed to reach diffusion equilibrium with these gases in the fluid. The composition of the bubble can then be analysed.

Gasometric methods involve the expulsion of the gases from the fluid and their subsequent analysis. Van Slyke and Neill (1924) extracted the gases under vacuum and the quantities of both oxygen and carbon dioxide can be determined (Kay, 1964). Only the oxygen and nitrogen can be measured by the method of Scholander

Fig. 14.4 Analysis of dissolved gases for completely aquatic species. (a) Closed system (after Omori & Ikeda, 1984). (b) Continuous-flow respirometer (after Brafield & Llewellyn, 1982).

et al. (1955), in which the gases are expelled from the sample by the addition of acid and a carbonate.

14.3 The energy budget, efficiencies, and transfer coefficients

The final energy budget may be presented in various forms; emphasis may be placed on the population studied (e.g. Lawton, 1970, 1971; McNeill, 1971), on trophic links (e.g. Gyllenberg, 1970; Shure, 1973), or on biomass turnover and features of the ecosystem (e.g. Teal, 1962; Jonasson & Kristiansen, 1967; Burky, 1971). Budgets are generally expressed in joules (J) or kilocalories (kcal) (1 kcal = 4186 J) or carbon (1 g C ~ 0.5357 kcal).

14.3.1 The energy budget of a population (or trophic level)
A graphic representation of the energy budget of a caddis-fly population is given by

Fig. 14.5 The energy budget of a population of the caddis-fly, *Potamophylax cingulatus*, calculated on a monthly basis in terms of cal/m² (after Otto, 1975).

Otto (1975) (Fig. 14.5). This budget is noteworthy as it emphasizes that a system is seldom without 'imports' (e.g. imagines from downstream and other streams) and 'exports' (e.g. predation of the adults by terrestrial animals that move away from the area). Furthermore, trophic links may cross habitat barriers (e.g. the predation of the larvae by crows).

Energy budgets can usefully be summarized and compared if the efficiencies of various processes are calculated. There have been many terms applied to these (Kozlovsky, 1968), but the following, which largely follow Wiegert (1964a), are now recognized as standard (Petrusewicz & Macfadyen, 1970) (the notation follows equation (14.1) at the start of this chapter):

Assimilation/consumption	*A/C*
Production/'assimilation' or 'net' production efficiency	*P/A*
Production/consumption efficiency	*P/C*
Production/respiration ratio	*P/R*

As Wiegert (1964a) emphasized, the terms (numerator and denominator) in these ratios should be precisely defined and the ratios described by the symbols (e.g. *P/R* ratio) or by a name that indicates these. The efficiencies, which represent energy transfers within organisms, are usually expressed as percentages. Wiegert

Table 14.4 Some energy efficiencies for populations of different invertebrates showing inter- and intra-specific variations.

Organism		Efficiencies			Source
		A/C%	P/A%	P/C%	
Predators					
Pyrrhosoma (Odonata)	Year 1	90	52	47	Lawton, 1971
	Year 2	88	48	42	Lawton, 1971
Phonoctonus (Het.)				c.45	Evans, 1962
Herbivores					
Leptopterna (Het.)	Year 1	29	58	17	McNeill, 1971
	Year 2	36	51	18	McNeill, 1971
	Year 3	36	52	19	McNeill, 1971
	Year 4	34	59	20	McNeill, 1971
	Year 5	28	55	16	McNeill, 1971
Rhynchaenus (Col.)	Larva I	24		20	Grimm, 1973
	Larva II	24		17	Grimm, 1973
	Larva III	20		14	Grimm, 1973
Melasoma (Col.) la.		c.72	35	24	Hågvar, 1975
Danaus (Lep.)			45	21	Schroeder, 1976
Euchaetias (Lep.)			49	20	Schroeder, 1977
Potamophylax (Trich.)		19	45	8	Otto, 1975
Saprophytic/microbial feeders					
Aphodius (Col.)		8	51	4	Holter, 1975
Pelodera (Nematoda)		60	38	22	Marchant & Nicholas, 1974
Brachionus (Rotifer) adult		19	57	11	Doohan, 1973

and Evans (1967) published an interesting table of P/C efficiencies; some values for other invertebrates are given in Table 14.4. The year-to-year variations (Gyllenberg, 1970; McNeill, 1971; Lawton, 1971) in the same population, which give an indication of the 'condition' of the population, its food supply, and environment, show that too much stress must not be placed on small differences. However, not surprisingly, herbivores have lower P/C values than predators and homotherms than poikilotherms. The distinction between the two latter groups is clearly shown when known P/R ratios are plotted; there are two linear relationships, the homotherms having a higher respiratory expenditure for the same production (McNeill & Lawton, 1970; Wiegert, 1976).

14.3.2 Energy transfer across trophic links

The pathways of energy transfer in an ecosystem are logically considered against a flow model of the type proposed by Wiegert and Owen (1971) (Fig. 14.6). The energy flow within the 'boxes' is described by the efficiencies or ratios listed above.

The rate at which energy moves between 'boxes' is best described as:

Production turnover rate $P_t/B = \Theta_P$

Elimination turnover rate $E_t/B = \Theta_E$

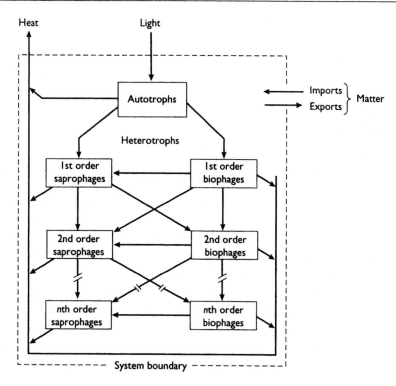

Fig. 14.6 Energy transfer in an ecosystem: dioristic model of Wiegert and Owen (1971).

where B = the average biomass over time unit t and P and E are the total production and elimination values over time. These two rates (which are, of course, related; see equation (14.4)) express the rate at which the production of the population (or trophic level) is being passed on to the rest of the ecosystem. The transfer of energy across particular links between boxes (e.g. between the autotroph and the first-order biophage (or herbivore)) may be expressed as:

Harvest/Production transfer coefficient $= E_t'' / P_t'$

Yield/Production transfer coefficient $= C'' / P_t'$

where E_t'' = the total biomass of the 'donor' (prey) eliminated by the next-level biophage during time t, P_t' = total production of the donor during this time period, and C_t'' = total consumption by the biophage or saprophage of the donor during period t. It will be noted that, for the links to the biophages (the boxes on the right of Fig. 14.6), there are two transfer coefficients. The 'harvest transfer coefficient' represents the removal of production to the biophage, i.e. the kill or cull by a predator. However, as Dagg (1974) found, a predator, especially if its prey is large relative to it, may only consume about half the energy it harvests. Its actual consumption, the yield to it—the flow across one of the links between the boxes in Fig. 14.6—is given by the 'yield transfer coefficient'.

It is important to make these distinctions between the efficiencies within organisms, the rates at which production is passed on through trophic links, the impact (in energy terms) of one trophic level on that below it (harvest transfer coefficient), and the actual energy flow along a particular link (yield transfer coefficient). Ratios other than those given here may be useful in a particular context, but it should always be made clear whether the energy flow described is within a trophic level or between levels, or spans both levels and links.

14.4 Identification of ecological pathways using stable isotopes

Both animals and plants can show selective uptake for the different isotopes of a number of elements and the characteristic isotope ratios produced can be transferred to their consumers. While unable to give precise information on the food sources used by an individual, isotope ratios can be used to infer the major sources of energy for an ecological group. An example of the application of these methods is the study of Forsberg *et al.* (1993) on the carbon sources of Amazonian fish. Flood-plain fish may enter the forest during the high-water season, so that it is unclear how much of their carbon was obtained from forest sources and how much from the extensive floating grasslands. These grasses (principally *Echinochloa polystachya* and *Paspalum repens*) use the C_4 photosynthetic pathway and contribute an estimated 52% of total primary production in the flood-plain. The forest trees, periphyton, and phytoplankton, which together produce the remaining 48% of production, use the C_3 pathway. Plants using the C_3 and C_4 pathways differ considerably in their carbon isotope ratio, because C_3 plants select the lighter ^{12}C isotope. Using the $^{13}C/^{12}C$ ratio of 35 fish species as measured using a mass spectrometer, it was found that C_4 macrophytes accounted for only 2.5–17.6% of the carbon in fish. The major source of fish carbon was phytoplankton, which only contributed 8% to the total flood-plain primary production but, surprisingly, 32% of the carbon in fish. This counter-intuitive finding would have been almost impossible to obtain by other means, because of the difficulty of identifying detritus from fish guts and because of the logistic problems posed by the spatial extent, instability, and biodiversity of the habitat.

14.5 Assessment of energy and time costs of strategies

Ecologists and evolutionary biologists have become increasingly interested in the quantitative comparison of different strategies used by animals for activities such as foraging or reproduction. Energy must be used in order to acquire food and the efficiency of this investment is the subject of optimal foraging theory (e.g. Townsend & Winfield, 1985). Strategies can be compared between species and, given polymorphism, within a species. Stam *et al.* (1996), for example, compared the strategies of different clones of the parthenogenetic collembolan *Folsomia candida*. The ultimate criterion of strategic success, of course, is reproductive success, but valuable indications of the relative costs and benefits can be obtained from the consideration of time and energy budgets (Emlen, 1966; Schoener, 1971; Van Valen, 1976; Norberg, 1977; Wells & Clarke, 1996). The methods utilized will often be similar to those described above, e.g. measurement of respiratory rate, energy value of 'harvest', but these will have to be related to particular activities. In essence the following are usually required.

1 A time budget of the various activities.

2 The rate of energy expenditure characteristic of each type of activity.

3 The rate of energy-gain food harvested or usually yielded (food ingested) for each type of activity.

Studies of this type have been made on territoriality and resources (e.g. McNab, 1963; Smith, 1968; Gill & Wolf, 1975a, b), on speed and pattern of movement in relation to resource availability (e.g. Calow, 1974), and on foraging strategies (e.g. Mukerji & Le Roux, 1969; Heinrich, 1972; Charnov, 1976; Davidson, D.W., 1977).

References

Allee, W.C. & Oesting, R. (1934) A critical examination of Winkler's method for determining dissolved oxygen in respiration studies with aquatic animals. *Physiol. Zool.* 7, 509–41.

Allen, K.R. (1951) The Horokiwi Stream, a study of a trout population. *NZ Mar. Dep. Fish. Rep. DIV Bull.* 10, 1–238.

Allen, M.D. (1959) Respiration rates of worker honeybees of different ages and at different temperatures. *J. Exp. Biol.* 36, 92–101.

Annis, P.C. & Nicol, C.R. (1975) Respirometry system for small biological samples. *J. Appl. Ecol.* 12, 137–41.

Arlian, L.G. (1973) Methods for making a Cartesian diver for use with small arthropods. *Ann. Entomol. Soc. Am.* 66, 694–5.

Arnold, D.I. & Keith, D.E. (1976) A simple continuous-flow respirometer for comparative respirometry changes in medium-sized aquatic organisms. *Water Res.* 10, 261–4.

Atkinson, E.G. & Wacasey, J.W. (1983) Caloric equivalents for benthic marine organisms from the Canadian Arctic. *Can. Tech. Rep. Fish. Aquat. Sci.* 1216, 32.

Bailey, C.G. & Riegert, P.W. (1972) Energy dynamics of *Encoptolophus sordidus cortalis* (Scudder) (Orthoptera: Acrididae) in a grassland ecosystem. *Can. J. Zool.* 51, 91–100.

Baird, D. & Milne, H. (1981) Energy flow in the Ytham Estuary, Aberdeenshire, Scotland. *Estuar. Coastal Shelf Sci.* 13, 455–72.

Benke, A.C. (1976) Dragonfly production and prey turnover. *Ecology* 57, 915–27.

Bennett, A.F. & Nagy, K.A. (1977) Energy expenditure in free-ranging lizards. *Ecology* 58, 697–700.

Beyers, R.I. & Smith, M.H. (1971) A calorimetric method for determining oxygen concentration in terrestrial situations. *Ecology* 52, 374–5.

Boyd, C.E. (1969) The nutritive value of three species of water weeds. *Econ. Bot.* 239, 123–7.

Boyd, C.E. & Blackburn, R.D. (1970) Seasonal changes in the proximate composition of some common aquatic weeds. *Hyacinth Contr. J.* 8, 42–144.

Boyd, C.E. & Goodyear, C.P. (1971) Nutritive quality of food in ecological systems. *Arch. Hydrobiol.* 69, 256–70.

Brafield, A.E. (1985) Laboratory studies of energy budgets. In: Tytler, P. & Calow, P. (eds) *Fish Energetics: New Perspectives.* Croom Helm, London, pp. 257–81.

Brafield, A.E. & Llewellyn, M.J. (1982) *Animal Energetics.* Chapman & Hall, London.

Brey, T., Rumohr, H., & Ankar, S. (1988) Energy content of macroinvertebrates: general conversion factors from weight to energy. *J. Exp. Mar. Biol. Ecol.* 117, 217–78.

Briggs, C.T. & Post, J.R. (1997) Field metabolic rates of rainbow trout estimated using electromyogram telemetry. *J. Fish Biol.* 51, 807–23.

Brody, S. (1945) *Bioenergetics and Growth.* Hafner, New York, 1023 pp.

Brown, D.S. (1961) The food of the larvae of *Chloeon dipierum* L. and *Baetis rhodani* (Pictet) (Insecta, Ephemeroptera). *J. Anim. Ecol.* 30, 55–75.

Burky, A.J. (1971) Biomass turnover, respiration

and interpopulation variation in the stream limpet *Ferrissia rivularis* (Say). *Ecol. Monogr.* **41**, 235–51.

Burnison, B.K. & Perez, K.T. (1974) A simple method for the dry combustion of ^{14}C-labelled materials. *Ecology* **55**, 899–902.

Cahn, R. (1956) *La Régulation des processus métaboliques dans l'organisme*. Hermann, Paris.

Calow, P. (1974) Some observations on locomotory strategies and their metabolic effects in two species of freshwater gastropods, *Anylus fluviatilis* and *Planorbis contortus* Linn. *Oecologia* **16**, 149–61.

Calow, P. & Fletcher, C.R. (1972) A new radio tracer technique involving ^{14}C and ^{51}Cr for estimating the assimilation efficiencies of aquatic primary consumers. *Oecologia* **9**, 155–70.

Carey, F.G. & Teal, J.M. (1965) Responses of oxygen electrodes to variables in construction, assembly and use. *J. Appl. Physiol.* **20**, 1074–7.

Caspers, N. (1977) Seasonal variations of caloric values in herbaceous plants. *Oecologia* **269**, 379–83.

Chaabane, K., Loreau, M., & Josens, G. (1996) Individual and population energy budgets of *Abax ater* (Coleoptera, Carabidae). *Ann. Zool. Fennici* **33**(1), 97–108.

Chapman, S.B. (1976) Production ecology and nutrient budgets. In: Chapman, S.B. (ed.) *Methods in Plant Ecology*. Blackwell Scientific Publications, Oxford, pp. 157–228.

Charnov, E.L. (1976) Optimal foraging: attack strategy of a mantid. *Am. Nat.* **110**, 141–51.

Coleman, D.C. (1968) Food webs of small arthropods of a broom sedge field studied with radio-isotope-labelled fungi. In: *Proceedings of IBP Technical Meeting on Methods of Study in Soil Zoology*. UNESCO, Paris, pp. 203–7.

Comita, G.W. & Schindler, D.W. (1963) Calorific values of microcrustacea. *Science* **1409**, 1394–6.

Corbett, J.L., Greenhalgh, F.D., McDonald, I., & Florence, E. (1960) Excretion of chromium sesquioxide administered as a component of paper to sheep. *Br. J. Nutr.* **14**, 289–99.

Cossins, A.R. & Bowler, K. (1987) *Temperature Biology of Animals*. Chapman & Hall, London.

Crossley, D.A. (1963a) Movement and accumulation of radiostrontium and radiocaesium in insects. In: Schultz, V. & Klement, A.W. (eds) *Radioecology*. Rheinhold, New York, pp. 103–5.

Crossley, D.A. (1963b) Consumption of vegetation by insects. In: Schultz, V. & Klement, A.W. (eds) *Radioecology*. Rheinhold, New York, pp. 427–30.

Crossley, D.A. (1963c) Use of radioactive traces in the study of insect–plant relationships. In: *Radiation and Radioisotopes Applied to Insects of Agricultural Importance*. ST1/PUB 74, International Atomic Energy Ag., pp. 43–54.

Crossley, D.A. (1966) Radio-isotope measurement of food consumption by a leaf beetle species. *Chrysomela knabi* Brown. *Ecology* **47**, 1–8.

Crossley, D.A. & Hoglund, M.P. (1962) A litter-bag method for the study of micro-arthropods inhabiting leaf litter. *Ecology* **43**, 571–3.

Dagg, M.J. (1974) Loss of prey body contents during feeding by an aquatic predator. *Ecology* **55**, 903–6.

Davidson, D.H. (1977) Assimilation efficiencies of slugs on different food materials. *Oecologia* **26**, 267–73.

Davidson, D.W. (1977) Foraging ecology and community organization in desert-seed eating ants. *Ecology* **58**, 725–37.

Davies, P. (1966) A constant pressure respirometer for medium-sized animals. *Oikos* **179**, 108–12.

Degan, A.A., Kam, M., Hazan, A., & Nagy, K.A. (1986) Energy expenditure and water flux in three sympatric desert rodents. *J. Anim. Ecol.* **55**, 421–9.

Dixon, M. (1951) *Manometric Methods as Applied to the Measurement of Cell Respiration and Other Processes*. Cambridge University Press, Cambridge.

Doohan, M. (1973) An energy budget for adult *Brachionus plicatilis* Muller (Rotatoria). *Oecologia (Berlin)* **13**, 351–62.

Dowdeswell, W.H. (1959) *Practical Animal Ecology*. Methuen, London.

Edmondson, W.T. & Winberg, G.G. (eds) (1971) *A Manual on Methods for the Assessment of Secondary Productivity in Fresh Waters*. IBP Handbook, Blackwell Scientific Publications, Oxford.

Edmunds, P.J. & Davies, P.S. (1989) An energy budget for *Porites porites* (Scleractinia), growing in a stressed environment. *Coral Reefs* **8**(1), 37–44.

Edwards, C.A., Reichle, D.E.C., & Rossley, D.A.J. (1969) Experimental manipulation of soil invertebrate populations for trophic studies. *Ecology* **50**, 495–8.

Emlen, J.M. (1966) The role of time and energy in food preferences. *Am. Nat.* **100**, 611–17.

Engelmann, M.D. (1961) The role of soil arthropods in the energetics of an old field community. *Ecol. Monogr.* **31**, 221–38.

Evans, D.E. (1962) The food requirements of *Phonoctonus nigrofasciatus* Stal (Hemiptera, Reduviidae). *Entomol. Exp. Appl.* **5**, 33–9.

Fagerström, T. (1977) Bodyweight, metabolic rate and trace substance turnover in animals. *Oecologia* **29**, 99–116.

Fewkes, D.W. (1960) The food requirements by

weight of some British Nabidae (Heteroptera). *Entomol. Exp. Appl.* **3**, 231–7.

Forsberg, B.R., Araujo-Lima, C.A.R.M., Martinelli, L.A., Victoria, R.L., & Bonsassi, J.A. (1993) Autotrophic carbon sources for fish of the central Amazon. *Ecology* **74**, 643–52.

Frolander, H.F. (1957) A plankton volume indicator. *J. Cons. Perm. Int. Explor. Mer* **229**, 278–83.

Fry, F.E.J. (1947) Effects of environment on animal activity. *Univ. Toronto St. Biol. Ser.* **55**, 1–62.

Gallucci, V.F. (1973) On the principles of thermodynamics and ecology. *Ann. Rev. Ecol. Syst.* **4**, 329–57.

Gill, F.B. & Wolf, L.L. (1975a) Economics of feeding territoriality in the golden winged sunbird. *Ecology* **56**, 333–45.

Gill, F.B. & Wolf, L.L. (1975b) Foraging strategies and energetics of cast African sunbirds at mistletoe flowers. *Am. Nat.* **109**, 491–510.

Golley, F.B. (1960) Energy dynamics of a food chain of an old-field community. *Ecol. Monogr.* **30**, 187–206.

Golley, F.B. & Buechner, H.K. (eds) (1968) *A Practical Guide to the Study of the Productivity of Large Herbivores.* IBP Handbook, Blackwell Scientific Publications, Oxford.

Golley, F.B. & Gentry, J.B. (1964) Bioenergetics of the southern harvester ant, *Pogonomyrmex badius. Ecology* **45**, 217–25.

Gosler, A.G. (1996) Environmental and social determinants of winter fat storage in the great tit *Parus major. J. Anim. Ecol.* **65**, 1–17.

Griffiths, D. (1977) Caloric variation in Crustacea and other animals. *J. Anim. Ecol.* **465**, 593–605.

Grimm, R. (1973) Zum energieumsatz phytophager Insekten im Buchenwald I. *Oecologia* **11**, 187–262.

Grodzinski, W., Klekowski, R.Z., & Duncan, A. (eds) (1975) *Methods for Ecological Bioenergetics.* IBP Handbook, Blackwell Scientific Publications, Oxford.

Gyllenberg, G. (1969) The energy flow through a *Chorippus parallelus* (Zett.) (Orthoptera) population in a meadow in Tyürmine, Finland. *Acta Zool. Fenn.* **123**, 1–74.

Gyllenberg, G. (1970) Energy flow through a simple food chain of a meadow ecosystem in four years. *Ann. Zool. Fennici* **7**, 283–9.

Hack, M.A. (1997) The effects of mass and age on standard metabolic rate in house crickets. *Physiol. Entomol.* **22**(4), 325–31.

Hagvar, S. (1975) Energy budget and growth during the development of *Melasoma collaris* (Coleoptera). *Oikos* **26**, 140–6.

Hanna, H.M. (1957) A study of the growth and feeding habits of the larvae of four species of caddis flies. *Proc. Roy. Entomol. Soc. Lond. A* **329**, 139–46.

Hartman, K.J. & Brandt, S.B. (1995) Estimating energy density of fish. *Trans. Am. Fish. Soc.* **124**, 347–55.

Heinrich, B. (1972) Energetics of temperature regulation and foraging in a bumblebee, *Bombus terricola* Kirby. *J. Comp. Physiol.* **77**, 48–64.

Henderson, P.A. & Walker, I. (1986) On the leaf-litter community of the Amazonian blackwater stream Tarumazinho. *J. Trop. Ecol.* **2**, 1–17.

Henen, B.T. (1997) Seasonal and annual energy budgets of female desert tortoises (*Gopherus agassizii*). *Ecology* **78**(1), 283–96.

Hinton, J.M. (1971) Energy flow in a natural population of *Neophilaenus litlealus* (Homoptera). *Oikos* **22**, 155–71.

Holter, P. (1973) A chromic oxide method for measuring consumption in dung-eating *Aphodius* larvae. *Oikos* **24**, 117–22.

Holter, P. (1974) Food utilization of dung-eating *Aphodius* larvae (Scarabaeidae). *Oikos* **25**, 71–9.

Holter, P. (1975) Energy budget of a natural population of *Aphodius rupes* larvae (Scarabaeidae). *Oikos* **26**, 177–86.

Hubbell, S.P., Sikora, A., & Paris, O.H. (1965) Radiotracer, gravimetric and calorimetric studies of ingestion and assimilation rates of an Isopod. *Health Phys.* **11**, 1485–501.

Humphreys, W.F. (1975) The food consumption of a wolf spider, *Geolylcosa godeffroyi* (Aracridae: Lycoridae), in the Australian Capital Territory. *Oecologia (Berlin)* **18**, 343–58.

Itô, Y. (1964) Preliminary studies on the respiratory energy loss of a spider, *Lycosa pseudoannulata. Res. Popul. Ecol.* **6**, 13–21.

Iversen, T.M. (1974) Ingestion and growth in *Sericostoma personatum* (Trichoptera) in relation to the nitrogen content of the ingested leaves. *Oikos* **25**, 278–82.

Ivlev, V.S. (1939) Transformation of energy by aquatic animals. *Int. Rev. ges. Hydrobiol. Hydrogr.* **38**, 449–558.

Ivlev, V.S. (1945) The biological productivity of waters. [In Russian.]. *Usp. Sovrem. Biol.* **19**, 98–120.

Jenderedjian, K. (1996) Energy budget of Oligochaeta and its relationship with the primary production of Lake Sevan, Armenia. *Hydrobiologia* **334**(1–3), 133–40.

Jonasson, P.M. & Kristiansen, J. (1967) Primary and secondary production in Lake Esrom: growth of *Chironomus anthracitius* in relation to seasonal cycles of phytoplankton and dissolved oxygen. *Int. Rev. ges. Hydrobiol.* **52**, 163–217.

Jones, J.D. (1959) A new tonometric method for the determination of dissolved oxygen and carbon dioxide in small samples. *J. Exp. Biol.* **36**, 177–90.

Kam, M. & Degen, A.A. (1997) Energy budget in free-living animals: a novel approach based on the doubly labelled water method. *Am. J. Physiol.* **272**, 1336–43.

Karo, J. (1973) An attempt to estimate the energy flow through the population of Colorado beetle (*Leptinotarsa decemlineato* Say). *Ekol. Polska* **21**, 239–50.

Kay, R.H. (1964) *Experimental Biology: Measurement and Analysis.* Chapman & Hall, London.

Kleiber, M. (1961) *The Fire of Life: an Introduction to Animal Energetics.* Wiley, New York.

Kormondy, E.J. (1965) Uptake and loss of zinc-65 in the dragonfly *Plathemis lydia. Limnol. Oceanogr.* **10**, 427–33.

Kozlovsky, D.G. (1968) A critical evaluation of the trophic level concept. 1. Ecological efficiencies. *Ecology* **49**, 48–60.

Kuenzler, E.J. (1961) Structure and energy flow of a mussel population in a Georgia salt marsh. *Limnol. Oceanogr.* **6**, 191–204.

Lawton, J.H. (1970) Feeding and food energy assimilation in larvae of the damselfly *Pyrrhosoma nymphula* (Sulz.) (Odonata: Zygoptera). *J. Anim. Ecol.* **39**, 669–89.

Lawton, J.H. (1971) Ecological energetics studies on larvae of the damselfly *Pyrrhosoma nymphula* (Sulz.) (Odonata: Zygoptera). *J. Anim. Ecol.* **40**, 385–419.

Lawton, J.H. & Richards, J. (1970) Comparability of Cartesian diver, Gilson, Warburg and Winkler methods of measuring the respiratory rates of aquatic invertebrates in ecological studies. *Oecologia (Berlin)* **4**, 319–24.

Ledbetter, M.C. & Flemion, F. (1954) A method for obtaining piercing–sucking mouth parts in host tissues from the tarnished plant bug by high voltage shock. *Contrib. Boyce Thompson Inst.* **17**, 343–6.

Le Febvre, E.A. (1964) The use of D2018 for measuring energy metabolism in *Columba livia* at rest and in flight. *Auk* **81**, 403–16.

Lifson, N. & McClintock, R. (1966) Theory of the use of the turnover rates of body water for measuring energy and material balence. *J. Theor. Biol.* **180**, 803–11.

Lindeman, R.L. (1942) The trophic–dynamic aspect of ecology. *Ecology* **23**, 399–418.

Lovegrove, T. (1966) The determination of dry weight of plankton and the effect of various factors on the values obtained. In: Bames, H. (ed.) *Some Contemporary Studies in Marine Science.* Hafner, New York, pp. 429–67.

Lucas, M.C., Johnstone, A.D.F., & Priede, I.G. (1993) Use of physiological telemetry as a method of estimating metabolism of fish in the natural environment. *Trans. Am. Fish. Soc.* **122**, 822–33.

Lyytikainen, T. & Jobling, M. (1998) The effect of temperature fluctuations on oxygen consumption and ammonia excretion of underyearling Lake Inari arctic char. *J. Fish Biol.* **52**, 1186–98.

McClintock, R. & Lifson, N. (1958) Determination of the total carbon dioxide output of rats by the D2018 method. *Am. J. Physiol.* **192**, 76–8.

Macfadyen, A. (1961) A new system for continuous respirometry of small air breathing invertebrates under near-natural conditions. *J. Exp. biol.* **38**, 323–43.

Macfadyen, A. (1963) *Animal Ecology: Aims and Methods.* Pitman, London, New York.

McGinnis, A.J. & Kasting, R. (1964a) Chromic oxide indicator method for measuring food utilization in a plant-feeding insect. *Science* **144**, 1464–5.

McGinnis, A.J. & Kasting, R. (1964b) Digestion in insects: calorimetric analysis of chromic oxide used to study food utilization by phytophagous insects. *J. Agric. Food Chem.* **12**, 259–62.

McNab, B.K. (1963) Bioenergetics and the determination of home-range size. *Am. Nat.* **97**, 133–40.

McNeill, S. (1971) The energetics of a population of *Leptopterna doiabrata* (Heteroptera: Miridae). *J. Anim. Ecol.* **40**, 127–40.

McNeill, S. & Lawton, J.H. (1970) Annual production and respiration in animal populations. *Nature* **225**, 472–4.

Malone, C.R. & Nelson, D.J. (1969) Feeding rates of freshwater snails (*Goniobasis clavaejormis*) determined with cobalt 60. *Ecology* **50**, 728–30.

Marchant, R. & Nicholas, A.L. (1974) An energy budget for the free-living nematode *Pelodera* (Rhabditidae). *Oecologia (Berlin)* **16**, 237–52.

Mathavan, S. & Pandian, T.I. (1974) Use of faecal weight as an indicator of food consumption in some lepidopterans. *Oecologia (Berlin)* **15**, 177–85.

Miles, P.W. (1972) The saliva of hemiptera. *Adv. Insect Physiol.* **9**, 183–225.

Moore, S.T., Schuster, M.F., & Harris, F.A. (1974) Radioisotope technique for estimating lady beetle consumption of tobacco budworm eggs and larvae. *J. Econ. Entomol.* **67**, 703–5.

Morista, M. (1959) Measuring of the dispersion of individuals and analysis of the distributional patterns. *Mem. Fac. Sci., Kyushu Univ., Ser. E (Biol.)* **2**, 215–35.

Mukerji, M.K. & Le Roux, E.J. (1969) A study of

energetics of *Podisus maculiventris* (Hemiptera: Pentatomidae). *Can. Entomol.* **101**, 449–60.

Mulkern, G.B. & Anderson, J.F. (1959) A technique for studying the food habits and preferences of grasshoppers. *J. Econ. Entomol.* **52**, 342.

Muthukrishnan, I. & Delvi, M.R. (1974) Effect of ration levels on food utilisation in the grasshopper *Poecilacerus pictus*. *Oecologica (Berlin)* **16**, 227–36.

Nagy, K.A. (1975) Nitrogen requirement and its relation to dietary water and potassium content in the lizard *Sauromalus obesus*. *J. Comp. Physiol.* **104**, 49–58.

Nagy, K.A. (1994) Field bioenergetics of mammals: what determines field metabolic rates? *Aust. J. Zool.* **42**, 43–53.

Neess, I. & Dugdale, C. (1959) Computation of production for populations of aquatic midge larvae. *Ecology* **40**, 425–30.

Norberg, R.A. (1977) An ecological theory on foraging time and energetics and choice of optimal food-searching method. *J. Anim. Ecol.* **46**, 511–29.

Odum, E.P. & Golley, F.B. (1963) Radioactive tracers as an aid to the measurement of energy flow at the population level in nature. In: Schultz, V. & Klement, A.W. (eds) *Radioecology*. Reinhold Pub. Corp., New York, pp. 403–10.

Odum, E.P. & Kuenzler, E.J. (1963) Experimental isolation of food chains in an old field ecosystem with the use of phosphorus-32. In: Schultz, V. & Klement, A.W. *Radioecology*. Reinhold Pub. Corp., New York, pp. 113–20.

Odum, H.T. (1957) Trophic structure and productivity of Silver Springs, Florida. *Ecol. Monogr.* **27**, 55–112.

Odum, H.T. & Odum, E.P. (1955) Trophic structure and productivity of a Windward coral reef community on Eniwetok Atoll. *Ecol. Monogr.* **25**, 291–320.

Omori, M. & Ikeda, T. (1984) *Methods in Marine Zooplankton Ecology*. J. Wiley & Sons, New York, 332 pp.

Onuf, C.P., Teal, L.M., & Valiela, I. (1977) Interactions of nutrients, plant growth and herbivores in a mangrove ecosystem. *Ecology* **58**, 514–26.

Otto, C. (1975) Energetic relationships of the larval population of *Potamophylax cingulatus* (Trichoptera) in a South Swedish stream. *Oikos* **26**, 159–69.

Ovington, J.D. & Heitkamp, D. (1960) The accumulation of energy in forest plantations in Britain. *J. Ecol.* **48**, 639–46.

Paine, R.T. (1964) Ash and calorie determination of sponge and opisthobranch tissues. *Ecology* **45**, 384–7.

Paine, R.T. (1971) The measurement and application of the calorie to ecological problems. *Ann. Rev. Ecol. Syst.* **2**, 145–64.

Paine, R.T. (1980) Food webs: linkage interaction strength and community infrastructure. *J. Anim. Ecol.* **49**, 667–85.

Paine, R.T. (1992) Food-web analysis through the analysis of per capita interaction strength. *Nature* **355**, 73–5.

Paris, O.H. & Sikora, A. (1965) Radiotracer demonstration of isopod herbivory. *Ecology* **46**, 729–34.

Paul, A.J., Paul, J.M., & Smith, R.L. (1988) Respiratory energy requirements of the cod *Gadus macrocephalus* Tilesius relative to body size, food intake, and temperature. *J. Exp. Mar. Biol. Ecol.* **122**, 83–9.

Pendleton, R.C. & Grundmann, A.W. (1954) Use of phosphorus-32 in tracing some insect–plant relationships of the thistle, *Cirsium undulatum*. *Ecology* **35**, 187–91.

Petrusewicz, K. & MacFadyen, A. (1970) *Productivity of Terrestrial Animals: Principles and Methods*. Blackwell Scientific Publications, Oxford.

Phillipson, J. (1960) The food consumption of different instars of *Mitopus morio* (F.) (Phalangiida) under natural conditions. *J. Anim. Ecol.* **29**, 299–307.

Phillipson, J. (1962) Respirometry and the study of energy turnover in natural systems with particular reference to harvest spiders (Phalangiida). *Oikos* **13**, 311–22.

Phillipson, J. (1964) A miniature bomb calorimeter for small biological samples. *Oikos* **159**, 130–9.

Plaut, I., Borut, A., & Spira, M.E. (1996) Lifetime energy budget in the sea hare *Aplysia oculifera*. *Comp. Biochem. Physiol. A* **113**(2), 205–12.

Pollard, D.G. (1973) Plant penetration by feeding aphids (Hemiptera: Aphoidea): a review. *Bull. Entomol. Res.* **62**, 631–714.

Raffaelli, D.G. & Hall, S.J. (1996) Assessing the relative importance of trophic links in food webs. In: Polis, G.A. & Winemiller, K.O. (eds) *Food Webs: Integration of Patterns and Dynamics*. Chapman & Hall, London.

Reichle, D.E. (1969) Measurement of elemental assimilation by animals from radioactive retention patterns. *Ecology* **50**, 1102–4.

Richman, S. (1958) The transformation of energy by *Daphnia pulex*. *Ecol. Monogr.* **28**, 273–91.

Ricker, W.E. (1968) *Methods for Assessment of Fish Production in Fresh Waters*. Blackwell Scientific Publications, Oxford.

Rogers, S.C. & Weatherley, A.H. (1983) The use of opercular muscle electromyograms as an indicator of fish activity in rainbow trout, *Salmo gairdneri* Richardson, as determined by radio telemetry. *J. Fish Biol.* **23**, 535–47.

Ross, L.G., Watts, W., & Young, A.H. (1981) An ultrasonic biotelemetry system for continuous monitoring of tail-beat rate from free-swimming fish. *J. Fish Biol.* **18**, 479–90.

Schoener, T.W. (1971) Theory of feeding strategies. *Ann. Rev. Ecol. Syst.* **2**, 369–404.

Scholander, P.F., Vandam, L., Claff, C.L., & Kanwisher, J.W. (1955) Microgasometric determination of dissolved oxygen and nitrogen. *Biol. Bull. Woods Hole* **109**, 328–34.

Schroeder, L.A. (1977) Energy, matter and nitrogen utilization by larvae of the milkweed tiger moth *Euchretias egle. Oikos* **28**, 27–31.

Shaw, I. & Beadle, L.C. (1949) A simplified ultra-micro Kjeldahl method for the estimation of protein and total nitrogen in fluid samples of less than 1.0 μm *J. Exp. Biol.* **26**, 15–23.

Shure, D.J. (1970) Limitations in radio-tracer determination of consumer trophic positions. *Ecology* **51**, 899–901.

Shure, D.J. (1973) Radionuclide tracer analysis of trophic relationships in an old-field ecosystem. *Ecol. Monogr.* **43**, 1–19.

Singh, J.S. & Yadava, P.S. (1973) Caloric values of plant and insect species of a tropical grassland. *Oikos* **24**, 186–94.

Slobodkin, L.B. & Richman, S. (1961) Calories/gm in species of animals. *Nature* **191**, 299.

Smith, A.H. & Douglas, J.R. (1949) An insect respirometer. *Ann. Entomol. Soc. Am.* **42**, 14–18.

Smith, C.C. (1968) The adaptive nature of social organization in the genus of three squirrels *Tamiasciurus. Ecol. Monogr.* **38**, 31–63.

Soofiani, N.M. & Hawkins, A.D. (1982) Energetic costs at different levels of feeding in juvenile cod, *Gadus morhua* L. *J. Fish Biol.* **21**, 577–92.

Stam, E.M., Van De Leemkule, M.A., & Ernsting, G. (1996) Trade-offs in the life history and energy budget of the parthenogenetic collembolan *Folsomia candida* (Willem). *Oecologia* **107**(3), 283–92.

Stanford, J.A. (1973) A centrifuge method for determining live weights of aquatic insect larvae, with a note on weight loss in preservative. *Ecology* **54**, 449–51.

Stiven, A.E. (1961) Food energy available for and required by the blue grouse chick. *Ecology* **42**, 547–53.

Strong, F.E. & Landes, D.A. (1965) Feeding and nutrition of *Lygus hesperus* (Hemiptera: Miridae). II. An estimation of normal feeding rates. *Ann. Entomol. Soc. Am.* **58**, 309–14.

Sushchenya, L.M. (1962) Quantitative data on nutrition and energy balance in *Artemia salina* (L.). [In Russian.] *Doklady Akad. Nauk. SSSR* **143**, 1205–7.

Teal, J.M. (1957) Community metabolism of a temperate cold spring. *Ecol. Monogr.* **27**, 283–302.

Teal, J.M. (1962) Energy flow in the salt marsh ecosystems of Georgia. *Ecology* **43**, 614–24.

Townsend, C.R. & Winfield, I.J. (1985) The application of optimal foraging theory to feeding behaviour in fish. In: Tyler, P. & Calow, P. (eds) *Fish Energetics: New Perspectives.* Croom Helm, London, pp. 67–98.

Utter, J.M. & Le Febvre, E.A. (1973) Daily energy expenditure of purple martins (*Progne subis*) during the breeding season: estimates using D2018 and time budget methods. *Ecology* **54**, 597–604.

Van Hook, R.I. (1971) Energy and nutrient dynamics of spiders and orthopteran populations in a grassland ecosystem. *Ecol. Monogr.* **41**, 1–26.

Van Slyke, D.D. & Neill, J.M. (1924) The determination of gases in blood and other solutions by extraction and manometric measurement. *J. Biol. Chem.* **61**, 523–73.

Van Valen, L. (1976) Energy and evolution. *Evol. Theory* **1**, 179–229.

Verdun, J. (1972) Caloric content and available energy in plant matter. *Ecology* **53**, 982.

Vollenweider, R.A. (1969) *A Manual on Methods of Measuring Primary Production in Aquatic Environments.* Blackwell Scientific Publications, Oxford.

Walker, I., Henderson, P.A., & Sterry, P.S. (1991) On the patterns of biomass transfer in the benthic fauna of an Amazonian black-water river, as evidenced by ^{32}P label experiment. *Hydrobiologia* **215**, 153–62.

Wells, M.J. & Clarke, A. (1996) Energetics: the costs of living and reproducing for an individual cephalopod. *Phil. Trans. Roy. Soc. Lond. B* **351**, 1083–104.

Whittaker, R.H., Bormann, F.H., Likens, G.E., & Siccama, T.G. (1974) The Hubbard Brook ecosystem study: forest biomass and production. *Ecol. Monogr.* **44**, 233–52.

Wiegert, R.G. (1964a) Population energetics of meadow spittle bugs (*Philaenus spumarius* L.) as affected by migration and habitat. *Ecol. Monogr.* **34**, 225–41.

Wiegert, R.G. (1964b) The ingestion of xylem sap by meadow spittle bugs, *Philaenus spumarius* (L.). *Am. Midl. Nat.* **71**, 422–8.

Wiegert, R.G. (1965) Intraspecific variation in calories/g of meadow spittle bugs (*Philaenus spumarius* L.). *BioScience* **15**, 543–5.

Wiegert, R.G. (1974) Litterbug studies of microarthropod populations in three South Carolina old fields. *Ecology* 55, 94–102.

Wiegert, R.G. (ed.) (1976) *Ecological Energetics. Benchmark Papers in Ecology*. Dowden, Hutchinson, & Ross, Stroudsbergh, Pennsylvania.

Wiegert, R.G. & Coleman, D.C. (1970) Ecological significance of low oxygen consumption and high fat accumulation by *Nasutitermes costalis* (Isoptera: Termitidae). *BioScience* 20, 663–5.

Wiegert, R.G. & Evans, F.C. (1967) Investigations of secondary productivity in grasslands. In: Petrusewicz, K. (ed.) *Secondary Productivity of Terrestrial Ecosystems*. Polish Academy of Science, Warsaw, Cracow, pp. 499–578.

Wiegert, R.G. & Owen, D.F. (1971) Trophic structure, available resources and population density in terrestrial vs. aquatic ecosystems. *J. Theor. Biol.* 30, 69–81.

Williams, E.C. & Reichle, D.F. (1968) Radioactive tracers in the study of energy turnover by a grazing insect (*Chrysochus amatus* Fab.: Coleoptera Chrysomelidae). *Oikos* 19, 10–18.

Wissing, T.E. & Hasler, A.D. (1971). Intraseasonal change in caloric content of some freshwater invertebrates. *Ecology* 52, 371–3.

Wood, T.G. & Lawton, J.H. (1973) Experimental studies on the respiratory rates of mites (Acari) from beech-woodland leaf litter. *Oecologia (Berlin)* 12, 169–91.

Yammamoto, T., Ueda, H., & Higashi, S. (1998) Correlation among dominance status, metabolic rate and otolith size in masu salmon. *J. Fish Biol.* 52, 281–90.

Zeuthen, E. (1950) Cartesian diver respirometer. *Biol. Bull. Mar. Lab. Woods Hole* 98, 139–43.

15 Large Scale Spatial and Temporal Studies and the Classification of Habitats

Technological advances over the last 30 years have greatly increased the opportunities for ecologists to study population and community processes over large spatial and temporal distances. Two important developments have made this possible. First, a variety of sensors on board satellites have produced a massive amount of remote-sensed data. This has allowed analysis of spatial and temporal variation in both physical variables, such as temperature, rainfall, or the area of open water, and biological variables, such as the distribution of closed forest, open grassland, or phytoplankton blooms. Secondly, the rapid reduction in the cost and size of powerful computers has made widely available to ecologists the computational power to analyse large data sets. Often this may now be done on personal computers. The final process aiding large-scale studies is the gradual accumulation of long-term data. Ecologists now have time series for a far greater number of populations or communities than were available to earlier generations of researchers (Leigh & Johnston, 1994). In addition, the time length of established long-term studies is gradually increasing to the point where time-series analysis techniques used by the physical sciences will become more applicable. All of these developments offer new opportunities to ecologists seeking to understand the way populations are organized and change through time. They also offer great advantages to conservation biologists seeking to identify, map, and protect vulnerable species or ecosystems. In this chapter, we do not aim to give a full account of the wide range of techniques used for the organization and analysis of large-scale spatial and temporal data sets. Rather, we aim to introduce the methods available and direct attention to some of the numerous books and articles that cover these topics in detail. Our purpose with respect to spatial data is to introduce, via some examples, the ways in which geographical information systems and remote-sensing data can be used by ecologists and conservationists. For temporal data, we introduce some elementary techniques for the analysis of the variability in population size and community structure.

The definition of large scale requires reference to the generation time and range of the organism. A long-term (large temporal scale) population study of an annually reproducing insect would be expected to include annual population estimates for at least 10 years. An equivalent study of an amoeba, which can reproduce daily, might be completed in a few weeks. However, if the focus of a long-term study is the role of

Table 15.1 The frequency at which the water level declines below selected water levels during the annual low-water season between 1902 and 1984 in the Reservá Mamirauá flood-plain between the Rios Japurá and Solimões in the Amazon basin, Brazil (from Henderson et al., 1998). The values are given for an arbitrary 0 m datum, which is the depth of the shallow lakes and channels found in the region. The frequencies were calculated using PORTOBRAS data.

Water depth (m)	Frequency of occurrence
−2	Once every 25 years
−1	Once every 10 years
0	Every 4–5 years
1	Every 2–3 years
2	Every 2 years
3	8 out of 10 years
4	9 out of 10 years

seasonal variation in determining population number, then it is likely that a study will need at least 15 years' data, irrespective of the size of the organism. Climatic variation and many biological processes almost invariably produce autocorrelated time series. As the primary requisite for the detection of any pattern is independent observations, any autocorrelation increases the length of time series that must be studied. Gaston and McArdle (1994) give, as a rule of thumb, that, if a time series has n observations and an autocorrelation coefficient of r, then the effective sample number is $(1 - r^2)n$. As an illustration of the typical levels of variability experienced by animals, Table 15.1 gives the frequency at which different low-water levels have been recorded in the Amazonian flood-plain this century (Henderson et al., 1998). Levels <−2 m will have a profound effect on the aquatic fauna, as most of the lakes and channels in the flood-plain would be dry. Such an event occurs, on average, once every 25 years. Thus, the fauna must have evolved to deal with conditions that may require a very long time series to actually observe. Even for the most temperate climate, anything shorter than 15 years is unlikely to include a representative selection of the possible seasonal patterns and events that the population could experience. Large spatial scales can be defined either in terms of the range of the population or by the extent of a particular ecosystem, such as a flood-plain. Large-scale studies of single species usually aim to analyse distribution or density over the entire population or metapopulation range (Hanski, 1981, 1982, 1987; Hanski & Gilpin, 1991; Hassell et al., 1991; Gruttke & Engels, 1998).

15.1 Remote sensing

Remote sensing is the measurement of reflected, emitted, or back-scattered electromagnetic radiation from the earth's surface, using instruments placed at a distance, most often on a satellite, although aircraft are used. General books introducing remote-sensing applications are Cracknell and Hayes (1991), Barrett and Curtis (1992), and Lillisand and Kiefer (1994). Table 15.2 lists the satellite systems offering data to researchers. It is normal to collect the information digitally as a stream of data representing the picture elements (the pixels) of the final image. The area of the earth's surface, represented by a single pixel, corresponds to the maximum resolution for the detector. The spatial resolutions of the receiving instruments differ widely and the value assigned to each pixel is the average for the corresponding land area. Pixels correspond to areas ranging from square kilometres for meteorological satellites to a few square centimetres for aircraft-borne sensors. For satellite-borne sensors, the temporal resolution—the time between repeated measurements at one locality—depends on the orbit. Polar orbiting satellites pass over different parts of the earth's surface on each orbit, so that there may be a number of days between repeat measurements of the same locality. Geostationary satellites orbit at the equator at the same speed as the earth, so that they remain above the same locality, allowing relatively small temporal resolutions.

Digital data derived from sensors must be processed to obtain useful images that display to best effect the features of interest. Specialist image-processing software, such as ERDAS and ILWAS (ITC, Enschede, the Netherlands; web site www.itc.nl/home.html), are available for processing the images. A frequent problem is cloud cover, which, if images for an area are available over a relatively long period, may frequently be removable by combining the images. The wavelengths chosen for study will depend on the variables of interest. The radiation at different wavelengths received from a pixel may be used to calculate an index that detects an important attribute. For example, the NOAA–AVHRR satellite data have been used to

Table 15.2 Remote-sensing satellites and their resolutions.

Name	Spectral resolution (μm)	Spatial resolution (m)	Temporal resolution (h)
SPOT/HRV	Channel 1 0.51–0.73 Channel 2 0.51–0.59 Channel 3 0.61–0.68 Channel 4 0.79–0.89	Channel 1 10 Channels 2–4 20	624
Landsat TM	Channel 1 0.45–0.52 Channel 2 0.52–0.60 Channel 3 0.63–0.69 Channel 4 0.76–0.90 Channel 5 1.55–1.75 Channel 6 2.10–2.35 Channel 7 10.40–12.50	Channels 1–6 30 Channel 7 120	384–432
NOAA/AVHRR	Channel 1 0.58–0.68 Channel 2 0.72–1.10 Channel 3 3.55–3.93 Channel 4 10.30–11.30 Channel 5 10.50–11.50	1100	12
Meteosat/HRR	Channel 1 0.40–1.10 Channel 2 10.53–12.50 Channel 3 5.70–7.10	Channel 1 2400 Channels 2–3 5000	0.5

measure vegetation cover and forecast potential famine by calculating a vegetation index. This is based on the fact that plants absorb light in the channel 1 waveband and reflect at channel 2 wavelengths (see Table 15.2). The normalized-difference vegetation index, NDVI, which compares the radiation received by channel 1 (Ch 1) and channel 2 (Ch 2), is:

$$NDVI = \frac{(Ch2 - Ch1)}{(Ch2 + Ch1)}$$

The internet is a particularly valuable source of remote-sensing information. A useful site that lists other geographical and remote-sensing sites is ftp://ftp.census.gov/pub/geo/gissites.txt.

Remote sensing can be a powerful tool for studying vegetation, particularly in inaccessible areas, such as tropical forest habitats in the Amazon basin, which are still poorly understood with respect to the distribution, patchiness, and area of different habitat types. These habitats present a considerable challenge for traditional survey methods, because of the high species richness, the large area to be studied, and the general inaccessibility of the habitats. For example, Tuomisto et al. (1994) analysed Landsat Multi Spectral Scanner (MSS) images from the Peruvian Amazon to detect and delimit vegetation types and geological formations. The authors used Landsat MSS images rather than those from the higher-resolution Landsat Thematic Mapper (TM), because of their lower cost per unit area and reduced computational requirements. A variety of methods were tried to give the best resolution of vegetation types using ERDAS and ILWAS software and satisfactory results were obtained using colour composites and histogram equalization. They also applied pixel-by-pixel classification techniques, which were unsatisfactory, because of the patchiness of the habitat, which often results in quite different electromagnetic spectra from adjacent pixels. The resulting images showed all the major features of Amazonian flood-plain systems, including scrole–swale topography caused by lateral channel migration, back swamps, inundation, and terra firma forest. Using these images, it was possible to calculate the amount of the different types of major habitat available and to appreciate its patchiness. The resolution, however, is insufficient to measure the amount of flood-plain water habitat, much of which, during the low-water season, is in the form of narrow channels or small pools. Tuomisto et al. (1994) noted the need for local knowledge and ground-truth surveys, in order to check that the different vegetation types were being distinguished. They concluded that some habitats could not be distinguished by their electromagnetic spectra alone. Ferrier and Anderson (1996) show how remote sensing can be used to quantify polluted areas and assess dispersal of pollutants.

Animal distribution is often closely linked to that of vegetation; Arseneault et al. (1997) used Landsat TM imagery to study the influence of caribou grazing on lichen biomass and percentage ground cover, and buzzard, Buteo buteo, nesting areas were predicted with the aid of vegetation-cover indices derived from satellite imagery

(Austin *et al.*, 1996). Vegetation indices may be particularly useful for predicting habitat suitability for animals in arid regions (e.g. Verlinden & Masogo, 1997). Such techniques are likely to become commonly used tools in wildlife management (Foster *et al.*, 1997). The measurement of deforestation (e.g. Hagner & Rigina, 1998), erosion (e.g. Dwivedi *et al.*, 1997), seaweed and sea-grass distribution (e.g. Ferguson & Korfmacher, 1997), and other forms of conservation monitoring are often aided by satellite imagery.

Climatic variables may also be the basis of a study as most animals are sensitive to them. The abundance and distribution of tsetse flies (*Glossina* spp.) in West Africa are sensitive to climatic variables, such as temperature, humidity, and rainfall (Bursell, 1957; Rogers & Randolph, 1991; Hay *et al.*, 1996; Rogers *et al.*, 1996). Unfortunately, there are insufficient meteorological data available from African ground stations to allow the spatial distribution of tsetse flies to be predicted. Using images collected by the NOAA and Metosat meteorological satellites, it was possible to create variables that were correlated with temperature, rainfall, and saturation deficit, as would be measured at ground-based meteorological stations. Because vegetation both responds to and changes atmospheric moisture, the NDVI (see above) was used to predict saturation deficit. Temperature close to the earth's surface was estimated from infrared channels 4 and 5 of the NOAA/AVHRR satellite, using the relationship of Price (1984). The estimation of land temperature using NOAA/AVHRR imagery has recently been discussed by Andersen (1997). Rainfall is obviously related to cloud cover and this can be estimated using a variable called cold-cloud duration, CCD, which measures the amount of convective cloud present. A Fourier series was used to summarize the seasonal climate for each pixel and the mean, annual, biannual, and triannual amplitudes and phases used in a discriminant analysis to predict tsetse density. Rogers *et al.* (1996) found that the presence–absence of tsetse fly was predicted with accuracies ranging from 67 to 100% and the thermal data were the best predictor of tsetse fly distribution. Similar success in the prediction of tsetse flies was found by Kitron *et al.* (1996) for Kenya, where Landsat TM band 7, which is associated with moisture content of soil and vegetation, emerged as being consistently highly correlated with fly density. Other examples of the use of satellite remote sensing to study important insects include: forecasting locust swarms (Hielkema, 1990); using rainfall estimated by remote sensing to predict Senegalese grasshopper, *Oedaleus senegalensis*, outbreaks (Burt *et al.*, 1995), monitoring the abundance of the mosquito *Aedes albifasciatus* (Gleiser *et al.*, 1997), and the identification of urban breeding sites of the mosquito *Culex annulirostris* (Dale & Morris, 1996). Washino and Wood (1994) review the application of remote-sensed imagery to the study of arthropod disease vectors.

Remote sensing can also be undertaken from aircraft, using photography, video, or direct observation. Wint (1998) discusses the situations where this technique has an advantage over satellite remote sensing. High-altitude surveys (5000–10 000 feet (1500–3000 m) above ground level) are useful for mapping, particularly vegetation; with high-resolution photography (conventional and digital), the condition of the plants may be recognized, damage levels evaluated, and indirect assessment made of pest populations (Franz & Karafiat, 1958; Aldrich *et al.*, 1959; Klein, 1973; Harris *et al.*, 1976; Wallen *et al.*, 1976; Everitt *et al.*, 1997). Airborne sensors can also be used for measuring lake primary production (see Chapter 14; George, 1997) and hence may be useful for predicting fish biomass or eutrophication.

Numerical information on populations of large animals is obtained from low-level surveys (400–2000 feet (120–160 m) above ground level). As Wint (1998) points out, visual observations are more reliable (and often cheaper) than automatic methods, which may fail to detect animals, e.g. those sheltering under trees. He recommends a combination of direct visual counting with direct photography of any group of >10 individuals. The sampling strategy may be stratified or systematic. Wint (1997) also indicates the conditions that determine the optimal choice for a given situation and describes methods for calculating the standard errors of the estimates.

Besides gathering primary information, aircraft surveys may be used to ground-truth the interpretation of satellite remote sensing.

15.2 Long-term studies

Most long-term studies have been of pests, food animals, or species that give valuable products, such as fur; perhaps the most famous examples are the Hudson's Bay Company's records of mammals from northern Canada (Elton, 1924; Elton & Nicholson, 1942). Time series of fish catches have been widely used to understand the exploitation of stocks, but these have, in a particularly acute form, a problem that affects most long-term data sets, namely, the efficiency of sampling and/or its methods changing with time. Fishing boats have tended to become more powerful and fishing more efficient as modern sonar and navigation aids became available. A standardization of sampling effort throughout the length of the time series is desirable but may be impossible to achieve. Numerous studies of long-term changes in marine ecosystems are described in Bachelet and Castel (1997).

Starting with the classical work of Schwerdtfeger (1941), there are many long-term studies on forest pests or other insects on long-lived vegetation. Klimetzek (1990) describes a 180-year series for certain defoliators in pine forests of southern Germany, whilst the Rothamsted insect survey (Woiwod & Harrington, 1994) provides faunal information on a national scale. Bird surveys, such as counts of migratory birds and ringing studies, have also produced many time-series. Recent examples of the application of such data are the analysis of variability in North American migratory bird counts (Keitt & Stanley, 1998) and the study on bird-ringing data from the Wash, England (Rehfisch *et al.*, 1996).

Through international cooperation, a global database has been established;[1] it comprises approximately 5500 data sets on 2000 different species from 1100 sites. The vast majority of data were collected in the northern hemisphere and they are biased in favour of insects, mammals, fish, and birds. Curiously, there are few long-

[1] Web address http://cpbnts1.bio.ic.ac.uk/gpdd/

term studies of higher plants. These time series may be used by ecologists to compare against their own time series. They can also be used to gain insight into the range of dynamic behaviour shown by animal populations and to identify whether different taxa display characteristic dynamic properties.

Time series may also be used for forecasting, but great caution is required. The simplest approach is to extrapolate a trend. While this may well be successful in the short term, it will almost certainly result in ridiculous long-term forecasts, because real populations can neither decline nor rise indefinitely. A more sophisticated approach is to fit an autoregressive integrated moving average (ARIMA) model. This is often referred to as the Box–Jenkins method, as it was advocated in Box and Jenkins (1976). While such models may give an impressive fit to the data, they are based solely on the existing time series. If conditions alter so that different biological processes become dominant, then the forecasts can be misleading. For example, Henderson and Corps (1997) showed that bass, *Dicentrarchus labrax*, juveniles of 2 and 3 years of age will cannibalize fish in their first year of life. During the 1980s, bass numbers were small because of low water temperatures, so that the suppression of the new recruits by the older year classes was undetectable. After the exceptional recruitment following the warm summer of 1989, the following 2-year classes were suppressed by cannibalism. An ARIMA model based on the 1980s time series would have greatly overestimated recruitment in the early 1990s. Population forecasts over time-scales greater than that at which reproduction and death will have replaced the entire population should only be made where there is considerable biological understanding.

Time series can also be reconstructed using fossils. Examples include the reconstruction of fish-abundance time series using scales preserved in the sediments and studies of changes in vegetation using palaeontological data (e.g. Lasen & MacDonald, 1998).

15.2.1 Planning spatial and temporal sampling

The guidance given in Chapter 2 on experimental design and planning should be consulted. In a typical study, sample sites are arranged over a spatial grid which is sampled at regular intervals in time. Generally, it is required to compare the abundance of animals over the grid at different points in time. A plot of abundance over the grid can be represented on a three-dimensional graph as a surface. Statistical comparison of such surfaces requires that sampling has the following features (Legendre & McArdle, 1997).

1 The same locations are used on each sampling occasion.

2 A regular grid is used. Stratified and random sampling are inferior for this purpose as changes at fixed points cannot be followed through time.

3 The samples are replicated at each sampling station on each sampling occasion.

4 If possible, these replicates should be from around the fixed station at positions chosen at random. If the replicates are too close together, they will not be independent.

5 A balance is needed between spatial resolution (more sampling stations) and the power to detect changes (more replication at each station) (see p. 7).

15.2.2 The classification of time series

The shape of abundance time series is likely to reflect the demographic strategy of the animal. For example, an opportunistic species able to exploit vacant space and capable of rapid reproduction will tend to produce sudden bursts in abundance when conditions are favourable. Ibanez and Fromentin (1995) describe a methodology for allocating abundance time series to one of the following six general forms.

1 Erratic series—shown by species that occasionally invade.

2 Series with periods of low abundance—exhibited by opportunistic species and those experiencing environmental stress.

3 Trending series—shown by species showing little seasonal recruitment and responding either to a favourable (positive slope) or unfavourable (negative slope) set of conditions.

4 Seasonal series—shown by species with clear seasonal recruitment but regulated to maintain an approximately constant long-term abundance.

5 Seasonal and trending series—species with seasonal recruitment and experiencing a change in conditions.

6 Spiky series—shown by permanently present species, which possibly display chaotic or limit-cycle dynamics plus stochastic variability.

Examples of these six types of dynamics are given in Fig. 15.1. While it may be possible to allocate time series to one of these six classes by visual examination, a rather less subjective method would be desirable. Given time series for a large suite of species, Ibanez and Fromentin (1995) used four synoptic attributes derived from the original time series in a flexible clustering method to divide the species into six groups. These attributes, which measured the trend, variability, and seasonality of the original series, were obtained using a technique called eigenvector filtering (Colebrook, 1978). This is, in effect, a principal-components analysis of the time series.

15.2.3 Time-series analysis

This section gives a brief introduction to time-series analysis from an ecological perspective; more detail about the mathematics and application of techniques introduced can be found in books by Diggle (1990), Tong (1990), Wei (1990), and Chatfield (1994). An ecological time series, X_t, normally comprises observations, X_1, $X_2 \ldots X_n$ of some population attribute, such as population size or density, obtained from data collected at discrete periods in time. Generally, the interval between observations, Δ, should, whenever possible, be held constant. When this is not possible, a constant-step time series can often be generated from the original set. Sometimes, this is done by removing some observations, but one of the commonest reasons for variable time steps is a missed observation, which then has to be estimated by interpolation. If it is possible to collect independent replicate time series,

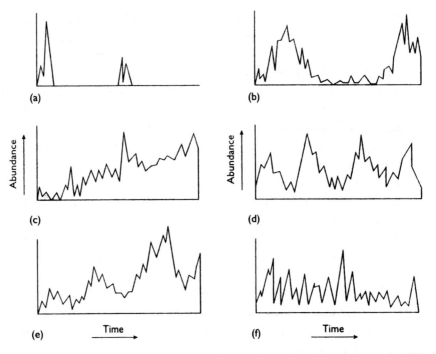

Fig. 15.1 Examples of the types of dynamics displayed by populations (after Ibanez & Fromentin, 1995). (a) Erratic series. (b) Series with periods of low abundance. (c) Trending series. (d) Seasonal series. (e) Seasonal and trending series. (f) Spiky series.

then this will be statistically advantageous. However, this can rarely be achieved; while cost may be a factor, it is often physically impossible, e.g. there may only be one population that can be studied. In the following account, we shall assume that only one time series is available for each attribute and that the time steps are constant.

At the outset, it is important to note that series analysis requires that the number of observations, n, in a series should be large if the conclusions are to be statistically valid. Probably $n = 100$ is a minimum requirement.

The analytical approach taken here assumes that the time series can be decomposed into a number of separate components, each of which contributes to the total variability of the observed series. These components can be classified as trends, periodic or cyclic patterns, and noise or random variation. An example of a time series broken down into these three components is given in Fig. 15.2. Quite often, the objective of an ecological study is to identify trends and any periodic changes and to assess their magnitude relative to the general noise in the time series. It is the teasing apart of these components that is the objective of time-series analysis.

The first stage of an analysis is almost always an investigation of the series for trends. This is because later stages of analysis require series to be stationary. In a stationary series, the mean and variance are constant with time. Thus, even if the trend is of no interest, it needs to be identified and removed before further analysis can

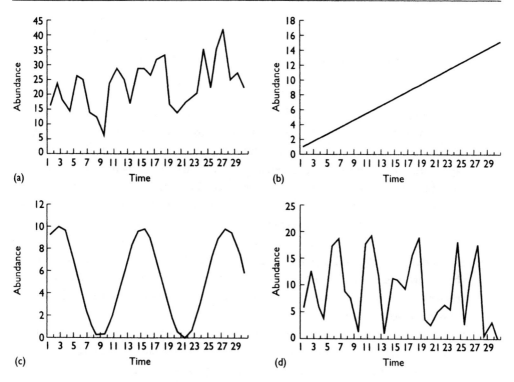

Fig. 15.2 The decomposition of the components frequently observed in biological time series. (a) Original series. (b) Trend. (c) Cyclic. (d) Noise or random component.

proceed. Trends are often defined as long-term changes in the mean level. In population studies, we often look for trends of increasing or decreasing number. However, it is clear that there is no way to distinguish between a long-term cycle, where only a small part of its cycle is covered by the period of study, and a trend. Indeed, Granger (1966) defines trend in the mean as all cyclic components whose wavelength exceeds the length of the observed series. There are two simple statistical procedures to test for the existence of a trend (Kendall *et al.*, 1983). The number of positive differences between successive observations in the series can be counted and a difference-sign test applied. Alternatively, the observations can be reordered from largest to smallest and this new series compared for resemblance against the original series, using Kendall's rank-order correlation coefficient.

Once a trend has been identified, it needs to be removed from the series. Quite probably, it is also of ecological interest, in which case an appropriate model may be found to describe it. The simplest model is a linear equation that can be fitted by regression. If the trend is exponential, as may occur with a growing population, then a linear model can be applied after initial logarithmic transformation. Trends need not be linear or exponential, and, more generally, a polynomial model may be fitted, although this may be difficult to interpret biologically. Once a model has been fitted, then the residuals obtained by subtracting the model from the original series are the detrended series.

Two other approaches frequently used to remove trend are the calculation of moving averages and differencing. Differencing provides a simple method to remove a trend. The first difference of a series x_t is defined as:

$$Dx_t = x_t - x_{t-1}$$

First differencing will remove a linear trend, as can be shown by calculating the first difference of the simple series showing a linear trend 1, 2, 3, 4, 5, for which the first differences are 1, 1, 1, 1. Higher-order differences can be calculated by repeated application of a first-order difference. Second-order differencing will remove the trend from a series which is accelerating at a constant rate. Generally, repeated differencing is undertaken until the series appears stationary, although it is rare to use higher than second-order differencing.

Differencing has a similar effect to subtracting a moving average from the original series. Moving averages are particularly useful for removing periodic patterns from data, such as seasonality. As shown in Fig. 15.3, if data are collected monthly, then a 12-month moving average will remove a strong seasonal signal, which may be obscuring other features within the series.

Not all series can be made stationary by differencing. In some cases, a series may have a constant mean (stationary in the mean) but variable variance. If the variance of a series is proportional to its level, then a square-root transformation will give a constant variance. The method of Box and Cox (see Chapter 2) to select a suitable transformation for stabilizing the variance can be applied.

The detrended and hopefully stationary series can then be analysed for periodic components. A series is defined as periodic with periodicity P if $x_t = x_{t+P}$ for all t. Typical examples of periodic pattern in biological time series are those caused by seasonality. Figure 15.3 shows strong seasonality in the abundance of dab, *Limanda limanda*, in the Severn Estuary, England. Cycles can also be caused by biological interactions such as delayed-density dependence or even cannibalism between age classes (Henderson & Corps, 1997). Classical examples of periodicity in animal abundance are those of northern temperate mammals (Elton & Nicholson, 1942; Finerty, 1980). Weak periodic signals may be difficult to detect if a series contains a lot of noise.

The autocorrelation coefficients of a series measure the correlation between observations at different times apart. They are calculated in a manner analogous to a simple correlation coefficient, except that the pairs of values both derive from the same series. The time or distance between the pairs of observations is normally termed the lag. The simplest approach to search for periodic components is to examine the correlogram. This is a plot of the autocorrelation of the series over a range of lags. The seasonality in the dab time series is shown in Fig. 15.4 by the clear 12-month periodicity in the correlogram. This correlogram also shows the normal time-series feature that observations collected close together (small lag) tend to be more similar (higher positive autocorrelation) than those further apart. Of course, with a lag of 0, the autocorrelation is +1, because each observation is being compared

Fig. 15.3 The effect of performing a 12-month moving average on a time series of monthly collected population abundance data which shows a strong seasonal variation. (a) Original series. (b) Series after the application of a 12-month moving average. The data are for the dab, *Limanda limanda*, in the Bristol Channel.

with itself. The autocorrelation at any particular lag is significantly different from zero at the 5% level if it lies outside the range $\pm 2/\sqrt{n}$. Chatfield (1994) recommends that this test be applied to the correlogram of the residuals of a time-series model to determine if the residuals are random and thus the model a good fit.

Some of the most powerful tools for investigating the properties of time series are based on the Fourier representation of a series:

Fig. 15.4 The correlogram of the dab time series plotted in Fig. 15.3, which shows a marked peak at 12 months because of the marked seasonality of abundance.

$$x_t = \sum_{k=0}^{k=n/2} (a_k \cos \omega_k t + b_k \sin \omega_k t) \tag{15.1}$$

where $\omega_k = 2\pi k/n$, $k = 0, 1, \ldots, n/2$, are Fourier frequencies. The terms a_k and b_k are the Fourier coefficients and are given by the expressions:

$$a_k = \frac{1}{n} \sum_{i=1}^{n} x_t \cos \omega_k t$$

for $k = 0$ and $n/2$ if n is even, and:

$$a_k = \frac{2}{n} \sum_{i=1}^{n} x_t \cos \omega_k t$$

for other values of k, and:

$$b_k = \frac{2}{n} \sum_{i=1}^{n} x_t \sin \omega_k t \tag{15.2}$$

for $k = 1, 2, 3, \ldots, (n-1)/2$.

Thus, in the Fourier representation, the original series is described as the summation of a series of cosine and sine terms. Using this approach, a time series may be analysed in terms of the sinusoidal behaviour at a range of frequencies. While analysis based on the correlogram is in the time domain, studies based on the Fourier transform are termed frequency-domain analyses.

Using Perseval's relationship, we can partition the total variance shown by the series between the different frequencies because:

$$\sum_{t=1}^{n} x_t^2 = na_0^2 + \frac{n}{2} \sum_{t=1}^{(n-1)/2} (a_k^2 + b_k^2)$$

if n is odd and:

$$\sum_{t=1}^{n} x_t^2 = na_0^2 + \frac{n}{2} \sum_{t=1}^{(n-1)/2} (a_k^2 + b_k^2) + na_{n/2}^2 \qquad (15.3)$$

if n is even.

This allows a time series to be analysed in a manner analogous to the more familiar analysis of variance. In tabulated form, the analysis of variance for a time series is as shown in Table 15.3.

The periodogram is formed by plotting the sums of squares (Table 15.3) against frequency. The periodogram is a direct estimate of the power spectrum and can be examined for information about the time series. However, it is an inconsistent estimate of the spectrum, because it uses n observations to estimate $n/2$ quantities and thus shows considerable noise. If, as is normal for ecological time series, we can assume that the underlying spectrum is a continuous function, then the periodogram can be smoothed in some way to produce an estimate of the spectrum. A number of different smoothing procedures are used, including a simple moving average of the periodogram.

In practice, the availability of time-series analysis software, which usually employs the fast Fourier transform algorithm, has made the calculation and plotting of the spectrum easy and rapid. A typical example of the power spectrum for a population time series is that for the pout, *Trisopterus luscus*, shown in Fig. 15.5. A visual examination of the spectrum shows the presence of two peaks: the first at a frequency of one cycle per year, which is the seasonal cycle, and a second at 0.33 cycles per year, which suggests a 3-year cycle. In this example, the data were collected monthly, so that the highest frequency, termed the Nyquist frequency, that can be studied is $1/2 \times 1 = 0.5$ cycles per month. In general, the Nyquist frequency is calculated as

Table 15.3 Analysis of variance for a time series.

Source	Degrees of freedom	Sum of squares
Frequency ω_0	1	na_0^2
Frequency ω_1	2	$n/2(a_1^2 + b_1^2)$
Frequency ω_2	2	$n/2(a_1^2 + b_1^2)$
Frequency $\omega_{(n-1)/2}$	2	$n/2(a_{(n-1)/2}^2 + b_{(n-1)/2}^2)$
Frequency $\omega_{n/2}$	1	$na_{n/2}^2$
Total	n	$\sum x_t^2$

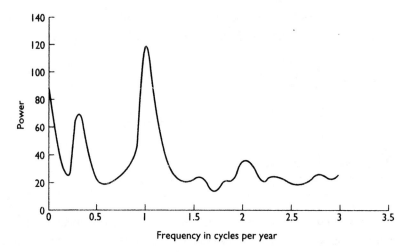

Fig. 15.5 A typical example of the power spectrum for a population time series. The data are for pout, *Trisopterus luscus*, in the Bristol Channel. The spectrum shows two peaks: the first at a frequency of one cycle per year, which is the seasonal cycle, and a second at 0.33 cycles per year, which suggests a 3-year cycle.

$1/2\Delta$, where Δ is the time between samples. The lowest frequency that can be studied is $2/n\Delta$. Tests to determine if the observed peaks in the power spectrum are significant were developed initially by R.A. Fisher, and useful worked examples of such tests are given in Chapter 12 of Wei (1990).

15.2.4 Detecting synchrony

While rarely detected, a number of important studies have investigated synchrony in the population dynamics between species. Well-known studies include those on microtine rodents and their predators (Henttonen *et al.*, 1987), some British butterfly populations (Pollard, 1991), and capercaillie, *Tetrao urogallus*, and grouse, *Lagpopus*, populations in Finland (Ranta *et al.*, 1995). The general analytical approach taken is to standardize all the time series to zero mean and unit variance, so that site- or species-specific differences in density are removed from the series, and then to calculate the cross-correlation (Diggle, 1990; Chatfield, 1994). Studies have also been made of synchrony between different populations of the same species (Saitoh *et al.*, 1998).

Methods have been developed to fit models to populations following a logistic growth pattern within a spatial framework within which the metapopulations may become synchronized (Dennis *et al.*, 1998).

15.2.5 Measuring temporal variability

The abundance of every animal changes through time and, while some species have remarkably stable populations, others change by orders of magnitude in a few generations. The study of the degree of variation shown by populations is of considerable theoretical and applied interest. In general, it would be anticipated that the more variable a population, the more likely it is to become extinct, because it will occasionally reach low numbers where stochastic processes dominate (May, 1973). Historically, biologists have used a surprising number of measures of variability, many of which are clearly inappropriate.

The variance of population time series usually increases with the length of the time series. Thus, care should be taken when comparing the variability of time series of different lengths. By far the safest approach is only to compare values obtained from contemporaneous time series of the same length. However, if the comparison is between species with greatly differing generation times, then each series should be adjusted for the same number of generations. The increase in variance with time can be caused by three main agents (Gaston & McArdle, 1994). First, there tends to be autocorrelation in the environment, so that, on average, population abundance in temporally close sampling periods is more similar than in those far apart. Secondly, autocorrelation is also generated within the population; small populations beget small populations and vice versa. Thirdly, in the short time series common in biological studies, there is often a trend caused by changes in the habitat. As yet, it is unclear whether, given time series of the length likely to be obtained, the variance will stabilize.

Most ecologists have considered that variability should measure proportional change, so that a change in density from 0.01 to 0.1 is equivalent to a change from 10000 to 100000 (Gaston & McArdle, 1994). If this is accepted, then variability is a measure of the deviation from the mean value expressed as a proportion of the mean. Gaston and McArdle (1994) consider that the two most appropriate measures of variation are the standard deviation of the log abundance values, $SD[\log(N)]$, or the coefficient of variation of the abundance values, $CV(N)$. If the time series contains no zeros, then $SD[\log(N)]$ is the more natural measure (Williamson, 1984). However, many abundance time series include zeros and the commonly used expedient of using $SD[\log(N + 1)]$ should be avoided, so for these cases $CV(N)$ should be used.

When the population shows a linear trend, it may be useful to measure the variation about the trend, for which the most suitable measure is $SD[\log(R)]$, where $R = N_{t+1}/N_t$ (Williamson, 1984). When such a series contains zeros, then no suitable measure exists.

Zeros arise either because the species is absent (termed structural zeros) or because it was not included in the samples (sampling zeros). It is normally impossible to distinguish between these possibilities. In principle, zeros caused by the absence of the species would not be included in the calculation of the variability, as they would tend to exaggerate the variability shown by the population when present. However, zeros arising from sampling variation should be included, otherwise the population variation will be underestimated. Unless there are strong biological reasons to believe that the species was not locally present, then zeros should be included and the coefficient of variation used as a measure of variation. As Gaston and McArdle (1994) argue, the sampling-unit size for a study should generally be chosen to be large enough to eliminate zeros. Such an approach is unavailable in long-term studies, where a single sampling method, such as grab sampling, is applied to a whole community. The pattern of relative species abundance is often such that millions of the commoner species would be caught in samples that were large enough to ensure that the rarer 50% of species were always to be present (see Fig. 13.4 for a typical aquatic-species abundance pattern).

15.2.6 Detection of density dependence

Where sufficient detail is available key-factor analysis (Chapter 11) is normally the most powerful method of detecting density dependence (Vickery & Nudds, 1991). However, many data sets of the type discussed in this chapter consist solely of a sequential series of estimates of one particular age or of the population at regular intervals. These investigations are extensive in time, whilst those permitting key factor analysis are intensive (Chapter 1). A number of approaches have been developed for the detection of density dependence in long time series (Rothery, 1998).

The underlying approach is to test the null hypothesis (H$_o$) that the data can be produced by a random walk with or without drift. A statistic must be calculated that will be capable of rejecting the null hypothesis in favour of the alternative hypothesis (H$_d$) that there is density dependence.

15.2.6.1 Bulmer's (1975) test
This is based on a Gompertz model of population growth and is appropriate for time series that display no obvious trend. The test statistic (R) is:

$$R = \frac{\sum_{i=1}^{k}(X_i - \bar{X})^2}{\sum_{i=1}^{k-1}(X_{i+1} - X_i)^2} \tag{15.4}$$

where X_i is log abundance in year i and k is the number of observations. The null hypothesis that there is no density dependence is rejected at the 5% level if R is smaller than about $0.25 + (k - 2)0.0366$. If the population is either increasing or decreasing this will increase the value of R and make density-dependence undetectable with this method and either of the following methods should be used instead.

15.2.6.2 Pollard et al.'s (1987) randomization test
A population subject to annual census for k years will yield a series $N_1, N_2, \ldots, N_k$ estimates of population size. If this series is transformed by taking the natural logarithm, $X_i = \log_e N_i$ ($i = 1$ to k) then the following three time series models for the population can be considered:

1 $X_{t+1} = X_t + e_i$
2 $X_{t+1} = r + X_t + e_i$
3 $X_{t+1} = r + bX_t + e_i$

In these equations r and b are parameters that determine the rate of change of the population and e_i are random, independent, variables. Model 1 is a simple random walk, model 2 describes a population that is changing without density-dependence (random walk with drift) and model 3 with $b \neq 1$, is a density-dependent model that is also termed the Gompertz equation.

For a density-independent population, the change in the $\log_e$ population between years, $D_i = (X_{i+1} - X_i)$, are just random fluctuations and could have occurred in any order. The procedure is therefore to calculate a test statistic, T, for the observed series and determine by Monte Carlo simulation if this value is so extreme that the null hypothesis of density-independence can be rejected.

To compare the density dependent model (3) against a random walk model (1) the test statistic $T(1,3)$ is

$$\sum_{i=1}^{k-1}(X_{i-1} - X_i)^2 \tag{15.5}$$

and to compare model 3 against a changing population model (2) the test statistic, $T(2,3)$ is

$$\frac{\sum_{i=1}^{k-1}(X_{i+1} - m_2)^2 - B\sum_{i=1}^{k-1}(X_i - m_1)(X_{i+1} - m_2)}{\sum_{i=1}^{k-1}(X_{i+1} - X_i)^2 - (X_k - X_1)^2/(k-1)} \tag{15.6}$$

where

$$B = \sum_{i=1}^{k-1}(X_i - m_1),$$

$$m_1 = \sum_{i=1}^{n-1}X_i/(k-1),$$

and

$$m_2 = \sum_{i=1}^{n-1}X_{i+1}/(k-1).$$

To test the null hypothesis of density independence a computer is used to undertake the following procedure:

1 Use the observed $X_1, X_2, \ldots, X_k$ to compute the test statistic (either $T(1,3)$ or $T(2,3)$).

2 Calculate the D_i values and either calculate all possible sets of X_i's that can be produced by a permutation of the sequence of D_i values given a fixed starting value of X_1 (there will be $(k-1)!$ sequences). If k is >7 then randomly permutate the D_i values to

produce a large number of sequences starting from X_1. 1000 such sequences is probably sufficient.

3 For each sequence of X_i values calculate the test statistic, $T(1,3)$ or $T(2,3)$.

4 If less than 5% of the computed T values are less than or equal to the observed T value calculated in step 1 then reject the null hypothesis that the population dynamics are density-independent.

15.2.6.3 Dennis and Taper's (1994) bootstrap approach

This can be viewed as a development of Pollard *et al*.'s (1987) randomization test. It makes more efficient use of the data to produce a more powerful test providing a stochastic logistic equation

$$N_{t+1} = N_t \exp(r + bN_t + \sigma Z_t) \tag{15.7}$$

is an appropriate model. N_t and N_{t+1} are the population sizes at times t and $t + 1$ respectively and r, b and σ are parameters to be estimated. The population growth rate in the absence of density dependence or random noise is given by r. For values of $b < 0$, b gives a measure of density dependence. The term σZ_t introduces random noise, as Z_t is a random variable with 0 mean and variance of 1. If density dependent regulation is acting then a model with $a > 0$, $b < 0$ will give a superior fit to the data than either a random walk ($a = 0$, $b = 0$) or random walk with drift ($a > 0$, $b = 0$) model. The test proceeds by fitting the random walk (only σ to be estimated), the random walk with drift (r and σ to be estimated) and density dependent (r, b and σ to be estimated) models to the data. The fit of these models is then compared using a likelihood ratio test for which the confidence limits are found using a bootstrap procedure. Dennis and Traper (1994) describe the computational steps required, which can be programmed on a personal computer. They also discuss extensions of this approach for situations where a logistic model is inappropriate.

Various simulation studies have been undertaken to compare the relative performances of these tests, but no clear conclusion emerges (Rothery, 1998). The underlying model must be considered for its appropriateness for the data in question. It is clear that the ability to detect density-dependence increases with the length of the time series and in some situations times series spanning 100 years or more may be required. Density-dependence can be detected in short time series for populations for which the Gompertz or logistic equations offer a good description of the dynamics.

15.3 Geographical information systems

A geographical information system (GIS) is a database system in which the data are referenced with respect to their geographical coordinates on our planet's surface. While this definition does not mention computers, in practice a powerful computer is at the core of a GIS, which will usually be expected to retrieve, display, and print quickly large amounts of data. GIS are becoming widely used in the fields of agriculture, conservation biology, environmental monitoring, epidemiology, and forestry (Burrough & McDonnell, 1998).

GIS are most useful for studying temporal variation in the spatial distribution of species. Typically, a GIS holds data derived from many different sources. For example, animal abundance at set locations may be estimated from field samples, while vegetation cover is estimated from satellite imagery. There is usually a temporal component to the database, so that changes in a physical or biological variable can be mapped through time. For the better-studied parts of the world, large amounts of spatially organized data are now becoming available, e.g. temperature records for Great Britain have recently become available on disk (Lennon & Turner, 1995).

A typical example of a GIS application in biology was the system developed for Projeto Mamirauá. The project objective was to create a management plan for a large reserve in the flood-plain of the Rios Solimões and Japurá in the upper Amazon basin. The multidisciplinary research team, comprising more than 50 specialists, worked on the project for more than 3 years. The reserve covers an area of about 1 250 000 ha and holds more than 1000 bodies of water. At the outset, a database was required to hold place names and to map community boundaries. As data accumulated on the flora and fauna within the reserve, it was natural to record data with reference to its geographical locality. Further, the regular seasonal flooding caused by the rise in flow of the main rivers created the need for time-series data. However, the creation of such a database is not easy and presents considerable organizational challenges. It is almost inevitable that the time needed to create, implement, and maintain the GIS will be underestimated. A frequent result is that fieldworkers are collecting data before the database is ready for use; this creates a backlog of data, which are often in an unsuitable format for rapid entry. Once completed, such a GIS offers the prospect that ecologists can rapidly gain access to huge amounts of information, justifying the cost. In the final stage, will a reserve management team in place, they can hopefully use the GIS to aid their implementation of the management plan and monitor the reserve.

15.4 Habitat classification

15.4.1 Qualitative

Zoologists frequently delimit their communities by reference to plants or environmental factors. The most universal classification of habitats is that of Elton and Miller (1954).

1 Terrestrial system.
  (a) Formations.
      Open-ground type—if any dominant plants, these not more than 15 cm high.
      Field type—dominant life form coincides with field layer, usually not more than 2 m in height.

Scrub type—dominant life form does not exceed a shrub layer, height generally not over 7.6 m.

Woodland type—trees dominant life form.

(b) Vertical layers.

Subsoil and rock.

Topsoil.

Ground zone, including low-growing vegetation, less than 15 cm.

Low canopy—up to about 7.6 m.

High canopy.

Air above vegetation.

2 Aquatic system.

(a) Formations.

See Table 15.4.

(b) Vertical layers.

Bottom light and dark zones, water-mass light and dark zones—free water not among vegetation.

Submerged vegetation.

Water surface—upper and under surface of film of floating leaves. Emergent vegetation—reed swamp and similar vegetation, the bases of which are in the water.

Air above vegetation.

3 Aquatic–terrestrial transition system—defined further by body type with which it occurs and by vegetational systems corresponding to the terrestrial system.

4 Subterranean system—caves and underground waters.

5 Domestic system.

6 General system.

Dying and dead wood.

Macrofungi.

Dung.

Carrion.

Table 15.4 Formation types in an aquatic system.

	Very small	Small	Medium	Large	Very large
Still	Tree hole	Small pond <17 m²	Pond <400 m²	Large pool or tarn <40 ha	Lake or sea
Slow	Trickle Gutter	Ditch Field dyke	Canal River-bank water		
Medium	Trickle	Lowland brook	Lowland river	Lowland large river	River estuary
Fast	Spring	Upland weir Small torrent or stream	Large torrent or stream		
Vertical or steep	Water drip Pipe outlet	Small weir Waterfall	Large weir Medium waterfall	Large waterfall	

Animal artefacts — nests, etc.

Human artefacts — fence posts, straw stacks, etc.

Further division of the habitat into communities may be made on the type of plant. In soil, marine benthic, and freshwater studies divisions are frequently based on the fauna itself.

15.4.2 Quantitative

The quantification of the characters of the habitat, especially plants, in general terms was pioneered by MacArthur and MacArthur (1961), who assessed foliage-height diversity in trees to relate it to bird diversity. MacArthur (1972) suggested the best measure of foliage height diversity was D of the Simpson–Yule index (p. 475), where the various values of p_i are the proportion of the vegetation at various heights or at the different layers in the Eltonian system (above). A quick method of assessing 15 interrelated variables in arboreal vegetation is described by James and Shugart (1970) and utilized by James (1971) to demonstrate the vegetational characteristics of bird habitats. Southwood et al. (1979) distinguished between structural diversity (the distribution of vegetation in three-dimensional space) and architectural diversity (the variety of plant parts — bark, leaves, flowers, etc.).

Numerical methods for classifying vegetation are reviewed by Kent and Coker (1992) and Jongman et al. (1995). One of the most popular vegetation classification methods is two-way indicator-species analysis (TWINSPAN), orginally developed by M.O. Hill. The analysis results in a hierarchical classification, in which different communities are distinguished by the presence or relative abundance of indicator species. The use of indicator species is particularly attractive for field ecologists, as it allows field sites to be quickly assigned to a predefined community. In marine benthic studies, TWINSPAN is now becoming more widely used for the classification of animal communities. In the study of plant communities, reciprocal averaging and its variant, detrended correspondence analysis (DECORANA), are widely used. These techniques, although applicable to animal communities, are little used. The fundamental difference between terrestrial and marine ecology below the photic zone is the absence of large plants in the marine habitat. Thus, the most important factor determining the presence and abundance of terrestrial animals — the species of plant that are present — is no longer acting.

In a review of habitat characteristics, Southwood (1977, 1988) stressed the need to define spatial heterogeneity, in relation to the trivial and migratory ranges of the animal, and temporal heterogeneity, in relation to generation time. Stability in space — the length of time the same location remains suitable for the species (durational stability) — and stability in time — the seasonal fluctuations in favourableness — should be distinguished; such temporal changes, together with the general level of unfavourableness, constitute the 'adversity axes' of Whittaker. It was proposed that most habitats could be characterized against the two axes of adversity and durational stability and that suites of bionomic adaptations were related to the position of the organism's habitat in this chart.

References

Aldrich, R.C., Bailey, W.F., & Heller, R.C. (1959) Large scale 70 mm color photography techniques and equipment and their application to a forest sampling problem. *Photogrammetric Eng.* **25**, 747–54.

Andersen, H.S. (1997) Land surface temperature estimation based on NOAA–AVHRR data during the HAPEX–Sahel experiment. *J. Hydrol.* **189**(1–4), 788–814.

Arseneault, D., Villeneuve, N., Boismenu, C., LeBlanc, Y., & Deshaye, J. (1997) Estimating lichen biomass and caribou grazing on the wintering grounds of northern Quebec: an application of fire history and Landsat data. *J. Appl. Ecol.* **34**(1), 65–78.

Austin, G.E., Thomas, C.J., Houston, D.C., & Thompson, D.B.A. (1996) Predicting the spatial distribution of buzzard *Buteo buteo* nesting areas using a geographical information system and remote sensing. *J. Appl. Ecol.* **33**(6), 1541–50.

Bachelet, G. & Castel, J. (eds) (1997) Long-term changes in marine ecosystems. *Oceanol. Acta.* **20**, 329 pp.

Barrett, E. & Curtis, L.F. (1992) *Introduction to Environmental Remote Sensing.* Chapman & Hall, London.

Box, G.E.P. & Jenkins, G.M. (1976) *Time Series Analysis: Forecasting and Control.* Holden-Day, San Francisco.

Bulmer, M.G. (1975) The statistical analysis of density dependence. *Biometrics* **3**, 901–11.

Burrough, P.A. & McDonnell, R.A. (1998) *Principles of Geographical Information Systems.* Oxford University Press, Oxford.

Bursell, E. (1957) The effect of humidity on the activity of tsetse flies. *J. Exp. Biol.* **34**, 42–51.

Burt, P.J.A., Colvin, J., & Smith, S.M. (1995) Remote sensing of rainfall by satellite as an aid to *Oedaleus senegalensis* (Orthoptera: Acrididae) control in the Sahel. *Bull. Entomol. Res.* **85**(4), 455–62.

Chatfield, C. (1994) *The Analysis of Time Series: an Introduction.* Chapman & Hall, London.

Colebrook, J.M. (1978) Continuous plankton records: overwintering and annual fluctuations in abundance of zooplankton. *Mar. Biol.* **84**, 261–5.

Cracknell, A.P. & Hayes, L.W.B. (1991) *Introduction to Remote Sensing.* Taylor & Francis, London.

Dale, P.E.R. & Morris, C.D. (1996) *Culex annulirostris* breeding sites in urban areas: using remote sensing and digital image analysis to develop a rapid predictor of potential breeding areas. *J. Am. Mosquito Contr. Assoc.* **12**, 316–20.

Dennis, B. & Taper, M.L. (1994) Density dependence in time series observations: estimation and testing. *Ecol. Monogr.* **64**, 205–24.

Dennis, B., Kemp, W.P., & Taper, M.L. (1998) Joint density dependence. *Ecology* **79**, 426–41.

Diggle, P.J. (1990) *Time Series: a Biostatistical Introduction.* Clarendon Press, Oxford.

Dwivedi, R.S., Sankar, T.R., Venkataratnam, L., et al. (1997) The inventory and monitoring of eroded lands using remote sensing data. *Int. J. Remote Sens.* **18**(1), 107–19.

Elton, C.S. (1924) Periodic fluctuations in the numbers of animals: their causes and effects. *Brit. J. Exp. Biol.* **2**, 119–63.

Elton, C.S. & Miller, R.S. (1954) The ecological survey of animal communities with a practical system of classifying habitats by structural characters. *J. Ecol.* **42**, 460–96.

Elton, C. & Nicholson, M. (1942) The ten-year cycle in numbers of the lynx in Canada. *J. Anim. Ecol.* **11**, 215–44.

Everitt, J.H., Richerson, J.V., Karges, J.P., & Davis, M.R. (1997) Using remote sensing to detect and monitor a western pine beetle infestation in west Texas. *Southw. Entomol.* **22**(3), 285–92.

Ferguson, R.L. & Korfmacher, K. (1997) Remote sensing and GIS analysis of seagrass meadows in North Carolina, USA. *Aquat. Bot.* **58**(3–4), 241–58.

Ferrier, G. & Anderson, J.M. (1996) The application of remote sensing data in the study of effluent dispersal in the Tay estuary, Scotland. *Int. J. Remote Sens.* **17**(18), 3541–66.

Finerty, J.P. (1980) *The Population Ecology of Cycles in Small Mammals.* Yale University Press, New Haven, 234 pp.

Foster, J.R., Roseberry, J.L., & Woolf, A. (1997) Factors influencing efficiency of white-tailed deer harvest in Illinois. *J. Wildl. Manag.* **61**(4), 1091–7.

Franz, J. & Karafiat, H. (1958) Eigen sich Kartierung und Scrienphotographic von Tannenläsen für Massenwechsel-studien? *Z. Anew. Ent.* **43**, 100–12.

Gaston, K.J. & McArdle, B.H. (1994) The temporal variability of animal abundances: measures, methods and patterns. *Phil. Trans. Roy. Soc. Lond. B* **345**, 335–58.

George, D.G. (1997) The airborne remote sensing of phytoplankton chlorophyll in the lakes and tarns of the English Lake District. *Int. J. Remote Sens.* **18**(9), 1961–75.

Gleiser, R.M., Gorla, D.E., & Almeida, F.F.L. (1997) Monitoring the abundance of *Aedes* (*Ochlerotatus*) *albifasciatus* (Macquart 1838) (Diptera: Culicidae) to the south of Mar Chiquita Lake, central Argentina, with the aid of remote sensing. *Ann. Trop. Med. Parasitol.* **91**(8), 917–26.

Granger, C.W.J. (1966) The typical shape of an economic variable. *Econometrica* **34**, 150–61.

Gruttke, H. & Engels, H. (1998) Metapolulation structure of *Carabus problematicus* in a fragmented landscape, significance of simulation results for conservation. In: Baumgärtner, J., Brandmayr, P., & Manly, B.F.J. (eds) *Population and Community Ecology for Insect Management and Conservation*. Balkema, Rotterdam 133–43.

Hagner, O. & Rigina, O. (1998) Detection of forest decline in Monchegorsk area. *Remote Sens. Environ.* **63**(1), 11–23.

Hanski, I. (1981) Coexistence of competitors in patchy environment with and without predation. *Oikos* **37**, 306–12.

Hanski, I. (1982) Dynamics of regional distribution: the core and satellite species hypothesis. *Oikos* **38**, 210–21.

Hanski, I. (ed.) (1987) Ecological significance of spatial and temporal variability. *Ann. Zool. Fennici* **25**, 1–106.

Hanski, I. & Gilpin, M. (1991) Metapopulation dynamics: brief history and conceptual domain. *Biol. J. Linn. Soc.* **42**, 3–16.

Harris, M.K., Hart, W.G., Davis, M.R., Ingle, S.J., & Van Cleave, H.W. (1976) Aerial photography shows caterpillar infestation. *Pecan. Quarterly* **10**, 12–18.

Hassell, M.P., Comins, H.N., & May, R.M. (1991) Spatial structure and chaos in insect population dynamics. *Nature* **353**, 255–8.

Hay, S.I., Tucker, C.J., Rogers, D.J., & Packer, M.J. (1996) Remotely sensed surrogates of meteorological data for the study of distribution and adundance of arthropod vectors of disease. *Ann. Trop. Med. Parasitol.* **90**, 1–19.

Henderson, P.A. & Corps, M. (1997) The role of temperature and cannibalism in inter-annual recruitment variation of bass, *Dicentrarchus labrax* (L) in British waters. *J. Fish Biol.* **50**, 280–95.

Henderson, P.A., Hamilton, W.D., & Crampton, W.G.R. (1998) Evolution and diversity in Amazonian floodplain communities. In: Newbury, D.M., Prins, H.H.T., & Brown, N.D. (eds) *Dynamics of Tropical Communities*. Blackwell Science, Oxford, 385–419.

Henttonen, H., Oksanen, T., Jortikka, A., &

Haukisalmi, V. (1987) How much do weasels shape microtine cycles in the northern Finnoscandian taiga? *Oikos* **50**, 353–65.

Hielkema, J.U. (1990) Satellite environmental monitoring for migrant pest forcasting by FAO: the ARTIMIS system. *Phil. Trans. Roy. Soc. B* **328**, 705–17.

Ibanez, F. & Fromentin, J.-M. (1995) Un typologie à partir de la forme des séries chronologiques (TFS). *Oceanol. Acta* **20**, 11–25.

James, F.C. (1971) Ordination of habitat relationships among breeding birds. *Wilson Bull.* **83**, 215–36.

James, F.C. & Shugart, H.H. (1970) A quantitative method of habitat description. *Audubon Fld. Notes* **24**(6), 727–36.

Jongman, R.H.G., Ter Braak, C.J.F., & Van Tongeren, O.F.R. (eds) (1995) *Data Analysis in Community and Landscape Ecology*. Cambridge University Press, Cambridge.

Keitt, T.H. & Stanley, H.E. (1998) Dynamics of North American breeding bird populations. *Nature* **393**, 257–60.

Kendall, M.G., Stuart, A., & Ord, J.K. (1983) *The Advanced Theory of Statistics*, Vol 3. 4th edn, Griffin, London.

Kent, M. & Coker, P. (1992) *Vegetation Description and Analysis: a Practical Approach*. John Wiley & Sons, Chichester.

Kitron, U., Otieno, L.H., Hungerford, L.L., *et al.* (1996) Spatial analysis of the distribution of tsetse flies in the Lambwe Valley, Kenya, using Landsat TM satellite imagery and GIS. *J. Anim. Ecol.* **65**(3), 371–80.

Klein, W.H. (1973) Beetle killed pine estimates. *Photogrammetric Eng.* **39**, 385–88.

Klimetzek, D. (1990) Population dynamics of pine-feeding insects: a historical survey. In: Watt, A.D., Leather, S.R., Hunter, M.D., & Kidd, N.A.C. (eds) *Population Dynamics of Forest Insects*. Intercept, Andover, UK, pp. 3–10.

Lasen, C.P.S. & MacDonald, G.M. (1998) An 840-year record of fire and vegetation in a boreal white spruce forest. *Ecology* **79**, 106–18.

Legendre, P. & McArdle, B.H. (1997) Comparison of surfaces. *Oceanologica Acta* **20**, 27–43.

Leigh, R.A. & Johnston, A.E. (eds) (1994) *Long-term Experiments in Agricultural and Ecological Sciences*. CABI, Wallingford, U.K. 428 pp.

Lennon, J.J. & Turner, J.R.G. (1995) Predicting the spatial distribution of climate: temperature in Great Britain. *J. Anim. Ecol.* **64**, 370–92.

Lillisand, T.M. & Kiefer, R.W. (1994) *Remote Sensing and Image Interpretation*. John Wiley & Sons, New York.

MacArthur, R.H. (1972) *Geographical Ecology:*

Patterns in the Distribution of Species. Harper & Row, New York and London, 269 pp.

MacArthur, R.H. & MacArthur, J.W. (1961) On bird species diversity. *Ecology* **42**, 594–8.

May, R.M. (1973) *Stability and Complexity in Model Ecosystems*. Princeton University Press, Princeton, New Jersey, 235 pp.

Morisita, M. (1959) Measuring of the dispersion of individuals and analysis of the distributional patterns. *Mem. Fac. Sci., Kyushu Univ., Ser. E (Biol.)* **2**, 215–35.

Pollard, E. (1991) Synchrony of population fluctuations: the dominant influence of widespread factors on local butterfly populations. *Oikos* **60**, 7–10.

Pollard, E., Lakhani, K.H., & Rothery, P. (1987) The detection of density dependence from a series of annual censuses. *Ecology* **68**, 2046–55.

Price, J.C. (1984) Land surface temperature measurement for the split window channels of the NOAA 7 advanced very high resolution radiometer. *J. Geophys. Res.* **89**, 7231–7.

Ranta, E., Lindstrom, J., & Linden, H. (1995) Synchrony in tetraonid population dynamics. *J. Anim. Ecol.* **64**, 767–76.

Rehfisch, M.M., Clark, N.A., Langston, R.H.W., & Greenwood, J.D. (1996) A guide to the provision of refuges for waders: an analysis of 30 years of ringing data from the Wash, England, *J. Appl. Ecol.* **33**, 673–87.

Rogers, D.J. & Randolph, S.E. (1991) Mortality rates and population density of tsetse flies correlated with satellite imagery. *Nature* **351**, 739–41.

Rogers, D.J., Hay, S.I., & Packer, M.J. (1996) Predicting the distribution of tsetse flies in West Africa using temporal Fourier processed meteorological satellite data. *Ann. Trop. Med. Parasitol.* **90**, 225–41.

Rothery, P. (1998) The problems associated with the identification of density dependence in population data. In: Dempster, J.P. & McLean, I.F.G. (eds) *Insect Populations*. Kluwer Academic Publishers, Dordrecht, pp. 97–133.

Saitoh, T., Stenseth, N.C., & Bjornstad, O.N. (1998) The population dynamics of the vole *Clethrionomys rufocanus* in Hokkaido, Japan. *Res. Pop. Ecol.* **40**, 61–76.

Schwerdtfeger, F. (1941) Ober die Ursachen des massenwechsels der Insekten. *Zanger. Ent.* **28**, 254–303.

Southwood, T.R.E. (1977) Habitat, the template for ecological strategies? *J. Anim. Ecol.* **46**, 337–65.

Southwood, T.R.E. (1988) Tactics, strategies and templates. *Oikos* **52**, 3–18.

Southwood, T.R.E., Brown, V.K., & Reader, P.M. (1979) The relationships of plant and insect diversities in succession. *Biol. J. Linn. Soc.* **12**, 327–48.

Tong, H. (1990) *Non-linear Time Series: a Dynamical Approach*. Oxford University Press, Oxford.

Tuomisto, H., Linna, A., & Kalliola, R. (1994) Use of digitally processed satellite images in studies of tropical rain forest vegetation. *Int. J. Remote Sens.* **15**, 1595–610.

Verlinden, A. & Masogo, R. (1997) Satellite remote sensing of habitat suitability for ungulates and ostrich in the Kalahari of Botswana. *J. Arid Environ.* **35**(3), 563–74.

Vickery, W.L. & Nudds, T.D. (1984) Detection of density dependent effects in annual duck census. *Ecology* **65**, 96–108.

Vickery, W.L. & Nudds, T.D. (1991) Testing for density-dependent effects in sequential censuses. *Oecologia* **85**, 419–23.

Wallen, V.R., Jackson, H.R., & MacDurmid, S.W. (1976) Remote sensing of corn aphid infestation, 1974 (Hemiptera: Aphididae). *Can. Ent.* **108**, 751–4.

Washino, R.K. & Wood, B.L. (1994) Application of remote-sensing to arthropod vector surveillance and control. *Am. J. Trop. Med. Hyg.* **50**(suppl.), 134–44.

Wei, W.W.S. (1990) *Time Series Analysis: Univariate and Multivariate Methods*. Addison-Wesley, Redwood City, California.

Williamson, M. (1984) The measurement of population variability. *Ecol. Entomol.* **9**, 239–41.

Wint, G.R.W. (1998) Rapid resource assessment and environmental monitoring using low level aerial surveys. In: *Drylands: Sustainable Use of Rangelands into the Twenty-First century. Proc. Int. Workshop on the Sustainable Use of Rangelands and Desertification Control*. International Fund for Agricultural Development, Rome; Ministry of Agriculture and Water, Jeddah, pp. 277–302.

Woiwod, I.D. & Harrington, R. (1994) Flying in the face of change. In: Leigh, R.A. & Johnston, A.E. (eds) *Long-term Experiments in Agricultural and Ecological Sciences*. CABI, Wallingford, U.K., pp. 321–42.

Index

Note: Page references in *italics* refer to figures; those in **bold** refer to tables

Abax ater (ground-beetle) 513
absolute population estimates 2, 187, 254–61
absolute population index 254, 298
absolute samples 270
abundance index 336
Acarina 224, 237, 240
ACCESS 27
accuracy 7–8, 20, 75
Acipenser transmontanus (sturgeon) 116, 118
Acmaea digitalis 380
Acomys cahirinus 523
Acomys russatus 523
acoustic detection 167
acoustic methods 263
activity level 249–50
Aculus schlechtendali (apple-rust mite) 154
Acyrthosiphon pisum (pea aphid) 39, 119, 151
Adelges piceae 281
Aedes 186, 286, 298, 441
Aedes aegypti 92, 93, 119, 366, 448
Aedes albifasciatus 544
Aeneolamia varia (sugar-cane froghopper) 62
aerial density 147
aerial population, total 147–8
aerial sweeps 270
age-dependent behavioural attributes 248–9
age-dependent capture and survival 91
age-grouping 437–47
   by developmental stage 438–9
   determination by using structures 439–47
   of insects 439–44
agglutination methods 365–6
aggregation indices 39–42
Agissiz trawls 197, *198*
air, sampling from 141 to 8
air — flight traps 270–3
airlift devices 206–8, *207*
Alces alces 335
Aleurotrachelus jelinekii 431
Allan grab 201, 206
Allen curve 515
Allobophora 239
Alnus incana (grey alder) 327

Alosa sapidissima (shad) 115
alpha index of diversity 463–4, 480–1
Ammodytes marinus (sand eel) 365
Amphimallon majalis (chafer beetle) 38
anaesthetics 207–8
analysis of variance 11
Anarsia lineatella (peach twig borer) 297
Anastrepha ludens 440
Anguilla anguilla (European eel) 248
annelids, age-grouping 439
Anopheles 127, 293, 440, 442
Antestiopsis (coffee bug) 255
Anthomonus 114
Aonidiella 370
Aonidiella ornatum (citrus scale insect) 36
Aphis pomi (apple aphid) 154
Aphodius 511, 520
Aplysia oculifera (sea hare) 513
Apodemus (wood mice) 86, 447, 491
apparent mortality 424
aquatic light traps 291–3, *292*
Arachnidomyia aldrichi 117
Archips (oak leaf-roller moth) 16
Archips argyrospilus (tortricid moth) 16, 18
Argryoploce 441
Argus 108
Argyoplace 444
Arianta arbustorum 444
arithmetic mean 14
Armadillidium 448, 517, 519
Artemia salina 520
artificial resting units 293
artificial substrate and sampling apparatus 194–5,
   195
Asphophondylia (gall-midge) 38
aspirator 268–9, *269*
assimilation rates 516–21
attraction, measure of 270
aural detection 264–5
autoregressive integrated moving average
   (ARIMA) 546
availability 253–4

Bactrocera cucurbitae 440

Lightning Source UK Ltd.
Milton Keynes UK
UKOW02f1603141113

221085UK00002B/22/P

Ecological Methods

01/06

UNIVERSITY OF
WOLVERHAMPTON